Earth and Solar System Constants

	Mass	Radius	Mean Orbital Radius
Earth	5.98×10^{24} kg	6.37×10^6 m	15.0×10^{10} m
Moon	7.34×10^{22} kg	1.74×10^6 m	3.84×10^8 m
Sun	1.99×10^{30} kg	7.00×10^8 m	—
Mercury	0.317×10^{24} kg	2.4×10^6 m	5.79×10^{10} m
Venus	4.87×10^{24} kg	6.26×10^6 m	10.8×10^{10} m
Mars	0.64×10^{24} kg	3.39×10^6 m	22.8×10^{10} m
Jupiter	1900×10^{24} kg	69.9×10^6 m	77.8×10^{10} m

$g = GM/R^2 = 9.8$ N/kg $= 9.8$ m/s^2

1 atmosphere (atm) $= 1.01 \times 10^5$ Pa $= (1.01 \times 10^5$ N/m$^2) = 76$ cm Hg

Solar constant (radiation from sun at earth's mean distance)   1.35 kW/m^2

seconds/year 3.15×10^7

PHENOMENAL PHYSICS

Clifford E. Swartz

The State University of New York
at Stony Brook

JOHN WILEY & SONS

New York
Chichester
Brisbane
Toronto

Cover: Photograph by Terry Lennon

To Barb, who is truly phenomenal

Copyright © 1981, by John Wiley & Sons, Inc.

All rights reserved. Published simultaneously in Canada.

Reproduction or translation of any part of
this work beyond that permitted by Sections
107 and 108 of the 1976 United States Copyright
Act without the permission of the copyright
owner is unlawful. Requests for permission
or further information should be addressed to
the Permissions Department, John Wiley & Sons.

Library of Congress Cataloging in Publication Data:

Swartz, Clifford E
  Phenomenal Physics.

  Includes index.
  1. Physics. I. Title.
QC21.2.S9 530 80-16690
ISBN 0-471-83880-2

Printed in the United States of America

10 9 8 7 6 5 4 3 2 1

PREFACE

Phenomenal physics? What can be so unusual or amazing about physics? Of course, physicists think physics is exciting, but that's because it's their life's work. There can be an enormous thrill in discovering a new subatomic particle, or in finding a new way to explain a familiar process. There can also be small but satisfying thrills in observing and understanding the daily natural events around us. Rainbows and hi-fi sets and spinning wheels and soap bubbles are more fun to deal with if you know their scientific origins. The success of physics in explaining everyday phenomena is truly phenomenal.

However, that's not why I named the book *Phenomenal Physics*. Phenomenal also means concerned with phenomena—phenomenological. In presenting the standard topics of physics, I first describe examples from the real world. Each chapter opens with a section called "Handling the Phenomena." You can usually do these activities at home or in a dorm with commonly available materials and in just a few moments. Of course, you can, if you wish, bypass these sections or just read about the activities. That would be a pity. Most people understand abstract ideas better if they first handle the phenomena involved. These activities are not a substitute for laboratory work. Instead, they supplement the demonstrations that many instructors present. You can learn more, however, if you play with these phenomena yourself.

This book contains far more information than you can cover in a one-year course. Different instructors, in different schools, in different years, choose to emphasize particular topics and skip others, but a solid core remains. I have tried to cover the core topics thoroughly, and in the standard sequence. The first ten chapters have few optional sections, matching the traditional treatment of this material in most colleges. From Chapter 11 on, however, there are many topics marked "Optional." Some of these involve treatment that is more sophisticated or requires more complicated math than the basic presentation. Other optional topics, such as the one on the physics of music, are not complicated, but might not be required in your particular course. You might enjoy such a section even if it is not required, or, in future years, you may find it useful for reference.

No calculus is used in the core topics, although some derivations in optional material require simple differentiation or integration. Nevertheless, the text appeals occasionally to geometrical arguments concerning the slopes of graph curves or the areas under them. I did not omit any topic in the standard introductory course because of the calculus restriction. Only simple algebra and a few trig-geometry facts are needed for the core topics. The text uses the international system of units (S.I. units) almost exclusively. Where translation to older units is needed, it is provided. There is a guide to S.I. units in the Appendix as well as a summary of the few math facts needed.

If you glance hastily at the table of contents, you might think that the book shortchanges the treatment of "modern" physics. There are no separate chapters on atomic, molecular, nuclear, particle, or astrophysics. Instead, from Chapter 1 on, the discussions use atomic models and examples that involve astronomical data. I point out many cases where the frontiers of research are only one short step beyond the introductory topic. For example, the nature of mass is a current (and recurrent) problem in cosmology, and I exploit this exciting situation in presenting Newton's laws. The final chapter serves both as a survey of atomic and subatomic physics and as a review of many topics studied earlier in the book. The same situation prevails with regard to biological and physiological applications. The human body and its parts are frequently used for examples in mechanics, fluids, sound, optics, and electricity.

The text contains many border diagrams and pictures. Arthlyn Ferguson skillfully and patiently drew the diagrams in a style meant to resemble the sketches that a professor might put on a blackboard while lecturing. These informal drawings describe, summarize, or sometimes comment on the discussions.

Every few pages the text interrupts itself with a question, often of the "yes, but" variety. Usually these follow some derivation or development where the conclusion should be challenged, or where there appears to be a paradox. Ideally, the reader should pause and try to answer such a question, or at least mull it over for a while. My suggested answer or comment about each question is at the end of the chapter. Students tell me that they have found this type of question useful in other books that I have written. If nothing else, the interruption caused by turning to the end of the chapter serves to waken the reader.

There are several features of the book designed to make your study easier and faster. Each chapter has both an Introduction and a Summary. Before studying a new chapter, read both. Then glance through the section headings. You'll get an idea of where you're heading and what to look for.

You can solve most of the homework problems without using a calculator. In fact, you should always work out the approximate size, or the order-of-magnitude, of an unknown quantity before plugging the detailed numbers into the calculator. Zero in on the answer! For order-of-magnitude calculations, $\pi \approx 3$, $\pi^2 \approx 10$, $1047 \approx 938 \approx 1 \times 10^3$, and so on. In the Appendix, there is a list of handy approximations.

Many people helped to produce this book. Naida Dewey faithfully typed and retyped the manuscript. At Wiley, Don Deneck as the College Physics Editor persuaded me to start the book. His successor, Robert McConnin, shepherded it through the final stages. Rosemary Wellner and Joan Knizeski prepared the manuscript for composition. Ann Renzi was the

PREFACE

designer of the book, and Kathy Bendo was in charge of photo research. The production supervisor was Nina R. West.

Several physicists read the first draft: Professor Alfred Romer of St. Lawrence University, Professor Arnold Strassenburg of the State University of New York at Stony Brook, and Professor Robert Bauman of the University of Alabama at Birmingham. They brought numerous errors to my attention and made useful suggestions for improvement that I tried to follow. Then Professor Bauman heroically reread a second draft and provided me with detailed criticism that was crucially important. No doubt some errors may remain. Blame me, or better yet, let me know and I will try to correct the error. Ater thirty years of teaching physics, I still find lots of things that I didn't know I didn't know before.

I hope that most of you who read this book are doing it because you want to learn more about our world. Some, I know, are taking physics only to fulfill some requirement. In either case, I hope you end up enjoying the book and the course work. We live in a mysterious and phenomenal universe, and, as far as we know, we're the only ones around who can comprehend it.

Clifford E. Swartz

CONTENTS

1. **THE SETTING AND THE DRAMA** — 1
   - Sizes and distances — 1
   - Handling the phenomena—distances — 7
   - Times — 8
   - Handling the phenomena—times — 14
   - Interactions and change — 15
   - Functional dependence — 18
   - Handling the phenomena—functions — 20
   - Summary — 21
   - Answers to questions — 22
   - Problems — 25

2. **MOTION ALONG A LINE** — 27
   - Handling the phenomena — 28
   - Description of an object standing still — 29
   - Description of an object with constant (but nonzero) velocity — 30
   - Instantaneous velocity — 33
   - Using a graph of v(t) to find distance traveled — 34
   - Definition of acceleration — 35
   - The special case of constant acceleration — 36
   - Constant acceleration formulas when v_o is not zero — 39
   - Units and dimensions — 40
   - Examples — 41
   - Simple problems and the real world — 44
   - Summary — 45
   - Answers to questions — 46
   - Problems — 48

3. **MOTION IN TWO DIMENSIONS** — 50
   - Handling the phenomena — 50
   - Position in two dimensions — 52
   - Vector quantities — 53
   - Combining velocities — 54
   - Flat earth trajectories — 56
   - On going in circles — 59
   - Summary — 62
   - Answers to questions — 62
   - Problems — 64

4. **FORCES IN EQUILIBRIUM** — 66
   - Handling the phenomena — 66
   - Kinds of forces — 68
   - Deformations and their measurements — 69
   - The relationship between force and distortion — 70
   - Pressures, stresses, and strains — 72
   - The directional nature of forces — 76
   - Equilibrium of forces — 78
   - Torques — 80
   - Equilibrium of torques — 82
   - Center of weight — 85
   - Summary — 87
   - Answers to questions — 88
   - Problems — 90

5. **MOTION PRODUCED BY FORCES** — 92
   - Handling the phenomena — 92
   - Experimental results of applying forces — 93
   - The constant of proportionality between force and acceleration: $F = ka$ — 95
   - Inertial and gravitational mass — 97
   - The unit of mass—the kilogram — 99
   - Weighing the earth — 100
   - Little g, the "acceleration due to gravity" — 101
   - Newton's second law and the real world — 102
   - Summary — 107
   - Answers to questions — 108
   - Problems — 110

6. **APPLICATIONS OF NEWTON'S SECOND LAW** — 112
   - Handling the phenomena — 112
   - Constant force in the direction of motion — 113
   - Circular motion—definitions and relationships — 116
   - Newton's second law for rotations — 119
   - Cases where centripetal force is perpendicular to gravity — 122
   - Cases where gravity provides the centripetal force — 123
   - The changing force that produces oscillations — 125

	The period of oscillation	128
	Simple harmonic motion—and other oscillations	130
	Summary	131
	Answers to questions	132
	Problems	134
7	**LINEAR MOMENTUM**	**136**
	Handling the phenomena	136
	The conservative nature of momentum	137
	Other properties of momentum	141
	Examples of the conservation of momentum	143
	Newton's third law and its relationship to Newton's second law	146
	Examples of impulses and kicks	148
	Center of mass and momentum conservation	152
	Summary	154
	Answers to questions	154
	Problems	156
8	**ANGULAR MOMENTUM**	**159**
	Handling the phenomena	159
	The forms of angular momentum	160
	Magnitudes of some angular momenta	161
	The conservation of angular momentum	162
	The directional nature of angular momentum	167
	The quantization of angular momentum	169
	Some special points of interest	171
	Summary	175
	Answers to questions	176
	Problems	178
9	**KINETIC ENERGY AND WORK**	**180**
	Handling the phenomena	181
	The second constraint on certain collisions	181
	The work necessary to produce kinetic energy	186
	Rotational kinetic energy	189
	Simple machines that save no work	191
	Summary	195
	Answers to questions	196
	Problems	198
10	**POTENTIAL ENERGY AND POWER**	**200**
	Handling the phenomena	201
	Storing energy by distorting systems	202
	Negative potential energy	205
	The earth's potential well	206
	Molecular potential wells	210
	Restoring forces and potential energy	211
	Molecular forces	213
	Centrifugal forces	215
	Power	216
	Units and ranges of energy and power	217
	Summary	218
	Answers to questions	219
	Problems	220
11	**REFERENCE FRAMES AND RELATIVITY**	**224**
	Handling the phenomena	225
	The Galilean transformation	225
	The local fields produced by accelerated reference frames	232
	Einstein's special theory of relativity	237
	A thought experiment	240
	The Lorentz transformation	244
	Proper time and the invariant interval	247
	The addition of velocities in moving reference frames	248
	The famous twin paradox	249
	Momentum, energy, and mass relationships	252
	Mass-energy and the Lorentz transformation	257
	Gravitation and the general theory of relativity	258
	Summary	259
	Answers to questions	260
	Problems	264
12	**THE INTERNAL ENERGY OF MATERIAL**	**267**
	Handling the phenomena	267
	Effects caused by the loss of mechanical energy	268
	Common thermometers	269
	Calibrating thermometers and temperature scales	271
	The behavior of gases as a function of temperature	275
	The effect of temperature on solids and liquids	278
	Phase changes in materials as a function of temperature	280
	The definition of temperature	282
	The first law of thermodynamics	284
	Heat capacities—the temperature change when heat is added	286
	Heat transfer	294
	Summary	297
	Answers to questions	298
	Problems	300
13	**THE MICROSTRUCTURE OF MATTER AND THE SECOND LAW OF THERMODYNAMICS**	**303**
	Handling the phenomena	303
	Assumptions about our atomic model	305
	Special assumptions for our model of gases	306
	The kinetic theory of gases	307
	Consequences of our model—average energy and speed	308

The distribution of molecular energy in a gas	309
Thermal effects in real gases, solids, and liquids	315
Specific heats according to the atomic model	318
Turning heat into work	322
The Carnot cycle	325
The second law of thermodynamics	327
Entropy	330
Summary	334
Answers to questions	335
Problems	339

14 FLUIDS — 341
Handling the phenomena	341
Fluid pressure	344
Archimedes' principle	347
Surface tension	350
"Ideal fluid" flow—a poor approximation	356
The real world of viscosity	362
Summary	370
Answers to questions	371
Problems	373

15 TRAVELING PULSES AND SIMPLE WAVES — 375
Handling the phenomena	376
Disturbance of an oscillator	377
Disturbance that propagates along a line	380
The velocity of a pulse	381
Wave trains	386
Reflection of pulses and waves	390
The variation of wave velocity with frequency	393
Power transferred by waves	395
The spread of wave energy	401
The dependence of wave reflection on impedances	402
Summary	404
Answers to questions	405
Problems	407

16 COMPLEX WAVES AND INTERFERENCE — 410
Handling the phenomena	410
A formal description of traveling waves	413
Traveling waves seen from a different reference frame—the Doppler shift	414
Phase effects and coherent waves	420
The linear superposition of waves	421
Beats between waves	421
Resonant lines	423
Standing waves	425
Fourier analysis of complex waves	429
The change of pulse shape during transmission	430
Interaction of waves in two dimensions	440
The interference produced by two wave sources	446
Interference produced by splitting and recombining a single beam	449
The grating—interference produced by multiple sources	451
Other interference patterns	452
Summary	454
Answers to questions	455
Problems	458

17 GEOMETRICAL OPTICS — 461
Handling the phenomena	461
The behavior of rays	462
Reflection from plane surfaces	462
Refraction	465
Dispersion and refraction	469
Model to explain reflection, transparency, and dispersion	471
Fermat's principle	472
Reflections from curved mirrors	474
Images formed by lenses	481
Optical instruments	489
Summary	498
Answers to questions	499
Problems	502

18 ELECTROSTATICS — 504
Handling the phenomena	505
The isolation of electric charge	506
Our modern explanation of charge separation	507
The natural production of static electricity	508
Local movement or distortion of charges	509
A research application of electrostatics	510
Quantitative electrostatics—Coulomb's law	510
The magnitude of the charges that we have separated	512
The electrostatic field	513
The electrostatic field produced by point charges	514
Gauss's law relating E and q	516
Summary of the dependence of E on source geometry	522
Electric potential	523
Electric potential in common geometries	526
The storage of electric charge—electrical capacitance	528
Practical capacitors	530
The microstructure of a dielectric	530
Energy storage in a capacitor	530
Combinations of capacitors	532
The natural unit of electric charge	533
Summary	536

	Answers to questions	538
	Problems	540
19	**ELECTRIC CURRENT**	542
	Handling the phenomena	542
	Current and power in familiar circuits	543
	Model of charge movement in a wire	544
	Velocity of charge flow in a wire	547
	Current as a function of V	548
	The special—but important—case of Ohm's law	550
	Temperature dependence of ρ	554
	Power loss in circuits	556
	Practical resistors	557
	Resistors in series and parallel	559
	Internal resistance of the power supply	562
	Voltaic cells and batteries	563
	Measuring current and voltage	566
	Special circuits	569
	Kirchhoff's rules for circuit analysis	571
	Summary	573
	Answers to questions	574
	Problems	576
20	**MAGNETISM**	578
	Handling the phenomena	579
	The force between parallel currents	580
	The magnetic field	582
	Field patterns from permanent magnets and electric current configurations	584
	Quantitative treatment of magnetic field—Ampère's law	587
	Quantitative calculations of magnetic field—the Biot-Savart law	589
	The magnetic field on the axis of a current loop	591
	Magnetic field in a solenoid	593
	Quantitative calculation of the force on a current in a magnetic field	594
	The torque on a wire loop	599
	Magnetism in materials	602
	Ferromagnetism	606
	The relative nature of electric and magnetic fields	610
	Summary	613
	Answers to questions	614
	Problems	616
21	**CURRENTS AND FIELDS THAT CHANGE WITH TIME**	618
	Handling the phenomena	618
	The decay of charge on a capacitor	619
	Moving a conductor in a magnetic field	622
	Effect of induced current—Lenz's law	624

	Faraday's law—the nonobvious generalization	626
	An emf is not a conservative potential	629
	Eddy currents	631
	Electric motors	633
	Electric generators	634
	Inductance	635
	Charging an inductor	637
	The energy in an inductor	638
	Inductance and capacitance in series	640
	Alternating current	642
	Transformers	647
	Summary	649
	Answers to questions	650
	Problems	653
22	**ELECTROMAGNETIC RADIATION**	655
	Handling the phenomena	655
	Review of the basic equations of electromagnetism	657
	The implications of Maxwell's equations	659
	Generation of electromagnetic fields from accelerated charges	662
	The energy radiated in electromagnetic fields	664
	Momentum carried by electromagnetic radiation	667
	Polarization of electromagnetic radiation	669
	The experimental confirmation of Maxwell's prediction	671
	The electromagnetic spectrum	672
	Summary	681
	Answers to questions	682
	Problems	683
23	**THE MICROSTRUCTURE OF THE WORLD**	685
	The electron	686
	Heisenberg's uncertainty principle	696
	Atomic structure	698
	Molecules	704
	The solid state	706
	Electrical conductivity in solids	707
	Spectra produced by hot dense sources	709
	The duality problem	712
	The atomic nucleus	714
	Fusion	716
	Fission	716
	Radioactivity	717
	Radioactive sequences	718
	Other particles	719
	Epilogue	721
	Answers to questions	722
	Problems	724
	Index	733

THE SETTING AND THE DRAMA 1

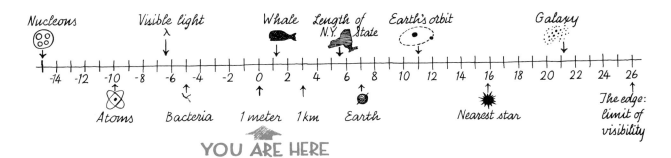

In fact, that's where we all are. The chart covers the size of everything that exists, from the smallest subatomic distance that has been measured, to the very edge of the universe. We stake out this entire realm as the subject of our study. Humans, living in the middle range of sizes, have discovered regularities and laws that link the behavior of atoms and the nature of galaxies. In probing both the microworld and the macroworld, we also learn more about ourselves.

SIZES AND DISTANCES

In order to map the whole universe and put it on a single page, we had to use a power-of-10 scale. As you go from left to right, each unit is ten times larger than the one before it. The basic unit of length in this map is the meter (m), a little longer than the English yard. The meter is shown at point 0 on the scale, since 10^0 is equal to 1. Another way to think of the scale markings is to visualize them as being on a logarithmic scale. "One meter" is at the 0 point, since log 1 is equal to 0. Take a look at some of the other familiar points on the map. A kilometer (about $\frac{5}{8}$ miles) is noted at point 3. A kilometer (km) is 1000 meters; the log of 1000 is 3, and 10^3 is 1000. Similarly, a millimeter is at the point -3 since 10^{-3} equals $(\frac{1}{10})^3$.

Question 1-1

We interrupt the text from time to time to challenge or emphasize what has just been said. The questions we raise will not always

> have definite answers. You can find our answer or opinion at the end of each chapter, but you should wrestle with each question for a while before turning to our solution.
>
> The question at this point concerns the location of humans on the map of the universe. They are placed very close to zero. Does that make them only a little taller than 1 meter?

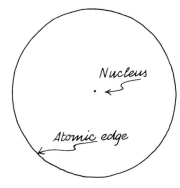

For an actual scale drawing, the atomic diameter would have to be at least 10,000 times the nuclear diameter. If the dot representing the nucleus is 0.1 mm, the atomic sphere should be 1 m in diameter.

Living creatures occupy a very small region of this map. The largest whale is no more than 30 meters long, and so is pictured at a point less than 2 on the map. Biological cells can be seen with microscopes and are larger than a millionth of a meter, shown on the scale at -6. (Since 10^6 is 1 million, 10^{-6} is $1/1,000,000$, or $0.000,001$, or 1 millionth.)

Marching down the scale into the microworld, we find atoms at -10. Most atoms are about the same size. From hydrogen to uranium, few of them differ by more than a factor of two in diameter. Note that if each atom is 10^{-10} meters across, you could line up 10^{10} of them, shoulder to shoulder, in a distance of 1 meter. Since 10^9, a thousand million, is one billion, that would be 10 billion atoms. The atom itself is huge compared to its tiny nucleus. Depending on the element, the nuclear diameter is smaller than the atomic diameter by a factor of 10,000 to 100,000. That would put the nucleus between -14 and -15 on the universe map. It is possible to measure quite accurately the diameters of the protons and neutrons that make up the atomic nucleus. As we will see in a later chapter, sizes much smaller than this are not meaningful in current research. We put the lower limit of our universe at 10^{-15} meters.

Does it make sense to talk about sizes this small? Only if we can measure them in some way. With your unaided eyes, using visible light, you can see the shapes of objects that are one millimeter (1 mm) across. A magnifying glass can provide a magnification of up to about 10, and a child's microscope may have a magnification of 100. Even the best research microscopes do not have magnification greater than 1000 to 2000. Since your unaided eye can see an object that has a size of 1 mm, with a research microscope it can see objects that are $\frac{1}{1000}$ of 1 mm, or 10^{-6} meters, which is called a micrometer. (The older name is micron.) The limit of magnification is caused by the fact that light has a wavelength, or a size, of its own. You cannot "see" something if the probe is larger than the detail you are trying to see. Electron microscopes use electrons as probes. These can be smaller than the wavelengths of visible light. Magnification by a factor of over 10^5 has been achieved, making it possible to see objects that are only 10^{-8} meters across.

For sizes below the electron microscope range, we have to scatter electrons or other subatomic particles and analyze their scattering patterns. We'll examine these methods in later chapters. Note, however, that in seeing anything, we shoot or probe with something, and then detect what happens to the probe. As you look at this book, for instance, light is being scattered from the print and the page. A small fraction of the reflected light enters your eye, where the information is processed, sent on to the brain, and recognized. In much the same way, an airplane can be seen by scattering

SIZES AND DISTANCES

radar waves from it, and detecting the waves that are reflected to the receiver. You could not read this book by scattering radar waves off it, however. The radar waves are about the size of the book itself, and so are too broad a probe to detect the printed details.

If it only makes sense to talk about sizes that we can measure, how can we justify using some of those distances on the right side of the universe map? You can measure a whale, or even a country, by fairly straightforward methods. Surveying is still done by measuring baselines and angles. The length of the meter was originally calibrated in a complicated and rather misguided surveying project. At the time of the French revolution in 1791, the French National Assembly voted to make the unit of length equal to 1/10,000,000 of a quadrant of the earth's surface. The only reason for prescribing such an exact ratio for an arbitrary standard was so that the standard could always be reproduced—presumably the earth would never change. A surveying party actually measured the earth's arc all the way from Dunkerque on the English channel to Mont-Juoy near Barcelona, Spain.

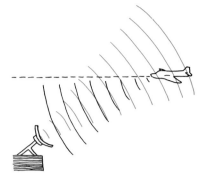

A radar scanner sees an airplane. The probes are electromagnetic waves much longer than those of visible light.

> ### Question 1-2
> What good did that measurement do them? Even if you know the length of France, how can you tell what fraction that is of the earth's circumference?

The trouble with using surveying techniques is that you need a measured baseline. To measure the distance to the moon, the largest baseline available is the diameter of the earth itself. Actually, a much smaller distance must be used since there are problems if telescopes are sighted too close to the horizon. Suppose you use two observatories 1000 km apart (about 630 miles). Both telescopes have to aim at the same point on the moon, and *at the same time*. (The moon is a moving object.) The geometry of the situation is shown in the diagram. By measuring the angle between the vertical and the moon at each location, the subtended angle can be found. That angle is approximately equal to the distance between the two telescopes divided by the distance to the moon.

$$\theta = (\text{baseline})/(\text{distance}) = 1000 \text{ km}/384{,}000 \text{ km} = 1/384 \text{ radians}$$

$$= \frac{1}{384} \text{ radians} \left(\frac{57°}{\text{radian}}\right) = 0.15 \text{ degree of arc}$$

An angle of only 0.15 degree is small, but easy enough for one telescope to measure. It's a little more difficult to determine that small an angle by having two telescopes synchronize their observations and subtract individual measurement.

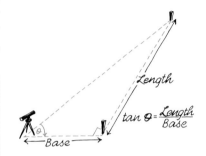

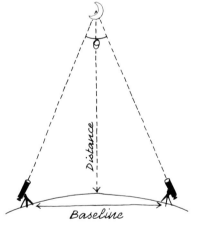

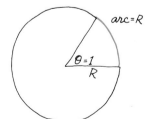

θ = arc length/radius When arc length equals R, θ equals 1 radian. There must be 2π radians in a circle or in 360°. Therefore, 1 radian = 57°.

> ### Question 1-3
> Isn't it necessary to use trigonometry to solve this problem? The distance between moon and earth is the long leg of a right triangle.

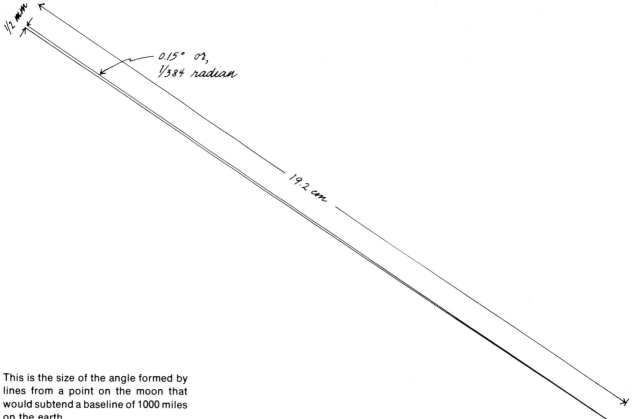

This is the size of the angle formed by lines from a point on the moon that would subtend a baseline of 1000 miles on the earth.

Half the baseline is the short leg. Therefore, the tangent of one half the subtended angle is equal to the ratio of one half the baseline divided by the moon-earth distance. Isn't this method right and the other wrong?

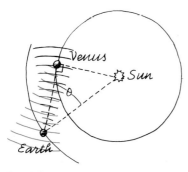

Earth-Sun distance can be measured by measuring Earth-Venus distance with radar, and knowing the angular relationships at that time of the Earth-Venus-Sun triangle.

The distance to the sun is 1.5×10^{11} m. If a baseline on the earth of 1000 km (1×10^6 m) were used, the subtended angle would be only 6×10^{-6} radians, or about one second of arc. That's too small an angle to measure for a wide target like the sun. The earth-solar distance is actually measured by a comparison of other solar system distances and angles. The earth-moon and earth-Venus distances are measured these days by timing the flight of radar pulses from earth to object and back again. Since the velocity of light—which is the same as that of radar—is known very precisely, the distance measurement can be as precise as the measurement of time between sending the pulse and receiving the echo.

The distance to the near stars is also determined by surveying techniques. In this case the baseline is unearthly! It is the diameter of the earth's orbit around the sun. Even though that baseline is 3×10^{11} m long, the distance to the nearest star makes the angular measurement tricky.

SIZES AND DISTANCES

> **Question 1-4**
>
> If the distance to the nearest star is 3.8×10^{16} m, what angle is subtended by the baseline?

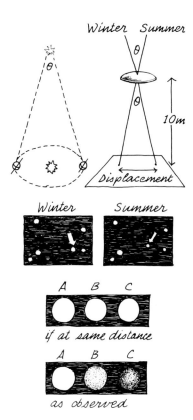

To make measurements using the orbit baseline, the same telescope photographs a star field at six-month intervals. When the two photographs are superimposed, most of the star images lie exactly on top of each other. However, the fall and spring images of a very near star will be slightly displaced from each other. The ratio of that slight displacement to the focal length of the telescope is the same angle as the ratio of baseline to stellar distance. For one second of arc and a focal length of 10 m, the image displacement on the photographic plate is found as follows:

$$\theta = 1 \text{ sec} \approx 5 \times 10^{-6} \text{ rad} = (\text{displacement})/(10 \text{ m})$$

The image displacement is about 50×10^{-6} m, or 50 micrometers. The distance to the stars must be measured with a microscope!

The nearest star to our sun is about four light-years away. A light-year is a unit of distance, not time. It's the distance that light travels in one year and is equal to 9.56×10^{15} m. (In the same way, we might say that the moon is 1.3 light-*seconds* from the earth.) Triangulation, or surveying techniques, can be used to measure distances as great as one hundred light-years or so. However, this is hardly a step out into our galaxy, which has a diameter of 100,000 light-years. To measure those distances, we must appeal to conclusions that follow from a whole set of observations about certain types of stars whose brightness varies in a periodic fashion. The argument goes like this: if you know how bright a star really is, you can tell how far away it is by measuring the brightness you see here on earth. For instance, suppose that all stars were really the same brightness but that we observe star A to be four times brighter than star B and nine times brighter than star C. Then B would be twice as far away as A, and C would be three times as far. The intensity of light from a point source falls off as the *square* of the distance. If you can measure the distance to A by triangulation, you can calculate the distances to B and C. Of course, stars are *not* all the same size and brightness. There are relationships, however, between the intrinsic brightness of certain stars and such characteristics as their color and frequency of changing brightness.

Relative brightness methods are also used to measure distances to the near galaxies. These galaxies are island universes, each containing millions or billions of stars. Our own galaxy contains about ten billion (10^{10}) stars, and looks something like our near neighbor, the galaxy that can just be seen with the naked eye in the Andromeda constellation. That galaxy, as you can see in the illustration, has a pinwheel shape—just like our own galaxy. Galaxies exist in a variety of forms, including spherical. Our solar system is about two-thirds of the way out on one of the arms of our galaxy. When we look toward the hub we see a high concentration of stars that appear in

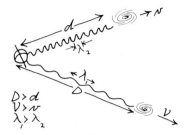

The galaxy at distance D is much farther from the earth than the galaxy at distance d. Because the universe is expanding, the velocity away from us of the far galaxy is greater. Therefore, the light of a particular element in that galaxy has a longer wavelength.

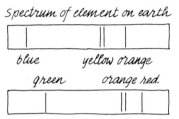

If a galaxy is traveling away from us, the pattern of its spectral lines shifts toward the red.

the night sky as the Milky Way. From detailed counts of small samples of the sky, we know that there are about ten billion (10^{10}) galaxies in the universe.

There is another method of measuring distances to the far galaxies. The velocity of a galaxy with respect to us can be measured by its Doppler shift. If a source of waves is moving toward an observer, the frequency of the waves seems higher. A train whistle, for instance, has a higher pitch as the train approaches a listener. As the train passes and races away, the pitch abruptly lowers. In the same way the characteristic patterns of light produced by various elements in distant galaxies are seen on earth to be shifted to lower frequencies. The amount of this "red shift" to long wavelengths tells us how fast the galaxy is fleeing from us. It turns out that all the galaxies are flying apart from each other. The universe is expanding. Furthermore, the further away the galaxy, the faster its speed away from us. Since this rule is followed by the close galaxies whose distances we can measure by brightness methods, we use the rule to find distances of the far galaxies by measuring their red shift velocities. In this way galaxies have been observed that are ten billion light-years away from us. They are traveling at speeds over half that of light, and are close to the edge of our visible universe.

If, as seems likely, the universe started in an immense explosion about fifteen billion years ago, and has been expanding ever since, there is, indeed, a limit to the universe. Many of the details of the origin and of the far reaches of the universe are still shrouded in mystery. Regardless of the details, however, there is a limit to *our* universe at about 10^{26} meters. Beyond that distance, any object would be traveling away from us so fast that any information sent to us (such as the characteristic patterns of the light produced by different elements) would be red-shifted into the background noise of low energy radiation. Blue light, for instance, would not just shift to red light, but might have the much longer wavelength of radio waves.

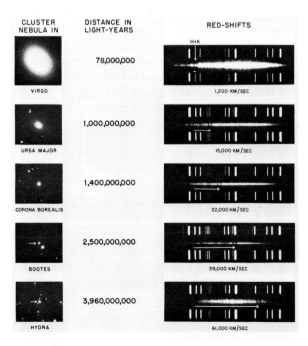

Note the horizontal white arrow drawn on each spectrum, showing the position of one prominent spectral line, appearing here just as a dot. The lines above and below the galaxy's spectrum show an earth spectrum.

> **Question 1-5**
>
> How long would it take something traveling at the speed of light (3×10^8 m/sec) to go 1×10^{26} meters?

HANDLING THE PHENOMENA—DISTANCES

Physics is not math or logic. All our theories and models are useful only if they predict real events in the real world. Futhermore, understanding physics is not just an intellectual exercise. The phenomena must be handled. When you talk about velocity, you should have some familiarity with various velocities either by experiencing them or by measuring them. When you talk about forces, you should feel them in your muscles. Such a goal is limited, of course. In our study we want to go far beyond the normal human experience, both to the very large and the very small. Nevertheless, there are benchmarks along the way that can be experienced and comprehended.

In every chapter we propose ways to handle some of the phenomena described in that chapter. Sometimes this will be done through the use of models or analogies. These are not laboratory exercises, but are simple things to do using materials available at home or in the dormitory. The precision required and the time to be spent are usually very small.

How on earth can you experience the beginning topic of this chapter? Perhaps you could build your own universe! Better yet, why not construct a scale map of something close to home, such as the solar system? The map of the universe that we presented was built on a logarithmic scale. To better understand the nature of that scale, build your solar system on a *linear* scale. That's the kind used in ordinary road maps. A state map, for instance, might use a scale of 20 miles to the inch (or 10 km to the cm). We list below the distances of the planets from the sun, and also the diameters of the planets and sun. Choose a scale so that Pluto can be included in whatever room or hall you use, but also a scale so that the sun is more than a point. Note the problem of trying to use a scale that represents the size of each planet and also fits the whole system into a reasonable indoor space. If the scale is such that the earth diameter is 1 cm, the earth-sun distance will be over 100 m. One suggestion, which still requires a corridor or good-sized room, is to use a scale of 1 cm to 5,000,000 km. The model of the sun itself will then be small, but more than a point. The other planets will be just points, but their locations can be marked with arrows on paper tabs. If you want a solar system that you can roll up and put in your pocket, make the scale model on a long paper tape, such as a cashier's tape. The diagram shows how to make a marker tab that will fold down when rolled.

Why go to the trouble of making such a map? Perhaps you will see something about the nature of the solar system that is not apparent from the

Planet	Average Distance from Sun (in 10^6 km)	Mean Diameter (in km)
Mercury	57.9	4840
Venus	108.1	12520
Earth	149.5	12740
Mars	225.8	6780
Jupiter	777.8	139800
Saturn	1426	115000
Uranus	2868	47400
Neptune	4494	43000
Pluto	5908	5800
Sun	—	1393000

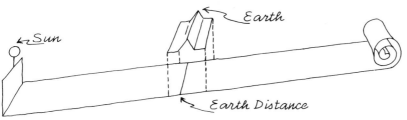

table of numbers. As you look out at the night sky and see Venus, Mars, Jupiter, and Saturn, it's hard to imagine where they are in relationship to the sun and to us. With the model, most people are impressed to see how close the inner planets are compared with the ranges of the outer planets. It's also surprising to see on this scale how such a small sun can control planets as far away as Neptune and Pluto.

Of course, we can't promise that any particular activity or view is going to intrigue some other particular person. Some people do not appreciate sunsets; some people don't like violin music. We each have our problems. At any rate, there's something else to be learned from the linear scale model. Observe how difficult it is on a linear scale to represent a large range of values. With the suggested scale, the planets shrink to points. Yet on this scale the next closest star would be at a distance of about 100 km, about 60 miles away.

TIMES

[Logarithmic time scale diagram with labels: Nuclear Interactions (−23); Lifetime of "strange" particles (−10); Period of A.C. (−2); Human heart beat (0, 1 sec); 1 day (5); Human life (7, 1 year); Human existence (9); (13); (15); Age of Earth (17, Age of Universe)]

Age of universe	5×10^{17} s
Age of Earth	1.5×10^{17} s
Earliest life	7.5×10^{16} s
Reptiles	7.5×10^{15} s
Mammals	5×10^{15} s
Humans	1×10^{14} s
One Year	3×10^{7} s

You've seen where we are in space; here's where we are in time. Our human lives are an insignificant dot in the history of the universe. Yet we are able to measure processes that take billions of years, and we can also detect events that occur in the time that it takes light to cross an atomic nucleus. As with distances, our increasing knowledge of events that take place in very short times enables us to measure the distant past.

The unit of our logarithmic map of time is the second, which is the same in the English system of units as in the metric. Like the other basic units, it is of human size. The period of a heartbeat is about one second. On the time map one second is shown at 0.

We all know what time is, yet there are some peculiar problems in setting up a standard unit of time. In the nineteenth century, the meter was defined, for practical purposes, as the distance between two scribe marks on a platinum-iridium bar. The bar, made of this very stable alloy, was kept in a temperature-controlled vault. Any other bar could be put next to it and compared. However, how do you take a standard second of time and put it in a velvet-lined box? The best you can do is define and preserve some system that performs some event in the unit time. Then you must make sure, or have a theoretical basis for assuming, that the system will always work in the same way that it did originally.

You could choose the time between your pulse beats, for instance, as the standard unit of time. As you can tell with a watch, however, your pulse rate is different when you are running from when you are sitting. Then why not use the watch as the standard? Even good watches have to be reset occasionally, usually with a radio time signal. What clock does the radio station set its clock against?

> **Question 1-6**
> What are some natural systems that might be relied on always to take the same time for the same event?

Time is involved in two kinds of events: repetitive ones like the swing of a pendulum, or processes of continuous change, such as radioactive decay or motions of objects. Whatever process is chosen to set the standard, the unit of time must keep pace with other repetitive processes and also produce simple and consistent relationships for motion. For instance, we have good reason for thinking that the speed of light (in a vacuum) is a universal constant. Velocity measurements usually involve time measurements (so many meters *per second*). If, in the future, a measurement of the speed of light produced a value different from normal, we would suspect that something had happened to our standard of time. In other words, we judge the reliability of any of our standards in terms of their mutual consistency.

Until 1964 the international unit of time was based on the repetitive rotation of the earth. Unfortunately for this purpose, the length of the day has all sorts of variations, some of them large enough to be detected with an ordinary electric clock. The length of the day depends on our position in orbit around the sun. That orbit is almost a circle, but is slightly elliptical. The earth also has a number of short-term and long-term wobbles as it rotates on its axis. Furthermore, the earth's rotation is gradually slowing down because of energy loss to tidal motion. To smooth out all these irregularities, an average solar day in 1900 was chosen as the standard. A solar day is about four minutes longer than a stellar day, the time that it takes the earth to turn 360°. To line up with the sun again, the earth must turn an extra degree each day, since it is moving around the sun about one degree per day (360° in 365 days).

In the United States, astronomers at the Naval Observatory are responsible for timing the passage of certain stars across the meridian (center line of the sky running from north to south) and thus determining sidereal time. They then make corrections for various earth motions and compute solar time accurate to one-thousandth of a second. Time signals accurate to 0.001 second are broadcast from several stations. Through satellite relays, clocks have been calibrated with those in other countries to within one-tenth microsecond (10^{-7} s).

Atoms and molecules are also systems that maintain particular frequencies. They can emit or absorb light of only certain sharply defined frequencies. After World War II, a number of devices were made that maintain resonance between electromagnetic radiation at radar frequencies and beams of isolated atoms or molecules. The constant frequency of the atomic vibrations keeps the radar signal frequency constant. As long as a constant frequency can be maintained and counted, the system acts like a

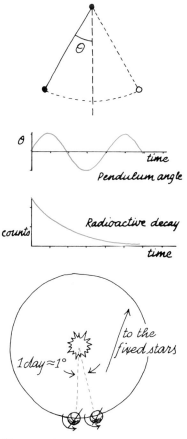

After rotation of 360°, one sidereal (stellar) day has passed. Solar day is slightly longer.

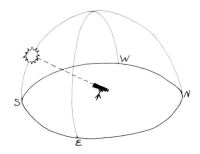

clock. Since 1964 the international standard of time has been defined in terms of the constant frequency produced by an atomic clock using the element cesium. The second is officially the time that it takes for 9,192,631,770 oscillations of the cesium signal. Atomic clocks of various types have been compared with each other for long periods of time and are precise (that is, they agree) to within one part in 10^{11}. That precision is over 100 times better than could be maintained with astronomical methods.

> Question 1-7
>
> In an Alpine village there was a clockmaker's shop with a sign reading, "This clock keeps perfect time." When a visitor asked the clockmaker how he knew this was true, the clockmaker told him that the clock was calibrated every day against the bell from the monastery observatory. When the visitor visited the monastery he asked to see how the monks determined the exact time from their observations of stars. "Oh," the monks said, "we no longer have to do that because we can now coordinate our clocks with one down in the village which keeps perfect time."
> If time itself were slowing down, affecting all repetitive motions equally, would we ever be able to detect the effect?

Our perception of time in everyday life seldom extends lower than 0.1 second. Human response times of hands or feet are about $\frac{1}{5}$ of a second. From personal experience we have a feeling for the length of time of a day, a year, and even a lifetime.

> Question 1-8
>
> Note on the time map the approximate number of seconds in a year or a human lifetime (taken to be threescore years and ten). Check the accuracy of these numbers using *order of magnitude* calculations. For instance, in one day there are
>
> $$\frac{60 \text{ seconds}}{\text{minute}} \times \frac{60 \text{ minutes}}{\text{hour}} \times \frac{24 \text{ hours}}{\text{day}}$$
>
> Note that the units of minutes and hours cancel out, leaving some number of *seconds*. The number is $60 \times 60 \times 24$. Another way of writing this is
>
> $$(6.0 \times 10^1) \times (6.0 \times 10^1) \times (2.4 \times 10^1) = 6 \times 6 \times 2.4 \times 10^3$$
>
> To one significant figure, the arithmetic can now be done in your head: $6 \times 6 = 36$. That has to be multiplied by 2.4, which is about $2\frac{1}{2}$: 2×36 is 72. Add on another half of 36, which is 18, and you get 90, which is close to 100. To one significant figure, there are 100

Translation of Units
Always express units in words or symbols. It is meaningless to say that your speed is 60, but your speed may be 60 mph, which can be written 60 miles/hr.

To translate units, always multiply by a fraction equal to one. For instance, there are 5280 feet in one mile; therefore, for this purpose,

$$\frac{5280 \text{ ft}}{1 \text{ mi}} = 1$$

To change mph to ft/s:

$$\frac{60 \text{ mi}}{\text{hr}} \times \left(\frac{5280 \text{ ft}}{1 \text{ mi}}\right) \times \left(\frac{1 \text{ hr}}{3600 \text{ s}}\right)$$
$$= \frac{5280 \text{ ft}}{60 \text{ s}} = 88 \text{ ft/s}$$

× 10³, or 1 × 10⁵ seconds in one day. Of course, if you need more precision, you can use a slide rule or calculator or even multiply the figures long hand. However, it's always a good idea to zero in on a numerical problem by getting the order-of-magnitude solution. Do this for the numbers of seconds in a year and in a lifetime.

Several frequencies used in everyday life are familiar to us in one way or another. The standard frequency for household electricity is 60 cycles per second. This means that in $\frac{1}{60}$ of a second, one of the prongs in an electric outlet goes from maximum positive to maximum negative and back to maximum positive again. The voltage goes through zero every $\frac{1}{120}$ second. A 60 cycle hum is a low note that sometimes can be heard in hi-fi sound systems.

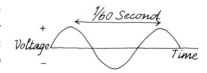

A higher note with which orchestra players are familiar is 440 cycles per second, or middle A, used for tuning. The vibrating system that produces the note may be a double reed (oboe), or string (violin), or lips (trumpet). Ordinary AM radio has frequencies from 540 kilocycles per second to 1480 kilocycles per second. (Cycles per second is now given the name *hertz*, with the symbol Hz, and so AM radio extends from 0.54×10^6 Hz to 1.48×10^6 Hz.) If your radio is tuned to the middle of the dial, the electric charge in the circuits is vibrating at 10^6 Hz, which means that it goes through a full cycle, or one period, in one microsecond (1×10^{-6} s). Television signals are higher in frequency by a factor of 100, and so the period of their oscillation is 10^{-8} s. Times that short, incidentally, can be seen on oscilloscopes. The radar signals used in the cesium atomic clock have a frequency—by defintion—of 9,192,631,770 Hz. The period in seconds can be found simply by taking the reciprocal of the frequency.

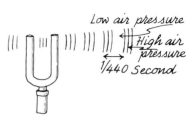

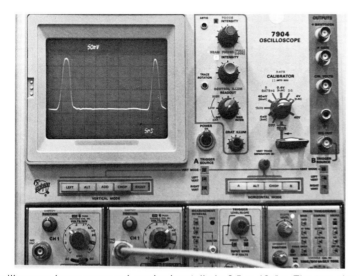

The oscilloscope beam sweeps 1 cm horizontally in 0.5×10^{-8} s. The two signals are 4×10^{-8} s apart.

> **Question 1-9**
>
> What is the period of the cesium clock signal? There is no need to calculate it more precisely than to one or two significant figures.

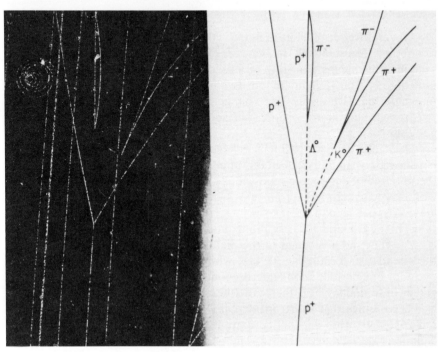

This is a bubble chamber picture of tracks of a proton-proton collision. Two neutral particles (Λ_0 and K_0) were produced, flew forward almost at the speed of light, and then decayed. Their decay times of about 10^{-10} s can be measured by measuring the decay distances: $t = x/v$

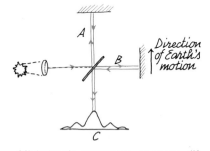

Michelson's attempt to measure difference in speed of light, parallel and perpendicular to Earth's rotation. Light coming from lens went through half-silvered mirror. Half went by A; half by B. After reflection from mirrors, light met at C. If light had been out of step at C by $\frac{1}{10}$ wavelength, Michelson could have seen the effect.

Processes other than periodic ones are used for timing purposes, in all ranges. If you drive on an expressway at a steady 60 mph (the speed limit of 55 plus 10% grace), you may well measure your progress in hours instead of miles. The 120 miles between A and B will take you 2 hours. Similarly, if you know that a subatomic particle is traveling close to the speed of light (3×10^8 m/s), and goes 6 cm from the point where it was produced to the point where it decays, it must have lived for 2×10^{-10} s.

$$\text{Lifetime} = \frac{0.06 \text{m}}{3 \times 10^8 \text{ m/s}} = \frac{6 \times 10^{-2} \text{m}}{3 \times 10^8 \text{m/s}} = 2 \times 10^{-10} \text{s}$$

It may seem as if times of millionths or billionths of seconds are beyond comprehension, and that their measurement can be done only with ultra-modern apparatus. Actually, lots of engineers and technicians in the electronics industries routinely make measurements in these ranges. Comprehension is easy and comes with familiarity. Way back in the 1890s, in

Cleveland, Ohio, a physicist named Michelson was comparing the flight times of light over two perpendicular paths. He could detect differences of these flight times to about $\frac{1}{10}$ the period of visible light vibrations—to about 10^{-16} s.

The shortest time interval on our chart of times is 10^{-23} s. Times shorter than this are part of theories, but this is the shortest time showing up in reasonably direct experimental results. It is the time taken for light, or a subatomic particle traveling at close to the speed of light, to travel across the diameter of a proton or neutron.

> **Question 1-10**
>
> Check that time. The diameter of a proton is about 2×10^{-15} m.

Since World War II there has been a great revolution in our ability to measure long time intervals. The prime new method available (although there are others) is the use of various radioactive decay schemes. Actual historical records can be pieced together extending back about 5000 years, to the days of the early pyramid builders in Egypt. Even within this period, many archaeological findings have been dated by determining the amount of radioactive carbon still in the material. We will study the details of this method in a later chapter. The general idea is this: living things take in carbon dioxide from the air. A small but definite fraction of the carbon is radioactive, and so any sample of tissue or wood or cloth made of materials that were recently alive has that same fraction of radioactive carbon. In the right type of detector, fresh materials produce 16 counts per minute per gram of carbon. In 5600 years, the counts will be down to 8 counts per minute per gram, and in another 5600 years the counting rate will be down to 4. By analyzing the counting rate from a baked sample of the wood from some ancient boat (to reduce the organic material to pure carbon), the number of years can be computed back to the time when the wood was a living tree and still absorbing and giving off carbon dioxide. Because eventually the counting rate gets too small to measure accurately, the carbon dating method is limited to about 25,000 years.

Other dating schemes, many of them depending on very complex geological sequences, have been used to push back the date for the origin of humanlike creatures to several million years ago. The ages of rocks and crustal features of the earth are determined by a variety of radioactivity measurements. For example, take a look at the list of naturally occurring radioactive materials and their half-lives. (Notice that carbon is not among them. Radioactive carbon, with its relatively short half-life, is continuously created by cosmic radiation in the upper atmosphere.) The radioactive elements listed are the ones left over from the origin of the earth. In a time of one half-life, half of any such element is transformed to a different element. In another half-life, half of the remaining amount is transformed, leaving only one-quarter of the original.

Fresh wood yields counting rate of 16 counts/min/g of carbon. Wood from funeral boat from pyramid yields rate of 8. Wood from cave fire yields rate of 4.

Naturally Occurring Radioactive Isotopes (that are not products of other decays)

Element	Half-life in Years
$_{92}U^{238}$	4.5×10^9
$_{92}U^{235}$	7.1×10^8
$_{90}Th^{232}$	1.4×10^{10}
$_{71}Lu^{176}$	1.0×10^{10}
$_{37}Rb^{87}$	5.0×10^{10}
$_{19}K^{40}$	1.3×10^9

> **Question 1-11**
>
> At that rate, it looks as if we could never get rid of radioactive material. What fraction is left after 5 half-lives? After 10 half-lives?

The shortest half-life of those radioactive elements that still exist on the earth is 7×10^8 years, for the form of uranium known as 235. That's a half-life of almost one billion years, and yet U^{235} is very scarce compared with U^{238}, which has a half-life six times as long. Why aren't there any elements with half-lives of 10^6 or 10^7 or 10^8 years? Such forms exist and can be made with atom smashers. They must also have been created when the other elements in our solar system were formed. If our solar system were only 10^6 or 10^7 or 10^8 years old, those radioactive elements with comparable half-lives would still be here. Since they aren't, we conclude that the solar system is older than 10^8 years. In fact, a detailed analysis of the ratios of the amounts of the existing elements (amount of U^{238} to amount of U^{235}, for instance), leads to an age for the solar system of about 5×10^9 years.

Our own solar system is at least a second generation star. All the present planets and sun—and humans—were once part of another star that exploded violently. The scattered dust from that star and others eventually swirled together, attracted by mutual gravitational pull, and started over again. This cycle has not been going on forever. For one thing, our own star is probably too small to go through an explosive end. In another five billion years or so it will expand greatly, consuming the inner planets, but then contract and turn eventually into a small star, gradually cooling off.

The universe itself is only two or three times the age of the sun. Everything—stars, galaxy, and space itself—started as a highly concentrated, very energetic collection of something. We don't know where it came from, or how it started, but the system exploded, creating our present subatomic particles and sending them out to gather into the galaxies and stars of our present universe. That was about fifteen billion years ago and the system has been expanding ever since. Earlier we described how the velocities of the galaxies could be measured in terms of the red shift of their characteristic light. Each galaxy is traveling with the right velocity to have reached the place where it is in a travel time of 15 billion years. There are many other pieces of evidence that lead to the same conclusion. On our time map, the age of the universe corresponds to 5×10^{17} s.

HANDLING THE PHENOMENA—TIMES

In describing times shorter than one second, we claimed that human response times for hands or legs are only about $\frac{1}{5}$ second. Are yours any better? Here's a simple way to measure hand response time. Have someone else hold a meter stick or a ruler, vertically, with your thumb and index finger poised to clamp the stick as soon as it drops. Start out with your fingers over a particular mark on the stick, and then see how many centimeters the stick drops before your fingers grab it. The graph shows the distance dropped

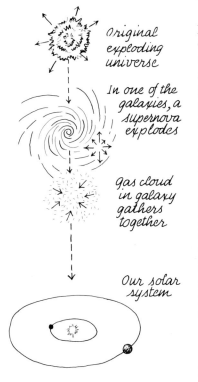

Original exploding universe

In one of the galaxies, a supernova explodes

gas cloud in galaxy gathers together

Our solar system

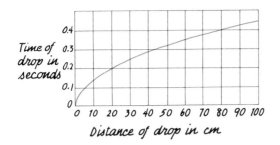

as a function of the time of dropping. See if you can get down to a response time of $\frac{1}{10}$ second.

On page 8, we used a logarithmic scale for time because of the vast range. Make a *linear* scale of the history of the universe by using this book to represent that whole history. Let the total number of pages represent 5×10^{17} seconds. Put paper markers in to represent great moments in the universe. Halfway through, for instance, you should put a marker labeled, "Solar system formed." Note such events as the beginning of life on earth, and the earliest reptiles, mammals, and humans. Make special note of the page and the place on the page where the written history of humans appears, about five thousand years ago.

INTERACTIONS AND CHANGE

So far we have described our universe in space and in time. If these realms and events were all known and static, models of them could be pinned in exhibit cases for our study and marvel. It would be a fantastic museum, but, finally, dull. That's not what the universe and its study are like, of course. The essence is not the length and breadth and age, but rather the complex interactions of all the parts and the way they change.

We know the world only through interactions with it. When we see an object our eyes receive light that has scattered off its surface. There is an interaction of the light and object that causes certain colors to be absorbed and others to be reflected. Within our eyes there are interactions that turn the light into electrical pulses in the optic nerves. A completely isolated object, with no interactions possible, is unknowable.

> ### Question 1-12
>
> Interactions between the "knower" and the object are not instantaneous. After the light and the object interact, it may take a while for the light to get to our eyes. The light is a messenger containing the information. If it takes four years for light to reach us from the nearest next star, how do we know at any given time that the star now exists?

The messengers that bring us information are the primitive ones of sight, sound, taste, touch, and smell. Actually, these are all basically electromagnetic, though in very complicated forms. In later chapters we will see that electric forces and magnetic forces are just different aspects of the same phenomenon. Furthermore, visible light is electromagnetic radiation of a particular range of frequencies, and the atomic forces (though not the forces of the atomic nucleus) are electromagnetic. But if the atom is held together with electromagnetism, and the same interaction is responsible for all of molecular binding, then all chemical and biological effects are electromagnetic in nature. The electromagnetic interaction between two objects is a function of their electric charge distributions, the distance between them, and their relative motion.

If electromagnetism accounts for the common messengers to which the human body responds, is it the only interaction between objects? No, so far as we know now, gravity is a separate form of interaction. Electromagnetism dominates in the atomic and molecular realms, and in the exchange of radiant energy, but on the cosmic scale it is the gravitational interaction that holds the planets in their courses and controls the motions of galaxies. Within the nuclei of atoms, two other interactions are important. We will study the Strong and Weak nuclear interactions later. They are responsible both for the stability of nuclei and for their transformations.

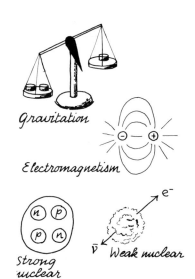

Question 1-13

We have named only four kinds of interactions. How simple this universe is! Can you think of an interaction that is not one of these four?

We keep speaking of interactions, instead of forces. Doesn't one object interact with another by exerting a force on it? That's frequently a useful way to look at the way things happen, but there is another way. It involves looking at objects as a whole system before and after some change. For instance, suppose that you hold a stone high in the air. Everyone knows that there is a force on it that will pull it to the earth as soon as you let go. We can describe that force in terms of the mass of the stone and the size of the earth, and we are going to learn how to describe the subsequent motion as a function of that force. From another point of view, the earth and stone (and you) are part of a system. To begin with, the stone and earth are separated. Afterwards, the stone and earth are together, some sound was created in the air, the earth is dented, and to a very slight extent the earth and stone are hotter. The original state was transformed into a final state. The interaction between these two situations, through which the first was transformed to the second, was gravitational to begin with. After the stone struck the ground, the electromagnetic interaction was responsible for the transformations that produce the sound, the denting, and the rise in temperature.

Question 1-14

A firecracker explodes. How would you describe the initial and final

INTERACTIONS AND CHANGE

> states, and the interaction by which one was transformed into the other?

So many different kinds of changes can take place during an interaction that it might seem hopeless to derive general rules. During the explosion of the firecracker, its color changed, its shape changed, and the location of its parts changed. There is another approach to studying interactions. Instead of asking, "What changes?," find out what stays the same. That problem may also seem hopeless. So many things change when a firecracker explodes; has anything stayed the same? As a matter of fact, in every interaction a number of properties cannot change—they are *conserved*. These conservation laws greatly restrict the number of possibilities for change. The conserved properties include energy, momentum, rotational momentum, and electric charge. We will study each of these, and their conservation laws, in detail. In the meantime, note that you would not expect all the pieces of the firecracker to fly off in the same direction. A change to that final state cannot take place because of the conservation of momentum. You would not expect to see the powder ooze quietly to ash and gas without the noise and scattering of the casing. The chemical energy of the initial state is transformed without loss into sound, light, and motion energy. The total energy is conserved.

There is another virtue in thinking of events as transformations between states of a system. If there is an interaction between some initial state and a variety of final states, the event will take place in such a way as to increase the probability of the situation. For instance, why couldn't we reverse the direction of all the pieces flying out of the exploding firecracker, and put the original system back together again? Momentum and energy could be conserved. It would require more than good aim, however. All the heating effects would have to be reversed, too, so that the thermal energy as well as the motion energy would be recaptured into the compact form of chemical energy. This is so highly unlikely that it just never happens. Highly concentrated forms of energy or complicated arrangements of materials tend to become dissipated, disorganized.

> ### Question 1-15
> The same effect is apparent in parking a car. It is easier to get out of a parking space than to get into it. In terms of the probabilities of being in one state or another—parked or unparked—why should this be?

There are ways to reverse this natural tendency for order to go to disorder. When water molecules form ice they must fit together in very definite positions in a crystal. It is an ordered, unlikely situation. There are far more ways in which water molecules can arrange themselves in a liquid. Nevertheless, energy would be conserved if all the molecules in a glass of warm water were suddenly to arrange themselves in crystalline positions and give off their surplus energy to the walls of the glass, making it warmer. That never happens all by itself, of course. However, something like this happens in your refrigerator all the time. The extra energy of the liquid water is extracted and thrown out the back end of the refrigerator. Heat runs up a temperature hill! The chaotic situation of the water molecules becomes ordered. A price is paid, however. This interaction takes place only as long as the refrigerator is plugged in, and the electric bill is paid.

A similar transformation from disorder to order takes place with living things. The DNA molecules in our cells busily fashion ordered, complicated molecules out of randomly arranged atoms and smaller molecules. Once again, it takes energy to perform this defeat of probability. In the case of both the refrigerator and the living cells, if you include the source of energy, it is still true that the total system goes from a less probable to a more probable situation. The electricity to drive the refrigerator came from a generator where highly concentrated chemical energy turned into dissipated thermal energy, a large fraction of which had to be thrown away into the cooling water of the system. In the case of living cells, their energy ultimately came from light from the sun, a highly concentrated form of energy being generated at the expense of less probable arrangements of atomic nuclei transforming into more probable combinations. The universe as a whole is becoming more likely.

> ### Question 1-16
>
> What difference does it make if a power plant throws away a large fraction of its chemical (oil or coal) energy in the process of making electricity? After all, energy is conserved and so it isn't lost.

FUNCTIONAL DEPENDENCE

Every variable is a function of other variables. As *they* change, *it* changes. You are a function of many variables, one of which is time. As the years roll by, you change. You are also a function of your heredity, of the newspapers you read, and of the temperature of your surroundings. In our descriptions of people or of things, we are concerned not only with what they are but also with how they will change.

The moment-by-moment position of a stone that you drop is a function of time. If we call the distance that it has fallen, y, then the shorthand for functional dependence is $y = F(t)$ or, more simply, $y(t)$. y is a function of time, t. But y is a function of other variables, too, some of them obvious and some not so obvious. Perhaps the distance, y, depends on the mass of the stone, the location on earth, the size of the earth, the color of the stone, the initial velocity with which you launched it (up or down), the air pressure, the power of your positive thinking, and so on.

> ### Question 1-17
>
> Which of these factors are important, and which can be ruled out? Are there other variables that must be taken into account?

FUNCTIONAL DEPENDENCE

Functional dependence of one variable on another can take many forms. Suppose you drop the stone and measure the distance that it has fallen after one second and then after two seconds. The time has doubled. Has the distance doubled? Whatever the functional dependence, we can model it by graphing one variable versus the other. In many simple cases, the dependence can also be described with a simple algebraic formula. Here are some graphs showing possible dependence of the distance that the stone has fallen as a function of the time. Each graph has its own algebraic formula.

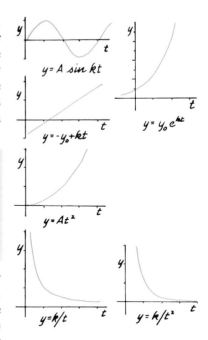

Question 1-18

The actual functional dependence, $y(t)$, for a stone starting at rest and then dropping in air may not follow any of these forms. If you are assured, however, that one of them is a good approximation, which one would you choose?

Fortunately, almost all the phenomena that we study in introductory physics can be modeled with the functions shown, plus one or two others. We can either conclude that the universe behaves in a very few simple ways, or that we choose to handle only phenomena that require simple functions. Regardless of how you look at it, it is surprising that so many interactions follow these simple rules. There are reasons for this, as we will see. Even when the simple functions do not completely describe the way one variable depends on another, they are often good first approximations. Let's take a look at an example of this situation.

When an object first starts falling in air, the distance traveled is proportional to the square of the time. That's because it is falling faster and faster, so that as time goes by it is traveling a greater distance each second. The graph of $y(t)$ is a rising curved line called a parabola, described by the formula, $y = At^2$. The value of the constant, A, depends on the units chosen, and also is a function of the other variables discussed in Question 1-17. If you drop a stone, that simple function, $y = At^2$, describes quite accurately what happens during the first few seconds. However, it does not describe what happens if you drop a feather or a piece of paper. These cases are obviously dominated by air friction. Even a falling stone's velocity is eventually affected by air friction. At first its velocity increases rapidly, but then the rate of increase gets less and less until finally the stone is traveling at a constant "terminal" velocity. Depending on the weight and shape of the stone, the terminal velocity might be about 60 m/s (130 mph). It takes only about 10 seconds for a falling stone to reach that velocity, and in that time it would have fallen about 400 meters, about the height of the Empire State Building.

Look at the graph of $y(t)$ that portrays the actual fall of a stone. It starts out following one simple function and ends up following another. At first $y = At^2$, but eventually the curve becomes a straight line described by $y = -y_0 + kt$. During the transition period, the functional form is quite complex.

As a matter of fact, even the curve as shown is only an approximation to a complete description. The drag effect of the air depends on its density,

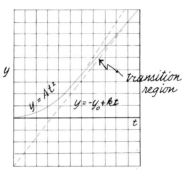

Distance traveled as a function of time for a stone dropping in air.

and that depends on the height. Furthermore, the weight of the stone depends on its distance from the earth, an effect that becomes vital in the launch of space rockets.

What good does it do us to describe the world with a few simple functions when the real world is more complicated? Isn't physics supposed to be the science of precision? Nonsense! That is, that would not make sense. Physics is preeminently the science of *common* sense. When you take a measurement, get only the precision required for that particular purpose. Precision is expensive; to get too much would be wasteful. Similarly, when describing functional dependence, zero in on the problem. Use simple models, good over limited ranges. There are only rare occasions when seven significant figures will tell you something that would be hidden with six. There are situations, such as shooting rockets at the moon, where you must take into account the effect of many small factors on very complex functions. Even in these cases, the computers zero in on the problems by calculating closer and closer approximations. As you will see, practically nothing is beyond our grasp in introductory physics. We deal with galaxies and subatomic particles, with power stations, and with the theory of special relativity. But in describing the mundane and the exotic we usually stick to first approximations and need only simple functions.

HANDLING THE PHENOMENA—FUNCTIONS

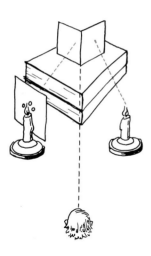

One of the simple functions that we will be using over and over again describes the way that various influences spread out from point or spherical sources. The influence may be the force exerted by an electrical charge or a gravitational mass. The same function also describes the intensity of light at different distances from a small source.

You can explore the nature of this function by using the simple photometer shown in the diagram. A photometer is a device for measuring illumination, usually by comparing the intensity of light arriving from two different sources. One source serves as the standard, or constant reference.

The two lights should be identical, but can be 60 watt light bulbs, flashlight bulbs, or candles. The experiment should be done in a room without any other lights, or at least the background illumination should be low and even. Line up the two sources of light at the same distance from the cardboard and at the same height, and sight along the bisector line of the bent cardboard. When the illumination is even on both sides, the crease almost seems to disappear. Notice that if you move one source further away or closer, the cardboard on that side gets darker or brighter than the other side.

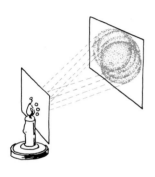

Suppose that you could reduce the intensity of the standard source on the left by a factor of two. Then the right side of the cardboard would appear brighter. You could match the two sides again by moving the right-hand source further away. If you could change the intensity of the control source by factors of 2, 3, 4, and 5, you could measure the factors by which you have to move the right-hand source to keep the illumination balanced. You would be measuring the intensity of light as a function of distance from the source.

One way to change the intensity of the left-hand light by known factors is to make the light shine through holes punched in a cardboard, as shown in the detailed diagram. Make sure that each hole is the same size, and that the light from each hole spreads out enough to cover the cardboard screen. Use five of these cardboard masks. One of them should have five identical holes clustered together, the next should have four holes, and so on. The holes should be small compared with the size of the light source, so that the same amount of light shines through each one. Start out with the five-hole mask, and balance the illumination on the two sides of the screen by moving the left-hand source. Then leave that source stationary, and change its intensity by changing the masks in front of it. The illumination of the left side of the screen will thus change from 5 units to 4 units to 3 units to 2 units to 1 unit. To balance the illumination initially, it may be necessary to put a duplicate five-hole mask on the right-hand source. If so, leave it on for the rest of the experiment.

For each value of illumination of the left side, find the distance for which the right-hand source balances the illumination. Measure this distance in terms of the original distance of the right-hand source. It would be a good idea to keep that original distance small—perhaps 10 cm. If you move it back to 40 cm, you have thus moved it from 1 unit distance to 4 unit distances. See if your results correspond to any of the functions shown on page 19. One way to find out if your data fits a function such as $I = k/r^2$ is to take the product Ir^2 for all values of r. If the data follow the function, all values of Ir^2 will equal the same constant.

Illumination in Arbitrary Units	Distance of Right-Hand Source in cm	Distance in Units of Initial Difference
I	r	r
5		1
4		
3		
2		
1		

SUMMARY

This introductory chapter stakes out the territory for our course of study. It might appear that we intend to cover just about everything under the sun and beyond. We will, indeed, explore the microworld of particles and the macroworld of the stars, paying special attention to the common features of both. In doing so, we will have to become familiar with time intervals much smaller than a microsecond and much larger than a million years.

While we are interested in objects and events that span vast ranges of space and time, our prime concern is with the interactions among systems.

There are only a few types of interactions and they are constrained by a small number of conservation laws.

This first chapter also introduces a number of techniques and ideas that are more important for future study than the facts themselves. First, in discussing the sizes of things and the times for various events, we felt it necessary to sketch the method used for each measurement. Some of these methods are covered in detail in later chapters, but note the prejudice: a quantity should be defined and explained in terms of how it can be measured. If, in principle, a quantity cannot be measured, then it is meaningless to talk about it.

The second aspect of the hidden agenda of this chapter involves an attitude toward solving problems. Zero in on the answer! First find an approximate solution, perhaps good to an order of magnitude. Then, when necessary, go to the trouble of computing the answer to one or two significant figures—or more if there is good reason. In order to do this sort of free-wheeling calculation, it is necessary to use power-of-ten notation. Some explanation and examples of the rules for this system were given in the chapter, and more drill is provided in the problems at the end of the chapter. For success in using the rest of this book, you really should be familiar with the method.

The third technique that we introduced was the way to represent functional dependence of variables. Physics is largely a study of functional dependence. In introductory physics we need only a few simple functions, such as the sine, log, exponential, linear, quadratic, and inverse powers. In future work we refer frequently to these functions and their graphs.

Answers to Questions

1-1: As the value on the log scale goes from 0 to 1, the corresponding distances go from 1 meter to 10 meters. Two meters, a little taller than the average human, is at the point 0.3 on the log scale, a little less than one-third the way between 0 and 1.

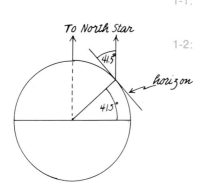

1-2: The French surveyors measured an arc of the earth's surface. The angle subtended can be found by astronomical sighting. For instance, the latitude of a place north of the equator is equal to the angle between the horizon and the line to the north star. The southern end of their line was at 41.5°N and the northern end was at 51°N. The angle subtended by their measured arc was thus 9.5°.

Having a permanent reference standard for a basic unit is useful only if comparisons with it can easily be made. No one would want to go to the trouble of surveying an arc of the earth again, and even comparisons with the standard meter defined by a platinum bar are hard to make. Since 1960 the official meter has been defined as 1,650,763.73 wavelengths of the orange-red line in the spectrum of krypton-86. Such a standard is not subject to theft or wear and tear, and the optical measuring devices used for comparison are fairly inexpensive to produce and use.

1-3: The radian measure of an angle is the ratio of the arc to the radius. The diagram shows the relationship between the angle in radians and the sine and tangent of the angle. Notice that for small angles the ratio of arc to radius is almost the same as the ratio of opposite leg to radius, or opposite leg to adjacent leg. In other words, for small angles the measure of the angle in radians is approximately equal to the sine or the tangent of the angle.

This approximation is quite good even for an angle as large as 30°. An angle of 30° is equal to $\frac{30}{57}$ radian or 0.525 radians. The sine of 30° equals 0.500 and the tangent of 30° equals 0.577.

For smaller angles the approximation rapidly gets better. For the fractional degree angles of astronomical surveying, it is inappropriate to use the sine and tangent relationships.

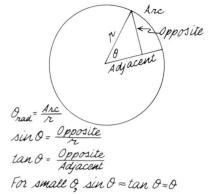

1-4: The subtended angle is equal to the ratio of baseline to radius.

$$\theta = (3 \times 10^{11} \text{ m})/(3.8 \times 10^{16} \text{ m}) = 0.8 \times 10^{-5} \text{ rad} = 8 \times 10^{-6} \text{ rad}$$

That is a little larger than one second of arc.

1-5: Velocity = (distance traveled)/(time taken). In this case:

$$3 \times 10^8 \text{ m/s} = (1 \times 10^{26} \text{ m})/(\text{time})$$

$$(\text{time}) = (1 \times 10^{26} \text{ m})/(3 \times 10^8 \text{ m/s})$$

$$(\text{time}) = 3 \times 10^{17} \text{ seconds} = 1 \times 10^{10} \text{ years}$$

(This time just happens to be about the age of the universe, as we will soon see.)

1-6: Sand, falling through the hole in a 3-minute egg timer? A spring with a weight on it, bobbing up and down? The time for a stone to fall a certain distance? The time between full moons? The time between annual returns of the swallows to Capistrano?

1-7: This question is not completely farfetched. With our expanding universe it is possible that certain basic constants are slowly changing. In this case the problem is whether or not *all* motions would be affected by the slowing down of time. For instance, if repetitive motions were affected, but others were not, the time required for a ball to drop a certain distance would be changed.

1-8: In one year there are

$$\frac{60 \text{ seconds}}{\text{minute}} \times \frac{60 \text{ minutes}}{\text{hour}} \times \frac{24 \text{ hours}}{\text{day}} \times \frac{365 \text{ days}}{\text{year}}$$

$$= 60 \times 60 \times 24 \times 365 \text{ seconds / year}$$

$$= (6.0 \times 10^1) \times (6.0 \times 10^1) \times (2.4 \times 10^1) \times (3.65 \times 10^2)$$

$$= 6 \times 6 \times 2.4 \times 3.65 \times 10^5 \text{ s/y}$$

$$\approx 3 \times 10^7 \text{ s}$$

The number of seconds in a scriptural lifetime is $(3 \times 10^7 \text{ s/y}) \times (70 \text{ y/lifetime}) \approx 200 \times 10^7 \text{ s/l} = 2 \times 10^9 \text{ seconds/lifetime}$

It often helps to translate the powers of ten to words. Note that your allotment is two billion seconds, and about 6×10^8 are already gone!

1-9: $1/(9{,}192{,}631{,}770 \text{ cycles/s}) \approx 1/(9 \times 10^9 \text{ cycles/s}) = \frac{1}{9} 10^{-9} \text{ s} = 1.1 \times 10^{-10} \text{ s/cycle}$. In words, the period is one-ninth of a billionth of a second.

1-10: Average velocity = (distance)/(time taken) (time) = (distance)/(velocity) = $(2 \times 10^{-15} \text{ m})/(3 \times 10^8 \text{ m/s}) \approx 1 \times 10^{-23}$ s. Consider how improper it would be in this type of calculation to divide 2 by 3 and get $\frac{2}{3}$ instead of 1.

1-11: After 5 half-lives, the fraction remaining would be $\frac{1}{2}\frac{1}{2}\frac{1}{2}\frac{1}{2}\frac{1}{2} = \frac{1}{32}$. After 10 half-lives there will be only $(\frac{1}{32}) \times (\frac{1}{32}) \approx \frac{1}{1000}$, or 1×10^{-3}. After 20 half-lives, only 1×10^{-6} would remain—one-millionth.

1-12: We don't, unless as a matter of philosophical principle we define "the present" to be anytime along a path connected by the passage of light.

1-13: How about love? A physicist with a wife, six children, and a large dog is surrounded by it. If you care to analyze love, it's probably electromagnetic. Of course, there are always psychics and dowsers looking for a sixth sense. The mainstream of science is concerned with reproducible results of experiments that have valid statistics. Not all the psychics are charlatans, but too many of their followers are deluded by the hope of finding a royal road to scientific learning.

Anyway, since, as far as we know, the electromagnetic interaction is responsible for all the chemical and biological processes, there doesn't seem to be a need for any more types of interactions.

1-14: In the initial state, there was a single solid object at rest. In the final state, there were many small pieces flying in all directions, with accompanying noise and flame. While forces were involved during the explosion, it would be very difficult to trace the detailed effects.

1-15: There are only a few acceptable positions for the parked car compared with the large number of positions and angular headings that the car may have once it is out in the road. In parking, you are going from a state with many possibilities—high probability—to one with smaller probability.

1-16: Indeed, energy is conserved, but the highly concentrated energy of chemical binding in the oil or coal is dissipated into the low temperature thermal energy of a much larger volume of water. That energy is so spread out that there is no efficient way to recapture it to produce useful work.

1-17: Aristotle, and most other right-thinking people until the time of Galileo, thought that the position (and velocity) of a falling object obviously depended on its mass. In the next few chapters you will be experimenting with such situations, and will find that to a first approximation, the position of a falling object does *not* depend on its mass. This simple fact has astonishing implications, as we will see. You will also find out the limitations of the approximation. In air, for instance, the position of a falling object certainly does depend on the object's mass.

The velocity and position of a falling object also depend slightly on the location on earth (faster at the north pole than at the equator); depend strongly on the size of the earth (faster on the earth than on the moon); do not depend at all on the color of the stone; do depend on the initial velocity; depend very slightly on air pressure; and do not depend on your thoughts except, perhaps, in very close tennis matches. How about dependence on the shape of the stone?

1-18: The sinusoidal function describes a motion where the distance from a starting point repetitively gets larger and smaller. It describes the motion of a pendulum bob, but not that of a falling stone.

The exponential function requires that the object has an initial separation from the origin at $t = 0$. As time proceeds, the separation, y, increases very rapidly. The first point is hard to reconcile with the observed motion, although the second point is the sort of description we want.

The linear function could be arranged so that $y_0 = 0$, thus starting the stone off with zero separation at zero time. However, if the distance covered is proportional to the time taken, the velocity must be constant. (2 m in 1 second, 4 m in 2 s, 6 m in 3 s—the constant velocity must be 2 m/s) Take a close look at a falling object and you can see that its velocity doesn't remain constant; it speeds up.

The squared function, which is parabolic and is also called quadratic, fits the behavior we just described. It might be a good approximation for the motion of a falling stone.

The next two graphs show inverse relationships. As time proceeds, the separation distance decreases. They certainly do not describe the motion of a falling stone.

Problems

1. $\dfrac{8253 \times 0.684}{3.4 \times 10^4 \times 4.8 \times 10^{-6}} = ?$
2. A bedroom is $9' \times 12' \times 8'$. What is its volume in m^3?
3. On a girl's eighteenth birthday, how many seconds has it been since she was born?
4. What percent of the nation's age have you lived?
5. What is the ratio of the radius of the universe to the radius of a nucleon?
6. How many atoms are in the earth? Assume that each atom has a diameter of 2×10^{-10} m, and that there is negligible volume lost in the packing of the spheres.
7. What is the angle in radians and in degrees that you see subtended by the fingernail on your little finger when held at arm's length?
8. For small angles, $\sin \theta \approx \theta$, when θ is in radians. What is the percentage error in using this approximation for $\theta = 30°$?
9. What is the angle in radians and in degrees subtended by the moon, as seen from earth?
10. What is the ratio of intensity of sunlight (energy per square meter) at the earth to that at Jupiter?
11. The nearest star to our sun is about 4 light-years away. What would happen to its apparent brightness if it were 100 light-years away?
12. How long does it take a radar signal to travel to the moon and back to earth?
13. How long does it take light to travel to you from a bulb 1 meter away?
14. How many years does it take light to reach us from the other side of our galaxy—about 10^{21} m away?
15. If you wanted to make a linear chart of times for our universe, allowing 1 mm for every thousand years, how long must the chart be?
16. Find your normal walking speed in miles per hour, and then convert the units to m/s, and light-years per year.
17. A furlong is 220 yards; a fortnight is 14 days. How fast is 1 furlong per fortnight in cm/s?
18. The half-life of Cs^{137} is 2.7 minutes. What fraction is left after one hour?
19. Every living thing contains carbon. A small fraction of the carbon is radioactive (C^{14}), producing 16 disintegrations per gram of carbon per minute. The half-life of C^{14} is 5600 years. If a wooden tool from a cave burial yields 4 counts per minute, when did the burial take place?
20. Which of the four main interactions is responsible for:
    - (a) rusting
    - (b) melting
    - (c) rolling downhill
    - (d) carbon-14 beta decay
    - (e) "atomic" bomb energy release
    - (f) satellite orbit
    - (g) thunder
    - (h) animal cell division
    - (i) TV
    - (j) stellar energy source
21. Which of the graphs shown on page 19 represents the position of a pendulum bob as a function of time? Which graph represents a car moving with constant speed?
22. If the smallest displacement of star images that can be measured on a photographic plate is 1 micron (1×10^{-6} m), what is the largest stellar distance that can be measured using the earth's orbit as baseline? Assume the focal length of the telescope is 20 m.
23. If the heavy elements were created in a stellar explosion about 10 billion years ago, what fraction of the original $_{92}U^{235}$ is now left?

24. The text describes why a solar day is longer than a stellar day. If the revolution period (stellar period) of our moon is 27 days, how long is the period from new moon to new moon?

25. On page 9 there is a graph of the angular position of a pendulum bob as a function of time. Real pendulums swing with smaller and smaller amplitude and eventually stop. Sketch a graph showing θ as a function of time for a real situation where the amplitude is nearly zero after five oscillations.

MOTION ALONG A LINE 2

Things move. A river of ice may flow with the glacial slowness of a meter per week. Southwestern California, including Los Angeles, is moving northwest along the San Andreas fault at an average of about 5 cm per year. If an electron or a baseball is dropped in vacuum for a distance of one meter, it will accelerate to a speed of almost 10 m/s. If the electron, but not the baseball, is put between metal plates in a vacuum, with a $1\frac{1}{2}$ volt flashlight battery connected to the plates, it will accelerate to a speed of almost 10^6 m/s. That's fast, especially if you compare it with the speed of a passenger jet at about 300 m/s. If the voltage across the plates is increased by a factor of 100, to 150 volts, the electron's speed will be increased by a factor of 10, to 10^7 m/s. (The final speed is proportional to the square root of the voltage.) With 15,000 volts across the plates, the electron would be going nearly 10^8 m/s. By using special, but readily available, techniques, a voltage of 1,500,000 volts could be put across the plates. Then the electron would accelerate to nearly 3×10^8 m/s.

Question 2-1
Are those numbers consistent with each other?

Velocity is a very special kind of variable in our universe. We usually refer to it in combination terms of distance and time: miles per hour, feet per second, meters per second, furlongs per fortnight. It might appear that distance and time are the two basic, or primitive, units, and that velocity is derived from them. On the other hand, as far as nature is concerned (so far as we know) there are no *universally natural* units of distance and time. There is, however, a natural unit of velocity—an ultimate speed limit. No object (whether baseball or subatomic particle) can be accelerated to go faster than the speed of light in a vacuum. That speed is 3.0×10^8 m/s. Not only has this prohibition law been experimentally confirmed in a direct fashion, it is also the foundation stone of the special theory of relativity. That theory has spectacular consequences (such as the release of nuclear

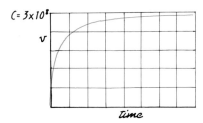

Speed of protons being accelerated in a synchrotron.

energy), all of which have been confirmed. In nuclear physics research, particle speeds are usually given in terms of β, the ratio of the speed of the particle to the speed of light. β can never be greater than 1.

Question 2-2

What is your β when you are running as fast as you can?

We have been using the words *velocity* and *speed* interchangeably. There is a technical difference between the words, emphasized more in school than in the real world. *Speed* tells how fast something is going; *velocity* tells how fast, *and* in what direction. In this chapter we deal only with motion along a line. That motion can be to the right or to the left (or up and down). We will distinguish these directions with + and − signs, and not become fussy about the words—at least not until the next chapter.

There's more to the subject of motion than velocity, of course. On the instrument panel of your car there's a speedometer, but on the floor there's an accelerator. *Velocity* is the measure of how fast your *position* is changing; *acceleration* is the measure of how fast your *velocity* is changing. In this chapter we describe relationships among position, velocity, acceleration, and time.

HANDLING THE PHENOMENA

We are in motion or observing motion all the time and so it might seem that we are all familiar with the phenomenon. Some of the details are not so easy to observe, however, and are sometimes confusing to describe. Why not start out with a simple motion that is easy to measure—slow, constant velocity? One way to achieve this is to roll a marble or ball on a tabletop. If the rolling object is slow enough, you can time its movement through a small distance with a stopwatch or by counting steadily. (In fact, counting rapidly and steadily works surprisingly well for this sort of timing. Do the counting out loud at a rate so that you count from 1 to 20 in about 5 seconds.) For marking the small distances, use something about 20 cm wide, like the width of a piece of typing paper. If the velocity remains constant, the time taken to go past a piece of paper at the beginning of the trip should be the same as the time to go past another piece of paper further on. Of course,

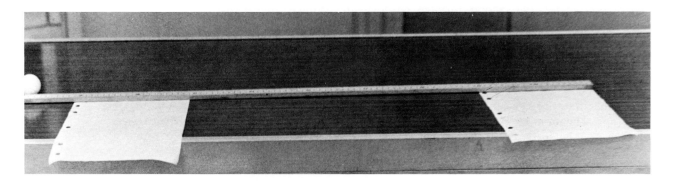

unless the table is tilted, the ball will slow down. You can defeat the slowing down by deliberately tilting the table. Shim up two of the legs with paper or cardboard. If the surface is smooth and clean, and the ball reasonably smooth and round, it should take only a very slight tilt to produce constant velocity. (Sight along the edge of the table to make sure that it is straight and doesn't have a dip in it. You may have to put a straight stick along the path, to guide the ball as it rolls.)

When you have achieved constant velocity, *tilt the table a slight amount more so that the ball speeds as it rolls.* Now the problem is, how does the time taken to roll through a short distance depend on the distance and time from starting? Here's a way to find out.

First Arrangement

Put one of the marking papers at the end of the table, and another so that its center point is halfway from the starting point to the center of the end paper. To reach the end paper, the object must roll twice as far as it does to reach the middle paper. Is it going twice as fast? That is, is the time taken to cross the end paper only half the time taken to cross the middle paper? Keep a record of the times (or counts) that you get.

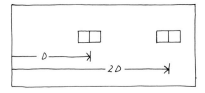

Second Arrangement

Find the time, T, that it takes the object to roll down the complete length of the table, and also find the time it takes to roll down the first half. Record these times for future use in Question 2-16. Now find the point to which it rolls in $T/2$. Is it the center point of the table? Record the length of the table and the distance to the half-time point. Place the center of one of the marking papers at the *half-time* point, and then repeat the first part of the experiment. Is the object rolling twice as fast when it crosses the end paper as when it crosses the paper at half-time?

If, in the first arrangement, the time to cross the middle paper is twice the time for the end paper, then the velocity under these conditions is proportional to the *distance* traveled; if you double the distance, you double the velocity. If, in the second arrangement, the time to cross the paper at the half-time point is twice the time for the end paper, then the velocity is proportional to the *time* of rolling. If you double the time, you double the velocity. Which of these two conclusions do you find?

You can see some of these effects if you throw an object into the air and watch closely. The action is too fast to see many details, but you can improve the observation by tossing something that is not round, and giving it a twist. A shoe does very nicely. The shoe will twirl at an almost constant rate, giving you an object with a built-in timer. Observe the vertical velocity of the object as it rises and falls. Where does it slow down and speed up? Is it gradual or abrupt?

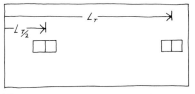

Time (counts) to cross end paper
_____
Time (counts) to cross middle paper
_____
Time (counts) to roll length of table (to middle mark of end paper)
_____
Time (counts) to roll half-length of table _____
Length of table to middle mark
_____
Length of table to half-time mark
_____

DESCRIPTION OF AN OBJECT STANDING STILL

In everyday life we describe motion in terms that are familiar to everyone. If you have to go 180 miles on an expressway, you know that it will take you at least 3 hours.

$$\frac{180 \text{ miles}}{60 \text{ mph}} = 3 \text{ hr}$$

In common words,

$$\frac{\text{distance}}{\text{speed}} = \text{time}$$

Similarly, if it takes you half an hour to go 15 miles in heavy traffic, your average speed must have been 30 mph.

$$\text{speed} = \frac{\text{distance}}{\text{time}}$$

The technical description of motion is not much more complicated but does require some shorthand symbols that save space and permit easy analysis. We introduce these symbols and techniques by describing very simple motions. We want to be able to describe the position of an object as a function of time. If the object can move only back and forth along a line, (not necessarily a straight line) we can define its position as the distance, x, from an origin. We describe its motion in terms of the function, $x(t)$. The graph of the function will show the object's position, x, along the vertical axis as a function of time, t, along the horizontal axis.

If the object is standing still, its velocity must be zero. The graph of the function is a horizontal straight line, since for any time, t, $x = x_o$.

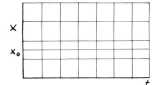

$x(t)$ for an object stationary at x_0.

> **Question 2-3**
>
> As time proceeds, the graph point traces out a line to the right. How can a line represent an object standing still?

DESCRIPTION OF AN OBJECT WITH CONSTANT (BUT NON ZERO) VELOCITY

If an object is moving with a constant velocity of 2 m/s, then in one second it will have traveled two meters, and in five seconds it will have traveled ten meters. The algebraic description of this motion is $x = vt$, where x is the distance traveled, v is the constant velocity, and t is the time taken. The graph of the function is a straight line going through the origin.

That simple algebraic formula, as it stands, can cause confusion. If you are on an expressway going from west coast to east coast, the milestones that mark position usually give the distance from some particular origin. Reno, for instance, may be at the 325 km mark from San Francisco. If you start out from Reno at nine o'clock, going 75 km/hr for one hour, you could not describe your new position by using $x = vt$, and substituting 10 (10 o'clock) for t. The t in the equation should refer to the time *interval*. To be more complete, we should describe the time interval as $(t_f - t_o)$, the difference between the final time and the original time. In the example given, $(t_f - t_o) = (10:00 - 9:00) = 1$ hr.

There is still a complication. If in this example we use the formula, $x = v(t_f - t_o)$, then we would get $x = (75 \text{ km/hr}) \times (1 \text{ hr}) = 75$ km.

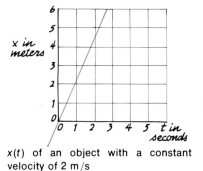

$x(t)$ of an object with a constant velocity of 2 m/s

Question 2-4

What's wrong with this solution? Isn't 75 km the distance traveled?

To eliminate all possible confusion, the formula for constant velocity should be written $(x_f - x_o) = v(t_f - t_o)$. Instead of writing all the subscripts, we frequently use another common notation. The difference between two variables is indicated by Δ, the symbol for the Greek letter delta. Our formula would then become: $\Delta x = v\Delta t$. In words, the formula is *not* "position equals the product of velocity and time." Instead, the formula is distance traveled equals the product of velocity and the travel time.

The traditional algebraic formula for a straight line for $y(x)$, is
$$y = mx + b$$
As you can see on the graph, b is the intercept on the y-axis, when $x = 0$. This equation has the same form as the one for constant velocity:
$$x_f = v(t_f - t_o) + x_o$$
If we start our clock at zero as we pass x_o, then $t_o = 0$, and the constant velocity equation reduces to
$$x_f = vt + x_o$$
This is called a linear equation, because its graph is a straight line.

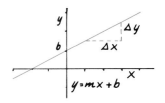

Question 2-5

Does linearity imply proportionality? Is x proportional to t?

In the algebraic formula for a straight line, the constant, m, is the value for the slope of the line. The slope of the $x(t)$ graph is the velocity, v. Algebraically, it is given by
$$v = \frac{(x_f - x_o)}{(t_f - t_o)} = \frac{\Delta x}{\Delta t}$$

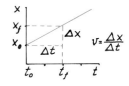

The words to match these symbols are the familiar ones that we use for velocity: distance traveled divided by time of travel. For instance, velocity might be given in meters per second.

The significance of these symbols for the graph is that the slope is equal to the ratio of the rise to the step. Since the velocity is constant, and the graph is a straight line, it doesn't make any difference how big you make the triangle for the rise and step. The ratio, $\Delta x/\Delta t$, remains the same. Notice, however, that $\Delta x/\Delta t$ is not necessarily the same as x/t.

Let's see how we can use an $x(t)$ graph to describe a journey along the road shown on the map on the next page.

Starting at A, a driver traveled at 80 km/hr until he got to B. He then had to travel on a detour between B and C at 20 km/hr. At C he stopped for half an hour for coffee, and then drove on city streets to D at 40 km/hr. After a half hour stop at D, he drove back for four hours on an expressway along the same route at 80 km/hr. As you can see, he passed A, going to the left on the map, and to negative values of x on the graph. The usual

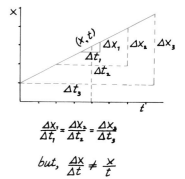

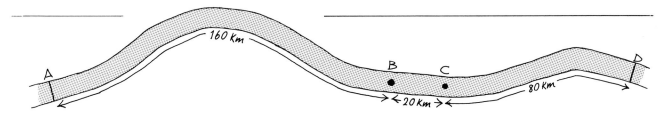

convention is that distances to the right of an origin are labeled positive and those to the left are labeled negative.

Dotted triangles are drawn on the graph so that values for $\Delta x/\Delta t$ can be calculated for the various parts of the trip. On the homeward journey, Δx is negative. Note that this is true both for regions where x itself is positive and where it is negative. During this return trip, the *differences* in values of x, $\Delta x = (x_f - x_o)$, are negative.

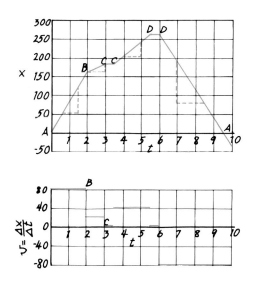

Question 2-6

In the negative x region, wouldn't $(x_f - x_o)$ be positive?

Directly underneath the graph of $x(t)$, there is the corresponding graph of $v(t)$, the velocity as a function of time. It is a graph of the *slopes* of $x(t)$. The steeper the slope, the faster the speed. When the graph of $x(t)$ is horizontal, the slope is zero, corresponding to zero speed. When the slope is negative, the speed is negative, describing motion to the left.

We did not connect the line segments in the graph of $v(t)$. These abrupt changes in speed correspond to the abrupt changes in slope shown in the $x(t)$ graph. Such abrupt changes are mathematical fictions, of course. As we warned in the first chapter, simple functions are only good first approximations in describing real events. A real car would take time to accelerate from zero to 80 km/hr, and from each subsequent speed to the next. On the $x(t)$ graph the real motion would be shown as a rounded connection between two straight line segments. In the $v(t)$ graph, the horizontal lines would be connected. Actually, the record of a real journey might look more like this next graph.

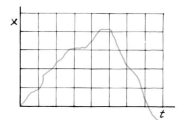

Question 2-7

It looks as if the real $x(t)$ graph closely follows the idealized one. If straight lines were drawn over the wiggly lines of the $x(t)$ graph, there would not be large deviations. If you did the same thing in the $v(t)$ graph, however, there would be sharp peaks outside the lines. Why should this be?

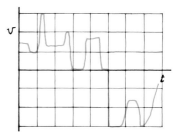

INSTANTANEOUS VELOCITY

Question 2-8

After traveling half a mile in one minute, a motorist was arrested for speeding in a 30 mph zone. The police officer said, "I am sorry to do this, since you are my son, although I am not your father." How can this be?

The definition of velocity that we have used gives the *average* velocity during the time interval, Δt. If we want the instantaneous velocity at a particular instant, we should include that instant in the time interval, presumably right in the middle. The question is, how big should we make Δt? If we want the velocity at $t = 1.0$ seconds, should Δt go from 0.5 to 1.5, or from 0.9 to 1.1, or from 0.99 to 1.01? If the velocity is constant, the size of Δt doesn't matter. But if the velocity is changing, we might get a different answer for each interval chosen.

The graph shows a plot of $x(t)$ for an object whose velocity is steadily increasing according to the function, $x = kt^3$, with k having the numerical value of 1. You can find $\Delta x/\Delta t$ for several different intervals centered about 1.0 second, as shown by the dotted triangles. Values for $\Delta x/\Delta t$ are given in the table for smaller and smaller values of Δt. You can see on the graph that the hypotenuse of the smallest triangle is nearly tangent to the curve at the $t = 1.0$ mark. The slope of an $x(t)$ curve at a particular time is the velocity at that time. For values of Δt too small to show on the graph, we computed Δx by calculating $(x_f - x_0)$. For instance, for $\Delta t = 0.10$ s, $x_0 = (0.95)^3 = 0.857375$, and $x_f = (1.05)^3 = 1.157625$. Then, $\Delta x = 0.30025$, and $\Delta x/\Delta t = 3.0025$ m/s.

As Δt gets smaller, $\Delta x/\Delta t$ tends toward a limiting value of 3.0000. That limiting value is called the instantaneous velocity at the point, $t = 1$. In symbols, the instantaneous velocity is defined as:

$$v = \text{limit}_{\Delta t \to 0} \Delta x/\Delta t$$

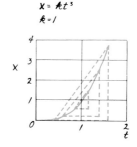

Δt	Δx	$\Delta x/\Delta t$
2	8	4
1	3.25	3.25
0.5	1.531	3.063
0.1	0.30025	3.0025
0.01	0.03000025	3.000025

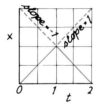

From $t = 0$ to $t = 1$, slope $= 1$.
From $t = 1$ to $t = 2$, slope $= -1$.
At $t = 1$, slope is not defined.

There are mathematical functions, such as those with abrupt changes in values, where $\Delta x/\Delta t$ does not approach a limit as Δt shrinks around certain points. We do not have to be concerned with such problems because we need only functions that describe physical phenomena where there are not abrupt changes in values.

Question 2-9

Many laboratory methods of measuring velocity, such as with strobe pictures, rely on measuring positions at various times and calculating average velocities between times. In practice we don't actually perform the limiting process. What common devices really measure instantaneous velocity?

USING A GRAPH OF $v(t)$ TO FIND DISTANCE TRAVELED

The velocity of an object at any time can be found by taking the slope of the $x(t)$ graph at that particular time. In this way, as we have seen, you can generate a $v(t)$ graph from an $x(t)$ graph. Knowing the position as a function of time allows you to find the velocity as a function of time. Does it work the other way around? Can you get the $x(t)$ graph out of the $v(t)$ graph?

Here's an almost trivial example of how you can get $x(t)$ information from a graph of velocity as a function of time. The graph describes constant velocity of 5 m/s. The distance traveled is equal to $v\Delta t$. After one second, the object has gone 5 m; after two seconds, it has gone 10 m; after three seconds, it has gone 15 m. Numerically, these distances are equal to the numerical value of the areas of the rectangles shown by crosshatching. The area of each rectangle is equal to the product of its two sides, which is $v\Delta t$, and this is equal to the distance, Δx.

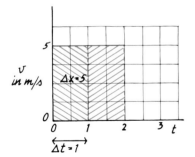

Question 2-10

Something must be wrong. According to the preceding argument, a distance (a length) is equal to an area (a length squared). How can this be?

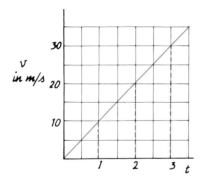

Let's try an example that is not so trivial. Suppose that the velocity of an object is not constant, but is steadily increasing. The graph shows such a case, where $v = 10t$. After one second, starting from rest, the object has a velocity of 10 m/s. The actual curve can be approximated by the step functions shown. Each curve is composed of a series of moves made at constant velocity. The distance traveled during each move is equal to $v\Delta t$, where v is the constant velocity during the time interval, Δt. In the first graph, the approximation is made that the velocity during each time interval is the velocity at the beginning of the interval. This choice produces a total distance traveled that is clearly too low. In the second graph, the approximation is that the velocity during each time interval is the velocity at the end of the interval. That assumption will produce too large a total distance. However, as the time intervals get shorter and shorter, those two estimates converge toward the same value. In this particular case (but not in general), the limiting value is the same as would have been obtained if the *average* velocity had been chosen for each interval—instead of the maximum or minimum.

Graphically, the step functions converge to the original straight line. The area bordered by the step functions becomes the area under the line. We could have used the same type of approximations, and the same arguments, about any curve showing $v(t)$ on a graph. The conclusion is general that the area under the $v(t)$ curve is equal to the distance traveled during the time interval. In the particular case of the straight line, describing the function $v = 10t$, the area under the curve is the area of a triangle. That area, for a time interval of one second, is $\frac{1}{2}$(base)(height) = $\frac{1}{2}$(1 s)(10 m/s) = 5 m. Note that this value is between the values of the two approximations. It is also the distance traveled during one second at an average speed of 5 m/s.

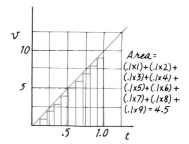

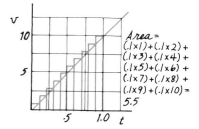

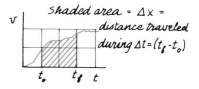

Question 2-11
How do you calculate the average speed?

At constant velocity, the distance traveled by an object is proportional to the time of travel. If you go 5 m in one second, you go 10 m in two seconds. However, that's not true when the velocity is changing. To find the distance traveled in the first two seconds when $v = 10t$, we calculate the area under the $v(t)$ curve from 0 to 2 seconds. That area is $\frac{1}{2}$(base)(height) = $\frac{1}{2}$(2 s)(20 m/s) = 20 m. Compare this distance with the distance traveled during the first second, which was 5 m. When the velocity of an object is proportional to the time, the distance it moves increases at a rate faster than simple proportionality. Remember the experiment that you did with the ball rolling down a tilted table. The velocity of the ball, as determined by the time it took it to cross a sheet of paper, was proportional to the time since it started rolling. The mark for the half-time position was much closer to the starting point than to the end. During the second half of the rolling time, the ball went much further than it did during the first half.

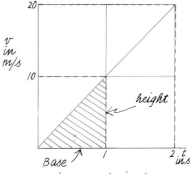

DEFINITION OF ACCELERATION
When you step on the accelerator of a car, your speed increases, but not at the same rate under all conditions. The effect seems to depend on the speed you start with and on the gear the car is in. To compare such effects we need a standard definition of rate of change of velocity. A useful definition is not so obvious. For instance, when you rolled a ball down a tilted table, the speed of the ball kept increasing. At the 20 cm mark it was going faster than at 10 cm, and at the 40 cm mark it was going much faster.

$$a \stackrel{?}{=} \frac{\Delta v}{\Delta x}$$

$$a = \frac{\Delta v}{\Delta t}$$

Question 2-12

Why not define acceleration as the *space rate of change of velocity?* In symbols, this definition would be $\Delta v/\Delta x$. It would tell us how velocity changes as a function of distance traveled: $v(x)$.

Since velocity is also a function of time, we can define the *time* rate of change of velocity: $\Delta v/\Delta t$. This is the standard definition of acceleration. The units for acceleration are those of change of velocity per second, or meters per second per second. The acceleration of a car is usually given in terms of the time that it takes to go from 0 to 30 mph or 0 to 60 mph. A sports car, for instance, might be able to go from 0 to 60 mph in 9 seconds. The rather confusing units in this case would be miles per hour per second.

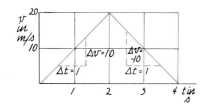

All three variables—position, velocity, and acceleration—are functions of time. To describe a particular motion, it is frequently revealing to plot the graphs of $x(t)$, $v(t)$, and $a(t)$. We have already seen two relationships between the graphs of $x(t)$ and $v(t)$. A point on the graph of $v(t)$ is the slope of the corresponding point on the graph of $x(t)$. The area under the $v(t)$ curve between two times represents the distance traveled during those times. There is also a relationship between the graphs of $v(t)$ and $a(t)$. Since $a = \Delta v/\Delta t$, the acceleration at any time is the slope of the $v(t)$ graph at that time. For the function that we used before, $v = 10t$, we plot $v(t)$, and, directly underneath, $a(t)$. Since the slope of $v(t)$ is constant for each of the two-second intervals, the acceleration during each interval is constant. Notice that in this graph we let $v = +10t$ for the first two seconds, and let $v = (20 \text{ m}) - 10t$ for the next two seconds. The slope for the latter part is negative, and so the acceleration is negative. Throughout the entire 4 seconds, the velocity is positive, to the right. During the last two seconds, however, the negative acceleration is reducing the speed. The function describes an object steadily moving to the right, speeding up for the first two seconds, and then slowing down to zero during the next two seconds.

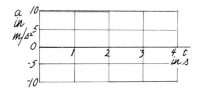

Question 2-13

How far does the object travel during the four seconds?

THE SPECIAL CASE OF CONSTANT ACCELERATION

We need only a few simple functions to describe most of the motions that we will study. Even when the math functions aren't perfect models, they are frequently good approximations. One such common type of motion involves constant acceleration: $a = \Delta v/\Delta t = $ constant. As we will see, this function is only a rough approximation to the motion of your car when you step on the accelerator. On the other hand, when objects are being pulled by gravity, either straight down or rolling down a tilted table, their acceleration is almost constant—at least over a useful range.

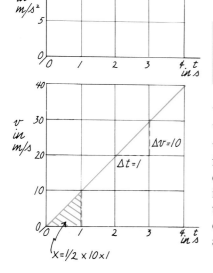

Let's see what sort of functions of velocity and position we must have

THE SPECIAL CASE OF CONSTANT ACCELERATION

if the acceleration of an object is constant. First, we plot the function, $a(t)$, where a = k. To use a number that will be useful later, let $k = 10$ m/s². On the graph the function is represented by a horizontal straight line. For all values of t, $a = 10$ m/s². The graph of the velocity function, $v(t)$, is drawn directly below the first graph. The curve of $v(t)$ must be a straight line with the constant slope of magnitude 10 m/s². The straight line could intercept the v-axis at any value at time $t = 0$. Whenever we start our clock, the object could have an initial velocity and still have the necessary acceleration. For simplicity, however, let's take the special case of zero initial velocity. Then we get the graph of $v(t)$ as shown.

To derive the graph of $x(t)$, the position as a function of time, we can measure the areas under the $v(t)$ line for various times. That calculation is shown under the graphs, and the data produce the third curve shown. We assume that at $t = 0$, $x = 0$.

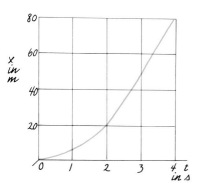

After 1 s, $x = 1/2 \times 10 \times 1 = 5$
After 2 s, $x = 1/2 \times 20 \times 2 = 20$
After 3 s, $x = 1/2 \times 30 \times 3 = 45$
After 4 s, $x = 1/2 \times 40 \times 4 = 80$

Question 2-14

Compare the motion described by the three graphs with the motion you produced by rolling a ball down the tilted table, and by letting an object drop from a height. What characteristics that you can actually observe show that these motions are described, at least approximately, by the graphed functions?

Here is a stroboscopic picture of a ball dropping freely, starting from rest. Each flash of light from the strobe lamp lasts only $\frac{1}{1000}$ second. For this particular picture, the lamp was flashing at the rate of 30/s. In other words, the time between flashes, Δt, was $\frac{1}{30}$ second. There is a meter scale beside the falling ball. As you can determine with a ruler, the picture has been printed with a scale reduction of 1:10. One centimeter on the picture represents ten centimeters of the original.

Get a centimeter ruler, measure $x(t)$ for the ball, and enter your data on the next page. (Don't forget the scale factor.) Now measure Δx for each interval of time, and enter that information in the next column. Since the average velocity during Δt is equal to $\Delta x/\Delta t$, and in this case each Δt equals $\frac{1}{30}$ second, you can calculate the average velocity during each time interval. Enter those numbers in the next column, and then calculate Δv from one interval to the next. Divide that number by $\frac{1}{30}$ second, and in the final column you can enter the value for $\Delta v/\Delta t$, which is the acceleration. Finally, you have all the data you need to plot graphs of $x(t)$, $v(t)$, and $a(t)$. Mark the data points on the same graphs that we worked out theoretically above, and see if your experimental data matches the theoretical curves. The actual value of the constant acceleration that we used in the graphs was chosen to be the same as the acceleration in free-fall.

Question 2-15

It is impossible to measure exact values for the positions, x, and the values for Δx are even more uncertain. By the time that you take

t	x	Δx	$\Delta x/\Delta t$	Δv	$\Delta v/\Delta t$
0	0				
1/30					
2/30					
3/30					
4/30					
5/30					
6/30					
7/30					
8/30					
9/30					
10/30					
11/30					
12/30					
13/30					

the difference of two uncertain values of Δx, the value for the acceleration is very uncertain. How can you record these uncertainties on the graphs?

For constant acceleration there are four simple relationships between x, v, a, and t.

1. One of them we have already seen. For constant acceleration (but only in this case), the average velocity during a time interval when the velocity is changing is

$$\bar{v} = v_{\text{ave}} = \frac{v_o + v_f}{2}$$ Eq. 2-1

For instance, if the speed of an object increased with constant acceleration during one second from 10 m/s to 20 m/s, then the average velocity during that second was 15 m/s. The plausibility of this relationship is shown graphically. The accompanying graph shows an example of why this simple relationship is not true when the acceleration is not constant.

2. Starting from rest, and accelerating constantly for a time, t, the final velocity attained is

$$v_f = at$$ Eq. 2-2a

This relationship is simply the one we have used graphically for the case when v is proportional to t. The formula also follows from the definition of acceleration, namely, $a = \Delta v/\Delta t = (v_f - v_o)/(t_f - t_o)$. If $v_o = 0$, and $t_o = 0$, then $a = v_f/t_f$, which is the same as Equation 2-2a.

3. If you start from rest at $t = 0$, and at $x = 0$, then with constant acceleration the graph of $v(t)$ is a straight line going through the origin. The distance traveled at time t is represented by the area under the line up to that time, t. The area of that triangular region is equal to $\frac{1}{2}(\text{base})(\text{height}) = \frac{1}{2}(t)(v_f) = \frac{1}{2}(t)(at) = \frac{1}{2}at^2$. For the special conditions of (a) constant acceleration; (b) original velocity equal to zero; (c) original position equal to zero; (d) starting at $t = 0$, the relationship between x, t, and a is

$$x = \tfrac{1}{2}at^2$$ Eq. 2-3a

Notice how this equation agrees with the graphs we have plotted showing $x(t)$ for constant acceleration. The squared, or quadratic, function describes a parabola. As time progresses, the distance traveled increases as the square of time.

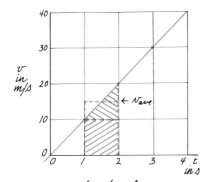

$\Delta x = $ area of rectangle + area of triangle
$= (10 \times 1) + (\tfrac{1}{2} \times 10 \times 1)$
$= 15$ m

$v_{\text{ave}} = \frac{\Delta x}{\Delta t} = \frac{15 \text{ m}}{1 \text{ s}} = 15$ m/s

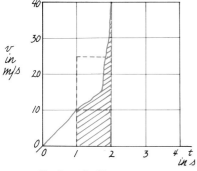

$v_{\text{ave}} \neq \frac{v_o + v_f}{2} = \frac{10 + 40}{2} = 25$ m/s

Question 2-16

When you rolled a ball down a tilted table, you recorded the times for the ball to roll halfway and the full length. Now substitute those times in Equation 2-3a and see if your data is satisfied by the formula. You also measured the distance traveled at half-time. Test that data with the formula also.

4. Of the four variables—x, v, a, and t—Equation 2-3a links x, a, and t for constant acceleration. There is another useful formula linking v, a, and t. We find the new relationship by combining two others.

$$v_f = at \qquad x = \tfrac{1}{2}at^2$$

Square the first formula and then eliminate t^2 with the second formula:

$$v_f^2 = a^2 t^2 \qquad t^2 = \frac{2x}{a}$$

$$v_f^2 = a^2 \frac{2x}{a}$$

$$v_f^2 = 2ax \qquad \text{Eq. 2-4a}$$

Notice that the final velocity with constant acceleration is proportional to the time, but proportional to the square root of the distance.

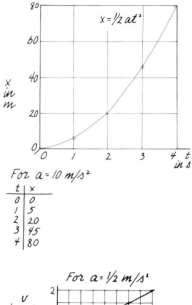

Question 2-17

Does the square root relationship agree with the data you obtained by rolling the ball down a tilted table? Compare the time it took to cross a paper width halfway down the table with the time it took to cross the paper at the end of the table.

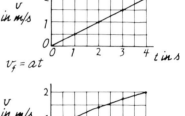

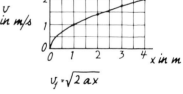

CONSTANT ACCELERATION FORMULAS WHEN v_o IS NOT ZERO

In the formulas derived above, we assumed that $t = (t_f - t_o) = \Delta t$. This assumption saves using messy subscripts and is easy to satisfy in practice simply by starting our clock at the beginning of the action being described. In the same way, we can usually define the origin so that $x_o = 0$. In our formulas, therefore, unless there is some special reason to do otherwise, $x = (x_f - x_o) = \Delta x$. There are many practical needs, however, for describing motions where the original velocity is not zero. For instance, a ball may be given an initial velocity when it is thrown, and be subject to the constant acceleration produced by gravity.

Equation 2-1 has the original velocity, v_o, built in. Equation 2-2a is easily expanded by going back to the definition of acceleration. Since $a = (v_f - v_o)/(t_f - t_o)$, and we are letting $(t_f - t_o) = t$, then

$$v_f = v_o + at \qquad \text{Eq. 2-2b}$$

To expand Equation 2-3a, calculate the area under the $v(t)$ graph shown,

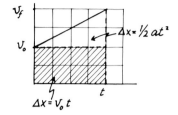

where there is an initial velocity. If there had been no acceleration, the distance traveled would have been $x = v_o t$, which is just the area of the crosshatched rectangle in the graph. The additional distance traveled because of the acceleration is represented by the triangular area. The total area, and thus the total distance traveled, is

$$x = v_o t + \tfrac{1}{2} at^2 \qquad \text{Eq. 2-3b}$$

To expand Equation 2-4a, we will use an algebraic derivation similar to the one used to derive it in the first place. In this case

$$v_f = v_o + at \qquad x = v_o t + \tfrac{1}{2} at^2$$

Square the first formula, and multiply the second by 2a:

$$v_f^2 = v_o^2 + a^2 t^2 + 2at v_o \qquad 2ax = 2a v_o t + a^2 t^2$$

The transformed second equation can be substituted into the squared first equation:

$$v_f^2 = v_o^2 + 2ax \qquad \text{Eq. 2-4b}$$

Consider the reasonableness of this equation. If there were no acceleration, then, of course, $v_f^2 = v_o^2$.

UNITS AND DIMENSIONS

Whenever formulas are available to solve a problem, there is a great temptation to substitute in the available data and get out a number. Resist that temptation until you make sure that the units of all your data are consistent. You cannot get velocity in miles per hour directly out of an equation where acceleration is in meters per second squared and time is in minutes. The safest routine to adopt is always to write down the units while setting up the problem. For instance,

$$v_f = \frac{10 \text{ m}}{\text{s}^2} \times (0.2 \text{ min}) \times \frac{60 \text{ s}}{1 \text{ min}} = 120 \text{ m/s}$$

Notice that the units were treated as algebraic quantities, to be multiplied and canceled until only the desired units were left.

The *dimensionality* of a variable is not the same as the unit in which it is expressed. By dimensionality, we describe how a variable is made up of the basic dimensions of length (L), time (T), mass (M), temperature (θ) and electric charge (Q). So far we have used only quantities that are composed of length and time. A distance, for instance, has the dimensions of L. An area, whether expressed in units of acres or square meters, has the dimensions of L^2. Velocity has the dimensions of L/T, or LT^{-1}.

You could add 3 meters to 2 feet, if you were careful with the numbers, but you cannot find the sum of 3 cows and 2 horses. You also cannot add together quantities that have different dimensions. Equations 2-3b and 2-4b involve additions of various combinations of *x*, *v*, *t*, and *a*. Does each term of one of those equations have the same dimensions as the other terms?

UNITS AND DIMENSIONS 41

Take the case of Equation 2-3b. We write the equation and underneath it the dimensions of each term:

$$x = v_o t + \tfrac{1}{2} a t^2$$

$$L = \frac{L}{\cancel{T}}\cancel{T} + \frac{L}{\cancel{T^2}}\cancel{T^2}$$

As you can see from the cancellation marks, the dimensions of the terms are consistent.

> Question 2-18
>
> Test the dimensional consistency of Equation 2-4b.

1. Instruction manuals for beginning drivers usually contain a chart showing how far a car will go after you step on the brakes. Of course, there are assumptions about the type of car, the kind of brakes, and the distance the car goes during your reaction time of moving your foot into place. One generalization can be made accurately, however. There is a standard relationship between original velocity and stopping distance. If the original velocity is v_o, and the desired final velocity is zero, then $0 = v_o^2 + 2ax$. The acceleration provided by the brakes is in the opposite direction from x, and so a is negative. In terms of the magnitudes of the numbers: $x = (\tfrac{1}{2}a)v_o^2$. The distance traveled after the brakes are applied is proportional to the square of the original velocity. The stopping distance at 60 mph is four times the stopping distance at 30 mph.

   The state of Florida driving manual lists 270 feet as the maximum distance your car should go after you apply the brakes starting at 60 mph. What acceleration are they assuming? (Notice that we avoid using the word "deceleration." Instead we use + or − acceleration.) Since Florida has not yet changed to the metric system, let us use their units, but convert 60 mph to 88 ft/s.

   $$(88 \text{ ft/s})^2 = 2a(270 \text{ ft})$$

   To find the acceleration with suitable precision, proceed as follows:

   $$a = \frac{88 \times 88}{2 \times 270} \approx \frac{90 \times 44}{270} = \frac{44}{3} \approx 15 \text{ ft/s}^2$$

   By itself, that number may not mean much to you. Compare it to the acceleration of an object in free fall, which is 32 ft/s². In terms of "g," the acceleration due to gravity, the Florida motor bureau is assuming that your brakes can exert, and you can withstand, an acceleration of about $\tfrac{1}{2}$g. That's reasonable.

2. Suppose that you are on the roof of a four story physics building and *throw* a ball down at the same time that a colleague *drops* a ball from a third story window. Your height from the ground is 20 meters; hers is 10 meters. What initial velocity must you give your ball if it is to reach the ground at the same time as the one that is dropped?

Examples

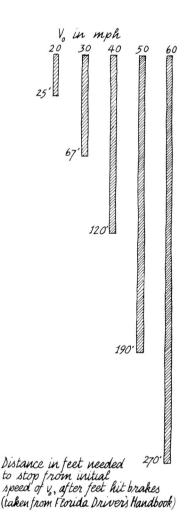

Distance in feet needed to stop from initial speed of v_o, after feet hit brakes (taken from Florida Driver's Handbook)

The motion of your ball is described by: 20 m = $v_o t + \frac{1}{2}at^2$. The motion of the other ball is given by 10 m = $\frac{1}{2}at^2$. The requirement of the problem is that the time, t, should be the same in both cases. The acceleration is produced by gravity and on the earth's surface has the value of 9.8 m/s², or 980 cm/s², or 32.2 ft/s². Gravitational acceleration is always down, a direction that by convention is called negative. On the other hand, in this problem all distances, velocities, and acceleration are in the same direction, and so we can leave all the terms positive. For purposes of this problem—since no one is really interested in the detailed answer—we can approximate 9.8 m/s² by 10 m/s². Our two equations then become

$$20 \text{ m} = v_o t + \tfrac{1}{2}(10 \text{ m/s}^2)t^2$$

and

$$10 \text{ m} = \tfrac{1}{2}(10 \text{ m/s}^2)t^2$$

Solve the second equation for t, which turns out to be 1.4 s. Substitute into the first equation to find v_o:

$$20 = v_o(1.4) + \tfrac{1}{2}(10)(2)$$

$$v_o = 10/1.4 = 7 \text{ m/s}$$

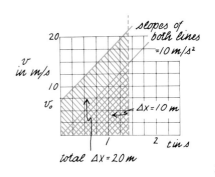

The problem can also be solved graphically. On a $v(t)$ graph, the motion of the dropped ball is represented by a straight line through the origin with a slope of 10 m/s². The motion of the thrown ball is shown by a line with the same slope but with an intercept at some value, v_o. At some particular time, t, the area under the first line up to that time is equal to 10 m. That's the time it takes the *dropped* ball to fall 10 m. Our problem requires that in that same time, the *thrown* ball travels 20 m. Therefore, v_o must be chosen so that the total area under the upper line up to the time, t, is equal to 20 m.

3. If you throw a ball straight up with an initial velocity of 20 m/s, how high will it rise, and how long will it be before it hits the ground, which is two meters below the height from which the ball is launched?

The problem can be solved in two parts, or all at once. To do it in two steps, calculate the distance and time for the upward flight first. The second part just involves free fall from the height found in the first part.

Upward flight: $v_f^2 = v_o^2 + 2ax$

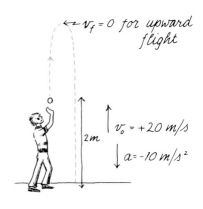

We use this formula instead of "$x = \tfrac{1}{2}at^2$" because we know the original velocity (20 m/s), the final velocity (0), and the acceleration (−9.8 m/s²). We do not, however, know either x or t. No matter what method we use, in this problem we must be careful to use + to indicate upward position or velocity, and − to indicate downward directions. We solve the equation for the upward flight, using the approximation that 9.8 ≈ 10.

$$0 = (20 \text{ m/s})^2 + 2(-10 \text{ m/s}^2)x$$

$$x = +20 \text{ m}$$

This is the distance above the launch point. The final height above the ground is thus 22 m. To find the time of the upward flight, we use

$$v_f = v_0 + at$$

$$0 = 20 \text{ m/s} + (-10 \text{ m/s}^2)t$$

$$t = (20 \text{ m/s})/(10 \text{ m/s}^2)$$

$$t = 2 \text{ s}$$

The time of the downward flight is a little longer because the ball falls 22 m to the ground. Since we know x and a, and want to find t, we use

$$x_f - x_o = \tfrac{1}{2}at^2$$

$$0 - 22 \text{ m} = \tfrac{1}{2}(-10 \text{ m/s}^2)t^2$$

$$t = \sqrt{4.4 \text{ s}^2} = 2.1 \text{ s}$$

Question 2-19

What is the velocity of the ball as it passes its launch point on the way down?

Here's a way to find the total time of the flight in one complicated step. The complete formula that includes initial position and initial velocity is

$$x_f - x_o = v_o t + \tfrac{1}{2}at^2$$

$$0 - 2 = (+20 \text{ m/s})t + \tfrac{1}{2}(-10 \text{ m/s}^2)t^2$$

$$5t^2 - 20t - 2 = 0$$

$$t = \frac{20 \pm \sqrt{(20)^2 - 4(-2)(5)}}{2 \times 5} = \frac{20 \pm \sqrt{440}}{10}$$

$$t = \frac{20 \pm 21}{10} = 4.1 \text{ s or } -0.1 \text{ s}$$

Question 2-20

The positive solution of the quadratic equation is the same number that we got in the two-step method. But what does it mean to get a negative number of seconds?

Here are the graphs of $x(t)$, $v(t)$, and $a(t)$ for the problem of the ball thrown upward. Although you can use the formulas and values that we have just developed to draw each of the graphs by itself, the easiest

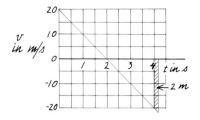

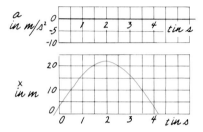

way to draw the graphs is to start with the $v(t)$ graph and derive the other two from it. At $t = 0$, $v_o = +20$ m/s. The slope of the line is a constant -10 m/s². Knowing the initial point of the line, and its slope, we can draw the line and get the value of t when $v = 0$. The velocity continues into the negative region, meaning that the ball has stopped rising and has started to fall. The total negative area between the line and the axis must be equal to the positive area plus the extra 2 meters that the ball fell past the launch height to hit ground. Since the slope of the $v(t)$ curve is constant, the $a(t)$ graph is a constant, horizontal line with the value of -10 m/s². The $x(t)$ graph, representing the area under the $v(t)$ graph, is made up of parabolas. At $t = 0$, $x_1 = 2$ m, and at $t = 4.1$ s, $x = 0$. Notice the other value of t for which $x = 0$, if the graph is continued to the left.

Simple Problems and the Real World

If we actually measure the motion of a ball thrown upward at 20 m/s, we find that it does not go quite so high as predicted, and its velocity at any point is slightly less than that given by the simple formulas. For an initial speed of 20 m/s, the effects of air friction are measurable but small. For a ball shot upwards with an initial velocity of 100 m/s, however, the simple formulas would be a very poor approximation. Short of doing an experiment, how can we know whether or not the simple formulas are good enough?

In making the simple mathematical model of the ball's motion, we deliberately ignored certain functional dependencies. Nothing was said about the ball's surface size or texture, or about its mass. We assumed that the acceleration caused by gravity is constant regardless of height, and that the earth's rotation does not affect the ball's motion. In other words, we assumed that the ball was being thrown in a vacuum on a flat, nonrotating earth. Certainly the earth can be considered flat for flights of a ball up to a few hundred meters. You can get a feel for the relative importance of this factor by comparing the rise distance to the radius of the earth, which is 6400 km. The trajectories of artillery shells that rise to several kilometers are sensitive to the change in acceleration and are also affected by the rotational motion of the earth. To determine whether or not certain factors can be ignored in a calculation, you either have to know the functional form of these factors and work out the theory, or you may be able to use rules of thumb from past experience.

Here's an example of using a rule of thumb to determine whether or not air friction can be ignored for an object thrown through the air. If an object is dropped from a great height, its speed will not increase indefinitely. Air friction will get greater and greater until there will be no more acceleration and the object will fall with a constant velocity. For a human body this *terminal velocity* is about 60 m/s. It depends on the size of the human, the clothing, and so on. For a parachutist the terminal velocity should be less than 7 m/s (15 mph). For a baseball the terminal velocity is about 75 m/s. If an object's speed is less than half its terminal velocity, air effects can usually be ignored for school-type trajectory problems. It does depend on the problem, however. This rule of thumb should not be applied to trajectories of spinning Ping-Pong balls, which are affected by air in another way.

Question 2-21

Using the simple formulas, from what height can you drop a baseball without having to take into account appreciable air friction effects?

SUMMARY

In this chapter we concentrated on describing motion in one dimension, along a line. For a complete description we must be able to give position, velocity, and acceleration of an object as functions of time—$x(t)$, $v(t)$, and $a(t)$. The average velocity during a time interval, Δt, is $\Delta x/\Delta t = (x_f - x_o)/(t_f - t_o)$. The instantaneous velocity at time t is equal to the limiting value of $\Delta x/\Delta t$ as Δt shrinks to zero about the time t. This definition of instantaneous velocity corresponds to the definition of the slope at a point on a graph of $x(t)$. The graph of $v(t)$ can be generated by measuring the slopes at various points on the graph of $x(t)$. Acceleration is defined as $\Delta v/\Delta t$. The graph of $a(t)$ can thus be generated by taking the slopes at various points on the graph of $v(t)$.

The graphs of $x(t)$ and $v(t)$ are also related in the following way. The area between the $v(t)$ curve and the axis represents the distance traveled during the time intervals marking the boundaries of the area.

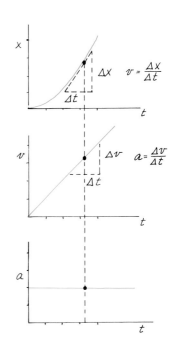

For the special case of constant acceleration, we developed several relationships between x, v, a, and t. These are

$$v_{\text{ave}} = \frac{v_f + v_o}{2}$$

$$x = x_o + v_o t + \tfrac{1}{2} a t^2$$

$$v_f^2 = v_o^2 + 2ax$$

Any equation must have terms with consistent units and dimensions. In the metric system, velocity is measured in m/s, and acceleration is given in m/s². The dimensions of these quantities describing motion are compounds of length (L) and time (T). The dimension of acceleration, for instance, is LT^{-2}.

The simple formulas describe many motions only approximately. They do not take into account such real factors as friction or variable acceleration. Nevertheless, many aspects of this introductory study of motion, such as the graphical representation, have general applicability.

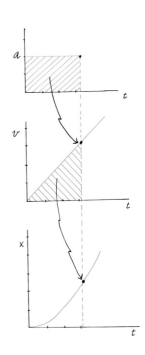

Question 2-22

There is one aspect of describing motion that we did not mention. Throughout this chapter it was assumed that you, the observer, were standing still and that you were measuring the motion of objects with respect to your position. How could the various descriptions change if you were in a car moving at 30 m/s? If you dropped a ball, would its path be a straight line? If you fired a gun with a muzzle velocity of 300 m/s, would its velocity depend on whether you fired it forward or backward?

Answers to Questions

2-1: Starting with 1.5 volts, each increase of voltage by a factor of 100 produces an increase of velocity by a factor of 10. But the change from 15,000 volts to 1,500,000 volts produces a velocity change of only 3. As we will see, the electrons cannot go faster than the speed of light, which is 3.0×10^8 m/s. The simple rules change in the relativistic region.

2-2: You might be able to dash 100 meters in 10 seconds. Your β would then be equal to $(10 \text{ m/s})/(3.0 \times 10^8 \text{ m/s}) = \frac{1}{3} \times 10^{-7}$. This speed is not in the relativistic range.

2-3: A horizontal line on a graph of $x(t)$ does not mean that the object is traveling to the right. The object stands still at x_o, but time marches on.

2-4: The distance traveled from Reno is 75 km, but that is not the position, x. The position, x, is equal to 75 km plus the original position, x_o, of Reno. Since x_o is 325 km, x would be 400 km.

2-5: Linearity does not imply proportionality. x is not proportional to t. It is $(x_f - x_o)$ that is proportional to t. If you double the time of travel, you double the *distance* traveled.

2-6: An example of $(x_f - x_o)$ in the negative region might be: $[-50 \text{ km} - (-25 \text{ km})] = (-50 + 25) \text{ km} = -25 \text{ km}$. For a negative slope, Δx is negative.

2-7: There cannot be sudden large changes in $x(t)$ since x is a cumulative value, dependent on all the distance covered up to time t. However, $v(t)$ can change value rapidly, regardless of the prior history of the motion. For instance, a brief stop at a stoplight would be represented by a very short horizontal line on the $x(t)$ graph, and the line would be at the value of x already reached. In the $v(t)$ graph, this brief stop becomes a sharp dip down to zero.

2-8: The policewoman had to arrest her son because although his *average* velocity was 30 mph, he was clocked at 60 mph in the middle of the block. It is a common experience of drivers on a busy highway that although they pass cars at 65 mph and try to press the speed limit of 55 mph, they still end up going only 45 miles in one hour.

2-9: The velocity meter that you use most makes use of a phenomenon that depends on instantaneous velocity. Most car speedometers employ a direct cable drive to a rotating magnet. The moving magnet induces a magnetic pull on a concentric cylinder. The magnitude of the pull depends directly on the speed of the magnet. This pull is balanced by a spring on the cylinder. The cylinder's position, indicating the pull and thus the speed, is given by a needle attached to it.

Weather forecasters and sailors have a variety of ways of measuring wind speed without measuring $\Delta x/\Delta t$. On the wind scale devised by Sir Francis Beaufort (1774–1857), and used ever since, a Beaufort "force number" of 5 corresponds to a "fresh breeze" of from 17 to 21 knots. Instead of measuring $\Delta x/\Delta t$, you observe the beginnings of whitecaps on the waves and know that the wind is about 19 mph.

The measurement of speed with radar does not appear to involve a calculation of $\Delta x/\Delta t$, but an argument can be made that the radar waves are doing just that. You can judge for yourself when we take a closer look at the nature of radar and the Doppler shift.

2-10: The actual surface of the graph paper is an area, but the axes of the graph may represent any kind of variable. In this case, the area of the graph represents the product of velocity and time.

2-11: There can be different kinds of averages. The simplest and most common average is calculated by adding the value of each component, and then dividing by the number of components. In this case the average of the values at 0 and 1 seconds is $(0 + 10)/2$. As we will see, the average velocity calculated this way is not always the average velocity determined by $\Delta x/\Delta t$.

2-12: Galileo worried about this question almost 400 years ago. There are circumstances where it is useful to know the velocity as a function of position, and how that velocity changes with position. Galileo demonstrated, however, that the mathematical representation of the motion of a falling object is particularly simple if the position and velocity are described as functions of *time*.

2-13: The distance traveled during the four seconds is represented by the area under the $v(t)$ curve for the four second interval. The area of the triangle is 40 m. The distance traveled in the first two seconds is equal to the distance traveled in the last two seconds.

2-14: From qualitative observations it is not possible to determine what happens to the acceleration. The speed, however, clearly increases during the time of fall, which agrees with the graph of $v(t)$. The measurement of the position of the rolling ball at half-time should also agree qualitatively with the $x(t)$ graph. The ball went much further during the second half-time than it did during the first.

2-15: If you record your data as actual *points* on the graph, you are stating that at exactly t_1 the value of x was exactly x_1. Suppose, however, that you can measure x only to plus or minus $\frac{1}{2}$ mm. Then on the graph, that value should be indicated by a vertical bar, centered on your best guess for the number and extending $\frac{1}{2}$ mm up and $\frac{1}{2}$ mm down. When you draw the graph line, it should be a smooth curve going through all the vertical bars, but not necessarily going through the center points.

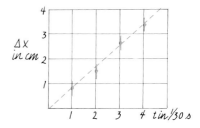

2-16: According to Equation 2-3a, the distance is proportional to the square of the time. If the ratio of distances is 1/2, the ratio of times must be $\sqrt{1/2}$, which is 1/1.4.

$$\frac{\text{Time for half distance}}{\text{Time for whole distance}} = \frac{1}{1.4}$$

$$\frac{\text{Distance in half time}}{\text{Distance in whole time}} = \frac{(\frac{1}{2}T)^2}{T^2} = \frac{1}{4}$$

2-17: Here is the relationship that you might expect:

$$\frac{\text{Velocity at half distance}}{\text{Velocity at whole distance}} = \frac{\sqrt{2a(L/2)}}{\sqrt{2aL}} = \sqrt{\frac{1}{2}} = \frac{1}{1.4}$$

2-18: $v_f^2 = v_o^2 + 2ax$. The dimensions of this equation are

$$(L/T)^2 = (L/T)^2 + (L/T^2)(L)$$
$$= (L/T)^2 + (L/T)^2$$

2-19: According to these equations, the speed at every position, x, is the same when the ball is going up as when it is coming down.

2-20: The mathematical model has no way of knowing that prior to t_o, the ball was in your hand. The initial conditions would be satisfied by a ball shot up from the ground and passing the 2 meter mark with a speed of 20 m/s. The ball would have left the ground (starting with a speed of 21 m/s) at 0.1 seconds before the t_o of our problem.

2-21: Suppose we set 40 m/s as the velocity at which we begin to notice air friction effects. To attain that speed, a ball would have to fall a distance, x, deduced from the formula

$$v_f^2 = 2ax$$
$$(40 \text{ m/s})^2 = 2(10 \text{ m/s}^2) x$$
$$x = 80 \text{ m}$$

2-22: These questions are explored in Chapter 11. In the meantime, note that you already are on a moving system, namely, the earth. If you drop a ball while you are in a moving car, you will see it drop straight down. An observer beside the road would see it move in a parabolic trajectory. With respect to you, the speed of the bullet would be 300 m/s regardless of the direction you fire it. However, an observer beside the road would measure the bullet's speed to be 330 m/s if you fired it forward, and 270 m/s if you fired it backward.

Problems

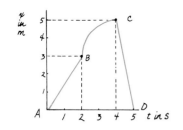

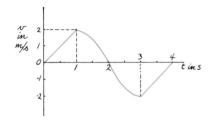

1. What is β (ratio of speed of an object to the speed of light) for a supersonic airplane flying at 1000 m/s?
2. The location of a particular car along the New York State Thruway is given by $x = (60 \text{ mph})t + 120$ miles, where x is the distance from New York City, and t is travel time, starting at 9:00 A.M. How far from New York City was the car at 9:00 A.M.? 12:00 noon? 8:00 A.M.?
3. In the first figure, what is the distance from the origin at B? At C? At D?
4. In the first figure, what is the distance traveled from A at B? At C? At D?
5. In the first figure, what is the velocity from A to B? What is the average speed from B to C? What is the velocity from C to D? What is the average velocity from A to D?
6. If $x = (60 \text{ mph})t + 120$ miles, show that $\Delta x/\Delta t = 60$ mph. (For instance, let $\Delta t = 1$ hr. Find x_o when $t_o = 0$ and x_f when $t_f = 1$).
7. In the second figure, how far has the object traveled during the first second? How far is the object from the origin at the end of three seconds? At the end of four seconds?
8. In the second figure, what is the acceleration at $t = \frac{1}{2}$ s? At $t = 3\frac{1}{2}$ s?
9. In the second figure, what is the distance traveled during the time between $t = 1$ s and $t = 2$ s?
10. What is the speed of a ball, dropping freely from rest, after one-tenth second? After one-half second? After one second?
11. How far does a ball fall, starting from rest in the first one-tenth second? In the first half second? In the first second?
12. What is the speed of a ball, dropping freely from rest, after it has fallen 1 cm? 1 m? 10 m?
13. If you throw a ball upward with an initial speed of 10 m/s, how long will it take to reach zero speed? How far will it rise? What is its velocity when it reaches your hand again?
14. A car traveling at 30 m/s undergoes constant acceleration of 2 m/s² for 5 s. What is its final speed? How far did it travel during this time?
15. A car traveling at 30 m/s undergoes constant acceleration of 2 m/s² for 100 m. What is its final speed? How long did it take?
16. The half-life of a muon is 2×10^{-6} s. Many of these are created by primary cosmic rays at the top of the atmosphere. They travel at nearly the speed of light. What fraction of those produced at a height of 30 km would you expect to reach us on the earth's surface?
17. To measure the height of your physics building, you could time the fall of a stone dropped from the roof. Is this feasible for a measurement good to 5%? (Assume a reasonable height; calculate the expected time of fall; judge the uncertainty you would have with a stopwatch measurement.)
18. A parachutist often lands at about 15 mph. If you wanted to practice landing at such a speed by jumping from a platform, what height should the platform be?

19. In the last figure the (idealized) velocity of a car is graphed as a function of time. Assume the car starts directly in front of you at the origin. Describe qualitatively its motion during the four seconds: what direction is it going; what happens to its speed; what is its acceleration; where does it go? Then draw the corresponding graphs for $x(t)$ and $a(t)$.
20. If a baseball is dropped from a balloon at a height of 1 km, about how long does it take to fall to earth?
21. The electrons in a television tube travel at constant speed 0.3 m (from electron gun to tube face) in 1×10^{-5} s. What is their speed?
22. If "up" is positive, what are the signs of velocity and acceleration for (a) an elevator rising and coming to rest; (b) an elevator starting to descend; (c) an elevator descending and coming to rest; (d) an elevator rising at constant speed?
23. In order to provide the comforts of home, a space ship accelerates in free space with the same acceleration produced by gravity on earth. How long will it take to reach 1/10 the speed of light? How far will it have gone in this time?
24. Superman is standing in a window 30 m above the street. A baby hurtles past, having been dropped from a window 15 m higher. With what constant acceleration (starting from rest) must Superman descend to catch the baby just before it is too late?
25. From a height of 60 m, a man throws a rock straight down with an initial speed of 10 m/s. Neglecting air resistance, how long will it take the rock to strike the ground? How fast will it be going when it hits?
26. You have a car that can accelerate from 0 to 60 mph in seven seconds. However, the speed limit is 60 mph. A car going a steady 40 mph passes you just as you start up from a stoplight. Obeying the law, how long will it take you to pass the other car? How far will you have gone? (It would be best to change units to m/s. One method of solution is to graph $x(t)$ for both cars on the same graph.)
27. An object moving at -30 m/s changes its velocity to -15 m/s in 5 seconds. What is the average acceleration?

3 MOTION IN TWO DIMENSIONS

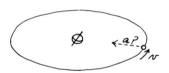

Is the moon falling down?

There are several problems involved in describing motion in two or three directions. To begin with, there is a problem in defining the distance traveled. Is the distance between two hilltops the distance you would have to walk to go from one to the other, or is it the distance between them as the crow flies? Another problem is whether motion in one direction affects motion in a second direction. If you throw a ball horizontally, it has an initial velocity parallel to the ground and meanwhile is accelerating downward because of the gravitational pull. Is the acceleration down reduced by the horizontal velocity? It certainly is in the case of airplanes! Finally, in this chapter we must take seriously the difference between speed and velocity. As we will see, there is a way an object can move with constant speed while at the same time it has a constant magnitude of acceleration. If the constant acceleration is always perpendicular to the velocity, the object will travel in a circle.

HANDLING THE PHENOMENA

1. In describing position and motion in two dimensions, we will continually refer to diagrams showing distances along the x- and y-axes and angles between lines. It is not necessary to be an artist to draw these diagrams, and the math is simple. All you need is a centimeter ruler, a protractor, and a set of trig tables giving sine, cosine, and tangent.

Here are some exercises in drawing and measuring distances in two dimensions. Consider these to be Question 3-1. Some comments about the exercises appear at the end of the chapter in Answers to Questions.

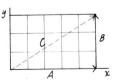

By actual measurement, does $C^2 = A^2 + B^2$?

a. Draw a line 5 cm long on the x-axis, starting at the origin. Draw another line 3 cm long in the y-direction, starting at the end point of the first line. Complete the triangle with a line between the two end points. Measure the length of this hypotenuse with your ruler. Do the sides of the triangle satisfy the Pythagorean theorem that $C^2 = A^2 + B^2$?

b. Draw a line 10 cm long, starting at the origin, at 30° from the x-axis. From the end of the line, drop dotted lines perpendicular to the x- and

y-axis. Do the measured lengths of the dotted lines satisfy the trig relationships for sine, cosine, and tangent?

c. Draw a full scale diagram of the lines shown here, and measure the distances x_A, y_A, x_B, y_B, x_C, and y_C. Compare

$(x_A + x_B)$ with x_C, and $(y_A + y_B)$ with y_C.

2. Does an object shot horizontally fall at the same rate as one that is just dropped? If you compare the fall of a stone with the fall of a paper glider launched from the same height, you know that the horizontal motion of the glider can keep it aloft long after the stone has hit the floor. Obviously, in this case it is the air that is responsible for the dependence of the vertical velocity on the horizontal velocity. To test the original question, you must either eliminate the air, or use objects whose motion is relatively independent of the air.

One way to test the independence of vertical and horizontal velocity is to drop one object at exactly the same time that another one is shot horizontally. Then compare the time that it takes them to fall. Both objects must be dense enough and slow enough so that the effect of air resistance is negligible. An easy way to launch two objects simultaneously is to balance a coin on the edge of a table. Shoot another coin along the table so that it gives a glancing blow to the first one. Both will leave the table at the same time, with one going horizontally faster than the other. You can hear them hit the floor, and determine whether or not the two sounds occur at the time time. The test is very sensitive because you can hear a difference in the arrival time of two sounds of at least 1/30 second.

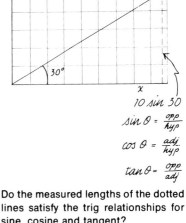

Do the measured lengths of the dotted lines satisfy the trig relationships for sine, cosine and tangent?

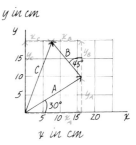

Does the sum of the components equal the component of the vector sum?

Question 3-2

The implication of the results of this experiment is that a baseball hurled horizontally will strike the ground as quickly as if it were dropped. How long does it take a coin to drop from a table to the floor? How far can a baseball be thrown in this time? (Assume a reasonable pitching speed.) Would the same sort of calculation apply to the flight of a rifle bullet?

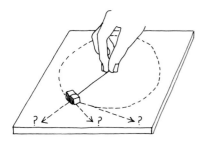

3. We claimed in the Preface that an object going at constant speed in a circle is undergoing acceleration with constant magnitude. If that seems like a paradox, see what happens to a rotating object when you remove the constraint that is forcing it to travel in a circle. Tie a string to a rubber eraser or soft ball, and twirl the object in a horizontal circle on the floor or a table. When it is traveling smoothly at constant speed, let go of the string. Note carefully the subsequent path of the object. Is the path radial? Tangential? Or something in between? It would be a good idea to have someone else observe the path also and compare notes.

POSITION IN TWO DIMENSIONS

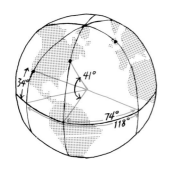

To locate a point on a line, whether the line is straight or twisted, only one number is needed—the distance from some origin. To locate a point on a surface, two numbers are needed. On the surface of the earth, for instance, 40.7° N, 74° W specifies the location of New York City. The first number is the *latitude,* the number of degrees north of the equator. The second number is the *longitude,* the number of degrees, along the equator, separating the meridian passing through New York from the zero meridian, which passes through Greenwich, England. The position numbers of Los Angeles are 34° N and 118.2° W. If you know those two sets of numbers, and the radius of the earth, you can find the distance between New York and Los Angeles by using spherical trigonometry, but the formulas are complicated.

On a flat plane, position can be specified by giving the x- and y-coordinates of the point, or by giving r and θ, the length and angular position of a line from the origin to the point. For most of the work in this book, we will use the so-called Cartesian coordinates, x, y. (These are called Cartesian coordinates in honor of René Descartes, a French mathematician and philosopher (1596–1650). Sometimes physicists are accused by classicists of putting Descartes before Horace.) With these coordinates, it is particularly easy to calculate the distance between two points.

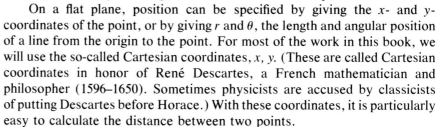

The distance between two points along a wandering path may be much longer than the straight line distance between the points. We call the straight line distance the *displacement.* The easiest way to calculate its length is to break the displacement into projections along perpendicular axes, such as north-south and east-west, or x and y. These projections are found by dropping perpendicular lines to the axes, as shown in the diagram. They are called the *components* of the displacement. The length of the displacement is related to the length of the components by the Pythagorean theorem: $D^2 = D_x^2 + D_y^2$.

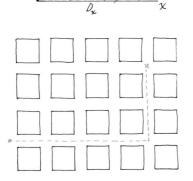

Here's an example of a displacement and its components in two dimensions. Suppose that you walk four blocks east and two blocks north. How far are you from your starting point? Of course, for most purposes in a city, you are six blocks from where you started. Not even a crow would go in a diagonal line across a block with skyscrapers. Your displacement, however, is along such a diagonal. If the east component is 1200 m, and the north component is 600 m, then

$$\text{length of displacement} = \sqrt{(1200 \text{ m})^2 + (600 \text{ m})^2} = 1340 \text{ m}$$

VECTOR QUANTITIES

A series of displacements can be added together to find the net displacement. The easiest way to do this is to break each of the separate displacements into components, and then add together those components that are along the same direction. If these are *x*- and *y*-components, for instance, then you end up with the *x*- and *y*-components of the net displacement, which is called the *resultant*.

Here's an example. Suppose that you walk 10 m in a direction 30° north of east; then 6 m in a direction 45° west of north; then 5 m due west; and finally, 8 m at 60° south of west. What is your net displacement? The first thing to do with any problem involving this much data is to draw a diagram. The addition of vectors in a diagram is shown by drawing the arrows head to toe, each to scale and at the specified angle. In adding vectors graphically like this, an arrow may be moved, and added in any sequence, as long as its magnitude and direction are not changed.

A sketch is satisfactory as long as some care is taken with the approximation of angles and scale. Each of the separate displacements should be broken into components, and the magnitudes of each line should be computed and labeled. Next, draw up a table of the N-E or *x-y* components. A component of 5 m N should be listed as +5 in the N column, and a component of 5 m S should be listed as −5. Add the components in the two separate columns and compare the results with the qualitative appearance of the components of the net displacement, or resultant, in your diagram. The length of the resultant can be found by using the Pythagorean theorem, and the direction of the resultant can be found in terms of the tangent relationship shown in the diagram.

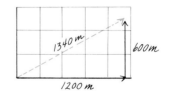

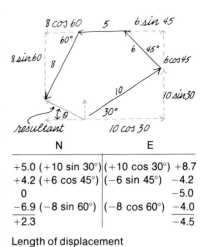

N	E
+5.0 (+10 sin 30°)	(+10 cos 30°) +8.7
+4.2 (+6 cos 45°)	(−6 sin 45°) −4.2
0	−5.0
−6.9 (−8 sin 60°)	(−8 cos 60°) −4.0
+2.3	−4.5

Length of displacement

$$= \sqrt{(2.3)^2 + (4.5)^2} = 5.1$$

$$\tan \theta = \frac{2.3}{4.5} = 0.51$$

$$\theta = 27° \text{ N of W}$$

Question 3-3

Where would you end up if you made the four moves in the opposite order—starting out with a displacement of 8 m at 60°S of W? Where would you end up if you made each of the moves in exactly the opposite direction—starting out with a displacement of 10 m at 30°S of W?

VECTOR QUANTITIES

A displacement is different from a simple distance, since it involves a particular distance *and* a particular direction. Furthermore, as we have seen, displacements do not add like distances. To add displacements you have to break them into components and then reassemble them. Quantities that follow these rules are called *vectors*. In the diagrams we indicate vector quantities by putting an arrow over the symbol for the quantity; in the text we use **boldface** type to indicate vectors. Displacement along the *x*-axis, for instance, might be labeled $\overrightarrow{\Delta x}$ in the diagram and Δx in the text. If only the magnitude of the vector is being considered, "absolute value" bars are used around the symbol. For instance, the magnitude of $\Delta \mathbf{x}$ is $|\Delta \mathbf{x}|$.

Every time that we take up a new topic in this study, we will define new quantities—mass, energy, momentum, temperature, electric potential, and so on. Besides describing how to measure each quantity, we must choose a mathematical representation that behaves like the quantity. When we measure time, for instance, are we apt to get +1 second if our clock is lined up north-south, and −1 second if it is lined up south-north? If so, we might have to use vector mathematics to describe time. To be represented by a vector, a quantity must have magnitude, direction, and combine in the same way that displacements combine.

If a quantity does not have a direction associated with it, it is called a *scalar*. Time is a scalar even though time can be positive or negative and always seems to be progressing in the same direction. Actually, a negative time is negative only with respect to an arbitrary zero point. During a rocket launch countdown, for instance, there is ignition at −3 seconds. Similarly, we could date Hannibal's crossing of the Alps as −218, although the date is more commonly listed as 218 B.C.

Can velocity be described by scalar mathematics? In recording motion along a road, scalar math usually describes the events of interest. Regardless of the displacement between two cities, your travel time is determined by the actual road distance and your speed as measured by your speedometer. On the other hand, an airplane or boat can move in two dimensions and the directional properties of its motion are important. Vector mathematics describes the important features in navigation. Following the standard usage of the words, we will call *speed* the distance traveled per time, without regard to the direction. Speed is a scalar quantity. *Velocity* is the displacement per time, and takes into account both speed and direction. Velocities combine in the same way that displacements do, and so velocities are vector quantities.

Question 3-4

If you are on the outer edge of a merry-go-round that is 10 m in diameter and is rotating one revolution every 10 s, what is your average speed during one revolution? What is your average velocity?

COMBINING VELOCITIES

When you add displacements, you usually make one move and then another move. In adding velocities, however, you must deal with two motions that are happening at the same time. Take a look at the problem shown in the first diagram. What is the velocity of the ball with respect to the ground? The dotted arrow found by vector addition shows both the resulting magnitude and direction.

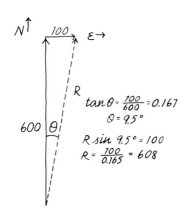

Here's another example. Consider a plane that has an air speed of 600 km/hr and is heading due north through a jet stream blowing at 100 km/hr from west to east. After one hour of such flight, the plane would be 600 km north and 100 km east of its starting point. Its velocity with respect to the ground was 608 km/hr at an angle of 9.5° east of north. The vector diagram for such a combined motion looks just like a displacement diagram. To solve any problem like this, sketch the vectors roughly to scale, label them, break them into perpendicular components, label the components, add together the x-components and then the y-components, find the magnitude and direction of the resultant.

Here's another example of combining velocities. Suppose that you want to cross a river that is flowing due south at 6 km/hr. In still water your boat can travel at 10 km/hr. You want to go straight across the river from the west bank to the east bank. To do that you will have to keep the boat heading at some angle upstream all the way across. (If you head due east, the current will carry you downstream.) Note that in this case the direction of the resultant is known (due east) but its magnitude must be determined. The magnitudes of two velocities are known (6 km/hr and 10 km/hr), but the direction of the boat heading and the resultant velocity across the river must be found. This is a typical situation in navigation. To solve such a problem, it is useful not only to sketch a vector diagram, but also to draw up a table of what is known and not known about the separate vectors. Three vectors are involved: the velocity of the stream, the velocity of the boat with respect to the water, and the resultant velocity of the boat with respect to the land. The table must give two quantities for each of the three velocities: the magnitude and the direction. Of the six quantities in any problem like this, four are invariably given and two remain to be found.

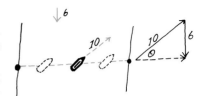

	Angle	Magnitude
Stream	S	6
Boat with respect to water	? θ	10
Boat with respect to land	E	? v

$$\tan \theta = \frac{6}{10} \quad \theta = 31°$$
$$v = \sqrt{10^2 - 6^2} = 8 \text{ km/hr}$$

> **Question 3-5**
>
> What two quantities were unknown in the problem of the plane heading north through the jet stream heading east?

In real navigation problems, the required directions may not be so convenient as due north or east. Suppose that a pilot wants to fly along a straight line between Atlanta and Chicago, but the jet stream is blowing *from* an angle of 30°N of W, and therefore the direction of its velocity is 30°S of E. (The direction of a "north" wind is to the south.) Chicago is on a bearing 17° W of N from Atlanta. The standard rules for treating vector problems still apply. Draw a sketch, break vectors into components, draw up a table of quantities. In this case, the heading of the plane is not known, although you know that the plane has to head at an angle larger than 17° W of N in order to buck the jet stream that is trying to push the plane east. As you can see in the worked-out solution to this problem, the components of the plane's velocity are given in terms of the unknown angle, θ, in which the plane must be headed. Note that this angle shows up in an equation in which it is the only unknown variable. In most school exercises, an algebraic equation with just one unknown can be solved immediately. In this case, however, the variable is embedded in both a sine and cosine function. The easiest way to solve such an equation is to zero in on the answer by trying a series of values for the unknown. We know from the sketch that θ must be larger than 17°. Without using the trig tables you can see that 30° is too large, since $(600 \sin 30°) = 300$, and $(180 \cos 30°)$ cannot be greater than 180. Therefore, $(600 \sin 30° - 180 \cos 30°) > 120$. The numerical results of using several other choices for θ are shown in the problem solution.

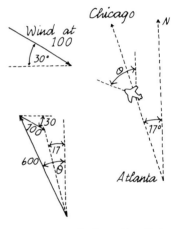

	Angle	Magnitude
Jet stream	30° S of E	100
Plane—air	? θ	600
Plane—ground	17° W of N	? v

$v \cos 17° = 600 \cos \theta - 100 \sin 30°$
 (North component)
$v \sin 17° = 600 \sin \theta - 100 \cos 30°$
 (West component)
$\tan 17° = 0.3 = \dfrac{600 \sin \theta - 87}{600 \cos \theta - 50}$
$180 \cos \theta - 15 = 600 \sin \theta - 87$
$600 \sin \theta - 180 \cos \theta = 72$
$\theta = 20° \quad 205 - 169 = 36$
$\theta = 25° \quad 254 - 163 = 91$
$\theta = 24° \quad 244 - 164 = 80$
$\theta = 23° \quad 234 - 165 = 69$
$\theta = 23.3°$

(Note that in demonstrating the method, we have found more significant figures than are justified by the original data.)

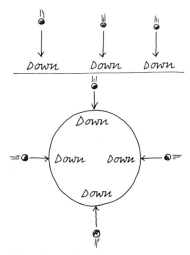

"Down" on a flat earth.
"Down" on a spherical earth

FLAT EARTH TRAJECTORIES

As long as you aren't watching ships sail over the horizon, the earth appears flat as far as your eye can see. In the flat earth approximation, objects that are dropped fall perpendicular to the earth, and thus their paths are parallel to each other. Actually, the paths of dropped objects lie along radii of the earth and so are not parallel to each other. To plot the orbits of earth satellites you must assume that the earth is spherical and that there is a radial acceleration of the satellite. To plot the trajectories of objects tossed for distances of less than a kilometer or so along the earth's surface, you can simplify the problem by assuming that the earth is flat and that any object free to move will accelerate downward in a direction parallel to all other downward directions.

In working out the trajectories of objects moving above a flat earth we must make one other assumption, one that you tested in "Handling the Phenomena." We assume that we can resolve the velocity of an object into vector components, parallel and perpendicular to the earth, and that there is no interaction between the component motions. The assumption is good as long as air effects are negligible. This means that regardless of whether a ball is thrown up or down or horizontally, as long as it is free to move, it is accelerating straight down at the rate of 9.8 m/s². Meanwhile, its horizontal motion continues independently of the vertical fall—at least until it hits the earth.

With these assumptions, every trajectory problem can be turned into two separate problems, each concerned with motion along a straight line. The initial velocity of a thrown object can be broken up into a vertical and a horizontal component. The vertical motion can be calculated with the formulas for one-dimensional motion with constant acceleration such as ($y = v_o t + \frac{1}{2} g t^2$) and ($v_f^2 = v_o^2 + 2gh$). All these v's refer to the vertical component of the object's velocity. The horizontal motion is simply one of constant velocity as long as the object is in the air. The constant velocity is the horizontal component of the initial velocity. The horizontal distance traveled is given by $x = v_H t$, where v_H is the horizontal component of the object's velocity.

1. Suppose that you pitch a ball horizontally from a height of 1.6 m with a speed of 30 m/s. The initial vertical component of velocity is zero, and so the vertical motion consists of a drop from a height of 1.6 m with constant acceleration.

$$y = y_o + \tfrac{1}{2} a t^2$$
$$0 = 1.6 \text{ m} + \tfrac{1}{2}(-9.8 \text{ m/s}^2) t^2$$
$$1.6 \text{ m} = \tfrac{1}{2}(9.8 \text{ m/s}^2) t^2$$
$$t^2 = 0.326 \text{ s}^2$$
$$t = 0.57 \text{ s}$$

During this time, the ball's horizontal velocity carries it a distance
$$x = v_H t = (30 \text{ m/s})(0.57 \text{ s}) = 17 \text{ m}$$

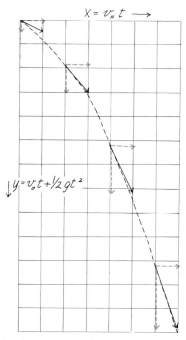

(y, x) trajectory of a falling object with an initial horizontal velocity.

This horizontal distance is less than the regulation baseball distance of $60\frac{1}{2}$ ft between pitcher and batter. The assumed speed of 30 m/s is about 67 miles/hr, which is a fast pitch for an amateur, though a professional may throw as fast as 100 miles/hour. Evidently, most pitches should not start out horizontally but must have a slight upward component of velocity.

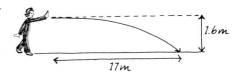

Question 3-6

Can a fast ball pitcher with a speed of 50 m/s throw horizontally and still get the ball in the strike zone? The pitcher's mound is 15 inches above the rest of the field.

2. An outfielder, trying to return the ball to the catcher, does not throw the ball horizontally. Instead, he aims the throw up at an angle and with a speed that experience has taught him will produce the right trajectory. We can analyze the motion by breaking it up into vertical and horizontal components. Suppose that the original velocity is 30 m/s at an angle from the horizontal of 30°. Then the vertical component of the original velocity is (30 m/s)(sin 30°) = 15 m/s straight up. The horizontal component is (30 m/s)(cos 30°) = 26 m/s. As long as the ball is in the air, it will travel 26 m/s parallel to the ground. The time that the ball is in the air is determined by how long it takes to rise and fall with an initial upward speed of 15 m/s. That time can be found by using the appropriate formula for constant acceleration:

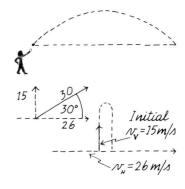

$$y = y_o + v_o t + \tfrac{1}{2}at^2$$

$$0 = 0 + (15 \text{ m/s})t + \tfrac{1}{2}(-9.8 \text{ m/s}^2)t^2$$

$$5t = 15 \text{ (dividing by } t, \text{ and assuming } 9.8 \approx 10)$$

$$t = 3 \text{ s}$$

The time that it takes the ball to rise to its maximum height could also be found by using the formula relating final velocity, acceleration, and time:

$$v_f = v_o + at$$

$$0 = +15 \text{ m/s} + (-9.8 \text{ m/s}^2)t$$

$$t = 1.5 \text{ s (once again, letting } 9.8 \approx 10)$$

Discounting air friction, a ball takes as long to fall as it does to rise. Therefore, the total time of the flight is 3 s, which is the answer we got using the first method. The horizontal distance traveled by the ball—its range—is

$$x = v_H t$$

$$x = (26 \text{ m/s})(3 \text{ s}) = 78 \text{ m}$$

Compare this range, obtained with a small upward component of the initial velocity, with the 17 m range obtained in Example 1 with the same initial speed.

Question 3-7

The comparison is not quite fair! In Example 1, we assumed that the ball started out 1.6 m above the ground (a reasonable height for a thrower's hand at the moment of release of the ball). In Example 2, we calculated the range covered by the ball during its flight up and down to the initial height. Does the ball then travel an additional 17 m in falling the additional 1.6 m?

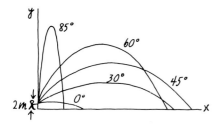

3. For a given speed, a ball thrown at 30° from the horizontal has a longer range than a ball thrown exactly horizontal. Can the range be made greater yet by throwing a ball nearly vertically? Of course, if a ball is thrown exactly vertically, it will come right back down where it started from and its range will be zero. Evidently, there must be some optimum angle between 0° and 90° that will produce maximum range. One way to find this optimum angle is to substitute a number of angles into the equations and zero in on the approximate solution. In this particular problem there is a nice alternative method that yields an exact solution. We will deal with the (original) vertical and horizontal components of velocity in terms of their angular relationships: $(v_V)_o = V \sin \theta$, and $v_H = V \cos \theta$. The two separate equations for x and y then become

$$y = y_o + v_o t + \tfrac{1}{2} a t^2$$

$$x = v_H t$$

$$0 = 0 + (V \sin \theta)t + \tfrac{1}{2}(-10 \text{ m/s}^2)t^2$$

$$x = (V \cos \theta)t$$

Solve the first equation for t, and substitute that value for t into the second equation:

$$t = \tfrac{1}{5}(V \sin \theta)$$

$$x = \tfrac{1}{5} V^2 (\sin \theta \cos \theta)$$

It appears that we have ended up with one of those equations that contains what we want to know in terms of only one variable (θ), but in two different functions of the variable. Once again, we could solve for maximum x, which is what we want to know, by trying different values of θ and zeroing in on the largest value of x. In this case, there is a simplifying trick. One of the trigonometric identities is

$$2 \sin \theta \cos \theta = \sin 2\theta$$

The formula for the range, x, can thus be written

$$x = (1/10) V^2 (\sin 2\theta)$$

The greatest value that x can have occurs when $(\sin 2\theta) = 1$, since the sine of an angle cannot be greater than 1. When $(\sin 2\theta) = 1$, $2\theta = 90°$. Therefore, for maximum x, $\theta = 45°$. Try it for yourself with a baseball!

Question 3-8

For an initial speed of 30 m/s, how does the maximum range compare with the ranges computed earlier for angles of 0° and 30°?

4. In solving these trajectory problems, we treated the vertical and horizontal motions separately. The vertical position, y, and the horizontal position, x, are independent functions of time. Of course, for every point, x, along the trajectory, the object was at a particular height, y. Instead of describing the motion as a function of time, we could describe the trajectory in terms of y as a function of x, or x as a function of y. In order to do this, we just eliminate the time, t, from the two separate equations for x and y.

$$y = v_V t - 5t^2$$
$$x = v_H t \quad \text{or} \quad t = x/v_H$$
$$y = (v_V/v_H)x - (5/v_H^2)x^2$$

The vertical distance traveled is a function of the square of the horizontal distance. The formula represents a very specific geometric curve, called a parabola. *All* trajectories above a flat, airless earth are parts of parabolas. In the diagram we contrast the parabolic path with a circular path and an elliptical path going through the same three points.

Question 3-9

The orbits of the moon and all the artificial satellites around the earth are ellipses, which in many cases are nearly circles. Yet the claim has just been made that the trajectories of objects above the earth are parabolic. How does an object that is launched near the surface of the earth know whether to follow a parabolic or an elliptical path?

Elliptical and parabolic trajectories on the earth.

ON GOING IN CIRCLES

In real orbits of satellites around the earth, the acceleration caused by gravity depends on the distance from the earth, and is radial. The resulting trajectory depends on the satellite's initial velocity, both on its speed and its direction. One special case of this type of motion is easy to solve, and that is when the satellite is going in a circle. Not only is circular motion important in its own right, because many objects travel exactly in circles, but it is also a good approximation to the actual orbits of many satellites.

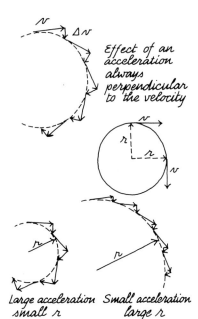

Effect of an acceleration always perpendicular to the velocity

Large acceleration small r Small acceleration large r

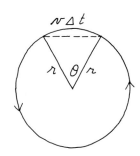

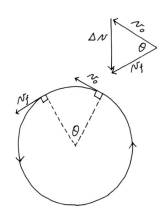

The moon's orbit around the earth, and the earth's orbit around the sun, are, to first approximation, circles.

Parabolic trajectories are produced when the acceleration is constant in both magnitude *and* direction. Circular trajectories are produced when the acceleration on an object is constant in magnitude, *but* the direction of the acceleration continually changes so that it is always perpendicular to the velocity. It might seem at first thought that if you continually accelerate a moving object, you will eventually warp its path so that it will move mostly in the direction of the acceleration. That's what happens in the flat earth trajectories. On the other hand, if the acceleration is always perpendicular to the velocity, the velocity can never increase or decrease in magnitude. The speed remains constant, but its direction continually changes.

The instantaneous velocity of an object moving in a circular path is along a direction tangent to the circle. You could see that this is the case when you released the rotating object in the "Handling the Phenomena" exercise. In a circle the radius is always perpendicular to the tangent. Therefore, an acceleration that is always in a radial direction will always be perpendicular to the tangential velocity of an object traveling in that circle. The amount of acceleration needed to force the object to travel in a particular circle depends on the velocity of the object and the radius of the circle. Only a small radial acceleration is needed to keep a slowly moving object following a circular path with a large radius. On the other hand, it would take a very large radial acceleration to warp the path of a rapidly moving object into a small circle.

The radial acceleration needed to produce circular motion is called *centripetal* acceleration. The word literally means "center seeking." (In Chapter 11 we will distinguish between centripetal and centrifugal acceleration—they're the same thing, but seen from different reference frames.) In the following derivation, we will find how a_c depends on v and r. Note that we work with two diagrams, one for the circular path, and the other for the velocities. The triangles that are formed are similar because the instantaneous velocities are tangent to the circle, and tangents are perpendicular to radii. During a small interval of time, Δt, the object moves through the arc length, $v\Delta t$. The figure formed with that arc as a base is not exactly a triangle, but if the time Δt is short enough, the arc is almost equal to the chord. To form the velocity triangle, simply move the two vector velocities so that their directions are maintained. The original velocity is turned into the final velocity by the change of velocity, Δv. For very short times, Δt, the direction of Δv must be radial.

$$\mathbf{v}_f = \mathbf{v}_o + \mathbf{\Delta v}, \text{ but in magnitude,}$$

$$|\mathbf{v}_o| = |\mathbf{v}_f| = |\mathbf{v}|$$

Since the velocity triangle is similar to the circular path triangle,

$$\frac{\Delta v}{v} = \frac{v\Delta t}{r}$$

Cross multiplying some of these terms, we get

$$\frac{\Delta v}{\Delta t} = \frac{v^2}{r}$$

By definition, $\Delta v/\Delta t$ is acceleration. Its magnitude is proportional to the square of the velocity, and inversely proportional to the radius of the circle. Since, as the diagram shows, it is in a radial direction, it always remains perpendicular to the tangential velocity of an object moving in a circular path.

> **Question 3-10**
>
> Does this formula for centripetal acceleration agree with our common sense expectations? The required acceleration is small for a large radius. On a merry-go-round, isn't there greater centripetal acceleration at the rim than near the center?

Examples

1. Let's calculate the magnitude of centripetal acceleration required to maintain a circular orbit close to the earth. The first earth satellites orbited in the late 1950s traveled at heights of only a couple of hundred kilometers, barely above the atmosphere. For a rough calculation, assume that the radius of orbit was about the same as the radius of the earth—6400 kilometers. The period for such an orbit is a little less than an hour and a half—5100 seconds. The speed of the satellite is equal to the circumference of the orbit, divided by the time it takes to go around.

$$v = \frac{2\pi r}{T} = \frac{2 \times 3.14 \times 6.4 \times 10^6 \text{m}}{5.1 \times 10^3 \text{ s}} = 7.9 \times 10^3 \text{ m/s}$$

The necessary acceleration, therefore, must be

$$a_c = \frac{v^2}{r} = \frac{(7.9 \times 10^3 \text{ m/s})^2}{6.4 \times 10^6 \text{m}} = 9.7 \text{ m/s}^2$$

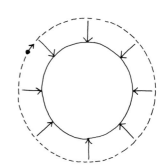

A satellite in orbit continually falls toward the center of the earth.

That calculated centripetal acceleration is just about equal to the free-fall acceleration on the surface of the earth. Apparently the satellite is continually falling down—radially—and therefore is able to maintain its circular orbit around the circular earth. If it did not accelerate downward, it would go sailing off tangent to the earth.

2. In one of the large particle accelerators—or atom smashers—protons travel at almost the speed of light in an evacuated horizontal circular path with a radius of 100 meters. Let's compare the centripetal acceleration that produces such an orbit with the vertical acceleration of the protons due to gravity. The centripetal acceleration is

$$a_c = \frac{v^2}{r} = \frac{(3 \times 10^8 \text{m/s})^2}{1 \times 10^2 \text{m}} \approx \frac{10^{17}}{10^2} = 10^{15} \text{ m/s}^2$$

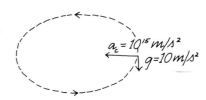

The required centripetal acceleration is 10^{14} times greater than the vertical acceleration caused by gravity. Evidently, the falling of the protons is a minor matter compared with their horizontal acceleration.

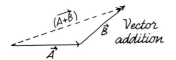

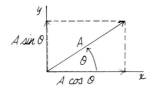

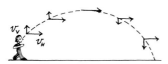

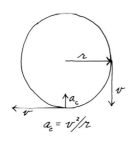

SUMMARY

In order to deal with motion in two dimensions, it is necessary to use vector algebra. Displacements are vectors, and their method of combining serves as the defining behavior of vectors. If a variable must be described in terms of both magnitude and direction, and if its rules of combination are the same as those of displacements, then it is a vector quantity. Velocity and acceleration are vectors; time is not. Since the magnitude of velocity is a frequently used quantity, it is called *speed*.

Vector problems are best analyzed by sketching the situation roughly to scale, and then breaking all the vectors into perpendicular components. All the components in the same direction can then be added algebraically, and recombined with the total of the components in the perpendicular direction to form the resultant.

When dealing with problems of motion in a moving medium, such as a boat crossing a stream, a foolproof method of solution is to draw up a table of values of magnitude and direction for each of the three relevant velocities. Usually two of the six variables are unknown. These can be found by writing the two equations linking the two perpendicular components.

There are many very complicated two-dimensional motions. We examined only two special cases: (1) constant horizontal velocity and constant vertical acceleration; (2) acceleration with constant magnitude and direction always perpendicular to the velocity. The first case gives us the trajectories of objects moving over a flat earth. The predicted paths are good approximations to the actual trajectories of objects thrown for short distances near the surface of the earth and not influenced strongly by the air. The second case introduces us to some of the properties of circular motion, more of which we must study later. Although we applied the analysis to a simple case of an orbiting satellite, real orbits are more often elliptical than circular. Furthermore, in gravitational orbits, the acceleration is not generally perpendicular to the velocity, and, for that matter, is not constant in magnitude. Nevertheless, the simple cases that we analyzed are good first approximations to many real motions. In later chapters we will see how to extend the approximations with simple, but powerful, conservation laws.

Answers to Questions

3-1: (a) The hypotenuse should be a little more than 5.8 cm long. Notice the way that the squaring of the terms causes the dominance of the larger number. 5.8 is not much larger than 5. (b) The values for sine and cosine of 0°, 30°, 45°, 60°, and 90° are so frequently needed that it will save you time if you memorize them. Note also that the side opposite 30° was written as (10 sin 30), and the adjacent component was listed as (10 cos 30). It is useful to be able to think of the components in these terms, rather than having to work out the expression each time by starting out with

$$\sin \theta = \frac{\text{opp}}{\text{hyp}}$$

(c) You can make *A* and *B* any lengths you like, thus determining the length of *C*.

When you drop the dotted lines perpendicular to the axes, you will find that

$$(x_A + x_B) = x_C \quad \text{and} \quad (y_A + y_B) = y_C$$

3-2: The standard height for a table is 76 cm. The time for an object to fall from that height to the floor is given by: $0.76 \text{ m} = \frac{1}{2}(9.8 \text{ m/s}^2)t^2$. $t = 0.39$ s. With a pitching speed of 30 m/s (about 67 mph), a ball pitched from that height (maybe underhand?) would travel a distance, $x = vt = (30 \text{ m/s})(0.39 \text{ s}) = 12$ m. (Notice what a short pitch that is.) Whether or not the same sort of calculation is justified in the case of a rifle bullet depends on the effects of the air on a high-speed spinning object. Probably the simple calculation is only a very rough first approximation. The initial speeds of rifle bullets can be faster than the speed of sound, which is about 330 m/s.

3-3: The order of the moves makes no difference in the final displacement. In mathematical terms, addition of displacements is commutative. $A + B = B + A$. If each of the moves were made in exactly the opposite direction, then each of the components would have the opposite sign. A NE displacement would become a SW displacement. The resultant would have the same magnitude, but be in the opposite direction.

3-4: On this merry-go-round, the distance that you travel in one revolution is $\pi d = 31.4$ m. Your average speed is 3.14 m/s. Your average velocity, however, is not the distance traveled divided by the time, but is the *displacement* divided by the time. In one revolution, the displacement is zero, since you are right back where you started from. Therefore, the average velocity is zero.

3-5: Both the magnitude and direction of the wind and the plane's velocity with respect to the wind are known. Neither the direction of the plane with respect to the ground nor its speed with respect to the ground are known to begin with.

3-6: With an additional height of $15'' = 38$ cm above baseline level, the pitcher can release the ball at a height of about 2m. At a speed of 50 m/s, the ball will reach the batter in a time, $t = (60.5 \text{ ft})(0.30 \text{ m/ft})/(50 \text{ m/s}) = 0.36$ s. In that time the ball will have fallen a distance $\Delta y = \frac{1}{2}(-9.8 \text{ m/s})^2 (0.36 \text{ s})^2 = -0.64$ m. It will cross the plate at a height of 1.4 m, high but fairly in the strike zone.

3-7: When the ball passes its original level, but now heading down, the component of its velocity down is equal in magnitude to that of the original upward component. The additional flight time can be calculated as follows:

$$y = y_o + vt + \tfrac{1}{2}at^2$$

$$-1.6 \text{ m} = 0 + (-15 \text{ m/s})t + \tfrac{1}{2}(-10 \text{ m/s}^2)t^2$$

$$5t^2 + 15t - 1.6 = 0$$

This equation can be solved with the quadratic formula

$$t = \frac{-15 \pm \sqrt{15^2 - 4 \times 5 \times (-1.6)}}{2 \times 5} = 0.1 \text{ s}$$

Actually, it is not necessary to go to the trouble of using the quadratic formula in this case. Note that the ball has only 1.6 m to go with an initial speed of 15 m/s. Evidently, the time taken will be about 0.1 second. The term containing t^2 will thus be less than 10% of the term containing t, since the factor in the t^2 term is only 5, while the factor in the t term is 15. For our purposes in this problem, we could have dropped the t^2 term and saved ourselves some work. In a time of 0.1 second, the ball will travel an additional distance in the x direction of $x = (26 \text{ m/s})(0.1 \text{ s}) = 2.6$ m.

3-8: $x = (0.1) V^2 (\sin 2\theta) = (0.1)(30 \text{ m/s})^2$ for $\theta = 45°$. $x = 90$ m. In comparing this distance with the range for 0°, remember that we assumed that the horizontal throw started from a height of 1.6 m. The range formula that we have derived only applies to a throw that rises and falls to the initial height.

3-9: Trajectories of objects launched near the surface of the earth are always elliptical. (The earth is not flat!) The parabolic path is just an approximation that is good for baseballs and similar low flying objects.

3-10: A person at the rim of a merry-go-round is at a position of greater r, but also has greater velocity, v. The speed is equal to the circumference divided by the period of revolution: $v = 2\pi r/T$. Since the period is the same for all points on the merry-go-round, the centripetal acceleration can also be written

$$a_c = v^2/r = 4\pi^2 r^2/T^2 r = (4\pi^2/T^2)r$$

Note how this situation changes the apparent functional dependence on r. For constant period of revolution, the centripetal acceleration is proportional to r.

Problems

1. A displacement is 8 meters at 30° north of west. What are the north and east components?
2. A person walks 10 meters east, 5 m at 30° N of E, 6 m at 45° N of W, and 8 m west. What is the final displacement?
3. A mountain can be assigned a magnitude (its mass or its height), and a direction (up). Are mountains vectors? If not, why not?
4. A person on the equator travels 3000 miles due east and then 6000 miles due north. How far is she from her starting point?
5. A fighter plane flying at 1000 km/hr fires a rocket with a muzzle speed of 1000 km/hr. How fast is the rocket going with respect to the ground if it is launched in the forward direction? The backward direction? Sideways?
6. An airplane with an air speed of 200 km/hr maintains a bearing due east in a wind of 100 km/hr blowing from the north east (45° N of E). What is the resultant ground velocity of the plane?
7. What heading should the plane in Problem 6 maintain in order to fly due east? What is the resultant ground speed?
8. A bicycle is going at 20 km/hr due north, while a wind blows from the east at 10 km/hr. What is the apparent velocity of the wind on the cyclist?
9. A ball is thrown sideways at 10 m/s in a railroad car traveling at 30 m/s. What is the velocity of the ball with respect to the ground?
10. At what angle must a boat be pointed to go straight across a river if the current is 3.5 km/hr and the boat can go 8.4 km/hr in still water?
11. A boat that can go 4 km/hr must cross a river flowing at 3 km/hr. In order to reach a point directly across from the starting place in the shortest possible time, should the boat be headed straight across the river, carried downstream until it reaches the opposite side, and then proceed directly upstream to its destination, or should the boat be headed so that it goes straight across? Is some other path faster?
12. An object is heading east at 20 m/s. Three seconds later it is heading north at 10 m/s. What are the magnitude and direction of the average acceleration during the 3 seconds? Before you start, draw a good diagram.
13. You throw a baseball at an angle upward and it comes down again past the height from which you threw it in 2.6 seconds, 30 meters away. (a) What is the horizontal acceleration? (b) What is the vertical acceleration? (c) What was the initial horizontal component of velocity? (d) What must have been the vertical component of initial velocity? (e) What was the angle at which the ball was thrown?

SUMMARY 65

14. A tennis ball launcher shoots a ball from a height of 2 m above ground. If the initial velocity is 20 m/s at an angle of 30° above the horizontal, how far from the launcher is the ball when it hits the ground?
15. Suppose you can throw a ball with a speed of 30 m/s. Compare the ranges (distance along ground) produced by throws at 30° above the horizontal, 45°, and 50°. Does the fact that the ball leaves your hand at about 1.6 m above ground make much of a difference in your calculations? (See answer to Question 3-7).
16. On a navigation exercise, a ship sails four miles north, five miles east, and two miles south. How far is it from the starting point? Another ship, starting from the same point, sails two miles south, five miles east, and four miles north. Where is it with respect to the first ship?
17. The simplest formula for a parabola is $y = kx^2$. Compare this formula with the one given on page 59 for a trajectory of a thrown object. Why the difference?
18. You can swing a pail of water in a vertical loop without spilling the water, providing the centripetal acceleration required is equal to the gravitational acceleration of free-fall. If the radius of the loop, provided by your arm and the pail handle, is 1 m, what is the necessary speed? What would be the time for one revolution?
19. Since the moon is 60 times further from the center of the earth than we are, we should expect its centripetal acceleration to be $(1/60)^2 g = 2.7 \times 10^{-3}$ m/s². Calculate the period of the moon's rotation, assuming the approximation that the orbit is circular with a radius of 384,000 km. Does your answer agree with the time from new moon to new moon? If not, why not?
20. On a merry-go-round, what is the ratio of the centripetal acceleration at $R/2$ compared with R?
21. A swimmer is caught in a narrow rip tide going out from shore at a speed of 1.5 km/hr. If she can swim steadily at 2.0 km/hr, will she get back to shore 300 m away faster by bucking the current or by swimming sideways for five minutes to escape the current and then swimming in? Don't forget that in the second maneuver she is being carried away from shore for five minutes. What are the respective swimming times?
22. What is the magnitude and direction of the acceleration of a person on a merry-go-round at a radius of 5 m with a revolution period of 20 seconds?
23. What is the acceleration of a car going at 40 km/hr around a curve with a radius of 40 m?
24. A ball is thrown from a roof 20 m above ground. It has an initial velocity of 25 m/s at 30° above the horizontal. What is the horizontal range?

4 FORCES IN EQUILIBRIUM

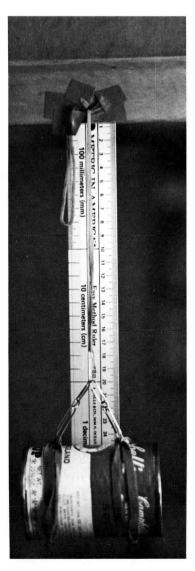

In everyday life we might speak of a *forceful* personality, or of troops being drawn up in *force,* or of being *forced* to make a decision. In the technical use of the word *force,* however, we'll be talking basically about pushes or pulls. Force, therefore, is not the same as power, or strength, or determination.

We can make an operational definition of what we mean by force by describing how to measure it. The measuring method depends on the effects produced by forces. There are two of these effects. First, a force applied to an object can produce a distortion in the object—a change in shape, such as a stretch or a twist or a dent. Second, a force applied to an object may change its velocity—speeding it up, slowing it down, or changing its direction of motion. In this chapter we concentrate on the first of these cases.

HANDLING THE PHENOMENA

Since life is usually filled with pushing and shoving, we should all be familiar with the phenomena produced by forces. As usual in physics, however, the subject has some interesting features that are not obvious. You can explore these features with your hands and a few simple things such as a ruler, rubber bands, light springs, and paper clips. If you can get hold of a force-measuring scale, so much the better.

   1. *Force and Stretch.* What we would like to do is to apply a known force to a spring or a rubber band and then measure the stretch. If we double the force, will the amount of stretch double? The trouble is, we haven't defined force yet. A reasonable thing to do is to make a preliminary definition and then see if it gives us simple and consistent descriptions of experimental facts.

   Hang a spring with a ruler beside it so that you can measure the amount of stretch. If necessary, hang a small object on the spring, which allows the spring to begin to stretch. Call this the starting condition. Get several identical objects, such as washers or cans of food. Choose the objects so that when you hang four of them together they stretch the spring a distance about equal to the starting length of the spring. Hang each of the identical objects by itself on the spring to make sure that each one produces the same stretch. Then measure the stretch for one hanging object, then two together, three together, and so on. Enter the data and sketch a curve on the graph.

Do the same experiment with a rubber band instead of a spring, but don't expect exactly the same results.

Note the basic assumptions that we are making if we try to draw conclusions from these simple experiments: (a) The hanging objects are "weights" and the weight of each one exerts a force on the spring. There are some profound and complicated features to that assumption, and we must pursue those problems in Chapter 5. (b) We assume that hanging two identical objects together exerts on the spring twice the force of hanging one alone. This seems like a reasonable assumption, but not everything adds up that way as you will see in the next experiment.

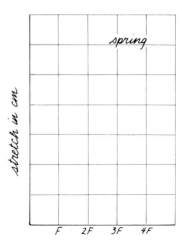

Question 4-1

Describe with words and algebraic formula the results of the experiment with the spring. Do the different results with the rubber band mean that our definitions and assumptions about weights (assumptions a and b above) are no good?

2. *1 unit of force + 1 unit of force = 2 units of force? (not necessarily).* See for yourself that the effect of a force depends on the direction in which you apply it. Use three force scales, or three springs, or three rubber bands. Use two of them to pull against the third, as shown in the picture. It will be easier if you have an extra hand, presumably from another person. Keep the third force the same by maintaining the length of the stretch producing it. Change the angle between the other two all the way from being in line with each other to being almost opposite each other. Observe qualitatively the forces that you have to apply to balance the third force.

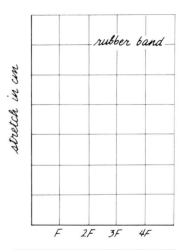

Question 4-2

What is the angle between the two forces when all three forces are equal?

3. *Handling Known Forces.* The unit of force in the International System of Units is the *newton*. We will be dealing with forces in newtons throughout the rest of the book. The unit will be meaningful to you only if your muscles are familiar with it. If you have a force scale calibrated in newtons, play with it until you know whether your little finger can pull as hard as one newton (1 N), or 10 N, or 100 N. The weight of a liter (which is about a quart) of water or milk is about 10 N. A pound is about the same as $4\frac{1}{2}$ N.

Question 4-3

What is the maximum force that you can exert with your body?

FORCES IN EQUILIBRIUM

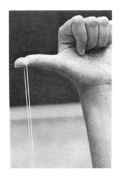

KINDS OF FORCE

It seems as if there are many different kinds of forces. Rubber bands and springs exert forces, and so do muscles, steel clamps, and compressed gases. Forces can also be exerted in an invisible sort of way, even through solid materials or empty space, by gravity, electricity, and magnetism. Regardless of what produces the force, we can deal with its effects in very general terms. We know that a force must exist if something gets distorted or accelerated.

There is an even greater simplification. All of the forces that we list above, plus practically all others, are really produced either by gravity or electromagnetism. As we will see much later in this study, electric and magnetic effects are just different versions of the same phenomenon. The combined electromagnetic effect is responsible for atomic and molecular binding, and thus is responsible for all the ordinary physical, chemical, and biological forces. As a matter of fact, besides gravity and electromagnetism, there are only two other kinds of interactions: the strong nuclear and the weak nuclear. These last two are responsible for the nuclear binding of atoms and for certain types of radioactivity. As a practical matter we continue to talk about spring forces and muscle forces, even though each of these *can* be described as the group results of an enormous number of electromagnetic interactions of atoms and molecules.

Although we may be primarily concerned with what a force does to one particular object, a force can't appear out of nowhere and act only on that object. A force can exist only as an interaction between two objects. If you pull on a rubber band with your finger, the rubber band stretches and the flesh on your finger is squeezed. Your finger feels a force in the direction away from your hand. As far as the rubber band is concerned, there is the same amount of force in the opposite direction. It is all one and the same interaction. If there is a pull of 10 N on the rubber band, there is a pull of 10 N on your finger.

Suppose that you tie one end of a rope to a wall and pull on it using a force scale to measure the force that you exert. The rope will become taut and will stretch a bit all along its length. The rope is under tension, which is another way of saying that a force is being exerted on it. The amount of the tension can be measured by inserting a force scale in the rope. If you pull with a force of 36 N, the force scale in the rope will read a force or tension of 36 N. Now change the conditions so that someone else is pulling on the other end of the rope. If the rope is not to start moving, she must pull with a force of 36 N also. What is the tension in the rope now? Still 36 N, as you can see in the pictures. As far as the rope is concerned, it is standing still but is slightly stretched because equal forces (of 36 N each) are acting in opposite directions on every piece of it.

Question 4-4

If a weight of 4 N is hung from a force scale with a string, the tension in the string will be 4 N. Suppose, however, that the string is looped over the force scale hook so that it is free to slide, and then the string is pulled down on the other side to balance the 4 N weight. This is a pulley arrangement, except that the upper force scale is measuring the total force being exerted by the string. What is the tension in the string? What is the total force measured by the upper scale? Try it and see for yourself.

DEFORMATIONS AND THEIR MEASUREMENT

It's easy to see what happens if you pull on a small, light spring. The deformation, or stretch, is in the direction of the pull and can be measured with a ruler. If you press down on a bed spring, however, your downward force produces distortions that spread out for some distance. Some of the internal springs are distorted sideways. More confusing still, if you press down lightly on a heavy car spring, you may not see any distortion at all. Similarly, there is no obvious distortion of a table that is holding up a book resting on it. In the case of a weight held by a spring, the spring is obviously stretched by the pull of the earth, but there are no apparent distortions in the earth itself.

Distortions occur in all these cases, although the effects may be small and also may be complicated in terms of spread and direction. The heavy car spring may not appear to change when a book is put on it, but certainly a car compresses it. The table may not appear to be dented by a book, but the effect can easily be seen with optical tools. Engineers use a variety of mechanical and electronic devices to measure expansions or compressions of floor and girders, sometimes down to distances smaller than the diameter of an atom. Of course, if the forces are large enough they may produce sizable distortions even in very hard and big objects. The pictures show two cases where solid objects were distorted. In the case of the golf ball, the material could spring back and form a sphere again. In the case of the car, the distortion went beyond the spring limits of the material. We will have to develop an atomic model of matter that can account for both cases.

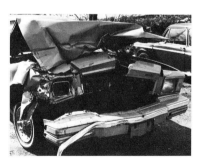

We might hope that at least the solid earth and massive foundations fastened in it would be immovable. That's not true, of course, as anyone knows who has experienced an earthquake. Distortions of the earth caused by titanic forces are also experienced daily by anyone who lives near the seashore. The tides are caused by distortions of the oceans as they are pulled by the gravitational attractions of the moon and sun. There are similar tides, though much smaller, in the solid crust of the earth itself.

Question 4-5

If atoms are too small to be seen (at least with our eyes), how can you measure a distortion smaller than the diameter of an atom?

THE RELATIONSHIP BETWEEN FORCE AND DISTORTION

In "Handling the Phenomena," you hung identical objects from a spring. The assumptions were that weight is a force, and that such forces add in an ordinary, arithmetic way as long as they are pulling in the same direction. Using such assumptions and ordinary springs, precision experiments—and perhaps yours—demonstrate a remarkable simplicity. *The distortion, or stretch, is proportional to the applied force.* If you double the force, you double the stretch.

$$F \propto x$$

This relationship is called Hooke's law. Robert Hooke (1635–1703) was a contemporary of Newton. His reputation is overshadowed by that of the great Newton, although he had many independent discoveries to his credit and almost anticipated some of Newton's work.

There are two important points to be noted here, almost contradictory ones. First, as you probably discovered when you hung weights on the rubber band, not all materials are consistent with the assumptions and/or Hooke's law. In fact, if all materials behaved like rubber bands, we wouldn't know whether our assumptions were useless or whether Hooke's law was no good.

> **Question 4-6**
>
> Suppose that for most springy materials, two weights produced four times the distortion of one weight, and three weights produced nine times the distortion. What would we conclude about our assumptions or how would we write Hooke's law?

The second important point is that since Hooke's law does hold for many materials, we can use the relationship for an operational definition of force. We could say that a particular spring is the "international standard spring," and that unit force is exerted by that spring when it is stretched unit distance. Note that we could say that the spring exerts unit restoring force when stretched unit distance, or that it takes unit force to stretch it that much. The magnitude of the restoring force is the same as the magnitude of the stretching force, but the forces are in opposite directions.

The standard unit of force, the newton, is actually defined in another way. For that we make use of the other property of force—an unbalanced force produces an acceleration. We will propose that definition in the next chapter, but in the meantime we can use the stretches of calibrated springs to exert known forces.

Hooke's law can be written as an equation instead of a proportionality, simply by inserting a proportionality constant.

$$F_{\text{restoring}} = -kx$$

The restoring force of a spring is equal to a constant times the stretch, and

THE RELATIONSHIP BETWEEN FORCE AND DISTORTION

is in a direction opposite to the stretch. The proportionality constant, k, must depend on the type of material in the spring, its size, and the way that it is made. A heavy lead spring might have a very small k. In fact, a lead spring might not obey Hooke's law at all. It would depend on what kind of lead alloy was used, how the spring was made, and the amount of distortion. Of course, you might not make a spring out of lead, but there are good and useful springs that also don't obey Hooke's law. Consider the springs in seat-belt reels or in measuring-tape windup cases. If the restoring forces in these were proportional to the stretch, you would be strangled in your seat and the measuring tape would have dangerous backlash.

Question 4-7

What should be the law relating $f_{\text{restoring}}$ and x for these devices?

A graph of the function, $F = -kx$, is shown below. Note that when x is positive, F is negative, and vice versa. The restoring force is in the opposite direction from the distortion. If this graph represents the behavior of a real spring, the zero point of x must be chosen so that both expansion (positive x) and compression (negative x) are possible. The slope of the line is $\Delta F / \Delta x$, which equals $-k$. In this case, the spring constant has a value of 200 N/m.

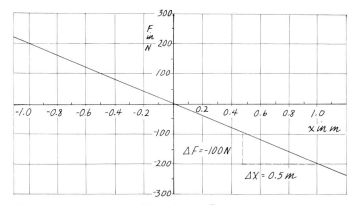

$F(x)$ for a spring obeying Hooke's law; $F = kx$

If this spring is 10 cm long to begin with and is pulled with a force of 6 N, how long will it become?

$\quad x = F/k$, where the negative sign is left out because F is the pulling force
$\qquad$ instead of the restoring force.
$\quad x = (6 \text{ N})/(200 \text{ N/m}) = 3 \times 10^{-2} \text{ m} = 3 \text{ cm}$

The length of the stretched spring will thus be 13 cm.

> **Question 4-8**
>
> How long will this spring become if it is pulled with a force of 200 N?

PRESSURE, STRESSES, AND STRAINS

Most people live under constant *pressure*, and the *stress* of modern living produces a lot of *strain* on all of us. Those everyday terms have technical meanings that are not far different from the common ones. Pressure and stress are similar quantities, although pressure is used in describing effects with gases or liquids, while stress refers to solids. In both cases the term is defined as *force per unit area*.

The concepts of pressure and stress are clearly needed to explain common phenomena. For instance, you can easily exert a force of 10 N with your thumb while pushing in a thumbtack. Try turning the tack over, however, and exerting the same force on the point. In the first case the pressure was about 10 N per square centimeter, or 10^5 N/m². In the second case the pressure might be 10 N per square tenth of a millimeter, or 10^9 N/m². Same force, but the pressure is greater by a factor of 10,000.

We live at the bottom of a great sea of air. The atmosphere presses on us equally on all sides and from within. If there were a vacuum inside us, we would be squashed. Under these circumstances it is meaningless to ask about the *force* exerted by the atmosphere unless we specify the area on which it is acting. Instead, we specify the atmospheric *pressure*. Normal atmospheric pressure at sea level is 1.013×10^5 N/m². (Compare this with the pressure you might exert on a thumbtack.)

A standard demonstration of the effect of atmospheric pressure is to produce a vacuum in a rectangular gallon can. This can be done with a vacuum pump or by boiling water in the can until the steam has driven all the air out, and then capping the can. When the steam condenses, the pressure inside the can drops almost to zero. The effect is shown in the picture.

PRESSURE, STRESSES, AND STRAINS

The wide side of the can has an area of 20 cm by 30 cm, or 600 cm², or 0.06 m². If the atmospheric pressure on the outside of the can is 1×10^5 N/m², and the inside pressure is essentially zero, then the can must withstand a force equal to the pressure times the area.

$$F = P \cdot A = (1 \times 10^5 \text{ N/m}^2) \cdot (6 \times 10^{-2} \text{ m}^2) = 6 \times 10^3 \text{ N}$$

A metric ton, which is about the same as our familiar American ton, is about equal to 10^4 N. It's no wonder that the can collapses.

The atmospheric pressure is the weight of a column of air divided by the cross-sectional area of the column. The height of the column is indeterminate because some traces of the atmosphere extend for several hundred kilometers above the surface of the earth. Most of the air, however, is contained in the first 30 kilometers, and the pressure decreases all the way up. The atmospheric pressure is frequently given in terms of the height of columns of mercury or water that can be supported by the atmosphere. The standard mercury barometer consists of a long tube sealed at one end and filled with mercury. If the tube is turned upside down with its lower end in a bowl of mercury, the mercury in the tube will flow out until the pressure at the bottom level caused by the weight of the mercury is just equal to the atmospheric pressure. A column of mercury 76.0 cm high can be supported by normal atmospheric pressure at sea level. (At Denver, Colorado, with an elevation of 1.6 km, normal atmospheric pressure supports a column of 64 cm.) Since the density of mercury is 13.6 times that of water, standard atmospheric pressure can support a column of water that is 76.0×13.6 cm high, or 10.3 m. That's a column as high as a three story house. It's also the maximum depth from which you can draw water from a well with a simple vacuum pump. For practical purposes, single-stage vacuum pumps are not used to draw up water from depths greater than about 8 meters.

Note that it is not necessary to give the cross-sectional area of the mercury or water columns. It doesn't make any difference whether the column has an area of 1 m² or 1 cm². The *pressure* is the same. The total force exerted by the weight of a column of mercury 76 cm high with an area of 1 m² is 1×10^5 N. That's a lot of mercury! Atmospheric pressure also produces that force over an area of 1 m². The total force exerted by a column of water 10.3 m high but with an area of only 1 cm² is only 10.1 N. That's a pressure of 1×10^5 N/m², which is atmospheric pressure also. (Incidentally, the volume of that column of water is about 1 liter, and the mass is about 1 kg.)

The pressure exerted by a liquid or gas acts on all sides of a submerged object. For instance, the atmosphere doesn't just press down on you. If you hold your hand out flat, there's almost the same force up on the palm as there is down on the back of your hand. It appears that pressure in a liquid or a gas has no direction. A force can be applied to a surface in any direction, but in a liquid or gas a sideways force along a surface would make the liquid or gas move. In the stationary situation in a liquid or gas, there can only be forces perpendicular to surfaces.

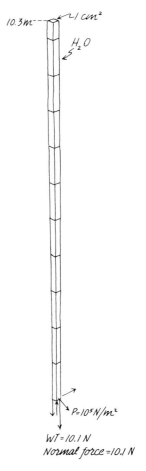

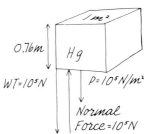

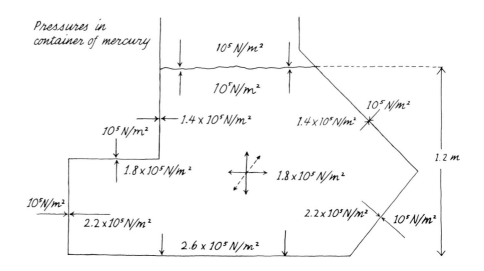

> **Question 4-9**
>
> If you have a balloon of hydrogen gas, and there is the same pressure from the atmosphere pressing on all sides of it, why does it rise?

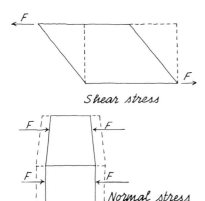

If a force is exerted at an angle on a solid surface, the part of the force that is *perpendicular* will compress the solid. The part of the force that is *parallel* to the surface will twist or "shear" the surface. Instead of being called pressure, the perpendicular force per unit area on a solid is called the *normal stress*. ("Normal," in this context, means perpendicular.) The parallel force per unit area on a solid is called the *shear stress*.

When stress is applied to real materials, there can be very complicated results. Real gases and liquids can weakly support shear stresses without appreciable motion, at least for short times, and real solids can flow under the influence of shear stresses. Furthermore, the molecular structure of most solids has directional regularities—such as crystalline structure or the grain of wood. The response of the material to stress is different in one direction than in another.

For certain types of geometries, it is convenient to give special names to the response of material to stress. For instance, if the stress is along the length of a wire or rod, then the response is relatively simple: there is a stretch. The stress is given by the ratio of applied force to cross-sectional area: stress $= F/A$. If we are told that a stress of 10^{10} N/m² produced a stretch of 1 cm in a wire, we have no way of knowing whether that's a little or a lot. If the original length of the wire was 10 cm, the strain on the wire was enormous. If the original length of the wire was 10 km, the strain was insignificant. It's the *fractional* stretch that's important, and that is the definition of strain: strain $= \Delta L/L$. Hooke's law can be expanded to a more sophisticated form to describe the relationship between stress and strain for simple stretch along one direction.

PRESSURE, STRESSES, AND STRAINS

Stress is proportional to strain: $F/A \propto \Delta L/L$

The proportionality constant is given a special name, Young's modulus. [Thomas Young (1773–1829) was an Englishman with many interests. He is best known for his demonstrations of the wave nature of light, and also for being the first person to decipher Egyptian hieroglyphics.]

$$F/A = Y(\Delta L/L), \text{ so that } Y = \frac{F/A}{\Delta L/L}$$

The units of Young's modulus are newtons per square meter, since the denominator is a ratio of two lengths and so has no units. Young's modulus may look a lot more complicated than the spring constant, k, in the first form of Hooke's law. Remember, though, that k applies to one particular spring. It depends on the type of material in the spring, the length, the cross section of the wire, and the geometry of the winding. Young's modulus, on the other hand, is characteristic of the particular material only. It applies only to geometry of stretch along one direction, and the length and cross-sectional area are taken into account. The table gives Young's modulus for a number of materials. Notice that they are all the same order of magnitude. Later we demand that our atomic model of matter explain this fact and we also relate the size of Young's modulus to the strength of atomic binding.

Young's Modulus (Y) and Bulk Modulus (B)

	Y	B
aluminum	70×10^9 N/m²	70×10^9 N/m²
copper	91×10^9 N/m²	61×10^9 N/m²
steel	190×10^9 N/m²	100×10^9 N/m²
nickel	210×10^9 N/m²	260×10^9 N/m²
tungsten	360×10^9 N/m²	200×10^9 N/m²
gold	16×10^9 N/m²	8×10^9 N/m²
glass	55×10^9 N/m²	37×10^9 N/m²
nylon	5×10^9 N/m²	
bone	$\sim 20 \times 10^9$ N/m²	
water		2×10^9 N/m²
alcohol (ethyl)		0.9×10^9 N/m²
mercury		27×10^9 N/m²

Question 4-10

Young's modulus for copper is 1×10^{11} N/m². Suppose that you produce a stress of 10^{11} N/m² on a copper wire only 0.1 mm in diameter—that's threadlike. The cross-sectional area of such a wire is about $(10^{-4})^2 = 10^{-8}$ m². A force of 10^3 newtons would produce a stress of 10^{11} N/m². But a force of 10^3 newtons is about $\frac{1}{10}$ of a ton! Can such a thin wire really support such a stress?

There are other moduli like Young's that are used to describe the stress and strains of matter. The shear modulus is the ratio of stress to angular or twisting strain. The bulk modulus, which we will want to use later, is used to describe what happens to materials when they are squeezed from all sides (instead of being stretched in just one direction). It is the ratio of pressure change to *fractional* change of volume: $B = \dfrac{-\Delta P}{(\Delta V/V)}$. The minus sign indicates that an increase in pressure produces a decrease in volume. The table gives values for the bulk modulus for a number of materials. Notice the relationship between the values for solids and liquids.

If you drop an aluminum cube into the ocean, it will sink because its density is 2.5 times that of salt water. If it drops into one of the deep trenches in the ocean, is it possible that it will be compressed so that its density will be greater? Or will the salt water compress even more so that the water density will be as great as the aluminum, in which case the aluminum would stop sinking?

From the table of bulk moduli, it appears that the compressibility of water is about 35 times that of aluminum. As a first approximation, we can ignore the effect of increasing pressure on the volume of the aluminum and consider only the effect on the water. To a first approximation, the pressure under water is given by

$$P = P_0 + (\text{weight density}) \times (\text{depth})$$

where P_0 is atmospheric pressure.

If the compression of the sea water produces an appreciable change of its density, this formula would not be valid because the assumption here is that density is a constant. Let's test that assumption by seeing what this formula predicts for a depth of 10 km. The weight density of salt water is about 1×10^4 N/m³. At a depth of 10 km, the increase in pressure would be $(1 \times 10^4 \text{ N/m}^3) \times (1 \times 10^4 \text{ m}) = 1 \times 10^8$ N/m². That's about 1000 times atmospheric pressure. The fractional change in volume (and hence in density) of the sea water would be

$$\Delta V/V = \Delta P/B = (1 \times 10^8 \text{ N/m}^2)/(2 \times 10^9 \text{ N/m}^2) = 0.05 = 5\%$$

A 5% increase in density is certainly measurable, but it would make no difference in whether or not aluminum would sink. Furthermore, the effect is small enough to justify using the approximation that pressure is given by a formula that assumes constant density.

THE DIRECTIONAL NATURE OF FORCES

While handling the phenomena you saw that two forces of the same strength acting on the same object don't necessarily produce twice the effect of one force. Forces can reinforce each other or cancel each other. Forces have both magnitude and direction. This is a vector-type property, but that doesn't necessarily mean that forces behave like vectors. There are other quantities that have magnitude and direction but which are not vectors. A more

THE DIRECTIONAL NATURE OF FORCES

advanced description of stress, for example, is given by a quantity called a tensor. Tensors have magnitude and direction but combine with other tensors in ways much more complicated than those of vectors. The rules of combination define whether or not a quantity is a vector. If two forces combine in the same way that two displacements combine, then they are vectors. Experimentally, this is found to be the case for forces—but it's an experimental fact, not one derived from logic.

> **Question 4-11**
>
> How do displacements combine? If you move 8 meters east and 6 meters north, how far are you from your origin?

In working with displacements we always chose two perpendicular directions as axes (such as north and east), and resolved all the vector displacements into components along those axes. Then we added all the components along each axis separately and found the resultant of those total components. Here's an example of the same process with forces.

What is the resultant (or combined) force in the following diagram? It is

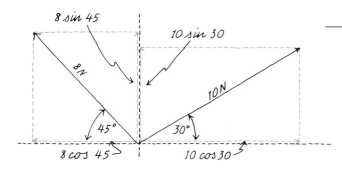

	East	North
	8.7	5
	−5.6	5.6
	3.1	10.6

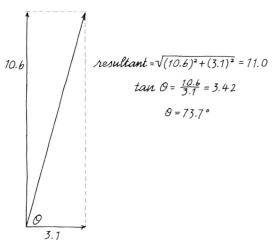

$$\text{resultant} = \sqrt{(10.6)^2 + (3.1)^2} = 11.0$$

$$\tan \theta = \frac{10.6}{3.1} = 3.42$$

$$\theta = 73.7°$$

useful, and usually timesaving, to sketch an approximately scaled drawing of such a problem. Label all the vectors and angles. Draw dotted lines to indicate the components of each vector. Label the magnitudes of the components. Once you produce a properly labeled diagram, your problem is nearly solved. Add the components separately along the two perpendicular axes. Recombine them, using both a scaled sketch and the Pythagorean theorem.

Sometimes vectors should be resolved into components that are perpendicular to each other but *not* vertical and horizontal. In the next example, the force vector representing the weight of the block on the slide is already vertical. What we want to know is how much of that force is acting perpendicular to the slide, and how much is acting parallel to the slide.

To find out which of the angles in the vector triangle is 30°, you can match the perpendicular and parallel lines in the similar triangles. Notice also that a sketch drawn with reasonable attention to scale is a great help in doublechecking the results.

EQUILIBRIUM OF FORCES

So far we have been talking only about situations where forces balance each other. The distortion in a stretched spring is caused by a force that pulls out on the spring, but it is balanced by the restoring force of the spring that pulls back with equal magnitude. The spring itself is motionless under such conditions. A book on a table is pulled toward the earth by gravity. The slight compression of the table indicates that there is an upward force on the book being exerted by the tabletop. Once again the book is motionless because the two forces balance. The first of Newton's three laws of motion describes these situations: *If the vector sum of all the forces acting at each point on an object is zero, then the object will maintain constant velocity.* In particular, if the original velocity is zero, it will remain zero. Later we will consider cases where the forces are in equilibrium but the velocity is not zero. It's not always easy to observe distortions that define forces in order to determine that the vector sum of the forces on an object is zero. We can turn Newton's first law around and use it as a definition of "zero net force." *If we observe that an object is maintaining constant velocity (with respect to us) then we can conclude that the net force acting on it is zero (with respect to us).* We'll see a little later why it's necessary to qualify that definition with the parenthetical clauses.

Question 4-12

Does this definition really make sense outside a physics class? Doesn't it take a sizable force pushing on a car to keep it going at constant speed? Therefore, if you see a car moving at constant speed you know that the motor or something is exerting a force on it. Why use a definition for zero force that contradicts common observations?

EQUILIBRIUM OF FORCES

The condition for no change in velocity can be written like this:

$$\Sigma \mathbf{F} = 0$$

The symbol, Σ, stands for "the sum of." The sum of the forces at every point of an object must be zero. Remember, however, that this is a *vector* sum. Since each vector force can be resolved into components along three perpendicular axes, Newton's first law requires that the sum of each group of components be zero:

$$\Sigma F_x = 0 \qquad \Sigma F_y = 0 \qquad \Sigma F_z = 0$$

1. The diagram shows a bow and arrow, with the arrow being pulled back with a force of 200 N. The string is resisting this pull, and the bent bow is resisting the tension in the string. It looks as if there are forces acting all over the place! We can greatly simplify the analysis, however, by considering only the arrow, and in particular, only the notched end of the arrow. Its velocity is constant, namely, zero. Therefore, the sum of the forces acting on it must be zero. Here is the first and foremost rule when analyzing forces in a system: ISOLATE THE OBJECT. Actually draw a circle around the point or the object, and then draw and label the forces acting on that particular point. That's what we've done in the second part of the diagram. We have also followed the other rules about breaking every vector into components and labeling their magnitudes. In this particular two-dimensional case, there are no components in the z-direction. Sum up all the component forces in the x-direction and set that sum equal to zero. Do the same thing for the y-components. In this example, that determines the components of the tension in the string, and from those you can find the tension itself.

    Incidentally, notice how large the tension is compared with the pull on the arrow. This situation is sometimes known as the clothesline problem. Turn the geometry by 90° so that the rope is horizontal and serves as a clothesline. Make the rope very taut so that θ is very small, and then hang some heavy wet clothes in the middle. The tension may be so great that the rope breaks, spilling the clothes and demonstrating the laws of physics. This can happen even though the rope could withstand a tension considerably greater than the actual weight of the clothes.

2. Here's a very simplified "mast and boom" problem. The simple feature is that we will assume that the boom itself has no weight and can exert only a compression force along its length. In a more realistic situation, which we soon take up, the boom would have weight that must be partially supported by the mast at the point where they meet. Furthermore, if the boom were solidly fastened to the mast in a cantilever arrangement, it might be able to support the load without any rope. First, however, let's solve the simple case.

    Note that the force diagram has been drawn to a reasonable scale, the one vector not in the x, y direction has been broken into components, all forces have been labeled, and the point of equilibrium has been circled to make sure that all of the forces acting there have been

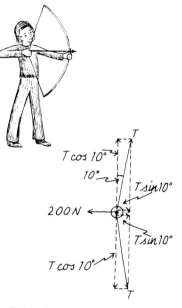

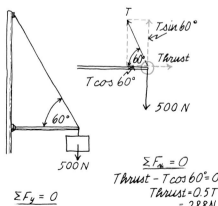

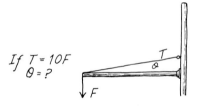

accounted for. Once again, the tension in the rope is greater than the weight supported.

Question 4-13

Could you make use of a device like this to exert a large tension in the rope to extract something fastened in the mast? What would be the angle between the rope and the boom in order to multiply your effective force by a factor of 10?

TORQUES

It's possible to have all the forces on an object nicely balanced, and still have the object accelerate! Here's how.

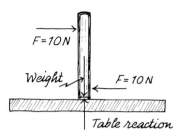

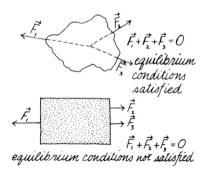

Take another look at the wording of Newton's first law on page 78. An object is in equilibrium only if all the forces cancel out at *each* point of the object. If the forces on an object all act through the same point and their sum is zero, the condition is automatically satisfied. However, if the forces are being exerted at various points of the object and in various directions, there has to be a second condition for equilibrium. The turning moments, or *torques*, must cancel each other.

If you apply a force to twist or turn an object around an axis, you are applying a torque on the object. When it comes to opening doors, you already know all about torques.

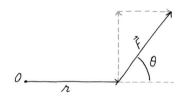

Question 4-14

Suppose that you want to open a heavy door. Do you shove near the hinge or near the handle? Do you shove parallel to the door or perpendicular to it?

The effect of a torque depends on the strength of the force, the length of the radius arm, and the angle between the radius arm and the force. Only the force component *perpendicular* to the radius arm is effective in turning the object. The magnitude of a torque is the product of the length of the radius (or lever) arm and the perpendicular component of the force. Any component *parallel* to the radius arm just exerts a push or pull on the axis.

TORQUES

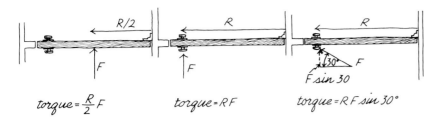

torque = $\frac{R}{2}F$ torque = RF torque = $RF \sin 30°$

In the metric system, the magnitude of a torque is given in meter·newtons. The unit has no special name. (As we will see in Chapter 9, work and energy also have a unit that is the product of force and distance. The newton·meter, or joule, is a completely different quantity, however.)

The symbol for torque is τ, the Greek letter, tau.

Torque not only has a magnitude, but also a direction. It tends to produce a rotation either clockwise or counterclockwise around an axis, which itself has a direction. It turns out that the most reasonable way to define the direction of the torque is to say that it is in line with the axis, in one direction or the other. In the case of torques exerted on a door, the direction of the torques would be either up or down along the line of the hinges. Whether the direction is up or down is determined by a standard convention: let the fingers of your right hand curl in the direction of the turning door. Your thumb will then point in the direction of the torque. This is shown in the diagram.

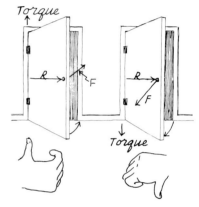

A torque is the product of two vector quantities: radius arm times a force. We defined vectors earlier in terms of the way they *add*. What does it mean to *multiply* two vectors? It can mean anything we want, as long as there is some consistency in the way we use the definition. Fortunately, there is a convenient form of mathematics that defines a *vector product* in a way that describes the way torques behave. The vector product, or cross product, of two vectors is itself a vector, with a direction perpendicular to the original two vectors and a magnitude equal to their numerical product times the sine of the angle between them. (By the angle between two vectors, we mean the angle between their directions.)

$$\boldsymbol{\tau} = \mathbf{r} \times \mathbf{F} \qquad |\tau| = |r||F| \sin \theta$$

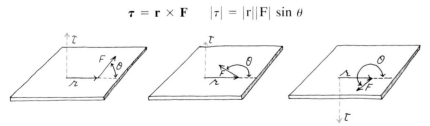

A vector product has some unusual properties. It is created by a noncommutative process: $\mathbf{A} \times \mathbf{B} \neq \mathbf{B} \times \mathbf{A}$. Instead, $\mathbf{A} \times \mathbf{B} = -\mathbf{B} \times \mathbf{A}$. This is because the direction of the vector product is defined to be in the direction that a right-hand screw advances if you turn the first vector into the second. The same direction is given by the right-hand rule as shown above.

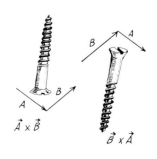

You may also be amused to observe that a vector product is not really a vector! If all of the components of a regular vector are reversed in

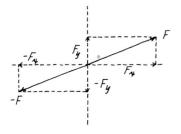

direction, then the vector itself is reversed in direction. If $F_x \to -F_x$ and $F_y \to -F_y$, then $\mathbf{F} \to -\mathbf{F}$. If the components of a vector product are all reversed, then $\mathbf{r} \to -\mathbf{r}$, $\mathbf{F} \to -\mathbf{F}$, but $\boldsymbol{\tau} = \mathbf{r} \times \mathbf{F} \to (-\mathbf{r}) \times (-\mathbf{F}) = \mathbf{r} \times \mathbf{F}$. The vector product is not changed. Because of this behavior, the vector product is technically known as a "pseudo-vector."

EQUILIBRIUM OF TORQUES

The second condition for equilibrium of an object is that the sum of all the torques about any point must be equal to zero. The condition is easy to test because if $\Sigma \mathbf{F} = 0$, and if the sum of the torques on an object about *any* particular point is zero, then the sum of the torques is zero for *any other* point. Several examples of this consequence are given in Example 1 below. To test for equilibrium, we draw a circle around the whole object and take the vector sum of all forces acting on the object from the outside. Then we choose *any* convenient point and add up the torques acting around that point as an axis. If both sums are zero, the object is in equilibrium.

$$\Sigma \mathbf{F} = 0 \qquad \Sigma \boldsymbol{\tau} = 0$$

Examples

1 *The Simple Seesaw.* Suppose two children of equal weight are on a seesaw. The forces are labeled in the diagram. The equilibrium condition concerning forces yields the obvious fact that the single support at the center must exert an upward force equal to the sum of the weights of the children.

The second condition, for rotational equilibrium, requires that the sum of the torques about *any* point be equal to zero. For this simple case, let's try a number of points.

$$\Sigma \boldsymbol{\tau} = 0 \text{ about center point}$$

$$\mathbf{r}_3 \times \mathbf{F}_3 + \mathbf{r}_2 \times \mathbf{F}_2 + \mathbf{r}_1 \times \mathbf{F}_1 = 0$$

$$2 \times 200 + 0 \times 400 - 2 \times 200 = 0$$
(clockwise) \qquad (counterclockwise)

$$\Sigma \boldsymbol{\tau} = 0 \text{ about end point}$$

$$\mathbf{r}_1 \times \mathbf{F}_1 + \mathbf{r}_2 \times \mathbf{F}_2 + \mathbf{r}_3 \times \mathbf{F}_3 = 0$$

$$0 \times 200 - 2 \times 400 + 4 \times 200 = 0$$
(counterclockwise) \qquad (clockwise)

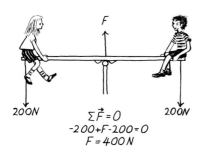

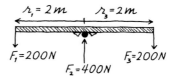

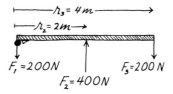

EQUILIBRIUM OF TORQUES

In each of these two cases we chose axis points through which one of the forces is acting. That eliminated one term from the equation, because if a force is acting in line with the axis point, the radius arm is zero and the torque is zero. Just to demonstrate that in equilibrium the torques about *any* point cancel, let's choose a very unlikely point, one that's not even on the seesaw.

$$\Sigma\tau = 0 \text{ about point in space}$$

$$\mathbf{r}_1 \times \mathbf{F}_1 + \mathbf{r}_2 \times \mathbf{F}_2 + \mathbf{r}_3 \times \mathbf{F}_3 = 0$$

$$2 \times 200 - 4 \times 400 + 6 \times 200 = 0$$
$$\quad\text{(cl)} \qquad \text{(ccl)} \qquad \text{(cl)}$$

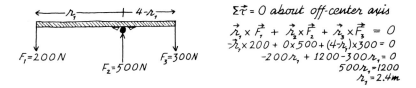

point around which torques are taken

2 *A More Complicated Seesaw.* Let's complicate the seesaw problem by making a balance with two children whose weights are not equal. However, let's also make an unreasonable assumption that the seesaw board has no weight. Where should this mythical board be placed on its fulcrum if it is to balance?

$$\Sigma\vec{\tau} = 0 \text{ about off-center axis}$$

$$\vec{r}_1 \times \vec{F}_1 + \vec{r}_2 \times \vec{F}_2 + \vec{r}_3 \times \vec{F}_3 = 0$$

$$-r_1 \times 200 + 0 \times 500 + (4-r_1) \times 300 = 0$$

$$-200 r_1 + 1200 - 300 r_1 = 0$$

$$500 r_1 = 1200$$

$$r_1 = 2.4 \text{ m}$$

3 *Mast and Boom with Weight.* We solved the earlier mast and boom problem using only the condition for equilibrium of forces. In that example, however, all the forces of interest were acting through a single point at the end of the boom. Let's complicate the problem by hanging a weight on the middle of the boom. Once again, we will assume that the boom itself has no weight. We start out by considering the stability of the whole boom as the object, and therefore circle it in the diagram and take into account all the forces acting on it. Notice that we cannot assume that the mast merely shoves out on the boom. There is also a force at the connecting point that helps to hold up that end of the boom. The resultant force of the mast on the boom is not in line with the boom. A rope can transmit forces only in its own direction, but a rigid object can sustain forces at any angle.

After isolating the object that is in equilibrium (the boom), we labeled all the forces and distances, and then wrote the equilibrium conditions for forces and torques. Although any point could have been chosen as axis for the torques, it is most convenient to choose a point through which forces are acting. The torques of those forces will then be zero. There are three equations and three unknowns—F_H, F_V, and T. In this particular example the actual length of the boom is not needed. Notice that it cancels out of the third equation.

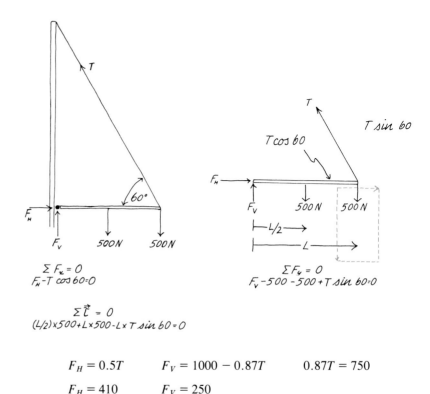

$$F_H = 0.5T \qquad F_V = 1000 - 0.87T \qquad 0.87T = 750$$

$$F_H = 410 \qquad F_V = 250$$

4 *Climber and Ladder.* Here's a case where the forces do not act perpendicular to the radius arm, and the radius arm is neither vertical nor horizontal. Nevertheless, forces should be resolved into components that are either perpendicular or parallel to the radius arm. The problem consists of finding the forces acting on a ladder as a person climbs up. Assume that the wall is smooth and therefore can only exert forces perpendicular to itself. Once again, assume that the ladder itself has no weight, but the climber weighs 500 N.

Not all of the forces had to be resolved into components perpendicular or parallel to the ladder. By choosing the axis at the ground point, the torques produced by the ground reaction were reduced to zero. In summing the forces, it was more convenient to take them as originally given instead of using their components. In computing the torques, however, only the force components perpendicular to the radius arm were considered.

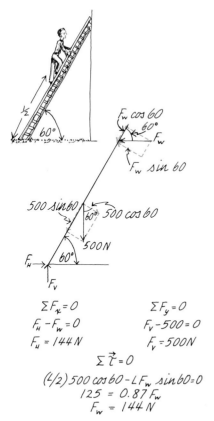

CENTER OF WEIGHT

> **Question 4-15**
>
> (a) In the problem, we placed the climber halfway up the ladder. What is the maximum force that he can produce on the wall by changing his position? (b) What is the direction of the resultant force exerted by the ground? Is it along the line of the ladder? How does this situation depend on the climber's position?

CENTER OF WEIGHT

In all the examples so far, we have left out the weight of the boom or the ladder. That's unreal, of course. The weight of the boom might be greater than the weight that it's holding. Fortunately, in any of these equilibrium problems the effect of the weight of each object can be calculated very simply. You can consider that the weight acts through one particular point, as if all of the object were concentrated at that point. In everyday situations, we make use of the center of weight concept without even thinking about it. If you are going to carry a ladder any distance, you pick it up in the middle instead of at one end. If the ladder is wider and heavier at one end than the other, you shift your balance point toward the heavier end.

Holding a ladder at its center of weight doesn't make it lighter, but it does make it easier to handle the ladder. The net torque about the center of weight is zero, and so your hands do not have to twist to exert a countertorque. The center of weight is defined to be that point around which all the weight torques of an object add to zero regardless of the orientation of the object.

Example

Where is the center of weight of a very light rod holding three heavy weights as shown in the diagram?

If the center of weight is at a distance x from the left side, a support at that point would balance the system. There would be zero net torque about an axis at x.

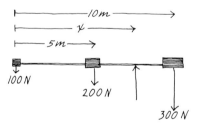

$$\sum \tau = 0$$

$$-x(100) - (x-5)200 + (10-x)300 = 0$$

$$-300x + 1000 + 3000 - 300x = 0$$

$$600x = 4000$$

$$x = 6\tfrac{2}{3} \text{ m}$$

If the weight of an object is distributed uniformly and the object has simple symmetry, the center of weight is in the geometric center point. The center of weight of a doughnut is right at the center of the doughnut—in mid-air!

For the general case of finding the center of weight of a continuous

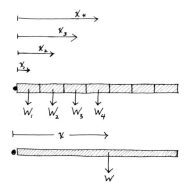

object, the weight can be divided into small sections. Find the sum of the torques of all these pieces around some particular point, such as one end of the object. This total torque should be the same as the torque produced by all the weight acting through the center of weight.

$$\text{Total torque} = x_1 W_1 + x_2 W_2 + x_3 W_3 + \cdots = xW$$

The weight of the third piece is called W_3 and the radius arm through which that force acts is x_3. The total weight of the object is the sum of all the individual weights and is called W. The distance x is the radius arm from the axis to the center of weight. In the general case:

$$x = \frac{\sum_{i=1}^{n} x_i W_i}{W}$$

Question 4-16

Does this general formula work for the simple case of a uniform boom? Try it and see. If the boom is divided into n equal pieces, each piece has a weight of W/n.

There is a very simple experimental method for finding the center of weight of any flat object. First of all, remember that if you hold an object at its center of weight it will not rotate. As a matter of fact, if you hold an object at any point vertically *over* the center of weight, there will be no rotation. Furthermore, if you hold an object at any point so that it *can* turn,

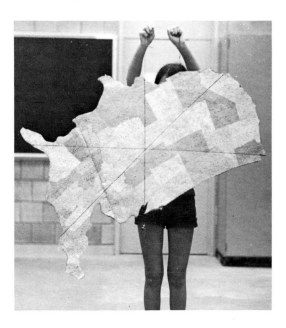

Finding the center of weight of the United States.

it will swing around until the center of weight is directly under the point of suspension. If you drop a line straight down from that point of suspension, the center of weight will be somewhere on that line. Next, hold the object from any other point. Once again the object will rotate until the center of weight is directly under the new point of suspension. Drop a line from there. Since the center of weight must be somewhere on both lines, and since they cross only at one point, that point must be the center of weight.

> Question 4-17
>
> Why will a suspended object rotate until its center of weight is directly underneath the point of suspension?

SUMMARY

When forces act on an object, they distort it or accelerate it. In this chapter we concentrated on the distortions. Indeed, one way to define force is to measure the distortion produced in standard devices. $F_{restoring} = -kx$, for certain objects such as springs, and for small x.

The size of the force unit, the *newton*, is small by human body standards. You can exert several newtons with your little finger.

The forces that we deal with in everyday life are either gravitational or electromagnetic.

Pressure is force per unit area. In a liquid or gas the pressure always produces effects perpendicular to the walls. A force exerted on a solid can produce a perpendicular, or normal, stress, and a parallel, or shear, stress. For many materials, the *fractional* change in length, called the strain, is proportional to the stress.

Forces always exist as an interaction between two objects. However, in studying the behavior of a particular system it is usually a good idea to isolate it mentally and consider the effect on it of all the outside forces. For the object to be in equilibrium, two conditions must be satisfied:

$$\sum \mathbf{F} = 0 \qquad \sum \tau = 0$$

These are vector equations. To use them, it is usually necessary to break all the vectors into perpendicular components.

On one level this subject of static forces is elementary and unexciting. After all, the rules have been known well for over 150 years and the general notions were understood 300 years ago. The stability conditions that we illustrated with simple booms and ladders also apply to complicated bridge and rocket structures. These applications to the real world, however, require an enormous jump in mathematical techniques and in the detailed knowledge of real materials. We already mentioned that stresses and strains are really tensor quantities. They usually have nine components each to describe the directional effects in materials that respond differently in one direction from another. Not only do wires and rods stretch in length if they are pulled; they also tend to pinch at a yield point. There are other complications. For instance, mechanical forces on certain crystals can produce thermal and electrical effects. The Young's modulus and other properties of materials can change with temperature, with previous strains, and with slight changes in chemical composition. Vibratory or short pulse forces may produce effects quite different from long-term stresses. All these effects must be taken into account in engineering practice and, in return, the effects are frequently useful tools for learning more about the microstructure of materials. The engineer's stumbling block has often been the physicist's clue.

Answers to Questions

4-1: The length of the stretch, x, is proportional to the applied force, F. $x \propto F$. The graph of stretch versus force is a straight line.

If we knew only the effect of hanging weights on rubber bands, we wouldn't know whether the assumptions are wrong or whether this relationship just doesn't apply to rubber bands. Perhaps adding one weight to one more weight does not produce twice the force. After all, adding one liter of water to one liter of alcohol yields less than two liters of liquid.

4-2: In this symmetric situation of equal forces, the angle between any two forces must be equal to the angle between any two other forces. Therefore each angle must be 120°.

4-3: You can at least exert your own weight. (How many newtons is that?) By pulling on a force scale across your chest with your two arms, you can probably exert at least 100 N. By bracing your legs and pulling up you may be able to exert close to 1000 N. Get a large force scale and see for yourself.

4-4: Only the string is holding up the 4 N weight, and so the tension in the string must be 4 N. However, the force scale must balance the sum of the downward weight of 4 N and the downward pull of 4 N on the other end of the string.

4-5: Distortions smaller than the diameter of an atom can be measured if they are repetitive or vibratory. Usually the small vibrations are turned into electric signals. The human ear is capable of measuring (hearing) vibrations of the ear drum that are this small.

4-6: We could conclude that Hooke's law holds and that weights can be added according to the law $W_{1+2} = (W_1 + W_2)^2$. Of course, this law might not agree with other weight experiments, such as what happens when they fall. Alternatively, we might create a different Hooke's law: $x \propto F^2$.

4-7: $F_{restoring}$ = constant. There should be no dependence on x.

4-8: If we used the formula blindly, the stretch would be 1 meter and the spring would then be 110 cm long. Very few springs can stretch to eleven times their original length, however. Here is an example of how formulas and rules are only models of physical phenomena. Models apply only within limited ranges and conditions. In this case, Hooke's law would probably not apply, long before the stretch got to be one meter.

4-9: The pressure of the atmosphere does depend on height. If the balloon is 1 meter tall, the pressure of the air on the bottom is greater than that on the top by the amount of pressure exerted by a column of 1 meter of air. Suppose that the column has a cross-sectional area of 1 m². The weight of 1 m³ of air is about 13 N. The difference of pressure between top and bottom of the balloon is therefore 13 N/m². That doesn't mean that the net upward force on the balloon is 13 N, however. One cubic meter of hydrogen weighs a little less than 1 N. The net upward force on such a balloon due to the air displacement would therefore be about 12 N. If the weight of the balloon itself is less than 12 N, the balloon will rise.

4-10: Here's another example of how we can apply formulas or models only within limited ranges. Since Young's modulus for copper is 1×10^{11} N/m², it would appear from the formula that a stress of 1×10^{11} N/m² would produce a strain of 1. The stretch would be equal to the original length. For most materials, strain is proportional to stress only in a range of a few percent of strain. Beyond that, the material either flows or breaks.

4-11: Displacements combine according to the rules of Euclidean geometry. A displacement of 8 meters east and then 6 meters north creates the two legs of a right triangle. The hypotenuse is found from the Pythagorean theorem: $C^2 = A^2 + B^2$. In this case, the distance to the origin (the hypotenuse) is 10 m.

4-12: The key word in answering this question is *net* force. As we will see later when studying Newton's second law of motion, friction dominates most of the motions in the world around us. When the driving force on a car just matches the retarding forces of road and air friction, the net force on the car is zero and the car cruises at constant speed.

4-13: To solve this problem, follow the suggested rules about making a sketch of the forces and writing down the equilibrium conditions. The required angle is 5.8°. (Note that this is 0.1 radian.)

4-14: If you don't know, consult your nearest door.

4-15: (a) The force on the wall is proportional to the fractional distance that the climber has reached. When he has reached the top of the ladder, the wall must produce a force of 288 N. There must be an equal horizontal force in the opposite direction provided by the floor. (b) The direction of the resultant force on the foot of the ladder is given by $\tan \theta = (L/x) \tan 60°$, where x is the distance of the climber up the ladder. When the climber is at the top, $x = L$, and the thrust of the ground is in line with the ladder. When $x = 0$, no horizontal component of thrust is needed. $\tan \theta = \infty$, corresponding to $\theta = 90°$.

4-16: If each $w_i = W/n$, then

$$x = \frac{\sum_{i=1}^{n} x_i (W/n)}{W} = \frac{\sum_{i=1}^{n} x_i}{n}$$

This expression is the ordinary arithmetic average, which yields the midpoint for the center of weight of a uniform boom.

4-17: If the center of weight of an object is not directly under the point of suspension, the weight of the object will produce a torque with a radius arm equal to the horizontal component of the distance between the center of weight and the point of suspension. When the center of weight is directly under the point of suspension, the horizontal component of the radius arm is zero.

Problems

1. About what is the weight in newtons of (a) a cup of water; (b) a pencil; (c) this textbook; (d) yourself; (e) a car?
2. One hundred people choose sides for a tug of war. If the average force exerted by each person is 200 N, what will be the tension in the rope?
3. For a particular spring, $k = 1000$ N/m. What force is required to stretch the spring 1 m?
4. About what value of spring constant would be appropriate for a coil spring supporting one quarter of a car's weight?
5. What is the pressure at the bottom of a swimming pool 3 m deep?
6. Atmospheric pressure at sea level ranges from about 77 cm of mercury in high pressure regions to 72 cm in the middle of hurricanes. What are these pressures in N/m²? What is the percentage change compared with standard atmospheric pressure?
7. During a severe wind storm air pressure might fluctuate by 1% in a sudden gust. If the air pressure in a closed house remains constant during the gust, what is the total force on a window that has an area of 1 m²? Express the force in tons.
8. If you pull with a force of 1000 N on a steel wire that has a diameter of 1 mm, what is the stress? What is the strain?
9. A closed cubic container, 10 cm on a side, is filled with air. When it is submerged in water with the top of the box level with the surface of the water, what is the extra force on the bottom tending to make it rise? (This is called the buoyant force.) If the box is pushed down 1 meter, what is the buoyant force?
10. A cord with a diameter of 2 mm and a length of 50 cm stretches 0.5 cm under an applied force of 500 N. What is the spring constant, k, for this suspension? What is Young's modulus for the material?
11. What is the resultant force on a post if one person pulls on it toward the east with a force of 100 N and another person pulls on it toward the north with a force of 200 N?
12. A car weighing 1×10^4 N rests on a 10° incline. What is the force exerted perpendicular to the road? What is the force parallel to the road, along the incline?
13. Three forces act on a small object in equilibrium. One is 20 N at 30° N of E, one is 20 N at 45° W of N. What is the third force?
14. An arrow, notched in a bow string, is pulled back with a force of 200 N. If the angle between the upper and lower part of the string is 150°, what is the tension in the string?
15. A weight of 1000 N is supported at the end of a horizontal boom that has negligible weight. What is the tension in the supporting rope: (a) if the rope is vertical? (b) if the rope is at 45° from the vertical? (c) if the rope is at 80° from the vertical?
16. A steel wedge, used to split wood, exerts forces perpendicular to its flat surfaces. If the wedge angle between the two faces is 12°, what is the force exerted by each face when the driving force at the blunt end of the wedge is 1000 N?
17. A thin clothesline 10 meters long supports a single suit that weighs 20 N. The hanger is in the middle of the line, and that point sags only 10 cm below the horizontal. What is the tension in the line?
18. The seesaw shown in the diagram is in equilibrium. Neglect the weight of the board.
    (a) What are the magnitude and direction of the force exerted by the center support?

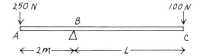

(b) Find the distance, *L,* by considering the torques about *B*.
(c) Confirm your result in (b) by considering torques about points *A* and *C*.

19. A chauvinist male tries to hold a door open for a woman by pushing at 45° at a point halfway between hinge and edge of door. What is the ratio of the force that he has to exert compared with that of the optimum way of opening the door? (Express each torque in terms of the applied force and width of door, *R*. F_a is the force he applied. F_o is the smaller optimum force that would produce the same torque. What is the ratio of F_a to F_o?

20. A force of 40 N is applied tangentially to the handle of a winch that is 30 cm from the center of the axle. If this torque is just sufficient to hold an object suspended by a rope fastened to the axle at 10 cm from the center, what is the weight of the object?

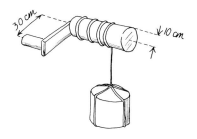

21. A uniform beam 5 meters long, with a weight of 500 N, is supported at its ends. Two men stand on it. One man, with a weight of 600 N, is 2 meters from the left end. The other, with a weight of 800 N, is 1 meter from the right end. What are the forces provided by the two supports?

22. A boom with uniform cross section has a weight of 1000 N. It is fastened to a wall at one end and is held by a cable at the other. The cable makes an angle of 30° with the boom, which is horizontal. If an object with a weight of 2000 N is supported at the end of the boom, what is the tension in the cable? What are the magnitude and direction of the force provided by the wall support?

23. A child weighing 300 N sits in a swing and is held stationary by a horizontal force of 200 N. Draw the vector diagram. What is the tension in each rope supporting the swing?

24. A picture with a weight of 100 N hangs by a single wire, which is attached to each side of the frame and passes over a nail in the wall. What angle does the wire make with the vertical if the tension in the wire is 100 N?

25. A girl with a weight of 500 N balances on one high heel. If the area of the heel touching the floor is 4 cm², what is the pressure on the floor?

26. The distance between the axles of a car is 2.3 m. When the car is weighed on a platform scale, it is found that the front wheels support 9000 N and the rear wheels support 6500 N. How far back of the front axle is the center of weight?

27. The three tubes of a telescope are each 20 cm long. The inner one weighs 0.8 N; the middle one weighs 0.9 N; and the outer one weighs 1.0 N. When the telescope is fully extended, how far is the center of weight from the big end?

28. A cube with uniform density weighs 100 N. What horizontal force applied to the top edge of the cube will just tip it over?

29. A ladder makes an angle of 70° with the ground and leans against a wall with negligible friction. When a 70 kg climber is two-thirds the way up the ladder, what are the forces exerted by the wall and by the ground?

5 MOTION PRODUCED BY FORCES

If you pull on a cart that is tied to a post, you may stretch the rope or bend the post but you won't change the cart's velocity. What happens if you untie the cart and then pull on it with a constant force? It will move, of course, but how? Will it move steadily with a speed that is proportional to the force? Will it go faster and faster with an acceleration proportional to the force? Does the choice between these two possibilities depend on how heavy the cart is, or whether it is being pulled along a road or through sand? As you may suspect, these simple questions have complex, but interesting, answers.

HANDLING THE PHENOMENA

In a laboratory situation, with equipment that is made for the purpose, you can apply a known force to an object and then record the motion in detail. However, you don't need elaborate apparatus to get a feel for the subject. To apply a constant force, attach a rubber band to an object and keep the rubber stretched to a constant length. (That's not always easy, because as the object moves you have to keep moving ahead of it in order to keep the rubber band stretched the same amount.) The best "object" to use is something with wheels that will reduce friction. If you use a small cart, a toy car, or roller skate, you can pile other objects on top to see how they affect the motion. If you use a large cart, you can use people for the load, but you will need metal springs instead of rubber bands to apply the force. If you can't get hold of a wheeled vehicle, just pull a block of anything hard along a smooth floor, using an ordinary rubber band.

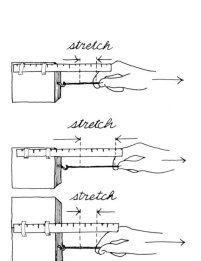

Tape a ruler to the side of the "object" so that you can watch the length of the rubber band and keep it stretched a constant amount. Try the setup and see what sort of motion is produced. Is it constant velocity? Does the object speed up? Whether or not there is constant acceleration may be hard to determine. To find out, you could use the same techniques we described in "Handling the Phenomena" of Chapter 2.

Double the stretch of the rubber band and compare the new motion with the old. Because of the nature of rubber bands, doubling the stretch does not necessarily produce double the force, but it's probably close.

Next, "double" the object. That is, pile on some load that is either identical to what you are pulling or that seems to have about the same weight. Pull the doubled object with the initial stretch and with the doubled stretch.

So far you have been pulling things on wheels or at least on smooth surfaces. What happens if you increase the friction so that it's a major part of the action? Try the same set of experiments on a rough surface, like a rug. You will have to pull harder than a particular "threshold" force, just to get the object moving. What happens then? Do you get constant velocity, or does the cart speed up?

After you have convinced yourself that you know how friction influences motion, take a tissue (like Kleenex), hold it spread out and horizontal, and then let it drop. It will waft around and dive eventually, but while it is still horizontal, is it drifting down with a steady speed or is it accelerating? For contrast, after you have watched it drift down a few times, wad it up and let it drop. Does it go down with constant speed? In both cases there is the same downward force applied, namely, its weight.

EXPERIMENTAL RESULTS OF APPLYING FORCES

With laboratory apparatus, the experiments described in "Handling the Phenomena" can be done with precision. The results are easiest to analyze if you can reduce friction almost to zero. To do this, you should obtain the conditions described by Newton's first law: without any applied force, the object should have constant velocity or be in such a situation that if put into motion its velocity will remain constant. Good approximations to these conditions can be obtained by floating gliders on an air track, or even by using carts with good roller-bearing wheels.

Air track glider with constant force being applied by weight hung from pulley.

Constant forces can be applied to the moving objects, either with springs stretched a constant amount, or by pulling with a thread that goes over a pulley and is attached to a falling weight. The position of the object as a function of time can be measured with a spark record on tape or by taking stroboscopic pictures of the object as it moves.

Strobe picture of air track glider being pulled with constant force. A cardboard arrow is mounted on the glider.

Question 5-1

The strobe picture makes it appear, qualitatively at least, that the glider is speeding up. How would you prove that the acceleration is really constant?

The quantitative results agree with the plausible conclusions that you probably drew from "Handling the Phenomena." Under these friction-free conditions, the application of constant force to an object produces constant acceleration in the direction of the force. Furthermore—and this is not so obvious—the acceleration is proportional to the force. If you double the force, you double the acceleration.

$$F \propto a$$

But why isn't this obvious? Most of you probably learned that fact in junior high school. Consider, however, the sophistication that it takes to analyze phenomena in this way. All around us we observe objects standing still unless a force is exerted on them. If you turn off the force applied to a car or a bike, they do not continue on with constant velocity. They slow down and stop. Furthermore, when you apply a constant force to a bike or a car (constant pedaling pressure or constant gas feed), you may accelerate briefly but then you cruise at constant speed. Do you think that you could convince an ox-cart driver that the constant pull of his ox will make his cart go faster and faster through the mud? No one thought about force and acceleration in our modern terms until about four centuries ago when Galileo described motion. For ages before then, the most intelligent of humans thought that it was obvious that objects moved so as to seek their natural positions—most things fall down, except for heavenly objects, which, of course, move in circles. To understand that constant acceleration is produced by constant force, and is proportional to it, takes a great deal of modern experience. It also requires the wit to simplify a situation by eliminating variables—such as friction. Note that simplifying the experiment takes not only the wit to think of doing it, but usually it also requires complicated apparatus.

There are two special points to note about the proportionality of acceleration and force. First, the law is true only under circumstances where Newton's first law holds. That law can be used to define a special set of conditions, namely, where an object will maintain constant velocity in the absence of external forces. (The technical description of such conditions is that they define an "inertial reference frame.") We don't really have those special conditions if we use the earth as our base of measurements. The earth is spherical and rotating, and that affects our observations of objects moving with respect to the earth and to us. Therefore, with respect to earth-based measurements, the rule about $a \propto F$ is just a good approximation for slow motions over short distances.

Question 5-2

In the northern hemisphere, if there is a low pressure region due north of high pressure air, in what direction will the winds blow?

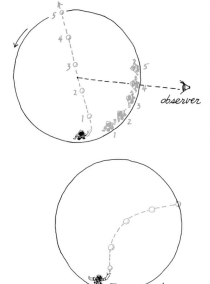

In the top diagram, the thrown baseball is observed from *above* the merry-go-round. Newton's first law holds.

In the bottom diagram, the thrown baseball is observed from a position *on* the merry-go-round. Newton's first law does not hold. This is not an inertial reference frame.

The second point about the rule that $a \propto F$ is simply that the proportionality really extends all the way to zero force. We have built into our language such expressions as "it takes a lot of force to get it moving." You might get the impression that a big object will remain motionless until a shoving force gets big enough, and then the object will start to move. Frequently, as a matter of fact, objects *are* stuck and do require a "threshold" force to start the motion. If friction is really zero, however, acceleration will be produced no matter how large the object or how small the force.

The Constant of Proportionality Between Force and Acceleration: $F = ka$

The proportionality constant, k, measures some property of an object having to do with its reluctance to change its velocity. If $F = ka$, then $k = F/a$. A large value of k means that a large force will produce only a small acceleration of the object. A system that is hard to change is said to have a lot of *inertia*. In this sense, the proportionality constant, k, is a measure of the inertia of the object. The problem is, how do you measure this inertia?

Perhaps k is equal to the volume of an object. It stands to reason, or at least to experience, that it is harder to accelerate a big object than a small one. That's true, however, only if the objects are made of the same material. What do you suppose the accelerations will be if you apply one newton of force steadily to each of the objects pictured here? The big chunk of Styrofoam in the picture will not necessarily have a smaller acceleration than the small piece of lead. It would not be a good rule to propose that $F = ka$, and $k =$ volume.

water   lead   styrofoam

Perhaps, instead of volume, the right measure to use for inertia is weight. The trouble with this idea is that weight is caused by an interaction of an object with some other object—usually with something big like the moon or the earth. The astronauts demonstrated the fact that hammers and other objects have as much inertia on the moon as they do on earth, even though everything the astronauts brought there weighed only one-sixth as much as it did on earth. We can see the same effect on earth. Both a bicycle and a train have their weight fully supported by the earth. If they were to roll slowly toward you, however, you could more easily change the velocity of the bike than the train. It wouldn't be the vertical force between earth and train that would bowl you over. You couldn't stop the train even in far outer space, where it would have no weight at all.

If the inertial property is not proportional to the volume of objects, and exists whether or not the objects have weight, perhaps it is proportional to the "amount of matter" in an object. The trouble with that suggestion is that there is no good way to measure the "amount of matter," except perhaps by measuring the amount of inertia. Even if you could count the number of atoms in an object, you would find that different atoms have different inertial responses. One idea, however, is to count the total number of protons, neutrons, and electrons in an object. These constituents of the atom are the same in every atom, but each atom has a different number of them. While it isn't practical to count every particle in large objects, we can actually compare the inertial response of subatomic systems where we can count the individual protons, neutrons, and electrons. One easily performed experiment (with a device called a mass spectrometer) allows us to measure the relative acceleration of a deuterium nucleus and a helium nucleus, when each is subjected to a known electric force. Deuterium is more popularly known as heavy hydrogen. Its nucleus contains one proton and one neutron. The nucleus of helium, sometimes called an alpha particle, contains exactly twice as much "matter" (or so it would seem). It consists of two protons and two neutrons. We might expect, therefore, that if the two nuclei were subjected to the same force, the acceleration of the deuterium nucleus would

Same number of atoms of hydrogen and uranium—but not the same inertia.

Deuterium nucleus	Helium nucleus
1 proton	2 protons
1 neutron	2 neutrons

Helium has twice as many particles as deuterium. Does it have twice the inertia?

be exactly twice that of the helium nucleus. But it's not quite true. The inertial property of the helium nucleus is a little less than twice that of the deuterium nucleus. We'll explore the reason for this strange effect in a later chapter.

It looks as if the only way to measure the inertial property of an object is to measure the inertial effect! The property is not necessarily linked to anything else. The property is called *inertial mass* and is given the symbol, *m*. The relationship between force and acceleration becomes

$$F = ma$$

This is the most important formula of classical mechanics. It is called Newton's Second Law.

The units of inertial mass could be defined to be those of force/acceleration—(newton)/(meter/second^2). As a matter of historical and technological fact, the units are defined the other way around. Mass was chosen to be a basic dimension, with a unit called the *kilogram* in the metric system. Force is then considered a derived property, with the newton defined as the force necessary to accelerate one kilogram at a rate of one meter per second squared: $1 \text{ N} = (1 \text{ kg})(1 \text{ m/s}^2)$.

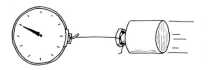

1 newton force applied to 1 kilogram produces acceleration of 1 m/s².

> **Question 5-3**
>
> Given a standard object with a defined mass of one kilogram, how would you measure the inertial mass of another object?

As we determined experimentally in Chapter 4, force is a vector quantity. We might expect then that since acceleration is also a vector, mass must be a scalar. That is certainly true if the acceleration produced by a force is always in the same direction as the force. The proportionality constant could not have any direction properties. Why should there be a different proportionality constant if you exert a force N-S instead of E-W? So far as we know, the proportionality is independent of direction, and mass is, indeed, a scalar. But logic can't tell us this. Only experiment can determine such a fact. For over a hundred years people have realized the possibility that the very existence of an inertial effect of an object may depend on all the rest of the mass in the universe. In particular, the mass of an object might be a little larger if the object were being forced to accelerate in the direction of high concentrations of mass. The highest concentration of mass for us is at the center of our own galaxy—the hub of the Milky Way we see in the night sky. Within the past decade, sensitive experiments have been done to see if the period of an oscillating system in line with the galactic center is different from what it is in a perpendicular direction. The results show that to within one part in 10^9, mass is a scalar. (If there were a difference, mass would have to be described with tensor algebra, which is like a more complicated form of vector algebra.) One kilogram plus two kilograms equals three kilograms no matter which way the apparatus is pointed.

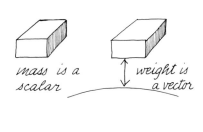

mass is a scalar   weight is a vector

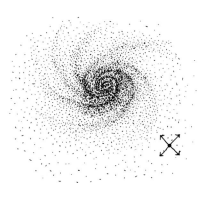

Does an object vibrating along a radial direction in the galaxy have a different period from one vibrating perpendicular to that direction?

INERTIAL AND GRAVITATIONAL MASS

There remains a very peculiar problem about the nature of mass. Surely mass is related to the property of weight in some way. We said that weight could not itself be the proportionality constant between force and acceleration because the weight of an object depends on some other gravitational body. An object can have inertial mass even if its weight is zero. Furthermore, mass is a scalar and weight is a force represented by a vector. Nevertheless, the larger the inertial mass of an object, the greater its weight.

Newton linked mass and weight over three hundred years ago. There is some historical basis to the legend that he thought of the theory of gravity after observing an apple fall from a tree. The grandeur of his revelation was more striking three hundred years ago than it would be today. Newton had been speculating about the way that the moon orbits the earth. There must be a centripetal acceleration, which means that in one sense the moon is continually falling toward the earth. The apple also falls toward the earth. Perhaps there is a common cause. Why not? It was a startling thought in those days because the common wisdom was that heavenly objects follow their own celestial rules, while earthly objects have more mundane behavior. Newton was proposing that the same rules apply throughout the universe.

Must the apple and the moon obey the same law?

Newton's law of universal gravitation is that any two objects attract each other with a force proportional to the product of their masses and inversely proportional to the square of the distance between them. (There is an important requirement if the statement is to hold in this form: the objects must either be spheres, or so far away from each other that their individual sizes are small compared with the distance between them. Also, if they are spheres, the distance between them must be taken to be the distance between centers.) In symbols, the equation is

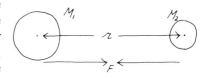

$$F_{\text{grav}} = G \frac{m_1 m_2}{r^2}$$

The distance between centers is r; G is a universal gravitational constant, having the same value in orchards as it does between here and the moon; m_1 and m_2 are the masses of the two objects.

Question 5-4

But what kind of mass should be used?

There are a number of profound points about Newton's simple formula. We have already emphasized that G is universal. Its numerical value depends only on the units being used. In our metric system of meters, seconds, and kilograms, $G = 6.67 \times 10^{-11}$ Nm²/kg². Notice, next, that the exponent of the distance, r, is 2. Is that so remarkable? Well, yes, especially since the exponent has been measured experimentally to be 2.000,000,000. The exponent of exactly 2 is a reflection of the Euclidean nature of space. (That is, points and distances in space are related by ordinary Euclidean geometry.) It shows up in this formula for the seemingly irrelevant reason that in a

Gravitational influence spreads out uniformly in space from centerpoint. Influence per unit area is proportional to 1/r^2 in Euclidean geometry.

Two definitions of "Mass"

1. Dynamic

$F = ma$

2. Static

$F = G \dfrac{m_1 m_2}{r^2}$

three-dimensional Euclidean world, the surface of a sphere is exactly proportional to the square of the radius. Finally, notice that the mass property of the objects serves as the source of the gravitational influence. Mass is the "charge" of the gravitational force, just as electric charge is the source or charge of the electric field.

The only mass that we defined in terms of an experiment was inertial mass. Now it appears that we could use Newton's gravitation law to define the quantity of matter in a different way. The definition of mass based on inertia makes use of the law: $F = ma$. A measurement of mass based on this definition requires a dynamic experiment—a known force is applied, and an acceleration must be measured.

The definition of mass based on gravity makes use of the law

$$F = G \frac{m_1 m_2}{r^2}$$

A measurement of mass based on this definition involves a static experiment—two objects are maintained at a set distance and the force between them is measured. Here are two entirely different definitions of mass. Can inertial mass be the same as gravitational mass?

The question was answered by Galileo, although he probably would not have recognized that he had done so. For three years Galileo was a professor of mathematics at Pisa in Italy. In one of his writings, he describes an experiment that might be done by dropping two objects at the same time from a tower. Ever since, there has been a legend that Galileo himself did such an experiment from the famous leaning tower of Pisa. He may have; he certainly described accurately what would happen. Even though the two objects have different weights, they will accelerate at the same rate and hit the ground at the same time. (Galileo realized that air resistance might affect the two objects differently if one were larger than the other, and he mentions this possibility in his argument. The purpose of Galileo's paper was not to prove that inertial and gravitational mass are proportional to each other. Instead, he was claiming that the traditional arguments of Aristotle about falling objects were wrong.) Nevertheless, the experiment does prove the proportionality of gravitational and inertial mass. We can follow the argument best by calling inertial mass, m_I, and gravitational mass, m_G. The weight of an object on the surface of the earth is equal to:

$$\text{weight} = F_{\text{grav}} = G \frac{m_G M_{\text{earth}}}{R^2}$$

where R is the radius of the earth. Since the quantity, $G(M_{\text{earth}}/R^2)$ is the same for all objects on the earth, let's call that quantity g. The weight of an object on the earth is, therefore, $m_G g$.

Now let's compare what happens when two objects are dropped from a tower at the same time. The weight of object 1 is $m_G g$.

The weight of object 2 is $m_G g$.

The acceleration of #1 is given by $F = m_I a$.

The acceleration of #2 is given by $F = m_I a$.

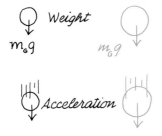

The force acting on each object is its weight

$$m_G g = m_I a \qquad m_G g = m_I a$$

The acceleration of each object downward is given by

$$a = \frac{m_G}{m_I} g \qquad a = \frac{m_G}{m_I} g$$

Since the value of g is the same for both objects, the ratio of their accelerations is

$$\frac{a}{a} = \frac{m_G}{m_I} \frac{m_I}{m_G}$$

The experimental observation is that both objects reach the ground at the same time. Therefore, $a = a$, and their ratio = 1. This can be true only if $m_G \propto m_I$. Gravitational mass must be proportional to inertial mass, and, if we choose our units properly, they are equal to each other.

This astonishing fact has been proved many times since with much more sophisticated experiments. Within the past decade, a group at Princeton showed that gravitational and inertial mass are equal to each other to a precision of one part in 10^{11}. It may seem strange that anyone would want to know such a thing to that precision, but a profound matter is at stake. The cornerstone of Einstein's general theory of relativity is the equivalence of inertial and gravitational mass. The nature of genius is clear in the simple matter of describing the equality of the two kinds of mass. Other people had thought how remarkable it was that gravitational mass was equal to inertial mass. It took an Einstein to explain that it was not remarkable that the two quantities are equal, for the simple reason that they are actually identical. Of course, out of this simple fact he then constructed a theory that successfully describes gravitation and the expansion of the universe.

If $m_G \propto m_I$, then $m_G = C\, m_I$

$$\frac{a}{a} = \frac{m_G}{m_I} \frac{m_I}{m_G} = C \frac{1}{C} = 1$$

Therefore, $a = a$

If gravitational mass is proportional to inertial mass, all free falling objects have the same acceleration.

$m_G = m_I$ Wisdom
$m_G \equiv m_I$ Genius

Question 5-5

Suppose that you lived in a universe where m_G were proportional to m_I^2. If you dropped a heavy object and a light object, which would hit the ground first?

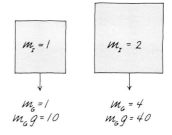

Conditions in a universe where $m_G \propto m_I^2$

THE UNIT OF MASS—THE KILOGRAM

When the metric system was established, during the French revolution, the unit of mass was chosen to be the mass of one cubic centimeter of water at its lowest density, which occurs at about 4°C. This much mass was named the *gram*. For practical purposes, the official standard of comparison was chosen to be a kilogram—a thousand grams. This is the mass of 1000 cc, or one liter, of water. A cylinder made of a special alloy of platinum was filed and polished until its weight balanced a liter of water on a pan balance. Notice that the standard was established in terms of the gravitational effect of mass.

Nations and their industries take standards of measurement very seriously. In 1821 two carefully calibrated copies of the international kilogram were sent to the United States. During the late 1800s, an international congress of distinguished scientists prepared revisions of all the standards. New duplicate kilograms were constructed and distributed to the participating nations. When the new kilogram was delivered to the United States, President Harrison officially received it at the White House and certified it as the United States standard mass.

It is unfortunate that the name of a basic unit is expressed in terms of a multiple of a subunit. The *kilo*gram is the unit—not the gram. During the 1960s several international meetings wrestled with the question of renaming the kilogram. Perhaps the gram should be called a millikilogram. We seem to be stuck with the present situation.

We gave the numerical value for G, the universal gravitational constant. It is not so obvious how you can obtain that number. You might suggest just putting two objects of known mass a known distance apart, and measuring the force between them. $F = G(m_1 m_2 / r^2)$. If m_1, m_2, and r are known, and F is measured, you can solve for G. The force between two ordinary objects is very small, however. You, yourself, are not gravitationally trapped by large buildings, for instance. Why not use an extremely large object for one of the masses? Use the earth. The catch is, how do you find the mass of the earth? Since we know the diameter of the earth, and thus its volume, we could estimate its mass by estimating its density.

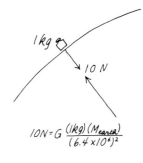

Question 5-6

Most rocks on the surface of the earth are compounds of silicon and oxygen, and have a specific gravity of about 2.5. Their density in metric terms is therefore about 2500 kg/m³. The radius of the earth is 6400 km. If the density of the earth were the same as the density of its surface rocks, what would be the mass of the earth? Once you have a value for the earth's mass, you can find G. Just remember that the force between the earth and a 1 kg mass is approximately 10 N. What is the value for G, based on that assumption about density?

The experiment to measure G was first performed by Henry Cavendish, an Englishman, in 1798. It involved measuring the gravitational attraction between two heavy, but human size, spheres. Two spheres were fastened at the ends of a rod, as shown in the diagram, and the rod was then suspended at its center by a very fine fiber. Two other spheres were brought up close to the suspended ones, so that the gravitational attractions rotated the suspended system until the torque was balanced by the twist in the fiber. Then the stationary spheres were moved to the other sides of the suspended ones, creating a rotation in the opposite direction.

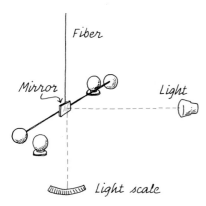

The Cavendish apparatus for measuring G.

Considerable engineering skill and planning are required to do such a delicate experiment. The larger the suspended masses, the greater the experimental effect. On the other hand, the larger the masses, the thicker the suspension fiber must be, and so the less sensitive the system will be to twisting. The whole system must be shielded against air currents and stray electrostatic fields. The rotation of the suspension is so small that it has to be detected by a light beam reflecting from a mirror on the suspended rod. Such an "optical lever" moves the light beam through a measurable arc even for a very small angular rotation.

Cavendish's value for G was within $\frac{1}{2}\%$ of the modern value, an astonishing feat considering the difficulty of the experiment and the state of technology 200 years ago. At the time, the measurement was widely publicized as a "weighing" of the earth. To be sure, once the gravitational constant is known, the mass of the earth and its average density can be found. The average density turns out to be 5500 kg/m³. Evidently the silicate rocks are only on the surface; most of the earth must be composed of higher density materials such as iron.

> **Question 5-7**
>
> What is the strength of gravitational attraction between a 50 kg girl and a 60 kg boy, 10 meters apart?

LITTLE g, THE "ACCELERATION DUE TO GRAVITY"

As we have seen, on the surface of the earth an object with mass m has a weight given by $F = G\,(mM/R^2)$, where M and R are the mass and radius of the earth. The quantity, GM/R^2 is approximately a constant, and is given the symbol g. Actually, g is not a constant, partly because the radius of the earth varies from pole to equator, and partly because the centrifugal effect caused by the earth's rotation depends on latitude. The first effect reduces g at the equator (compared with the poles) by 0.18%; the second effect (which we will study in Chapter 6) reduces g by 0.34%. Let's evaluate g using an average radius:

$$g = (6.67 \times 10^{-11}\,\text{Nm}^2/\text{kg}^2)(5.98 \times 10^{24}\,\text{kg})\,(6.37 \times 10^6\,\text{m})^2$$

$$= 9.8\,\text{N/kg}$$

The weight of an object with mass m on the earth's surface is equal to mg. For instance, a 1 kg mass has a weight of $(1\,\text{kg})(9.8\,\text{N/kg}) = 9.8\,\text{N}$.

This number, 9.8, first appeared when we described objects falling freely at the surface of the earth. Any such object, regardless of mass, accelerates at a rate of 9.8 m/s². Little g (as opposed to capital G, the universal gravitational constant) is usually called "the acceleration due to gravity," and is assigned the units of m/s².

What's the relationship between these two uses and two sets of units—N/kg and m/s²? An object free to fall toward the earth is subject to a force equal to its weight: $F = \text{weight} = mg$. The force will produce an acceleration

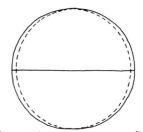

The earth has a very slight equatorial bulge. At the poles, an object is closer to the dense core of the earth, and so weighs slightly more.

according to the formula: $F = ma$. Therefore, the acceleration is given by $ma = mg$, or $a = g$. This can be true only if 9.8 m/s² = 9.8 N/kg.

> **Question 5-8**
>
> Show that these two expressions are equivalent, first in dimensions, and then in units.

One of the confusions of calling g the acceleration due to gravity is that frequently we use the value to find the weight of some object that is tied down and not accelerating anywhere. Alternatively, we can call g the value of the gravitational field. When you multiply the strength of the field, g, with the gravitational source charge, m, you get the gravitational force, mg, which is the weight. This way of looking at things is completely analogous with the way of describing the electric field, as we will see in a later chapter.

NEWTON'S SECOND LAW AND THE REAL WORLD

It may seem that Newton's second law applies only in special laboratory situations. In the first place, it applies only under circumstances where Newton's first law applies: "When no force is acting on an object, the velocity of the object does not change." This first law is really a definition of a reference frame from which the observations are made. It requires that the observer should not be accelerating. Unless appropriate corrections are made, strict applications of Newton's laws do not hold on the surface of our spherical, rotating earth.

There is another small problem with the law $F = ma$. Even in an inertial reference frame, it's only an approximation. Remember that there is a speed limit in the universe—the velocity of light. A constant force cannot produce constant acceleration endlessly, or else the speed limit would be passed. Furthermore, there are a few subtle problems about the nature of mass that we have not yet mentioned. It turns out that the mass of an object is a function of its speed. We'll explore these interesting features in later chapters.

Even without the complications of reference frames or speeds close to that of light, it may seem that Newton's second law does not describe the events of everyday life. In physics laboratories, which are filled with ideal machines and frictionless planes, objects may travel at constant velocity without any applied force. In the outside world, however, they slow down and stop. Everyone knows why. In the real world, there's lots of friction.

Friction acts as a force opposing motion. In using Newton's second law we must remember that the force described by the left side of the equation must be the *net* force acting on the object. That net force is equal to the effective applied force minus any retarding friction force.

$$F_{\text{applied}} - F_{\text{friction}} = ma$$

In some cases the friction force is a constant, and in other cases it depends on the speed. The effects of these two cases are radically different.

1. *Dry Sliding Friction.* When two dry objects slide on each other, the retarding force of friction is almost independent of the velocity, at least for low velocities. In fact, the friction force depends only on the nature of the surfaces and the force pressing the two objects together. It isn't obvious, at least on first thought, why such an approximation should be so good.

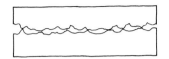

Imagine the surface boundary between two objects. On the atomic scale there are bumps, valleys, and hills. You might think that a considerable force would be required to slide one object past another, but that this force would decrease as the objects moved faster, riding along the tops of the bumps. Furthermore, you might expect that the larger the surface area touching, the greater the friction force. It doesn't work that way. The friction is not caused by the meshing of bumps and valleys into each other, but by actual molecular bonding between the two surfaces. One proof of this explanation is the fact that very smooth surfaces usually have large sliding friction between them. Wires with smooth insulation are hard to pull through electrical conduits; special rough insulation is used for wires that have to be fed through long distances.

The molecular bonding model of friction also explains why the amount of friction does not depend on the apparent area of contact. When one object is placed on another, only a small percentage of the mutual area is actually in molecular contact. The bumps on the surfaces of both objects are touching. There is just enough surface contact to support the weight of the upper object, or to resist whatever force is pressing the two objects together. In the case of the rectangular block in the diagram, there is the same amount of actual contact area regardless of how the block is placed. That area is determined only by the weight, and the weight of the block is the same no matter how it is turned. If another weight is placed on top of the block, the block will sink a little lower so that the actual contact area increases, even though the apparent surface area remains the same.

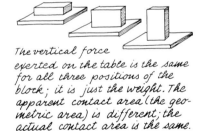

The vertical force exerted on the table is the same for all three positions of the block; it is just the weight. The apparent contact area (the geometric area) is different; the actual contact area is the same.

We can summarize with a formula the way that friction between dry sliding surfaces depends only on the force between the objects and the nature of the surfaces:

$$F_{\text{friction}} = \mu F_N$$

F_N is the "normal" force—normal in the mathematical sense, meaning perpendicular. If one object is sliding on top of another, the normal force is usually the weight of the object on top. However, there may be extra forces pressing the objects together. The Greek letter, μ (mu), is used for the "coefficient of friction." The coefficient is a function of the combination of materials sliding past each other. It has a different value for hardwood and iron than for hardwood and hardwood. Engineering handbooks give values for many different combinations. On the next page there is a table of some of them.

Surfaces	Static μ	Kinetic μ
steel on steel	0.74	0.57
steel on aluminum	0.61	0.47
steel on brass	0.51	0.44
glass on glass	0.94	0.40
glass on copper	0.68	0.53
steel on Teflon	0.04	0.04

Notice that there is a column of values for the coefficient of static friction, and that these values are higher than for the coefficient of moving, or kinetic, friction. We see here a demonstration of the fact that friction between dry surfaces is not completely independent of velocity. At least when the velocity is zero, in a static situation, the frictional bonding is greater. As one object begins to move past the other, the original molecular bonds are broken and the coefficient of friction falls to its almost constant value. Rolling friction is much less than sliding friction. When one object rolls over another, the molecular bonds are broken by a lifting motion, whereas a sliding motion requires a shearing effect.

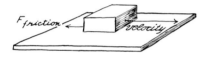

Question 5-9

We have described the magnitude of friction forces, but haven't said anything about their direction. If one object is resting on another, and there is a friction force between them, why don't the objects accelerate in response to this force?

Let's examine several examples of how sliding friction affects simple motion:

1. An aluminum cube, 10 cm on a side, has a mass of 2.7 kg. If it is resting on a steel table, and a force of 10 N is applied, what is the acceleration? The acceleration is zero. The retarding force produced by static friction can get as large as $\mu_s F_N$ before the static bonds are broken. The normal force, F_N, is provided by the weight of the cube, which is equal to mg. The threshold force is

$$F = (0.61)(2.7 \text{ kg})(9.8 \text{ N/kg}) = 16.1 \text{ N}$$

Until the external horizontal force gets that large, the cube will stand still.

2. Take the same case of an aluminum cube on a steel table, and now exert a horizontal force of 20 N. What is the acceleration? Once the cube is in motion, the retarding force is

$$\mu_k F_N = (0.47)(2.7 \text{ kg})(9.8 \text{ N/kg}) = 12.5 \text{ N}$$

Notice that this is less than the threshold force needed to get the cube moving. Newton's second law applied to this case is

$$F_{\text{external}} - F_{\text{friction}} = ma$$

$$20 \text{ N} - 12.5 \text{ N} = (2.7 \text{ kg})a$$

$$a = 2.8 \text{ m/s}^2$$

3. Suppose that you push a trunk across the floor as shown in the diagram, with the force of 300 N applied at an angle of 30° from the horizontal. If the coefficient of friction between trunk and floor is 0.3, and the trunk has a mass of 50 kg, what is the acceleration? In this case, the normal force is the sum of the weight of the trunk and the downward component of pushing.

$$F_N = mg + F_{\text{applied}} \sin 30°$$

$$= (50 \text{ kg})(9.8 \text{ N/kg}) + (300 \text{ N})(0.5)$$

$$= 490 + 150 = 640 \text{ N}$$

The effective force in the direction of motion is $F \cos 30° = (300 \text{ N})(0.87) = 260 \text{ N}$. The acceleration produced is given by

$$F_{\text{effective}} - F_{\text{friction}} = ma$$

$$(260 \text{ N}) - (0.3)(640 \text{ N}) = (50 \text{ kg}) a$$

$$a = 1.4 \text{ m/s}^2$$

Question 5-10

What would be the acceleration if the trunk were pulled with a rope with the same force but at an upward angle of 30°?

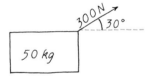

Optional

Case 2: Friction that Depends on Velocity

As much as possible in the real world, we avoid dry, sliding friction, unless we deliberately want to keep things from slipping. Of course, in the case of sideways motion of tires on the road, or of shoes on the sidewalk, we want to keep friction as large as possible. When you want to reduce friction, you arrange to have objects roll over each other, or you lubricate them. Ancient pyramid drawings show the Egyptians pouring milk under the runners of sledges used to drag stone blocks. Since the butterfat content of bottled milk is low, we usually use oil nowadays. The lubricant keeps the molecular bonds between the moving objects from forming, and provides a solid-fluid boundary. The behavior of friction forces is very different for a solid moving through a fluid than it is for a solid sliding on a solid. For one thing, the retarding force depends on the velocity.

The consequences of velocity-dependent friction are familiar to everyone. As you saw in "Handling the Phenomena," a tissue that is dropped floats gently as long as it is spread out. A parachutist falls, not with constant acceleration, but with constant velocity. If you floor the accelerator pedal of your car, you accelerate only to some maximum velocity and then continue at that speed. The friction forces in all these cases become as large as the driving forces. When the net force becomes zero, the acceleration becomes zero, and the velocity remains constant. That constant velocity is usually called *terminal* velocity.

Question 5-11
Why doesn't dry sliding friction produce terminal velocity? Apparently all that is required is for the friction force to be as large as the driving force.

When an object moves through a fluid, such as air or water, the friction force increases with velocity. For human size objects moving in air, for example, $F_{friction} = kv^2$. When an applied force first starts accelerating an object through the air, the velocity is still small and so the friction term is small.

$$F_{applied} - kv^2 = ma$$

remains approximately

$$F_{applied} = ma$$

As the object accelerates, however, the velocity gets greater and greater. Eventually, $kv^2 = F_{applied}$. Then, $(F_{applied} - kv^2) = 0$. With no net force, the acceleration is zero, and the velocity remains constant.

The exact dependence of the friction force on the velocity depends on the nature of the fluid and the size of the object. For dust-size particles drifting in air, the friction force is approximately proportional to the first power of the velocity. For boats moving in water, the friction force is proportional to the square of the velocity. In the case of boats, however, greater complications arise when the waves thrown up by the boat's movement begin interfering with each other. There are more details about fluid friction in Chapter 14.

The constant, k, is a function of geometrical size and surface shape. "Streamlining" can often reduce the value of k, for a particular object. There are peculiar complications that affect the value of k. For small boats, under general conditions, the maximum practical speed is proportional to the square root of the length of the boat. For spheres (such as golf balls) traveling in air, k is less if the spherical surface is rough rather than smooth.

The terminal velocity of a large object falling in air is proportional to the square root of its weight. This relationship arises because at terminal velocity $(F_{applied} - kv^2) = 0$. Since the applied force is the weight of the object, and v is the terminal velocity, $v_{terminal} = \sqrt{weight/k}$. The geometric factor, k, is roughly proportional to the surface area of the falling object. For a human body, without a parachute, the terminal velocity in air is a little more than 50 m/s, or about 120 mph. Parachutes are usually designed for a terminal velocity—and thus a landing speed—of 6.5 m/s, or 15 mph.

Question 5-12
If you drop two objects of the same shape, but of different size and weight, which will achieve the larger terminal velocity? If you dropped an elephant and a mouse from an airplane at the same time, which would land first?

The friction affecting automobile motion is partly due to the road and partly due to the air. For most cars, depending on tires, load, and streamlining, the road friction dominates for speeds under 30 mph (48 km/hr). Since the road friction is relatively independent of velocity, and since air drag is proportional to the square of the velocity, air effects rapidly take over the performance characteristics as speed increases. Each car has its own upper speed limit, or

$v(t)$ for an object subject to a constant force and to friction that increases as velocity increases.

terminal velocity. Increasing the power of the engine of a car does not increase that limit by much. The power needed is actually proportional to the cube of the terminal velocity. If you doubled the power, the terminal velocity would rise by only 26%.

SUMMARY

This chapter is concerned with just one equation: $F = ma$. The equation is simple, but the implications of Newton's second law are profound. We analyzed and questioned every aspect of the symbols and their relationship.

1. A force applied to an object produces an acceleration of the object in the direction of the force and proportional to the force. This claim appears to be contrary to most everyday experience. The law applies only in inertial reference frames—defined by Newton's first law—and the force involved must be the net force.
2. The constant of proportionality between force and acceleration is called the inertial mass. It is not the same as the volume, or the weight, or the number of subatomic particles in an object. The amount of inertial mass in an object can be determined by applying a known force and measuring the acceleration.
3. Inertial mass is a scalar quantity, although this is a fact subject to experimental challenge. The inertial property of an object may be caused by the influence of all the rest of the mass in the universe.
4. Gravitational mass is determined by another one of Newton's discoveries. The gravitational attraction between two spherical (or distant) bodies is given by $F_{grav} = G(m_1 m_2)/r^2$. Gravitational mass plays the role of the source charge of the gravitational field. So far as we know, the gravitational charge of all objects is positive, always producing attraction. Gravitational mass is defined in terms of weight measurements.
5. Newton's gravitational formula displays important facts about the universe. The constant, G, is universal. It has the same value on earth as it does in space. The exponent of the distance in the denominator is apparently exactly 2, corresponding to the properties of three-dimensional Euclidean space. The value of G was determined in a laboratory experiment by Cavendish almost 200 years ago. With G determined, the mass of the earth could be calculated. The average density of the earth turns out to be much greater than the density of surface rocks.
6. On the surface of the earth, (GM/R^2) is almost a constant and is equal to about 9.8 N/kg. The constant is called g. On the surface of the earth, the weight of an object with mass m, is mg.
7. It has been demonstrated experimentally with great precision that gravitational mass is proportional to inertial mass. Einstein declared that this fact could not be a coincidence, and that the two properties are actually identical. Therefore, the same property, m, can be used in the two unrelated formulas: $F = ma$, and $F = G(m_1 m_2)/r^2$.
8. The unit of mass is a human choice and arbitrary. It is called the kilogram and was originally defined as the mass of one liter of water.

The official standard in each country is a platinum cylinder whose mass has been calibrated against the international standard kilogram.

After worrying about the meaning of Newton's second law, we warned that it is only an approximation good for velocities much less than the speed of light. Nevertheless, the law applies to all sorts of everyday phenomena, providing that opposing friction forces are taken into account.

We considered two types of friction, each of them producing forces that oppose motion. Dry, sliding friction does not depend on velocity, at least to a first approximation. This type of friction force is equal to $F = \mu F_N$, where μ is the coefficient of friction, a characteristic of the two surfaces, and F_N is the normal force pressing the two objects together. The force does not depend on the apparent surface area between the objects.

The other type of friction does depend on the velocity of the object. This situation occurs when objects move through a fluid, whether liquid or gas. For most objects moving through air, the retarding force is approximately proportional to the square of the velocity. The actual dependence varies with surface boundary conditions and geometric size. If the retarding force increases with velocity, and a constant or maximum external force is applied to an object, terminal velocity will be reached. Cars at their top speed, and parachutes drifting down are examples of terminal velocity situations.

Answers to Questions

5-1: Using the techniques of strobe picture analysis developed in Chapter 2, measure the differences in position, Δx. Since the time intervals, Δt, are constant, the differences in position are proportional to the speed, $\Delta x/\Delta t$. The plot of Δx versus t should be a straight line if the acceleration is constant.

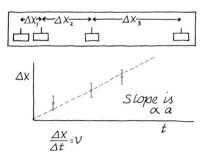

5-2: Take a look at a weather map if you don't remember. Winds moving in the northern hemisphere veer to the right because of the earth's rotation. (In the southern hemisphere they veer to the left.) Therefore, if the pressure regions are such as to force a wind from south to north, the wind will swing around toward the east. As the winds swirl into a low pressure region, they move in a counterclockwise rotation, as viewed from a satellite.

5-3: With the known mass, you could calibrate a spring so that you could exert a known force. For instance, you could use the spring to give the one kilogram mass an acceleration of 1 m/s². For that much stretch, the spring must be exerting a force of 1 N. With the calibrated spring you could then produce a measured acceleration of the unknown mass. You would know F and would measure a. Find m from $F = ma$.

SUMMARY

5-4: The only kind of mass that we have defined so far is inertial mass. It's not obvious, however, that the inertial property of an object resisting a change in its motion should have anything to do with the phenomenon of attracting another object at a distance.

5-5: In comparing the accelerations of the two objects, we concluded that

$$\frac{a}{a} = \frac{m_G}{m_1}\frac{m_1}{m_G}$$

If $m_G \propto m_1^2$, then $\dfrac{a}{a} = \dfrac{m_1^2}{m_1}\dfrac{m_1}{m_1^2} = \dfrac{m_1}{m_1}$

The acceleration of objects would be proportional to their inertial mass.

Another way to see this effect is by substituting simple trial numbers in the equation, as done in the diagram accompanying Question 5-5. If the inertial mass of the first object is 1, and the inertial mass of the second object is 2 (whatever units are used), then the gravitational mass of the first one is proportional to 1, and the gravitational mass of the second object is proportional to 4 (not 2). Then the weight of the second object would be 4 times the weight of the first one, but its inertial mass would be only twice the inertial mass of the first one. Since weight = $F = m_I a$, the heavier one would have twice the acceleration of the lighter one.

5-6: The volume of the earth is $(4/3)\pi r^3$. The mass of the earth is equal to the density times the volume = $(4/3)\pi(6.4 \times 10^6 \text{ m})^3 (2.5 \times 10^3 \text{ kg/m}^3) = 2.5 \times 10^{24}$ kg.

$$G = FR^2/m_1 M = (10 \text{ N})(6.4 \times 10^6 \text{ m})^2/(1 \text{ kg})(2.5 \times 10^{24} \text{ kg})$$

$$= 1.6 \times 10^{-10} \text{ Nm}^2/\text{kg}^2$$

The value is high, by over a factor of 2. Still, it gives the right order of magnitude.

5-7: $F = (6.67 \times 10^{-11} \text{ Nm}^2/\text{kg}^2)(50 \text{ kg})(60 \text{ kg})/(10 \text{ m})^2 \approx 2 \times 10^{-9}$ N. To the extent that the question is reasonable at all, it is reasonable to use the inverse square law as a good approximation. The boy and girl are not spherical, but the distance between them is large compared with their size. Evidently, any attraction between them is scarcely physical.

5-8: The dimensions of m/s² are L/T^2. The dimensions of N/kg are those of force/mass = (mass)(acceleration)/mass = acceleration = L/T^2. The same analysis shows that N/kg = (kg)(m/s²)/kg = m/s².

5-9: Friction force always acts in the opposite direction to the velocity, or in the opposite direction to the net applied force to prevent a stationary object from moving.

5-10: If the trunk is pulled with an upward force instead of pushed with a downward force, there is the same component of force along the horizontal direction. The vertical component, however, reduces the normal force between trunk and floor, thus reducing friction. $F_N = mg - F_{\text{applied}} \sin 30° = 490 - 150 = 340$ N. $F_{\text{effective}} - F_{\text{friction}} = ma$. $(260 \text{ N}) - (0.3)(340 \text{ N}) = (50 \text{ kg})a$. $a = 3.2$ m/s²

5-11: A delicate balance between a constant driving force and a constant friction force is theoretically possible, but is unlikely to happen in practice. The crucial point is that when the friction force is constant, the left side of Newton's second law becomes the difference between two constants: $F_{\text{applied}} - F_{\text{friction}}$. Unless those two constants are the same and remain the same, there will be acceleration. If, however, the friction term increases with velocity, a completely different situation arises. The friction term will increase until it just cancels the applied force, and then it will stay at just that required amount.

5-12: The terminal velocity is proportional to the square root of the ratio of weight and the constant, k. The constant, k, depends on the boundary layer between object and fluid, and on the geometry. As long as a comparison is made between two objects whose drag depends on the same function of velocity, it is approximately true that

k is proportional to the *area* of that falling object. For objects of the same general shape and density, the weight is proportional to the *volume*, which is proportional to the cube of some average length. The area creating drag is proportional to the square of that length. Therefore, the ratio of weight to k will be proportional to the ratio of L^3 to L^2. The terminal velocity will thus be proportional to the square root of the average length. Elephants will fall faster than mice.

Problems

1. List four situations where your *applied* force does not produce motion that satisfies the equation $F \propto a$.
2. Give two examples of what's wrong in defining mass as "the amount of matter" in an object.
3. Write the defining equation for inertial mass and the defining equation for gravitational mass. Explain each term in each equation. Explain why the equations do not require that inertial mass equals gravitational mass.
4. By appealing to the result of Galileo's "leaning tower" experiment, demonstrate algebraically that inertial mass must be proportional to gravitational mass.
5. If you lived in a universe where $m_G \propto m_I^3$, what would be the ratio of acceleration for objects with $(m_G)_1 = 1$ and $(m_G)_2 = 2$?
6. On the surface of the moon, 1 kg weighs 1.7 N. Using the assumption that the surface density of moon rocks (2500 kg/m³) is the average density throughout, and knowing that the radius of the moon is 1.7×10^6 m, what is the approximate value for G? (Compare this value with the one obtained in Question 5-6. Why the difference?)
7. All the gold in the world (that has been mined) would make a sphere only 22 m in diameter. The density of gold is 19,300 kg/m³. If this gold sphere could be constructed, what would be your gravitational attraction to it if you were at its surface?
8. What is "g" on the surface of the moon, which has a radius of 1.7×10^6 m and a mass of 7.3×10^{22} kg?
9. In the English engineering system of units, the pound is a unit of force (not mass). A kilogram weighs 2.2 pounds. How many newtons equal one pound?
10. What is your mass in kilograms? What is your weight in newtons?
11. A rectangular block has sides of 10 cm, 2 cm, and 5 cm. When the block is resting on the 10 × 5 face, the area of molecular support is A_1. When the block rests on the 10 × 2 face the molecular support area is A_2. What is the ratio, A_1/A_2?
12. A spring has a constant of 50 N/m. If it is used to pull a 2.0 kg box along the floor at constant velocity, and its length increases from 10 cm to 15 cm, what is the coefficient of friction between box and floor?
13. A block slides at constant speed down a plane inclined at 30°. What is the coefficient of friction? (Why is it not necessary to know the mass of the block?)
14. The coefficient of sliding friction between skis and snow depends on the type of snow and the wax on the skis. For a typical μ of 0.04, what is the angle from the horizontal of a slope that will provide constant velocity?
15. A parachute is designed to land at 15 mph with a woman whose mass is 50 kg. If, by mistake, a man with a mass of 100 kg uses this parachute, what will be his landing speed?
16. A 100 kg box is pulled along the floor at constant speed with a horizontal force of 100 N. What is the coefficient of friction? If an identical box is put on top of the first one, what force would be required to pull both of them at constant velocity? Would it take less force to pull them if both were on the floor?

17. The coefficient of kinetic friction between a 100 kg box and the floor is 0.2. A rope is tied to a harness that goes through the center of weight of the box. What force is required to pull the box at constant velocity if the rope makes an angle of 30° with the horizontal?
18. In Shakespeare's *Julius Caesar,* Cassius says, "the fault, dear Brutus, is not in our stars, but in ourselves, that we are underlings." Demonstrate the plausibility of this assertion by calculating your gravitational attraction to the sun and also to the next nearest star, Proxima, which is about 4 light-years away. Assume Proxima has the mass of our sun ($M = 2.0 \times 10^{30}$ kg). Compare both of these values with your gravitational attraction to the earth.
19. In 1959 a stewardess fell to her death from an airplane one mile above the earth. About how long did the fall last? (Calculate how long it would take in free fall to reach terminal velocity. Then make appropriate approximations.)
20. An approximate formula for the viscous drag on a mammal falling through air is $F = 0.2 \, (N \cdot s^2/m^4) A v^2$, where A is the area presented in the direction of travel and v is the speed. Find the terminal velocity of a 5 kg cat with an area of about 0.1 m².

6 APPLICATIONS OF NEWTON'S SECOND LAW

In Chapter 5 we questioned, marveled, and worried about every feature of the law: $F = ma$. Now let's apply the law to a variety of situations. First, we will study cases where the applied force is constant and the motion is in the direction of the force, or at least in the direction of a component of the force.

> **Question 6-1**
>
> Doesn't the motion always have to be in the direction of a component of the force?

Next, we will take another look at circular motion. All the variables of length, velocity, acceleration, mass, and force have their counterparts in rotation. Transforming one set into the other yields a form of Newton's second law that applies to rotation. We will learn how to describe rolling motion.

There are some interesting phenomena on earth where centripetal force is perpendicular to the force of gravity. There are even more interesting phenomena in space where gravity *is* the centripetal force.

Finally, we will begin the exploration of a type of motion caused by forces that do not remain constant, but that change according to a simple rule. The vibrating motions produced appear in bridges, violin strings, clocks, atoms, and stars.

HANDLING THE PHENOMENA

1. All objects, large and small, fall with the same acceleration, except when air friction is important. That's because a heavy object also has a lot of inertia. Here is a way to produce a falling system that has an *effective* weight that is not proportional to its mass. Get a thread or a string and hang it over a smooth rod, like a plastic pen. Tie each end to a paper clip, which is bent to form a hook. On each hook you can hang more paper clips, all of

the same type. If you have a total of ten paper clips on each side, the system will not move. If you then take one paper clip off one side and hang it on the other, you have the same total mass, but the weight of one side is more than that of the other. Observe how the system moves, with various combinations of paper clips on one side or the other. Is there a threshold weight difference before motion can start? Can constant velocity be produced if the weight difference is small enough? How does the acceleration compare with g as you change the total mass and the weight difference?

2. You need a record player to see this phenomenon, but it's worth it even if you have to borrow one. Get an orange-juice-size can, two paper straws, scotch tape, thread, and paper clips—the standard scientific materials used in physics. Stick one straw a short distance into the other one, and tape them to the top of the can, as shown in the diagram. When you place the can on the center of the turntable, the straws will form a diameter over the table. Hang paper clips at the 5 cm and 10 cm radial positions on each side. They act like plumb bobs, which show the "down" direction. Watch what happens to the plumb bob directions when the turntable is rotating. As closely as you can, estimate the angle from the vertical of each suspended weight. How does the angle for the 5 cm plumb bob compare with the angle for the one at 10 cm? If your turntable has variable speed, increase the frequency from 33 rpm to 45 rpm or 78 rpm. In each case, estimate the angles from the vertical of the plumb bobs.

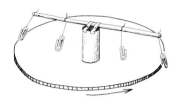

3. One of the easiest ways to produce oscillating motion, which we will be studying, is to make a pendulum. Just hang a weight from a string that is free to swing from a point. Choose a support that can't move, and tie the string to it tightly so that the string is swinging from a knot rather than slipping on the support. Any weight will do but, if you can, get something like a salt shaker and suspend it upside down. Either enlarge a center hole so that salt trickles out freely, or replace the cap with a paper or plastic cover that has a center hole.

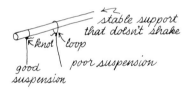

Cover up the hole so that the salt isn't coming out, and see how the pendulum behaves for small swings and large swings. Compare the time for 10 small swings and 10 large swings. Compare the time for 10 swings with a string only 25 cm long and with a string 1 m long. Compare the time for 10 swings with the salt shaker full and empty.

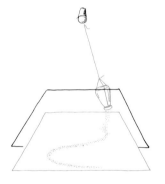

But the best part of playing with the pendulum is still to come. Tape several sheets of paper together and place it under the pendulum. Open the hole so that the salt comes out, depositing a line on the paper as the pendulum swings. Now slowly and steadily move the paper perpendicular to that line while the pendulum is swinging. The salt will trace out a pattern described by a familiar math function.

CONSTANT FORCE IN THE DIRECTION OF MOTION

Here are some examples of applications of Newton's second law. In these cases, the total applied force is not necessarily in the direction of the resulting motion, but an effective component of the applied force is in that direction and remains constant.

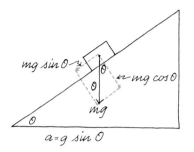

1. In the diagram we show a block sliding down an incline that has an angle θ. The weight of the block is mg, represented by a vector arrow pointing straight down. Here is one of those cases where a vector should be broken into perpendicular components that are not vertical and horizontal. The components of the weight that are of interest are the ones parallel and perpendicular to the inclined plane. Following the usual rule with vector problems, we draw a sketch of the geometry, label the variables, and break the vectors into components and label them. The force component perpendicular to the inclined plane will increase friction. We will pursue that problem next, but assume here that the plane is frictionless. The force component parallel to the plane will cause the block to accelerate along the same direction. Newton's second law becomes $mg \sin \theta = ma$. Notice that the mass of the block cancels out, as it does for a free-fall.

You should always check to make sure that you have used the right trig function (sine or cosine in this case) by asking whether the equation would be reasonable for extreme angles, such as 0° or 90°. If $\theta = 90°$, for instance, the block would fall straight down, and a would equal g. That's what this formula calls for, since $\sin 90° = 1$.

Free-fall acceleration is too large for many study purposes. Objects fall so fast that it is hard to record their position as a function of time. Back in Chapter 2, in "Handling the Phenomena," we made use of a plane that was inclined at a slight angle to slow down the effects of falling.

Question 6-2

What angle of incline should you use if you want to make a cart roll with an acceleration of 0.1 m/s²?

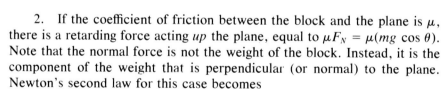

2. If the coefficient of friction between the block and the plane is μ, there is a retarding force acting *up* the plane, equal to $\mu F_N = \mu(mg \cos \theta)$. Note that the normal force is not the weight of the block. Instead, it is the component of the weight that is perpendicular (or normal) to the plane. Newton's second law for this case becomes

$$mg \sin \theta - \mu mg \cos \theta = ma$$

Once again, the value for the mass drops out of the equation.

Question 6-3

If $\mu_{kin} = 0.1$, about what is the minimum angle that you could use to keep the block sliding down?

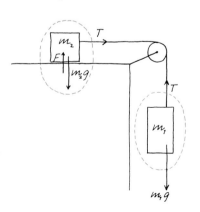

3. In this example, we have another way to dilute the effect of gravity on a falling object. In the diagram we show a pulley system arranged so that the falling object is pulling another one that slides along the table. Once again, we assume that there is no friction in the sliding object or in the pulley.

CONSTANT FORCE ON THE DIRECTION OF MOTION

There are two ways of solving this problem to find the motion. First, we can assume that the two blocks are really part of the same system, and just write Newton's second law for the whole system. The net force is produced by the weight of m_1; the total mass is the sum of m_1 and m_2.

$$m_1 g = (m_1 + m_2)a$$

If $m_1 = m_2$, the acceleration is just one-half of g.

There is a more general method of solution for this problem—one that will be useful and even necessary for more complicated problems. We appeal to that great slogan we used in the previous chapter: When dealing with forces, ISOLATE THE SYSTEM. In this case, we have drawn isolation rings around each of the blocks in the problem. Consider all of the forces acting on each system.

On block 1, the weight, $m_1 g$, is acting downward. The tension in the string, T, is acting upward. Newton's second law for this system is

$$m_1 g - T = m_1 a$$

So far we have only one equation, and two unknowns: T and a.

On block 2, the upward force exerted by the compression of the table balances the downward force of the block's weight. In the horizontal direction, there is only one force—and that is the tension in the string. In the idealized case of no inertia and no friction in the pulley, there is the same tension, T, all along the string; if there were not, the pulley wheel would turn until the tension became uniform. Therefore, Newton's second law for the second block is

$$T = m_2 a$$

Once again there are two unknowns, but we now have two equations. Use the value of T from the second equation and substitute that value into the first equation:

$$T = m_2 a \qquad m_1 g - m_2 a = m_1 a$$

By grouping terms, we end up with the same equation that we got from the first method:

$$m_1 g = (m_1 + m_2)a$$

Aside from the more general applicability of this second method, it also provides us with information about the tension in the string.

4. The apparatus shown in this example bears a classic name: Atwood's machine. (The real nineteenth-century device was far more sophisticated.) It is a slightly more complicated form of the simple apparatus used in "Handling the Phenomena." Once again, we could treat the system as a whole, noting that the total mass is $(m_1 + m_2 + m_3)$, all of which must accelerate together at the unknown rate, a. The effective force is the difference between the two hanging weights. Newton's second law becomes

$$m_1 g - m_3 g = (m_1 + m_2 + m_3)a$$

If we isolate each object and treat each as a separate system, then we find three equations. The common acceleration, a, is unknown, and the two

ISOLATE THE SYSTEM

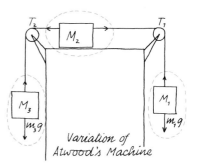

Variation of Atwood's Machine

tensions are unknown. T_1 must be different from T_2 or else object 2 could not accelerate. Newton's second law for the three objects yields these equations:

Equation 1 $\qquad m_1 g - T_1 = m_1 a$

Equation 2 $\qquad T_1 - T_2 = m_2 a$

Equation 3 $\qquad T_2 - m_3 g = m_3 a$

This set of equations can be solved in several ways by eliminating one variable or another. Perhaps the easiest method is to add all three equations, thus eliminating T_1 and T_2 at once:

$$m_1 g - T_1 + T_1 - T_2 + T_2 - m_3 g = m_1 a + m_2 a + m_3 a$$

This equation reduces to the original one:

$$m_1 g - m_3 g = (m_1 + m_2 + m_3) a$$

Question 6-4

In both methods of solving this problem, we assumed that the direction of a was such that object 2 would move to the right. Solve this problem for the following values and see if our general equations still hold: $m_1 = 5$ kg, $m_2 = 10$ kg, and $m_3 = 10$ kg.

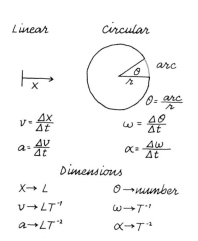

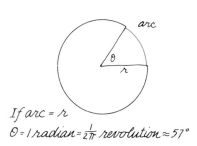

CIRCULAR MOTION—DEFINITIONS AND RELATIONSHIPS

Since so much of life is spent going in circles, we should learn the language of circular motion. Each of the variables used to describe motion along a line has its counterpart in circular terms. The position, x, becomes the angle, θ. The time-rate-of-change of position, $\Delta x/\Delta t$, is linear speed. Similarly, the angular displacement divided by the time, $\Delta \theta/\Delta t$, is angular or rotational speed. Angular speed is given a special symbol: ω (omega). The time-rate-of-change of the angular velocity, $\Delta \omega/\Delta t$, is the angular acceleration. It has the symbol α (alpha).

The dimensions of these circular units are different from their linear counterparts. An angle has the dimensions of just a number, since the measure of an angle is the ratio of an arc (a length) to the radius (a length). The dimensions of ω are therefore T^{-1}, and the dimensions of α are T^{-2}. Three kinds of units for angular velocity are used, depending on the unit chosen for θ. The angle might be given in degrees, radians, or revolutions. The angular velocity would then be given in degrees per second, radians per second, or revolutions per second. Conventional usage is that ω refers to radians per second, and frequency, f, is the name for revolutions per second. If a wheel has a frequency, f, of one revolution per second, it turns 2π radians during one second. Consequently, $\omega = 2\pi f$.

Velocity and acceleration are vectors. It is not obvious how to represent the direction of angular velocity and acceleration, except in terms of clockwise and counterclockwise. Of course, when we say "clockwise," we

CIRCULAR MOTION—DEFINITIONS AND RELATIONSHIPS

refer to a rotation around an axis that itself has a particular direction. The custom is to give the direction of rotation with the same system that we described for torques in Chapter 4. Let the fingers of your right hand curl in the direction of the rotational motion. Your thumb points along the axis in the direction assigned to the motion. However, this system works only for small rotational displacements or instantaneous rotational velocities and accelerations. If you use the system for large displacements or long-term average velocities, you will find that vector addition does not work: $A + B \neq B + A$. An example of the difficulty is shown in the diagram.

There are several other useful relationships to be noted before we start applying the language. The period of rotation, T, is just equal to the reciprocal of the frequency: $T = 1/f$. If a wheel is rotating at 10 revolutions per second, then the period for one revolution is just $\frac{1}{10}$ second. There is also a simple relationship between the tangential velocity of an object going in a circle and its angular velocity. With a tangential velocity, v, the object would move through an arc, $v\Delta t$, during a time interval, Δt, small enough so that the tangential distance and the arc are approximately the same. That arc length is equal to $r\Delta\theta$. Since $\Delta\theta/\Delta t = \omega$, $v = r\omega$. This formula holds only if ω is expressed in units of radians per second. The same argument gives a similar relationship between the tangential acceleration and the rotational acceleration: $a = \alpha r$

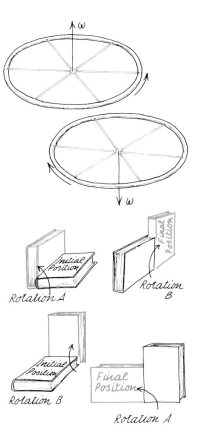

> **Question 6-5**
>
> There are two problems about these relationships for rotational velocity and acceleration. Are the dimensions and units consistent for α and ω? Is this angular acceleration the same thing as the centripetal acceleration?

The centripetal acceleration derived in Chapter 3 was in terms of the radius and tangential velocity: $a_c = v^2/r$. Let's translate this quantity into several other forms using the rotation variables. Since $v = \omega r$, $a_c = \omega^2 r^2/r = \omega^2 r$. Since $\omega = 2\pi f$, $a_c = 4\pi^2 f^2 r$. Since $f = 1/T$, $a_c = (4\pi^2/T^2)r$. Here are all the different forms:

$$a_{\text{cent}} = v^2/r = \omega^2 r = 4\pi^2 f^2 r = (4\pi^2/T^2)r$$

Notice that the original form for centripetal acceleration is *inversely* proportional to the radius. In rotation variables, the centripetal acceleration turns out to be *proportional* to the radius. The reason for this apparent paradox is that, for a rotating system with a constant period, the larger the radius the larger the tangential velocity.

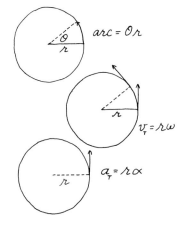

Let's use all the variables to analyze the turntable plumb bobs described in "Handling the Phenomena." On the standard slow-speed setting, a record player has a frequency of $33\frac{1}{3}$ rpm. In revolutions per second, $f = (33\frac{1}{3}$ rpm$)(1$ min$/60$ s$) = 0.555$ r/s. The period for one revolution is $T = 1/f = 1/(0.555$ r/s$) = 1.8$ s. The angular frequency in radians per second is $\omega =$

$2\pi f = 2(3.14)(0.555 \text{ r/s}) = 3.48 \text{ rad/s}$. The angular acceleration is zero unless the turntable has a wobble, in which case you have other troubles. All these variables—f, T, ω, and α—apply to every point on the turntable, regardless of radius.

> **Question 6-6**
>
> What is the direction of the angular velocity of a standard turntable?

The tangential velocity of a point on the turntable is proportional to the radius of the point. For the 5 cm plum bob, $v = \omega r = (3.48 \text{ rad/s})(5 \text{ cm}) = 17.4 \text{ cm/s}$. The tangential velocity of the 10 cm plumb bob is just twice as great, or 34.8 cm/s.

Now let's use the values for all these variables to calculate the centripetal accelerations of the 5 cm and 10 cm plumb bobs.

5 cm
$$a_c = v^2/r = (17.4 \text{ cm/s})^2/(5 \text{ cm}) = 60.5 \text{ cm/s}^2$$
$$a_c = \omega^2 r = (3.48 \text{ rad/s})^2(5 \text{ cm}) = 60.5 \text{ cm/s}^2$$
$$a_c = 4\pi^2 f^2 r = 39.4 \, (0.555 \text{ r/s})^2(5 \text{ cm}) = 60.5 \text{ cm/s}^2$$
$$a_c = (4\pi^2/T^2)r = [39.4/(1.8 \text{ s})^2](5 \text{ cm}) = 60.5 \text{ cm/s}^2$$

10 cm
$$(34.8 \text{ cm/s})^2/(10 \text{ cm}) = 121 \text{ cm/s}^2$$
$$(3.48 \text{ rad/s})^2(10 \text{ cm}) = 121 \text{ cm/s}^2$$
$$39.4 \, (0.555 \text{ r/s})^2(10 \text{ cm}) = 121 \text{ cm/s}^2$$
$$[39.4/(1.8 \text{ s})^2](10 \text{ cm}) = 121 \text{ cm/s}^2$$

As you can see, all the calculated values are consistent with each other. Notice that if you double the radius on the turntable, you double the centripetal acceleration. However, if you double the frequency, you quadruple the centripetal acceleration.

> **Question 6-7**
>
> What are the ratios of the three standard turntable speeds: $45/(33\tfrac{1}{3})$; $78/(33\tfrac{1}{3})$; $78/45$? What are the ratios of the centripetal accelerations?

We have been talking about centripetal *acceleration*, but, of course, an object accelerates centripetally only if there is a centripetal *force*. The force may be provided by a string fastened to a whirling ball, by tires preventing a car from slipping away from its turn, or by gravity pulling on a planet. The strength of the force required is given by Newton's second law: $F = ma$. In this case,

$$F_{\text{cent}} = mv^2/r = m\omega^2 r = m 4\pi^2 f^2 r = m(4\pi^2/T^2)r$$

NEWTON'S SECOND LAW FOR ROTATIONS

For an object to move in a circle, a centripetal force must continually pull the object out of its tangential path. Since the centripetal acceleration produced is always perpendicular to the velocity, the object does not speed up or slow down. The angular velocity remains unchanged. In order to change the angular velocity, there must be an angular acceleration. A force must be applied along the direction of the tangential velocity.

Let's apply Newton's second law to the point object with mass, m, shown rotating in the diagram. We will exert a force that is constant in magnitude, but continually changing in direction so that it is always in line with the tangential velocity. If it were not for this changing direction, we could just use Newton's second law in its usual form: $F = ma$. In the case of static forces, we had an expression for the result of a force exerted on a radial arm—it is *torque*. $\tau = \mathbf{r} \times \mathbf{F}$. The cross mark signifies a vector product, which means that the largest effect occurs when the force is perpendicular to the radius arm. That's just the way that we propose to push on the object traveling in a circle. Therefore, in this geometry, $|\mathbf{r} \times \mathbf{F}| = rF$. To emphasize this geometry, we will multiply both sides of Newton's second law by r, and then call the left side the torque:

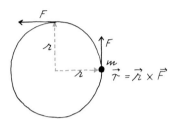

$$rF = rma$$
$$\tau = rma$$

Remember that the tangential acceleration is equal to the radius times the rotational acceleration: $a = \alpha r$. Substitute this change into the above equation:

$$\tau = (mr^2)\alpha$$

This transformed equation is Newton's second law for rotational motion. Notice how parallel it is to the original. Force becomes torque; tangential acceleration becomes angular acceleration; and mass becomes a combination of mass and its location. This particular combination (mr^2) is called the *moment of inertia*. It is usually given the symbol I. The moment of inertia plays the same role in rotational motion as mass does in linear motion. The greater the moment of inertia of an object, the larger the torque needed to produce a change in its angular velocity. However, it's not just the mass that determines the moment of inertia; the radial location of the mass is even more important since I depends on the *square* of r.

Although we derived Newton's second law for rotational motion in terms of a point object at radius r, the same formula would apply to a thin hoop or rim with radius r. The moment of inertia of each little piece on the hoop would be mr^2. The moment of inertia of the whole hoop is Mr^2, where M is the sum of the masses of all the individual pieces. For any other kind of rotating object, each piece with mass m has its own particular radius and contributes its own particular value, mr^2, to the total moment of inertia.

If two rotating systems have the same mass, but the mass of the first is mostly around the rim while the mass of the second is concentrated at the

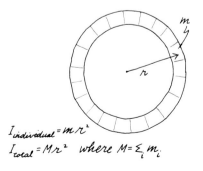

hub, the first one will be much more sluggish in responding to torques speeding it up or slowing it down. Have you ever experienced this phenomenon? If not, here's an easy remedy. Take four full soup cans (or any other convenient weights) and place them symmetrically on a record player turntable. With the turntable disengaged and free to rotate, try spinning it slowly with your hand—first while the cans are near the axis, and then while the cans are near the rim. The mass is the same in both cases, but the torque required to start or stop the spinning is noticeably different.

Question 6-8

Since in both cases all of the mass is accelerated up to the same angular velocity, why should different torques be required?

Let's calculate how much torque is necessary to accelerate a bicycle wheel in 5 s up to a speed corresponding to 15 mph. Our example is unreal because we are not proposing to accelerate the whole bicycle, but just one wheel, spinning freely. (The problem of accelerating the whole bike can be done most easily in terms of energy, which we will discuss later.) A typical front wheel from a standard bike has a mass of about 2 kg and a radius of 33 cm. We will assume that all of the mass is concentrated at the rim, or approximately at $r = 30$ cm. At 15 mph the metric speed is 6.7 m/s, and this is the tangential speed of a point on the tire. The angular velocity is therefore $\omega = v/r = (6.7 \text{ m/s})/(0.33 \text{ m}) = 20$ rad/s. The acceleration needed to get up to this velocity can be calculated from a formula that is parallel to the linear case for constant acceleration: $\omega_f = \alpha t$, which is similar to $v_f = at$; $\alpha = \omega_f/t = (20 \text{ rad/s})/(5 \text{ s}) = 4$ rad/s². The torque necessary to provide this acceleration is: $\tau = I\alpha = [(2 \text{ kg})(0.3 \text{ m})^2](4 \text{ rad/s}^2) = 0.7$ mN. The force necessary to yield this torque depends on where the force is applied. If it is applied at the rim, then $F = \tau/r = (0.7 \text{ mN})/(0.33 \text{ m}) = 2$ N.

Notice the crucial assumption in this example that all the mass of the wheel was located at the rim. This is a reasonable assumption for a bicycle wheel, which has wire spokes of low mass. However, for a uniform disk, it would not be a good assumption. If the mass is distributed over the whole radius of a rotating object, the moment of inertia must be some fraction of MR^2, where M is the total mass and R is the outside radius. The appropriate fraction for each geometric shape can be calculated with the aid of the calculus.

The moment of inertia also depends on the location of the axis of rotation. In many cases objects rotate on an axis through their center of mass. But when an object is rolling on its rim, it may be convenient to consider the moment of inertia with respect to an axis going through the momentary contact point on the edge.

Here is a table of moments of inertia for various shapes and axis locations.

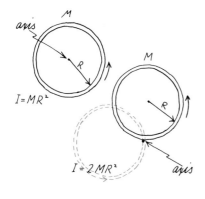

NEWTON'S SECOND LAW FOR ROTATIONS

	Moments of Inertia	
	About Center Axis	About a Point on the Edge
Ring	MR^2	$2MR^2$
Washer (inner R_1, outer R_2)	$\frac{1}{2}M(R_1^2 + R_2^2)$	$\frac{1}{2}M(R_1^2 + 3R_2^2)$
Solid cylinder	$\frac{1}{2}MR^2$	$\frac{3}{2}MR^2$
Spherical shell	$\frac{2}{3}MR^2$	$\frac{5}{3}MR^2$
Solid sphere	$\frac{2}{5}MR^2$	$\frac{7}{5}MR^2$
Uniform thin rod	$\frac{1}{12}ML^2$ (about axis through center, $\perp$ to length)	$\frac{1}{3}ML^2$ (about axis through end, $\perp$ to length)

Optional

There is a simple method for calculating the approximate moment of inertia of a solid cylinder rotating about its axis. Divide the cylinder into thin rings. Each ring has a moment of inertia of $m_i r_i^2$. Add together the separate moments of inertia.

$$\Sigma m_i = M \quad \Sigma I_i = I$$

For instance, suppose $r_1 = \frac{1}{2}$; $r_2 = \frac{3}{2}$; $r_3 = \frac{5}{2}$; $r_4 = \frac{7}{2}$; $R = 4$. The mass of a ring is approximately equal to (density) × ($2\pi r$) × (radial thickness) × (length). Choose rings with a radial thickness of 1 unit.

$$m_1 = \rho 2\pi(\tfrac{1}{2})(1)L \quad m_2 = \rho 2\pi(\tfrac{3}{2})(1)L$$
$$m_3 = \rho 2\pi(\tfrac{5}{2})(1)L \quad m_4 = \rho 2\pi(\tfrac{7}{2})(1)L$$

The moment of inertia of the ith ring $= m_i r_i^2 = I_i$.

$$I_1 = \rho L \pi(\tfrac{1}{4}) \quad I_2 = \rho L \pi 3(\tfrac{9}{4}) \quad I_3 = \rho L \pi 5(\tfrac{25}{4}) \quad I_4 = \rho L \pi 7(\tfrac{49}{4})$$

$$I = (\rho L \pi/4)(1 + 27 + 125 + 343) = \rho L \pi(124)$$

Since $M = \rho L \pi R^2 = \rho L \pi(16)$, $I = [\rho L \pi(124)] \times [M/(16 \rho L \pi)] = M (7.8)$.

The actual formula for the moment of inertia of a solid cylinder is $I = \frac{1}{2}MR^2$. For a cylinder with radius $R = 4$, $I = 8M$. Compare this value with that obtained from our approximate calculation: $I = 7.8M$. The approximation would get better as the rings chosen became thinner.

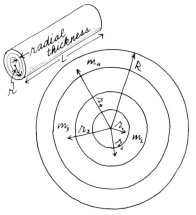

If an object is rolling downhill, its weight provides a torque as shown in the diagram. Notice that the choice of $R \sin \theta$ for the lever arm of the torque is plausible, since if $\theta = 0$ the weight is acting through the contact point, and the lever arm is zero. The torque produces an angular acceleration, making the object roll downhill with increasing speed.

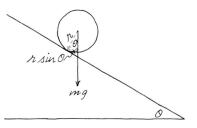

Question 6-9

Suppose that you have a downhill rolling race between three objects with the same mass and the same radius: a hollow cylinder, a solid sphere, and a solid cylinder. Which one would win? Would it make any difference if they had different masses? Would it make any difference if they had different radii?

CASES WHERE CENTRIPETAL FORCE IS PERPENDICULAR TO GRAVITY

On turntables or merry-go-rounds, a plumb bob does not hang straight down; it flies out at an angle. You examined a situation like this in "Handling the Phenomena." Two forces acted on the plumb bob. One was vertical, pulling the bob down. The other was exerted by the thread, which slanted obliquely so as to provide a centripetal component, pulling the bob in radially.

Why should it be necessary to pull the bob *in* radially? Is there a force tending to pull the bob *out*? Certainly the string slants as if the bob were being pulled out. There is another term used to describe the radial force involved in circular motion: the centrifugal, or center-fleeing, force. Which description you choose depends on which frame of reference you prefer. If you are *outside* the rotating system, you observe that a *centripetal* force is necessary to force an object to move in a circular path. If you are *in* the rotating system, you observe that there is a *centrifugal* force acting on the object. We are going to examine this and other reference frame problems in a later chapter.

For now we will look at the turntable from the outside. The thread holding the bob exerts a force that balances the weight of the bob and pulls it in radially with the required centripetal force. The geometry of the vertical and horizontal components is shown in the diagram. There are several things to notice in the formula for the plumb bob angle. First, the mass cancels out. That seems to happen in quite a few formulas where gravity and mass are concerned. In this case, the mass disappears even though centripetal force is mixed in. There's more to this than meets the eye, as we will see in the chapter on reference frames.

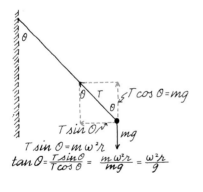

The next thing to notice about the plumb bob angle is the dependence on r and ω. The tangent of the angle is proportional to the radial position of the bob. Is that what you observed in "Handling the Phenomena"? It's hard to make precise measurements on a rotating system, but at $33\frac{1}{3}$ rpm the angle for the 10 cm bob should have been about twice the angle for the 5 cm bob. The formula is in terms of the tangent of the angle, but for small angles the tangent is about equal to the angle (measured in radians). Since the centripetal force varies with the square of the angular velocity, the angles for 45 rpm and 78 rpm may be so large that the small-angle approximation no longer applies. Let's actually find the expected angles for the several cases with the turntable. First, we calculate the angle for a bob at 5 cm with the turntable set at $33\frac{1}{3}$ rpm, which is 3.48 rad/s (see page 118).

$$\tan \theta = \frac{\omega^2 r}{g} = \frac{(3.48 \text{ rad/s})^2 (0.05 \text{ m})}{(9.8 \text{ m/s}^2)} = 0.062$$

CASES WHERE GRAVITY PROVIDES THE CENTRIPETAL FORCE

$$\theta = 3.5° = 0.062 \text{ rad}$$

The *tangent* of the angle for a bob at 10 cm will be just twice the value for 5 cm. Since the angle is so small, the approximation is good that the *angle* will also be twice the angle for the 5 cm case: $\theta = 7°$.

If the turntable is rotating at 78 rpm, the tangent of the angle will be greater by a factor of $[78/(33\frac{1}{3})]^2 = 5.5$ The approximations that we used before are not quite so good in this case. For $\tan \theta = (0.062)(5.5) = 0.34$, $\theta = 18.8° = 0.33$ rad. The tangent of the angle for the 10 cm bob is $(0.34)(2) = 0.68$, and $\theta = 34.2° = 0.60$ rad. Another approximation that we have made all along is that the bobs are at the same radius as their support ties. As they fly out radially, they move to positions where the centripetal force required is stronger than we have calculated.

One practical application of the analysis that we have made is in the construction of roads that are banked on turns. If you try to take a curve at high speed on a flat road, you may skid off along the tangential direction. Where a road is banked at the correct angle for a particular radius of curvature and a particular speed, the road surface exerts exactly the right force both to balance the weight of the car and to provide the centripetal force that the car needs to round the curve. The diagram shows the geometry of the situation.

Let's calculate the correct banking angle for a speed of 60 mph on a curve with a radius of curvature of $\frac{1}{10}$ mile. In metric units, $v = 27$ m/s, and $r = 160$ m. Then,

$$\tan \theta = \frac{v^2}{gr} = \frac{(27 \text{ m/s})^2}{(9.8 \text{ m/s}^2)(160 \text{ m})} = 0.46.$$

The required angle for complete banking would be almost 25°. While that angle would be acceptable for a roller coaster, it would be a little too dramatic for a public road. An actual road with such a sharp curvature would probably have smaller banking and be posted with a 30 mph speed limit.

Question 6-10

Without calculating the formula all over again, and without using a trig table, find the banking angle required for 30 mph.

CASES WHERE GRAVITY PROVIDES THE CENTRIPETAL FORCE

Popular legend has it that before 1492 everyone thought that the earth was flat, and that before the discoveries of Copernicus, Galileo, and Kepler in the sixteenth century everyone thought that the earth stood still at the center of the universe. Not so. We have the word of Archimedes in the *Sand Reckoner,* written in the third century B.C., that Aristarchos of Samos had found from his measurements that the moon and sun were far away from the earth, that the sun was very large, and that the most reasonable explanation was that the earth rotated on its axis and revolved around the sun, as did the other planets. It took a while for that idea to get accepted.

The actual orbits of the planets around the sun, and of the satellites around the earth, are ellipses. The math needed to demonstrate this fact and to analyze the motions is complicated. Since the orbits of most of the planets are almost circles, we can calculate good approximations for the orbits and laws with the simple arithmetic of circular motion. Gravitational force, which depends on the orbital radius, provides the centripetal force needed to keep the planets in their courses:

$$G\frac{mM}{r^2} = \frac{4\pi^2 m}{T^2} r$$

gravitational force = centripetal force

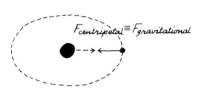

The first thing to notice is that once again, in this formula, the mass of the moving object cancels. We are assuming that the mass of the planet, m, is much smaller than the mass of the sun, M, and that therefore only the planet moves. This is just an approximation, of course, but a good one, even in the case of the moon going around the earth.

The equation applies to all of the satellites of the controlling body. There are only two variables, the radius and the period, and these are linked by the formula. Let's separate them by cross-multiplying:

$$T^2 = \left(\frac{4\pi^2}{GM}\right) r^3$$

The square of the period of a planet is proportional to the cube of its orbital radius. This is one of the laws for the planets that Kepler deduced in the early 1600s. He did not derive the equation from theory, but figured out the relationships from the experimental data that Tycho Brahe, a Danish astronomer, had collected during the previous half-century. The table shows the values of T and r for each of the planets, not all of which were known in Kepler's day. As you can see, the ratio T^2/r^3 is the same for all of the planets.

Question 6-11

The period of the moon is about 27 days and its orbital radius is 3.8 × 10^8 m. Does it have the same ratio of T^2/r^3 as the planets?

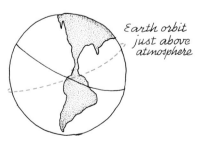

In Chapter 3 we calculated the acceleration of an earth satellite with an orbital radius only slightly larger than the radius of the earth itself. At that point we asserted that the period for such an orbit is 5100 s = 85 min. Here's an easy way to calculate the period. Since both moon and satellite are going around the same controlling body, they must have the same ratio of T^2/r^3.

Planet	T Period (days)	r Orbital Radius (miles)	T^2/r^3 (days²/miles³)
Mercury	0.88×10^2	0.36×10^8	1.66×10^{-19}
Venus	2.25	0.67	1.66
Earth	3.65	0.93	1.66
Mars	6.87	1.42	1.66
Jupiter	43.3	4.85	1.66
Saturn	107.6	8.90	1.66
Uranus	307	17.8	1.66
Neptune	602	28.1	1.66
Pluto	909	36.7	1.66

$$\frac{(T_{\text{moon}})^2}{(r_{\text{moon}})^3} = \frac{(T_{\text{sat}})^2}{(r_{\text{sat}})^3} \rightarrow \left(\frac{T_{\text{moon}}}{T_{\text{sat}}}\right)^2 = \left(\frac{r_{\text{moon}}}{r_{\text{sat}}}\right)^3$$

The ratio of the orbital radius of the moon to the radius of the earth is

$$(3.8 \times 10^8 \text{ m})/(6.4 \times 10^6 \text{ m}) = 6.0 \times 10^1$$

The cube of this number is 2.16×10^5, and the square root of that is 465. Therefore, $(T_{\text{moon}})/(T_{\text{sat}}) = 465$. The period of the moon is (27 days) × (24 hr) × (60 min) = 3.9×10^4 min. The period of the close earth satellite must therefore be $(3.9 \times 10^4 \text{ min})/465 = 84$ min.

Question 6-12

According to the calendar, new moon to new moon is about 29 days. Why did we say that the period of the moon's revolution is 27 days?

THE CHANGING FORCE THAT PRODUCES OSCILLATIONS

If you pull a pendulum bob to one side and let it go, a force drives it back toward its center position. The bob overshoots and goes beyond the center position to the other side. On that side, some force slows it down and pulls it back again, acting in the opposite direction from the first force. When the bob is on the right side of the equilibrium point, the force is to the left; when the bob is on the left, the force is to the right. What sort of force is it that can automatically change direction?

We have already seen one kind of force that depends on position. For many springs, the restoring force when they are stretched is proportional to the stretch, and in the opposite direction.

$$F_{\text{restoring}} = -kx$$

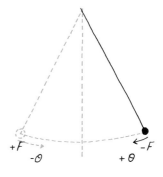

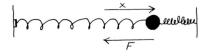

We used this property of springs to define force in the first place. If a bob on a spring is given a long stretch in the positive x direction, the restoring force is large and in the negative direction. As the bob is restored to the equilibrium position where x equals zero, the restoring force becomes zero, but the bob is traveling fast. It overshoots into the negative x region of compression. The restoring force is now toward the positive direction and increases as the compression increases. Then the action repeats itself all over again.

Question 6-13

At what position is the speed of the oscillating bob greatest? At what position is the speed zero? Is the velocity always positive when x is positive?

Let's analyze the motion of oscillation in the same way that we did for objects in free-fall. We plotted graphs of $x(t)$, $v(t)$, and $a(t)$. We can do this in two ways: first, by making some plausibility claims and, second, by a geometric argument. Of course, the test of our analysis is to compare the results with actual measurements of $x(t)$, $v(t)$, and $a(t)$ in an oscillating system. That's a project that you ought to do in the laboratory.

First Analysis: A Plausibility Argument

Here are three sets of possible graphs for the oscillating motion produced by a bob on a spring. In all three cases, we start out with the assumption that at $t = 0$, the moving bob is at $x = 0$, the equilibrium position. A short time later the bob reaches a maximum positive value of x; an equal time

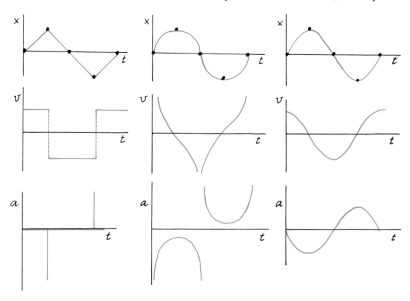

THE CHANGING FORCE THAT PRODUCES OSCILLATIONS

after that the bob is back at $x = 0$; another equal time and the bob has reached maximum negative x; after a fourth equal time period the bob is back at $x = 0$. The question is: What is the actual graph of $x(t)$ as it goes between those established points? We offer three possibilities—however, there are some constraints on the choices. The slope of a curve of $x(t)$ is the velocity, $\Delta x/\Delta t$. The velocity graph given by the slopes of $x(t)$ must agree with our observations and be reasonable. Furthermore, the slope of the velocity curve is the acceleration, $\Delta v/\Delta t$. Not only must that curve agree with our observations, but the force formula for this motion requires a very special relationship between acceleration and position. The motion must obey the law: $F = -kx$. Since $F = ma$, it must be that $ma = -kx$. Therefore, the acceleration must be proportional to negative x: $a = -(k/m)x$. Whatever the shape of the curve of $x(t)$, the curve of $a(t)$ must have the same shape, but be upside down (because it is negative). Let's see which of the three sets of motion graphs satisfies these constraints for oscillating motion.

In the first set, we connected the fixed data points of $x(t)$ with straight lines. That's the simplest assumption. However, such a graph requires that the speed of the bob have a constant value—first positive and then negative. Casual observation of a pendulum or vibrating spring can rule out this possibility. The moving bob slows down before turning around at the end points. Furthermore, notice what the curve of $v(t)$ would require of the acceleration. The acceleration (and, hence, the force) would be zero all during the swing except at the end points. Then it would have to be infinite in order to change the velocity abruptly from positive to negative. Clearly, the first set of graphs cannot be right.

In the second set of graphs, we connected the fixed points of $x(t)$ with semicircles. Such a graph has a slope of zero at the end points of the motion. That conclusion agrees with our observations that the bob slows down as it approaches the end points and momentarily comes to rest before starting back. Unfortunately, the slope of a semicircle at the $x = 0$ position is infinite. To be sure, the bob is traveling rapidly as it passes through the equilibrium point, but its speed is not infinite.

In the third set of graphs, we compromise between the first two possibilities. The curve of $x(t)$ has a reasonable slope as it passes through zero; it also is gently curved in the regions where x is maximum. The slopes of the curve have the correct properties to describe the observed velocities of an oscillating bob. Furthermore, the slopes of the velocity curve yield an acceleration curve that is just an upside-down reflection of the curve of $x(t)$. Notice that all three curves have identical shapes, but pass the origin at different times of their swings. The function described by such a curve is the sine function.

Question 6-14

How can we call the graph of $v(t)$ a sine function when it appears to be a cosine curve?

Second Analysis: A Geometric Argument

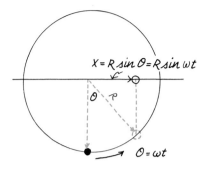

The second method of exploring oscillatory motion is to use a geometric argument. Instead of putting a bob on a spring or a string, and letting it bob back and forth, fasten it to the rim of a rotating wheel. At first glance the rotational motion does not appear to be related to an oscillation, although note that in both cases the bob keeps coming back to the same point and repeats this motion with a particular frequency. The similarity of the two motions becomes even more evident if we look at the perpendicular projection of the rotating bob's motion.

In this diagram, we show the geometry of projecting the rotational motion onto the horizontal axis. As the bob goes around the circle, the projection point goes back and forth on the x-axis. The angular position of the bob is given by $\theta = \omega t$, where ω is the angular velocity in radians per second. The position of the projection on the x-axis is $x = R \sin \theta = R \sin \omega t$. Notice that this geometric argument leads directly to a sinusoidal expression for the position as a function of time: $x(t) = R \sin \omega t$.

The tangential velocity of the rotating bob always has the same magnitude but changes in direction. The projection of that vector velocity on the x-axis is just the x component of the velocity. That component is

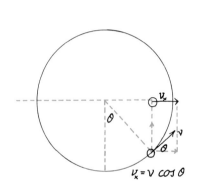

$$v_x = v \cos \theta = v \cos \omega t = R\omega \cos \omega t.$$

(Remember that $v = R\omega$.) Notice how this function agrees with our plausibility arguments for the third set of graphs. When $\theta = 0$, the projection of the bob on the x-axis is at the center of the oscillation, where $x = 0$. At that same time, v_x is a maximum because $\cos 0 = 1$. A quarter of a cycle later, when $\theta = 90°$, the x-projection is a maximum with $\sin 90° = 1$; the velocity at that turnaround point of the oscillation is zero, since $\cos 90° = 0$.

Question 6-15

We still have to determine the acceleration of the projection point. The bob is rotating at constant angular velocity. What acceleration can there be?

THE PERIOD OF OSCILLATION

Whenever an object is subject to the special force relationship, $F = -kx$, it will undergo oscillation. The period of the oscillation—the time that it takes for one complete cycle—can be calculated from the geometric relationship between rotational motion and oscillatory motion. As you found in answering Question 6-15, the acceleration of the oscillating point is given by $a = -\omega^2 R \sin \omega t$. The angular velocity is ω, which is also equal to $2\pi f$ or $2\pi/T$. The radius of the rotational motion is R; for an oscillation, the maximum amplitude is usually called A. Therefore, the *acceleration* of the oscillating bob can be written

$$a = -\omega^2 A \sin \omega t$$

THE PERIOD OF OSCILLATION

The *position* of the oscillating bob has the same form: $x = A \sin \omega t$. Let's substitute the expressions for x and a into the force law:

$$F = ma = -kx$$
$$m(-\omega^2 A \sin \omega t) = -k(A \sin \omega t)$$

The common term, $-A \sin \omega t$, cancels out, leaving

$$m \omega^2 = k$$
$$\omega = 2\pi f = 2\pi / T = \sqrt{k/m}$$

The frequency of an oscillating system that obeys the law, $F = -kx$, is

$$f = (1/2\pi)\sqrt{k/m} \quad \text{cycles/second}$$

The period is

$$T = 2\pi \sqrt{m/k} \quad \text{s}$$

In the formula for the period, note that the value for the mass of the bob is in the numerator. The larger the mass, the more sluggish the bob is, and the longer it takes to oscillate through one period. The spring constant, k, is in the denominator in the formula for T. A strong spring has a large value of k, and will restore the bob toward equilibrium more quickly than a weak spring.

Let's calculate the period in a couple of cases.

1. If you hang a kilogram weight on a typical laboratory spring, it might stretch the spring 20 cm. The spring constant is

$$k = -F/x = -(1 \text{ kg})(9.8 \text{ N/kg})/(-0.2 \text{ m}) = 49 \text{ N/m}$$

   If you stretch the bob a little more and then let it oscillate, the period is

$$T = 2\pi \sqrt{m/k} = 6.3 \sqrt{(1 \text{ kg})/(49 \text{ N/m})} = 0.9 \text{ s}$$

2. If you hang a kilogram weight from a string and let it swing as a pendulum, what is the period? There is an immediate problem. Not only do we not have a spring constant, we do not have a spring. To be sure, the motion of a pendulum is oscillatory, but there is no guarantee that the motion satisfies the prime requirement that $F = -kx$. The diagram shows the forces involved on the bob of a pendulum. The weight of the bob is broken into two components—(1) radial, and thus along the line of the string, and (2) tangential, and thus along the line of motion. The tangential component is the one that produces the restoring acceleration. It is equal to

$$F_{\text{restoring}} = -mg \sin \theta = -mg \sin(x/L)$$

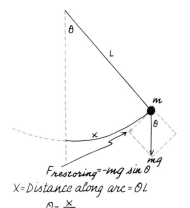

This is not the force relationship that we need. Instead of having the restoring force proportional to the negative of the displacement, we have ended up with the restoring force being proportional to the negative of the *sine* of the displacement. However, for small angles, $\sin \theta \approx \theta$. In that case,

$$F_{\text{restoring}} \approx -mg(x/L) = -(mg/L)x$$

Now we have the force law in the familiar form, with the proportionality constant, $k = (mg/L)$. The period of a pendulum for small swinging angles is therefore

$$T = 2\pi \sqrt{m/k} = 2\pi \sqrt{m/(mg/L)} = 2\pi \sqrt{L/g}$$

Notice that the period of the pendulum does not depend on the mass of the bob. As we have seen before, the acceleration produced by gravity is independent of mass. For a given location on earth, and thus for a particular value of g, the period depends only on L, the length of the string. In particular, to the extent that the approximation is good that $\sin \theta \approx \theta$, the period does not depend on the amplitude of the swing. The period given by this simple formula is good to within 1% for swings up to 23°; it is too small by only 4% when the maximum θ is 45°.

Question 6-16

How long is the string of a simple pendulum with a period of 1 second? How long would the string have to be for a period of 10 seconds?

SIMPLE HARMONIC MOTION—AND OTHER OSCILLATIONS

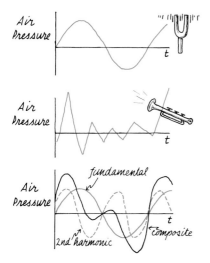

Whenever a system can move back and forth past an equilibrium point, and the restoring force is proportional to the displacement, the resulting motion is sinusoidal. It is called Simple Harmonic Motion, or SHM. It is simple because of the simple force law and because only one frequency of oscillation is produced. It is harmonic in the sense that a fundamental note in music can be the base for harmonic overtones. Not all oscillations are simple or harmonic. In musical tones, as we will see in a later chapter, the air pressure sequences of the sound follow a sine wave only for tuning fork notes or when strings are allowed to vibrate in very special ways. Most musical tones consist of a combination of many different sine waves, with frequencies having special relationships to each other. These are the fundamental and harmonic frequencies. The instrument that produces the sound is oscillating in a very nonsimple way, although usually harmonically. A cymbal or a bell oscillates in a way that is neither simple nor harmonic.

If there are many different types of oscillation, why take up the special case of SHM? There are two reasons. First, many systems—mechanical, electrical, chemical, even economic—are in dynamic equilibrium. If they are disturbed, there is built-in stability to come back to the equilbrium position. *For small displacements,* the restoring force is proportional to the displacement. Simple harmonic motion results. For large displacements, SHM is usually a good first approximation. The second reason for studying SHM is that even in cases where the restoring force is not simply proportional to the displacement, the actual motion can be described in terms of a combination of sine waves with different frequencies. The method is the same as the one used to combine a fundamental and harmonics to produce a complex musical tone.

SUMMARY

In solving practical problems involving Newton's second law, the most important step is to isolate the object. Decide on one object that is going to accelerate, and then write down all the forces acting on that object. Frequently, one or more of the forces must be resolved into components parallel and perpendicular to the direction of the allowed motion. The parallel force is effective in acceleration, but sometimes the perpendicular component contributes to the friction force, which is always in the direction opposed to the velocity.

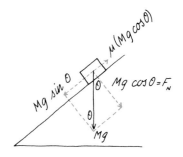

Circular motion is described with variables and formulas that are similar to those for motion along a line. These are summarized in the diagrams and formulas below.

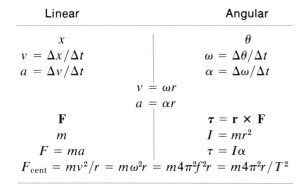

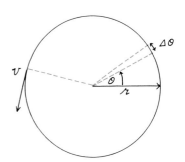

When an object on earth is traveling in a horizontal circular path, the gravitational force on the object is perpendicular to the centripetal force. A plumb bob in such a situation will hang at an angle from the vertical given by

$$\tan \theta = \frac{\omega^2 r}{g}$$

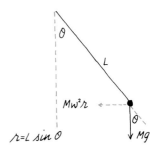

For orbital motion of satellites, gravity provides the centripetal force:

$$G \frac{mM}{r^2} = m\omega^2 r$$

This formula leads to Kepler's law connecting the periods and radii of all the satellites of a controlling gravitational body:

$$T^2 = \frac{4\pi^2}{GM} r^3$$

The orbits of all the sun's planets satisfy Kepler's law, with M, in this case, being the mass of the sun. Similarly, the orbits of the moon and all the artificial satellites that humans have placed around the earth are linked by this formula when M is the mass of the earth. Although we derived the law for circular paths, it also holds for elliptical orbits, if r is the semimajor axis.

If an object can move back and forth past an equilibrium point, and if the restoring force on the object is proportional to the displacement, the motion produced is Simple Harmonic. The functions $x(t)$, $v(t)$, and $a(t)$ are all sinusoidal. The period of oscillation is given by $T = 2\pi\sqrt{m/k}$, where k is the proportionality constant between force and displacement. A pendulum's motion is approximately Simple Harmonic. If the amplitude of the swing is under 23°, the period is within 1% of $T = 2\pi\sqrt{L/g}$.

In terms of the study of physics, one of the intriguing features of learning about these various types of motion is that we will make repeated use of them in future work. Not only do springs exert forces on "objects," but electric fields exert forces on electrons. In both cases, $F = ma$. Not only does gravity maintain planets in orbit, but magnetic fields force protons to travel in circles. In both cases,

$$F_{\text{cent}} = m\omega^2 r$$

Not only is the restoring force on a spring bob proportional to the displacement, but the same law applies to atoms bound in a crystal. In both cases, $F = -kx$, and the resulting motion is described by $x = A \sin \omega t$. Apparently, the basic laws of the universe are few and simple. The applications are legion, and, as we all know, complex.

Answers to Questions

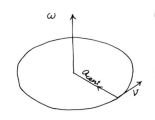

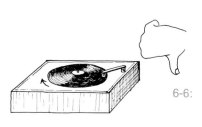

6-1: In circular motion, the acceleration—and hence the force—is always perpendicular to the velocity. Force, acceleration, and velocity are vectors. An applied force can change the direction of velocity without changing its magnitude.

6-2: An acceleration of 0.1 m/s² is about 1% of $g = 9.8$ m/s². Therefore, $\sin \theta = 0.01$. Since for small angles, $\sin \theta \approx \theta$, the angle itself must be 0.01 radians = (0.01 rad)(57°/rad) = 0.57°.

6-3: If the sliding block is in motion, but not accelerating, then $\sin \theta - \mu \cos \theta = 0$. This leads to $\tan \theta = \mu = 0.1$. For small angles, $\tan \theta \approx \theta$. Approximately (and very closely), $\theta = 0.1$ radian = 5.7°.

6-4: (5 kg − 10 kg)(9.8 m/s²) = (25 kg)a $a = -2$ m/s²

According to the formula, the acceleration is negative. This means that the system is accelerating to the left as we expect, opposite to the original assumption. The arithmetic faithfully gives us the right answer as long as we are consistent.

6-5: The dimensions of linear velocity must be L/T, and the dimensions of linear acceleration must be L/T^2. The dimensions of ω and α are $1/T$ and $1/T^2$. If $a = \alpha r$, and $v = \omega r$, the dimensions of α and ω are each multiplied by a length, L, making the dimensions consistent. As long as ω and α are given in units of radians per second and radians per second squared, the units of v and a will be m/s and m/s² if r is given in meters.

Angular acceleration is not at all the same as centripetal acceleration. An object rotating at constant angular velocity, and thus with zero angular acceleration, still is subject to centripetal acceleration, which keeps it going in the circular path. The centripetal acceleration is radial; the direction of angular acceleration is axial.

6-6: As you look down on a record player turntable, the disk is turning clockwise. If you curl the fingers of your right hand in that direction, your thumb will point down. The direction assigned to the angular velocity is therefore down the spindle.

6-7: $[45/(33\frac{1}{3})] = 1.35$ a_c ratio $= \omega^2$ ratio $= 1.8$

$[78/(33\frac{1}{3})] = 2.35$ a_c ratio $= 5.5$

$(78/45) = 1.73$ a_c ratio $= 3$

6-8: The weights are accelerated to the same *angular* velocity regardless of where they are on the turntable. However, for a given angular velocity, the weights have a greater *tangential* velocity at the larger radius.

6-9: According to the diagram, the torque on the rotating object is equal to $mgr \sin \theta$. Newton's second law becomes

$$mgr \sin \theta = I\alpha = Kmr^2\alpha = Kmr^2 a/r$$

The constant, K, depends on the shape of the rolling object and the location of the axis. The constants for the three shapes, in this case rolling around points on their surfaces, are given in the table on page 000. Canceling factors in the equation above leads to

$$g \sin \theta = Ka$$

The acceleration of the objects is independent of mass and radius, and depends only on the effective component of gravity (and thus on the angle, θ), and on the shape of the rotating object. For a hollow cylinder, $K = 2$. For a solid sphere, $K = \frac{7}{5}$. For a solid cylinder, $K = \frac{3}{2}$.

Since $a \propto 1/K$, the sphere, which has the smallest value of K, will win the race, followed by the solid cylinder, and then the hollow cylinder. Try it for yourself and see. Since the effect is independent of mass or size of the objects, you can race any kind and any size of solid ball against tin cans or solid pencils, and so on.

6-10: At 30 mph, the tangent of the required banking angle would be only one-fourth the value calculated for 60 mph. Therefore, $(\tan \theta)_{30 \text{ mph}} = 0.46/4 = 0.11$. For that small a value, $\tan \theta \approx \theta$. Therefore, $\theta = 0.11$ radians $= 6.3°$.

6-11: No! The ratio, T^2/r^3, has the same value for the satellites of a *particular* central gravitational body. The planets are circling the sun. The value of T^2/r^3 for the moon is the same as for the other satellites of the *earth*.

6-12: As you can see in the accompanying diagram, after 27 days the moon has made one complete revolution, but meanwhile the earth-moon system has traveled 26° along the orbit around the sun. The moon has to travel that extra angular distance to get back to the new moon position. It takes a couple of extra days.

6-13: The speed of the oscillating bob is greatest when the bob is moving through its equilibrium position. The speed is zero at the turn-around points, where the displacement is greatest. During the first quarter cycle, after passing through the zero point into the $+x$ region, the velocity is positive. After going to zero, the velocity becomes negative for the next half cycle. Therefore, during half the time that x is positive, v is negative.

6-14: The sine curve and cosine curve look just alike except that the zero time chosen for the origin is different in the two cases. The zero time is a quarter of a cycle ($T/4$) different for the two. For the sine curve, $t = 0$ when the function is passing through zero and increasing. For the cosine curve, $t = 0$ when the function is at its maximum positive and is starting to decrease.

6-15: For an object rotating at constant angular speed, there must be a centripetal acceleration. The direction of this acceleration is radial inward—just the opposite direction from the position vector for the object, which is radial outward. The magnitude of this acceleration is $\omega^2 R$. The component along the x-axis is $-\omega^2 R \sin \theta = -\omega^2 R \sin \omega t$. Notice that this value of the acceleration is proportional to the value of the x position: $x = R \sin \omega t$; $a_c = -\omega^2 R \sin \omega t = -\omega^2 x$

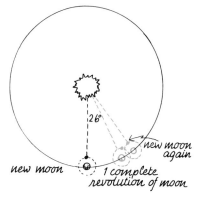
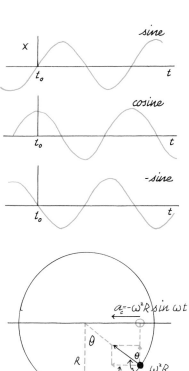

6-16: If $T = 1$ s, $\quad 1 = 2\pi \sqrt{L/g} = 6.28 \sqrt{L/9.8}$

$$0.16 = \sqrt{L/9.8} \quad 2.6 \times 10^{-2} = L/9.8$$

$$L = 2.5 \times 10^{-1} \, \text{m} = 25 \, \text{cm}$$

For a period 10 times as long, the length would have to be 100 times as large as for the 1 s pendulum: $L = 25$ m

What is the period of a 1 meter pendulum?

Problems

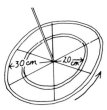

1. A car rolls down an incline of 5° with little friction. What is the acceleration?
2. If the coefficient of rolling friction is 0.02, how long will it take a car, starting from rest, to roll 100 m down an incline of 2°?
3. On page 114, $M_1 = 5$ kg and $M_2 = 5$ kg. Assume zero friction between M_2 and the table, and assume that the pulley wheel has no friction and insignificant inertia. What is the tension, T?
4. Suppose, in Problem 3, that the coefficient of kinetic friction between M_2 and the table is 0.2. What is the acceleration and what is the tension, T?
5. In the first exercise of "Handling the Phenomena," I suggested that you hang 10 paper clips on each side. If you remove one from one side, and if friction is negligible, what acceleration is produced?
6. On page 115, $M_1 = 10$ kg, $M_2 = 10$ kg, $M_3 = 8$ kg. There is negligible friction between M_2 and the table, and the pulleys have negligible friction and inertia. What are the values for a, T_1, and T_2?
7. Suppose, in Problem 6, that the coefficient of kinetic friction between M_2 and the table is 0.2. What are the values for a, T_1, and T_2?
8. For the minute hand of a clock, what are the values of T, f, and ω?
9. The sweep second hand of a clock goes around once each minute. If the radius is 10 cm, what is the angular velocity of the point, what is its tangential velocity, what is its angular acceleration, and what is its centripetal acceleration? What is the direction of each of these vectors?
10. Two hoops are connected by lightweight spokes as shown in the diagram. The mass of the outer one, which is made of aluminum, is 1 kg. Its radius is 30 cm. The mass of the inner one, which is made of brass, is 5 kg. Its radius is 20 cm. What is the moment of inertia of the combination about their common axis?
11. Why should the moment of inertia about a point on the edge of a rolling ring be greater than the moment of inertia about its axis?
12. What is the torque on a sphere rolling down an incline with angle θ? Write Newton's second law for this situation in order to find the angular acceleration, α. Now, since α and a are related, find the linear acceleration of the sphere rolling down the incline. Compare this acceleration with that of a block sliding without friction down the same incline.
13. A carousel with a radius of 5 m has a rotational period of 10 s. If you suspend a plumb bob at the rim, what angle does it make with the vertical?
14. If you go around a curve on an unbanked road, what exerts the necessary centripetal force on the car? How large is that force for a 1500 kg car going at 100 km/hr around a curve with a radius of $\frac{1}{2}$ km?
15. Space colonies are being planned in which thousands of people would live for long periods. In order to provide a simulation of normal gravity, the spherical

or cylindrical or doughnut-shaped colony will rotate. If the radius of the structure is 500 m, what is the period of rotation that will provide 1 g at the perimeter?
16. There are now many geosynchronous satellites around and above the equator, with periods of 24 hours. These satellites are particularly useful as relay stations since they hover over fixed locations. What is the height of these satellites *above the surface of the earth*?
17. Try twirling a weight with a rubber band in a horizontal circle. The restoring force of the rubber provides the centripetal force. For small distortions, x, the restoring force is approximately equal to $-kx$. The radius of the swing is $(L + x)$, where L is the unstretched length of the rubber band. What is the formula for the period of rotation as a function of m of the weight, k of the rubber, L, and x? Do your experimental results agree qualitatively with the formula?
18. Assume that you start observing a pendulum bob as it swings through its lowest point. Call that position $x = 0$, and the time $t = 0$. The bob is swinging to the right into the region defined as positive. (a) Describe velocity, v, at $t = 0$. (b) Describe the acceleration, a, at $t = 0$. (c) Describe v at $t = T/4$. (d) Describe a at $t = T/4$. (e) Describe v at $t = T/2$. (f) Describe a at $t = T/2$. (g) Describe v at $t = 5/8T$. (h) Describe a at $t = 5/8T$.
19. SHM is defined to be oscillatory motion produced by a restoring force that is proportional to the displacement from equilibrium. Explain why SHM is characteristic of a bob on an ordinary spring, a pendulum bob, and the shadow of a bob on the rim of a wheel rotating at constant velocity.
20. A particular slinky has a spring constant of 10 N/m. What mass must be suspended from this slinky to produce an oscillation period of 5 s?
21. A car spring may have a spring constant of about 2×10^4 N/m. What would be the period of oscillation if 500 kg (about a quarter of a car's mass) were to bounce on the spring?
22. How long is a simple pendulum with a period of 2 seconds?
23. What is the period of a simple pendulum with a length of 10 cm? 100 cm? 10 m?
24. The value of g on the moon is about $\frac{1}{6}$ the value of g on earth. If a simple pendulum has a period of 1 second on earth, what will be its period on the moon? Does your answer depend on the mass of the bob?
25. A superball bouncing from the floor executes oscillations with amplitude that decreases slowly. Show that even if the amplitude stays constant, the motion is not simple harmonic.
26. A bob on a spring oscillates through a total vertical distance of 30 cm, with a period of 1 second. What is the maximum velocity? What is the maximum acceleration?
27. It was pointed out in Chapter 5 that g varies on the earth's surface. The largest factor results from the earth's rotation. Part of the gravitational attraction of the earth provides centripetal force to keep objects from flying off. What is the percentage reduction of weight (or of g) on the equator compared with the poles? (Compare the centripetal force on one kilogram at the equator with its weight of 9.8 N.)

7 LINEAR MOMENTUM

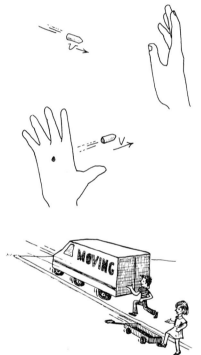

A .22 caliber short bullet has a mass of 2 grams. If somebody tossed one to you, you could catch it easily, with or without gloves. If you tried to catch the bullet at its muzzle speed of 300 m/s, wearing gloves would again not help. However, the results would be different from the first case. There is something about the high velocity of a moving body that makes the object hard to stop.

If a toy cart is rolling toward you, you can stop it with your toe. If a truck is rolling toward you, you should keep your toes out of the way. There is something about the large mass of a moving body that makes the object hard to stop.

The product of mass and velocity of an object is given a special name—*momentum*. Momentum is a measure of how much force, applied for a specific length of time, is needed to stop a moving object or to get it up to its original speed in the first place. More importantly, momentum is one of a very few quantities that are always conserved during interactions. The total momentum of a system before an interaction—such as an explosion—must be the same as the total momentum after the interaction. In this respect, and in many experimental situations, momentum appears to be a more basic quantity than either mass or velocity separately.

HANDLING THE PHENOMENA

A good pool player is already familiar with some of the properties of momentum. For instance, it's possible for the cue ball to collide head-on with another ball and stop dead, sending the target ball on in the same direction and with the same velocity that the cue ball had. Actually, the interactions of pool balls are complex because of their spin and because they interact with the table as well as with each other. As you can see in the accompanying strobe picture, the bombarding ball glanced off to its left after the collision and thus had to start rolling around a different axis than it had to begin with. To change its spin axis, it dug into the table a bit and skidded into the new path. You can see in the picture the way that the ball traveled in a slightly curving path before settling down into the new straight-line path.

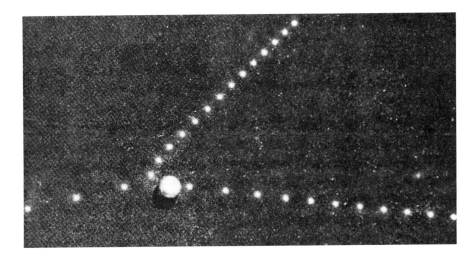

In spite of the problems caused by rolling and table contact, you can still get a feel for some of the effects of momentum by rolling one ball against another. If you can, get several balls the same size but with one of them having greater mass. Try two golf balls or squash balls and a ping-pong ball, or two tennis balls and a hard ball. Roll one of the identical balls against the other, both in head-on and glancing collisions. See if you can get the target ball to take on the direction and speed of the bombarding ball. Note the range of angles that you get between the paths of the two balls after a glancing collision.

Next, roll a heavy ball into a light one. Once again, try both head-on and glancing collisions. Can you make the heavy ball stop dead? In a head-on collision, does it keep going, and with what speed compared to the original speed and the speed of the target ball? What range of angles between the paths do you get in glancing collisions? Try the same experiments with the light ball striking the heavy one.

THE CONSERVATIVE NATURE OF MOMENTUM

When an egg hits a fan, it splatters. When you lean on a friend, you both give a little. When a rocket is fired, the gases shoot out in one direction and the rocket moves in the opposite direction. In all of these cases, the interactions between objects produce change in complex ways.

Up until now, we considered the change of velocity of an isolated object when outside forces acted on it. For that purpose, we didn't even have to know where the forces came from. Now we want to consider the whole interaction. When two or more objects exert forces on each other, all sorts of changes can occur. Take the case of a firecracker exploding. It might seem hopeless to try to analyze the resulting changes and motions. The force acting on any particular fragment changes rapidly in a way that is very hard to predict. In principle we could calculate the changes; in practice it is very difficult.

There is a way out. Instead of trying to predict the changes, see what stays constant. But this advice may not seem very helpful. For example, when a firecracker explodes, it appears that nothing stays the same. The color changes, the shape changes, the locations and velocities of the pieces change. A few things, however, don't change. Because these properties must remain constant, the number of possibilities for change is greatly reduced. A conservation law for one property of an interaction is a law of constraint on the other properties that *do* change.

Let's look at a very simple interaction. The strobe photo shows an air track glider bumping into one just like it that had been standing still. The first one stops; the target glider takes off with the same speed that the first one had. In spite of the complexity of the actual interaction of the springs when they bumped, something did stay the same. The momentum of the first glider before the collision was mass × velocity or (mv). Since the second glider was identical and had the same mass, and moved away with the same velocity, v, the final momentum was (mv).

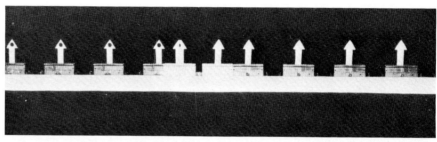

Target glider (with plain arrow) was standing still in center of air track. A glider with equal mass (with spotted arrow) came in from left. After collision by way of their spring bumpers, the moving glider stopped, and the target glider moved off to the right with the speed of the original glider.

Question 7-1

Momentum evidently was transferred from the first glider to the second, but the momentum of the *first* glider was not conserved. What good is the conservation law?

The second strobe photo shows a different sort of interaction between air track gliders. Two of them had been tied together with the spring between them compressed. When the tie was cut, the gliders shot apart, each of them evidently having the same speed. It looks as if our attempt to find a conservation law really breaks down in this case. The original momentum of the system was clearly zero; nothing was moving. The interaction occurred and suddenly there was momentum.

Question 7-2

How can you preserve a conservation law under these circumstances?

THE CONSERVATIVE NATURE OF MOMENTUM

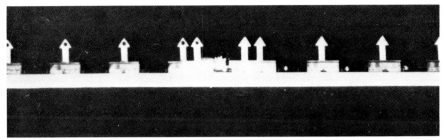

The two gliders were motionless in the center of the track. There was a compressed spring between them. Note the flame of the burning string that held them together. After the string burned, the gliders shot apart from each other with equal speeds.

The third strobe photo shows what happens when two nonidentical gliders spring apart from each other. The mass of the glider on the right was one-half that of the glider on the left. The speeds are not the same. The glider with mass 2 picked up only half the speed of the glider with mass 1. Momentum, however, is conserved with these values. If we use the customary notation that motion to the right is positive and to the left is negative, then the momentum before and after the interaction can be equated as follows:

Momentum of system before = momentum of system after

$$0 = (m_2 v_2) + (m_1 v_1) = (2m)(-\tfrac{1}{2}v) + (m)(+v) = 0$$

The fourth and fifth strobe photos show what happens when the glider with double mass runs into the one with single mass, and vice versa.

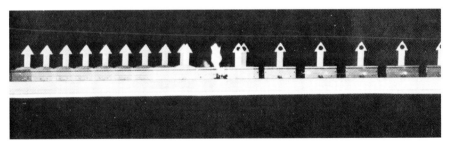

The glider on the left has twice the mass of the glider on the right. When they shoot apart from each other, the speed of the glider on the left is half that of the glider on the right. (Measure and see.)

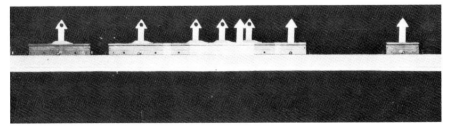

Glider on left (with spotted arrow) has twice the mass of glider on right. Single-mass-glider was originally standing still. After collision, both gliders moved to right, but with different speeds.

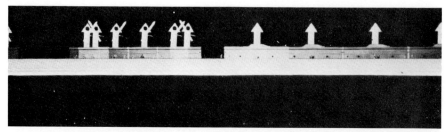

Glider on left (with spotted arrow and "bayonet" over its shoulder) has half the mass of glider on right. Double-mass glider was originally standing still. After collision, double-mass glider moved to right; single-mass glider bounced back to left (with its "bayonet" swung by collision so that it is over its shoulder as it heads back.)

Question 7-3

Use a ruler to measure Δx for each of the motions. Since Δt, the strobe light interval, is always the same in all of the illustrations, measuring Δx gives you a value proportional to $\Delta x/\Delta t$, the speed. See if momentum is conserved in these two collisions.

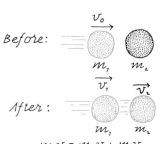

With the special provision that (mv) can be positive or negative, it looks as if momentum is conserved in the special cases that we showed. Maybe this fact is neither unusual nor useful after all. As for the relationship being unusual, maybe any combination or product of mass and velocity is conserved in the same way. When one glider hits an identical one and sends the target glider on with the original velocity, there has been no change in (mv) or (mv^2), or (m^2v), or (m^2v^3), or any other combination of m and v, including just v. However, some of these combinations are ruled out in the case of the repulsion of the two identical gliders that are at an original position with zero velocity. While v can be positive or negative, v^2 is always positive. Before the explosion, (mv^2) for the system was zero. After the interaction, the total value of this product is the positive quantity:

$$\left| (m_1v_1^2) + (m_2v_2^2) \right|$$

Therefore, the combination of (mv^2) is not a conserved quantity.

Question 7-4

Try the other combinations on the data from the interactions involving a single-mass and double-mass glider. Is any combination of m and v conserved, except for (mv)?

Here is a second question to ponder: How useful is it to know that the momentum of a system is conserved? Does that constraint rule out all other types of interactions except the ones that actually happen? For instance, when one glider strikes an identical one at rest, why can't they both go gliding off together at only half the original speed? Momentum would be conserved:

$$(m_1 v) = (m_1 \tfrac{1}{2} v) + (m_2 \tfrac{1}{2} v) = (m_1 v) \quad \text{if } m_1 = m_2$$

This sort of collision is exactly what does happen if the two gliders stick together, as you can see in the strobe photo of this event. When the gliders interact through springs, however, nature allows only the original results. It appears that there is yet another constraint on the interaction. We'll pursue that subject in Chapter 9.

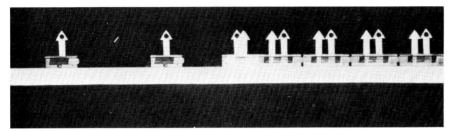

Identical mass gliders. Target glider was motionless in center of track. Instead of a collision with bumper springs, this collision resulted in the two gliders sticking together afterwards. They moved off to the right with half of the original speed.

OTHER PROPERTIES OF MOMENTUM

In spite of the importance of momentum, no special name has been given to its units. The dimensions are just those of the product of mass and velocity: MLT^{-1}. In the S.I. metric system, the units are kg·m/s.

We defined momentum to be the product of mass and velocity. In later chapters, we will see that this simple definition is good only at velocities considerably smaller than the speed of light. Photons, the particles of light, carry momentum in spite of the fact that they always travel at the speed of light and have no mass of the form assumed in the simple definition. As we will see, the complete formula for momentum reduces to (mv) when v is small compared with the velocity of light.

Since mass is a scalar and velocity is a vector, their product is a vector. Experimentally, momentum behaves like a vector; therefore, the components of momentum have to be treated separately. The cloud chamber picture

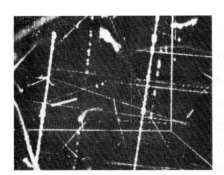
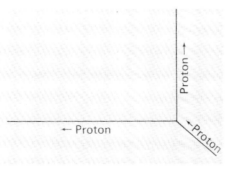
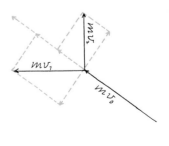

shows the track of a proton striking another proton. Protons are positively charged particles that, along with neutrons, make up the nuclei of atoms. For our purposes here, they are acting like identical marbles colliding without friction. Before the collision, all the momentum of the system resided in the bombarding proton. It struck the nucleus of one of the hydrogen atoms in the gas of the cloud chamber and knocked it forward in a glancing collision. The diagram beside the picture shows the momentum vectors of the particles. We have broken up those vectors into components parallel and perpendicular to the original direction. Both components of momentum must be conserved. Since there was no momentum perpendicular to the original direction to begin with, the perpendicular components afterward must cancel each other. The two parallel components of the momentum afterward must add up to the original forward momentum of the bombarding particle.

Question 7-5

After the collision, can the two protons each go off at 90° to the original direction and thus be going exactly opposite each other? If the two protons after the collision each go off at 45° from the original direction, what is the velocity of each of them if the original proton velocity was 2×10^7 m/s?

In the one-dimensional air track collisions, the requirement of momentum conservation yields just one equation, which links the momenta before and after the interaction. For example, if the glider with mass m_1 had an initial velocity of v_0 and ran into a stationary glider with mass m_2, the equation is

$$m_1 v_0 = m_1 v_1 + m_2 v_2$$

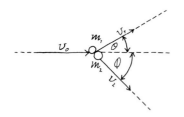

In the case of the proton-proton collision, or any other interaction of two objects taking place in two-dimensional space, momentum conservation yields two equations. For the collision angles shown in the diagram, momentum conservation in both the x-direction and y-direction produces these equations:

$$\text{Momentum in } x: m_1 v_0 = m_1 v_1 \cos \theta + m_2 v_2 \cos \phi$$

$$\text{Momentum in } y: 0 = m_1 v_1 \sin \theta - m_2 v_2 \sin \phi$$

The first equation says that the momentum in the original direction must remain unchanged. The second equation says that the components of momenta perpendicular to the original direction must cancel each other.

Question 7-6

In the case of one-dimensional air track collisions, we found that momentum conservation did not give us enough information to predict the final velocities. Of course, the final velocities had to satisfy the momentum equation, but there were two unknown velocities, v_1 and v_2, and only one equation. Now with two-dimensional collisions, we get two equations. Does that mean that if one marble glances off another one at some observed angle—say, $\theta = 45°$—that we now have enough information to predict the final velocities of the marbles?

EXAMPLES OF THE CONSERVATION OF MOMENTUM

1. Earlier we raised the question of what would happen if you tried to stop a bullet. The detailed solution of that problem can be messy, but let's see what happens if you try to shoot a bullet in the first place. A .22 slug with a mass of 2 grams can be shot with a muzzle velocity of 340 m/s. (The speed of sound in air at room temperature is 343 m/s; ordinary jet liners travel at no more than 270 m/s.) Just before firing, the gun-bullet system has zero momentum. Immediately afterward, the lead slug is carrying off momentum equal to $(2 \times 10^{-3} \text{ kg}) \times (3.4 \times 10^2 \text{ m/s}) = 0.68 \text{ kg} \cdot \text{m/s}$. If the gun were floating by itself and had a mass of 3 kg, it would recoil with a velocity of 0.23 m/s. Momentum would be conserved:

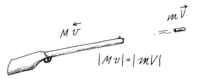

Momentum before = momentum after

$$0 = (3 \times 10^{-3} \text{ kg})(3.4 \times 10^2 \text{ m/s}) + (3 \text{ kg})(-0.23 \text{ m/s})$$

The gun, of course, is not floating; it is braced against your shoulder. In a later section of this chapter, we will figure out how large a kick you have coming.

2. Is momentum conserved when a tennis ball bounces from the ground? A new tennis ball, with a mass of 56 grams, when dropped from 1 meter rises to a height of 63 cm. Its speed just before and after striking the ground can be found by using one of the formulas for constant acceleration:

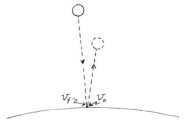

Before: $v_f^2 = 2gy = 2(9.8 \text{ m/s}^2)(1 \text{ m})$
$= 19.6 \text{ m}^2/\text{s}^2 \quad v = 4.4 \text{ m/s}$

After: $v_0^2 = 2(9.8 \text{ m/s}^2)(0.63 \text{ m})$
$= 12.4 \text{ m}^2/\text{s}^2 \quad v = 3.5 \text{ m/s}$

Question 7-7

What was the momentum just before the ball hit the ground? What was its momentum just afterward? Look out! There's a catch to the answers.

At first thought, it appears that momentum is not conserved for several reasons. The ball starts out from rest with zero momentum. The momentum increases steadily, but during the collision with the ground the momentum abruptly goes to zero, then suddenly becomes large in the opposite direction, but not so large as it had been. Momentum appears to be increasing, decreasing and changing direction. How can this behavior satisfy a conservation law? The answer is that we never claimed that the momentum of a single object is conserved. The *conservation law* is concerned with the *momentum of a whole system,* free from outside influences. In this case, the system includes both the ball and the earth. While the ball is falling toward the earth and picking up momentum in one direction, the earth is "falling" towards the ball, picking up equal momentum in the opposite direction. The total momentum is zero at all times. When the ball and earth bounce, the ball goes one way, the earth the other, each with equal magnitude of momentum but in opposite directions so that the vector quantities always add to zero.

> ### Question 7-8
> Isn't momentum lost during the bounce when the ball's speed changes from 4.4 m/s to 3.5 m/s?

Surely it is fantastic to think of the earth falling up toward a tennis ball, and then bouncing back. Let's calculate the speed of the earth just before and after it bounces off the ball.

In order to conserve the momentum of the earth-ball system, it must be true at all times that $MV = -mv$. The capital letters refer to the earth; the lowercase letters refer to the ball.

Before bounce: $(6 \times 10^{24} \text{ kg})(V) = -(0.056 \text{ kg}) \times (-4.4 \text{ m/s})$
$V = 4.1 \times 10^{-26}$ m/s

After bounce: $(6 \times 10^{24} \text{ kg})(V) = -(0.056 \text{ kg}) \times (+3.5 \text{ m/s})$
$V = -3.3 \times 10^{-26}$ m/s

The change of velocity is -7.4×10^{-26} m/s, which is not enough to make anything spill!

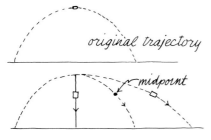

3. When bombs are bursting in air, the fragments shoot out in such a way that the original momentum of the bomb remains unchanged. That condition imposes severe constraints on the relative speeds and directions of the fragments. Let's study a simple example of a shell exploding in midtrajectory into two equal pieces. Furthermore, take the special case where one piece is shot horizontally backward at a relative speed just equal to that of the shell when it exploded. That piece is suddenly left with zero horizontal velocity, with respect to the ground, and will start dropping straight down. The other half of the shell must be kicked in the forward direction with twice the horizontal velocity that it did have. In that way, the momentum of the shell before explosion, which was horizontal with the value (mv), remains unchanged since the momentum of the fragment is $(\frac{1}{2}m)(2v) = (mv)$.

As the two fragments fall, their vertical heights are identical at all times; their heights are also the same as the height that the whole shell would have been at those same times. The first fragment falls straight down, and the

EXAMPLES OF THE CONSERVATION OF MOMENTUM

horizontal distance covered by the second fragment is just twice the horizontal distance that the original bomb would have traveled. (Of course, all these statements ignore air resistance.) Note on the diagram the location of the point that is halfway between the two equal pieces.

> Question 7-9
>
> That point keeps changing as the pieces fall. What is the trajectory of that midpoint?

4. Here is an example of how the law of momentum conservation can tell us that something must have happened even when we can't see it. The bubble chamber picture shows the tracks of one subatomic particle decaying into a second one; the second particle is then shown decaying into a third one. In a later chapter, we will study these subatomic particles and the methods used to produce them. For now, all we have to know is that the tracks in the picture are trails of bubbles in a tank of liquid hydrogen. The subatomic particles shot through the tank just as the hydrogen reached an unstable condition and was about to boil (at a temperature of 21° above absolute zero—that's −252°C). At that crucial moment, a bright light flashed through a window in the side of the chamber, and a camera took the picture looking down through a glass window in the top. The actual chamber had a diameter of about 30 cm.

The track of the original particle, coming in from the bottom of the picture, is labeled π^- (pi minus; the minus means that it has negative electric

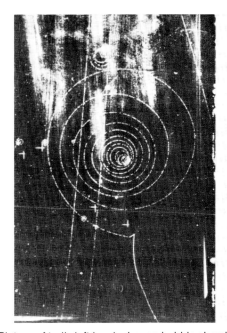
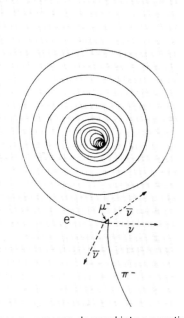

Picture of trails left in a hydrogen bubble chamber as a π^- meson decayed into a negative muon (μ^-), and the muon decayed into an electron (e^-).

charge). The particle must have been slowing down, as is indicated by the thickening trail. It came to rest, and then decayed. From other experiments, we know that the decay time must have been about 10^{-8} seconds. The short, thick trail of the decay product is labeled μ^- (mu minus). It also came to rest and then, after about 10^{-6} seconds, decayed into an electron that produced a very sparse, curved trail. The fact that the trail of the final decay product is very thin indicates that the electron was traveling very fast—close to the speed of light. From the thickness of the trails of the other particles, we know that they were traveling considerably slower. All three particles had a negative electric charge and followed curved paths because they were traveling in a magnetic field. The electron's spiral path shows that the electron continually lost momentum going through the hydrogen liquid and thus that the sideways force exerted by the magnetic field had more and more effect on its path.

So far we have described most of the things that you can see in the picture. Some of the most interesting events cannot be seen, but the law of momentum conservation tells us that they must have taken place. Note that the π^- particle slowed down, stopped, and then decayed with the emission of a single particle going off to the side. What if you saw a hand grenade roll to a stop, explode, and shoot just one fragment to the side? It couldn't happen. Something must have shot out in the opposite direction in order to conserve momentum. In the case of the subatomic particles, we know that the invisible recoil particle was an anti-neutrino. Because the anti-neutrino has no electric charge, it leaves no trail in a bubble chamber.

The same argument can be used to describe what we do not see when the μ^- particle decays. Once again, there must have been recoil particles. The evidence of many similar pictures is that the μ^- track always has the same length, regardless of the angle between it and the π^- track. On the other hand, the electron track is sometimes long and sometimes short. In the case of the π-μ decay, the μ can get the same momentum each time only if it is recoiling against one—and only one—particle. In the μ-e decay, the momentum of the electron can vary from one time to the next only if it is recoiling from two or more particles. From other evidence, we know that when the μ decays it turns into an electron, a neutrino and an anti-neutrino. We cannot see the neutrino and anti-neutrino, but we know that the vector sum of their momenta must be equal and opposite to the momentum of the electron.

Question 7-10

Why couldn't the electron always get the same amount of momentum if two other particles are involved?

NEWTON'S THIRD LAW AND ITS RELATIONSHIP TO NEWTON'S SECOND LAW

The law of conservation of momentum says that the momentum of an isolated system after an event is the same as it was before the event. If the momentum of a firecracker is zero before the explosion, the vector sum of

NEWTON'S THIRD LAW AND ITS RELATIONSHIP TO NEWTON'S SECOND LAW

the momenta of all the fragments is still zero after the explosion. The *change of momentum*, $\Delta(mv)$, is zero. Newton's second law, as we first presented it, is also concerned with a change, Δv. The force exerted on an object with mass, m, produces a change of velocity, Δv, during the time interval, Δt:

$$F = \frac{m \Delta v}{\Delta t}$$

If there is no change in the mass during the time interval, $m \Delta v$ is the same as $\Delta(mv)$. In that case, Newton's second law could be written

$$F = \frac{\Delta(mv)}{\Delta t}$$

Force is equal to the time rate of change of momentum. As a matter of fact, this is the general form of Newton's second law, which is always good whether or not the mass changes during the time interval. We will soon see a case where the mass of the object does change. The usual formula, $F = ma$, is just a special form of the more general law.

A special name is given to the product of the force and the time interval: *impulse* $\equiv F \Delta t = \Delta(mv)$. Two special rules follow from this form of Newton's law. First, if the net *external* force, F, on a system is zero, then the system's momentum will remain constant. If $F = 0$, then $\Delta(mv) = 0$. This is just a shorthand way of saying that a system's momentum is conserved if no external forces are acting on it.

The second special rule about impulse is concerned with the *internal* forces in a system. For the sake of simplicity, let's analyze the interaction between two parts of the same system. The part with mass m_1 is pushing on part 2 with a force equal to F_{12}. (This is not F sub twelve; it is the force that object 1 exerts on object 2.) Meanwhile, the part with mass m_2 is exerting a force, F_{21}, on part 1. Since the two parts are being shoved apart, each of them will suffer a change of momentum. However, the vector sum of the two changes of momenta must equal zero, because there are no external forces:

$$\Delta(mv)_1 + \Delta(mv)_2 = 0$$

The change of momentum of each part is caused by the force exerted on it by the other part:

$$F_{21} \Delta t = \Delta(mv)_1 \quad \text{and} \quad F_{12} \Delta t = \Delta(mv)_2$$

Therefore, $F_{21} \Delta t + F_{12} \Delta t = 0$. Since the time interval, Δt, is the same for both sides, it must be that $F_{21} = -F_{12}$. The force that one object exerts on a second object is equal and opposite to the force that the second exerts on the first. *This is one of the forms of Newton's third law.*

There are several ways of stating the law that we have just derived. Sometimes it is said that "action equals reaction." The trouble with that statement is that "action" is not technically defined, and reaction sounds like something out of politics. Sometimes the fact that two objects exert equal but opposite forces on each other is given as the primary statement of Newton's third law. Momentum conservation then becomes a derived consequence. The best way to view the situation is to think of two objects interacting with each other, not that object 1 is exerting a force on object

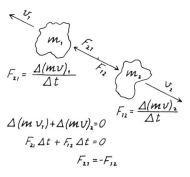

2, and object 2 is exerting a force on object 1. From the point of view of a single interaction, of course the forces on the individual parts *must* be equal and opposite.

> ### Question 7-11
> The weight of a tennis ball near the surface of the earth is 0.55 N. What is the weight of the earth in the gravitational field of the tennis ball?

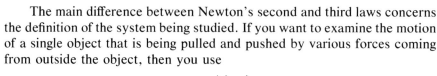

The main difference between Newton's second and third laws concerns the definition of the system being studied. If you want to examine the motion of a single object that is being pulled and pushed by various forces coming from outside the object, then you use

$$F = \frac{\Delta(mv)}{\Delta t}$$

In adding all the external forces, you isolate the object. On the other hand, if you are considering a whole system composed of parts that can interact with each other, then Newton's third law tells you that the sum of the changes of the individual momenta will be equal to zero and that, therefore, the internal forces between pairs of the parts are equal but opposite to each other:

$$\text{If } \Delta(mv)_1 + \Delta(mv)_2 = 0 \quad \text{then } F_{12} = -F_{21}$$

or vice versa.

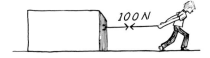

> ### Question 7-12
> If you pull on a box with a force of 100 N, then the stretch in the material of the box must be exerting a force of 100 N on you in the opposite direction. Apparently the forces cancel. How can anything move?

EXAMPLES OF IMPULSES AND KICKS

1. Earlier we pointed out that you would get a "kick" out of firing a gun. The recoil momentum of the .22 rifle was 0.68 kg·m/s. That's $\Delta(mv)$, which must equal $F\Delta t$. The size of the force exerted on the gun evidently depends on the length of the time interval, Δt. That time interval depends on the length of the gun barrel, the muzzle velocity, and the nature of the explosion. Actually, the recoil force is greatest shortly after the explosion begins and falls off as the gases expand and the slug shoots down the barrel. However, because the slug starts out slowly, it is close to the explosion point during most of the time that it spends in the barrel. For our purposes, a good approximation is that the force is constant during the time that it takes the slug to leave the gun. If the muzzle velocity is 340 m/s, the average

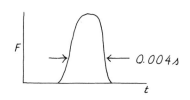

velocity with constant acceleration is 170 m/s. If the gun barrel is $\tfrac{2}{3}$ m long, the time interval is $\Delta t = (\tfrac{2}{3}\text{ m})/(170\text{ m/s}) = 0.004$ s. That's 4 milliseconds. Now we can find the force on the gun:

$$F = \Delta(mv)/\Delta t$$

$$F = (0.68\text{ kg}\cdot\text{m/s})/(0.004\text{ s}) = 170\text{ N}$$

> ### Question 7-13
>
> Is that much of a force? How does it compare with the weight of the gun? (The gun has a mass of 3 kg.)

2. The force on a tennis ball as it bounces is certainly not constant. Whether the ball bounces against the ground or against the strings of a racket, the restoring force must be approximately proportional to the distortion. That requirement would produce a force-versus-time curve something like the dotted one shown in the diagram *if* the bounce were symmetrical. (It is not, since the upward velocity is less than the downward velocity.) For an easier calculation, however, let's use the triangular approximation. The time it takes for the ball to come to rest during the first half of the bounce can be estimated by assuming that the ball slows down with constant acceleration in a distance equal to some fraction of its diameter. Let's assume that when the ball is dropped from a height of 1 m, it has stopped by the time it is squashed one-fourth of its diameter—about 1.5 cm. From the earlier calculation, we know that just before it hits the ground, a ball dropped from a height of 1 m has a speed of 4.4 m/s. If it were to slow down with constant acceleration, which is a very rough approximation, its average speed during the bounce would be 2.2 m/s. The time it would take to come to rest would be

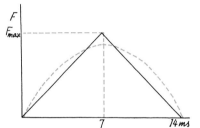

$$\Delta t = (1.5 \times 10^{-2}\text{ m})/(2.2\text{ m/s}) = 6.8 \times 10^{-3}\text{ s}$$

which is about 7 milliseconds.

We can now calibrate the graph of force versus time. The time scale must be such that it takes 14 milliseconds for the complete bounce. The product, $F\Delta t$, must be equal to the change of momentum, $\Delta(mv)$. We know what the change of momentum must be

$$\begin{aligned}\Delta(mv) &= (mv)_{\text{after}} - (mv)_{\text{before}} \\ &= (0.056\text{ kg})(+3.5\text{ m/s}) - (0.056\text{ kg})(-4.4\text{ m/s}) \\ &= +0.44\text{ kg}\cdot\text{m/s} = 0.44\text{ N}\cdot\text{s}\end{aligned}$$

The trouble is, we cannot simply set $F\Delta t = 0.44$ N·s.

> ### Question 7-14
> Why not?

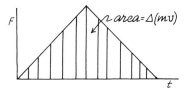

The area under the force-time curve is equal to the impulse, which is the change of momentum. In order to simplify the calculation, we assume that the curve of $F(t)$ is triangular. The area under such a curve is equal to: $\frac{1}{2}F_{max}t = \frac{1}{2}F_{max}(14 \times 10^{-3} \text{ s}) = 0.44 \text{ N} \cdot \text{s}$. Therefore, $F_{max} = 60$ N. The ball itself, with a mass of 56 grams, has a weight of only 0.55 N. Apparently, even for a drop of only 1 m, the restoring force on the tennis ball becomes more than 100 times the weight of the ball. In the language of test pilots of jets or rockets, the ball is subject to more than 100 "g's" ("gravities").

3. In case you feel no empathy for an accelerated tennis ball, consider the forces involved when you yourself stop suddenly in a car. Assume that you drive at 40 mph into a concrete wall. The car will stop in a distance of less than 1 m, that distance being taken up by the space where the motor and things used to be. Assuming that the car slows down with constant acceleration from an initial speed of 40 mph, which is about 18 m/s, the time it takes is equal to

$$\Delta t = \Delta x / v_{ave} = (1 \text{ m})/(9 \text{ m/s}) = 0.11 \text{ s}$$

If your seat belts are well adjusted and hold, you will slow down in that same time. If you are not wearing seat belts, you will feel no strain until your body hits the steering wheel. The wheel will not take long to collapse, giving you some more free time before you hit the dashboard and windshield. Then you will stop in a hurry unless you make it all the way through the windshield. On the graph of force versus time for this accident, we have drawn three curves: one for the car frame, one for a driver with seat belts, and one for the driver without. The time scale of events is realistic in terms of automobile manufacturers' tests with dummies. Actual events vary drastically, of course, depending on whether your stomach gets caught in the wheel and on which portion of your head hits the glass.

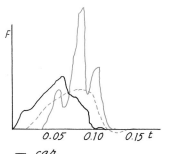

— car
--- passenger with seatbelt
— passenger without seatbelt

There is one important thing to keep in mind during the tenth of a second that all this action takes place. The change of momentum of your body will be the same regardless of whether you are wearing a seat belt. You were traveling at 18 m/s, and about a tenth of a second later your speed is zero. Therefore, the total area under the $F(t)$ curve must be the same:

$$\Delta(mv) = (60 \text{ kg})(0 - 18 \text{ m/s}) = -1.1 \times 10^3 \text{ kg} \cdot \text{m/s} = -1.1 \times 10^3 \text{ N} \cdot \text{s}$$

If you could stretch the impulse out evenly over the time of the collision, the force would be $F = (-1.1 \times 10^3 \text{ N} \cdot \text{s})/(0.11 \text{ s}) = -1 \times 10^4$ N. The minus sign just indicates that the direction of the force is opposite to your original direction.

Question 7-15
How many "g's" would you experience?

As you can see from the graph, the actual effect of the seat belt is not so ideally provident, but the maximum g produced is tolerable. Besides, the seat belt exerts its restraining forces on sections of the body that are more flexible than the skull, spreading the force out over a larger area of the body.

EXAMPLES OF IMPULSES AND KICKS

Notice that it is not exactly the car collision that kills the driver. Instead, the fatal blow comes from the second collision when the driver strikes the inside of the car. The car has pretty much come to rest before the unbelted driver hits. The driver's stopping distance, and thus the stopping time, is shorter than that of the car by a factor of about 10. Consequently, the stopping force on the body is about 10 times what it would have been had the driver been fastened to the car.

■ Optional

4. In each of the examples so far, it made no difference whether we described the change in momentum, $\Delta(mv)$, or the product of mass and change of velocity, $m\Delta v$. The mass of the bullet, or the tennis ball, or the human body did not change during the interaction. When a subatomic particle travels at speeds close to that of light, its mass changes and, experimentally, momentum becomes more meaningful than mass. There is another case, perhaps less exotic, where the mass of an object changes during an interaction. A rocket accelerates by shooting away some of its own mass. The exhaust gases have a fairly constant velocity with respect to the rocket. If the rate of ejection of gas is constant during firing, the rate of change of momentum will be constant. However, the acceleration of the rocket will not be constant because the mass of the rocket will be decreasing. Not only will the velocity of the rocket increase, its acceleration will increase. If you watched the television reports of the Apollo shots, you may remember that the astronauts underwent the largest g forces toward the end of the firing of the first stage. Let's take a look at the arithmetic of this situation.

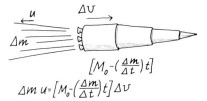

In free flight, or neglecting gravitational influences, the momentum of the rocket-gas system must be conserved. If the rocket's mass at the beginning of a firing is M_0, and if it loses mass at the rate of $\Delta m/\Delta t$, then after a time, t, its mass will be: $[M_0 - (\Delta m/\Delta t)t]$. During the next interval of time after t, the rocket will pick up an additional velocity, Δv, and will lose an additional mass, Δm. Meanwhile, an amount of gas with mass, Δm, has been shot out with an exhaust velocity of u with respect to the ship. The forward momentum gained by the ship must be equal in magnitude to the backward momentum of the gas.

$$[M_0 - (\Delta m/\Delta t)t - \Delta m]\,\Delta v = \Delta m u$$

Question 7-16
If the rocket ship is traveling to the right, in what direction are the exhaust gases traveling? The answer is not quite simple.

If the amount of mass shot out during this interval is small compared with the mass of the ship at that time, we can make the approximation that

$$[M_0 - (\Delta m/\Delta t)t - \Delta m] \approx [M_0 - (\Delta m/\Delta t)t]$$

Rearrange the variables and divide both sides by the time interval, Δt:

$$\frac{\Delta v}{\Delta t} = \frac{\Delta m}{\Delta t}\,\frac{u}{[M_0 - (\Delta m/\Delta t)t]}$$

This formula is not so formidable as it may appear. The left side is simply the acceleration of the rocket ship. From Newton's second law, we expect that an acceleration is equal to a force divided by a mass. In this case, the force is constant and equal to the product of a constant velocity, u, and a constant mass loss, $\Delta m/\Delta t$. The mass in the denominator of the right side is a variable mass because the rocket ship itself is continually getting lighter. Even with a

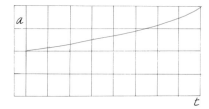

Acceleration of a rocket with constant thrust but with decreasing mass.

constant force exerted on the ship by the constant exhaust, the acceleration increases because the mass decreases.

Let's try some actual numbers in the rocket equation. Consider the lift-off of a 100 ton rocket. At $t = 0$, the mass is simply M_0. The thrust of the rocket must be at least equal to its weight or the rocket will never get off the pad. The thrust of the rocket is equal to the product of its mass and acceleration: $M_0 (\Delta v/\Delta t)$. From the rocket equation, at $t = 0$,

$$\text{Thrust} = M_0 \frac{\Delta v}{\Delta t} = u \frac{\Delta m}{\Delta t}$$

The exhaust velocity of the gas is about equal to the average thermal velocity of the molecules of gas at the combustion temperature. The average velocity of oxygen and nitrogen molecules at room temperature is about 340 m/s. On the absolute temperature scale, the combustion temperature of rocket gas is about 10 times higher than room temperature. The molecular velocities are proportional to the square root of the temperature and inversely proportional to the square root of the mass of the molecule. Since hydrogen is part of the exhaust, and since the temperature is very high, the exhaust velocities can be as high as 3000 m/s. If we require a thrust of exactly 100 tons, just enough to balance the initial weight of the rocket ship, then the rate of mass loss must be:

$$(\Delta m/\Delta t) = (100 \text{ tons})(9 \times 10^3 \text{ N/ton})/(3 \times 10^3 \text{ m/s}) = 300 \text{ kg/s}$$

To produce an upward acceleration of g, the rocket will have to shoot out twice this much, or 600 kg/s. (Incidentally, note that an English ton is about equal to 900 kg.) The rocket cannot lose mass at the initial rate for very many seconds.

The exhaust velocity of 3000 m/s is about as high as can be obtained with chemical burning and thermal molecular speeds. One way to avoid having to carry along so much mass and then throwing it away is to shoot a smaller amount of mass out at much higher velocities. Protons, the nuclei of hydrogen atoms, can be accelerated to speeds of 10^7 m/s with fairly small high-voltage accelerators. A propulsion device making use of such an accelerator could be used only after the rocket ship were out of the atmosphere and in the vacuum of space. Incidentally, note that the rocket exhaust does not push against the air in order to provide propulsion. On the contrary, the air is a nuisance, creating very considerable drag.

Question 7-17
If you are working with your tool kit outside a space station and become separated from the tether holding you, how could you make your way back to the station?

Center of Mass and Momentum Conservation
In Chapter 4 we defined the center of weight of an object. It is that point around which all the weight torques of an object add to zero. For a system of n point objects fastened together, each with weight W_i, and each located a distance, x_i, from some axis, the distance of the center of weight from that same axis is

$$x = \frac{\Sigma x_i W_i}{W}$$

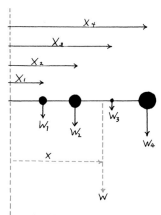

Center of weight of a system.

If the gravitational force on each part is parallel to that of every other part (the flat earth approximation), then the summation of torques, $x_i W_i$, is just ordinary arithmetic instead of vector addition. We can substitute $m_i g$ for W_i and rewrite the equation for the center point:

$$x = \frac{\Sigma x_i m_i g}{Mg} = \frac{\Sigma x_i m_i}{M} \quad \text{where } M = \Sigma M_i$$

With this change, x becomes the distance along the x-axis from some origin to the *center of mass*. Similarly, the y-component of the distance from the origin to the center of mass is

$$y = \frac{\Sigma y_i m_i}{M}$$

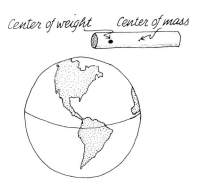

The distances, x and y, are "weighted averages" of the distances from the origin to the individual objects. It may seem as if the center of mass is the same as the center of weight. For small systems of objects on earth that is generally true, but it isn't true for a system of objects on the celestial scale. As an extreme example, consider the difference between center of mass and center of weight for the uniform rod shown in the diagram.

The idea of the center of mass is vital to the law of conservation of momentum. If you follow the path of one point object moving all by itself without external forces, it is simple to assert that momentum is conserved. If the object bumps into another point object, it is not so obvious that the momentum of the system must be conserved, but we have seen that this is the case. If there is a whole system of objects, exerting all sorts of internal forces on each other, it is not even clear what momentum is being conserved. Certainly the momentum of any individual part changes. The claim of the law of momentum conservation is that the vector sum of all the individual momenta will not change. Therefore, the sum of the x-, y-, and z-components of the individual momenta must not change. Let's write out that sum for the x-components:

$$\Sigma m_i (v_x)_i = \Sigma m_i \left(\frac{\Delta x}{\Delta t}\right)_i = \frac{1}{\Delta t}[\Sigma m_i (x_f - x_0)_i]$$
$$= \frac{1}{\Delta t}[\Sigma m_i (x_f)_i - \Sigma m_i (x_0)_i]$$

We wrote the x-component of the velocity of object i as

$$\left(\frac{\Delta x}{\Delta t}\right)_i$$

Since the time interval, Δt, is common to all the objects, we took it outside the summation sign. The displacement, Δx, is equal to $(x_f - x_0)$, the difference between the final x and the original x.

Now the definition of the center of mass enters the derivation. The weighted sum of the x-components is $[\Sigma m_i x_i] = Mx$. Let's substitute this expression into our statement about the sum of momenta of the objects in the system:

Sum of the x components of momenta $= \frac{1}{\Delta t}[(Mx)_f - (Mx)_0]$
$$= \frac{1}{\Delta t} M \Delta x = M \frac{\Delta x}{\Delta t} = Mv_x$$

It appears that the sum of the x-components of momenta of all the objects in a system is equal to the total mass times the x-component of the velocity of the center of mass. The same statement is true for the y- and z-components.

The law of momentum conservation for a system of objects states that the momentum of the center of mass remains constant. Furthermore, if the system is subject to external forces, the motion of the center of mass is the same as if the vector sum of all the external forces acted on the total mass situated at the center of mass.

SUMMARY

The greater the mass of an object and the greater its velocity, the harder it is to start or stop the object. The product of mass and velocity is given a special name: $mv \equiv$ momentum. The units of momentum have no special name, but are simply kg·m/s.

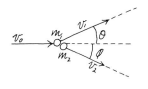

We demonstrated that in two-body interactions, the total vector momentum is a conserved quantity. Although the momentum of each of the two interacting objects usually changes, the vector sum of the momenta of the objects remains unchanged during the interaction. For the whole system, $\Delta(\Sigma m_i \mathbf{v}_i) = 0$. In the case of collision of a moving object with mass m_1 with a stationary object of mass m_2, momentum conservation requires that $m_1 \mathbf{v}_0 = m_1 \mathbf{v}_1 + m_2 \mathbf{v}_2$. The vector equation yields one equation for a one-dimensional collision and two equations for a two-dimensional collision. Although the equations must be satisfied by the data from any collision, in neither case is there enough information to predict the complete results. Another equation, representing another conservation requirement, is needed.

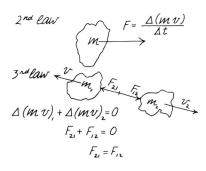

The fact that momentum of a system must be conserved is one statement of Newton's third law. Our original expression for Newton's second law was $\mathbf{F} = m\mathbf{a} = m(\Delta \mathbf{v}/\Delta t)$. The more general form of Newton's second law is that force equals the time rate of change of momentum: $\mathbf{F} = \Delta(m\mathbf{v})/\Delta t$. It might appear that Newton's second and third laws are the same thing, since if there are no external forces acting on an object, $\mathbf{F} = 0$, and according to the second law, $\Delta(m\mathbf{v}) = 0$. This means that momentum of the object is conserved. However, the emphasis of the second law is on an isolated object and the external forces acting on it. The emphasis of the third law is on the interaction between objects or systems. In particular, we showed that if the momentum of a system of two objects is conserved, then the internal force between them must be such that the force that object 1 exerts on object 2 is equal but opposite to the force that object 2 exerts on object 1. This is another, and popular, way of expressing Newton's third law.

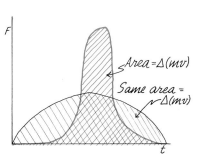

Sometimes it is convenient to solve problems in terms of the impulse, or change of momentum, given to the system. During an interaction, $\Delta(mv)$ may be known, but the way the resultant force changes with time may not be easily calculated. The total impulse, $F\Delta t$, is represented by the area under a curve of $F(t)$. During collisions, the peak force on an object depends crucially on the collision time.

As an example of the usefulness of defining F as $\Delta(mv)/\Delta t$, we showed how rockets are propelled by a thrust produced by gases emitted at constant relative velocity. The thrust is proportional to the time rate of loss of mass.

For a system of many objects, there is a particular point called the center of mass. Regardless of the motions of the separate parts of the system, the motion of the center of mass is the same as if all the mass of the system were concentrated there, and all the external forces were applied there.

Answers to Questions

7-1: Momentum is conserved only for a complete system uninfluenced by external forces. The first air track glider was subject to external forces when it hit the second one. However, since the two of them together are a complete system, the total momentum of the pair must stay the same.

SUMMARY

7-2: Momentum, like velocity, must have direction. It is true that each of the gliders has momentum after the explosion, where none existed before. However, if we call the momentum to the right, positive, and momentum to the left, negative, then the sum of the momenta remains zero.

7-3: When the double glider struck the single one with a velocity of v, the single one took off with a velocity of $\frac{4}{3}v_0$. Meanwhile the double glider continued on with a velocity of $\frac{1}{3}v_0$. The original momentum was $(2)(v_0)$. The final momentum was $(1)(\frac{4}{3}v_0) + (2)(\frac{1}{3}v_0) = 2v_0$.

When the single glider struck the double glider with a velocity of v_0, the double glider took off with a velocity of $\frac{2}{3}v_0$. The single glider bounced back with a velocity of $-\frac{1}{3}v_0$. The original momentum was $(1)(v_0)$. The final momentum was $(2)(+\frac{2}{3}v_0) + (1)(-\frac{1}{3}v_0) = v_0$.

7-4: See if the quantity (mv^2) is conserved when the double glider struck the single one: $(2)(v_0)^2 = (1)(\frac{4}{3}v_0)^2 + (2)(\frac{1}{3}v_0)^2$. $2v_0^2 = \frac{16}{9}v_0^2 + \frac{2}{9}v_0^2 = \frac{18}{9}v_0^2 = 2v_0^2$. For this particular collision, it appears that (mv^2) is a conserved quantity, although it is not conserved for the explosion-type interactions. Is (mv^6) conserved when the single glider strikes the double one? How about (m^2v) or (m^2v^2)? In Chapter 9 we will take a closer look at the quantity (mv^2).

7-5: If the two protons went off at 90° to the original direction, there would be no remaining momentum in the forward direction. If the two protons go off at 45° from the forward direction, then the component of velocity of each in the forward direction is $(v \sin 45°) = (\frac{1}{2}\sqrt{2}\,v)$. The combined components of momentum in the forward direction are $2m_p(\frac{1}{2}\sqrt{2}v) = \sqrt{2}m_pv$. The final momentum must equal the original momentum, which was $m_p(2 \times 10^7 \text{ m/s})$. Therefore, $v = 1.4 \times 10^7$ m/s. Since the two protons have equal speeds and equal angles from the original direction, the components of their momenta perpendicular to the original direction cancel out.

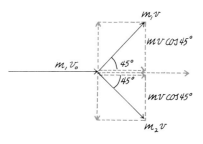

7-6: In this two-dimensional collision of marbles, or protons, we assume that we know the masses and the original velocity and that we want to know the resulting velocities if one of the particles comes out of the collision at 45°. It is true that there are just two unknown speeds, v_1 and v_2, and that we now have two equations. Unfortunately, there is now a third unknown—the other angle, ϕ. Our equations would be satisfied by a large number of solutions; nature allows only one. It must be that there is still another constraint, or conservation law, that we can use. We will find out what that is in Chapter 9.

7-7: The momentum of the tennis ball just before it hit the ground was $(0.056 \text{ kg})(-4.4 \text{ m/s})$. Note the negative sign for the velocity; the ball is headed down. After the bounce, the momentum was $(0.056 \text{ kg})(+3.5 \text{ m/s})$. The *change* of momentum was $(\text{Mom}_f - \text{Mom}_0) = (0.056 \text{ kg})(+7.9 \text{ m/s})$.

7-8: There is, indeed, a momentum change of the ball itself when its velocity changes from -4.4 m/s to $+3.5$ m/s. Note that this change is not just because the ball slowed down, but because it turned around. The fact that the ball lost some speed has nothing to do with momentum conservation. The original momentum of the ball-earth system was zero; the final momentum of the ball-earth system was zero. Presumably, if the ball recoiled with smaller speed than before the collision, so did the earth.

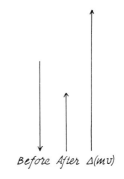

7-9: The original bomb was at the midpoint of its flight and from that point would have dropped with the same vertical acceleration as the two fragments did. Since one fragment dropped straight down, and the horizontal velocity of the other fragment was twice that of the original bomb, the midpoint of the two fragments has a horizontal velocity of half that, or just the original horizontal velocity of the bomb. Therefore, the midpoint will follow the trajectory that the original bomb would have had if it had not exploded.

7-10: If an object explodes into three fragments, there is usually no reason why the angles between fragments must have particular values. In the diagram we suggest several

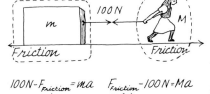

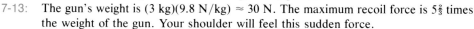

ways that the three fragments might come off. In all cases, the total momentum must remain zero, but the electron can get more or less momentum depending on the directions of the two neutrinos. On the other hand, if only two fragments are emitted in an explosion, the two must come off exactly opposite to each other—each carrying the same magnitude of momentum. As we will see in Chapter 9, the fragments in such an explosion always share the same amount of energy, which must be divided up in such a way that the momentum constraints are satisfied.

7-11: Since there is just one interaction between earth and the tennis ball, its magnitude can have only one value. The pull of the earth on the tennis ball is down, toward the earth; the pull of the tennis ball on the earth is up, toward the tennis ball. The weight of the earth in the gravitational field of the tennis ball is 0.55 N.

7-12: The equal but opposite forces described by Newton's third law always act on *different* objects; the law says nothing about equal but opposite forces on the *same* object. For instance, a book has a gravitational interaction with the earth. The book pulls up on the earth; the earth pulls down on the book, with an equal but opposite force. The forces are acting on different objects—the book and the earth. The book and earth fall toward each other. If the book is on a table, the book does not fall, because the compression of the table exerts an upward force on the book. This compression force is indeed equal to the weight of the book, and opposite in direction, but it arises from a different interaction. The book is subject to two forces—a gravitational force down, and the table compression up.

In the case of pulling on the box, there is a tension of 100 N between puller and box. As far as the box is concerned, however, there is a force of 100 N pulling it in one direction, and probably some friction forces acting in the other direction. The equal but opposite force of 100 N is not acting on the box; it is acting on the puller. The puller is subject to other forces, most importantly those of friction between his feet and the ground. Those friction forces, which the puller brings into existence by pressing against the ground, are actually shoving the puller forward. In this case, the friction force does not slow down motion, but makes it possible. The forward force of friction (which is equal but opposite to the backward push of the feet) is greater than the backward pull of 100 N acting on the puller.

7-13: The gun's weight is $(3 \text{ kg})(9.8 \text{ N/kg}) \approx 30 \text{ N}$. The maximum recoil force is $5\frac{2}{3}$ times the weight of the gun. Your shoulder will feel this sudden force.

7-14: We know that the restoring force is not constant. Either we must assume some average F during Δt, or we must assume how F depends on t. The maximum force will surely be greater than the average force.

7-15: With a mass of 60 kg, your weight is about 600 N. The ratio of acceleration force to weight is the value of the "g's" to which you are subjected: $(1 \times 10^4 \text{ N})/(600 \text{ N}) = 16\,g$.

7-16: The exhaust velocity, u, is the velocity of the gases *with respect to the ship*. As seen by an observer at rest on the ground, when acceleration first starts, the rocket ship goes to the right and the gases shoot out to the left. After the rocket ship reaches a speed greater than u, the gases as seen from the ground are traveling to the right also. Their velocity, however, is always less than that of the ship by the amount u.

7-17: If you throw your tools away from the ship, you will recoil toward the ship. Hang on to your tether. The technique might take a bit of practice.

Problems

1. An ice skater (or roller skater) stands still facing another skater gliding toward her with a speed of 2 m/s. Her mass is 50 kg; his is 70 kg. If they embrace (and do not stop their motion by digging their skates into the ice), what is their resultant joint velocity?

2. On an air hockey table, a floating disk with a speed of 4 m/s strikes another disk with equal mass. The first disk glances off at an angle of 30° from its original direction with a speed of v_1; the second disk is knocked at an angle of 60° from that original direction with a speed of v_2. What are the values of v_1 and v_2?

3. In a head-on collision, a proton (relative mass 1) traveling at 3×10^7 m/s strikes the nucleus of a helium atom (relative mass 4) at rest and bounces directly backward with a speed of 1.8×10^7 m/s. What is the velocity of the helium nucleus?

4. In the radioactive decay of uranium, an alpha particle (helium nucleus), with a mass of about 4 on the atomic scale, is emitted with a velocity of 1.5×10^7 m/s. What is the recoil velocity of the remaining atom of thorium, which has a mass of about 234 on the same scale?

5. A rocket ship emits hot gases at a speed of 2000 m/s with respect to the ship. If it burns and shoots out 100 kg of mass every second, what is the resultant thrust, or force, on the rocket?

6. A child with a mass of 22 kg running at 2.5 m/s jumps onto a 12 kg wagon from the rear. What is the resultant velocity of wagon and child?

7. A 1.8 kg ball hits a wall straight on at 6.5 m/s and bounces off at 4.8 m/s. What impulse is experienced by the ball?

8. A 12 kg wagon is pushed with a force of 7.0 N for 1.5 s, then with 4.5 N for 1.2 s, then with 10.0 N for 2.0 s. (a) What is the total impulse applied to the wagon? (b) What is the wagon's change of velocity (assuming that rolling friction is negligible)?

9. Suppose that these are graphs of the force as a function of time exerted by a karate expert and by a boxer.
   (a) What is the impulse delivered in each case?
   (b) Speculate about which would be the more dangerous blow. Which would send you flying across a room? Which would be more apt to break a bone?

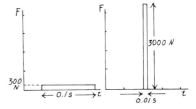

10. A 50 g tennis ball strikes a racquet perpendicular to its surface. Its speed is 10 m/s and it rebounds with the same speed. The force applied to the ball increases linearly to a maximum in 0.01 s and then decreases linearly to zero in the next 0.01 s. (a) What is the change of momentum of the ball? (b) What is the *average* force exerted by the racquet on the ball? (c) What is the *maximum* force exerted by the racquet on the ball?

11. What thrust is required to accelerate a 30 kg rocket sled to a speed of 30 m/s in 5 s, starting from rest?

12. A 5.0 kg rifle, suspended on strings, fires a 4.0 g slug with a muzzle velocity of 520 m/s. What is the recoil velocity of the rifle?

13. Four people line up on a long, giant skateboard. Each one, in turn, makes a running jump off the back end of the skateboard, with a speed of 2 m/s relative to the board. Each person has a mass of 60 kg. The skateboard has a mass of 10 kg. What is the final velocity of the skateboard? (The problem must be worked in steps. At each jump, momentum of the system must be conserved.)

14. Suppose that you are coasting on your bike at 5 m/s with a combined mass of 70 kg. You lean down and snatch up a 15 kg knapsack that was on the ground. What is your resulting speed? If the pickup takes 0.1 s, what is the average force exerted by your arm? (Don't forget, you have to lift the knapsack as well as drag it forward.)

15. Suppose you are coasting on your bike at 5 m/s, as in Problem 14, but this time you are holding the 15 kg knapsack. Your combined mass is therefore 85 kg. If

you now drop the knapsack (which hits the road and stops), what is your resulting speed?

16. The thrust of a particular rocket just after lift-off is 1.8×10^6 N. If the exhaust velocity is 2500 m/s, at what rate (mass per unit time) are the combustion products being expelled?

17. Machine gun bullets are fired with a velocity of 800 m/s at a steel target. Each bullet has a mass of 20 g. They are fired at the rate of 20 per second. If each bullet bounces back with a speed of 200 m/s, what is the time average force exerted on the target by the burst of bullets?

18. A skyrocket, traveling straight up with a speed of 100 m/s, bursts into three pieces. Two pieces, each with a mass of 0.5 kg, end up traveling horizontally—one to the east and the other to the west. What is the velocity of the third piece, which has a mass of 1.0 kg?

19. A skyrocket, traveling straight up with a speed of 100 m/s, bursts into three pieces. One piece, with a mass of 0.5 kg, ends up traveling horizontally to the east with a speed of 20 m/s. Another piece, with a mass of 0.5 kg, ends up traveling at 45° to the horizontal and to the east. (All three pieces are thus in the east-west vertical plane.) What are the velocities of the second and third pieces?

20. You are sitting at rest in an airplane seat. Suddenly you start exerting a force of 210 N on the back of your seat without exerting any forward forces. If your mass is 70 kg, what is your velocity after 15 s?

21. For the passenger in a car accident, the *change* of velocity is a crucial factor. Suppose that on an icy road a station wagon with a mass of 2000 kg, traveling at 30 mph, runs head-on into a 1000 kg compact that was traveling at 30 mph in the opposite direction. The cars lock together. Assume that on the icy road they slide without friction. What is the change of velocity of each car?

ANGULAR MOMENTUM 8

A comet in its elongated orbit moves very slowly until it approaches the sun. Then it goes faster and faster, dashes around the sun, and finally loses speed as it retreats to the far reaches of the solar system. A cyclist, riding no-hands, *leans* to the right, and instead of tipping over, *turns* to the right. An ice skater extends her arms and starts twirling on the point of one blade. She lowers her arms and suddenly spins much faster. All these things happen because a quantity called *angular momentum* must be conserved.

Angular momentum plays the same role for rotational motion that linear momentum plays for motion in a line. However, angular momentum is responsible for some special, startling effects. Spinning objects, such as tops, seem to move in directions perpendicular to what you would at first expect. The conserved nature of angular momentum permits certain interactions among subatomic particles and forbids others. Strangest of all, angular momentum is quantized, existing only in multiples of small units, much like the situation with electric charge. A rotating system cannot increase or decrease its angular momentum continuously, but only in unit amounts that are integral multiples of a basic unit.

HANDLING THE PHENOMENA

1. If you are a good ice skater or ballet dancer, you may already have experienced the phenomenon of angular momentum conservation. If you can't spin on your toes, get a rotating chair or stool. If you can't get a rotating chair, look at the illustration and see how easy it is to start rotating and then spinning on one toe. In any case, get two sizable weights—one for each hand. Try to get something with a mass of at least 1 kg; a quart of almost anything will do. On your seat or on one toe, start spinning with the weights held in your outstretched hands. Then, abruptly bring your hands and the weights down close to your body. You have to experience the effect to believe it; watching someone else is no good. As we will learn later in this chapter, your angular *velocity* increased because your angular *momentum* was conserved!

2. There's another way to feel angular momentum. Get a bicycle and hold it by the front handle so that the front wheel is off the floor. Have

someone else spin the wheel fast, and then feel what happens as you try to turn the wheel to the left or right or tilt it. In particular, make a note of the direction of rotation of the wheel (right-hand rule), and what direction the effect is in when you try to turn the wheel one way or the other. Angular momentum is a vector quantity. The conservation law applies not only to its magnitude, but also to its direction.

THE FORMS OF ANGULAR MOMENTUM

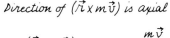

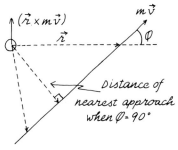

An object barreling along in a straight line has momentum (**mv**). With respect to some axis that it is passing, it also has *angular* momentum equal to **r** × (**mv**). The cross product of the two vectors simply yields the product of (mv) and the distance of nearest approach, as shown in the diagram. So far this may seem like a useless definition. Just because an object is going past an axis does not mean that any rotational motion exists. However, if the object ran into a paddle of a wheel extending into its path, it would start the paddle wheel spinning. Therefore, it is consistent to say that angular momentum already existed in the complete system—whether or not it gets transferred from the projectile to the paddle wheel. Note that if the object is headed straight for the axis, it has no angular momentum about that axis.

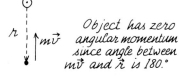

> ### Question 8-1
>
> Doesn't the angular momentum change continuously as an object approaches and passes an axis? Note that the mass and velocity stay constant, but the distance, r, to the axis changes steadily.

In the case of a spinning bicycle wheel, the magnitude of the angular momentum about its own axis is approximately (mvr). To get this value, we make the approximation that all of the mass is at the radius, r. The tangential velocity, v, of each part of the rim is automatically perpendicular to the radius. In Chapter 6, we saw alternative ways of expressing rotational motion. The angular velocity, ω, is equal in magnitude to v/r, and the moment of inertia of a system, where all the mass is at the same radius, is $I = mr^2$. Therefore, the angular momentum becomes

$$mvr = m(r\omega)r = (mr^2)\omega = I\omega$$

This last form is valid, regardless of how the mass of the rotating system is distributed.

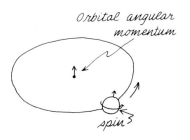

There is a fundamental difference between the angular momentum of an object rotating around its center of mass, and the angular momentum of the object as it goes past some axis point. The former is frequently called *spin*, to differentiate it from the angular momentum caused by rotation around or motion past an external axis. An object's spin is a feature of the object itself, without regard to any outside axis. In general, however, the magnitude of any angular momentum depends on the axis point chosen. For instance, the angular momentum of a bicycle wheel around its axis—its spin—depends only on the radius of the wheel, the mass and mass distribution of the wheel, and the speed. For some purposes, however, it is convenient to consider the angular momentum of the wheel around the momentary axis where it touches the road. That angular momentum is larger than the spin, because it includes not only the spin but also the angular momentum of the center of mass of the wheel going past the momentary axis.

In Chapter 6, Newton's second law for rotational motion was given as $\tau = I\alpha$. In the same way in which the more general form of [$\mathbf{F} = m\mathbf{a}$] is [$\mathbf{F} = \Delta(\mathbf{mv})/\Delta t$], the more general form of Newton's second law for rotation is $\tau = \Delta(\mathbf{I\omega})/\Delta t$. The torque applied to a rotating system is equal to the time rate of change of angular momentum.

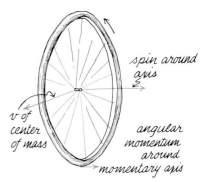

MAGNITUDES OF SOME ANGULAR MOMENTA

Objects appear to be spinning or in orbit at every range of magnitude, from the subatomic realm to the domain of the galaxies. Let's calculate the angular momenta of objects in three different ranges of size. We should compare these angular momenta in terms of some common unit. Although, as we will see, nature has a basic unit of angular momentum, no special name is given to the S.I., or humanmade unit. In terms of kilograms, meters, and seconds, the unit angular momentum is $1 \text{ kg} \cdot \text{m}^2/\text{s}$. These same units appear whether you define angular momentum as (mvr) or ($I\omega$).

1. Let's start with an object that is human size. An ordinary bicycle wheel has a mass of about 2 kg and a radius of 30 cm. If we make the approximation that all of the mass is at the rim, then a wheel traveling at 10 mph ($4\frac{1}{2}$ m/s) has a spin (an angular momentum about its axis) of $mvr = (2 \text{ kg})(4.5 \text{ m/s})(0.30 \text{ m}) = 2.7 \text{ kg} \cdot \text{m}^2/\text{s}$.

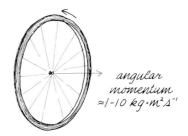

2. In the Bohr model of the atom, proposed by Niels Bohr in 1913 and still very useful, the valence electron in an atom travels in a circular orbit around the central nucleus. The mass of an electron is about 1×10^{-30} kg; the radius of an atom (and therefore the radius of the orbit of the outer electron) is about 1×10^{-10} m; and the speed of the electron is about 1×10^6 m/s. (Later in the text we will show how to calculate or measure these quantities.) *On the basis of this model,* the angular momentum of such an electron is

$$mvr = (1 \times 10^{-30} \text{ kg})(1 \times 10^6 \text{ m/s})(1 \times 10^{-10} \text{ m}) = 1 \times 10^{-34} \text{ kg} \cdot \text{m}^2/\text{s}$$

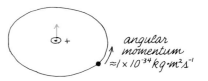

This seems like an unimaginably small quantity. Compare it with the angular momentum of a bicycle wheel. As we will soon see, however, angular momentum in such tiny amounts plays a controlling role in the behavior of atoms.

3. Now let's calculate the angular momentum of something big. The earth spins on its axis with a frequency of 1 revolution per day. *If* the earth's density were uniform (it isn't), the moment of inertia would be $I = (2/5)MR^2 = (0.4)MR^2$. The actual moment of inertia is $(0.3444)MR^2$. (The earth is more dense toward the center.) The angular momentum is $I\omega = (0.344)(5.98 \times 10^{24} \text{ kg})(6.38 \times 10^6 \text{m})^2(2\pi \text{ rad})/(24 \text{ hr})(1 \text{ hr}/3600 \text{ s}) = 6.1 \times 10^{33}$ kg·m²/s.

This quantity of angular momentum is about as large in one direction as the size of angular momentum of the valence electron is small in the

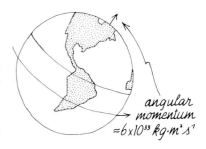

other direction. Actually, as we will see, this angular momentum of the spinning earth is small compared with other angular momenta in the solar system.

THE CONSERVATION OF ANGULAR MOMENTUM

For a closed system, with no external torques, the change of angular momentum, $\Delta(\mathbf{I}\omega)$, must equal zero.

> **Question 8-2**
>
> If you are sitting on a rotating stool, why can't you shove against the rim of the seat, thus exerting a torque to get you and the seat into rotation?

Angular momentum is a conserved quantity. That does not mean, however, that isolated rotating objects always have the same angular *speed*. Here are four examples that illustrate this peculiar nature of rotational motion.

1. The orbits of the planets (and of most earth satellites) are not really circles, but ellipses. The eccentricity of the earth's orbit around the sun is so small that the elliptical nature can hardly be shown in a page-size drawing. The ratio of the largest diameter to the shortest diameter differs from unity by only $1\frac{1}{2}$ parts per 10^4. Nevertheless, the difference between aphelion radius, when we are furthest from the sun, and perihelion radius, when we are closest, is about 3%. This is because the focal point of the ellipse, which is the sun's location, is off center by about $1\frac{1}{2}\%$ of the radial length. What happens to the angular momentum in a noncircular orbit? If the radius is not always perpendicular to the velocity, perhaps the sun exerts a torque on the earth and changes its angular momentum.

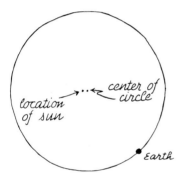

To simplify this problem, let's assume that the effect of the other planets is negligible, and that sun-earth system can be considered to be an isolated two-body system. The sun does, indeed, exert a force on the planet, but it is a gravitational force directed along the radius, connecting the planet to the sun. The torque on the planet with respect to the sun as axis is $|\boldsymbol{\tau}| = |\mathbf{r} \times \mathbf{F}| = |\mathbf{r}| |\mathbf{F}| \sin \theta$. Remember that θ is the angle between the radius arm and the force vector. Torque is maximum when θ is 90°; it is zero when θ is 0° or 180°. For any *radial* force, $\mathbf{F}$ lies along $\mathbf{r}$, either in the positive or negative direction, and therefore θ is 0° or 180° and the torque is zero. Consequently, there can be no change in angular momentum. This is true for the orbits controlled by any central force. In the case of gravitational force, which is proportional to $1/r^2$, the allowed orbits can be circles, ellipses, parabolas, or hyperbolas.

As you can see in the next diagram, which shows a very eccentric ellipse, the angular momentum of a planet is not just (mvr), where v is the tangential velocity and r is the radial distance between sun and planet. The definition of angular momentum is $\mathbf{r} \times (\mathbf{mv})$, and that is the product of the radius and the component of the tangential momentum that is perpendicular to the radius. The amplitude of the angular momentum = $m |\mathbf{v}| |\mathbf{r}| \sin \phi$. At all points of the orbit, $\mathbf{v}$, $\mathbf{r}$, and ϕ must change in such a way that angular momentum stays constant. At aphelion and perihelion, the velocity and radius are perpendicular to each other: $\phi = 90°$. In those cases, the angular momentum is simply (mvr). Since that product must be the same at both points, at aphelion the tangential velocity of the earth must be 3% smaller than it is at perihelion.

A geometrical consequence of the conservation of angular momentum was discovered by Kepler back in 1609. He not only figured out that the orbits of the planets are ellipses, but he also found from his observations that the radius arm from sun to planet sweeps out equal areas in equal times. The derivation of this rule (which had to wait until Newton) is shown in the diagram. As you can see, when a planet is far away from the sun, its velocity is smaller than when it is close. In the case of comets, this effect is very dramatic. The eccentricity of their orbits is usually very large. For most of its period a comet moves slowly at great distances from the sun. When it does finally approach the sun, its speed gets greater and greater, thus conserving its angular momentum.

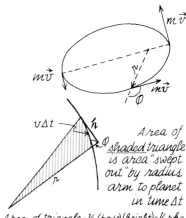

Area of shaded triangle is area "swept out" by radius arm to planet in time Δt

Area of triangle = ½(base)(height) = ½rh =
= ½$r(v\Delta t)\sin\phi$ = $\frac{\Delta t}{2m}[r(mv)\sin\phi]$ =
= $\frac{\Delta t}{2m}$ [angular momentum]

Area swept out in unit time = $\frac{area}{\Delta t}$ =
= ½m[angular momentum] = constant

2. Now we must explain how skaters can twirl so fast, and what happened to you when you tried to spin on one foot while holding the quart jars or books. The person starts turning with arms outstretched. Once on tiptoe, the whirler is not subject to much external torque; therefore, angular momentum must be nearly conserved. The diagram shows a crude model of the situation. The trunk of the body has become a solid cylinder with a mass of 60 kg and an outer radius of 15 cm. The arms and weights turn into lumps at a radius of 80 cm, with a total mass of 4 kg. The total angular momentum of the model is

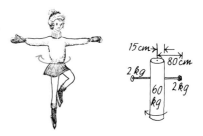

$$[\tfrac{1}{2}(60 \text{ kg})(0.15 \text{ m})^2]\omega + [(4 \text{ kg})(0.80 \text{ m})^2]\omega = [0.68]\omega + [2.6]\omega$$

Note that the angular momentum of the much lighter arms and hands is about four times as large as the angular momentum of the core body. If the original frequency is 1 revolution every 2 seconds, $\omega = 3.1$ rad/s. The total angular momentum is $(3.3 \text{ kg} \cdot \text{m}^2)(3.1 \text{ rad/s}) = 10 \text{ kg} \cdot \text{m}^2/\text{s}$.

Next the twirler brings her arms and the weights down beside her body. The angular momentum stays the same, but the radius of the arms and weights is now greatly reduced.

$$10 \text{ kg} \cdot \text{m}^2/\text{s} = (0.68)\omega + [(4 \text{ kg})(0.15 \text{ m})^2]\omega$$
$$= (0.68)\omega + (0.09)\omega = (0.77 \text{ kg} \cdot \text{m}^2)\omega$$
$$\omega = 13 \text{ rad/s}$$

The angular frequency has increased by over a factor of 4. The twirler is now going over 2 revolutions per second.

Question 8-3

Does this explanation of a skater's twirl also explain how a diver can do a somersault, or how a cat can always land on its feet? Where do the diver and the cat get their original angular momentum?

3. Rotational motion is one of the most common and striking features of the universe. Planets, moons, and stars spin on their axes; the planets have moons and the sun has planets revolving around it; a large fraction of stars appear to be binaries—rotating double stars with a complete range of size of partners; the stars and their satellites revolve around the centers of their galaxies; and many galaxies are parts of swirling clusters of other galaxies. On a more homely scale, you have probably seen dust devils of wind pick up dust or leaves and swirl them into the air as if there were a miniature tornado. In rowing a boat, or in swimming, you may have noticed tiny whirlpools spinning off in the wake. Most of these rotations start with motion of material along straight lines. What makes everything swirl?

We gave an example earlier of why an object hurtling past an axis really has angular momentum with respect to that axis. If the object hits the paddle of a wheel centered on the axis, the wheel will start turning. If two currents of fluid, either air or water, are flowing side by side in opposite directions, they have angular momentum with respect to each other. If the conditions of viscosity and relative velocity are sufficient, whirlpools will be generated at the boundary between the two currents.

On a cosmic scale, if atoms were uniformly distributed in space and motionless to begin with, they would all fall toward the center of mass along straight, radial lines. Eventually they would collide and form a star with no rotational motion at all. However, the original condition of uniform distribution of the raw materials just never occurs. In a region of the galaxy where a star is going to form, the atoms are distributed with varying density and varying initial velocities. The motions of the atoms are not radial toward the center of mass and, therefore, angular momentum exists. As the atoms condense into a star, the radius of the system shrinks from a distance of a light-year or so to the star's final size of about 10^9 m, a reduction by a factor of 10^7. You have felt what happens as you rotate and change the radial distribution of a small fraction of your mass. If a new star had no companion or satellite system to absorb the angular momentum, in most cases it would be left spinning so fast that it would not be stable.

Let's see where the angular momentum in our solar system resides. We have already calculated the spin angular momentum of the earth. The sun is also spinning on its axis, although since it is not a rigid body the spin period varies with latitude. The period for the equatorial region is $24\frac{2}{3}$ of our days, and at a latitude of $75°$ the rotation period is about 33 days. The assumption that the sphere has uniform density is even less accurate for the sun than for the earth. The density of the sun as a whole is 1.4×10^3 kg/m^3; the density at the center must be about 100 times greater. If we assume that for such a body the moment of inertia is $I = (0.1)MR^2$, then the angular momentum of the sun is about $I\omega = (0.1)(2 \times 10^{30}$ kg$)(7 \times 10^8$ m$)^2$ $(2\pi$ rad$)/(25$ days$)(1$ day$/8.6 \times 10^4$ s$) = 3 \times 10^{41}$ kg·m^2/s. The absolute value of this enormous quantity is unimportant; just compare it with the angular momentum of the spinning earth, which is smaller by a factor of 5×10^7. But these quantities are still much smaller than the *orbital* angular momentum of the planets.

THE CONSERVATION OF ANGULAR MOMENTUM

Question 8-4

To one significant figure, calculate the orbital angular momenta of earth and Jupiter.

$$M_{earth} = 6 \times 10^{24} \text{ kg} \qquad M_{Jupiter} = 1.9 \times 10^{27} \text{ kg}$$

$$\text{Orbital radius}_{earth} = 1.5 \times 10^{11} \text{ m}$$

$$\text{Orbital radius}_{Jupiter} = 7.8 \times 10^{11} \text{ m}$$

Period of Jupiter's revolution = 11.9 earth years
1 earth year = 3×10^7 s

As you can see, Jupiter in its orbit possesses much more angular momentum than the spinning sun itself. The masses and orbital radii of the other planets are given in the table. Is it possible that some other planet, perhaps Neptune with its great orbital radius, carries more angular momentum than even Jupiter?

Planet	Distance from Sun (10^6 km)	Mass (10^{24} kg)	Period in Days
Mercury	57.9	0.3	88
Venus	108.1	4.9	225
Earth	149.5	6.0	365
Mars	227.8	0.6	687
Jupiter	777.8	1900	4333
Saturn	1426	569	10760
Uranus	2868	87	30690
Neptune	4494	103	60190
Pluto	5908	5.4	90740

Question 8-5

There is an easy way to compare the angular momenta of the planets without doing the detailed calculation. According to one of Kepler's laws that we derived in Chapter 6, the relationship between the period and the orbital radius of a satellite in a system is $T^2 = K(r^3)$. Use this relationship, and data from the table, to compare the angular momenta of Neptune and Jupiter.

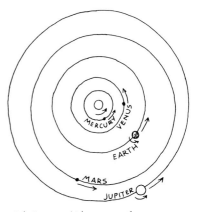

Neither orbits nor planets are drawn to scale in this diagram. Compare the scale with that of the solar system map you made in Handling the Phenomena in Chapter 1.
(Or, if drawn to scale with $r_{Jupiter} = 5$ cm, $r_{Mars} = 1.46$ cm, $r_{Earth} = 0.96$ cm, $r_{Venus} = 0.69$ cm, $r_{Mercury} = 0.37$ cm)

As viewed from our North Pole, all of the planets revolve about the sun in a counterclockwise direction. Their orbits are all approximately in the same plane, and the orbits of most of their moons are also in that plane. Pluto's orbit is the most out of line. It is inclined at an angle of 17° to the plane going through the sun's equator, and Mercury's orbit is inclined at an angle of 7°. The axes of most, but not all, of the planets are perpendicular to the rotation plane, and their spins are mostly in the counterclockwise direction. Most of the angular momenta therefore add up arithmetically. Presumably, the sum is the angular momentum of the original condensing cloud of atoms that formed the solar system.

The spin angular momentum and orbital angular momentum of a satellite can be linked through tidal forces with the parent body. The tidal motions (within the solid sphere as well as in any surface oceans) drain energy from the spinning body, reducing the spin. In order to conserve angular momentum, the distance between satellite and parent increases, thus increasing the *orbital* angular momentum. Apparently this happened long ago to our moon, because now the period of the moon's spin is the same as its period of revolution. The motions are locked so that we always see the same face of the moon. The earth's spin continues to be drained by the friction of the tides. The day is lengthening by 0.0016 second per century and the moon-earth distance continues to increase in order to conserve angular momentum.

4. The prime evidence for the conservation of linear momentum came from the collisions of two objects. Is there an equivalent form of rotational collision? If you have two old records that are already scratched, you can easily arrange such a collision. Get a thread or string, at least a meter long, and tie one end to a large paper clip. Thread the other end of the string through the center holes of the two records, and then suspend the array from some overhead support. Tape a small piece of paper to the rim of the bottom record, so that you can time its rotation. Most likely the records will not hang in a horizontal plane, but will tip. Get someone else to hold one of the records some distance up the thread, while you give the bottom record a spin in the horizontal plane. Immediately you will see one effect of angular momentum. The spinning record will remain in the plane in which you spun it. You have a gyroscope. (With only the one record on the thread, you can let the whole system swing back and forth like a pendulum. The spinning record will remain flat—which looks remarkable.)

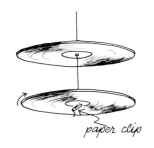

Time several revolutions of the bottom record with a watch or by counting steadily. Then have your assistant drop the upper record a short distance onto the bottom one. There will be some grinding of the grooves, but then the two will spin together. Immediately time the rotation period for the two records together. Regardless of what else is lost during the collision (and, as we will see in the next chapter, energy is lost), angular momentum must be conserved. Since the two records have equal moments of inertia, there must be a definite relationship between the original and final periods.

Question 8-6

What is the relationship?

If the records rotate too many times in one direction, they will wind up the string so much that the restoring torque of the string will interfere with your observations. To avoid this, use a long string, and reverse direction of the spin each time.

THE DIRECTIONAL NATURE OF ANGULAR MOMENTUM

When we defined the direction of angular velocity in Chapter 6, it was more a matter of convenience than for actual use. After all, for most purposes you can describe a rotation perfectly well by calling it clockwise or counterclockwise as seen from a particular direction. Now we will see a practical use for saying that the appropriate direction for torque, angular velocity, and angular momentum is along the axis of rotation. Without such a description, it's hard to explain the motion of ordinary tops.

You can create a dramatic example of a kind of top, or gyroscope, with a bicycle wheel. If you can liberate the front wheel from a bike, support one end of the axle with a string, as shown in the illustration. Let go of one side of the axle and let the wheel be supported only by the string on the other side. Immediately there is an unbalanced torque acting on the wheel. Gravity exerts a force equal to the weight of the wheel, which acts downward through the center of the wheel. The radius arm of the torque is the distance between the center of the wheel and the suspension point of the string.

> Question 8-7
>
> What is the direction of the torque and what happens to the wheel?

Now start out the same way, but this time get somebody else to spin the wheel before you let go. There will still be the same unbalanced torque due to gravity, but the wheel won't tip over and fall. Instead, it will remain essentially vertical, and the whole wheel will rotate slowly in a horizontal circle around the string. Why doesn't it tip over, and where did it get the angular momentum to go in a horizontal circle?

For a qualitative explanation, consider what happens to the point P at the top of the wheel. As seen from above, the top of the wheel is tipping over to the right. Consequently, P receives an impulse to the right. However, P is also moving rapidly forward (as we look down on the wheel). The tipping impulse provides a small momentum to the right that adds vectorially to the main forward momentum of P. The resultant momentum is in the direction of a turn to the right for the whole wheel. Instead of tipping over, the wheel, and its plane of rotation, turns to the right. (Try analyzing the similar reactions for the point at the bottom of the wheel, as well as for the case where the wheel is rotating in the opposite direction.)

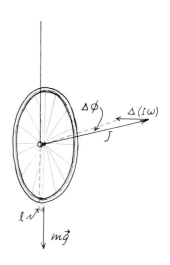

In the diagram we represent the angular momenta and torques with arrows along the axial directions. The largest angular momentum in the system is that of the spinning wheel, and that is in the direction of the horizontal axle. The torque due to gravity is in the horizontal plane also, but is perpendicular to the axle. During each interval of time, the torque produces a change of angular momentum: $\tau(\Delta t) = \Delta(I\omega)$. That change of angular momentum is perpendicular to the main angular momentum of the wheel and so does not change its magnitude. But the torque does change the *direction* of the angular momentum of the wheel. The axle moves in the horizontal plane and continues to do so as long as the gravitational torque continues. This type of rotation is called *precession*.

Let's find the angular velocity of the bicycle wheel's precession. So as not to get confused with the angular velocity of the spinning wheel itself, call the wheel's angular momentum, J. We'll assume that the magnitude of J stays constant for the duration of our experiment—no friction in the axle bearing or from the air. Then, as the diagram shows, a small change in the precession angle is given by

$$\Delta\phi = \frac{\Delta(I\omega)}{J} = \frac{\tau(\Delta t)}{J} = \frac{mgl(\Delta t)}{J}$$

The precession frequency is

$$\Delta\phi/\Delta t = (mgl)/J$$

Question 8-8

To a first approximation, the angular momentum of the spinning wheel is just equal to $(mr^2)\omega$. The approximation proposes that all of the mass, m(2 kg), is at the outer radius, r(30 cm). Assume some reasonable value for ω, and for the support length on the axle, and calculate the precession frequency. Does it reasonably agree with what you observe with your own bicycle wheel?

The phenomenon of precession is actually more complicated than the simple explanation we have given. Note that the wheel also has angular momentum in the *vertical* direction, because it is precessing around a vertical axis. Where did *that* component of angular momentum come from? You can see the answer qualitatively by noticing that, even though the axle starts in the horizontal plane, it dips below it as it goes around. Consequently, part of the angular momentum of the axle is vertical.

Spinning objects react to torques in directions perpendicular to what we would naively expect. On the other hand, an experienced bike rider makes use of the effects almost intuitively. If you are riding no-hands, and want to turn right, you *lean* to the right. Instead of tipping the front wheel over to the right, that direction of torque *turns* the wheel to the right. This gyroscopic effect must be taken into account in flying a single-engine plane. If you try to climb abruptly, the resultant *horizontal* torque applied to the propeller will make the plane swing sideways.

Some years ago a popular song called prematurely for the dawning of the Age of Aquarius. Aquarius is one of the twelve astrological "houses," or constellations, marking the ecliptic (sun's path) in the sky. Each 24 hours the sun, moon, and stars appear to rotate once around the earth, all moving from east to west. However, as the earth makes its annual trip around the sun, the sun appears to move *eastward* with respect to the stars— about 1° per day. In the course of one year the sun moves through each of the twelve regions, or "houses," along the ecliptic. The astrological "age" depends on which house the sun is in at the beginning of spring (the vernal equinox). That position changes slowly from one year to the next. Starting some hundreds of years from now, each spring the sun will be entering the ill-defined region called Aquarius.

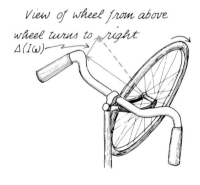

View of wheel from above
wheel turns to right
$\Delta(I\omega)$

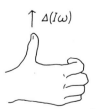

↑ $\Delta(I\omega)$

Tipping torque caused by leaning to right produces torque in horizontal, forward direction.

The reason the sun's position on a given date keeps changing is because our spinning earth precesses. The tilt of the axis remains the same, but the direction of the axis slowly changes. When spring comes now, we are always on a particular side of the sun (with respect to the stars). Eventually spring will come when we are on the opposite side of the sun. The axis now points toward Polaris, our North Star, but it didn't point in that direction when the pyramids were built. The period of precession is about 26,000 years—about 2200 years per house.

Precession requires a torque. In the case of a spinning toy top, the torque comes from the weight of the top acting through the lever arm that exists because the top is tilted. In the case of the earth, the torque must come from the pull of the sun trying to tilt the earth in some way. If the earth were truly spherical, there would be no lever arm and no torque—you can't "straighten up" a sphere. However, the earth is not quite spherical. There is a slight bulge around the equator, no doubt produced by the centrifugal force as the spinning earth formed. The equatorial radius is 6378.2 km, and the polar radius is 6356.7 km. This slight difference of 21.5 km is enough to provide a girdle, which the sun can twist. There is a greater force on the bulge toward the sun than on the side opposite. Instead of straightening up the earth, the resulting torque makes the spinning earth precess.

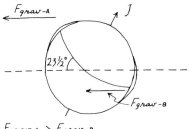

$F_{grav-A} > F_{grav-B}$

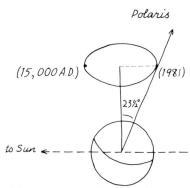

Axis precesses through circle on celestial sphere with radius of 23½°.

THE QUANTIZATION OF ANGULAR MOMENTUM

So far as we know, the length of an object can be as large or as small as we choose. No minimum length is known, not even for the subatomic particles. The same condition holds for mass and time. There are no basic units, m and t, such that some other mass or time must be exactly $1083m$ or $2t$. As we will see, our measured values of length, mass, and time for objects depend on our velocity with respect to those objects. The standard kilogram itself would have a larger mass if it went hurtling by us, and we could change that mass by an infinitesimal amount just by changing the velocity slightly. Consequently, length, mass, and time cannot be integral multiples of some small basic unit; they are not quantized.

On the other hand, the angular momentum of a system is quantized. The basic, ultimate unit of angular momentum is called *Planck's constant*. When the frequency is given in terms of cycles or revolutions per second, Planck's constant is

$$h = 6.626 \times 10^{-34} \text{ kg} \cdot \text{m}^2/\text{s}$$

We will use Planck's constant in this form when we study the energy of photons. For now, however, we have been describing angular velocity in terms of radians per second. In those units, Planck's constant is called *h-bar*:

$$\hbar = h/2\pi = 1.054 \times 10^{-34} \text{ kg} \cdot \text{m}^2/\text{s}$$

That's a very small amount of angular momentum. We found a quantity about that size when we calculated the orbital angular momentum of a valence electron in an atom. The order-of-magnitude calculation on page 161 yielded the value 1×10^{-34} kg·m² s. Our calculation was based on a very crude, and now rather old-fashioned, model of the atom. The remarkable thing is that such a model, picturing electrons in planetlike orbits, should work so well. In many cases, the subatomic particles act as if they were just tiny versions of the large-scale objects that we can handle. In many other cases, however, such mechanical models break down, and we have to describe the particles in more exotic terms.

As an example of how a mechanical model for subatomic particles sometimes works and sometimes doesn't, consider the analogy between electrons in atoms and planets in orbit. The planets not only revolve in orbit; many of them rotate on their own axes. It turns out that many of the subatomic particles also act as if they were spinning on their own axes. However, their spin angular momentum can only be zero, or $\frac{1}{2}\hbar$, or $1\hbar$, or $\frac{3}{2}\hbar$, or $2\hbar$, and so on. Immediately there appears to be a contradiction! How can a particle have half of the basic unit? Perhaps the basic unit is $\frac{1}{2}\hbar$. Actually, the quantization rule is that angular momentum must be either integral or half integral, but cannot *change* except in integral multiples of $\hbar$.

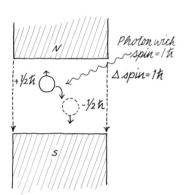

The angular momentum of a particle or a system of particles must have a particular direction, usually determined by the direction of electric or magnetic fields. For instance, an electron may have its spin aligned with a magnetic field and then suddenly flip so that it is lined up in the opposite direction. At all times the electron keeps the magnitude of its angular momentum of $\frac{1}{2}\hbar$, but the *change* of angular momentum is one whole unit of $\hbar$—from $+\frac{1}{2}\hbar$ (up) to $-\frac{1}{2}\hbar$ (down).

We know that an electron or an atom can't just change its angular momentum without something else happening. Some other particle must be given a recoil angular momentum. Angular momentum must be conserved. Frequently when an atomic system changes angular momentum, the recoil is carried off by a photon—the particle that is the messenger of the electromagnetic interaction. Ordinary visible light is composed of photons. They not only carry linear momentum, but also angular momentum. Both effects have been demonstrated in very direct mechanical experiments. If an intense light is reflected from one side of a tiny vane suspended in a high vacuum, the vane will recoil, thus demonstrating that light carries linear momentum.

In an ingenious experiment in 1935, Richard Beth produced rotation in a disk by shining polarized light along the axis of the disk. (Technically, the light was circularly polarized, which means that the photons were selected so that the direction of the angular momentum of all of them was the same. In ordinary, untreated light, there is a 50% chance that the angular momentum of a photon will be in the direction of travel, and a 50% chance that it will be in the opposite direction.)

Question 8-9

If the angular momentum of any rotating system must equal $n\hbar$, where $\hbar$ is Planck's constant and n is an integer, about what is the value of n for a bicycle wheel traveling at 10 mph ($4\frac{1}{2}$ m/s)? Assume that the wheel has $m = 2$ kg and $r = 30$ m.

SOME SPECIAL POINTS OF INTEREST

Evidently, the quantization of angular momentum has no effect on the motion of things that we handle in everyday life. If Planck's constant were larger by a factor of 10^{34}, a bicycle wheel would behave very strangely. It could rotate only at a certain speed, or twice that speed, or three times that speed, and so forth, and its axis could change direction only abruptly and only by certain amounts.

SOME SPECIAL POINTS OF INTEREST

Radius of Gyration

The vector sum of all the forces acting on a system of objects will accelerate the center of mass of the system as if all the mass were located there and the resultant force were applied there. However, as we saw on page 80, there may also be rotation of the system, and that is not described by assigning all the mass to the center-of-mass point. For instance, the center of mass of a wheel rotating on its axis is right at the center of the wheel. If you throw the wheel, no matter how it tumbles, the center of mass will move in a parabolic trajectory. In order to describe the rotation around the axis, however, the mass of the wheel must be considered to be distributed at various radii.

It is sometimes convenient to describe a radius for a rotating object where all the mass could be concentrated without affecting the angular momentum. This distance is called the *radius of gyration, j*.

For any rotating object, the angular momentum is given by $I\omega$, where I is the moment of inertia. For a system composed of parts at various radii, $I = \Sigma m_i r_i^2$. As we saw on page 120, the moment of inertia of common objects can be written as $I = KMR^2$. The total mass is M, the outside radius is R, and K is a constant that depends on the geometrical distribution of mass of an object with a particular shape. We can produce the same moment of inertia by rotating a point object or a hoop with mass M at a distance j from the axis. To find this distance j, the radius of gyration, we equate the moments of inertia of the two systems:

$$KMR^2 = Mj^2$$

Evidently, $j = \sqrt{K}\, R$.

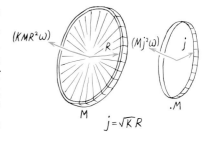

Question 8-10

What is the radius of gyration of a solid disk, 10 cm in radius, rotating around its axis?

Parallel Axis of Rotation

On page 121, there is a table of values for the moments of inertia of some common objects. Notice that there is a difference between the moment of inertia of a wheel around its axis and its moment of inertia as it turns around the instantaneous axis, where the rim is touching the ground. Of course, the latter moment of inertia is larger because some of the mass is much farther away from the axis than in the first case. Frequently, it is easy to calculate the moment of inertia around an axis going through a point of symmetry, such as the center of mass, but harder to do the calculation for any other axis. There is a simple formula for finding the moment of inertia around any axis parallel to the center-of-mass axis and at a distance ℓ from it:

$$I = I_{cm} + M\ell^2$$

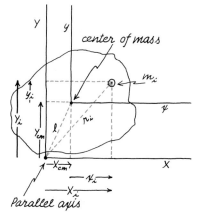

Parallel axis

■ Optional

Here is the proof of the "parallel axis" theorem. The moment of inertia of the object shown in the diagram is

$$I = \Sigma m_i r_i^2 = \Sigma m_i (X_i^2 + Y_i^2)$$

The x- and y-locations of particle i can be written in terms of the x- and y-location of the center of mass:

$$X_i = X_{cm} + x_i \qquad Y_i = Y_{cm} + y_i$$

Substitute these values into the equation for I:

$$I = \Sigma m_i (x_i^2 + y_i^2) + \Sigma m_i (X_{cm}^2 + Y_{cm}^2) + 2X_{cm}\Sigma m_i x_i + 2Y_{cm}\Sigma m_i y_i$$

Each of the last two terms is zero, because the sum of the $m_i x_i$ to the right of the center of mass cancels the sum of the $m_i x_i$ to the left of the center of mass. The same argument applies to the y-components. In the second term in the equation, $(X_{cm}^2 + Y_{cm}^2)$ is just the distance squared from the center of mass to the parallel axis around which the rotation is taking place. Call that distance l. The first term in the equation is the moment of inertia around an axis through the center of mass. Therefore, the moment of inertia of an object around an axis parallel to the center of mass, and a distance of l from it, is

$$I = I_{cm} + M\ell^2$$

■

Question 8-11

What is the moment of inertia of a bicycle wheel with respect to the point of contact with the road? Assume that all of its mass, m, is located along the circumference at a radius, r, from the center.

The Physical Pendulum

The *simple* pendulum, with which we have worked so far, consists ideally of a bob with its mass concentrated at a point, swinging on a massless string. Any solid object supported from a point above its center of mass will also swing back and forth with a pendulum motion. However, the relationship between its period and its length is different from that of the simple pendulum.

SOME SPECIAL POINTS OF INTEREST

The moment of inertia of a solid swinging object, called a "physical pendulum," would be the same if all of its mass were concentrated at its radius of gyration. Perhaps its period is the same as the period of a simple pendulum with a string as long as the radius of gyration? It's not that simple. The trouble is, placing the mass at the radius of gyration produces the right moment of rotational inertia, but the restoring torque provided by gravity acts through the *center of mass*—which is at a different radius. Here is the proper analysis:

simple pendulum   Complex pendulum

$$\tau = I\alpha$$
$$(mg \sin \theta)\rho = I(a/\rho)$$

In the substitution, the component of the weight—acting perpendicular to the line through the axis and the center of mass—is ($mg \sin \theta$). The distance from the axis to the center of mass is ρ (the Greek letter, rho). The acceleration of the center of mass along the arc is a, which equals $\alpha\rho$. The distance along the arc from the equilibrium line is called, s. Remember that for small angles, $\sin \theta \approx \theta = s/\rho$. The pendulum equation then becomes

$$mgs = I(a/\rho) \rightarrow a = mgs\rho/I$$

The restoring force on the center of mass is $F = ma = -(m^2 g\rho/I)s$. The negative sign indicates that the restoring force is opposite in sign to s, the displacement along the arc.

We have arrived at the condition for simple harmonic motion. The restoring force is proportional to the negative of the displacement. For $F = ma = -kx$, the period of oscillation is

$$T = 2\pi \sqrt{m/k}$$

Therefore, the period of the physical pendulum is $T = 2\pi \sqrt{m/k} = 2\pi \sqrt{m/(m^2 g\rho/I)} = 2\pi \sqrt{I/(mg\rho)}$. Remember that the period of a simple pendulum is $T = 2\pi \sqrt{L/g}$. For a simple pendulum to have the same period as a physical pendulum, the length of the simple pendulum would have to be $L = I/m\rho$.

> ### Question 8-12
> Does this last formula give the right length for a physical pendulum that is almost like a simple pendulum?

In terms of the radius of gyration, j, the moment of inertia is $I = mj^2$. The length of the simple pendulum with the same frequency is therefore $L = mj^2/m\rho = j^2/\rho$. Evidently, the length of the equivalent simple pendulum is not the same as the radius of gyration.

> ### Question 8-13
> What is the period of a meter stick swinging from one end? What is the length of the simple pendulum with that same period?

Radius of Percussion

There is a very familiar type of collision involving angular momentum of one object and linear momentum of the other: a bat hits a ball. The very least that we can do in a physics course is to explain the dynamics of the great American pastime. As every schoolboy knows, a bat has a diamond imprint about $\frac{4}{5}$ the distance from the handle to the end. If the bat strikes the ball at that point (but on the opposite side), there will be no kick or lashback to the hands. How does the bat maker know where to put that imprint?

Before reading the analysis of this problem, handle the phenomena in the following simple way. Get a ruler, or better yet a meter stick, and place it flat on a table or other smooth surface. Put your finger or some marker at one end, and snap the stick at the far end. Notice what happens to the end near the marker. Now line up the stick again and snap it close to the marker end. The end at the marker will move again, but in the opposite direction from the first time. Clearly, there must be some point along the stick where you could snap it and not make the marker end move. Find that point. Is it at the half mark? The two-thirds mark? The three-fourths mark? Notice that a blow at this special point rotates the stick around the end point as an axis, but does not make the end point move forward or backward. If the stick were a bat, and you were holding the bat at the end point, your hands would feel no pain. Now let's calculate where that point is. The distance from the axis to that point is called the *radius of percussion*.

We will assume that the only force acting on the stationary bat comes from the blow when the ball strikes it. In other words, we assume that the bat is only loosely held by the hands, and we require a condition such that the hands do not have to exert any force as a result of the blow. The ball is traveling perpendicular to the bat and strikes it at a distance, x, from the pivot point held by the hands. The force delivered to the bat is $F = \Delta(mv)/\Delta t$. The change of momentum of the ball is $\Delta(mv)$. That force will produce a torque on the bat around the pivot point: $\tau = \mathbf{x} \times \mathbf{F} = I\alpha$. The moment of inertia of the bat around the pivot point is I.

The force delivered by the ball will also make the center of mass of the bat move. *If* no force is exerted by the hands, then the center of mass must accelerate in the same direction as the force applied by the ball. That direction is perpendicular to the bat. The only way that that can happen (if the pivot point doesn't move) is for the center of mass to start moving on an arc around the pivot point, since momentarily that direction is perpendicular to the bat. The equation of motion for the center of mass must be $F = Ma = M\alpha\rho$, where M is the mass of the bat. The angular acceleration of the bat around the pivot point is α, and the distance from that pivot point to the center of mass is ρ. Now let's equate the force expressions from the torque equation and the force equation.

$$M \alpha \rho = I \alpha / x$$

The special distance, x, which allows the bat to swing around the pivot point without any reaction force by the hands, is $x = I/M\rho$. We can change this expression one step further, by writing the expression for the moment of inertia in terms of the radius of gyration: $I = Mj^2$. Then, $x = Mj^2/M\rho = j^2/\rho$.

This is a remarkable result! The radius of percussion of an object is the same as the length of a simple pendulum with the same period. To find the radius of percussion of a bat (or any other solid object), let it swing from the pivot point and measure the period of oscillation. (In the case of a bat or golf club, the pivot point is midway between your two hands.) Then, find (or calculate) the length of a simple pendulum that has the same period. That length is j^2/ρ, which is the same as the radius of percussion.

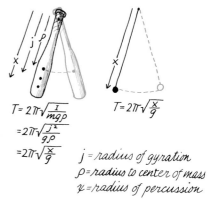

Question 8-14

You found the center of percussion of a ruler or meter stick by hitting it at various points. Now calculate the radius of percussion and compare your results. Also compare your answers with the results of your experiment and calculations for Question 8-13.

SUMMARY

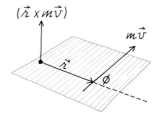

The angular momentum of an object with respect to an axis is $\mathbf{r} \times \mathbf{mv}$, where $\mathbf{r}$ is the radius vector from the axis to the object that has linear momentum $\mathbf{mv}$. The vector product of $\mathbf{r}$ and $\mathbf{mv}$ has a direction along the axis, perpendicular to $\mathbf{r}$ and to $\mathbf{mv}$. The magnitude of the angular momentum is equal to $|r| |mv| \sin \phi$, where ϕ is the angle between $\mathbf{r}$ and $\mathbf{mv}$.

The angular momentum of an object around an internal axis is usually called spin. The angular momentum of an object in a closed orbit is usually called orbital angular momentum.

In terms of rotational variables, the angular momentum of an object about an axis is equal to $I\omega$, where I is the moment of inertia about that axis and ω is the angular velocity in radians per second.

Angular momenta of systems of atomic size are in the range of 10^{-34} kg·m²/s. The angular momentum of a bicycle wheel in normal motion is a few units of kg·m²/s. The orbital angular momenta of the planets is in the range of 10^{40} kg·m²/s.

Newton's second law becomes, for a rotating system, $\tau = \Delta(\mathbf{I\omega})/\Delta t$. The torque exerted on a system is equal to the time rate of change of angular momentum. If there is no external torque on a system, its angular momentum will remain constant.

Constant angular momentum does not mean constant angular speed. If there is a change of the geometric distribution of mass with respect to the axis, the moment of inertia of the object changes and consequently its angular speed changes. For example, planets or comets in elliptical orbits travel faster when they are closer to the sun. Similarly, as skaters pull their outstretched arms close to their bodies, they twirl faster. The ingathering of galactic dust to form a new star usually involves angular momentum so great that a planetary system or a double star is formed in order to accommodate the angular momentum without whirling apart.

The direction of angular momentum is along the axis and in the direction given by the right-hand rule. If a torque is applied to a rotating system in such a direction that the resulting change of angular momentum is perpendicular to the original angular momentum, then the rotating system will precess. The precession frequency is $\Delta\phi/\Delta t = \tau/J$, where τ is the applied torque and J is the angular momentum of the spinning system. The reaction of a rotating wheel or a spinning top to gravitational torque appears at first thought to be perpendicular to the expected result. These effects are called gyroscopic.

The angular momentum of a rotating system is quantized; in other words, the angular momentum must be an integral (or half-integral) multiple of a basic unit. That basic unit is about the size of the angular momentum carried and exchanged by subatomic particles. It is equal to the value of Planck's constant, which, when angular frequency is expressed in radians per second, is given by $\hbar = 1.05 \times 10^{-34}$ kg·m²/s. While the quantization of angular momentum dominates atomic and subatomic reactions, the effect is completely negligible in the domain of human sizes.

There are several special points on a rotating object that can be used to characterize its motion or to simplify calculations concerning the motion. The center of mass was defined in the previous chapter as a point where all the mass of an object could be concentrated, and then the momentum of that point would respond as if all the forces on the body acted at that point. The distance from the axis of rotation to the center of mass of an object was called ρ. The center of gyration is the point on a rotating object where all of the mass could be concentrated and the moment of inertia would remain unchanged. The distance from the axis to the center of gyration is called the radius of gyration, j. If the moment of inertia of an object about an axis is KMR^2, then $j = \sqrt{K}\,R$. The point of percussion of a rotating object is the point where the object can be struck without any reaction force at the axis point. The radius of percussion is $x = j^2/\rho$.

The period of a physical pendulum is $T = 2\pi\sqrt{I/(mg\rho)}$. The length of a simple pendulum with the same period is $L = I/m\rho = j^2/\rho$. This is the same length as the radius of percussion.

These distances apply to a uniform stick, like a meter stick, swinging around an axis at one end.

—50.0—
—57.8—
—66.7—

Answers to Questions

8-1: The definition of angular momentum is not (mvr) but, instead, the vector quantity, $\mathbf{r} \times (\mathbf{mv})$. The magnitude of this vector is $|r|\,|mv|\sin\phi$. The angle, ϕ, is the angle between the radius vector, $\mathbf{r}$, and the linear momentum vector, $(\mathbf{mv})$. Multiplying r by $\sin\phi$ yields a constant that is the distance of closest approach. Since ($r\sin\phi$) is a constant, the angular momentum with respect to the axis is a constant.

8-2: Any part of a system, such as yourself, can exert a torque on the rest of the system. The situation is the same as for linear momentum. Parts of a system can exert mutual forces on each other. However, the sum of the internal forces is zero. Similarly, the sum of the internal torques of a system must be zero. In order to exert a torque in one direction, you must produce a recoil torque in the opposite direction. That's why helicopters have a small tail propeller with its axis horizontal. If the tail propeller didn't provide a countertorque, the body of the helicopter would rotate in the opposite direction from the overhead propeller.

If you were inside a squirrel cage in space and started running, the cage would rotate backward. You would be going around the axis in one direction, and the cage would be rotating in the opposite sense, thus conserving angular momentum. *On earth,* real hamsters or gerbils can spin their cages while still staying down near the bottom. They apparently can generate angular momentum of the cage even though they themselves do not go around. How can they do this?

8-3: Divers create a torque that produces angular momentum as they spring off the board. At the moment their feet are thrown into the air by the board, their center of mass is already beyond the edge of the board. From that point on, their angular momentum is approximately constant, although air resistance can exert a torque in a high dive. How fast divers spin depends on how they distribute their mass. If they stay stretched out, their feet will rise relative to their center of mass and they will enter the water after half a revolution—more or less, depending on the height of the board. If divers tuck their arms and legs together, they will go into a fast spin, still conserving angular momentum, but with a smaller moment of inertia.

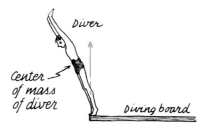

If a cat starts falling without angular momentum, and if air effects are negligible, the cat cannot generate net angular momentum. However, the cat can rotate one part of its body one way and its legs in the opposite direction and thus can change its angular orientation. Even a Manx cat, which has no tail, can land on its feet.

8-4: For earth: $I\omega = (6 \times 10^{24} \text{ kg})(1.5 \times 10^{11} \text{ m})^2(2\pi \text{ rad})/(3 \times 10^7 \text{ s}) = 3 \times 10^{40}$ kg·m²/s. For Jupiter: $I\omega = (1.9 \times 10^{27} \text{ kg})(7.8 \times 10^{11})^2 (2\pi \text{ rad})/(11.9 \text{ yr})(3 \times 10^7 \text{ s/yr}) = 2 \times 10^{42}$ kg·m²/s.

8-5: The angular momentum due to orbital rotation of a planet is $mr^2\omega = mr^2 2\pi/T = mr^2 2\pi/\sqrt{Kr^3} = 2\pi m \sqrt{r}/\sqrt{K}$. It appears that the orbital angular momentum of a planet in a system is proportional to the mass of the planet and to the square root of its distance from the controlling body. Comparing the angular momenta of Neptune and Jupiter, we see that the orbital radius for Neptune is 45×10^{11} m, and the orbital radius for Jupiter is 7.8×10^{11} m. The ratio of these distances is 5.77, and the square root of the ratio is 2.4. On the other hand, the ratio of the mass of Jupiter to the mass of Neptune is $1900/103 = 18.4$. Therefore, the orbital angular momentum of Jupiter is greater than that of Neptune by a factor of $18.4/2.4 = 7.7$.

8-6: The original moment of inertia is I; the final moment of inertia is $2I$. If the original period is T, the final period must be $2T$. The original angular momentum, $I(2\pi/T)$, must remain constant.

8-7: The direction of the torque is horizontal and parallel to the face of the wheel. The wheel, of course, will tip over and hang from the string.

8-8: The angular frequency depends on how fast you spin the wheel. You can determine the revolutions per second by making a mark on the wheel and timing its passage ($\omega = 2\pi f$). If $\omega = 10$ rad/s, then $J = (2 \text{ kg})(0.3 \text{ m})^2(10 \text{ rad/s}) = 1.8$ kg·m²/s. If you tie the support string on the axle so that $\ell = 6$ cm, the precession frequency will be $\Delta\phi/\Delta t = (mgl)/J = (2 \text{ kg})(9.8 \text{ N/kg})(0.06 \text{ m})/(1.8 \text{ kg·m}^2/\text{s}) = 0.65$ rad/s. The period of the precession will be $T = 1/f = 2\pi/\omega = 9.7$ s.

8-9: The spin angular momentum of the bicycle wheel is $(MR^2)\omega = (MR^2)(v/R) = (2 \text{ kg})(0.3 \text{ m})^2(4.5 \text{ m/s})/(0.3 \text{ m}) = 2.7$ kg·m² s. This quantity is equal to $n\hbar = n(1 \times 10^{-34}$ kg·m² s). Therefore, n is equal to some integer with 35 digits, approximately equal to 2.7×10^{34}.

8-10: The moment of inertia of a solid disk rotating around its axis is $\frac{1}{2}mr^2 = mj^2$, where j is the radius of gyration. Therefore, $j = \sqrt{\frac{1}{2}} r = 0.7r = 7$ cm.

8-11: Since the moment of inertia of a bicycle wheel around its axle is approximately mr^2, the moment of inertia about the point of contact with the road is $mr^2 + mr^2 = 2mr^2$.

8-12: The moment of inertia of a physical pendulum that is almost like a simple pendulum is $I = mL^2$. For such a pendulum, the distance from the axis to the center of mass would be the same as the length: $\rho = L$. Therefore, the length of the equivalent pendulum would be $L = I/m\rho = mL^2/mL = L$. The formula for the length of the equivalent simple pendulum gives the right answer for the case where the physical pendulum is simple.

The cat was dropped one-third of a second before this picture was taken. It had been held upside down by its legs, and was released without protest or angular momentum.

8-13: The moment of inertia of a meter stick, which is a uniform rod, swinging about one end, is $I = \frac{1}{3}ML^2$. Its period of pendulum motion is $T = 2\pi\sqrt{I/(Mg\rho)} = 2\pi\sqrt{\frac{1}{3}M(1\text{ m})^2/M(9.8\text{ N/kg})(0.5\text{ m})} = 1.6$ s.

The length of the simple pendulum with that period is $L = I/M\rho = (\frac{1}{3}M)(1.0\text{ m})^2/M(0.5\text{ m}) = \frac{2}{3}$ m. To check this answer, substitute it into the formula for the period of a simple pendulum:

$$T = 2\pi\sqrt{L/g} = 2\pi\sqrt{(\frac{2}{3}\text{ m})/(9.8\text{ m/s}^2)} = 1.6\text{ s}.$$

Note that in all these cases, the mass of the pendulum cancels out.

8-14: The radius of gyration of a uniform rod swinging around one end is given by $\frac{1}{3}ML^2 = Mj^2$. Therefore, $j^2 = \frac{1}{3}L^2$. The radius of percussion is $x = j^2/\rho = \frac{1}{3}L^2/\frac{1}{2}L = \frac{2}{3}L$. This result agrees with the calculations of Question 8-13. The radius of percussion is the same as the length of a simple pendulum, which has the same pendulum period as the meter stick.

Problems

1. On an air hockey table two identical disks approach each other. Each has a speed of 2 m/s, and the mass of each is 0.2 kg. The paths on which they move are such that they will miss each other, with their centers being 10 cm apart at closest approach. What is the angular momentum of the pair with respect to a point midway between them at closest approach?
2. Does the angular momentum of the system described in Problem 1 depend on the axis point chosen? Explain.
3. A particular $33\frac{1}{3}$ rpm record has a mass of 112 g. Its radius is 15 cm. What is its angular momentum while it is being played?
4. Two wheels with masses concentrated on their rims are mounted on a common axle. The lower wheel has a mass of 5 kg and a radius of 0.75 m and is spinning at 5 rev/s. The upper wheel is at rest and has a mass of 12 kg and a radius of 0.30 m. The upper wheel is dropped onto the lower one so that they spin together. What is their resulting frequency?
5. What happens if you stand on a low-friction turntable holding a spinning bicycle wheel with its axis vertical, and then turn the wheel over? Assume that you and the turntable can be approximately modeled by a solid cylinder with a mass of 75 kg and a radius of 15 cm. Assume that the 2 kg mass of the wheel is concentrated in the rim at a radius of 30 cm, and the frequency is 2 rev/s. If you are motionless to begin with, what is the magnitude of your resulting motion?
6. At aphelion a particular comet is 2.5×10^9 km from the sun; at perihelion its radial distance is only 2×10^7 km. If its speed is 1×10^3 m/s at aphelion, what is it at perihelion?
7. What is the direction of the gyroscopic effect on the front wheel of a bike when you turn left or right? To begin with, what is the direction of angular momentum of the wheel?
8. As seen from the pilot's seat, the propeller of a particular single-engine plane spins counterclockwise. What is the direction of its angular momentum? If, in flight, the plane starts to climb, what is the direction of the gyroscopic effect?
9. In the diagram on page 145, there is a bubble chamber picture of sequential decays of subatomic particles. From other experiments, we know that the original particle (π^-) has spin 0. The μ^- and the e^- each have spin $\frac{1}{2}\hbar$. The unseen particles are neutrinos. What must be the spin of a neutrino? How is angular momentum conserved in each of the two decays?

10. What is the radius of gyration of a sphere rotating around its axis?
11. In the parallel axis theorem on page 172 there are two terms. Define each in words. What is the total *angular momentum* of a bicycle wheel with respect to the instantaneous axis where it touches the road? Express your answer in terms of the mass of the wheel, M, its radius, r, and its velocity, v. What would be the *angular momentum* of the wheel with respect to that instantaneous axis if it were *sliding* past with velocity v? Describe in words the significance of the two terms for the total angular momentum of the rolling wheel.
12. Try twirling a bob in a horizontal circle as shown in the illustration. Any tube can be used as the holder; make the bob out of an eraser or something soft. Produce a frequency of 1 rev/s, and then pull the string down in the tube until the radius is reduced to one-half its original value. Did you exert any torque on the rotating system by pulling the string? What is the new frequency? Calculate it, and observe it.

13. A solidly packed cylinder of apparatus in space is rotating about its axis at 10 rev/s. Its mass is 100 kg and its radius is 15 cm. Two radial arms are then extended, each carrying a small detector with a mass of 10 kg to a distance of 1 m from the axis. What is the new frequency of revolution of the system?
14. Two wheels are mounted concentrically. The bottom one is made like a bicycle wheel with its mass concentrated in the rim. It has a moment of inertia of 0.2 kg·m² and is spinning in a horizontal plane at 3 rev/s. The other wheel is motionless at first and then is dropped onto the lower one so that they spin together. The resulting frequency is 1 rev/s. What is the moment of inertia of the upper wheel?
15. A typical baseball bat has an overall length of 83 cm. When held in the usual fashion, the distance from the end to a point between the two hands is 70 cm. When suspended from that point, the bat swings with a period of 1.5 s. What is the distance from the end of the bat to the center of percussion—the point where the ball should be hit?
16. In an experiment to demonstrate the angular momentum of photons, circularly polarized light (the spin of each photon in the same direction) might be absorbed by a tiny disk 1 cm in diameter, with a mass of 2 mg (2×10^{-6} kg). How many photons would it take to give the disk a spin of 1/10 revolution per second?
17. The moment of inertia about its axis of a particular car wheel and tire is $0.7Mr^2$. The mass of the wheel is 15 kg and its radius is 30 cm. What is the total angular momentum of the wheel with respect to a point on the road if the car is moving at 100 km/hr?
18. A stainless steel tennis racquet, suspended from the center of the normal grip, swings with a period of 1.3 s. What is the distance from that point to the center of percussion? (Get data from your own racquet, and see if the center of percussion is where you hit the ball.)
19. When suspended from one corner, a particular book swings with a period of 0.9 s. The book is 16 cm wide and 25 cm long. What are the values for $\rho, j,$ and x? (Measure the periods of various books of your own.)
20. In each example worked out in the book, the radius of gyration was greater than the radius to the center of mass: $j > \rho$. Consequently $x > j > \rho$. Do these inequalities necessarily hold for any physical pendulum? One way to explore this problem is to analyze a pendulum consisting of two point masses fastened together: M is at radius r_1, and m is at radius r_2.

9 KINETIC ENERGY AND WORK

The conservation law of interest to most people is the one concerned with energy. We hear that the rate of using energy is steadily increasing, and we know that recent shortages of energy affect both home life and international politics. Energy appears to be a term meaning oil, coal, falling water, and uranium. Energy not only makes our cars go and warms our houses, but it is also necessary for the production of things like metals and fertilizers. Indeed, in a very direct way, all living things eat energy in order to keep alive. It is well known from advertisements that certain breakfast foods will give you "instant energy" with which to start the day.

Considering the current importance of energy, it is surprising that the concept was not understood until the middle of the nineteenth century. For all their sophistication, Galileo, Newton, and Franklin did not know that a quantity called energy could be defined in such a way that it is always conserved.* Perhaps they didn't discover the idea because the concept is not at all obvious. Energy appears in many different forms. A rolling car has energy. A motionless flashlight battery has energy. A stone at the top of a cliff has energy. A pat of butter has energy. A kettle of boiling water has energy. Sunlight has energy. Energy in all these different forms can be defined in such a way that the total energy is conserved as one system transforms into another. However, so long as a system never changes in any way, its energy content is meaningless. It is only during transformations, from one shape to another or from one place to another, that the concept of energy becomes useful as a bookkeeping device.

Question 9-1
Bookkeeping device! Energy costs money. In the form of oil or coal it is sent through pipelines or carried in railroad cars. How can it be just a bookkeeping device?

---

* In the first edition of the *Encyclopaedia Brittanica,* 1771, the entire entry under "Energy" was: "Energy, a term of Greek origin, signifying the power, virtue, or efficacy of a thing. It is also used, figuratively, to denote emphasis of speech."

In this chapter, we will concentrate only on the form of energy involving motion—*kinetic energy*. In the next chapter, we will see how to define energy of position or distortion. Then in Chapter 11, we will be forced to the conclusion that energy and mass are really the same thing—a concept that is not obvious at all. In Chapter 12, we will define heat energy in such a way that the energy conservation law can be preserved even when there is friction. Throughout the rest of our studies of such topics as optics, electricity, and microstructure, the analysis of the energy transformations will be a unifying theme.

HANDLING THE PHENOMENA

If an object in motion has energy, it must lose the energy if it stops. As we will see, the energy is transferred to some other object in a single collision, or it is frittered away in friction. In the latter case, does a car with an initial speed of 20 mph coast twice as far as a car that starts coasting with a speed of 10 mph? If you have a car, or a bike with a speedometer, see for yourself. You will need a stretch of smooth, flat road.

There's an easy table-top simulation of the road test. Get a ruler or flat stick, some paper clips, and several projectiles such as aspirin pills or thumbtacks. Place one projectile at twice the radius of the other along the ruler. Then swing the ruler around one end, through an arc of about 30°, so that it hits a finger and stops abruptly. The pills or tacks will be swept along until the ruler stops, and then they will continue along the table in a straight line. Since the radial distance of the outer projectile is twice that of the inner, the outer one will be given twice the velocity of the inner one. Does the outer one slide twice as far before it stops? Try the game several times and find the approximate ratio of stopping distances. Then make the ratio of radius arms 3:1, and find the ratio of stopping distances for projectiles that have an initial ratio of velocities of 3:1.

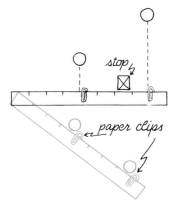

THE SECOND CONSTRAINT ON CERTAIN COLLISIONS

In Chapter 7, we found that in certain types of air track glider collisions, two different quantities of motion are conserved. The momentum, which is the product of mass and velocity, is always conserved. If the gliders interact through springs, and don't stick together, one other combination of mass and velocity is also conserved. That is the product, mv^2. For reasons that we will soon see, this particular quantity of motion is usually written as $\frac{1}{2}mv^2$ and is called *kinetic energy*, or E_{kin}.

There are several points to be noted about the nature of this defined quantity. First, kinetic energy must be a scalar and not a vector. The velocity, which *is* a vector, is squared in the expression, making the whole term just a directionless scalar. Second, the existence of the constant, $\frac{1}{2}$, implies that there must be some more basic definition of the unit of energy. That's right, and we will soon get to it. In the meantime, note that, if an object with a mass of 1 kilogram moves with a speed of 1 meter/second, then its kinetic energy is, by definition, equal to one-half of the basic unit of energy, or one-half *joule*. The unit is named after James Joule (1818–1889), an English scientist who first demonstrated the energy equivalence of heat.

The symbol for joule is J; the 1 kg object moving at a speed of 1 m/s² has an energy of ½ J. Finally, note that we now have two different quantities to define the quantity of motion—the momentum, **mv**, and the kinetic energy, ½mv^2. One is a vector; the other is a scalar. What is the relationship between these two? Why do we need them both? Three hundred years ago, people did argue about such questions; the Germans championed the preeminence of mv^2, which was called vis viva, or living force, and the English were sure that the quality of motion was best described in terms of the momentum, **mv**.

Question 9-2

In the collisions described in Chapter 7, momentum was always conserved, but the mv^2 quantity was not conserved in certain interactions. Which ones were these?

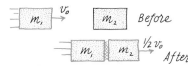

Momentum is conserved if $m_1 = m_2$ and objects stick together with final velocity of ½v_0.

The collisions that took place on the air track satisfied the law of momentum conservation, but that law by itself did not allow us to predict the results of any particular collision. For instance, when one glider coming in from the left struck a motionless glider with the same mass, the first one stopped completely and sent the target glider toward the right with the same speed. Momentum was conserved, but momentum could also have been conserved if the two gliders had stuck together and gone on with half the original speed. That is,

$$m_1 v_0 = (m_1 + m_2)(\tfrac{1}{2}v_0) \qquad \text{if } m_1 = m_2$$

Momentum conservation constrains the number of possible combinations of speeds after the collision, but not sufficiently. There is *one* equation which must be satisfied (momentum before = momentum after) and *two* unknowns (the velocities of the two objects after the collision). If kinetic energy must also be conserved, there are then two equations to be satisfied, and that is enough to determine completely the two unknown velocities. Here's how it works:

Conservation of momentum: $m_1 v_0 = m_1 v_1 + m_2 v_2$

Conservation of E_{kin}: $\tfrac{1}{2}m_1 v_0^2 = \tfrac{1}{2}m_1 v_1^2 + \tfrac{1}{2}m_2 v_2^2$

Since, for this case, $m_1 = m_2$, the m's can be canceled out and so can the factor of ½. The equations become

$$v_0 = v_1 + v_2 \qquad \text{and} \qquad v_0^2 = v_1^2 + v_2^2$$

Square the first equation and compare it with the second one:

First equation squared: $v_0^2 = v_1^2 + v_2^2 + 2v_1 v_2$

Second equation: $v_0^2 = v_1^2 + v_2^2$

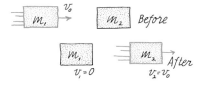

Momentum and kinetic energy are conserved in this elastic collision with $m_1 = m_2$.

These two equations are consistent with each other only if $2v_1 v_2 = 0$. There are two possibilities: either $v_1 = 0$, or $v_2 = 0$. What we actually saw happen

THE SECOND CONSTRAINT ON CERTAIN COLLISIONS

on the air track was that the first glider stopped after the collision. That's the case when $v_1 = 0$. The equations then require that $v_2 = v_0$, which is just what happens. The two conservation laws completely constrain the outcome of the collision so that only one combination of velocities is possible, and the prediction matches the experimental results.

> ### Question 9-3
>
> What about the other possibility? The mathematics would be satisfied if $v_2 = 0$, and $v_1 = v_0$. To satisfy these conditions, the original glider would have to continue on its way with the original velocity. How could that happen?

Let's apply both conservation laws to the collision of identical spheres in *two* dimensions. Once again, since the spheres are identical, the masses will cancel. Now we will have three equations, two for momentum and one for kinetic energy:

Momentum in x: $mv_0 = mv_1 \cos \theta + mv_2 \cos \phi$

Momentum in y: $0 = mv_1 \sin \theta - mv_2 \sin \phi$

E_{kin}: $\tfrac{1}{2}mv_0^2 = \tfrac{1}{2}mv_1^2 + \tfrac{1}{2}mv_2^2$

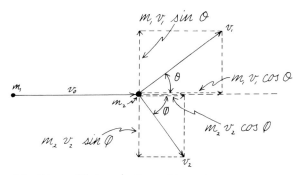

Elastic collision in two dimensions. $m_1 = m_2$

To solve these three simultaneous equations, eliminate the common mass, m; cancel the factor of $\tfrac{1}{2}$ in the third equation; and square the first and second equations:

$v_0^2 = v_1^2 \cos^2\theta + v_2^2 \cos^2\phi + 2v_1v_2 \cos \theta \cos \phi$

$0 = v_1^2 \sin^2\theta + v_2^2 \sin^2\phi - 2v_1v_2 \sin \theta \sin \phi$

Add these two equation, remembering that $\sin^2\theta + \cos^2\theta = 1$, for any θ.

$v_0^2 = v_1^2 + v_2^2 + 2v_1v_2(\cos \theta \cos \phi - \sin \theta \sin \phi)$

Compare this equation with the kinetic energy equation. They are consistent only if $2v_1v_2(\cos \theta \cos \phi - \sin \theta \sin \phi) = 0$. If either $v_1 = 0$ or $v_2 = 0$, the term would always be zero. Those conditions would correspond with either a head-on collision or a complete miss. For any other kind of collision, both spheres have nonzero velocities afterward. Consequently, the expression in the parentheses must be identically zero. That expression can be transformed by using the following trig identity:

$$(\cos \theta \cos \phi - \sin \theta \sin \phi) = \cos (\theta + \phi)$$

But if $\cos (\theta + \phi) = 0$, then $(\theta + \phi) = 90°$. The opening angle between the paths of the two spheres after the collision must be exactly 90°. Notice what a remarkable constraint is imposed by the conservation laws on the collision of identical particles!

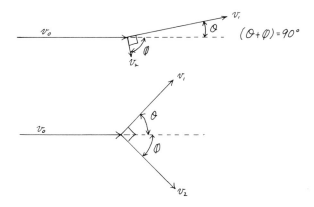

Question 9-4

Does the 90° condition mean that each of the spheres must travel away at an angle of 45° from the original direction? If not, aren't the final results of the collision undetermined? There are only three equations, but aren't there four unknowns—v_1, v_2, θ, and ϕ?

The 90° requirement is satisfied if, and only if, the colliding particles have the same mass and don't lose any kinetic energy in the collision. These conditions are satisfied by the proton-proton collision, shown in the cloud chamber picture on page 141. In the cloud chamber, the rapidly moving electrically charged particles ionized the gas molecules in their wake. In the supersaturated atmosphere of the chamber, a liquid droplet grew on each ion, making visible the ion trail left by the electrons or protons. Note that in the ball-bearing collision shown on page 137, the 90° requirement was not satisfied. In this case, the colliding objects were not isolated; the table was part of the system.

If two objects have equal mass and stick together in a head-on collision, kinetic energy is not conserved. Since there is only one final velocity after the collision (the two objects are stuck together), the single equation describing conservation of angular momentum is all that is needed.

THE SECOND CONSTRAINT ON CERTAIN COLLISIONS

$$m_1 v_0 = m_1 v_f + m_2 v_f = v_f(m_1 + m_2) = 2m_1 v_f \quad \text{since } m_1 = m_2$$

The final velocity is thus exactly half the original velocity: $v_f = \tfrac{1}{2}v_0$. Let's see what happens to the kinetic energy in such a situation. The original kinetic energy was

$$(E_{\text{kin}})_0 = \tfrac{1}{2} m_1 v_0^2$$

The final kinetic energy is

$$(E_{\text{kin}})_f = \tfrac{1}{2}(2m_1)v_f^2 = \tfrac{1}{2}(2m)(\tfrac{1}{2}v_0)^2 = \tfrac{1}{2}(\tfrac{1}{2}m_1 v_0^2) = \tfrac{1}{2}(E_{\text{kin}})_0.$$

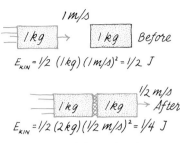

Half of the original kinetic energy disappears when the objects with equal mass collide and stick.

The final kinetic energy after the collision is exactly half of the kinetic energy of the bombarding object before the collision. Where did the kinetic energy go? At this point of the argument, we don't know.

Apparently kinetic energy is conserved only in certain types of collisions where there is no denting or sticking of the objects. Such collisions are called *elastic*. Therefore, while it is true that in elastic collisions kinetic energy is conserved, the only way you know that a collision is elastic is if the kinetic energy doesn't change. As a practical matter, collisions between very hard, spring materials are nearly elastic, while collisions that result in dents or an increase in temperature are inelastic. A tennis ball loses about a third of its kinetic energy when it bounces from a hard surface. On the other hand, steel ball bearings or glass marbles lose only a few percent of their energy when they collide gently with each other. Collisions of air track gliders can be made almost completely elastic by equipping them with steel spring bumpers or with opposing magnets.

We saw how the two conservation laws require certain results from the collision of two elastic objects with the same mass. The same laws can be used to derive a general expression for head-on collisions, where $m_1 \neq m_2$. As usual, we start out with m_2 at rest and m_1 coming in from the left with positive velocity.

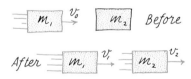

Momentum conservation: $m_1 v_0 = m_1 v_1 + m_2 v_2$

E_{kin} conservation: $\tfrac{1}{2} m_1 v_0^2 = \tfrac{1}{2} m_1 v_1^2 + \tfrac{1}{2} m_2 v_2^2$

Divide the first equation by m_1, and the second equation by $\tfrac{1}{2}m_1$:

$$v_0 = v_1 + (m_2/m_1)v_2 \quad \text{and} \quad v_0^2 = v_1^2 + (m_2/m_1)v_2^2$$

Square the first equation:

$$v_0^2 = v_1^2 + (m_2/m_1)^2 v_2^2 + 2(m_2/m_1)v_1 v_2$$

Subtract the second equation from this one:

$$0 = (m_2/m_1)v_2^2(m_2/m_1 - 1) + 2(m_2/m_1)v_1 v_2$$

Divide through by $(m_2/m_1)v_2$:

$$0 = (m_2/m_1 - 1)v_2 + 2v_1$$

Notice that after all this algebra, we have ended up with one equation for two unknowns. At least, however, we can solve for the ratio of v_1/v_2:

$$v_1/v_2 = \tfrac{1}{2}(1 - m_2/m_1) = (m_1 - m_2)/2m_1$$

> **Question 9-5**
>
> Does the formula for the general case agree with what you already know about the special case of $m_1 = m_2$? See what the general formula predicts for the ratio of v_1/v_2 for the cases where $m_1 = \frac{1}{2}m_2$; $m_1 = 2m_2$; $m_1 \gg m_2$.

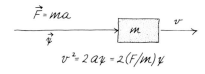

THE WORK NECESSARY TO PRODUCE KINETIC ENERGY

How does an object get to have kinetic energy? The energy can be transferred to the object in a collision, but it can also be produced by shoving the object until it has the required velocity. Starting from rest, an object will reach the velocity, v, if it is subject to a constant acceleration, a, over a distance, x, satisfying the condition $v^2 = 2ax$. The constant acceleration is produced by a constant force, given by $a = F/m$. If we substitute this formula for acceleration into the formula for final velocity, we get

$$v^2 = 2(F/m)x$$
$$\tfrac{1}{2}mv^2 = Fx$$

If we exert a constant force on an object for a distance, the kinetic energy produced is given by the above formula. The product of force, and the distance through which the force is exerted, is called *work*.

This simple definition of work raises all sorts of problems. Both force and displacement are vectors. Is their product a vector product, or a cross product, like torque? Does the simple defintion of work agree with the everyday meaning of the word? Is the definition any good if the applied force is not constant? Are work and energy part of the same conservation law—so that if you do some work on an object and produce kinetic energy, you can make the object exert a force and do the same amount of work? Finally, we must look into the relationships that exist among force, displacement, time, momentum, and kinetic energy.

First, what about that product of force and displacement? Certainly the maximum acceleration, and hence the maximum velocity, will be produced if the force is applied in the direction of the displacement. In fact, any force applied perpendicular to the displacement produces no useful effect at all—unless it reduces friction. In the case of the product of force and a lever arm to produce torque, the situation is exactly the opposite. Only the component of force *perpendicular* to the lever arm is effective. Apparently, we need another kind of product of two vectors. This one is called the *dot product*, or *scalar product*. The product sign is a dot between the two vectors.

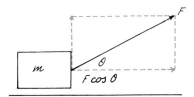

$$\text{Work} = W = \mathbf{F} \cdot \mathbf{x} = |\mathbf{F}|\,|\mathbf{x}|\cos\theta$$

The magnitude of the force component *in the direction of the displacement* is given by $|F|\cos\theta$, where θ is the angle between $\mathbf{F}$ and $\mathbf{x}$.

The *vector* product of two vectors yields another vector, perpendicular to the first two. The *dot product* of two vectors yields a scalar. Work and energy have no direction properties. The amount of work done or energy

THE WORK NECESSARY TO PRODUCE KINETIC ENERGY

produced is the same, regardless of whether the object is headed north, south, east, or west.

> Question 9-6
>
> According to this definition, no work would be done on a heavy box if you carried it in your arms across the room. The force that you exert to hold it is directed upward, but the displacement is horizontal. Therefore, the angle between force and displacement is 90°, and the work done on the box is zero. Can this be right?

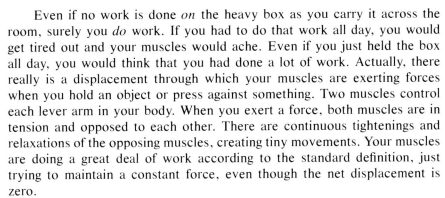

Even if no work is done *on* the heavy box as you carry it across the room, surely you *do* work. If you had to do that work all day, you would get tired out and your muscles would ache. Even if you just held the box all day, you would think that you had done a lot of work. Actually, there really is a displacement through which your muscles are exerting forces when you hold an object or press against something. Two muscles control each lever arm in your body. When you exert a force, both muscles are in tension and opposed to each other. There are continuous tightenings and relaxations of the opposing muscles, creating tiny movements. Your muscles are doing a great deal of work according to the standard definition, just trying to maintain a constant force, even though the net displacement is zero.

The unit of kinetic energy is the joule and, since work can produce kinetic energy, we will use the joule as the unit for work. *If a force of 1 newton is exerted on an object causing it to move a distance of 1 meter, then the work done is 1 joule.*

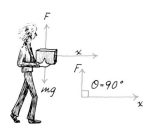

$$1 \text{ newton} \cdot 1 \text{ meter} = 1 \text{ joule}$$

> Question 9-7
>
> Does this agree with our earlier statement that an object with a mass of 1 kg traveling at a speed of 1 m/s only has $\frac{1}{2}$ J?

To get the relationship that we did between $\mathbf{F} \cdot \mathbf{s}$ and $\frac{1}{2}mv^2$, we made use of one of the formulas for constant acceleration: $v^2 = 2as$. What happens if the force, and hence the acceleration, is not constant? We can always make the approximation that the force is constant over a short distance. Our definition of work would then be

$$W = \mathbf{F} \cdot \Delta \mathbf{x}$$

We can also make a graph of the effective force exerted at every point: $F(x)$. If the force is constant, clearly the product of force and displacement is equal to the work done. It is also equal to the area under the curve of the $F(x)$ graph. As we saw in Chapter 2, in the case of the $v(x)$ graph, the area under the curve is equal to the product of the variables, even when the

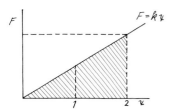

Area under $F(x)$ curve from $x=0$ to $x=2$ is four times area from $x=0$ to $x=1$.

dependent variable is not constant. For example, the diagram shows the graph of $F(x)$ for a spring that obeys Hooke's law: $F = kx$. In this case, there is no negative sign, because F is the force that we must exert in the direction of x in order to stretch the spring—as opposed to the restoring force of the spring, which is in the opposite direction. The work done in stretching the spring to a distance x is the triangular area under the $F(x)$ curve. That area, which is equal to the work, is

$$W = \tfrac{1}{2}(\text{base})(\text{height}) = \tfrac{1}{2}(x)(F_x) = \tfrac{1}{2}(x)(kx) = \tfrac{1}{2}kx^2$$

It takes four times the work to stretch a spring to 2 cm that it takes to stretch it to 1 cm.

Question 9-8

Applying a force to a spring will stretch it, and that takes work according to the definition, but the spring doesn't pick up any kinetic energy. Didn't we define work, on pages 186–187, in terms of kinetic energy?

If we count only the work done to produce kinetic energy, do we have a conserved property? Can we get the work back again? In principle, yes! For instance, if you exert a force through a distance to accelerate an air track glider, the work that you do is by definition equal to the kinetic energy of the glider. If the glider then runs into a spring at the end of the air track and stops momentarily, its kinetic energy has been transformed into the work of compressing the spring. Shortly thereafter, the spring expands, doing work on the glider and restoring its kinetic energy. With good elastic springs, the final kinetic energy is almost equal to the original. Evidently, we have expanded our energy conservation law by including work as an energy term.

Now let's reconsider that ancient question about whether the true quantity of motion is best described by mv or $\tfrac{1}{2}mv^2$, by momentum or by vis viva. If you exert a force, F, on an object for a *time* interval, Δt, then you have exerted an impulse that produces a change in momentum of the object:

$$\mathbf{F}\Delta t = \Delta(\mathbf{mv})$$

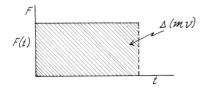

If you exert a force, $\mathbf{F}$, on an object, through a *distance* interval, $\Delta \mathbf{x}$, then you have done work on the object. If all the work goes into changing the kinetic energy:

$$\mathbf{F} \cdot \Delta \mathbf{x} = \Delta(\tfrac{1}{2}mv^2)$$

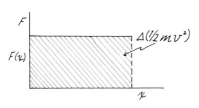

The area under the curve of the graph of $F(t)$ is equal to the change of momentum. The area under the curve of $F(x)$ is equal to the work done. If the object is free to accelerate, and F is the net force in the direction of displacement, then the area under the $F(x)$ curve is the change in *kinetic* energy of the object.

As an example of the different uses of momentum and kinetic energy in a problem, suppose that you want to accelerate an air track glider with a mass of 200 g to a velocity of 1 m/s by exerting a force of 0.1 N. How long will it take?

$$\Delta(mv) = F\Delta t \qquad (0.2 \text{ kg})(1 \text{ m/s}) = (0.1 \text{ N})\Delta t$$
$$\Delta t = 2 \text{ s}$$

How far does the glider go before its speed has reached 1 m/s?

$$\Delta(\tfrac{1}{2}mv^2) = \mathbf{F} \cdot \mathbf{\Delta x} \qquad \tfrac{1}{2}(0.2 \text{ kg})(1 \text{ m/s})^2 = (0.1 \text{ N})\Delta x$$
$$\Delta x = 1 \text{ m}$$

Notice that the two methods agree with each other. Since the acceleration is constant, the average velocity is just half the final velocity. At an average velocity of $\tfrac{1}{2}$(1 m/s), the glider will travel 1 meter in 2 seconds.

> **Question 9-9**
>
> To figure out the damage a car can do in a collision, and the force that it exerts while it stops, should you consider the car's momentum or its kinetic energy? Note that at 60 mph the car has twice the momentum that it does at 30 mph, but it has four times the kinetic energy.

While "Handling the Phenomena," you shot pills or tacks along a table. They slowed down and stopped because of the retarding forces of friction. As we saw in Chapter 4, dry, sliding friction is almost independent of velocity. Consequently, the pills or tacks were subject to forces that were approximately constant. If the launching system gave pill A twice the velocity of pill B, then A had four times the kinetic energy. Since both A and B were subject to the same frictional drag, A should have traveled four times as far as B. The same explanation applies to the car stopping distances taken from the Florida drivers' manual, as described in Chapter 2. Since the kinetic energy of a car is proportional to the square of its velocity, the distance required for stopping after the brakes are applied is also proportional to the square of the velocity. The constant force in this case is applied by the brake drums if all goes well.

ROTATIONAL KINETIC ENERGY

An object traveling in circles or spinning on its axis has kinetic energy. For a *point* object, at a distance r from the axis, the angular velocity and tangential velocity are related by $v = \omega r$. The angular velocity, ω, is in radians per second, and the moment of inertia, I, is equal to mr^2. The kinetic energy of the object is therefore

$$E_{\text{kin}} = \tfrac{1}{2}mv^2 = \tfrac{1}{2}m(\omega r)^2 = \tfrac{1}{2}I\omega^2$$

The same expression for rotational kinetic energy holds good for any kind

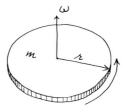

$I = \tfrac{1}{2} m r^2$ (for uniform disk)
$E_{kin} = \tfrac{1}{2} I \omega^2$

of rotational motion, whether or not it is a point object, and whether the object is circling an outside axis or spinning on its own.

In "Handling the Phenomena" of Chapter 8, you dropped a record onto another record that was spinning. Since angular momentum had to be conserved, the final angular velocity of the combination was only half that of the original spinning record:

$$\text{Conservation of } I\omega: \quad I\omega_o = 2I\omega_f \quad \omega_f = \tfrac{1}{2}\omega_o$$

However, the kinetic energy of the system is not necessarily conserved. As you can tell by the sound when the records mesh, the collision is not elastic. The final rotational kinetic energy is

$$E_f = \tfrac{1}{2}(2I)\,\omega_f^2 = \tfrac{1}{2}(2I)(\tfrac{1}{2}\omega_o)^2 = \tfrac{1}{2}(\tfrac{1}{2}I\omega_o^2) = \tfrac{1}{2}\,E_o$$

The result is just like that for the inelastic linear collision of two objects with equal masses. Exactly half of the kinetic energy disappeared.

Let's compare the rotational kinetic energy of a bicycle wheel with the translational kinetic energy of the wheel. As we did in Chapter 6, let's assume that the 2 kg mass of the bicycle wheel is concentrated on the rim at a radius of 33 cm. At 15 mph, the linear velocity is 6.7 m/s; the angular velocity is

$$\omega = v/r = (6.7\ \text{m/s})/(0.33\ \text{m}) = 20\ \text{rad/s}$$

The rotational kinetic energy of the wheel about its axis is therefore

$$E_{\text{rot. kin.}} = \tfrac{1}{2}\,I\omega^2 = \tfrac{1}{2}\,(2\ \text{kg})(0.33\ \text{m})^2(20\ \text{rad/s})^2 = 45\ \text{J}$$

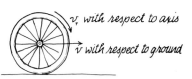

v_t of point on rim with respect to axis is equal to v of axle with respect to ground.

The translational kinetic energy of the wheel is

$$E_{\text{trans. kin.}} = \tfrac{1}{2}mv^2 = \tfrac{1}{2}(2\ \text{kg})(6.7\ \text{m/s})^2 = 45\ \text{J}$$

It appears numerically in this case that the rotational kinetic energy is equal to the translational kinetic energy. We could have deduced the same result from the general formulas. For any rolling wheel with the mass on the rim, the rotational kinetic energy is equal to the translational energy:

$$\tfrac{1}{2}I\omega^2 = \tfrac{1}{2}(mr^2)\omega^2 = \tfrac{1}{2}m(r\omega)^2 = \tfrac{1}{2}mv_t^2 = \tfrac{1}{2}mv^2$$

The total kinetic energy of the rolling wheel is $\tfrac{1}{2}I\omega^2 + \tfrac{1}{2}mv^2$. (Is the total kinetic energy also divided equally between $\tfrac{1}{2}I\omega^2$ and $\tfrac{1}{2}mv^2$ for a rolling disk or sphere?)

Serious research is being done on the design of super flywheels for the storage of very large amounts of energy. Since the energy is proportional to the square of the rotational velocity, obviously it pays to increase the rotational frequency of a flywheel. Unfortunately, the centripetal force needed to keep the outer rim going in a circle is also proportional to the square of the rotational velocity:

$$F_{\text{cent}} = m\omega^2 r$$

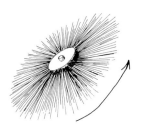

Let's calculate the energy storage in one proposed type of flywheel. Suppose that the wheel is made up of radial wires, anchored at a hub. Then the centripetal force must be exerted along the length of each wire. So long as the wires do not pull out of the hub, the only effect of the rotation is to stretch the wires. A giant circular wire brush of this kind is not the most efficient design for energy storage. With the wires packed closely together at the hub, they spread out at larger radii, occupying only a fraction of the volume. Furthermore, the part of the system most subject to stress is the anchor point of each wire. A better design consists of cylindrical layers of materials bonded together. Still, the order of magnitude of energy storage in a radial wire wheel is easy to calculate.

Assume that the radius of a wheel is $\frac{1}{2}$ m, so that it could fit easily into a car. If the thickness of the wheel were $\frac{1}{2}$ m, the volume would be about $\frac{1}{3}$ m^3. If the total volume were filled with a material of the density of steel, the mass would be 2000 kg. Since the wires occupy only a fraction of the volume, let's assume a mass of 500 kg, composed of individual radial wires, each with a moment of inertia equal to $\frac{1}{3}Mr^2$. (This is the moment of inertia of a thin rod rotating about an axis at one end.) Such devices can be spun in a vacuum at frequencies of 4000 rpm. The energy storage would be

$$E_{\rm rot} = \tfrac{1}{2}I\omega^2 = \tfrac{1}{2}(\tfrac{1}{3})(500 \text{ kg})(\tfrac{1}{2}m)^2 \, 4\pi^2 \, [(\tfrac{2}{3})(10^2) \text{ rps}]^2 =$$

$$\frac{500 \times 4\pi^2 \times 2 \times 2 \times 10^4}{2 \times 3 \times 2 \times 2 \times 3 \times 3} \approx 3.5 \times 10^6 \text{ J}$$

We can compare this energy with other types of storage by noting that 3.6×10^6 J = 1 kilowatt-hour (kWh), and that a standard car battery can store about 1 kWh. The flywheel would thus provide the equivalent of about one car battery.

Question 9-10

Can ordinary wires exert the necessary centripetal force without flying apart? The mass of one of the wires is equal to its density times the cross-section area times the length: $M = \sigma A r$. Exaggerate the effect of the force by assuming that all of that mass is located at the outer radius. Calculate the centripetal force required for a wire in the flywheel that we examined. Assume a density of 6×10^3 kg/m^3, but specify the cross-section area simply as A. When you find the force, substitute the value into the formula that links stress to strain. Young's modulus for steel is 1×10^{11} N/m^2. (If the strain of an ordinary steel wire is much greater than about 2×10^{-3}, it will break.)

SIMPLE MACHINES THAT SAVE NO WORK

What good is a machine if it doesn't save work? Unfortunately, no machine puts out more work than is put into it. In fact, most of them put out less. Machines are really devices allowing us to trade a smaller input force and a longer input displacement for a larger output force and a shorter output displacement. Or vice versa. Such a trade is frequently very useful. Let's examine a few simple cases.

Levers

To get simple rules for levers, we assume that the lever bar itself has no weight, and that the forces are exerted at right angles to the lever. Actually, in the case of a heavy crowbar dislodging a rock, the weight of the crowbar may add important leverage, and neither the weight nor the applied force may be perpendicular to the bar. At any rate, if the applied force is perpendicular to the lever, then the distance through which that force is exerted is just the arc through which that end of the lever moves. The arc is proportional to the length of the lever from the fulcrum to the point where the force is applied. The same conditions are true for the output force and the output distance.

The input work is equal to the product of input force and the input distance. The output work is equal to the product of output force and output distance. It is a matter of experimental determination that the output work is almost as great as the input work, since for most levers practically no work is lost because of friction.

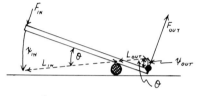

$$\text{Work input} = \text{work output}$$
$$(F_{\text{in}})(\text{arc}_{\text{in}}) = (F_{\text{out}})(\text{arc}_{\text{out}})$$
$$(F_{\text{in}})(L_{\text{in}})(\theta) = (F_{\text{out}})(L_{\text{out}})(\theta)$$
$$(F_{\text{in}})(L_{\text{in}}) = (F_{\text{out}})(L_{\text{out}})$$

Or

$$F_{\text{out}}/F_{\text{in}} = L_{\text{in}}/L_{\text{out}}$$

This ratio, $F_{\text{out}}/F_{\text{in}}$, is called the mechanical advantage of the lever.

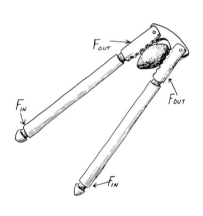

With most levers—nutcrackers, can openers, crowbars—we want to exert only a small force, even though for a longer distance, in order to apply a large force on some object. Usually the lever provides other conveniences as well, providing a coupling between large, soft hands and small, hard materials. The input force of your hand on a can opener is distributed over a much larger area than is the output force at the cutting edge. Not only is the input force smaller than the output force because of the leverage, but the input *pressure* is much smaller than the output pressure. It is well known (at least in Georgia) that it is hard to crack one pecan by itself in the hand, but it is easy to crack one pecan if two are pressed against each other in the hand.

We derived the force relationship for levers by requiring that work output be equal to work input. Note that the same relationship follows from equating torques. A torque is equal to the vector product of the radius arm and the applied force.

$$(\mathbf{r} \times \mathbf{F})_{\text{in}} = (\mathbf{r} \times \mathbf{F})_{\text{out}}$$

Consequently,

$$(F_{\text{in}})(L_{\text{in}}) = (F_{\text{out}})(L_{\text{out}})$$

since we have assumed that the forces are perpendicular to the lever arms.

It may seem as if most lever arrangements that we use provide a greater output force in exchange for a greater input distance. Actually, the levers that we use most of the time are rigged just the other way. Most of the joints of our body consist of levers arranged to deliver large output displacement—and consequently high output speed—in exchange for large input force. The diagram shows the main bones and muscles of the human arm. For every body part that can move, there have to be two muscles. Muscles can't push; they can only pull. When you raise your lower arm, the biceps contracts and the triceps expands. When you lower your arm, the opposite happens, as you can readily feel for yourself.

On a typical arm, the biceps muscle is fastened to the forearm at a point about 5 cm beyond the elbow. The hand is about 35 cm from the elbow. The biceps must act through a lever arm of 5 cm while anything being lifted in the hand has a lever arm seven times as long. If the point where the biceps is fastened moves upward 1 cm, the hand rises 7 cm. That's great! It means that your muscle can make a short movement, which makes your hand move much further and thus very rapidly. On the other hand (on either hand, for that matter), a lever that trades distance for force demands a great deal of force. If you hold 10 kg in your hand, the force down is 100 N, and your biceps must pull up with a force of 700 N. That's why biceps have to bulge.

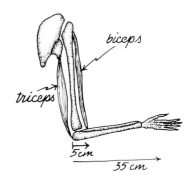

Wheel and Axle

This class of devices includes the screwdriver and the doorknob. The principle is really the same as that for the lever. In most wheel-and-axle arrangements a small input force exerted on a long lever arm moves through a large arc. The output force is larger, but is exerted through a smaller arc. Note, however, that the output torque and angular displacement are the same as the input torque and angular displacement. In the case of the screwdriver and doorknob, the devices also serve to match the size of human hands to the smaller size of mechanical systems.

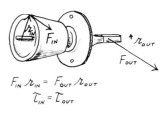

$F_{IN}\, r_{IN} = F_{OUT}\, r_{OUT}$
$\tau_{IN} = \tau_{OUT}$

Gears

Gears are toothed wheels that are not on the same axle. One gear may mesh with another directly or they may be connected with an endless chain. The important point about gears is that they provide a way for a shaft turning at a particular angular speed to drive another shaft at a different angular speed, and therefore with a different torque. A small gear with few teeth can rotate at high speed and low torque to drive a large gear with many teeth at low speed and high torque. The ideal mechanical advantage, the ratio of output torque to input torque, is the ratio of angular displacement of the drive wheel to the angular displacement produced in the final wheel.

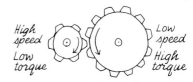

The gear ratio is 2:1. The smaller wheel (called the pinion) must rotate twice to make the larger gear rotate once. The torque delivered by the pinion is half that delivered by the larger gear.

> ### Question 9-11
>
> Take the case of a bike with several gear ratios. The drive system is a combination of gears and two wheel-and-axle arrangements (the pedal wheel and the back wheel). Why not measure the overall mechanical advantage by taking the ratio of the diameter of the circle in which the pedals travel to the wheel diameter?

Inclined Plane

If you want to lift shoppers from the ground floor of a store to the first floor, do you do more work with an elevator or an escalator? The elevator itself is counterbalanced like an Atwood's machine, but an extra upward force must be supplied equal to the weight of the passengers. The work done is thus equal to the product of weight and height raised.

> ### Question 9-12
> Doesn't the upward force have to be slightly larger than the weight in order to lift the load? If the upward force is exactly the same as the weight, won't the load just hang still?

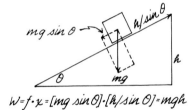

If the load is raised at an angle, less upward force is needed. The weight components are shown in the diagram. Neglecting friction, the force that has to be exerted along the plane is: $mg \sin \theta$. Unfortunately, the distance up the plane, to get to the height h, is $h/(\sin \theta)$. The work done to raise an object to a height, h, along an inclined plane is $W = \mathbf{F} \cdot \mathbf{x} = (mg \sin \theta) \cdot [h/(\sin \theta)] = mgh$. Apparently the work done in lifting a load with an inclined plane does not depend on the angle of inclination. What *does* change with the angle is the necessary applied force. There is the familiar trade-off between force and distance. A gradual slope permits a small force, but requires a long distance to reach the intended height.

Actually, the work done to move things up an inclined plane usually does depend on the angle. That's because there usually is friction between the load and the plane. The force required to keep the load moving up the plane is $(mg \sin \theta) + F_{\text{friction}}$. The ideal mechanical advantage of an inclined plane is the ratio: (force required to pull the load straight up)/(force required to push the load up the plane without friction). That is also the ratio: (distance up the plane to reach height h)/(height h). Evidently, that ratio is equal to $[h/\sin \theta]/h = 1/(\sin \theta)$. The *actual* mechanical advantage of an inclined plane is equal to the ratio: (weight of the load)/(the actual force required to push the load up the plane while overcoming friction).

Inclined planes were used to build the pyramids, and they are part of everyday life in roads and ramps. The principle is also used in any screw arrangement, from wood screws to jar covers. In these cases, the incline is wrapped around a central axis.

Pulleys

When a rope goes around a wheel that is free to turn without friction, the tension in the rope on one side of the wheel is the same as the tension on the other side. Hence it is possible with a pulley wheel on the ceiling to lift something by pulling down. By using another wheel traveling with the load, you can suspend the load with two strands of the same rope, each pulling up with a force equal to the tension. The person pulling on the rope is exerting a force equal to the tension in the rope; and thus the person's effective force is doubled. However, for every meter of rope that is pulled, the load moves through only half a meter. Once again, in an ideal case without friction, input work equals output work.

The ideal mechanical advantage of a simple pulley system can be calculated by counting the number of strands supporting the load. The real mechanical advantage is usually considerably less than the ideal, since pulleys under heavy load frequently have a lot of friction. To measure the real mechanical advantage, the ratio of output force (the load) to input force must be measured *while the load is actually being moved*.

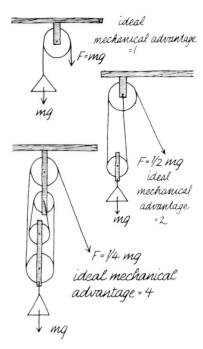

SUMMARY

Notice that we do *not* define energy as the ability to do work. Instead, we look for a quantity concerned with motion, with the special property that the quantity does not change in certain types of collisions. For collisions where springs bump and there are no dents or sticking, the quantity $\frac{1}{2}mv^2$ remains constant. This quantity is defined to be kinetic energy. In elastic collisions, both momentum and kinetic energy are conserved. This double constraint severely limits the possible results of the collision. For a head-on elastic collision between two objects with given masses and original velocities, only one combination of final velocities is possible.

An object obtains kinetic energy if it is subject to a force over a distance. The product of force and displacement is defined as work. This particular product of two vectors is called a scalar product, or dot product, and yields a scalar: work = $\mathbf{F} \cdot \mathbf{x}$. The magnitude of the work done is equal to $|\mathbf{F}||\mathbf{x}| \cos \theta$, where θ is the angle between the applied force and the resultant displacement.

The unit of work is the joule, abbreviated J. One joule is the work done by a force of 1 newton exerted through a distance of 1 meter. An object with a mass of 1 kg and a velocity of 1 m/s has a kinetic energy of $\frac{1}{2}mv^2 = \frac{1}{2}$ J.

Rotational kinetic energy is equal to $\frac{1}{2}I\omega^2$, where I is the moment of inertia of the rotating object and ω is the angular velocity in radians per second. Rotating systems can be used for energy storage, although they are limited by the stresses created by the centripetal forces necessary to hold the rotating object together.

Work and kinetic energy are defined so that they are part of the same conservation law. The work done in increasing the speed of an object is equal to the increase in the object's kinetic energy. Similarly, a moving or spinning object can turn its kinetic energy into work by exerting a force through a distance.

Simple machines are devices that can provide a trade-off between input force and input distance, and output force and output distance. The input work is always equal to or greater than the output work. If the output force is greater than the input force, the input displacement must be larger than the output displacement:

$$\mathbf{F}_{in} \cdot \mathbf{x}_{in} \geq \mathbf{F}_{out} \cdot \mathbf{x}_{out}$$

The interchangeability of work and kinetic energy has only limited application because the energy of objects seems to disappear when moving objects go uphill and stop, or compress springs and stop, or just simply slow down and stop. In the next chapter, we will extend the energy conservation law by defining new forms of energy.

Answers to Questions

9-1: Money, just like energy, is useful only in terms of exchange processes. Transformations occur when money is spent, not when it stays in the bank. It's the same with energy. A car battery doesn't start the car until a switch is thrown. A change in the internal structure of the battery takes place, and an electric motor starts turning. We can account for the transformations in terms of the battery losing energy and the motor gaining it.

9-2: The interactions in which mv^2 was not conserved involved permanent changes in the shapes of springs or in cases of friction or sticking. In particular, the interactions caused by the sudden expansion of a spring between two gliders did not conserve the quantity, mv^2.

9-3: The original glider could have continued on its way with $v_1 = v_0$, leaving the target glider with zero speed, if the original glider completely missed the second one. Of course, on an air track that's not very likely, but the mathematics doesn't know that it is describing an air track collision. The formulas provide a model for the real situation, and in this model a clean miss is possible.

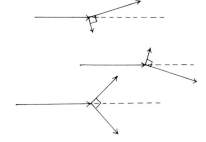

9-4: The 90° condition does not require each of the spheres to move off at 45° from the original direction. Depending on how close the collision is to head-on, or to a glancing collision that almost misses, the target sphere may go at any angle from almost in the forward direction to almost at right angles to the forward. However, in whatever direction the target sphere travels, the angle between the two paths of the identical spheres must be 90°.

It is true that the complete results of the collision are undetermined unless one more condition is imposed. However, if you measure just one of the four final variables, the other three are then determined. For example, if you observe that the target sphere goes off at 30°, then there is just one possible set of values for the speeds of the two spheres and the angle of the bombarding one. Furthermore, if you know the center-to-center separation of the spheres at the moment of collision, you know the angular momentum of the system. Since angular momentum must be conserved, there is a fourth constraint, or equation, so that the values of all four final variables could be predicted.

9-5: If $m_1 = m_2$, then $v_1 = 0$. The colliding ball stops dead in its tracks, as we have already seen.

When $m_1 = \frac{1}{2} m_2$, then $v_1/v_2 = -\frac{1}{2}$. In terms of v_0, as given by the momentum equation, $v_0 = -\frac{1}{2}v_2 + 2v_2$. Therefore, in this case where a glider strikes one with twice as much mass, $v_2 = \frac{2}{3} v_0$. The more massive target glider goes off with a velocity of two-thirds that of the original. The bombarding glider bounces back with a velocity of $v_1 = -\frac{1}{2}v_2 = -\frac{1}{3}v_0$.

When $m_1 = 2m_2$, then $v_1/v_2 = \frac{1}{4}$. In terms of the momentum equation, $v_0 = \frac{1}{4}v_2 + \frac{1}{2}v_2$. Therefore, when a glider strikes another one with one-half its mass, the target glider shoots forward with a velocity: $v_2 = \frac{4}{3} v_0$. The more massive bombarding glider continues in the forward direction with a velocity: $v_1 = \frac{1}{3} v_0$.

If $m_1 \gg m_2$, then $v_1/v_2 \approx \frac{1}{2}$. The target object goes shooting off with a velocity twice that of the bombarding object. As you can see from the momentum equation, the velocity of the bombarding object is approximately the same as its original velocity. (A car hitting a tennis ball does not slow down appreciably. Note that the tennis ball would be knocked forward with twice the velocity of the car.)

9-6: That's right! No work need be done on the box to move it across the room. In terms of our everyday meaning of the word, work is something that somebody gets paid for. Why should anyone pay you for carrying a box across a room when the job could be done just as well by sliding the box along a low-friction runway? Such a method would require no fuel, cost no money, and hence involve no work.

9-7: Let's calculate how much work it takes to accelerate an object with a mass of 1 kg, from rest to a speed of 1 m/s. Suppose that we exert a force of 1 N. The acceleration will be $a = F/m = 1$ m/s². The distance through which the force must be exerted in order to attain a speed of 1 m/s is given by $v^2 = 2ax$. $(1 \text{ m/s})^2 = 2(1 \text{ m/s}^2)x$. The distance x, is $\frac{1}{2}$ m. Therefore, the work necessary to accelerate an object with a mass of 1 kg to a speed of 1 m/s is $\mathbf{F} \cdot \mathbf{x} = (1 \text{ N})(\frac{1}{2} \text{ m}) = \frac{1}{2}$ J. This calculation agrees with the formula for the kinetic energy of an object: $\frac{1}{2}mv^2$.

9-8: The spring can expand against an air track glider, as shown in the strobe pictures of glider interactions, and produce kinetic energy of the glider. The spring appears to be a device for transmitting work to an object that can then take on kinetic energy. Of course, the spring can also store the work before transmitting it. We will explore this phenomenon more in Chapter 10.

9-9: Either momentum or kinetic energy can be used as a criterion for the damage done in a car collision. If momentum is used, then the *time* of stopping is the critical factor in finding the forces produced. If kinetic energy is used, then the *distance* of stopping should be used. With a constant force being exerted to stop the cars, a car starting at 60 mph will take twice the time to stop as a car at 30 mph, but the 60 mph car will require four times the stopping distance.

9-10: $F = m\omega^2 r = (\sigma A r)(4\pi^2 f^2)r$

$$F = [6 \times 10^3 A \tfrac{1}{2}][4\pi^2(\tfrac{2}{3} \times 10^2)^2]\tfrac{1}{2}$$

$$F/A = \frac{6 \times 10^3 \times 4\pi^2 \times 2 \times 2 \times 10^4}{2 \times 3 \times 3 \times 2} = 24 \times 10^7$$

$$F/A = Y\frac{\Delta L}{L}$$

$\Delta L/L = (24 \times 10^7)/(1 \times 10^{11})$

$\Delta L/L = 2.4 \times 10^{-3}$

It appears that the wire is at the breaking point. Actually, much stronger materials and better designs are available.

9-11: If the ideal mechanical advantage of a bicycle were the ratio of pedal-circle diameter to wheel diameter, then there would be no effect caused by gear changes. The way to measure the mechanical advantage, and see the effect of the gears, is to measure the distance traveled by a point on the wheel circumference, while the pedals go around once. The ratio of pedal-circle circumference to wheel arc is the ideal mechanical advantage. Actual mechanical advantage is less because there is friction in the gears and chain drive. Friction does not affect the ratio of arcs, but it does require one to push harder on the pedals. Incidentally, even for 10-speed bikes, the mechanical advantage is less than 1 in all gears.

9-12: If the upward force is exactly the same as the downward force, the elevator will be in equilibrium, but this does not mean that the elevator is necessarily standing still. It merely means that there is no acceleration; the velocity is constant. If a slight extra impulse upward is given to the elevator, it will start rising. If the upward force were then equal to the downward force, the elevator would continue up with constant velocity. As a practical matter, a real elevator would be subject to friction in the pulley system. The upward force would have to be large enough to overcome the friction.

Problems

1. An ice-hockey puck glides for 5 m when tapped to produce an initial speed of 2 m/s. How far will it glide if given an initial speed of 4 m/s?
2. If you are in a car traveling at 60 mph, what is the kinetic energy, in joules, of a half kilogram ball in your hand—with respect to you? With respect to a person standing beside the highway?
3. Roll a ball (for instance, a marble or tennis ball or golf ball) against an identical one at rest so that there is a glancing collision. Is the opening angle between the balls after the collision equal to 90°? Explain.
4. A 2 ton car traveling on an icy road at 50 km/hr crashes head-on with a 1 ton car traveling in the other direction at 50 km/hr. They stick together after the collision. What is the resulting speed? What is the change of velocity for an occupant in the heavier car? In the lighter car? How much kinetic energy was lost in the collision?
5. On an air track, a glider with mass of 300 g, traveling at 1 m/s, collides elastically with a 100 g glider at rest. What are the resulting velocities?
6. On an air track, a 100 g glider, traveling at 1 m/s, collides elastically with a 300 g glider at rest. What are the resulting velocities?
7. A golf club with an effective mass of 0.5 kg strikes a golf ball with a mass of 46 g at a speed of 30 m/s. Assuming that the collision is elastic (which it isn't—but it's a reasonable approximation), what is the resulting speed of the golf ball?
8. What is your kinetic energy when running as fast as you can?
9. What is the kinetic energy of a ball with a mass of 300 g when you throw it with a speed of 40 m/s?
10. In an electron collision with another electron at rest, the first has a velocity of 1×10^7 m/s. After the collision it was going at 60° to the original direction. What were the velocities of the two electrons afterward? What was the direction of the second one?
11. A compressed spring between two objects contains 100 J of stored energy. One object has a mass of 900 g and the other has a mass of 100 g. How is the energy distributed when the spring is released? Assume that the spring delivers all its energy to the two objects, and that they can move without friction. (Remember that momentum must be conserved.)
12. A woman with mass M stands still on ice skates holding a ball with mass m. She throws the ball with velocity v against a wall. It makes an elastic collision with the wall, and then the woman catches it. What is her final velocity, assuming that her motion is essentially friction free? Where did her momentum and kinetic energy come from, since she started out with neither?
13. A person lifts a box with a mass of 10 kg from the floor to a height of 1 m, then carries the box at that height for a distance of 10 m, and finally lowers the box to the floor again. How much work did the person do on the box in each step of the chore, and what was the total work done on the box by the person?

14. When a block with a mass of 2 kg is hung on a particular spring, the spring stretches 4 cm. How much work does it take to stretch the spring from 0 to 12 cm?
15. What is represented by the area under a curve of $F(x)$? What is represented by the area under a curve of $F(t)$?
16. A sled is pulled for a distance of 100 m with a force of 80 N exerted on a rope at an angle of 30° from the horizontal. How much work is done on the sled?
17. The moment of inertia of a solid ball about its diameter is $I = \frac{2}{5}Mr^2$. As the ball rolls downhill, what fraction of its kinetic energy is translational (of the center of mass), and what fraction is rotational?
18. A solid ball, a solid disk, and a hoop with its mass concentrated on the rim each has a mass of 2 kg and a radius of 10 cm. How much energy does it take to accelerate each to a speed of 10 revolutions per second about its axis?
19. What is the rotational kinetic energy of a $33\frac{1}{3}$ rpm record that has a mass of 100 g and a radius of 15 cm?
20. What is the rotational kinetic energy of the earth spinning on its axis? Assume that $I = 0.2Mr^2$. Compare this energy with the rotational energy of the moon going around the earth.

$$M_{earth} = 6.0 \times 10^{24} \text{ kg} \qquad r_{earth} = 6.4 \times 16^6 \text{ m}$$
$$M_{moon} = 7.3 \times 10^{22} \text{ kg} \qquad r_{moon\text{-}earth} = 3.8 \times 10^8 \text{ m}$$

21. To pry up a 200 kg rock, a farmer inserts a 2.0 m crowbar under the edge of the rock. The fulcrum is a log 25 cm from the end of the bar. What force must be exerted down on the other end of the bar to move the rock?
22. The handle of a winch in a sailboat is 30 cm from the center of the drum. The radius of the drum on which a rope is wound is 5 cm. If a force of 100 N is exerted on the handle, what is the tension produced in the rope?
23. The pedal gear of a bike has a diameter of 30 cm. It is connected by a chain to a gear with a diameter of 10 cm. If the pedal gear revolves at a frequency of 1 revolution per second, what is the rotation speed of the back wheel?
24. What is the ideal mechanical advantage of an inclined plane that makes an angle of 10° with the horizontal? What is the real mechanical advantage if the coefficient of friction between cart and plane is 0.1?
25. A pulley system consists of one wheel on the ceiling and one wheel attached to the load. A person pulls down on the rope which goes up to the ceiling wheel, down to the load wheel, and back up to the ceiling. What is the ideal mechanical advantage? If the efficiency is 80%, what force is needed to pull on the rope in order to lift 100 kg?

10 POTENTIAL ENERGY AND POWER

The word "energy" brings to mind images of waves crashing, cars racing, people jumping, and vigorous activity of all kinds. There's another kind of energy, though. It lurks quietly beneath the ground in pools of oil, or looms over canyons in dammed up lakes of water. The car battery and the motionless mousetrap are really coiled with stored energy, ready to lash out into the motion forms that we have already studied. These motionless forms of energy are called *potential,* as if to indicate that they may potentially be turned into motion energy. Actually, any form of energy can be turned into any other form, and so in one sense any form of energy could be called potential. Usually, however, the name *potential energy* refers to energy stored in the distortion of material or the displacement of objects in some electric or magnetic or gravitational force field. If the objects are moved away from a particular position, and then moved back, the system will regain its original potential energy.

In this chapter we are going to examine several different forms of potential energy. In each case, kinetic energy or work can be turned into the stored form of energy, and then recovered without loss. Furthermore, we will *define* the magnitude of the potential energy so that at all times the total amount of energy will stay constant. As work is done, or as kinetic energy disappears, the potential energy will increase. Energy will be conserved in such processes, which is hardly surprising since we define the potential energy just for that purpose. Actually, of course, in most systems both potential and kinetic energy disappear sooner or later. Then we define a new form of energy, in terms of the internal structure of material, and once again save the energy conservation law. We will see how to do this last step in Chapter 12.

Question 10-1

It sounds as if energy conservation is just a human invention. If energy disappears, we invent a new form and say that the energy turned into that new form, and so is conserved. Does the law represent a reality of nature, or a human convenience?

HANDLING THE PHENOMENA

1. Many toys involve energy storage systems. The most common system these days is the electric flashlight battery, or dry cell. In an earlier generation, rubber band motors were standard, particularly for driving model airplanes or small boats. A convenient feature of a twisted rubber band (as opposed to a stretched rubber band) is that the torque required to wind it up does not increase greatly as the number of turns increases. To a first approximation, if you twist a long rubber band twice as many times, you store twice as much energy. You can test this energy storage property for yourself by building a simple toy paddle boat. When you finish your scientific experiments with it, in the bathtub, you can give the boat to any lucky youngster.

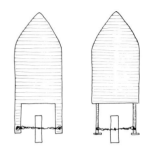

The classic shape of the boat is shown in the diagram. Staple the paddle to the rubber band so that it won't fall off every time the rubber band unwinds. Test the linearity of energy storage by measuring distance traveled as a function of number of turns of windup.

2. Another simple way of storing energy is to lift a weight to a higher position. If the weight falls from that height, the stored energy turns into kinetic energy. Pendulum motion is an example of a continuous interchange between gravitational potential energy and kinetic energy. At the highest point of its swing the pendulum bob has no kinetic energy. As the bob swings through its low point, the stored gravitational energy has turned into kinetic energy, and then, as the bob rises, the kinetic energy turns back into potential energy. As we will see, the gravitational potential energy is defined in such a way that the total energy of the swinging pendulum stays constant—except for the gradual dying down.

Here is a way to take energy, both potential and kinetic, away from one pendulum, feed the energy to another pendulum, and then get it back again in the first one—over and over again. Tie a string to the backs of two chairs as shown in the diagram. From the string suspend two pendulums, each having the same length, and so the same period. Start one pendulum swinging and see what happens to the other. By moving the chairs and so tightening or loosening the support string, you can change the speed with which energy feeds from one pendulum to the other. These are called "coupled pendulums." As you watch the system, you can almost see the energy move from one part to the other.

3. The time rate of doing work or using energy is called power. If you do one joule of work per second, your power output is one watt. An eighteenth century unit of power, still used in some engineering circles, is the horsepower. It was defined by James Watt in 1783 as the average amount of work per second that could be done by a strong English dray horse, working steadily all day. One horsepower equals 746 watts.

If you had to work steadily all day, you probably couldn't generate even one-tenth horsepower. During brief bursts of effort, however, you may be as powerful as a horse. Find out for yourself by lifting weights. The most convenient weight to use is your own. You can lift it by chinning yourself or by running upstairs. In either case, multiply your weight in newtons by the vertical distance through which you rise, and divide by the time it takes. For example, if you have a mass of 65 kg, your weight is about 650 N. Every time you chin yourself, you rise about 50 cm, and so do $(650 \text{ N})(\frac{1}{2}\text{m})$

= 325 J of work. If you can chin yourself at a rate of once a second, you are producing a little less than one-half horsepower. You can probably produce more power by using your leg muscles in running upstairs. Find out.

STORING ENERGY BY DISTORTING SYSTEMS

The kinetic energy of an air track glider disappears when the glider runs into a spring, or runs uphill on a tilted track, or approaches a stationary magnet while carrying a magnet that is repelled. In each case, most of the kinetic energy reappears when the glider is bounced back to its original position at the other end of the track. We can preserve an energy conservation law by saying that the spring, the earth-glider gravitational system, and the magnets can store energy. The amount of energy stored is just the amount of kinetic energy lost or the work done by the glider as it moves against the opposing force. Let's define the stored potential energy in each case by calculating the work done.

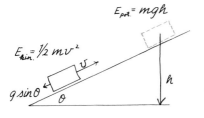

(a) A glider with an initial speed of v will move up a hill with constant slope through a distance x derived from

$$v_f^2 - v_o^2 = 2ax$$

The acceleration is $(-g \sin \theta)$, and the vertical rise of the glider is $h = x \sin \theta$. When the glider comes to rest:

$$0 - v_o^2 = 2(-g \sin \theta)(h/\sin \theta) = -2gh$$

Multiply both sides by $\frac{1}{2}m$:

$$\tfrac{1}{2}mv_o^2 = mgh$$

The amount of kinetic energy that disappears is $\frac{1}{2}mv_o^2$. If we define the gain in gravitational potential energy to be mgh, then there has been no energy loss. Kinetic energy has turned into gravitational potential energy—by definition.

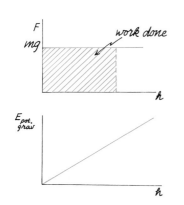

Look at the graph of F versus h for lifting an object with mass m. The graph is a horizontal line because the force is a constant equal to the weight of the object, mg. The work done in lifting the object is equal to the area under the $F(h)$ curve. That work goes into the separation of the object from the earth, and is being stored as gravitational energy. The graph of gravitational energy versus h is the same as the graph of work done in lifting the object. Notice that while the force is constant, the work done or energy stored is proportional to the height.

Work done in lifting object to height h
$$= \text{gravitational potential energy stored} = mgh$$

Question 10-2

The acceleration of a pendulum bob as it swings is certainly not constant. Nevertheless, according to the potential-kinetic energy transformation, how is the velocity at the bottom of the swing related to the maximum vertical height of the swing?

STORING ENERGY BY DISTORTING SYSTEMS

Gravitational potential energy can turn into rolling kinetic energy as well as kinetic energy of the center of mass of an object. As a round object rolls down hill, the two forms of kinetic energy share the released gravitational potential energy:

$$mgh = \tfrac{1}{2}mv^2 + \tfrac{1}{2}I\omega^2 = \tfrac{1}{2}mv^2 + \tfrac{1}{2}Kmr^2\omega^2 = \tfrac{1}{2}mv^2 + \tfrac{1}{2}Kmv^2 = (1+K)\tfrac{1}{2}mv^2$$

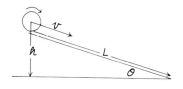

The geometric constant, K, in the moment of inertia, is for the moment of inertia about *the center of mass*. All three energy terms refer to the center of mass: the change in the potential energy as the center of mass falls through the vertical height, h; the translational kinetic energy of the center of mass; the rotational kinetic energy of the object about the center of mass.

For a hoop: $mgh \to (1+1)\tfrac{1}{2}mv^2 = mv^2$

Solid disk: $mgh \to (1+\tfrac{1}{2})\tfrac{1}{2}mv^2 = \tfrac{3}{4}mv^2$

Solid sphere: $mgh \to (1+\tfrac{2}{5})\tfrac{1}{2}mv^2 = \tfrac{7}{10}mv^2$

Question 10-3

We can now obtain the velocity of an object that has rolled downhill, without regard to forces or torques. Compare this method of calculation with the results of the calculation in Question 6-9.

(b) In Chapter 9 we calculated the work done in compressing or stretching a spring. The applied force necessary at any point is proportional to the stretch: $F = kx$. The total work required to distort the spring by a length, x, is therefore

$$W = \tfrac{1}{2}kx^2$$

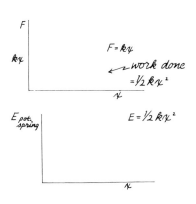

Since the spring can perform this same amount of work as it returns to its undistorted shape, we define the potential energy stored in the spring to be:

$$E_{\text{spring}} = \tfrac{1}{2}kx^2$$

The diagram shows the graph of $F(x)$ for a spring. Note that in this case we have plotted the *applied* force, which is in the same direction as the distortion. The restoring force exerted by the spring is in the opposite direction.

(c) Instead of using springs on air track gliders, you can use magnets with their poles opposed. The diagram shows a standard arrangement using cylindrical "Alnico" magnets. Collisions of gliders equipped with such magnets are completely elastic with no kinetic energy lost. The gliders never actually touch each other. For the geometry shown, the force necessary to shove the magnets together is *approximately*

$$F \propto 1/x^2$$

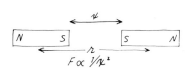

Note that r is the center to center distance between the two bar magnets.

In the next diagram we have plotted the function $F = C/x^2$, where C is a constant that depends on the strength of the magnets. The work done in shoving the two magnets together is defined as the potential energy stored

Note that x is the actual separation distance between the ends of the two bar magnets.

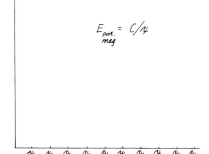

in the system. That energy is proportional to the area under the curve on the $F(x)$ graph. For any particular situation we could actually measure the area by counting squares on the graph paper. It is useful, however, to derive a general expression for the energy stored when the force is proportional to $1/x^2$. Remember that the gravitational force between two spheres is proportional to $1/x^2$. The same law applies to the force between spheres covered uniformly with electrostatic charges. The following derivation, therefore, applies to several very important phenomena.

The problem is to calculate the work done in moving the magnets together from a distance x_{10} to a distance x_1. The work is proportional to the area under the $F(x)$ curve from $x = 10$ to $x = 1$. The work is more than $F_{10}(x_1 - x_{10})$, but is less than $F_1(x_1 - x_{10})$. The average force is certainly larger than F_{10} but smaller than F_1. The force doesn't change much over a small distance, however. Let's break the action into nine small steps. In moving from x_{10} to x_9, for example, the average force is somewhere between $F_{10} = C/(x_{10})^2$ and $F_9 = C/(x_9)^2$. The geometric average of these two forces is $F = C/x_9 x_{10}$. Therefore, the work done in bringing the magnets closer through this small distance is approximately: $\text{Work}_{10-9} \approx (C/x_9 x_{10})(x_{10} - x_9)$. The work done in moving through all nine steps is

$$\text{Work}_{10\text{-}1} = C\left[\frac{(x_{10} - x_9)}{x_{10} x_9} + \frac{(x_9 - x_8)}{x_9 x_8} + \frac{(x_8 - x_7)}{x_8 x_7} + \cdots + \frac{(x_2 - x_1)}{x_2 x_1}\right]$$

$$= C\left[\left(\frac{1}{x_9} - \frac{1}{x_{10}}\right) + \left(\frac{1}{x_8} - \frac{1}{x_9}\right) + \left(\frac{1}{x_7} - \frac{1}{x_8}\right) + \cdots + \left(\frac{1}{x_1} - \frac{1}{x_2}\right)\right]$$

$$= C\left[\frac{1}{x_1} - \frac{1}{x_{10}}\right]$$

As you can see, the complicated algebra leads to a simple answer. All the terms of the intermediate steps cancel out. Furthermore, all the intermediate terms would cancel out even if we had divided the action into a hundred steps or a thousand steps.

If we want to know how much work is done in pushing the two magnets together until they are only a distance, x, apart, then we can let x_{10} be so large that the magnets are barely influencing each other. As $x_{10} \to \infty$, $1/x_{10} \to 0$. The work done in pushing the magnets together until they are a distance, x, apart, is therefore equal to C/x. Since the magnets can do work in shoving apart again, we define their magnetic potential energy to be

$$E_{\text{pot}} = C/x$$

for a system where the repulsion force = C/x^2. This relationship applies to gravitational and electrostatic systems where the objects are points or spheres. If they are spheres, the distance, x, is the distance between centers.

Question 10-4

The potential energy is positive if the objects repel each other. How can we define potential energy if the objects attract each other? (In a later section we will explore this situation as it relates to gravity.)

NEGATIVE POTENTIAL ENERGY

What is your gravitational potential energy where you are right now? Whether you are sitting or lying down, your center of mass is at some height, h, with respect to the floor. Is your gravitational potential energy therefore equal to mgh? But how high is the floor from the ground? If you are at ground level, is your potential energy zero, and if so what is your potential energy if you go downstairs to the cellar? The zero point of potential energy appears to depend on a rather artificial definition. Actually, only *differences* in potential energy are ever measured. When work is done to distort a system, the stored energy is the difference between the final and initial potential energy.

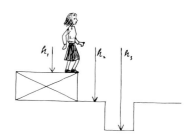

If your potential energy equals mgh, what is h?

Consider the situation shown in the diagram of the golf ball in the hole. The potential energy of the ball appears to depend on whether the zero point is chosen at the level of the fairway, the green, or the hole. Depending on this choice, the ball has a potential energy that is positive, negative, or zero. The physical significance of the definition, however, involves only the *change* of potential energy as the ball moves from one height to another. As the diagram shows, the change of potential energy of the ball if it is raised from the cup to the green is the same positive quantity regardless of how the zero point is defined.

While the definition of zero point of potential energy is arbitrary, there are some reasonable conventions. For instance, if a system like a spring is undistorted so that it will neither expand nor contract, then it is reasonable to assign zero potential energy to the condition. If two objects that influence each other through electric, magnetic, or gravitational forces are so far from each other that the influence is not measurable, then we might say that at that distance their potential energy is zero.

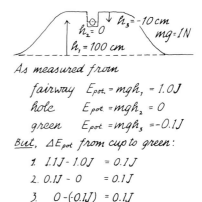

> ### Question 10-5
>
> According to this last convention, where is an object when its gravitational potential energy is zero with respect to the earth?

If objects *attract* each other, it takes no work to push them together. On the contrary it takes outside work to pull them apart. If magnets, aligned in such a way that they *repel*, have positive potential energy when they are close to each other, then two magnets that *attract* each other must have negative potential energy.

$$E_{\text{pot}} = -C/x \text{ for attracting magnet poles}$$

What can it mean to have *negative* energy? It's just a matter of bookkeeping again. If a system can do work as it returns to an undistorted geometry, then it has positive potential energy. If work must be done *on* the system, then the system has negative potential energy. In either case, the bookkeeping works out (in friction free cases) so that

$$\Delta E_{\text{kin}} + \Delta E_{\text{pot}} = 0$$

If E_{pot} is positive and decreases, then ΔE_{pot} is negative and ΔE_{kin} is positive.

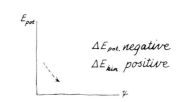

ΔE_{pot} positive
ΔE_{kin} negative

ΔE_{pot} negative
ΔE_{kin} positive

When a ball rolls down a hill, the potential energy decreases and the kinetic energy increases. For the opposite case, consider what happens when a magnet trapped by another one is given a shove to separate them. E_{pot} was negative but becomes less so; therefore ΔE_{pot} is positive. As the magnet pulls away, its speed decreases so ΔE_{kin} is negative.

If an object has negative potential energy, it is trapped in a *potential well*. If you fall into a water well, ΔE_{pot} is negative and as you fall ΔE_{kin} is positive. As you land in the water, your kinetic energy disappears into other forms of energy. Then you are trapped in a gravitational potential energy well. To furnish the positive quantity of E_{pot} to raise you to the surface, there would have to be an expenditure of energy from some other source. If you could jump out, your kinetic energy as you start the leap is greater than it is at the top: ΔE_{kin} is negative.

THE EARTH'S POTENTIAL WELL

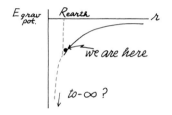

You don't have to fall into a water well to be trapped by the earth's gravitational field. All of us here on the earth are bound in a gravitational well. In the diagram we plot $E_{grav\ pot}$ versus r. Remember that if $E_{grav\ pot} = -(c/r)$, then $F_{grav} = -(c/r^2)$. The negative sign of the force means that the direction of the force is inward—toward a smaller radius.

As $r \to \infty$, $E_{grav\ pot} \to 0$. This situation agrees with our convention that potential energy should be zero where an object is beyond the influence of the source. Note, however, that as $r \to 0$, $E_{grav\ pot} \to -\infty$. This means that it would take an infinite amount of work to separate an object from the earth if it starts at $r = 0$—according to the model.

> **Question 10-6**
>
> How can this be? Can't we get rocket ships away from the earth without expending an infinite amount of energy?

If you could drill a hole through a diameter of the earth, starting in the United States, you would come out near Australia—not China. (After all, China is also north of the equator.) If the earth were not rotating, had uniform density, and weren't hot, and if you could drop a billiard ball down this hole without any air friction, what would happen? The gravitational force on the ball is equal to $-G(mM)/r^2$, but, inside the earth, where and what is the mass, M? There is a theorem proposed by Karl Gauss a hundred and fifty years ago stating that inside a uniformly distributed sphere, there is a cancellation of the gravitational effect on an object of all the mass at larger radius. The outward gravitational pull of the small amount of mass close to the object but at a larger radius is balanced by the inward pull of the larger amount of mass on the opposite side of the sphere. The geometrical explanation is shown in the diagram. The mass of a sphere of radius, r, is equal to $\rho \frac{4}{3}\pi r^3$, where ρ is the density. Therefore, at any radius, r, inside the earth, the gravitational force is equal to $F_r = -G(m/r^2)(\rho \frac{4}{3}\pi r^3) =$

$-Gm\frac{4}{3}\rho\pi r = -kr$. The gravitational pull in this geometry is not an inverse square force. Instead, it is proportional to r, and goes to zero as $r = 0$.

> Question 10-7
>
> If $F \to 0$ as $r \to 0$, does the billiard ball go slower and slower as it approaches the center of the earth?

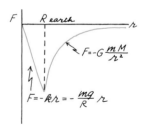

We have already explored a situation where the restoring force on an object is proportional to the displacement: $F = -kx$. Simple harmonic motion is produced, with a period equal to $T = 2\pi\sqrt{m/k}$. In this case, $k = \frac{4}{3}\pi m\rho G$. Since the mass of the earth is $M = \frac{4}{3}\pi\rho R^3$ and $g = G(M/R^2)$, another expression for k is $k = mG(M/R^3) = m(g/R)$. Therefore,

$$T = 2\pi\sqrt{\frac{R}{g}} \approx 2\pi\sqrt{\frac{(6.4 \times 10^6 m)}{\left(10\,\frac{N}{kg}\right)}} \approx 2\pi(0.8 \times 10^3) \approx 5 \times 10^3 s \approx 1.4\ hr$$

(So it would require only 42 minutes for a one-way trip to Australia.)

For a restoring force proportional to r, the potential energy is proportional to r^2, just as in the case of a spring. The gravitational potential energy inside the earth would not go to infinity as r goes to zero. Instead, the change of gravitational potential energy as an object moves from $r = 0$ to a radius, r, inside the earth is

$$\Delta E_{\text{grav pot}} = \tfrac{1}{2}kr^2 = \tfrac{1}{2}G\frac{mM}{R^3}r^2$$

The change of potential energy from the center to the surface where $r = R$ is

$$\Delta E_{\text{grav pot}} = \tfrac{1}{2}G\frac{mM}{R}$$

The potential energy at the surface is equal to

$$E_{\text{pot grav}} = -G\frac{mM}{R}$$

Therefore, at the center of the earth,

$$E_{\text{pot grav}} = -G\frac{mM}{R} - \tfrac{1}{2}G\frac{mM}{R} = -\tfrac{3}{2}G\frac{mM}{R}$$

The graph shows how the potential energy of the ball inside a planet with uniform mass is a quadratic function, merging at the surface with the inverse first power function. Actually, the mass inside the earth is not distributed uniformly. As we have seen, the core is much denser than the surface. Nevertheless, our argument here is a good first approximation.

Even if the billiard ball is resting quietly on the surface of the earth, it is still in a potential well at negative potential energy. To escape from the earth to where its potential energy would be zero requires a large amount of kinetic energy. To make $E_{\text{total}} = 0$, which means:

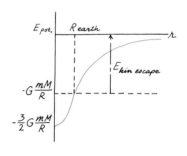

$$\tfrac{1}{2}mv^2 + E_{\text{grav pot}} = 0$$

$$\tfrac{1}{2}mv^2 - G\frac{mM}{R} = 0 \quad \text{and} \quad \tfrac{1}{2}mv^2 = G\frac{mM}{R}$$

The escape velocity is

$$v_{\text{escape}} = 2G\frac{M}{R} = 2gR \approx 2(9.8 \text{ m/s}^2)(6.4 \times 10^6 \text{m})$$

$$= 11.2 \times 10^3 \text{ m/s} = 25{,}000 \text{ mph}$$

Question 10-8

The final formula does not contain the mass of the billiard ball. Where did we go wrong?

How deep an energy well are we in, here on earth? Let's put numbers into the potential energy formula and compare the result with more familiar amounts of energy. For one kilogram bound on the surface of the earth, $m = 1$ kg, and $r = R = 6.4 \times 10^6$ m.

$$E_{\text{grav pot}} = -G\frac{mM}{r}$$

$$= -(6.7 \times 10^{-11} \text{ N} \cdot \text{m}^2/\text{kg}^2)(1 \text{ kg})(6 \times 10^{24} \text{ kg})/(6.4 \times 10^6 \text{ m})$$

$$\approx -6 \times 10^7 \text{ J}$$

Is that a lot of energy? It's about the amount that can be obtained by burning one-half gallon of gasoline. In terms of electrical units, it's about 17 kilowatt-hours, the amount of energy your clothes iron would use if you left it on for 17 hours. To free the kilogram object from the earth, you must shoot it upward with an initial kinetic energy of 6×10^7 J. Then its total energy will be zero, and it can escape. Note that although the escape *velocity* is independent of mass, the escape kinetic energy is proportional to mass.

Even if you escape from the earth, you are still bound in the sun's gravitational field. Let's compute the binding energy of a 1 kilogram object that is motionless, not on the earth but at the orbit of the earth.

$$E_{\text{grav pot}} = -G\frac{mM}{r}$$

$$= -(6.7 \times 10^{-11} \text{ N} \cdot \text{m}^2/\text{kg}^2)(1 \text{ kg})(2 \times 10^{30} \text{ kg})/(1.4 \times 10^{11} \text{ m})$$

$$\approx -9 \times 10^8 \text{ J}$$

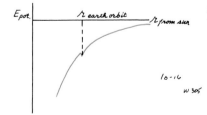

Here we have used the mass of the sun and the orbital radius of the earth. Evidently, we are in a hole due to the sun's pull that is 15 times deeper than the local well caused by the earth.

Throughout this chapter, we have been casually using two different expressions for gravitational force and gravitational potential energy. We started out saying that if you raise an air track glider through a height, h, the constant force required is the weight, mg. The work done, and the increase in gravitational potential energy, is mgh. On the other hand, we know very well that gravitational force caused by the earth is not constant,

THE EARTH'S POTENTIAL WELL

but decreases with distance as $1/r^2$. Furthermore, gravitational potential energy is really equal to $-GmM/r$, not mgh. Why are we justified in using wrong expressions?

Once again, remember that only *changes* in potential energy are measurable. Let's compare the change of potential energy given by the two different expressions for the same specific case. Let's calculate the change in potential energy of an object with mass, m, as it is raised through a height, h, from the earth's surface:

$$\Delta E_{\text{grav pot}} = -GmM\left[\frac{1}{R+h} - \frac{1}{R}\right] = -GmM\left[\frac{R-(R+h)}{(R+h)R}\right] = GmM\frac{h}{R^2 + Rh}$$

For small heights, hR is negligible compared with R^2 and can be ignored. Remember that $g = G\left(\dfrac{M}{R^2}\right)$. Then $\Delta E_{\text{grav pot}} \to mgh$ for small h.

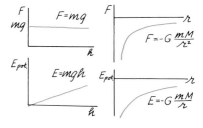

The expression mgh is a good approximation for small changes in height. Similarly, the weight of an object is not much different in an airplane 10 km above the earth's surface than it is on the surface. At 10 km, the distance to the center of the earth is greater by one part in 640 or about 0.15%. (6388 km compared with 6378 km.) The weight of an object at that height is less by 0.3%. (Since the weight is inversely proportional to r^2, an increase of $x\%$ in r leads to a decrease in $2x\%$ in weight, for small x.) Incidentally, note that if gravitational potential energy were equal to mgh for any height above the earth's surface, it would take an infinite amount of energy to free an object from the earth. We could never get those planetary probe rockets up and away.

If an object changes height in a gravitational field, there must be some other change in energy so that the total energy is conserved. In free fall through short distances, for example, $\Delta mgh = \Delta\frac{1}{2}mv^2$. For a satellite in a circular orbit, there is also a constant relationship between potential and kinetic energy. As shown on page 124, the centripetal force on an object in orbit is $mv^2/r = GmM/r^2$. Therefore the kinetic energy is $E_{\text{kin}} = \frac{1}{2}mv^2 = \frac{1}{2}GmM/r$. The gravitational potential energy of the satellite is $E_{\text{grav pot}} = -GmM/r$.

The potential energy of a satellite in circular orbit has twice the magnitude of the kinetic energy but is negative. The total energy is $E_{\text{tot}} = E_{\text{kin}} + E_{\text{pot}} = \frac{1}{2}GmM/r - GmM/r = -\frac{1}{2}GmM/r$. Notice that the total energy is negative, meaning that the satellite is bound, which, of course, it is. The smaller the radius of the orbit, the larger the magnitude of the negative total energy; that is, the satellite is deeper in the potential well and is more tightly bound. However, the smaller the orbital radius, the *larger* the kinetic energy. As a satellite descends, its velocity gets greater and greater. The velocity of a satellite close to the earth is greater than the velocity of the moon.

Question 10-9

Suppose that a satellite on the fringes of the atmosphere gradually loses energy through air friction. As the total energy of the satellite decreases, what happens to its speed?

MOLECULAR POTENTIAL WELLS

We can use the idea of potential energy curves even in cases where we don't know the exact nature of the energy or the detailed shape of the curve. For example, we know that certain combinations of atoms stick together to form molecules, and so there must be a force of attraction between the atoms. On the other hand, the atoms don't fall completely into each other, and so at close range there must be a force of repulsion. When an atom is in its normal molecular position with respect to its neighbors, it must be in a potential energy well. If the atom moves to a larger radius, or a smaller radius, there must be a restoring force to bring it back to its equilibrium position. In the diagram we sketch a plausible shape for such a potential energy well. The curve is a plot of $E(r)$ for a bound atom as a function of its distance, r, from a neighbor.

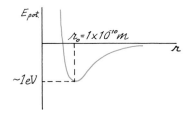

When an atom moves too close to its neighbor, at small r, the potential energy becomes positive because of the repulsion. When the atom moves beyond the equilibrium point, the potential energy becomes less negative and finally goes to zero, indicating that the atom has been freed. Note that the potential energy well is not symmetrical. The repulsion force that dominates when an atom gets too close to its neighbor usually rises more abruptly than the attraction force when the atom is pulled away. We will see the importance of this feature when we study heat.

Even without knowing the source of the molecular forces (they are electromagnetic), we can roughly calibrate the axes of the $E(r)$ curve. As we saw in Chapter 1, the radius of most atoms is about 1×10^{-10} m. In a molecule the atoms are essentially touching each other and so the equilibrium distance to the boundary between atoms is about 1×10^{-10} m. The binding energy of atoms (the depth of the potential well) has about the same value in most molecules. Within a factor of 3 or so this value is 1.6×10^{-19} J, a unit of energy that has a special name—an *electron volt* or eV. It is the amount of energy acquired by an electron that is accelerated through an electric potential of one volt. We will present more information about that definition in the chapter on electrostatics. In the meantime, all you have to know is that it is a very small amount of energy on a human scale, but is a natural size of unit for the atomic realm. Because the electron volt is defined in terms of electrons and volts this does not mean that it cannot also be used as a measure of mechanical energy or thermal energy, and for protons, atoms, or baseballs. After all, yardsticks are not used exclusively for measuring yards.

As an example of how the electron volt is a common magnitude of energy on the atomic scale, consider that most electrical cells (flashlight batteries, car batteries) yield between one and two volts. These potentials are created by the rearrangement of molecules, each one of which contributes one to two eV. It is also the case, as we will see in the chapter on light, that the energy of each photon, or particle, of visible light is from $1\frac{1}{2}$ to 3 eV, depending on the color. Since light can cause many chemical reactions—bleaching dyes, photosynthesis, tanning skin, darkening photographic film—it must be that the binding energy of chemical compounds is in the electron volt range.

Question 10-10

Suppose that a mole (6×10^{23}) of molecules combine or rearrange themselves, each yielding 1 eV. How many joules of energy are released?

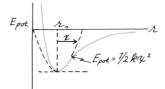

There is one more plausibility argument that we can make about the shape of the $E(r)$ curve for bound atoms. Although the whole well is asymmetrical, for small displacements from the equilibrium point any smooth-shaped well can be approximated by a parabola: $E_{pot} \approx \frac{1}{2}kx^2$, where x is the displacement (positive or negative) from the equilibrium point. That is the expression for the potential energy of a simple harmonic oscillator. If the bound atom is slightly disturbed, it should oscillate about the equilibrium point with simple harmonic motion. In the next section we will figure out an order-of-magnitude value for the constant, k, and so will be able to find an approximate frequency of oscillation of bound atoms.

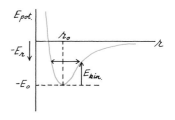

In the meantime, note the transformations of energy of an object vibrating in a potential well. Kinetic energy and potential energy change back and forth, much like the situation with a pendulum. The total energy is $E_T = E_{pot} + E_{kin}$, a quantity that is negative so long as the object is bound. The maximum kinetic energy exists as the object dashes through the equilibrium point, where $E_{pot} = -E_0$. Therefore, $E_T = -E_0 + E_{kin}$ (max). As the object slows down and finally stops at the turnaround point, the kinetic energy turns into potential energy that then has a smaller negative value. At all times, $\Delta E_{kin} + \Delta E_{pot} = 0$. As the kinetic energy decreases (ΔE_{kin} is negative), the potential energy increases (ΔE_{pot} is positive), and then the process reverses itself. If enough energy is fed into the system so that the maximum kinetic energy is greater than E_0, the total energy is positive and the object is no longer bound.

RESTORING FORCES AND POTENTIAL ENERGY

The amount of energy stored in a gravitational, or spring, or magnet system depends on the amount of distortion of the system. The distortion may be the displacement of a heavy object to a height, h; a stretch of a spring by an amount, x; or the forcing together of two repelling magnets by an amount, x. In the diagrams we have plotted the amount of stored energy as a function of the distortion, h or x.

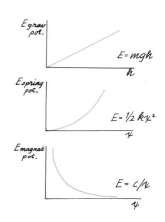

Question 10-11

Since the $E(x)$ graphs were derived from considerations of the forces necessary to cause the distortion, what good does it do to know $E(x)$? Why isn't $F(x)$ enough?

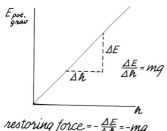

The potential energy of a system is a scalar quantity expressed in joules and by itself does not provide information about the future action of the system. Look at the sketches of $E(x)$ for three different spring systems, and notice the point for each of them where $E_{pot} = 1$ J. The first one apparently represents the situation for a weak spring that has been stretched a long distance. The second one must represent a strong spring, which had to be stretched only a short distance in order to store one joule. In the third case the spring has been compressed. Although the potential energy has the same value in each case, the actions of the springs if released will be quite different. The first one will pull back slowly to the left; the second one will snap back to the left; the third one will expand to the right. Although the single value of potential energy does not allow one to predict these different actions, the information is apparent in the shape of the whole $E(x)$ graph. It is the *slope* of the $E(x)$ curve at any point that is related to the restoring force *in the x-direction* that the system exerts at that point. Let's take some examples.

The graph of $E(h)$ for an object lifted away from the earth's surface (for small distances) has a constant slope. $\Delta(mgh)/\Delta h = mg$. The slope is equal to the weight of the object. There is a catch, however. The restoring force of gravity is pointed downward, and so is negative. The slope of $E(h)$ is positive. If we want the *restoring* force of the system, we must take the *negative* slope. $F_{restoring} = -\Delta E(h)/\Delta h$. The *outside* force that you would have to exert to store energy in the gravitational system is in the opposite direction, upward and positive.

The same situation applies to energy stored in a spring. The *restoring* force is given by

$$F_{spring} = -\Delta E(x)/\Delta x = -\Delta(\tfrac{1}{2}kx^2)/\Delta x = -kx$$

The restoring force obeys Hooke's law; it is proportional to the distortion and in a direction opposite to the distortion. Notice that this definition agrees with what you would expect qualitatively in the cases of the three springs that we described. In the first case the slope is small and positive; therefore the restoring force will be small and negative—toward smaller x. In the second case the slope is large and positive; the restoring force will be large and negative. In the third case the slope is negative; therefore the restoring force will be positive, making the spring expand.

■ Optional

There are several ways to see that the slope of the graph ($E = \tfrac{1}{2}kx^2$) is equal to kx. First, note the plausibility of this relationship by examining the accompanying graphs and the slopes that are drawn. The slope of ($\tfrac{1}{2}kx^2$) starts out horizontal at $x = 0$ and increases steadily. Second, remember the relationship between x and v for constant acceleration (page 38). The formula for x is ($x = \tfrac{1}{2}at^2$) and for v is ($v = at$), for $x_o = 0$ and $v_o = 0$. Since $v = \Delta x/\Delta t$, the slope of the curve ($x = \tfrac{1}{2}at^2$) is ($v = at$). Therefore, the slope of ($\tfrac{1}{2}kx^2$) must be (kx). Third, we can get the slope by performing the algebraic definition of slope:

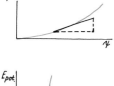
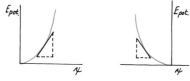

$$\frac{\Delta E}{\Delta x} = \frac{\Delta(\tfrac{1}{2}kx^2)}{\Delta x} = \frac{\tfrac{1}{2}kx_f^2 - \tfrac{1}{2}kx_o^2}{\Delta x} = \frac{\tfrac{1}{2}k\,[(x+\Delta x)^2 - x^2]}{\Delta x}$$
$$= \frac{\tfrac{1}{2}k\,[x^2 + 2x\Delta x + \Delta x^2 - x^2]}{\Delta x} = kx + \tfrac{1}{2}\Delta x$$

If we take the slope with smaller and smaller Δx, as Δx → 0, the slope becomes kx.

In the case of the magnets, where $E(x) = C/x$, $F_{magnet} = -\Delta(C/x)/\Delta x = +C/x^2$. The derivation of the slope in this case is shown in the box. Notice that the restoring force is positive; the magnets are shoving each other apart toward larger values of x.

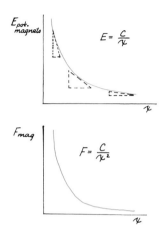

> Here is the algebraic derivation of the slope of $E = C/x$:
> $$F = -\frac{\Delta E}{\Delta x} = -\frac{C\left[\dfrac{1}{x+\Delta x} - \dfrac{1}{x}\right]}{\Delta x} = -\frac{C}{\Delta x}\left[\frac{-\Delta x}{(x+\Delta x)x}\right]$$
> $$= \frac{C}{(x+\Delta x)x} = \frac{C}{x^2 + x\Delta x} \to \frac{C}{x^2} \quad \text{as } \Delta x \to 0$$

Once again, notice the slopes that are drawn on the graph of $E(x)$. At small x, the slope is very steep and negative. Therefore, the force is large and positive. ($F = -\Delta E(x)/\Delta x$.) At large x, the slope is shallow and negative. Therefore the force is small and positive.

MOLECULAR FORCES

Any time that we have a potential energy system, we can find the restoring force associated with it. The force at any point is simply the negative of the slope of the $E(x)$ curve at that point. In a previous section we sketched a plausible graph of the potential energy of an atom bound in a molecule. From that graph we can sketch the qualitative nature of the force on the atom as a function of its distance from the equilibrium point. At distances smaller than the equilibrium point, the slope of $E(r)$ is large and negative. Therefore the restoring force on the atom is large and positive, toward larger r. At the equilibrium point, the slope of $E(r)$ is zero. Of course, the force should be zero when the atom is in equilibrium. For larger values of r, the slope of $E(r)$ is positive; the restoring force is negative, bringing the atom back toward equilibrium. This force of attraction gets larger as the atom gets further from equilibrium, but then, if the atom gets too far away, the force gets smaller and becomes zero.

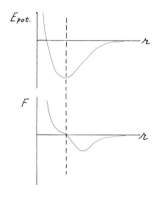

Since we have a rough value for the binding energy and radial separation of atoms in a molecule, we can derive a rough value for the forces involved. Let us assume that the half width of the potential well is equal to an atomic radius. In other words, when the atoms are touching, shoulder to shoulder, they are in their bound equilibrium position, but if they separate by more than a distance of the atomic radius, they are free. For $\Delta r = 1 \times 10^{-10}$ m, $\Delta E = 1$ eV $= 1.6 \times 10^{-19}$ J. Then the average restoring force during this separation is

$$F = -\Delta E/\Delta r = -(1.6 \times 10^{-19} \text{ J})/(1 \times 10^{-10} \text{ m}) \approx -2 \times 10^{-9} \text{ N}$$

That is a very small force, but then that's just the force on one atom. Let's

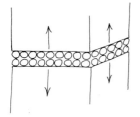

find the force that it would take to pull apart one square meter of such atoms. There are 10^{10} atoms along each meter, and so there are 10^{20} atoms in 1 m². The force exerted per square meter, F/A, is $\approx (2 \times 10^{-9})(1 \times 10^{20})$ $\approx 2 \times 10^{11}$ N/m².

In Chapter 4 we defined stress and strain and claimed that they are proportional: $F/A = -Y \Delta L/L$, where Y is Young's modulus. The values of Young's modulus for metals are in the range of 1×10^{11} N/m². The value of F/A that we obtained from the molecular forces was based on the assumption that the separation distance (in the direction of the force) between each atom was being doubled, from 1×10^{-10} m to 2×10^{-10} m. In other words, $\Delta L/L = 1$. It appears that Young's modulus is just the molecular force per unit area required to stretch a wire to double its original length.

Of course, Young's modulus does not have the same value for all materials, nor do all atoms have the same radius and the same binding energy. Still, it is apparent that our plausibility arguments about the nature of the potential energy curve for bound atoms must be qualitatively correct. As we saw before, a wire cannot be stretched until the strain, $\Delta L/L$, is equal to 1. As a matter of fact, most metals will fracture if the strain gets as large as 0.002. From the shape of the molecular $E(r)$ curve, there is no reason to expect breakdown for separation distances of only 2/1000 of the atomic radius. Fractures occur because of impurities or imperfections in the crystal structure of the metal.

We can also use the values for the size of the molecular energy well to determine the frequency of oscillation of a bound atom. If the well were parabolic all the way to a distance of an atomic radius, the potential energy would be

$$E_{\text{pot}} = -\tfrac{1}{2} kx^2$$

$$-1 \text{ eV} = -1.6 \times 10^{-19} \text{ J} = -\tfrac{1}{2} k(1 \times 10^{-10} \text{ m})^2$$

$$k \approx 30 \text{ J/m}^2$$

> ### Question 10-12
>
> Does this value of k agree with the value obtained from Hooke's law: $F = -kx$?

The frequency of a harmonic oscillator is given by $f = \dfrac{1}{2\pi}\sqrt{k/m}$. The mass of a carbon atom is about 2×10^{-26} kg. According to the model we have proposed, the oscillation frequency of a carbon atom bound in graphite or diamond should be in the range

$$f = \frac{1}{2\pi}\sqrt{(30 \text{ J/m}^2)/(2 \times 10^{-26} \text{ kg})} \approx \frac{1}{2\pi}(4 \times 10^{13})$$

$$\approx 10^{13} \text{ Hz (cycles per second)}$$

As we will see in the chapter on light, this frequency is smaller than that of visible light by a factor of about 100. It is in the far infrared range of frequencies. We will consider the effects of such oscillations in the chapter on heat, and will find that our simple model of vibrating atoms is useful. However, for a reason that we could not have anticipated, the model is a little too simple.

CENTRIFUGAL FORCES

If you swing an object around in a circle, it has rotational kinetic energy equal to

$$E_{\text{rot}} = \tfrac{1}{2} I \omega^2$$

If no external torques act on the system, the angular momentum, $I\omega$, must remain constant. To emphasize this condition, the rotational energy can be written as

$$E_{\text{rot}} = \tfrac{1}{2} (I\omega)^2 / I$$

If the rotating object is a point mass at the radius r, the moment of inertia is mr^2, and the rotational energy is

$$E_{\text{rot}} = \left(\frac{1}{2m}\right)(I\omega)^2 / r^2 = C/r^2$$

The constant, C, ties together the constants of mass and angular momentum. It appears that under these conditions (constant angular momentum and circular motion) the rotational energy is a function of position only. Energy that is a function of position can be considered potential energy. Each time that we define potential energy, however, we have to choose an appropriate zero point, and determine whether the energy is positive or negative. In this case, the convenient zero point is where $r \to \infty$.

> Question 10-13
>
> Wouldn't an object rotating with a very large radius have a large rotational energy—instead of zero energy?

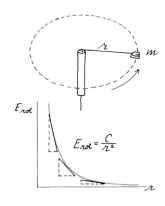

Whether we define the energy as being positive or negative depends on who does the work of changing position. If the system automatically tries to restore itself to zero energy, then the potential energy is positive. (This is like the situation with a compressed spring.) Take a look at the diagram showing a simple system of the type we are analyzing. Once the eraser has been set into rotation, the only torque on it is caused by the small amount of friction where the string rubs the edge of the tube. Since this torque can be made very small, the angular momentum remains constant for short periods of time. The only force that can be exerted within the system is radial. If you try spinning this device, you will find that the eraser tries to move toward a larger radius. To make it move in toward smaller radius, you must do external work on the system by pulling on the string. Therefore, according to our convention, the rotational energy is positive. The graph of $E(r)$ is shown in the diagram.

Every system with potential energy provides a restoring force equal to the negative slope of the potential energy curve. Since the energy of the rotating eraser is a function of r, the force must be radial. The slope of $E(r)$ is negative, and so the restoring force must be positive—in the direction of increasing r. The magnitude of the force is given by

$$F_{\text{restoring}} = -\Delta E/\Delta r = -\Delta(C/r^2)/\Delta r \to 2C/r^3 \quad \text{as } \Delta r \to 0$$

The proof of the last step in the derivation can be done algebraically with the same method used on page 213. By examining the graphs you can see how plausible it is that the function C/x should have a slope equal to $-C/x^2$, and that the function C/x^2 should have a slope proportional to $-C/x^3$.

Let's take a closer look at the value for the restoring force in this rotating system.

$$F = 2C/r^3 = 2\left(\frac{1}{2m}\right)(I\omega)^2/r^3 = (1/m)(m^2r^4)\omega^2/r^3 = m\omega^2 r$$

The restoring force has the same magnitude as the centripetal force, but is in the outward radial direction while the centripetal force is directed inward. The restoring force is called the *centrifugal* force. Whether the radial force in circular motion is outward or inward depends on the reference frame in which the force is measured. If you are outside a rotating system, you observe that it is necessary to exert an inward radial force to keep the object rotating in a circle instead of flying off at a tangent. If you are inside the rotating system, then you observe that there is a force tending to throw objects out radially. We will see other examples of reference frame problems in the next chapter. In the meantime, compare the derivations of the magnitude of this force as done here from energy considerations, and from a kinematic argument on pages 61 and 117.

> Question 10-14
>
> Suppose that you twirl the eraser with the device shown, and then pull down on the string, bringing the eraser in to a smaller radius. Are you doing work on the system or is the system doing work for you? Does the angular momentum of the system change? Does the angular speed change?

POWER

Power is the time rate of doing work: power = $\Delta E/\Delta t$. The unit of one joule per second is called the *watt*, symbolized by W. It is named after James Watt, a Scottish engineer and inventor who designed steam engines in the latter part of the eighteenth century. The watt is more familiar to most of us than is the joule, since electrical power is commonly measured in watts. A 60 watt bulb, for instance, uses 60 joules of electrical energy each second. We don't pay the electric company for watts—or power—however; we pay for energy. The common commercial unit of electrical energy is the kilowatt-hour (kWh). Notice that the unit is a product of power and time, yielding a unit of energy.

$$1 \text{ kilowatt-hour} \times \frac{3600 \text{ seconds}}{\text{hour}} \times \frac{10^3 \text{ watts}}{\text{kilowatt}}$$
$$= 3.6 \times 10^6 \text{ watt-seconds} = 3.6 \times 10^6 \text{ J}$$

In "Handling the Phenomena," we mentioned another unit of power, the one proposed by Watt. One horsepower = 746 watt. Don't feel inferior if you couldn't produce one horsepower while running upstairs. Most ordinary horses can't produce one horsepower either, particularly while running upstairs.

In many cases, the time during which a certain amount of energy is produced or absorbed makes a tremendous difference in the consequences. The amount of energy contained in a lightning bolt is in the range of 10^9 joules, or 300 kWh. That's about as much electrical energy as a household would use in one or two weeks. When it arrives in about one-third second, however, the narrow path down which it passes does not have time to dissipate the energy harmlessly, and instead vaporizes or explodes. The average power during a stroke is $(10^9 \text{J})/(\frac{1}{3}\text{s}) = 3 \times 10^9 \text{W}$, or 3000 megawatts (MW). Peak power, during flashbacks that last a few miscoseconds, reaches 10^{12} W.

> **Question 10-15**
>
> On page 150 we described a typical car collision. If the car has a mass of 1000 kg and stops in 0.1 second from a speed of 40 mph (19 m/s), what power must be dissipated, in watts and in horsepower?

UNITS AND RANGES OF ENERGY AND POWER

Many different units are used for energy and power, both because of the difference between English and metric (or S.I.) systems, and because of the enormous ranges of magnitude of energy dealt with in science and commerce. Let's take a look at some of the values and link them with familiar phenomena.

First, consider the size of the joule. It takes about 1 J to lift a one-kilogram book 10 cm. A standard two-cell flashlight uses about 1 J in one second. (The light uses about 1 W of power.) To raise the temperature of one gram of water one Celsius degree requires about 4 J. A ping-pong ball struck by a champion player may carry as much as 1 J of kinetic energy; most amateurs don't give a ping-pong ball more than about 0.1 J. If you burn a milligram of gasoline you produce about 40 J of heat energy, but only about 8 J of useful kinetic energy for your car. (The rest is wasted.) A milligram of sugar will provide your body with about 40 J of energy, which you use very efficiently to maintain body temperature and other activities.

The English unit of the foot-pound is the amount of work done in lifting a pound one foot. Note that it is roughly the same size as the joule, which is the work done in lifting a kilogram through 10 cm. Both are human-size units.

Chemists rate the energy of chemical reactions in terms of kilocalories per mole. A kilocalorie (which is the diet calorie) is 4184 J, or 2.6×10^{22} eV. Since a mole of molecular reactions is 6×10^{23} reactions, 1 kcal/mole is the same as $(2.6 \times 10^{22}$ eV$)/(6 \times 10^{23}$ reactions$) = 0.04$ eV reaction, or $\frac{1}{25}$eV/reaction. Most chemical reactions involve energy changes between 10 and 200 kcal/mole, or between $\frac{2}{5}$ and 8 eV per molecular reaction. The difference between chemical and physical changes, to the extent that the distinction is meaningful at all, is in terms of the energy involved per molecular transformation. For instance, it takes 2.7 eV per molecule of water to dissociate water into hydrogen and oxygen, clearly a chemical change. To melt ice into water, a physical change, takes about 0.06 eV/molecule.

SUMMARY

Potential energy is energy stored in the distortion of material or the displacement of material in electric or magnetic or gravitational force fields. We defined the magnitude of various forms of potential energy in such a way that in friction-free situations, $\Delta E_{pot} + \Delta E_{kin} = 0$. From air track experiments, letting gliders coast up hill or into springs or bar magnets, we define

$$E_{\text{grav pot}} = mgh \qquad E_{\text{spring pot}} = \tfrac{1}{2}kx^2 \qquad E_{\text{magnet pot}} \approx C/x$$

The zero point of potential energy is arbitrary. Only differences of potential energy are measured. If we say that an object has zero potential energy when it is "free" from the influence of some other object, then $E_{\text{grav pot}} \to 0$ as $r \to \infty$. $E_{\text{spring pot}} = 0$ when $x = 0$. Potential energy is usually considered to be negative if the object is bound in a system so that work is required to remove it. Thus, objects on the surface of the earth are bound in a gravitational potential well. The complete expression for gravitational potential energy for $r > R$ is $E_{\text{grav pot}} = -G(mM/r)$. For small changes of r, the complete expression for $\Delta E_{\text{grav pot}}$ reduces to mgh, where $h = \Delta r$. An object can escape from the surface of the earth if the positive kinetic energy it is given is equal in magnitude to its negative binding energy $\tfrac{1}{2}mv^2 = GmM/R$. The escape velocity of an object does not depend on its mass.

Given a graph of $E_{pot}(x)$, the restoring force of a system can be calculated

$$F_{\text{restoring}} = -\frac{\Delta E_{pot}}{\Delta x}$$

Reasonable assumptions about the shape of the potential energy for atoms bound in molecules yield a curve of restoring force that agrees with known facts about the strength of materials.

For circular motion with constant angular momentum, the rotational energy can be written as a function of r: $E_{rot} = \tfrac{1}{2}I\omega^2 = C/r^2$. For a point object where $I = mr^2$, the restoring force becomes $F = m\omega^2 r$, which is the equation for centrifugal force.

Power is the time rate of doing work: power $= \Delta E/\Delta t$. 1 J/s = 1 watt.

Answers to Questions

0-1: Philosophers (and physicists) love to argue about such questions. Physicists, however, usually take the practical approach that regardless of the true nature of reality, our business is to describe phenomena as consistently and efficiently as possible. The definition of a new form of energy may, in the first instance, be a human invention. The test of the usefulness of the invention, however, is the consistency with which the model describes a wide range of phenomena. If our definition of gravitational potential energy preserves an energy conservation law only for air track gliders, we should look further for a law of nature.

0-2: If h is the vertical height above the equilibrium point of the pendulum, then at all times there must be a constant value for the quantity $mgh + \frac{1}{2}mv^2$. At the top of the swing, where $h = h_{max}$, $v = 0$, and the total energy $= mgh_{max}$. The kinetic energy at the bottom of the swing is the total energy, which is equal to the gravitational potential energy at the top of the swing: $\frac{1}{2}mv_{max}^2 = mgh_{max}$. Therefore, the velocity at the bottom of the swing is equal to $v_{max} = 2gh_{max}$. Notice that this is just the velocity that the bob would have if it fell straight down through a height, h.

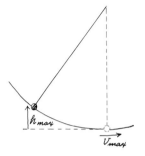

0-3: From energy considerations we can find the velocity of a round object after it has rolled down a hill through a vertical height,

$$v^2 = 2gh/(1 + K)$$

Note that the shape that has the smallest K yields the highest velocity. This fact agrees with the calculations in Question 6-9; the sphere will win the race. The earlier derivation gave the linear acceleration of the center of mass of each object: $a = (g \sin \theta)/(1 + K)$. (In this case, K is the geometrical factor for the moment of inertia about the axis that goes through the center of mass. The geometrical factor of $(1 + K)$ is for the moment of inertia about an axis on the rolling edge.) For constant acceleration down a distance, L, starting from rest, the final velocity can be found from $v^2 = 2aL = 2[(g \sin \theta)/(1 + K)](h/\sin \theta) = 2gh/(1 + K)$. The two different methods of calculation agree about the value of the final velocity.

0-4: If two objects attract each other, we would have to provide positive, external work to separate them. Therefore, it would be consistent to say that when the objects are close to each other they have negative potential energy.

0-5: Since the gravitational attraction to the earth is proportional to $1/r^2$, the only way to escape its influence is to go so far away that $1/r^2 = 0$, or is negligible. As $r \to \infty$, $1/r^2 \to 0$. Therefore, $E_{grav\ pot} \to 0$ as $r \to \infty$.

0-6: We don't shoot rocket ships up from the center of the earth where $r = 0$. We start them from the surface, where $r = R$, and $E_{grav\ pot} = -GmM/R^2$. The potential energy at the surface has a large, but not infinite, negative value. Can it really be true, however, that the potential energy at the center of the earth is infinite?

0-7: If the force goes to zero, the acceleration goes to zero—but that doesn't mean that the velocity goes to zero. Remember that the effective acceleration on a pendulum bob is zero at the equilibrium point, but its speed at that point is maximum.

0-8: In all cases where gravitational force produces acceleration, the result is independent of the mass of the accelerated object. The escape velocity is the same for rockets as it is for atoms of gas. Hydrogen and helium atoms near the top of the atmosphere eventually attain escape velocity due to chance bombardment by other gas molecules. There is a continual leakage of these gases from the atmosphere. But if escape velocity is independent of mass, why don't we also lose the heavier molecules and so lose all of our atmosphere? We'll have to take another look at this problem in Chapter 12.

0-9: Since $E_{tot} = -\frac{1}{2}G\ mM/r$, as E_{tot} decreases, the satellite sinks deeper into the gravitational potential well, toward greater negative values. The radius, r, decreases, giving E_{tot} a larger negative value. However, $E_{kin} = \frac{1}{2} G\, mM/r$. Therefore, as r decreases, E_{kin} gets larger, and the speed of the satellite increases.

10-10: $(6 \times 10^{23}$ interactions/mole$) \times (1.6 \times 10^{-19}$ J/interaction$) = 1 \times 10^5$ J/mole. For an order-of-magnitude comparison, note that a gallon of gasoline has a mass of about 3 kg and contains about 100 moles of molecules. If each molecule yields only 1 eV, the gallon of gasoline would furnish 1×10^7 J. The actual value is about 10 times greater.

10-11: The potential energy of a system, $E(x)$, is a function only of the final position or distortion of the system, regardless of how it got that way. A spring could be compressed rapidly or slowly, with a large force part of the way producing acceleration, and then a reverse force bringing the spring to rest. Extra force might also have been necessary to overcome friction. No matter what the history, once the spring is distorted by an amount, x, its potential energy is $E(x)$.

10-12: We found the *average* restoring force for an individual atom to be $\approx 1.6 \times 10^{-9}$ N. For the same simple model, the *maximum* force for $\Delta r = 1 \times 10^{-10}$ m would be 3×10^{-9} N. According to Hooke's law, $k = F/x = (3 \times 10^{-9}$ N$/(1 \times 10^{-10}$ m$) = 30$ N/m. This value for k is the same as the one obtained from the potential energy expression for a parabolic well.

10-13: The usual expression for rotational kinetic energy is $\frac{1}{2} I \omega^2 = \frac{1}{2} mr^2 \omega^2$. It certainly looks as if the rotational kinetic energy is greater for large r. However, if the angular momentum, $I\omega$, must remain constant, then the rotational kinetic energy of a system is $\frac{1}{2}(I\omega)^2/I = \frac{1}{2}(I\omega)^2/mr^2$. For such a system, without external torques, as r increases, the kinetic energy decreases.

10-14: To pull down on the string, thus bringing the eraser in to smaller radius, you must exert a force through a distance. You are doing work on the system, and providing energy to it. Since you are exerting a force only radially, you are creating no torque; therefore the angular momentum of the system must remain the same. If the angular momentum remains the same but the radius decreases, the angular speed must increase. Since for constant angular momentum, the energy of the system is equal to C/r^2, the system has greater energy at smaller radius. It gets this extra energy from the work that you do in pulling down on the string.

10-15: $E_{kin} = \frac{1}{2}(1000$ kg$)(18$ m/s$)^2 = 1.6 \times 10^5$ J.   Power $= \Delta E/\Delta t = (1.6 \times 10^5$ J$)/(0.1$ s$) = 1.6 \times 10^6$ W ≈ 2000 horsepower.

Problems

1. A stone with a mass of 2.0 kg is shot straight up with an initial velocity of 20 m/s. (a) What is the initial kinetic energy? (b) What is its potential energy at maximum height? (c) How high is it then? (d) What is its speed at half the maximum height?

2. If air friction is negligible, an object thrown up has a speed at every point equal to its speed on the way down. (a) Demonstrate that this must be so if the sum of kinetic and potential energy is conserved. (b) If there is energy loss due to air friction, what effect is there on the comparative speeds up and down at the same height, and what is the overall effect on the relative times of ascent and descent? It will help if you sketch the trajectories in the case without air resistance and with air resistance.

3. Do a Fermi type calculation (good to an order of magnitude) to find out how

much power is required to type a single symbol on a manual typewriter. (Force? Displacement? Time?)

4. Calculate and compare the gravitational escape energies and velocities for two objects, one on the surface of the moon and the other on the surface of the earth. Each has a mass of 1 kg.

$$M_{earth} = 6.0 \times 10^{24} \text{ kg} \qquad r_{earth} = 6.4 \times 10^{6} \text{m}$$

$$M_{moon} = 7.3 \times 10^{22} \text{ kg} \qquad r_{moon} = 1.7 \times 10^{6} \text{m}$$

5. Calculate and compare the gravitational escape velocity from the earth and the orbital velocity slightly above the earth's surface. To get the latter, recall what you learned about centripetal force and orbits. To compare newspaper accounts with your calculations, find the speeds in mph.

6. Do the calculations of Problem 5 for the moon, using the data in Problem 4. What is the period of a ship in close lunar orbit?

7. A rocket falling into the atmosphere from near-earth orbit attains a velocity so great that most of it will burn up because of air friction. Since we have to give a rocket the same velocity to get it into orbit, why doesn't it burn up as it leaves?

8. Write down algebraic expressions for total energy, kinetic energy, and potential energy of an object of mass m in circular orbit around the earth. Be sure that you have the signs correct. To *increase* total mechanical energy (sum of kinetic and gravitational potential) if you are in a spaceship you fire your rockets so that the exhaust is toward the rear. To *decrease* total mechanical energy you fire them forwards. If you increase your total mechanical energy, what happens to your distance from the Earth? What happens to your velocity? How would you go about catching up with something directly ahead of you?

9. Our atmosphere contains oxygen, nitrogen, and many other gases, but does not possess appreciable amounts of hydrogen or helium. Since these gases are continually fed into the atmosphere by various processes, they must be able to escape out of the gravitational potential well of the earth. Use the equations for kinetic energy and gravitational potential energy to explain how lighter gases might escape. After all, isn't escape velocity independent of the mass of the object? (At any given temperature, the *average* kinetic energy of any kind of gas molecule is equal to the average kinetic energy of any other kind of gas molecule. In other words, at any temperature where hydrogen and oxygen are mixed:

$$(\tfrac{1}{2} m_{H_2} v_{H_2}^2) = (\tfrac{1}{2} m_{O_2} v_{O_2}^2)$$

10. Suppose that you spin an object in a circle using a spring (or rubber band) as the radius arm. The system has two forms of energy: that due to rotation, and that from the stretched spring. Write down the equations for the sum of these two, which is the total energy of the system. Graph the two separate expressions, making sure that you use the proper sign for rotational energy. Note that the spring constant, k, is related to the value of the rotational constants. What happens to the dependence of the total energy on r? Does your result make sense? (Note that if you tried to walk in radially on a merry-go-round, you would have to be working against centrifugal force. If you were fastened to the center by a spring, the restoring force of the spring would always provide the correct force so that you could walk back and forth radially without exerting any other force.)

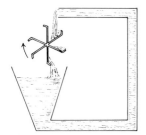

By "perpetual motion machine," we usually mean a device that produces more energy than is put into it. (There is also a second kind.) A diagram of such a machine is shown in the figure. The greater weight of water in the tank on the left forces water up the narrow tube, raising its potential energy. As the water falls through the water wheel, it converts its potential energy to kinetic energy and then into useful work in turning the wheel. What's wrong with this reasoning?

12. A 50 kg girl can dash up a flight of stairs 3 m high (about the standard distance between floors in a house) in 2.5 s. What is her horsepower during this effort?

13. Springs are frequently used for energy storage, for instance, in clocks. Compare the energy *density* (available energy/mass) of a spring with that of gasoline. A typical laboratory spring with a mass of 100 g will stretch 20 cm when holding a mass of 1 kg. The energy density of gasoline is about 4×10^7 J/kg. The spring might be stretched 20 cm before exceeding its elastic limits.

14. What is the escape velocity from the sun (and thus from the whole solar system) for a particle on the surface of the sun? ($R_{sun} = 7.0 \times 10^8$ m)

15. A roller coaster car is released with zero velocity at the highest point, which is 20 m above ground. It zooms down to 2 m and then back up to the top of the next hill, which is at a height of 15 m. If no energy is lost to friction, what are the speeds of the car at the 2 m valley and the 15 m hill?

16. A roller coaster has a loop-the-loop section with a radius of 10 m. If energy loss from friction is negligible, what is the minimum starting height, h, above the bottom of the loop, for the car to stay on the track? (For that height, the velocity of the car at the top of the loop will be just sufficient to require a centripetal force equal to the weight of the car.)

17. A crossbow shoots a bolt straight up with an initial speed of 60 m/s. (About how fast is this in mph?) Neglect the energy lost to air friction. If the bolt has a mass of 200 g, how high will it rise? How high will it rise if it has twice this mass?

18. A spring gun shoots a marble straight up to a height of 30 cm after the spring is compressed 1 cm. What is the initial (muzzle) speed of the marble? How high will the marble rise if the spring is compressed 3 cm?

19. Choose the zero of gravitational potential energy to be at the surface of the earth. In that case, the potential energy of an object in a well is negative. Suppose a one-kilogram rock drops from a point five meters down a well to the bottom, which is 30 meters beneath the surface. What is the final potential energy when the rock hits the bottom of the well? What is the initial potential energy when the rock is 5 meters down? What is the change in potential energy, and is this change positive or negative? What is the final speed just before the rock hits?

Two cylindrical magnets are placed in a smooth trough so that they will be pushed away from each other. Their magnetic potential energy in joules is $E_{mag\ pot} = 0.16/x$, where the separation distance, x, is in meters. They are prevented from separating too far by a light bronze spring arranged so that it is in its equilibrium, unstretched condition when the magnets are touching—that is, when $x = 0$. The spring potential energy is $E_{spring\ pot} = 1 \times 10^4\ x^2$, where x is in meters. Evidently the lowest *magnetic* potential energy is obtained by letting the magnets separate as far as possible. On the other hand, the lowest *spring* potential energy is obtained when the magnets are touching and the spring is unstretched.

On the same graph, plot the magnetic and spring potential energy. The potential energy of the system is the sum of these two. Add the two curves, point by point, to obtain the combined potential energy curve. The magnets will come to equilibrium at a separation distance that makes the potential energy a minimum. Use your graph to determine this distance.

In the figure there is a graph of $E_{\text{grav pot}}$ of an object as a function of horizontal displacement, x. What are the restoring forces for sections A, B, and C? In each case, what is the direction of the force?

Suppose that an atom is held in a potential well that is approximated by the graph shown here. Draw the corresponding $F(r)$ curve, labeling the axes with the proper units.

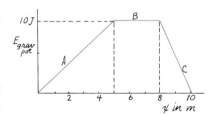

23. Calculate the energy required to raise 1 kg to a height of 100 km (above most of the atmosphere). Do the calculation two ways: (a) assume that g stays constant; (b) use the exact expression for potential energy in the earth's field. What is the percentage difference in the two calculations?

24. If your body needs 2000 kilocalories per day and could be fueled with gasoline, how much gasoline would you need each day? How much would it cost? If you could use electric power instead, how many kilowatt hours would you require? How much would that cost in your region?

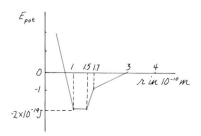

11 REFERENCE FRAMES AND RELATIVITY

It's hard to see things from someone else's point of view. It's particularly difficult if the other person is moving past you or circling around you. The other person has a different reference frame or, in terms of the physical world, a different coordinate system. If you are stopped at a red light beside another car, you might notice that the other car is rolling backward; the other person, in the other car, might claim that your car is rolling forward. In this case, a third person beside the road could tell you which car was moving *with respect to him*.

Suppose that while in the moving car you drop a coin; it will fall straight down into your lap, not into the seat behind you. According to you, the path of the falling coin is a straight line down. The observer beside the road sees both you and the coin traveling forward. Then he sees the coin drop along a path that we have proved to be a parabola. Is the true path of the falling coin a straight line or a parabola?

If the car in which you are sitting accelerates suddenly in the forward direction, you will be thrown back against the seat. From your reference frame you were evidently accelerated backward, presumably by some backward force. The observer on the side of the road will claim, however, that your car accelerated forward and so did you, presumably because there was a forward force acting on you. Is the true direction of the force on you forward or backward? If you go around a sharp curve while riding in a car, you will be thrown to the outside, presumably by an *outward* radial force. An observer watching from a roadway above would conclude that you started to travel along a line tangential to the car's curved path, but were then pulled into the circular path by an *inward* radial force. Is the force on you inward or outward?

Describing events from different reference frames is the subject of *relativity*. That word usually implies Einstein's theories of special and general relativity. Most of the special effects (but not all) of the *special* theory of relativity show up only when an object is traveling past us at nearly the speed of light. The *general* theory of relativity accounts for what happens in reference frames that are *accelerating* with respect to our own. But there is an older form of relativity that we will study first. It is named after Galileo, though he, himself, never used it in the modern form. A

THE GALILEAN TRANSFORMATION

Galilean transformation of coordinates allows us to simplify the description of many events in which objects are moving with respect to us. For instance, we will analyze cases where an object has a constant speed in one direction while undergoing acceleration in a perpendicular direction. We have already met such a case in describing trajectories, and we will study another application concerning simple harmonic motion. We will also see how the Galilean transformation can simplify the description of colliding objects.

> Question 11-1
>
> Does the earth go around the sun, or does the sun go around the earth?

HANDLING THE PHENOMENA

1. It may take some ingenuity to observe relativistic effects, although most of them can be seen or felt if you have access to a car, a driver, and a quiet road. First, note what happens to a ball that is dropped or thrown vertically upward while the car is rolling at slow, constant speed. Does the act of tossing it or catching it inside the car depend on your speed? Are the motions different from what they would be if the car were standing still? Station an observer outside the car. As you roll past her, lean out the window and lob the ball straight up, catching it as it comes straight back down again. What is the path of the ball according to the observer on the road?

2. While you are a passenger in a car, hold a plumb bob and watch it as the car starts up, goes around corners, and stops. The plumb bob can be any kind of small weight hung on the end of a string. When you and the plumb bob are motionless with respect to the earth, the plumb bob hangs straight down. In fact, the direction of the plumb bob can define "down." What direction does the plumb bob assume while you are traveling at constant speed? What direction is "down" when the car accelerates forward while starting up, or backward while braking? What does the plumb bob do when you go around corners to the right or the left? [If you ever get your hands on a helium balloon, try all of these car maneuvers (with someone else driving) while you hold the balloon on the end of a string. When you are motionless, the balloon is an anti-plumb bob; its direction defines "up." What will the balloon do as you accelerate forward, backward, or sideways?]

THE GALILEAN TRANSFORMATION

In spite of the formidable name, the problem that we face is very simple. Two people, with measuring instruments, observe the same phenomenon. It might be a ball shooting through the air, or the collision of two objects, or a weight on a spring moving up and down with simple harmonic motion. If the two observers are moving with constant velocity with respect to each other, what differences do they observe in the phenomenon? To make it easier to talk about, we will give each observer his own x, y, z coordinate system. One of them will move in the x direction at constant velocity, v,

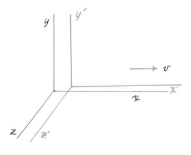

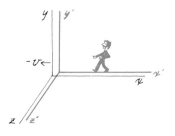

with respect to the other one. In the diagram, the x', y', z' frame is moving to the right with respect to the x, y, z frame. We must resist the temptation of thinking that the x, y, z frame is stationary while the x', y', z' frame is the one actually moving. All that is important is that they are moving with respect to each other. An observer in the x', y', z' frame assumes that his frame is stationary while the x, y, z frame is moving to the left with a velocity $-v$.

In order to transform a description of motion from one frame to the other, we must write down the coordinate transformation. Since the motion is only in the x-direction, the y- and z-coordinates in the two frames are the same. Furthermore, time is the same in both reference frames. After all, why should your clock run fast or slow just because you are traveling along a highway? (Before we are through with this chapter, we will see that your clock *does* change with respect to somebody by the side of the road, but for now we will use the Galilean assumption.) The Galilean transformations are

$$t = t' \qquad y = y'$$
$$z = z' \qquad x = x' + vt$$

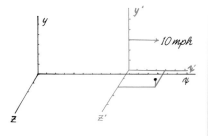

Question 11-2

Suppose that at 10 o'clock an object in the primed frame is at the point $x' = 4$ miles, $y' = 1$ mile, and $z' = 2$ miles. If the origins of the coordinate frames were coincident at 9 o'clock and the primed frame has been moving in the x-direction with a velocity of 10 mph, what are the coordinates of the object in the x, y, z frame?

Galilean Transformation: Example 1

Let's use the Galilean transformation to confirm the observation that you have already made in "Handling the Phenomena." If you drop a stone while you are inside a car that is moving with constant velocity, the stone appears to you to fall straight down. To an observer at the side of the road, however, the stone appears to be falling in a parabolic path. Let the observer by the side of the road take the unprimed reference frame and use the primed coordinate system for yourself. The falling stone satisfies the equation

$$y' = \tfrac{1}{2}g(t')^2$$

If we simply substitute the expressions for y and t, we find what we already know. The y-component of a moving object is unaffected by its motion in the perpendicular x-direction:

$$y = \tfrac{1}{2}gt^2$$

However, since the *time* is also related to x and v, let's make another substitution for t':

$$t' = t = (x - x')/v$$

Substituting this expression for t into the formula describing y, we get

$$y = \tfrac{1}{2}g\left(\frac{x - x'}{v}\right)^2$$

If we so choose, we can claim that the stone is dropping from the origin of our moving reference frame, which means that $x' = 0$. In that case,

$$y = \tfrac{1}{2}(g/v^2)x^2$$

The equation of motion in the unprimed frame describes a parabola ($y = Ax^2$). This is the path as seen by an observer at the side of the road. The formula is the same as the one we derived in Chapter 3, page 000. In that case we described a ball that had been thrown with horizontal velocity but that was also freely falling with an acceleration g. So we have learned no new physics by using the Galilean transformation. But note the different mental reference frame in considering the problem in this new way. Before, we threw a ball and knew very well that it was traveling independently in both the x- and y-direction. However, in terms of a reference frame problem, the ball's motion in one reference frame is straight down and in other reference frame it follows a parabolic path.

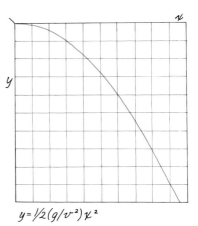

$y = 1/2(g/v^2)x^2$

Question 11-3

What is the true path of the falling ball?

Galilean Transformation: Example 2

Now let's transform a motion from the x, y, z frame into the x', y', z' frame. Suppose that the motion is simple harmonic, so that at $x = 0$ an object is oscillating up and down with a motion:

$$y = A \sin 2\pi ft$$

What does this motion look like in a reference frame that is moving along the x-axis with a velocity v? You can see the answer approximately by moving a pencil up and down on a piece of paper, using a motion that is as close to S.H.M. (simple harmonic motion) as you can make it. While you are doing this, have someone else slowly and steadily move the paper in a direction perpendicular to the pencil motion. You can also get the same effect all by yourself by moving a chalk up and down on a blackboard with S.H.M. while walking steadily beside the blackboard.

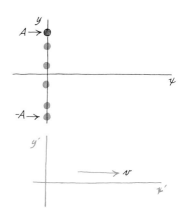

Question 11-4

Aren't these two demonstrations basically different? In one case, the paper is moving and, in the other case, you are moving along beside the blackboard.

To transform the equation for simple harmonic motion to the primed system (the paper or the blackboard), we use the substitutions:

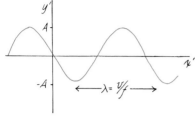

$$y = y' \quad t = t' = (x - x')/v$$

If the S.H.M. always takes place at the origin of coordinates where $x = 0$, the transformation is

$$y' = A \sin 2\pi f(-x'/v) = -A \sin 2\pi (f/v)x'$$

The transformation has changed a sinusoidal motion in *time* into a sinusoidal motion in *distance*. The wave length of this sine wave is λ (lambda), which is a function both of the frequency of the S.H.M. and also of the velocity between the two reference frames.

$$y' = -A \sin (2\pi/\lambda)x' \quad \text{where } \lambda = v/f$$

The reference frame change that we have just described changes an oscillation in time into an oscillation in space. The best known example of this transformation is the phonograph record. Both recording needle and pick-up needle oscillate along a single line with a motion that is vibratory, if not S.H.M. The grooves in the record, however, snake along with a sinusoidal-type wave in space. Note that the wave length along the groove in the record is $\lambda = v/f$. In this case, v is the velocity of the record groove past the recording needle, and f is the frequency of the recording needle. In the playback arrangement, the coordinate systems must be transformed again. The sinusoidal motion in space along the groove turns into a sinusoidal vibration in time of the playback needle. If the velocity of the playback is v_1 instead of v, the oscillations of the playback needle are

$$y = -A \sin (2\pi/\lambda)x' = -A \sin (2\pi/\lambda)(-v_1 t)$$

We obtained this substitution by choosing $x = 0$ in the expression: $x' = x - vt$. Since $\lambda = v/f$, our formula for y becomes

$$y = A \sin 2\pi(v_1 f/v)t = A \sin 2\pi f_1 t$$

The frequency that we hear is not necessarily the frequency that was recorded. The playback frequency is $f_1 = (v_1/v)f$.

Question 11-5

What happens if you play a modern $33\frac{1}{3}$ rpm record on an old turntable at a speed of 78 rpm?

Galilean Transformation: Example 3

We have seen that a change in reference frame can change the description of the path of a moving object and can also transform displacement as a function of time into displacement as a function of position. How do velocities transform? If you are walking at 2 m/s in a train that is moving at 30 m/s, how fast are you moving with respect to the ground? ("Nonsense," said the inspector as the train hurtled past the station, "I have been sitting motionless in my seat all this time.") We know intuitively that, if you are walking toward the front of the train, your speed with respect to the ground must be the sum of the train's speed and your speed with respect to the train. If you are walking toward the back of the train, your resultant speed with respect to the station is the difference between the train's speed and your speed. Let's prove this analytically.

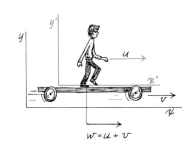

At time t: $x = x' + vt$

At time $(t + \Delta t)$: $(x + \Delta x) = (x' + \Delta x') + v(t + \Delta t)$

Subtracting the first equation from the second:

$$\Delta x = \Delta x' + v\Delta t$$

Dividing through by Δt:

$$\Delta x/\Delta t = \Delta x'/\Delta t + v$$

If we call the resultant velocity in the unprimed frame w and the velocity in the primed frame u, then

$$w = \Delta x/\Delta t \quad \text{and} \quad u = \Delta x'/\Delta t$$

$$w = u + v$$

Actually we have been dealing only with the component of velocities in the x-direction. The components of velocities in the other directions do not depend on the motion along the x-axis. In the direction of motion, however, the velocity in one reference frame is the algebraic sum of the velocity in the other reference frame plus the relative velocity between the two frames.

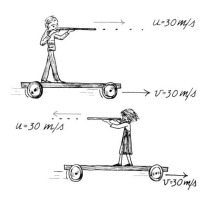

Question 11-6

Suppose that you have a BB gun with a muzzle velocity of 30 m/s and that you are on a train that is traveling at 30 m/s. If you shoot in the forward direction, what is the speed of the BB with respect to an observer at a station? If you stand on the back platform of the train and fire in the backward direction, what is the motion of the BB as seen by an observer at a station?

Galilean Transformation: Example 4

As usual, when we start probing into nature, simple and obvious rules sometimes lead to paradoxical conclusions. It seems reasonable that velocities should add algebraically in a change of coordinate systems. (Note, however, that in our derivation we assume that time is the same in both reference frames. As we will see at the end of this chapter, that assumption is not valid and, therefore, the simple addition rule that we have derived is only an approximation.) What happens to the kinetic energy of the BB as seen in the two reference frames? *Energy must also be relative.* In the moving train, the BB has the same kinetic energy—regardless of which direction it is fired: $E_{kin} = \frac{1}{2}m \ (30 \text{ m/s})^2$. The observer at the station, however, sees quite a different situation. If the BB is fired in the forward direction, then the observer measures the kinetic energy to be $E_{kin} = \frac{1}{2}m$ (60 m/s)². That is four times the kinetic energy that is measured inside the train. To make matters worse, if the BB is fired in the backward direction, the observer on the platform is correct in claiming that the kinetic energy is zero since its velocity is zero.

> ### Question 11-7
>
> Does the BB have kinetic energy or doesn't it? And, if so, how much? What happens if you measure that energy by firing the BB into a target and measure its penetration?

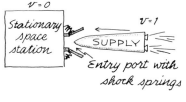

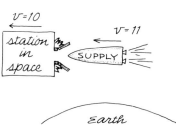

Collision problems such as the ones we have just proposed must be handled very carefully and very thoroughly as you transform from one coordinate system to another. In each reference frame you must set up the equations for conservation of energy and momentum within that reference frame, and you must include the motions and recoils of all of the participating objects. As an example of the problems encountered when you don't take these precautions, here is a simplified version of a problem that was described years ago in the magazine *Astounding Science Fiction*. The author claimed that he had discovered an insurmountable obstacle to the establishment of space stations. Suppose that a supply rocket must link with a space station that has a relative velocity of 1 (1 m/s or any other unit—it makes no difference). The kinetic energy of the supply rocket, relative to the station, is therefore $\frac{1}{2}m(1)^2$. That much energy must be absorbed by the shocks of the space station. These shock springs are tested on earth so that they can slow down and stop the supply rocket without damage. Now let's put the space station in orbit where it is traveling at a velocity of 10 with respect to the earth (10 of the same units, whichever you have chosen). The supply rocket must have a velocity of 11 units with respect to the earth if its relative velocity with respect to the space station is to remain 1 unit. Therefore, the kinetic energy of the supply rocket, with respect to the earth, is: $\frac{1}{2}m(11)^2$. If it slows down and stops in the space station, its final velocity will be 10, and its final kinetic energy will be $\frac{1}{2}m(10)^2$. Look at the disastrous consequences! The kinetic energy that must be lost to the shocks in the space station is $[\frac{1}{2}m(11)^2 - \frac{1}{2}m(10)^2] = \frac{1}{2}m(21)$. That's 21 times the energy that we calculated would be lost on earth. The actual situation with the space station would be much worse, because the actual velocity of a satellite in orbit around the earth is so large that taking the difference between two large squares yields an enormous difference. As the author of the note in *Astounding Science Fiction* said long ago, "The supply rocket will gently ease its way into the shock springs and slowly but inexorably push them aside and continue plowing its way through the space station and out the other side."

> ### Question 11-8
>
> Is the space program doomed? Where did we go wrong?

Optional

Galilean Transformation: Example 5

So far we have been comparing descriptions of events as seen from two different reference frames. The one containing ourselves we usually consider to be stationary; the other we assign to some system moving past us. Sometimes it is useful and revealing to analyze events from a reference frame attached to the center of mass of interacting objects. In Chapter 7, and again in Chapter 9, we learned how to describe the collision of two objects. These objects might be cars on a slippery road, in which case we can probably calculate only approximate results, or they might be subatomic particles where we can apply the conservation laws very precisely. (With the collisions of cars, there are problems concerned with the inelasticity of the collision, the need to conserve angular momentum even though the impact parameters are not well known, and the unknown road interaction.) The laws of conservation of energy and momentum are particularly simple in the reference frame of the center of mass. In the diagram, we compare the collision of two protons as seen from two reference frames. In the first, we assume that one of the protons is at rest with respect to us. We get the same geometry that we found when we analyzed the elastic collision on page 183. In the second part of our diagram, we see the same event from the point of view of the center of mass. Since the protons have identical mass, the center of mass must always be halfway between the two protons. From our point of view, if the bombarding proton is traveling to the right with velocity v_0, the center of mass is traveling to the right with velocity $v_0/2$. From the reference frame of the center of mass, however, the proton from the left is approaching it at $v_0/2$ and also the proton to the right is approaching it with a velocity $-v_0/2$. Necessarily, in this privileged reference frame, the net momentum of the participants is 0. At all times, including after the collision, the net momentum must remain 0. That can happen only if the protons in their glancing collision shoot out in directions opposite each other. The speed of each proton with respect to the center of mass will remain the same: $v_0/2$. Otherwise kinetic energy would not be conserved in the center-of-mass frame. The protons can shoot off at any angle as long as they leave opposite each other. In this reference frame, it is easier to calculate the probability of a proton leaving at any particular angle.

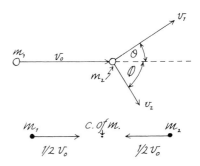

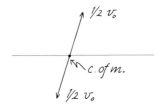

Question 11-9:
If the protons can leave at any angle, why isn't there the same probability for each angle?

The transformation back to the laboratory system is easy. Simply add vectorially the velocity of the center of mass to the velocity of each of the protons. That geometrical construction is shown in the diagram. The transformation is particularly easy in the case of identical particles and is not very difficult even when the particles have different masses. In the case of identical particles, note that the vector addition of velocities for each particle forms an isosceles triangle. The simple geometrical relationships demonstrate a fact that we derived algebraically in Chapter 9. In a collision between two particles with identical mass, one of which was originally motionless, the opening angle between their final directions must be 90°.

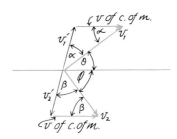

Because: $v_1' = \tfrac{1}{2} v_0 = v_{c.of\,m.}$
and: $v_2' = \tfrac{1}{2} v_0 = v_{c.of\,m.}$
∴ each triangle is isoceles
and $\alpha = \alpha$
$\beta = \beta$

Because of equality of interior angles:
$\theta = \alpha$ and $\phi = \beta$

Since: $\alpha + \theta + \phi + \beta = 180°$
$2\theta + 2\phi = 180°$
$(\theta + \phi) = 90°$

THE LOCAL FIELDS PRODUCED BY ACCELERATED REFERENCE FRAMES

As your jet plane races forward on takeoff, are you accelerated forward or backward? The observers at the airport would say that you are certainly accelerated forward. You yourself know very well, however, that you are thrown backward against your seat. If you are a passenger in a car careening around a corner to the left, are you thrown toward the door or toward the driver? If the door is loose, you might be thrown outward by the centrifugal force. To be sure, the observer from the helicopter above would then explain that the door was not able to provide sufficient centripetal force to keep you going in a circle. In both these cases, it appears that the direction of forces and the consequent accelerations depend on the reference frame in which they are described.

Newton was very concerned about the problems of describing motions from different reference frames. His three laws apply strictly only within the reference frame that he proposed—a reference frame at rest or moving with constant velocity with respect to the fixed stars. Such a frame is called *inertial*. Newton's first law holds strictly only within an inertial frame. Indeed, the first law can be used as a definition of an inertial reference frame.

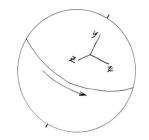

> ### Question 11-10
>
> What is Newton's first law, and how would you use it to test whether we are in an inertial frame of reference here on the surface of the earth?

Newton claimed that you could tell whether you were in an inertial reference frame by looking at the surface of liquids. For instance, when you are at rest or traveling with constant velocity, the surface of a liquid is horizontal and flat. Of course, there must be a uniform gravitational field perpendicular to the surface in order to keep it flat. Although you cannot measure absolute velocity, you can determine *acceleration* in the inertial reference frame simply by looking at the liquid surface. If you are accelerating forward, the water piles up toward the back, creating a flat surface at an angle to the horizontal. If you spin a pail of water, the surface becomes parabolic. If you were to station an observer on the surface of the rotating water and let her claim that she is at rest, then she would have to explain why all of the fixed stars were rotating rapidly around her. As a matter of fact, Ernst Mach (1838–1916) pointed out that it was not so far-fetched for an observer to assume that the universe rotating around her produces centrifugal force. The source of inertia itself is not understood. Perhaps the inertial mass of any object is produced by the gravitational influence of all of the rest of the mass in the universe. An object without any universe around it would have zero mass. Perhaps in the reference frame of an object, a rotating universe produces centrifugal force.

THE LOCAL FIELDS PRODUCED BY ACCELERATED REFERENCE FRAMES

Regardless of these philosophical problems, it is frequently convenient to describe phenomena from frames of reference that are not inertial. We are frequently in reference frames that are accelerating with respect to the fixed stars. We ride in elevators, go around corners in cars, and map the movement of the winds on our rotating earth. In all of these cases, we may find it convenient to deal with inertial forces that are caused by our acceleration with respect to inertial reference frames. These inertial forces cannot be described in terms of "source charges," such as gravitational mass, electromagnetic charge, or nuclear hypercharge. Instead, the inertial forces act on the mass of an object as if they were local gravitational fields. These inertial fields are functions of the position or motion of the object within the noninertial reference frame. Let's look at some examples.

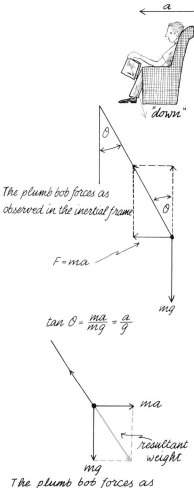

The plumb bob forces as observed in the inertial frame

$$\tan \theta = \frac{ma}{mg} = \frac{a}{g}$$

The plumb bob forces as observed in the accelerating frame.

Noninertial Reference Frames: Example 1

The direction of "down" is determined experimentally by a plumb bob. A plumb bob does not necessarily point toward the center of the earth. There are local anomalies created by nearby mountains. Besides, the earth is not quite spherical. For our present purposes, however, a plumb bob points "down" because that is the direction of the gravitational field. The gravitational force on a plumb bob with mass, m, is $F = mg = G(mM/R^2)$. The radius of the earth is R and the mass of the earth is M. If you hold a plumb bob in a car or airplane that is accelerating forward, the line is no longer vertical; it points slightly backward.

The effect is particularly easy to see as you take off in a jet liner. Instead of a regular plumb bob, you can use the magazine from the seat pocket in front of you. Simply hold the magazine at one corner between your thumb and index finger. You will see the corner that is diagonally opposite swing back toward you as the plane accelerates.

One way to describe this phenomenon is with the force diagram shown in the margin. The string attached to the bob exerts a force sufficient to balance the downward gravitational force on the bob and also to provide a horizontal force that produces horizontal acceleration. As you can see from the diagram, the tangent of the angle is equal to the ratio of the horizontal acceleration, a, to the gravitational field strength, g. However, suppose that your plane is completely sealed so that you do not know that the airport is rushing backward away from you. The direction of the plumb bob is the direction of "down." That direction is apparently produced by the vector sum of two gravitational-type fields. The gravitational field of the earth produces a force of mg; a local inertial acceleration field produces a force of ma. Their resultant is the net field. Not only does the plumb bob point in a new direction, but its weight is also slightly greater.

Instead of a plumb bob which defines "down," you could use a helium balloon whose direction defines "up." The reaction of a helium balloon in a closed car is exactly opposite to that of a plumb bob. As the car accelerates forward, the helium ballon stretches in the forward direction. If you turn a corner sharply to the left, the helium balloon will also swing left. In effect, a helium balloon is an anti-plumb bob.

> **Question 11-11**
>
> Can the action of the helium balloon also be analyzed in terms of the motion of the air in which the balloon floats?

Noninertial Reference Frames: Example 2

In an elevator, the acceleration is in line with the gravitational field, although its direction may be plus or minus. The direction of the gravitational field is always minus. You have experienced the result of these local fields in elevators that start or stop abruptly. When the elevator accelerates upward you feel heavier; just before it stops at a floor you feel lighter. The effect is easy to measure quantitatively by standing on a bathroom scale while riding an elevator. If other passengers get on, you can devise many interesting explanations of what you are doing.

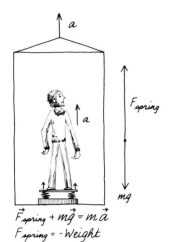

$\vec{F}_{spring} + m\vec{g} = m\vec{a}$
$\vec{F}_{spring} = -\text{Weight}$

Once again, we have two ways of describing the readings on the bathroom scales in the elevator. First, we can simply use Newton's second law on the person being weighed. As you can see in the diagram, there are two forces in the vertical direction that act on the person standing on the scale. The gravitational attraction of the earth is pulling down with a force equal to $m\mathbf{g}$. The springs on the scale are pushing up with a force that can be read from the dial. Since it is well known that bathroom scales never lie, that force is equal to the person's weight, W (but in the opposite direction). According to Newton's second law:

$$\mathbf{F}_{net} = m\mathbf{a}$$

$$\mathbf{F}_{spring} + m\mathbf{g} = m\mathbf{a}$$

If you know your mass, m, and the value of the local gravitational field, g, you can determine your upward acceleration by reading your weight on the scale.

> **Question 11-12**
>
> Some elevators can produce an upward acceleration of 1 m/s². If your weight while standing still is 500 N (about 110 pounds), what is your weight when the elevator first starts up? Do the bathroom scales continue reading this quantity as the elevator rises?

local field inside elevator is $(\vec{g} + \vec{a})$ ∴ $w = m(g+a)$

The other way of explaining the experimental results is to conclude that the local gravitational field changes from time to time. At the moment when the ground floor of the building leaves you and hurtles downward (as seen in your reference frame), the gravitational field increases. As the first floor arrives opposite you, your local gravitational field decreases. The net field acting on your mass is the sum of the earth's gravitational field and the acceleration field. If the outside world has an acceleration, a, downward, the resultant field is $(\mathbf{g} + \mathbf{a})$. Your weight in that case is $W = m[|\mathbf{g}| + |\mathbf{a}|]$.

As the outside world slows down opposite your elevator, its acceleration must be positive. According to the bathroom scales, and your own feeling, your weight is now only $W = m[|\mathbf{g}| - |\mathbf{a}|]$. Notice that the quantitative conclusions are the same regardless of which mental model we use to describe the elevator trip.

Noninertial Reference Frames: Example 3

Physics courses and physics textbooks frequently warn against the fallacy of thinking that there are centrifugal forces. Centrifugal forces are alleged to be merely fictitious forces. Indeed, you can describe the events in a rotating system by removing yourself from the action and standing above it all in an inertial reference frame. Take the case of the passenger in the car that is careening in a left-hand turn. The car door flies open, the passenger flies out. From your sublime and inertial vantage point, you can explain that the car door no longer provided sufficient centripetal force to keep the passenger going in a circle. From the passenger's point of view, of course, the door flew open and he shot out radially. You would have a hard time convincing the coroner that the death was caused by a fictitious force.

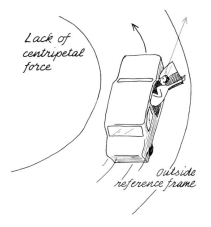

Lack of centripetal force

Outside reference frame

Which reference frames we use for our descriptions of phenomena is a matter of convenience. There is never any conflict between descriptions from different reference frames *as long as we don't mix them up*. We must not, for example, equate a centrifugal force in one reference frame to a centripetal force in another. If we are stationed in an inertial reference frame, observing events in an accelerating frame, then we have no need to use inertial forces such as the centrifugal force. On the other hand, if we are *in* an accelerating system, such as a merry-go-round, it may be more convenient to describe effects using the net local field produced by gravitation and any inertial forces. As a matter of fact, since we live on a giant merry-go-round, it is common practice to name and use the inertial forces just as if they were real in the Newtonian sense.

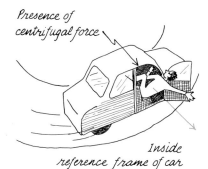

Presence of centrifugal force

Inside reference frame of car

> ### Question 11-13
>
> What fraction of your weight do you lose on the equator because of the centrifugal force produced by the earth's rotation?

Noninertial Reference Frames: Example 4

So far, we have only described one kind of inertial force produced by rotation—the centrifugal force. However, there is another important effect—the Coriolis force. This is the force that acts always to the right on an object moving along the surface of the earth in the northern hemisphere and always to the left on an object moving in the southern hemisphere. It is the Coriolis force that creates the great cyclonic motions of the atmosphere as the winds move from regions of high pressure to regions of low pressure. In the diagram, we show how winds moving from all directions toward a low-pressure region in the northern hemisphere produce a counterclockwise circulation; around a high-pressure region, a clockwise circulation is set up.

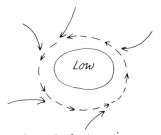

Counter-clockwise winds around a low pressure region

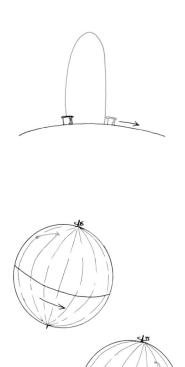

The smaller and more intense cyclonic phenomena of the hurricane and tornado also follow the same directional rule.

The origin of the Coriolis force can be understood in terms of the requirements of conservation of angular momentum. Suppose for instance that a cannon shoots a shell vertically upward from the equator. If the shell is lobbed up only a few feet, we would expect it to fall right back down into the mouth of the cannon. To be sure, the earth is rotating, but the shell shares in that tangential velocity as it leaves the mouth of the cannon. If the shell rises to a great height, however, another effect begins to be important. The shell must maintain its angular momentum with respect to the earth. That angular momentum is given by $mvr = mr^2\omega$. As the distance, r, from the axis *increases,* the angular velocity, ω, must *decrease.* Consequently, the angular velocity of the shell is less than that of the cannon throughout its entire trajectory. It will fall to the west of the cannon.

Now consider what happens when the wind blows due north in the northern hemisphere. Since the wind follows the curvature of the earth, it is getting closer to the axis of the earth. Therefore, the angular velocity must increase, since r is decreasing. This increase of angular velocity will divert the wind to the east, which is to the right of its northward motion. Similarly, if the wind is attempting to blow due south, it will follow the curvature of the earth and move to regions of greater r from the axis. Therefore, its angular velocity must decrease, making it slip to the west. Once again this direction is to the right of its main motion.

A complete analysis of the Coriolis force on the rotating earth shows that it is equal to

$$\vec{F}_{\text{Coriolis}} = -2m\boldsymbol{\omega} \times \mathbf{v}$$

where **v** is velocity (with respect to earth) and $\boldsymbol{\omega}$ is the angular frequency *of the earth.* Notice in this formula that the Coriolis force is zero for *tangential* velocities north or south on the earth's equator. For such motion, the velocity is parallel to the direction of rotation, $\boldsymbol{\omega}$. If the two vectors are parallel, their cross product is 0. At the poles, the Coriolis force is maximum, since any tangential velocity is automatically perpendicular to the direction of rotation. Heading northward in the northern hemisphere, there is a component of tangential velocity in the direction of the axis, and a component perpendicular to the axis that produces a force to the right. Similarly, in moving southward, the perpendicular component of velocity is away from the axis. The resultant force is once again toward the right.

There is a legend that wash bowls in the northern hemisphere drain counterclockwise while wash bowls in the southern hemisphere drain clockwise. Let's use the formula for Coriolis force and some reasonable estimates to see whether the force in a wash bowl is sufficient to produce such rotational motion. Let's compare the Coriolis acceleration with the acceleration that results from gravity. The ratio is $(2\omega \times v)/g$. The angular velocity ω is that which is due to the earth. Its magnitude is 2π radians for 24 hours, which is equal to 7.3×10^{-5} rad/s. The tangential velocity of water in a wash bowl in the United States is not perpendicular to the axis of rotation of the earth as it would be at the North Pole. But let's estimate that the perpendicular component is about one-half of the velocity. That takes care of the factor of 2 in the Coriolis force equation. Now we must

Satellite photo of hurricane Ava, May 3, 1973. (NOAA)

estimate the velocity of water in an average wash bowl as it moves in toward the center. Let's say 1 m/s? In that case, the ratio of the Coriolis force to the gravitational force on any chunk of water is about equal to

$$F_{\text{Coriolis}}/F_{\text{grav}} \approx (7 \times 10^{-5} \text{ rad/s})(1 \text{ m/s})/(10 \text{ m/s}^2) = 7 \times 10^{-6}$$

The relative importance of the Coriolis force in wash bowls is negligible. The drainage circulation, clockwise or counterclockwise, is determined by asymmetries in the shape of the bowl or by the long-term memory of the water concerning the direction in which it swirled as it filled. Actually, very careful experiments were done in the 1960s by a group from the Massachusetts Institute of Technology to observe the wash bowl effect. They used special hemispherical bowls that were machined to be very symmetrical. They also let the water rest for long intervals to eliminate any motions remaining from the filling process. Under these special laboratory conditions, they observed that the drainage circulation was generally counterclockwise in Boston. They shipped the device to Australia and, sure enough, the drainage there was clockwise.

Noninertial Reference Frames: Example 5

Serious proposals are being made to establish space colonies that can house many thousands of people. These colonies would be located in orbits between earth and the moon. Since the inhabitants would have to live in the space colonies for some years, if not permanently, living conditions would have to be pleasant and in many ways earthlike. In particular, humans and other living things adapted to the earth expect to live in a gravitational field equal to 9.8 N/kg. To produce this effect in the colonies, the habitations would be giant cylinders, or toruses (doughnuts), or spheres rotating at constant velocity. The centrifugal force experienced by the inhabitants—who would live on the inside of the shell—would seem to them like normal earth gravity.

An artist's conception of the "Stanford Torus", a possible space colony that would orbit in the region between earth and moon. Ten thousand people could live comfortably in the spinning torus (doughnut) which has a diameter of 1.5 km. Above the torus is a giant mirror to reflect sunlight into the living quarters. (Courtesy of the L.S. Society.)

Question 11-14

How fast would a cylinder with a diameter of 1 km have to spin to produce centrifugal acceleration equal to g?

One of the advantages of living in such a colony would be that you would have easy access to regions near the axis of the rotating system where "gravity" would be less. This feature opens up a whole new realm of sports possibilities. On the other hand, the Coriolis reaction would introduce other, larger complications that we are not used to on earth.

EINSTEIN'S SPECIAL THEORY OF RELATIVITY

The Galilean transformation depends crucially on the assumption that time intervals are the same in all reference frames: $\Delta t = \Delta t'$. This does not mean that if it is 10 o'clock in New York, it must be 10 o'clock in Milwaukee; but it does mean that, if an event takes 10 seconds in one reference frame, it takes 10 seconds in the other reference frame. This assumption is so obviously true that it would not be worth mentioning if it were not indeed false. We will soon see just why the assumption of equal time intervals is false. In the meantime, note that the assumption led directly to the conclusion that relative velocities add algebraically. It was this very point that first bothered Albert Einstein at the beginning of this century.

Einstein was disturbed about a problem with the laws of electromagnetism called *Maxwell's equations*. We will study these in Chapter 22. They predict that electromagnetic radiation travels with a very particular velocity, the speed of light. The equations say nothing, however, about the reference frame in which this velocity is to be measured. Apparently the equations hold in *any* reference frame and in *all* reference frames. Consequently, in a vacuum, light always travels at this particular velocity of 3.0×10^8 m/s.

> Question 11-15
>
> But how can this be? One of the distant galaxy stars, which are called "quasars," is traveling away from our earth at half the speed of light. It emits light which we detect. What is the velocity of that light with respect to us?

The truth of the matter is that this unique velocity of light is the same regardless of the speed of the light source or the speed of the observer who actually makes the measurement. Is this a hypothesis or is it an experimental fact? It is both.

First, many experiments have demonstrated that the speed of light is independent of the velocity between source and observer. For instance, we can see many double stars in the heavens. If twin stars rotate around each other at high speed, we may see them first as a double star, then as a single star when one passes behind the other, and then as a doublet again. In such cases, we have light sources swinging toward us and away from us at high velocities (as they go around each other), and these light sources are also turning off and on regularly (as they go behind each other). If the light from one of the stars swinging away from us travels slower, we would see a delay in its becoming obscured by the front star. If its light traveled faster as it swings toward us, the effect would be very strange indeed.

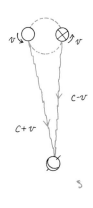

> Question 11-16
>
> What sort of effect would we see?

The explanation of the proper behavior of twin stars might be that light always travels at the same speed, independent of the speed of the *source*, because light is some sort of wave motion in stationary space. After all, that is the way that sound travels in air. Regardless of the speed of a train toward an observer or away from him, an observer measures a constant speed of sound from the whistle (but the pitch may change). If this explanation applied to light, an observer on earth could detect a change in the velocity of light caused by the *earth's* motion through space. If the light waves exist in the medium of space, like sound waves exist in the medium of air, then the velocity of light in the direction of the earth's motion should be different from the velocity of light perpendicular to that direction.

Suppose that we do an experiment to measure the effect of the earth's motion on light traveling in the same direction and also perpendicular to that direction. (To guard against the possibility that during the first measurement we were at the moment relatively motionless in space, we should do the experiment again six months later. At any given time, we on earth are traveling 120,000 mph (6×10^4 m/s) opposite to the direction we were traveling in six months before. If by chance we were motionless the first time, then surely we will be going 6×10^4 m/s with respect to space the second time.)

The experiment that answered this question was done almost 100 years ago in Cleveland, Ohio. This was the famous Michelson-Morley experiment. An optical interferometer was used to compare the time of flight of light parallel and perpendicular to the earth's motion. So sensitive was this device that Michelson could have detected a difference of flight time over the several meter path length of only 10^{-16} s (in 1881! No electronics, no computers, and no National Science Foundation).

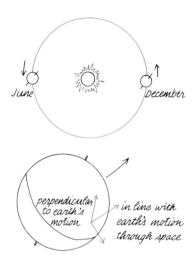

Question 11-17

How far can light travel in 1×10^{-16} s?

Optional

A schematic sketch of the interferometer that Michelson used is shown here. We will learn more about the theory of the device in Chapter 16. The speed of light is not measured directly, but the speed in one direction is compared with that in another. Light from a single source is divided by a half-silvered mirror, so that half continues in the same direction and half goes in a direction perpendicular to the original. Mirrors reflect both beams back. Where the two beams come together again on a screen, there is either constructive or destructive interference of the light wave. In the usual arrangement, a series of light and dark bands is seen. Movement of any of the mirrors by even a tenth of a wavelength of light (5×10^{-8} m) causes a detectable movement of light and dark bands. Similarly, even if the distances to the mirrors remain constant, a slight change in velocity of light in one of the perpendicular directions would make the bands shift. Michelson floated his apparatus on a pool of mercury and so could swing the whole interferometer 90° without disturbing the mirror spacings.

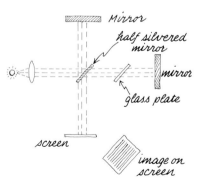

Question 11-18:
Why would he want to rotate the apparatus through 90°?

Observations of double stars show that the speed of light is independent of the speed of the source; the Michelson-Morley experiment shows that the speed of light does not depend on the speed of the observer through space. Indeed, this latter experiment implies that it is meaningless to talk about velocity through space. The velocity of one object with respect to another can be measured; but there apparently is no reference frame attached to space itself.

There are numerous other experiments that demonstrate that light has a unique speed. We also claimed that we could start out with this assumption as a theoretical hypothesis. Claiming that the speed of light is a constant and then seeing what conclusions can be drawn results in the *special theory of relativity*. This theory calls for many unexpected results—the slowing of clocks moving past us, the identity of mass and energy, and so on. These predictions have been experimentally confirmed; therefore, the plausibility of the original hypothesis is strengthened. In the next few sections, we will assume the uniqueness of the speed of light and then see what that simple assumption forces us to conclude.

A THOUGHT EXPERIMENT

If you are sitting down reading a book and turn on an overhead lamp, how long does it take the light to reach your book? Naturally we are not concerned with exactly this experimental situation. It usually takes much longer for the lamp to get bright than it takes for the light to travel from bulb to book. Still, light sources can be switched on rapidly enough to make this experiment feasible. Furthermore, modern electronic timing gear is quite capable of measuring such a short time.

> ### Question 11-19
> Approximately how long is the time interval?

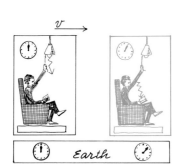

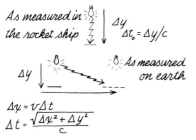

To calculate the answer to Question 11-19, you assumed some particular length for the distance from the lamp to your book (traveling down in a straight line); then you divided by c, the speed of light. Suppose now that you observe the same experimental situation going on in a rocket ship that is flying past you at the speed v. Just as the ship passes overhead, the light is pulsed and you observe the time on your clock. Some distance along the flight path the light strikes the book and this particular time is noted on a clock directly underneath the ship at that time. The chain of clocks on the ground has been previously synchronized so that the time interval as measured on the ground is equal to the difference in the two clock readings. Once again, the experiment is straightforward and perfectly feasible. You now have two measurements of the time it takes the light to reach the book—one was made in the rocket ship, just as you calculated in Question 11-19, and the other was made on the ground. The geometry of each measurement is shown in the diagram.

Question 11-20

How far did the light travel in each case?

Now the questions are: How far did the light *really* travel, and how long did it *really* take? In neither case, of course, does anyone see the light travel along the zig-zag path shown in the diagram. The only observations are that the switch is thrown to pulse the light and a short time later the book (perhaps a photocell) is illuminated. Both observers see the same event, and one set of measuring instruments is just as good as the other. In case you feel prejudiced in favor of the observer on the ground, remember that she and all her chain of clocks might be in another rocket ship with both ships out in space so that there is no way to tell which one is moving. Notice particularly the use of our basic hypothesis.

Question 11-21

What is the basic hypothesis and how is it used?

Since we observe the events in a ship going past us (regardless of whether *we* are moving with respect to the earth or the *ship* is moving with respect to the earth, *we* observe the ship moving past *us*), we must conclude that the light traveled farther than simply Δy in reaching the book. Incidentally, we assume that the perpendicular distance Δy appears the same to either experimenter. Therefore, the time interval that we observe, Δt, must be greater than the time interval Δt_0 measured in the rocket ship.

$$\Delta t_0 = \Delta y / c \qquad \Delta t = \frac{\sqrt{\Delta x^2 + \Delta y^2}}{c} = \frac{\sqrt{v^2 \Delta t^2 + \Delta y^2}}{c}$$

Since $\Delta y = c \Delta t_0$, we can substitute for Δy in the second equation and solve for Δt in terms of Δt_0.

Question 11-22

Carry out this solution.

The answer is that the measured time interval *between the same two events* is longer in our system than it is in the rocket ship: $\Delta t = \Delta t_0 / \sqrt{1 - v^2/c^2}$. The fact that time measured in one system is not the same as time measured in another system is contrary to all our everyday experience. After all, if you travel at high speeds on the highway, you do not expect to have your watch tick at a different rate from what it would if you were standing still. For that matter, the observer in the rocket ship does not notice his watch slowing down. In the ship, *everything* is going at a slower rate, including the biological processes such as the heartbeat. It is not just a case of clocks behaving strangely; time itself is different.

Only by comparison of time in two systems traveling past each other do we notice the effect. However, on a highway the effect will hardly be noticeable. The relationship between the two times depends on v^2/c^2. At 60 mph (27 m/s), the value of v/c is only 10^{-7}. The factor of v^2/c^2 is therefore only 10^{-14}, giving rise to a negligible difference between time intervals measured in a car and those measured on the road. Because of the dependence on the square of the velocity ratio, the factor remains negligible unless the velocity between the two observers is very close to that of light.

Question 11-23

What must be the speed between observers to produce a 10 percent difference between their two time scales? What speed would produce a factor of 2.00 in the length of time intervals?

■ Optional

An Apparent Paradox

Perhaps you have already noticed an apparent paradox in the situation that we have described. If rocket ship A goes by rocket ship B in space, B observes that the clocks in A are going at a slow rate. For instance, if the relative velocity is the same which you calculated in the second part of Question 11-23, B would claim that, while his clocks ticked off a full hour, the clock in A only advanced half an hour. On the other hand, as far as A is concerned, *she* is standing still and B is hurtling by her. Observer A also has a chain of clocks strung out in the direction in which B is going, and she has carefully synchronized all these clocks. Then the derivation applies to her observations as well, and she observes that B's clocks are going slow. Can they both be right? They certainly can!

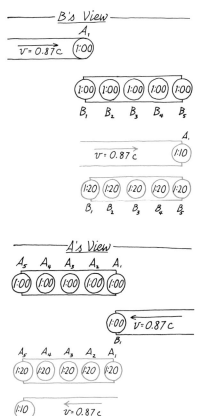

Let us suppose that just as A and B pass each other, both of their clocks (A_1 and B_1) read 1:00 o'clock (follow the action in the diagram). Now as far as B is concerned, he will note the times read by his chain of clocks as A_1 rushes past. Suppose that when A_1 passes the last clock in B's chain, B_5 reads 1:20. If the velocity with which A is passing is about $0.87c$, it will be found by the observer on B_5 that A_1 reads only 1:10. In other words, B observes that A's clock is going slower by a factor of 2. Now switch your thoughts over to A's system. She agrees that when A_1 passes B_1, both clocks read 1:00. In her system, B is rushing past to the left and, when B_1 passes A_5, A_5 reads 1:20 but B_1 reads 1:10. As far as A is concerned, B's clocks are running slow by a factor of 2. What is B's response to this? After all, he knows that he has synchronized his five clocks, which is easy to do since they are at rest with respect to him.

Question 11-24

How could B synchronize his clocks when they are not at the same place?

Observer B can also compare his clock B_1 when it passes A_5. Sure enough, he would agree that A_5 reads 1:20 while he reads only 1:10. Does he conclude that now A's clocks must be going fast compared with his? Not a bit. After all, as far as B is concerned, A_5 is not the clock which he is timing in his B system. (He is timing A_1). His explanation is very simple; A did not properly synchronize her clocks. Up in the A system, however, A is making the same complaint about B. Neither one is actually to blame. The system of clocks synchronized in one system is not synchronized in another system traveling past it. No error is involved; synchronization in both systems at the same time simply is not possible.

Question 11-25
Try using the synchronization method developed in the answer to Question 11-24 to synchronize clocks at the same time in the two systems moving past each other.

Experimental Evidence for Different Clock Rates
Look what has happened as a result of accepting one simple hypothesis (for which of course there is good experimental evidence). If the speed of light is the same for all observers independent of their velocity, then time itself must be different in systems moving with respect to each other. Do we have any experimental evidence of this difference in time? Certainly. The evidence is so commonplace that you are being affected by it right now. Within the last minute several dozen muons have passed through your body leaving trails of ionized atoms and destroyed molecules. These muons are short-lived heavy electrons produced near the top of the atmosphere by incoming cosmic rays. When they are at rest, they decay with a half-life of about 2×10^{-6} seconds.

Question 11-26
Traveling almost at the speed of light, how far could a muon go in 2×10^{-6} s?

Because the muons are traveling very fast with respect to us, their time goes more slowly. Therefore, most of them live long enough to come 20 miles through the atmosphere and so bombard us all. The experimental evidence for this effect is solid and quantitatively agrees with the time relation formula that we have deduced. In every high-energy physics laboratory the time relation formula must be used in dealing with the experimental effects of many other kinds of particles. It always works.

Another Apparent Paradox: Length Measurements
Like many other relativistic applications, there seems to be a paradox in this case of the muon. By realizing that the muon clock is ticking slowly, we can explain how it lives long enough to travel 20 miles. But in the muon's system it still lives only 2×10^{-6} s, and in both systems the velocity, close to the speed of light, is the same. How does the muon explain its ability to travel so far when it lives such a short time?

The apparent paradox with the muon could have been foreseen in our original example of the two clock chains passing each other. Instead of asking A and B to compare their clocks, let us see what they conclude about the lengths of each other's clock chains. If A wants to measure a length in B, she cannot hold a meter stick against B since B is racing past. However, both parties will agree that their relative velocity is v (in this case, equal to $0.87c$). If you want to measure the length of an object moving past you at constant velocity, simply time how long it takes to pass, and multiply by the velocity. Then B_1 will observe that the leading edge of A arrived at 1:00 and that the tail end at A_5 passed at 1:10 (to be sure, the clock at A_5 did not read 1:10, but that, according to B, is A's synchronization problem). Furthermore, B knows that the leading edge of A does not arrive at B_5 until 1:20. If it takes 20 minutes for the leading

B has synchronized all his clocks. In B's frame, when $B_5 = 1:20$, $B_4 = 1:20$, $B_3 = 1:20$, $B_2 = 1:20$ and $B_1 = 1:20$

A has synchronized all her clocks. In A's frame, when $A_1 = 1:10$, $A_2 = 1:10$, $A_3 = 1:10$, $A_4 = 1:10$ and $A_5 = 1:10$

When B observes A's clocks, she monitors only one of them — such as A_1.

When A observes B's clocks, he monitors only one of them — such as B_1.

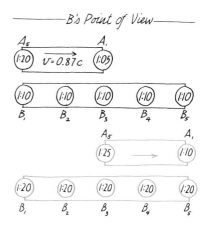

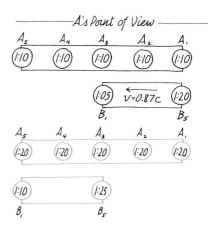

edge of A to travel the whole length of B but only 10 minutes for the whole object to pass a particular point, then A must be only half as long as B.

On the other hand, A could use the same arguments and would come up with the conclusion that B is only half as long as A. Can they both be right? Certainly! The logic is straightforward and inescapable. Once again, the resolution of the paradox is that there is no way to synchronize clocks between moving systems. To compare a length with a standard you must line up both edges of the standard *at the same time*. When B performs his measurement, he would say that he lined up A against marks at B_1 and B_3 at the very instant when the leading edge of A was at B_3 and the tail end of A was at B_1. The observer in A would say that B had used those marks all right, but had lined up the mark at B_3 long before A_5 and B_1 were together. According to A, when her clock at A_1 reads 1:10, B_5 is opposite A_1 (B agrees) and *at the same time* (B disagrees) B_1 was opposite A_3, which read 1:10. B agrees that when B_1 passed A_3, A_3 read 1:10, but B claims that it was really 1:05, which is what the clock at B_1 read.

Here is a summary of the times involved:

A_1 passes B_1 A_1 reads 1:00 B_1 reads 1:00
A_1 passes B_3 A_1 reads 1:05 B_3 reads 1:10
A_1 passes B_5 A_1 reads 1:10 B_5 reads 1:20

Both A and B agree about all of these readings. However, B claims that his clocks are synchronized so that the time read at B_1 is the same as the time in B_2, and so on. Observer A claims that B's clocks are not synchronized.

B_1 passes A_1 A_1 reads 1:00 B_1 reads 1:00
B_1 passes A_3 A_3 reads 1:10 B_1 reads 1:05
B_1 passes A_5 A_5 reads 1:20 B_1 reads 1:10

Both A and B agree about all of these readings. However, A claims that her clocks are synchronized. B claims that A's clocks are not synchronized.

B claims that at 1:10 (according to his clocks) A_1 was lined up with B_3 and A_5 was lined up with B_1. Therefore, B claims that A is only half the length of B. On the other hand, A claims that at 1:10 (according to her clocks) B_1 was lined up with A_3 and B_5 was lined up with A_1. Therefore, A claims that B is only half the length of A. Each is right in his own system. There is no such quantity as absolute length. If you measure a length to be Δx in a system passing by you, then an observer in that system will measure a longer length Δx_0. The relationship is

$$\Delta x = \Delta x_0 \sqrt{1 - v^2/c^2}$$

Remember now the problem of the muon lasting long enough to penetrate the atmosphere. Here on earth we explain that the muon lived long enough to go 20 miles, because time is slowed in a system moving very fast with respect to us. The muon, however, lasts on the average only 2×10^{-6} in its own system.

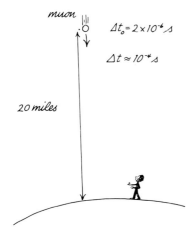

Question 11-27
How does the muon explain the fact that in such a short lifetime it can go from the top of the atmosphere to the earth?

The Lorentz Transformation

Clearly the Galilean transformations cannot describe events satisfactorily in a system moving past us with a velocity close to that of light. With the Galilean transformations, time intervals are the same in all reference frames, and the length of a meter stick in any system remains 1 m. The true transformations between moving coordinate systems are named for the Dutch physicist Hendrik Lorentz (1853–1928). The Galilean transformations are merely approximations that are good if $v/c \ll 1$.

To determine the Lorentz transformations, we will assume once again that the primed coordinate system is moving to the right along the x-axis with velocity v. The origins of the coordinate system touch when $t = t' = 0$. We can no longer assume, however, that $t = t'$ at any other time or place. However, since the relative motion is along the x-axis, we can argue because of symmetry that $y = y'$ and $z = z'$. Since the Lorentz transformations must reduce to the Galilean form at low velocity, it is a reasonable guess that

$$x' = \gamma(x - vt)$$

The constant γ must be some function of v and c such that when v/c goes to 0, γ goes to 1. In such a case, the Lorentz transformation would reduce to the Galilean transformation. The inverse relationship for x must be

$$x = \gamma(x' + vt')$$

To find the value of the constant γ, we now appeal to the basic hypothesis of the special theory of relativity: *the velocity of light is constant and equal to c in all reference frames.* If we start a burst of light at the origin when the origins of the two system coincide, then the law for the spread of the light must be the same in both reference frames:

$$x = ct \qquad x' = ct'$$

Since we have transformations for x and x', we can substitute these, eliminating x and x' and also getting rid of t and t'.

$$ct = x = \gamma(x' + vt') = \gamma(ct' + vt') = \gamma(c + v)t'$$

$$ct' = x' = \gamma(x - vt) = \gamma(ct - vt) = \gamma(c - v)t$$

$$ct = \gamma(c + v)\gamma(c - v)t/c$$

$$\gamma^2 = \frac{c^2}{c^2 - v^2} = \frac{1}{1 - v^2/c^2}$$

$$\gamma = \frac{1}{\sqrt{1 - v^2/c^2}}$$

Notice that the unknown constant of our transformation turns out to be the distortion factor that affects time and length intervals.

Question 11-28
From what we have seen so far, does the Lorentz transformation reduce to the Galilean transformation when $v/c \ll 1$?

Now we must find the transformation between t and t'. To find this, we use the original transformation for x and x' and solve for t or t'.

$$x = \gamma(x' + vt') = \gamma[\gamma(x - vt) + vt'] = \gamma^2 x - \gamma^2 vt + \gamma vt'$$

$$\gamma vt' = x + \gamma^2 vt - \gamma^2 x = x(1 - \gamma^2) + \gamma^2 vt$$

$$t' = \gamma t + \frac{1 - \gamma^2}{\gamma v} x = \gamma\left(t + \frac{1 - \gamma^2}{\gamma^2 v} x\right)$$

$$t' = \gamma\left[t + \left(\frac{1}{\gamma^2} - 1\right)\frac{x}{v}\right] = \gamma\left[t + \left(-\frac{v^2}{c^2}\right)\frac{x}{v}\right] = \gamma\left(t - \frac{v}{c^2}x\right)$$

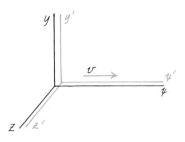

Galilean: $x' = x - vt$
$x = x' + vt$
$y' = y$
$z' = z$
$t' = t$

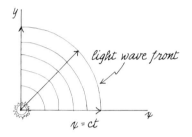

light wave front
$x = ct$

light wave front
$x' = ct'$

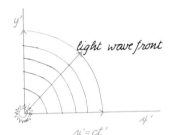

Lorentz Transformations

$x' = \gamma(x - vt) = \dfrac{x - vt}{\sqrt{1 - v^2/c^2}}$

$x = \gamma(x' + vt') = \dfrac{x' + vt'}{\sqrt{1 - v^2/c^2}}$

$t' = \gamma\left(t - \dfrac{v}{c^2}x\right) = \dfrac{t - vx/c^2}{\sqrt{1 - v^2/c^2}}$

$t = \gamma\left(t' + \dfrac{v}{c^2}x'\right) = \dfrac{t' + vx'/c^2}{\sqrt{1 - v^2/c^2}}$

Remember that

$$\gamma = \frac{1}{\sqrt{1 - v^2/c^2}}$$

and therefore

$$\frac{1}{\gamma^2} = 1 - \frac{v^2}{c^2}$$

A similar analysis leads to the corresponding formula for t:

$$t = \gamma\left(t' + \frac{v}{c^2}x'\right)$$

The Lorentz transformations for time show that clocks synchronized in one system are not synchronized in a moving system. For instance, when the origins of our two systems coincide, $x = x' = 0$ and $t = t' = 0$. Suppose that we have a whole string of clocks in our unprimed reference frame that we have carefully synchronized with the method described in the answer to question 11-24. When the origins of the two systems coincide, all of the clocks in the unprimed reference frame will read 0. Not so in the primed frame. Out along the x-axis, the recorders in *our* frame report that the clocks in the primed reference frame are reading earlier and earlier:

$$t' = \gamma\left(0 - \frac{v}{c^2}x\right)$$

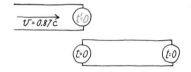

Our recorders along the *negative* x-axis find that the clocks in the primed frame register positive times. Yet at all the points along the x-axis where we make these observations, our clocks at that instant read 0.

Do the Lorentz transformations account for the space and time warping that we calculated earlier? To see that they do, we must apply them very carefully. Rather than dealing with the equations algebraically, it is a good idea to draw simple diagrams of the action. In the diagram, we show the two reference frames after the primed system has been moving to the right for a time, which according to our clocks is $t = 1$. All of the clocks along the x-axis read 1 at that instant since we can operationally define simultaneity within a system. The clocks in the primed system have various readings, but we are interested in the particular clock at the origin, since we know that it read 0 when $t = 0$. The time that it now reads is

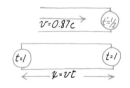

$$t' = \gamma\left(t - \frac{v}{c^2}x\right)$$

The distance that the origin of the moving system has gone is $x = vt$. Since $t = 1$:

$$t' = \gamma\left\{t - \frac{v}{c^2}vt\right\} = \gamma\left\{1 - \frac{v^2}{c^2}\right\} = \gamma\left\{\frac{1}{\gamma^2}\right\} = 1/\gamma$$

As we saw earlier, if $v/c = 0.87$, then $\gamma = 2$. In this case the clocks at the origin of the primed frame would read $t = \frac{1}{2}$ when $t = 1$. The clock in the moving frame appears to us to be slow.

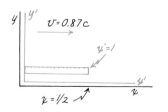

Now let's place a meter stick in the primed reference frame, with one end at $x' = 0$ and the other end at $x' = 1$. We will measure the length of this meter stick in the unprimed frame at the instant when the origins coincide. Since $x' = x = 0$, all we have to do is to find the position x of the other end at the same instant *in our reference frame*, when $t = 0$. The Lorentz transformation is $x' = \gamma(x - vt)$. When $t = 0$:

A THOUGHT EXPERIMENT

$$x' = 1 = \gamma x \quad x = 1/\gamma$$

For our familiar case when $v/c = 0.87$ $\gamma = 2$ and $x = \frac{1}{2}$ m. The meter stick in the moving frame appears to us to be shortened to only $\frac{1}{2}$ m. Evidently the Lorentz transformations yield the same shortening of length and slowing of clocks that we found with the earlier arguments.

Proper Time and the Invariant Interval

If the time and the distance between two events depend on the reference frame in which we measure them, it may seem impossible to describe the separation between two events. One person claims that two bolts of lightning struck at the same time, only 1 m apart. A person traveling past him would deny that the lightning struck at the same time and would measure a different distance between the two marks where the lightning hit.

There is a somewhat similar problem in describing the length of a stick in ordinary two-dimensional geometry. In the diagram, we have shown the stick with one end at the origin of the coordinate system and the other end defined by $x = 1, y = 2$. Then $\Delta x = 1$ and $\Delta y = 2$. If we rotate the coordinate system but leave the stick lying still, we change both the x- and y-coordinates of the end of the stick. For instance, with the proper rotation, we can make the x-component of the length 0. On the other hand, we know very well that the length of the stick itself is invariant and is given by the formula: $L = \sqrt{\Delta x^2 + \Delta y^2}$. We can describe this invariance of length by writing

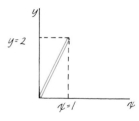

$$L^2 = \Delta x^2 + \Delta y^2 = (\Delta x')^2 + (\Delta y')^2$$

There is an analogous invariant interval in space-time. It must, of course, include variables of both space and time. This invariant interval is

$$s^2 = \Delta x^2 + \Delta y^2 + \Delta z^2 - c^2 \Delta t^2 = (\Delta x')^2 + (\Delta y')^2 + (\Delta z')^2 - c^2(\Delta t')^2$$

Question 11-29

If the velocity between two moving systems is in the x-direction so that we can ignore the y- and z-components, do the Lorentz transformations satisfy the invariance interval formula? Find out for yourself. For x and t in the invariance equation, substitute the Lorentz formulas for x and t in terms of x' and t'.

Let's take a specific example of how this invariant interval works. Suppose that the observer in the primed frame claims that lightning struck in two places at the same time. She observes that at $t' = 0$, the lightning struck at $x' = 0$ and $x' = 1c$ (one light-second). The space-time interval between the two events in her frame is given by

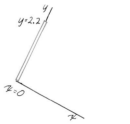

$$\Delta x'^2 - c^2(\Delta t')^2 = (1c)^2 - 0 = c^2$$

The observer in the unprimed frame does not agree about the sequence of events. He does agree that a bolt of lightning struck at the origin $x = 0$ at $t = 0$, but he claims that the second bolt of lightning struck later in time and further away. His coordinates for the second bolt of lightning are

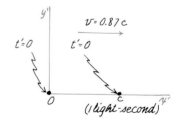

$$x = \gamma(x' + vt')$$
$$x = \gamma(c + 0) = \gamma c$$
$$t = \gamma\left(t' + \frac{v}{c^2}x'\right)$$
$$t = \gamma\left(0 + \frac{v}{c}\right)$$

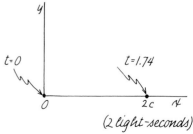

If we use our standard example where v/c equals 0.87, and $\gamma = 2$, then

$$x = \Delta x = 2c \quad \text{and} \quad t = \Delta t = 2\left(0 + \frac{0.87c}{c}\right) = 1.74$$

since $x_0 = 0$ and $t_0 = 0$.

The invariant interval in the unprimed frame is given by

$$\Delta x^2 - c^2 \Delta t^2 = 4c^2 - c^2(1.74^2) = c^2$$

The space-time interval is indeed the same in both reference systems. Notice how the clock times and distance times compare in the two systems. In the primed frame, the observer claims that the events were simultaneous. In the unprimed frame, they are separated by a considerable time. In the primed frame, the two marks left by the lightning are separated by a distance of $1c$. In the unprimed frame, the two marks are separated by a distance of $2c$. Do we have a paradox here? Shouldn't we in the unprimed frame observe that the distance separation in the primed frame is only $\tfrac{1}{2}c$? Actually, we do. It is in *our* frame that we measure the separation distance of $2c$. As we observed the primed reference frame, we agreed that there was a mark made by a lightning bolt at $x' = 0$ when $t = t' = 0$. However, according to us, the next lightning bolt struck at $t = 1.74$ and at $x = 2c$. We know that by that time the origin of the primed frame has traveled a distance of $ct = (0.87c)(1.74) = 1.5$. Since, according to us, the lightning struck at $x = 2$ and the origin of the primed system is at $x = 1.5$, we claim that the distance between the two marks in the primed frame must be 0.5. As we have seen before, distances in a system moving past us appear to shrink.

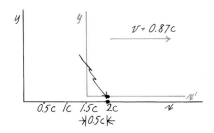

The spatial distance between any two events *in* a system is larger than it appears to be in any system moving past it. The time between any two events measured *in* a system is smaller than it is measured in any other system moving past. The space-time interval between two events is the same in any system.

The Addition of Velocities in Moving Reference Frames

In Galilean relativity it is easy to calculate the velocity, with respect to the ground, of a BB that is shot from a gun on board a moving train. If the muzzle velocity is 30 m/s and the train speed is 30 m/s, then, with respect to the ground, the BB travels at 60 m/s if fired straight ahead or with a speed of 0 if fired directly backward. If u is the speed of the BB with respect to the gun and v is the speed of the train with respect to the ground, then the speed of the BB with respect to the ground is

$$w = u + v$$

Obviously this simple formula cannot apply when the speeds are close to that of light. For instance, if a distant galaxy is moving away from us with a speed of $\tfrac{1}{2}c$ and sends light to us with a velocity of c with respect to itself, we still measure the velocity of that light to be c and not $\tfrac{3}{2}c$. The Galilean formula for the addition of velocities must be just an approximation to the true one given by the Lorentz transformations. Let's derive the true law.

The velocity of the BB in the primed frame is found by measuring the distance it travels, $\Delta x'$, during a time interval $\Delta t'$. Its velocity then is $u = \Delta x'/\Delta t'$. We will use the Lorentz equations to find Δx and Δt:

$$\Delta x = \gamma(\Delta x' + v\Delta t') \quad \text{and} \quad \Delta t = \gamma\left(\Delta t' + \frac{v}{c^2}\Delta x'\right)$$

The velocity that we measure in the unprimed frame is

A THOUGHT EXPERIMENT

$$w = \frac{\Delta x}{\Delta t} = \frac{\gamma(\Delta x' + v\Delta t')}{\gamma\left(\Delta t' + \frac{v\Delta x'}{c^2}\right)} = \frac{\Delta x'/\Delta t' + v}{1 + (v/c^2)(\Delta x'/\Delta t')} = \frac{u + v}{1 + uv/c^2}$$

Let's see how this true formula for the addition of velocities applies to specific cases. First, note that for the case of BBs the true formula reduces to the approximate Galilean formula. If $u \ll c$ and $v \ll c$, then $w \approx u + v$.

Let's see what happens with the galaxy that is flying away from us at a speed of $\tfrac{1}{2}c$. In its own reference frame, the speed of light is $u = c$. The velocity that we observe is

$$w = \frac{c - \tfrac{1}{2}c}{1 - (c)(-\tfrac{1}{2}c)/c^2} = \frac{\tfrac{1}{2}c}{1 - \tfrac{1}{2}} = c$$

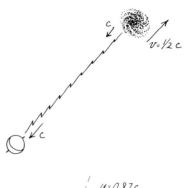

The velocity of light that we measure is c, even though it is being sent to us from a reference frame moving rapidly away from us.

Let's take one other example, which we can use later. We have already seen that in a reference frame moving past us with a velocity of $0.87c$ ($v^2/c^2 = \tfrac{3}{4}$) the Lorentz transformation constant is $\gamma = 2$. Suppose that we have a third system, C, moving with a velocity of $0.87c$ with respect to B. If B is moving with a speed of $0.87c$ with respect to us in A, what is the speed of C with respect to A? For ease of doing the calculation, we will work with fractions rather than decimals:

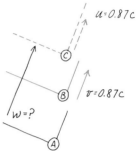

$$v/c = \sqrt{\tfrac{3}{4}}$$

$$w = \frac{u + v}{1 + uv/c^2} = \frac{2\sqrt{\tfrac{3}{4}}c}{1 + \tfrac{3}{4}} = \frac{\sqrt{3}c}{1.75} = 0.9897c$$

Note that both u and v are very close to the speed of light. Their sum is even closer, but is still less than c. While we are at it, let's compute γ for C as observed by A.

$$\gamma = \frac{1}{\sqrt{1 - w^2/c^2}} = \frac{1}{\sqrt{1 - 3/(7/4)^2}} = \frac{1}{\sqrt{1 - (48/49)}} = \frac{1}{\sqrt{1/49}} = 7$$

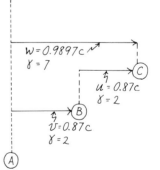

The factor of space warping and time dilation between B and C or A and B is simply 2; but the γ factor between C and A is 7.

The Famous Twin Paradox

First, there is no paradox. We will be able to calculate every detail of the event. Here is the story and the problem. There are two twins, G (girl) and B (boy). B stays at home, presumably on earth, while G hurtles off at constant velocity to a distant star. As soon as she gets there, she turns around and returns at the same speed to earth. Ignoring all details of acceleration time and turnaround time, it is clear that when she gets back, G is younger than B. Or is it? B argues that G has been traveling continuously at high velocity, except for the one short break at turnaround, and that therefore her clocks have been slow. Therefore, G herself has not aged as much as B. The seeming paradox concerns the symmetry of the situation. Why doesn't G observe that B went shooting in the opposite direction and then eventually turned around and came shooting back again? G would argue that B has been traveling at high velocity and therefore his clocks should be slower. Is there a physical difference between the sequence of events for G and those for B? Yes, there is. B stays in the same reference frame at all times. G, on the other hand, must switch reference frames. First,

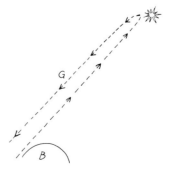

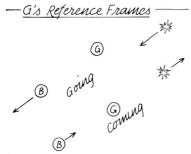

G's Reference Frames

(Note that G has to use two different reference frames. She transfers from the first to the second when the star reaches her.)

The origins of the three reference frames coincide at the beginning of the journey. At the common origin, $t_B = t_G = t_R = 0$

The view at the beginning as seen by B

she is in a reference frame that is traveling away from B, and then she must transfer to a reference frame that is moving toward B. The transfer between reference frames is a physical event (requiring acceleration, for instance) that can be measured. There is no symmetry between the experiences of G and B. B just sits still for the entire time.

Let's choose some simple numbers and actually calculate the various times and distances as seen by both G and B. Let us arrange for G to travel at a speed of about $0.87c$. so that the space-time warping constant is $\gamma = 2$. Therefore, as B observes events in G's reference frame, he will see that her meter sticks are shorter by a factor of 2 and that her clocks run slow by a factor of 2. Of course, G will also observe the same effects as she measures events taking place in B. We will require G to travel for a time of 20(years) to the distant star. The distance to that star as measured by B is therefore $x = ct = (0.87c)(20 \text{ yr}) = 17.4$ light-years. During this time G will only age by 10 years, since her clocks run slow by a factor of 2. If G turns around immediately, she will take another 20 years to return home as measured by B. In G's reference frame, however, she will age only another 10 years. At the end of the trip, G, will be only 20 years older, while her stay-at-home twin will be 40 years older.

To start the action, we will arrange for the origins of the coordinate systems to be at the same place at $t = t' = 0$. The situation at the beginning is shown in the diagram. We record the distances (in terms of an arbitrary unit: $1 = 17.4$ light-years) along the bottom of each reference frame. Along the top we place the reading of the clocks at each position. Notice that a third reference frame, R (return), also has its origins in common with the first two. This is the reference frame to which G must transfer when she starts her return journey home. The times in this reference frame are crucial because they will be compared with the earth clocks at the end of the journey. Both G and R are traveling with a velocity $v = 0.87c$ with respect to B, but they are traveling in opposite directions. In an earlier problem, we calculated that the relative velocity between G and R is $0.9897c$, and the γ factor between them is 7.

The next diagram shows positions and clock times for the three reference frames at the time when G reaches the distant star. The event is shown as viewed from each of the three different references frames. Note that they all agree about what the actual clock readings and position coordinates are *at the position of the star*. G says that her clock reads 10. She explains that it took her only 10 years to reach the star because from her reference frame the star's distance from earth has shrunk by a factor of 2. The clocks attached to the reference frame at rest, with respect to earth, read 20 when the rocket ship arrives at the distant star. The clocks in R read 70. Note that each observer claims that all of the clocks in his reference frame are synchronized, but the clocks in the other reference frames are not. Hence, as seen by B, his chain of clocks all read 20 at the time when G arrives at the distant star.

How do we know that a clock in R at the position of the star reads 70 when the rocket ship arrives? We could argue either through symmetry or by using the Lorentz transformation. As seen by B, the clock at the *origin of R* must read 10 when the rocket ship reaches the star. On the other hand, at the origin of B where $x = 0$, the time in either of the two moving frames must be $t' = \gamma[t + (xv/c^2)] = \gamma t$. Therefore, the clocks in both G and R passing the origin of B when $t = 20$ must read $t' = 2(20) = 40$. Since in R's reference frame as seen by B, the clock at the left end reads 10 and in the middle reads 40, then at the right end it must read 70.

A THOUGHT EXPERIMENT

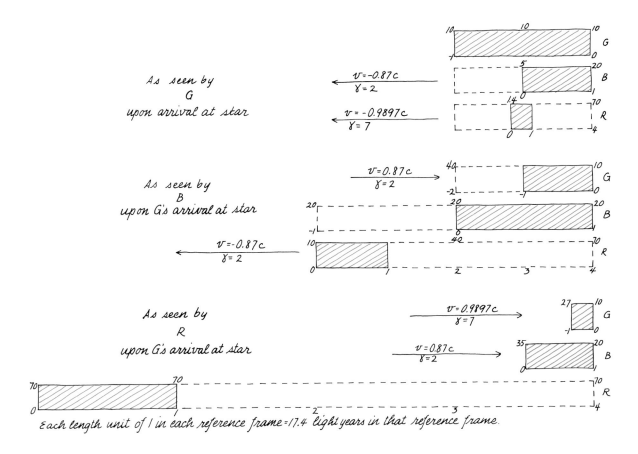

Each length unit of 1 in each reference frame = 17.4 light years in that reference frame.

Question 11-30
Does the Lorentz transformation also predict that R's clock at the star should read 70 when the rocket gets there? Try it and see, but remember that the appropriate distance unit for x is not 1 but rather 17.4 light-years.

The nature of the paradox is seen by the views of G and B when G arrives at the star. B knows that it took G 20 years to get there, but agrees that G's clock reads only 10 years. Now look at the scene as viewed by G. Ten years have passed and G is at the distant star. But according to the string of clocks in her own reference frame, the clock back on earth where $x = 0$ in B reads only 5 years. According to G, the fact that B's clock at the position of the star reads 20 is merely a matter of B's problem of synchronization. The resolution of the paradox comes when G starts home again. To do so she must transfer to the reference frame R. That reference frame had common origins with the others at $t = 0$. It is the clock system that G must use on the return flight. Note that there has been an abrupt shift of time as far as G is concerned. From a reading of 10, she suddenly jumps to a clock reading of 70. She can note with satisfaction, however, that at that same instant, as seen in her new reference frame R, the clock back on earth, where $x = 0$, reads only 35. Since earth seems to be hurtling toward her with a γ of 2, the rate of subsequent clock time should once again be only half of that in her new reference frame.

The arithmetic of the rest of the journey is simply a repeat of the first part. As seen by B, G will return in another 20 years at which time B will be 40 years older than at the start. As far as G is concerned, in her new reference frame R, she need travel only another 10 years to get home. At that time, her clock in R will read 80 as she observes the clocks on earth, which have advanced slowly from 35 to 40.

Although G changes reference frames and B does not, and although the necessary acceleration during that change creates a physical difference between the situations of G and B, nevertheless, it is not the acceleration itself that produces the sudden change of clock readings between G and R. Acceleration does indeed produce changes in time. This phenomenon is part of Einstein's general theory of relativity. However, in this case, we need appeal only to the rules of the special theory of relativity, which concern the relationships between inertial frames traveling at constant velocity with respect to each other. ∎

MOMENTUM, ENERGY, AND MASS RELATIONSHIPS

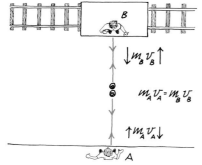

We have already seen with Galilean relativity that the energy and momentum of an object depend on the reference frame in which we measure them. If we are traveling beside an object, its velocity with respect to us is 0 and so are its momentum and kinetic energy.

We might expect that the Lorentz transformations for momentum and energy would produce some surprises, since time also depends on the reference frame. In the diagram, we show an experiment that would be difficult to perform even at low speeds, but that can be analyzed with both Galilean and Lorentz transformations. Two identical basketballs are chosen and thrown toward each other by two observers. One of the observers is on a train and the other is on the station platform. If the train is standing still, the two skillful experimenters can shoot the balls toward each other with equal speeds and therefore with equal momentum since the basketballs have the same mass by definition. (At rest they are identical.) The balls will meet halfway and bounce back with equal speeds since these are ideal basketballs. The same experiment can be done if the train is moving past the station at constant velocity. Of course, the timing is more crucial. Each experimenter still shoots the basketball with the same speed that he produced in the first case, but this time the observer on the platform sees a triangular trajectory for the ball coming toward him from the train, as shown in the upper diagram. We know that the observer on the train sees the inverse of the situation as shown in the lower diagram. Nevertheless, in Galilean physics the y-component of the ball's velocities both going and coming will be the same in each reference frame.

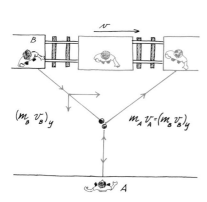

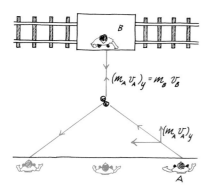

If the relative velocity between station and train is very great, however, time will be different in the two frames. The experimenter A on the platform observes that the clock of the moving experimenter runs slow. The y-component of the ball's velocity coming from B is therefore smaller since everything in B is moving more slowly. Nevertheless, the balls hit as before, and A's ball returns with its usual speed. Since momentum must be conserved, it must be that the relativistic mass of B's ball as measured by A has increased by the same factor by which its transverse velocity has decreased. The true momentum of an object is not (mv) but rather

$$\text{Momentum} = \gamma(mv) = \frac{(mv)}{\sqrt{1 - v^2/c^2}}$$

This relativistic increase of momentum is sometimes thought of as a relativistic increase in the mass of the object. However, the only way to measure the mass of a moving object is through a momentum exchange experiment. Experimentally, all that we can measure is the relativistic increase of momentum as a whole. And experimentally we do measure it—not, of course, by bouncing basketballs. Instead, we observe the effect in the equivalent experiments with subatomic particles. For instance, as electrons or protons are accelerated in circular guide-fields, their velocity rapidly approaches that of light. Energy and momentum continue to be fed into them, but their velocity can increase very little. Nevertheless, it is necessary to keep right on increasing the strength of the magnetic guide-field because the momentum of the particles continues to increase as γ increases.

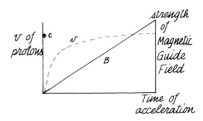

The particles in an accelerator not only increase in momentum but also in energy. Evidently the formula for kinetic energy cannot be simply $E_{kin} = \frac{1}{2}mv^2$. We cannot account for the increased energy simply by multiplying the kinetic energy by γ. Instead, we must make use of a new formula, which we will not derive, relating the energy of an object to its momentum and its rest mass. The rest mass, m, is just the ordinary mass that is measured in a static experiment with the object at rest. The energy-mass-momentum relationship is

$$E^2 = p^2c^2 + (mc^2)^2$$

where p is the momentum $= \gamma(mv)$.

The *form* of this equation is something like that of the Pythagorean theorem: $c^2 = a^2 + b^2$. It appears that momentum and the rest mass of an object are associated mathematically like two components of a vector. The magnitude of the resultant is the total energy of the object, including both its rest mass and its motion energy. Consider the consequences. According to our formula, an object has energy even if it is standing still with respect to us:

$$E = mc^2$$

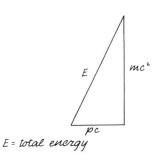

when p (momentum) is 0 with respect to us.

Here we have the statement of the equivalence of energy and mass. Note that it is not a case of being able to turn energy into mass or mass into energy. *Energy and mass are the same thing.* We could and we do measure energy in mass units of kilograms (or the equivalent), and we could and we do measure mass in energy units. The transformation equation is basically the same as the one transforming feet into yards: 1 yard = 3 feet. In the case of mass-energy, the conversion factor has the dimensions of velocity squared. However, there is nothing fundamental in this situation; it results because of the units originally chosen for mass and energy.

Let's calculate the energy equivalent of 1 kg.

$$E = mc^2 = (1 \text{ kg})(3 \times 10^8 \text{ m/s})^2 = 9 \times 10^{16} \text{ J (joules)}$$

> ### Question 11-31
> That's a lot of energy. The power complex at Niagara Falls produces about 10,000 megawatts. How long would it take this system to generate 9×10^{16} J?

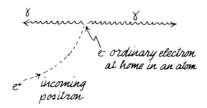

In an atomic bomb explosion less than 0.1% of the rest mass-energy turns into other forms of energy. But that is usually sufficient. In the diagram, we illustrate an interaction of subatomic particles where rest mass-energy is completely converted to other forms. A positron, which is a positive antielectron, comes to rest in the vicinity of an ordinary electron that is attached to an atom. The two annihilate each other, and their total mass-energy is converted to the energy of two gamma rays (also called X-rays). These energetic X-rays shoot out in opposite directions, thus conserving momentum. Each shares equally in the available energy from the electron annihilation. Let's calculate the energy of one of these gamma rays. The mass of an electron or a positron is 9×10^{-31} kg. The energy equivalent is

$$E = mc^2 = (9 \times 10^{-31} \text{ kg})(3 \times 10^8 \text{ m/s})^2 = 8 \times 10^{-14} \text{ J}$$

In joules that doesn't seem like much energy. But the appropriate unit of energy for atomic particles is the electron volt: 1 eV = 1.6×10^{-19} J. In electron volts the energy of each gamma ray produced in the annihilation is

$$E = (8 \times 10^{-14} \text{ J}) \left(\frac{1 \text{ eV}}{1.6 \times 10^{-19} \text{ J}} \right) = 5 \times 10^5 \text{ eV}$$

$$= \tfrac{1}{2} \text{ MeV } (\tfrac{1}{2} \text{ million electron volts})$$

We have calculated the mass, in energy terms, of one electron. This is also the energy of one of the gamma rays produced in an electron-positron annihilation. We must not conclude, however, that "mass has been converted to energy." Mass and energy are names for the same thing. The original mass-energy of the electron and positron was concentrated in two small regions of space. After the annihilation the same amount of mass-energy existed in the two gamma rays streaking away at the speed of light. To be sure, the gamma rays or photons or X-rays have no rest mass since they cannot exist at rest. Nevertheless, they carry momentum and energy and, hence, the original mass has not been lost. The same thing is true of an atomic bomb explosion. The small fraction of mass that turns into other forms of energy has not been lost but, instead, is scattered throughout the countryside, at first in the form of the kinetic energy of the exploding particles.

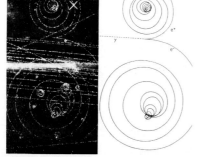

In the photo, there is a picture of the inverse process where a high-energy X-ray materializes into an electron-positron pair. In this case, the energy of the photon provided the rest mass-energy of the electron and positron and also their kinetic energy.

MOMENTUM, ENERGY, AND MASS RELATIONSHIPS

We have asserted that the total mass-energy of an object is composed of two parts—a term that depends on the momentum and a term that depends on the mass of the object measured when it is at rest. The total mass-energy can also be expressed in another way:

$$E = \frac{mc^2}{\sqrt{1 - v^2/c^2}} = \gamma mc^2$$

Let's demonstrate that the two formulas for total energy are equivalent.

$$\gamma^2(mc^2)^2 = E^2 = p^2c^2 + (mc^2)^2 = \gamma^2(mv)^2c^2 + (mc^2)^2$$

In the equation above, we have substituted on the left our alternative expression for E and on the right have made the substitution that the momentum of an object moving at high speeds is $\gamma(mv)$. Now let's gather terms and reduce our equation.

$$(\gamma^2 - 1)(mc^2)^2 = \gamma^2(mc^2)mv^2$$

$$\gamma^2(mc^2 - mv^2) = mc^2$$

$$\gamma^2(c^2 - v^2) = c^2$$

$$\frac{c^2 - v^2}{(1 - v^2/c^2)} = c^2$$

$$1 = 1$$

Evidently the two expressions for mass-energy are the same.

Question 11-32

A photon, whether visible light or an X-ray, has no rest mass since it can never be at rest with respect to any observer. Therefore, how can it have any energy if $E = mc^2 (1 - v^2/c^2)^{-1/2}$?

We can get some useful insight into the relative sizes of rest mass-energy and other forms of energy by performing an algebraic expansion on our second formula for E.

$$E = \gamma mc^2 = mc^2(1 - v^2/c^2)^{-1/2}$$

The algebraic procedure that we will use is the standard expansion of a binomial:

$$(1 - x)^{-n} = 1 + nx + \frac{n(n + 1)}{2}x^2 + \cdots \qquad x < 1$$

Question 11-33

How well does this approximation work if you use just the first three terms for $x = 0.1$ and $n = \frac{1}{2}$? Try it also for $x = 0.5$ and $n = \frac{1}{2}$.

When we use the binomial expansion for the mass-energy expression, we get

$$E = \gamma mc^2 = mc^2 + \tfrac{1}{2}mv^2 + \tfrac{3}{8}mv^2(v^2/c^2) + \cdots$$

The total mass-energy of an object consists of its rest mass, mc^2, plus its ordinary kinetic energy, $\tfrac{1}{2}mv^2$, plus other terms that depend on higher powers of (v^2/c^2). For ordinary velocities, we can neglect all of the kinetic energy terms except $\tfrac{1}{2}mv^2$. Let's compare the sizes of the first three terms for the case of a very fast jet plane with $v = 10^3$ m/s. For 1 kg of the plane, the first term is equal to: $mc^2 = (1 \text{ kg})(3 \times 10^8 \text{ m/s})^2 = 9 \times 10^{16}$ J. The value of the second term is:

$$\tfrac{1}{2}(1 \text{ kg})(1 \times 10^3 \text{ m/s})^2 = 5 \times 10^5 \text{ J}$$

The value of the third term is:

$$\tfrac{3}{8}mv^2(v^2/c^2) = \tfrac{3}{8}(1 \text{ kg})(1 \times 10^3 \text{ m/s})^2 \left(\frac{1 \times 10^3 \text{ m/s}}{3 \times 10^8 \text{ m/s}}\right)^2 = 4 \times 10^{-6} \text{ J}$$

As you can see, the third term is less than the ordinary kinetic energy by a factor of 10^{11} and so can be neglected. But notice that the ordinary kinetic energy is smaller than the rest mass-energy by a factor of 10^{11} also. Why don't we just neglect the kinetic energy in ordinary motions? The reason is that in ordinary interactions the rest mass-energy remains essentially the same for all of the participants before and after the interaction. Therefore, it always cancels out.

Actually, the rest mass-energy *does* change during ordinary interactions. There are both mechanical and chemical methods of feeding energy into objects. A stretched spring, for instance, has extra energy and therefore extra mass compared with its unstretched state. Let's calculate the extra mass. A good spring with a mass of 1 kg might require an *average* force of 2000 N to compress it by 0.1 m. The stored energy is $\Delta E = \mathbf{F} \cdot \mathbf{\Delta x} = (2 \times 10^3 \text{ N})(1 \times 10^{-1} \text{ m}) = 200$ J. The mass increase of the compressed spring is therefore

$$\Delta m = \frac{\Delta E}{c^2} = \frac{200 \text{ J}}{9 \times 10^{16} \text{ m}^2/\text{s}^2} \approx 2 \times 10^{-15} \text{ kg}$$

Clearly, no laboratory balance can measure such a slight increase in mass.

If you feed energy into an object and make it hotter, you can also increase its mass. As we will see in Chapter 13, when heat is fed into material, the molecules and atoms in their locked positions vibrate faster. We might expect therefore that their mass would increase, but note that such an increase affects the *first* term of the relativistic expansion for energy. The rest mass of the object as a whole is actually greater. For example, it takes about 2×10^5 J of thermal energy to make a 1 kg metal ball red hot. The extra mass of the ball is

$$\Delta m = \frac{\Delta E}{c^2} = \frac{2 \times 10^5 \text{ J}}{(3 \times 10^8 \text{ m/s})^2} \approx 2 \times 10^{-12} \text{ kg}$$

Once again, the increase in mass is too small to be measured.

Let's calculate the mass increase for one of the most energetic chemical interactions that we can produce. The energy provided in the production of 1 kg of water, by burning hydrogen and oxygen, is 1.6×10^7 J. This energy release corresponds to a loss (or dissipation) of 1.8×10^{-10} kg. Even this much of a mass change is too small to be detected by the best precision chemical balance.

MASS-ENERGY AND THE LORENTZ TRANSFORMATION

In space-time, as we have seen, there is an invariant interval between two events, regardless of which inertial reference frame you are in. The quantity $(\Delta x^2 + \Delta y^2 + \Delta z^2 - c^2 \Delta t^2)$ has the same value in your reference frame as it does in any other inertial frame. There is a similar invariant quantity concerning the momentum and energy of an object. This quantity is

$$E^2 - c^2 p^2 = (mc^2)^2$$

The invariant quantity is just the rest mass-energy of the object. That's the mass-energy that we measure when the object is motionless with respect to us. From a reference frame moving rapidly past us, the object would seem to be in motion. Its total energy E would be greater, but it would also have momentum p. If the other observer measured both E and p for the object and calculated the quantity $\sqrt{(E^2 - c^2 p^2)}$, he would find just the value that we measure to be the rest mass-energy.

Total energy and momentum for an object are related in the same way as the space-time coordinates for the object are related. If you measure E and p in one inertial frame, you can transform their values for any other inertial frame through the Lorentz transformations. The momentum variable, pc, transforms in the same way as the position coordinate, x. The energy variable, E/c, transforms similarly to the time, t.

For instance,

$$x' = \gamma(x - vt) \rightarrow p'c = \gamma\left(pc - v\frac{E}{c}\right)$$

$$p' = \gamma\left(p - v\frac{E}{c^2}\right)$$

The Lorentz transformations for the two sets of variables look like this:

$$x' = \gamma(x - vt) \qquad p'_x = \gamma[p_x - (v/c^2)E]$$
$$t' = \gamma[t - (v/c^2)x] \qquad E' = \gamma(E - vp)$$

There are other sets of variables that can be transformed together by the Lorentz transformation. One of the examples which we will study later is the set of variables for the electric and magnetic field. As we will see, it is possible for one observer to measure a pure electric field and no magnetic field with a particular electric charge arrangement while another observer in a reference frame moving by would observe both electric and magnetic fields.

GRAVITATION AND THE GENERAL THEORY OF RELATIVITY

In all of our examples with Galilean relativity or with Einstein's special theory of relativity, we have compared observations between observers moving with constant velocity with respect to each other. That's an unfortunate restriction since so many interesting events happen in accelerated reference frames. At the beginning of this chapter, we saw that pseudo or inertial forces arise in accelerated reference frames. With rotating systems these are the centrifugal and Coriolis forces. In acceleration along a straight line, there appears to be a gravitational-type force in the direction opposite to that of the acceleration. These inertial forces can be described in terms of local gravitational-type fields that can be added vectorially to any gravitational field of the normal Newtonian type.

Einstein proposed that these fields—gravitational and inertial—are completely equivalent in their action at any *point* in space. He went on from there to work out a theory in which gravitational mass warps space in such a way that gravitation becomes simply a consequence of geometry. In this model, planets go around the sun not because they experience a radial centripetal force but because the sun has distorted space in its vicinity. The planet is constrained like a marble rolling around on the inside of a bowl.

Einstein's union of geometry and gravitation reduces to Newtonian gravitation except for very large masses. Nevertheless, the differences in predictions concerning experimental effects are sufficiently large in several cases to be tested. The three original predictions that Einstein made have been confirmed with considerable precision. These are:

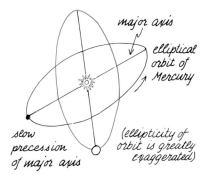

1. The elliptical orbit of Mercury precesses at a rate slightly different from the rate that could be accounted for by the known effects of other planets. The difference in precession rate as predicted by Einstein is only 43 seconds of arc per century. That small discrepancy is measurable and had been approximately known, but not understood, before Einstein made his prediction.

2. If space is warped near massive objects, then the path of light through such space should not be a straight line. One way to observe such an effect is to observe a star whose light has just grazed the surface of our sun. If the star's position seems to shift slightly as the sun moves close to the path, then the light from the star must have been deflected by the warped space near the sun. Until recently such an observation could be made only during a total eclipse of the sun. Shortly after Einstein predicted the effect, it was observed during the total eclipse of 1919. The measurement has been made many times since then with varying degrees of success and precision. Recently Einstein's prediction has been confirmed with great accuracy at radio wave lengths which can be used even when the sun is not eclipsed. It should be noted that, since the special theory of relativity requires that photons have mass-energy (but not rest mass), we would expect a gravitational deflection of the passing photon even with Newtonian gravitation. The general theory of relativity, however, predicts a deflection twice as great.

3. A third prediction of the general theory is that time depends on the

gravitational field or, in the Einstein model, the curvature of space. Clocks slow down in gravitational fields. If a particular element emits a characteristic pattern of different frequencies of light when it is incandescent in our laboratory, we would expect to see the same pattern from that element in a distant star. However, if time is slower in that star because of the strong gravitational field, each of the frequencies should be smaller because time is slower. We might expect to see the same *pattern* of frequencies, but shifted toward lower frequencies or longer wave lengths. This frequency shift is mixed up with other effects and has not served as a critical test of the general theory. However, there is a frequency-dependent phenomenon (the Mössbauer effect) so sensitive that the gravitational effect on clocks has been measured here on earth. A clock (an atomic process) in the basement of a building runs slow compared with a similar clock in the attic. The effect has been measured and agrees with Einstein's prediction.

For many decades after Einstein's publication of the general theory, very few new consequences were drawn concerning its implications. During the late 1970s there has been renewed interest in applications of the theory to astronomy and cosmology. Apparently the general theory of relativity is basic to our understanding of many events within the galaxy and of the universe as a whole.

SUMMARY

Both paths (kinematics) and forces (dynamics) look different to observers in different reference frames. The Galilean transformation relates motions in reference frames traveling at constant velocity with respect to each other and at speeds much less than the speed of light.

The Galilean transformation, for v in the x-direction, is:

$$x = x' + vt \qquad z = z'$$
$$y = y' \qquad t = t'$$

Note particularly that $t = t'$. The sum of two velocities, u and v, is $w = u + v$. The Galilean transformation shows how a ball dropping along a straight line in one system is seen as a ball in a parabolic trajectory in another system. Simple harmonic motion in one frame becomes sinusoidal motion in another. The magnitude of velocity, momentum, and kinetic energy of an object are functions of the relative velocity of the observing system. Frequently, the motion of a system can be described most simply by using the reference frame attached to the center of mass of the system.

An inertial reference frame is one moving with constant velocity with respect to the fixed stars. In a noninertial frame, the acceleration produces pseudo forces that have the characteristics of local gravitational fields. Vertical accelerations produce weight variations:

$$\mathbf{W} = m\mathbf{g} - m\mathbf{a}$$

Circular motion produces centrifugal and Coriolis forces, as seen from within the rotating system. The centrifugal force depends on angular frequency and radial position:

$$\mathbf{F}_{cent} = m\omega^2 \mathbf{r}$$

The Coriolis force depends on angular frequency and the velocity (within the rotating frame):

$$\mathbf{F}_{Coriolis} = -2m\boldsymbol{\omega} \times \mathbf{v}$$

For objects moving close to the speed of light with respect to us, Einstein's special theory of relativity must be used to describe phenomena. The cornerstone of the special theory is that the speed of light is constant regardless of the relative speed of source and observer. This condition implies that time and length intervals are different in different reference frames.

$$\Delta t = \Delta t_0 / \sqrt{1 - v^2/c^2} \qquad \Delta x = \Delta x_0 \sqrt{1 - v^2/c^2}$$

The Galilean transformation is a low-velocity approximation to the Lorentz transformation, which is correct at all speeds. For velocity in the x-direction, the Lorentz transformations are

$$x' = \gamma(x - vt) = \frac{x - vt}{\sqrt{1 - v^2/c^2}} \qquad t' = \gamma\left(t - \frac{v}{c^2}x\right) = \frac{t - vx/c^2}{\sqrt{1 - v^2/c^2}}$$

$$y' = y \qquad z' = z$$

There is an invariant space-time interval between events as seen from any reference frame:

$$\Delta x^2 + \Delta y^2 + \Delta z^2 - c^2 \Delta t^2 = (\Delta x')^2 + (\Delta y')^2 + (\Delta z')^2 - c^2(\Delta t')^2$$

The sum of two velocities u and v is

$$w = \frac{u + v}{1 + uv/c^2}$$

Momentum and energy must also be transformed by the Lorentz relationships:

$$p'_x = \gamma[p_x - (v/c^2)E]$$

$$E' = \gamma(E - vp)$$

The corresponding invariant quantity is $E^2 - c^2p^2$, which is $(mc^2)^2$, the square of the rest mass-energy of an object. Mass and energy are simply different names, for convenience often measured in different units, for the same quantity.

In Einstein's general theory of relativity, transformations between accelerating reference frames can be calculated. The effects are interpreted as gravitational, and gravitation itself becomes a property of space.

Answers to Questions

11-1: For most of the purposes of everyday life, it is convenient to assume that the sun goes around the earth. As you may have noticed, the sun rises in the morning and

SUMMARY

sets at night. The Ptolemaic system of astronomy, in which the earth stands still and the sun and moon and planets go around the earth, is successful and convenient for such purposes as ocean navigation and naked-eye astronomy. If you are going to send rockets off the earth or make some type of astronomical calculations, it will probably be more convenient for you to use Copernican astronomy in which the planets go around the sun. The choice between these two models is not a matter of one being true and the other therefore false. The choice is determined by the convenience of the reference frame that you use.

1-2: First, if it is 10 o'clock in the primed frame, it is also 10 o'clock in the x, y, z frame. Furthermore, since the coordinate frames were coincident at 9 o'clock and the motion is only in the x-direction, the y- and z-coordinates of the object are $y = 1$ mile and $z = 2$ miles. The x-coordinate is given by $x = x' + vt = (4 \text{ miles}) + (10 \text{ mph})(1 \text{ hr}) = 14$ miles. The coordinates of the object in the two reference frames at this time are shown in the diagram on page 226.

1-3: There is no such thing as a "true" path. The path depends on the reference frame.

1-4: The relative motion is the same in both of these cases. Therefore, the results will be the same.

1-5: The voices will sound like Donald Duck. All the frequencies or pitches will be higher by a factor of $78/(33\frac{1}{3}) = 2.3$. A factor of 2 in frequency raises the pitch of sound by one octave.

1-6: When you fire the gun in the same direction that the train is traveling, the BB travels at 60 m/s with respect to an observer at the station. If you fire in the backward direction, however, the observer sees the BB drop straight down.

1-7: Evidently the value of the kinetic energy of an object depends on the reference frame from which the measurement is made. The size of the hole in a target, which might be used as a measure of the kinetic energy lost by the BB when it stops, depends on the relative speed of the BB *and the target*. For instance, if the BB is fired from the train in the backward direction so that its velocity is 0 with respect to the observer at the station, it will not penetrate a target attached to the station. However, if the target is attached to the back end of the train, then the BB will enter it traveling at a speed of 30 m/s. In this case, the observer at the station might claim that the BB was standing still and the target ran into it.

1-8: We did not take into account the necessity for applying the law of conservation of momentum. The supply rocket cannot just enter the space station and come to rest. There must be a recoil of the merged system. The recoil velocity is small because the mass of the space station is much larger than that of the supply ship. Nevertheless, the kinetic energy associated with the recoil accounts for the extra energy that the supply ship appears to lose. The arithmetic is easy to work out. Try it! Remember that in the reference frame of the space station before collision: $m(1) = (m + M)v_{\text{final}}$. Assume $m = 1$ and $M = 10$.

1-9: Let us assume that there is equal probability for a proton to pass through any small region on the surface of a concentric sphere, as shown in the diagram. This would mean that there is equal probability for a proton to be emitted into any solid angle. In terms of the angle θ, however, there is a much larger solid angle available in the region of $\theta = 90°$ than there is in the region $\theta = 0°$. Therefore, more protons would be emitted at $90°$ in the center-of-mass system, which becomes $45°$ in the laboratory system.

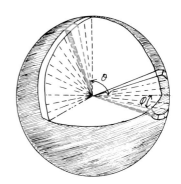

1-10: Newton's first law states that, if there are no forces acting on an object, the velocity of the object will remain constant. On the surface of the earth, we observe that objects fall down with increasing speed. We blame this action on the force of gravity. We also observe that in the northern hemisphere objects moving along the surface of the earth swing to the right with respect to straight lines on the surface of the earth. We blame the Coriolis force for this effect, although it is not a force between

11-11: The air that the balloon displaces acts like a plumb bob with respect to the balloon. The heavier air will force itself "down" with respect to the balloon, thus maneuvering the balloon to the direction that is "up."

11-12: According to the bathroom scales, your weight is $W = -F_{spring} = mg - ma = m(g - a)$. Since $a = 1$ m/s², $(g - a) = (-9.8 - 1)$ m/s² ≈ -11 m/s², which is about 10% greater than g. Therefore, the scales read 550 N (or 121 pounds). The scales read this extra amount only as long as the elevator accelerates. After the initial surge, elevators usually move at constant velocity. The bathroom scales would then read a normal 110 pounds.

11-13: On the equator, you experience a centrifugal acceleration: $a_c = v^2/R = \omega^2 R = 4\pi^2 R/T^2$. For the earth, $R = 6.4 \times 10^6$ m and $T = (24 \times 3600)$ s.

$$a_c = \frac{4\pi^2(6.4 \times 10^6 \text{ m})}{(24 \times 3600 \text{ s})^2} = 0.034 \text{ m/s}^2$$

This centrifugal acceleration should be compared with the acceleration due to gravity in free-fall: $g = 9.8$ m/s². Your loss of weight on the equator due to the rotation of the earth is about $\frac{1}{3}$%.

11-14:
$$\omega^2 r = 4\pi^2 f^2 r = g$$

$$4\pi^2 f^2 (500 \text{ m}) = 9.8 \text{ m/s}^2$$

$$f = 2.2 \times 10^{-2} \text{ revolutions/second}$$

$$T = 45 \text{ s}$$

11-15: The velocity of light that we observe is always 3.0×10^8 m/s. There is, however, a change in the color of the light from the quasar.

11-16: The effect would depend on the distance from us of the twin stars. (Why?) If the light from the star swinging toward us is faster than when it comes from the star swinging away from us, we might observe that the star spends only a short time in the eclipsed position. That could also happen if it were a large star and its twin were small. But then the twin's motion would not appear correct. We might end up with "ghost" stars—sometimes the doublet would become a triplet. Note also that experimental observations of twin stars demonstrate that in vacuum all colors of light travel at the same speed. Otherwise the eclipsing twin stars would be seen in rapidly varying colors—a remarkably pretty effect if it happened. It does not.

11-17:
$$\Delta x = c \Delta t = (3 \times 10^8 \text{ m/s})(1 \times 10^{-16} \text{ s})$$

$$= 3 \times 10^{-8} \text{ m} = 0.03 \text{ micrometers (microns)}$$

11-18: The apparatus is first lined up so that one of the light paths is parallel to the motion of the earth around the sun, and the other is perpendicular to it. If light took longer to go one way than the other, rotating the apparatus would reverse the situation and produce a shift in the interference fringes.

11-19:
$$\Delta t = \Delta y/c \approx (1 \text{ m})/(3 \times 10^8 \text{ m/s}) \approx 3 \times 10^{-9} \text{ s} = 3 \text{ nanoseconds}$$

11-20: *In* the rocket ship, the light traveled only Δy. Viewed from the ground, the light had to travel $\sqrt{\Delta x^2 + \Delta y^2}$.

11-21: The speed of light is the same for both observers. Therefore, as long as the two observers are moving at constant velocity with respect to each other, there is no way of telling which one is moving. We do not have to choose between the two observations because each is correct in its own system.

-22:
$$\Delta t = \frac{\sqrt{v^2 \Delta t^2 + \Delta y^2}}{c} = \frac{\sqrt{v^2 \Delta t^2 + c^2 \Delta t_0^2}}{c}$$

$$c^2 \Delta t^2 = v^2 \Delta t^2 + c^2 \Delta t_0^2$$

$$[(c^2 - v^2)/c^2]\Delta t^2 = \Delta t_0^2$$

$$\Delta t = \frac{\Delta t_0}{\sqrt{1 - v^2/c^2}}$$

-23: If $\Delta t / \Delta t_0 = 1.1$ (10% effect):

$$(1.1)^2 = 1/(1 - v^2/c^2)$$

$$1.21 - 1.21 v^2/c^2 = 1$$

$$1.21 v^2/c^2 = 0.21$$

$$v^2/c^2 = 0.17$$

$$v/c = 0.4$$

If $\Delta t / \Delta t_0 = 2.00$ then $4.00 = 1/(1 - v^2/c^2)$

$$4 - 4v^2/c^2 = 1 \quad \text{and} \quad 4v^2/c^2 = 3$$

$$v/c = 0.866$$

Notice how the effects are appreciable only when the velocity is very close to that of light.

-24: One way to synchronize clocks which are at rest with respect to you is simply to gather them together in one place, set them to read the same time, and then walk with them to their stations. Since everything is done at low velocities, you might expect that there would be no trouble. To check them when they are in place, establish yourself at the midpoint between any two clocks. Arrange for light signals to be sent from each clock every hour on the hour. If the signals from the two clocks arrive at your central location at the same time, the two clocks are synchronized.

-25: Suppose two observers are each at the midpoint of their respective systems. B_1 and B_5 emit light signals just at the moment when A_3 and B_3 are opposite each other. B_3 receives the two signals at the same time a short while later. B therefore says that B_1 and B_5 are synchronized. Meanwhile, however, A_3 has been traveling toward the right and so receives the signal from B_5 before she receives the signal from B_1. Therefore, A says that B_1 and B_5 are *not* synchronized. It does no good to say that A should know she is moving. As far as she is concerned, she is at rest and B is moving. Because light travels at the same velocity with respect to both A and B, both are equally good observers.

1-26: $\Delta x = c \Delta t = (3 \times 10^8 \text{ m/s})(2 \times 10^{-6} \text{ s}) = 600$ meters. Notice that 600 meters is only about $\frac{1}{3}$ mile. Most of the muons are produced at least 20 miles above the surface of the earth.

1-27: Since the muon in its rest system sees our length as being shorter than we do, the muon would claim that our atmosphere is very thin—of the order of 600 meters. As far as the muon is concerned, the earth and all its works are rushing up toward the muon.

1-28: When $v/c \ll 1$, then the denominator in the factor γ approaches 1 and so does the value of γ. Thus, the Lorentz transformation reduces to the Galilean transformation.

11-29: Here is the substitution to demonstrate that $\Delta x^2 - c^2 \Delta t^2 = (\Delta x')^2 - c^2(\Delta t')^2$. For simplicity, let $x_0 = 0, x'_0 = 0, t_0 = 0,$ and $t'_0 = 0$. The Lorentz transformations are

$$x = \gamma(x' + vt')$$

$$t = \gamma\left(t' + \frac{v}{c^2}x'\right)$$

$$x^2 - c^2t^2 = \gamma^2(x' + vt')^2 - c^2\gamma^2\left(t' + \frac{v}{c^2}x'\right)^2$$

$$= \gamma^2(x')^2 + \gamma^2 v^2(t')^2 + 2\gamma^2 x' vt' - c^2\gamma^2(t')^2 - \frac{\gamma^2 v^2 (x')^2}{c^2} - 2\gamma^2 t' vx'$$

$$= \gamma^2[(v^2 - c^2)(t')^2 + (x')^2(1 - v^2/c^2)]$$

$$= \gamma^2[-c^2(t')^2(1 - v^2/c^2) + (x')^2(1 - v^2/c^2)]$$

$$= \gamma^2[-c^2(t')^2(1/\gamma^2) + (x')^2(1/\gamma^2)]$$

$$= (x')^2 - c^2(t')^2$$

11-30: The Lorentz transformation for time is

$$t' = \gamma\left(t + \frac{v}{c^2}x\right)$$

If we express time in years, then $t = 20$, $v/c = 0.87$, and $x/c = 17.4$. Consequently, $t' = 2[20 + (0.87)(17.4)] = 70$.

11-31: To generate 9×10^{16} J, the power complex at Niagara Falls would have to work for $(9 \times 10^{16} \text{ J})/(10^{10} \text{ J/s}) \approx 10^7 \text{ s} \approx 4$ months.

11-32: In the form given, E is an indeterminate quantity since the numerator contains m, which is 0 for a photon, but the denominator is also 0 since $v = c$. The energy of a photon is best expressed by our first relativistic expression for the total energy: $E^2 = p^2c^2 + (mc^2)^2$. Since $m = 0$, the energy of a photon is simply $E = pc$. Note that this formula gives us the magnitude of the momentum of a photon: $p = E/c$.

11-33:
$$(1 - 0.1)^{-1/2} = \frac{1}{\sqrt{0.9}} = 1.05409$$

$$1 + \tfrac{1}{2}(0.1) + \frac{(\tfrac{1}{2})(\tfrac{3}{2})}{2}(0.01) = 1.0538$$

The approximation is good to three parts in ten thousand.

$$(1 - 0.5)^{-1/2} = \frac{1}{\sqrt{0.5}} = 1.4142$$

$$1 + \tfrac{1}{2}(0.5) + \frac{(\tfrac{1}{2})(\tfrac{3}{2})}{2}(0.25) = 1.3438$$

In this case, the approximation is good to seven parts in 140 for 5%. For a closer approximation more terms would be needed.

Problems

1. You are in a car driving north at 100 km/hr. Describe your motion in reference frames where the origins of coordinates are: (a) fastened to you, (b) fastened to the road, (c) fastened to a truck going south at 100 km/hr. Assume that in all cases you were at $x = 0$ at $t = 0$.

2. The coordinates of two Galilean reference frames were coincident at noon. One frame moves at 100 m/s in the negative x-direction. At 2 P.M. an object in that frame is located at the point $x' = 300$ m, $y' = 10$ m, $z' = 20$ m. What are the coordinates at that time in the x-frame?
3. The orchestral tuning note, A, has a frequency of 440 cycles per second (440 Hz). If this note is recorded on a record turning at $33\frac{1}{3}$ rpm and then is played back at 78 rpm, what is the resulting frequency?
4. A ball is hurled at a speed of 30 m/s inside a railroad car. The train is traveling at a speed of 50 m/s. What is the speed of the ball with respect to the ground if the ball is thrown: (a) forward, (b) backward, (c) perpendicular to the train's velocity?
5. A boat is sailing with the wind at a speed of 6 mph with respect to the water. The wind is blowing at a speed of 10 mph. What is the velocity of the wind with respect to the boat? What direction does the flag fly at the top of the mast?
6. In still air (with respect to the land) a boat floats down a river with the current at 4 mph. If the crew raises their sail, can they go faster downstream than just by drifting?
7. A pellet with a mass of 10 g is fired with a muzzle velocity of 50 m/s. If the pellet gun is fired from a flatcar of a train traveling at 30 m/s, (a) what is the kinetic energy delivered to a target beside the track if the gun is fired in the forward direction? (b) In the backward direction? (c) At a target fastened to the train?
8. A grenade with a mass of 2 kg traveling at 100 m/s explodes into two equal parts, each going off at 45° to the original direction. Describe the action from the center-of-mass coordinate system. (a) What is the original momentum in that system? (b) What is the final momentum in that system? (c) What is the observed direction of the pieces in that system? (d) What is the speed of each piece in that system? (e) What is the kinetic energy in that system after the explosion? (f) What is the kinetic energy of the system before and after the explosion in the original reference frame?
9. An elevator starts rising with an acceleration of 0.5 m/s². If a 150 lb man is standing on a spring scale, what weight does he read? What weight does he read while the elevator is rising with constant velocity? If the elevator then stops with an acceleration of 1.5 m/s², what does the scale read?
10. An elevator starts down with an acceleration of 1.0 m/s². If a 50 kg girl is standing on a spring scale, what weight does she read? What weight does she read when the elevator comes to an emergency stop with an acceleration of 2 m/s²? What weight does she read if the cable then breaks and the elevator falls freely?
11. What is the Coriolis acceleration for a river flowing due south at a speed of 4 mph at a latitude of 45°? [The component of the velocity perpendicular to the earth's axis is (0.7×4) mph.] Compare with g.
12. What is the Coriolis acceleration on a rifle bullet traveling with a speed of 600 m/s, when it is fired straight up on the earth's equator? Compare with g.
13. Habitats for space colonies must rotate in order to provide pseudogravity with centrifugal force on the people living inside. If a colony has a radius of 1 km, what must be the period of rotation in order to provide a field of 1 g at the rim?
14. One proposed form of space colony habitat is a giant torus (hollow doughnut). Assume that the outer radius is 1 km and that the torus temporarily contains a vacuum and is not rotating. A ball suspended in the tube will remain motionless. (The gravitational field in space is negligible because the space station itself is in orbit, and thus in free-fall.) Now the torus is set into rotation with an observer (in a space suit) held to the inside rim with a centrifugal field of 1 g. She observes

the ball rotating in the opposite direction, but it remains in orbit. As far as the internal observer is concerned, the ball can remain in orbit only if there is a *centripetal* force on it. How does the observer explain the nature of this centripetal force? (Remember that the observer knows that both she and the ball are in an outward field of 1 g.) Show that the magnitude is just right to keep the ball in orbit inside the torus. (Consider the Coriolis force on the ball, as seen by the internal observer.)

15. The half-life at rest for the decay of a π meson into a muon is 1.8×10^{-8} s. What is the half-life in the laboratory for π mesons coming from a target with a speed of 2.9×10^8 m/s?

16. What is the β (v/c) of a muon that observes that the earth's atmosphere has shrunk by a factor of 50? How much slower is this (in km/s) than the speed of light?

17. How far will a lambda (a neutral particle a little more massive than a proton) travel before decay if its lifetime in its rest system is 2×10^{-10} s and its γ is 10?

18. If a gas jet from a distant galaxy is shooting away from the galaxy, but toward us, at a speed of 1×10^8 m/s with respect to the galaxy, and the galaxy is moving away from us at a speed of 1.5×10^8 m/s, what is the velocity of the gas jet with respect to us?

19. If a K^+ meson decays in flight to a muon and a neutrino with the muon almost in the forward direction, what is the β of the muon with respect to us if the β of the muon with respect to the K^+ is 0.9 and the β of the K^+ with respect to us is also 0.9?

20. A gamma ray with an energy of 1.2 MeV creates an electron pair. If each electron (positive and negative) receives half the surplus energy (which does not usually happen), what is the speed of each? Use nonrelativistic formulas. Does their use turn out to be justified?

21. In the fission of uranium, about 1×10^{-3} of the rest mass is dissipated in the kinetic energy of the fragments. How many joules are dissipated per kilogram of uranium?

22. How fast is a proton going with respect to us when its mass is twice its rest mass?

23. Following the sequence of the twin's relativistic journey described in the text, how much older is each person when the girl reaches the distant star—according to the boy and to the girl? When the girl transfers to the return system, what time is it (in years) where the girl is according to the boy and to the girl? At that moment, what time is it on earth according to the boy and to the girl? What time is it according to each when she gets back home?

24. A spaceship traveling at 2.6×10^8 m/s past us fires two flares simultaneously in their system. The astronauts claim that the distance between the flare guns was 1000 m. What is the time interval between shots that we observe? How far apart are the flare guns according to us—in our system and in theirs?

THE INTERNAL ENERGY OF MATERIAL 12

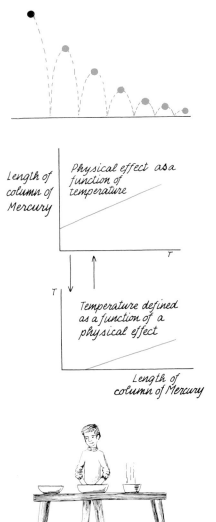

A glider on an air track eventually loses its kinetic energy; after a few bounces a tennis ball rolls to a resting place; even a pendulum finally stops swinging. So much for the law of conservation of energy! We must either conclude that the law is very limited, or else we must discover other forms of energy and invent appropriate definitions so that the missing energy is accounted for. The realization that energy could be defined in this way did not occur until about 140 years ago. Through a methodical series of experiments done to high precision, James Joule in England showed that the mechanical work done, and presumably lost, in stirring water was proportional to the temperature rise of the water. Apparently, the lost mechanical energy had changed the internal properties of the material. The results were the same as they would have been had the water been heated on a stove.

In this chapter, we will explore the changes that take place in solids, liquids, and gases as energy is fed into them. We will see that many of these changes can be defined as a function of temperature, which, in turn, will be defined in terms of one or more of the changes. These changes in materials occur in the same way whether the energy is fed in through friction effects or through exposing the material to a hotter object. The energy that flows between objects because one is hotter than the other is called *heat*. When heat enters a system, it can provide the energy to do work and can also raise the temperature or change other properties of the system.

HANDLING THE PHENOMENA

1. Most measuring instruments extend our own senses. Sensitivity to temperature is built into living systems. Indeed, the thermostat controlling human internal body temperature normally maintains constant temperature to within half a degree Fahrenheit and warns us dramatically of changes of more than one degree. But how good a thermometer are you when it comes to judging other temperatures?

Fill three pans of water: one, as hot as you can stand; the second as cold as you can get; the third at room temperature. For at least one minute, soak your left hand in the hot water and your right hand in the cold water. Then put both hands in the room-temperature water and feel how reliable your senses are for measuring temperature.

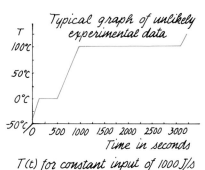

$T(t)$ for constant input of 1000 J/s to 1 kg of H_2O

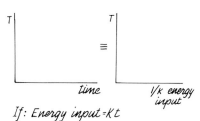

If: Energy input = Kt

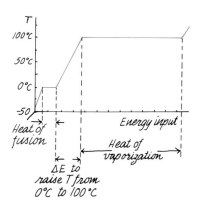

2. You can at least use your sense of touch to tell you if something has become hotter or colder. Waste some mechanical energy by rubbing your hands together, hard and briskly. You can produce an even larger effect by rubbing wood with sandpaper or by hammering a nail. Mechanical energy is lost; something gets hotter. With a power drill, the drill may get hot enough to burn your fingers if you touch it. A simple demonstration of the effect of mechanical energy loss can be done with a paper clip while sitting at your desk. With the paper clip partly unfolded, rapidly bend it back and forth a few times, and then quickly touch the bent region to your tongue. You will sense that the energy has not disappeared.

3. Very few people know how to cook ice properly. In most elementary texts, there is a graph showing the temperature of water as a function of time as it is heated from ice to water to steam. Here is a typical graph of this kind. It is almost impossible to duplicate such a graph with real experimental data, for reasons that you can easily find out.

Get a thermometer, either Celsius or Fahrenheit, a cooking pan, a hot plate, and some ice. Make sure that the hot plate is not controlled by a themostat. Once you start cooking, you want the hot plate to feed energy into the pan at a constant rate instead of turning on and off. Use the thermometer to get the starting temperature of the ice. Immediately you will face a problem. How do you know that the surface temperature of the ice cubes is the same as the interior temperature of the cubes? The solution to the problem is to use very finely chopped ice. Ideally, the ice should come from the freezer at below-freezing temperature, but fine enough and dry enough so that it can be stirred with the thermometer.

Stir the ice or water constantly and take temperature readings at regular intervals. If you assume that energy is flowing into the pan at a constant rate, then a graph of temperature, T, versus time, t, is the same as a graph of T versus energy input. Although you will not know what fraction of the energy goes into the water (compared with the energy lost into the air), you can assume that the fraction stays about the same throughout the entire cooking process. Therefore, you can make a rough measurement of the amount of energy it takes to melt the ice (the "heat of fusion")—compared with the amount of energy it takes to change the temperature from melting to boiling and also compared with the amount of energy it takes to turn all the water into steam (the "heat of vaporization"). Just compare the respective times for those processes. So that the final step won't take too long, start out with only 100 to 200 cc of ice—the equivalent of two or three ice cubes.

Question 12-1
Did you get the ideal graph shown for this experiment? If not, why not?

EFFECTS CAUSED BY THE LOSS OF MECHANICAL ENERGY

If you rub one stick against another—in the traditional scout manner—you can start a camp fire. The most obvious result of losing mechanical energy through friction is that things get hotter. Drill tips get hot enough to burn; brake drums get hot enough to smoke; match tips get hot enough to kindle.

What James Joule proved, back in the 1840s, was that the mechanical energy lost through friction produces the same effects as heat from a flame. Before that time, the general wisdom was that heat consisted of a fluid, called "caloric." A hot object had surplus caloric, which could flow over into another object that touched the hot one. A serious objection to this theory had been pointed out by Benjamin Thompson, a double agent in the American Revolution who fled to England and later served as Minister of War in Bavaria. While supervising the boring of cannon he performed a series of experiments to demonstrate that heat was produced as long as the boring continued, and yet neither the cannon nor the metal scraps seemed to have changed in any way after they cooled. It was a strange kind of fluid that could be continuously rubbed out of material without any change in the material itself. Thompson is better known as Count Rumford of the Holy Roman Empire, a title that he was given in Bavaria by a grateful prince.

For practical purposes, we can investigate the effects of lost mechanical energy by using either gas flames or electric stoves to introduce heat into materials. As long as the energy introduced does not change the gravitational potential or kinetic energy of the object as a whole, the effects are the same whether the energy comes from friction or from a heat source. In either case, materials change in many ways as energy is poured into them. They may change in volume or electrical resistance or color or hardness. They may melt or evaporate or change their chemical nature. All of these effects are used in thermometers. Let's take a look at several types of practical thermometers and then come back to the question of just what it is that thermometers measure. Of course, we all know what temperature is, but, as we will see, it is a surprisingly subtle thing to define.

COMMON THERMOMETERS—FAMILIAR AND UNFAMILIAR

Mercury and Colored Alcohol

The most common thermometers exploit the fact that liquids usually expand as they get hotter. The liquid used in laboratory or clinical thermometers is mercury. A less expensive and potentially less dangerous liquid used in household thermometers is a red-colored alcohol, or toluene. A liquid thermometer can be used only in the range above the freezing point of the liquid and well below its boiling point. In the case of mercury, this practical range is from −38 to 260°C.

The ordinary type of liquid thermometer is really a very clever instrument. If it consisted simply of liquid in a glass capillary tube, the temperature changes would be very hard to detect. Liquids expand as they get hotter, but the percentage change of volume for a reasonable temperature change is very small. For mercury, it is only 0.018% per degree Celsius. From the melting point of ice at 0°C to the boiling point of water at 100°C, the volume of a column of mercury (and therefore its height in a tube) would change by only 1.8%. The actual construction of this type of thermometer makes use of a clever technique of amplification. Most of the volume of the thermometer liquid is in the bulb at the end. The column of liquid, which indicates the temperature, is in a capillary tube with a very small volume. A 2% change in the total volume of the liquid can change the length of the column by a factor of 10 or more.

> Question 12-2
>
> As the temperature increases, the volume of the glass tube will also increase. Does the volume of the cavity inside get larger or smaller? What effect does this have on the reading of the thermometer?

The Thermocouple

Temperature changes produce electrical changes in many materials. One kind of thermometer that makes use of such changes is called a thermocouple. Two fine wires of *different* metals are welded or soldered together at one end. The other ends are connected to a voltage-reading device. The voltage across the wires is a function of the temperature difference between the welded end and the voltmeter. In this case, the microstructure of the wires at the junctions acts like an electrical generator. The standard metals used for most thermocouples are copper and constantan. The latter is an alloy of copper and nickel. A table giving the voltage of this combination as a function of temperature is readily available in many handbooks. One of the virtues of thermocouples is their small size—no larger than the junction point of the fine wires. The wires can be embedded deep in the material and the internal temperature read remotely. Using several combinations of metals, thermocouples can be used to measure temperature all the way from −269°C to almost 2300°C.

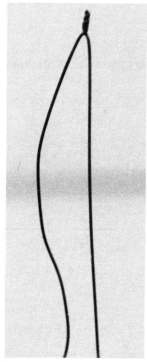

Thermocouple

Electrical Resistance Thermometers

The resistance of metals generally increases with increased temperature. This is a very reproducible effect and is used in many precision thermometers. Platinum wire wound in a small coil is usually used because the characteristics of platinum remain so stable. The temperatures can range from −258 to 900°C.

Thermistors

Thermistor is the name for a class of semiconductors that have a temperature-resistance response just opposite to that of metals. The electrical resistance decreases with increasing temperature. The thermistor materials, which are usually claylike metal oxides, can be made in the form of tiny beads. They can be much more sensitive than thermocouples or platinum-resistor thermometers, but are not so stable over long periods of time. Hospital thermometers with electrical readout use thermistors.

Bimetallic Strips

The thermometer in most home thermostats consists of two pieces of different metals bonded together. The metals are chosen so that one expands much more than the other as the temperature increases. The only way that this can happen with the two pieces fastened together is for the bimetallic strip to change its radius of curvature, as shown in the diagram. One of the metals usually used has the trade name Invar. It is an iron-nickel alloy whose volume changes very little over a large temperature range. Thermometers with long needle probes, such as those used in darkrooms or in kitchen roasts, have a coiled bimetallic strip in the probe. You can see the coiled bimetallic strip in a home thermostat simply by taking off the cover cap. Usually the coil moves a needle to indicate the temperature and also trips a mercury switch to turn the heating system on or off. Thermometers of this type are usually accurate only to 0.5°C and have a small, limited range.

Bimetallic coil in thermostat

Gas Thermometers

When heat is fed into a gas, the pressure of the gas increases or, if the pressure is kept constant, the volume increases. Either of these effects can be used to indicate temperature, but they are seldom used any more in actual thermometers. However, experiments done with constant volume gas thermometers, shown schematically in the diagram, helped to define the meaning of temperature. All of the other thermometers depend critically on the nature and purity of the materials used. As we will see, under the right conditions the behavior of gas thermometers is almost independent of the type of gas used. Furthermore, under these conditions the theory of the temperature response of gas is simple and well understood.

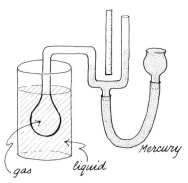

Constant volume gas thermometer

By raising or lowering the mercury reservoir, the volume of the gas can be maintained constant, regardless of temperature. However, the *difference* in mercury levels in the two columns measures the pressure of the gas.

Other Thermometers

Many other physical properties have been used to measure temperature. Skaters and skiers can judge temperature crudely in terms of the slipperyness of ice or snow. Generations of fudge makers know that when the candy has reached the "soft ball" stage (when a little of the hot syrup dropped into a cup of cold water clings together) it is ready to take off the stove. A novelty thermometer item makes use of liquid crystals that can be mixed to change color at a particular temperature. For industrial purposes, with hot ovens and furnaces, there is a whole range of waxes, each of which melts at a particular temperature. The color of glowing metals is a function of temperature. There are special instruments—optical pyrometers—that allow the brightness of glowing metals to be judged against various standards.

> Question 12-3
>
> What other physical evidence do you use in daily life to judge temperature?

CALIBRATING THERMOMETERS AND TEMPERATURE SCALES

We still haven't defined temperature in any formal way, but some of the difficulties in doing so will become apparent when we try to calibrate thermometers. Apparently no one measured temperature with instruments until Galileo designed an air thermometer in 1592. Many types of liquid-in-glass thermometers were made during the seventeenth century, each calibrated to the individual standard of its maker. In the early 1700s, Gabriel Fahrenheit standardized procedures for making mercury thermometers very much like our present ones. He also chose a scale that, with a few changes, we still use for household temperatures in English-speaking countries. The zero point of the scale was chosen to be the lowest temperature that Fahrenheit could conveniently produce in the laboratory. It is the temperature of salt and ordinary ice mixed together, the same mixture and process that is used in home ice cream makers. On this scale, 32° was assigned to the melting point of ordinary ice and 96° to normal body temperature. Our present Fahrenheit scale maintains the 32° as the melting point of ice and 212° as the boiling point of water at normal atmospheric pressure at sea level but, with those points fixed, normal body temperature turns out to be 98.6°. During the same years, Anders Celsius, a Swedish astronomer, proposed that the range of temperature between the melting point of ice and the boiling point of water be divided into 100 degrees. The scale derived from his proposal was originally called "centigrade," but since 1960 has been known officially as the Celsius scale. Celsius degrees are standard in the S.I. system of units, which we use in this book. The old-fashioned Fahrenheit degree does have the advantage of being about the smallest change of temperature to which the human body is sensitive.

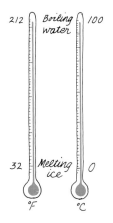

> ### Question 12-4
>
> How many Fahrenheit degrees are there between the melting point of ice and the boiling point of water? How many Celsius degrees are there in that same range? Then which is smaller, a Fahrenheit degree or a Celsius degree? How much larger is one than the other?

When translating temperature from one scale to the other, the first question to ask is: How many degrees is the temperature above the melting point of ice? If the original temperature is in Celsius, the answer is easy. A Celsius temperature of 20° is 20 Celsius degrees above the melting point of ice. A temperature of 68° Fahrenheit, however, is 68°F − 32°F = 36F° (Fahrenheit degrees) above the melting point of ice. To transform a number of small Fahrenheit degrees to the larger Celsius degrees, multiply the number of Fahrenheit degrees by $\frac{5}{9}$. For instance, 36F° × $\frac{5}{9}$ = 20C°. Evidently, 68°F—which is 36 Fahrenheit degrees above the melting point of ice, is also 20 Celsius degrees above the melting point of ice and is therefore equal to 20°C. Here are the formulas summarizing the transformations:

$$(\Delta T_C)C° = \tfrac{5}{9}(\Delta T_F)F°$$

$$T_C = \tfrac{5}{9}(T_F - 32)$$

$$T_F = 32 + \tfrac{9}{5}T_C$$

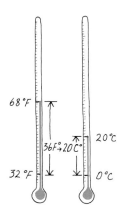

Let's use these formulas to change some familiar temperatures to unfamiliar temperatures. For instance, (1) normal body temperature is 98.6°F. Let's convert that temperature to the Celsius scale.

$$T_C = \tfrac{5}{9}(98.6 - 32) = \tfrac{5}{9}(66.6)$$

$$T_C = 37.0°C$$

(2) A temperature of 20° below zero is mighty cold on the Fahrenheit scale. Is −20°C so cold?

$$T_F = 32 + \tfrac{9}{5}(-20) = 32 - 36$$

$$T_F = -4°F$$

> **Question 12-5**
>
> What temperature has the same number in both the Celsius and the Fahrenheit scale?

Calibrating a thermometer should be easy in principle. If you have an unmarked thermometer, place it in a dish of melting ice and mark the indicator position on the thermometer 0°C. Then place the thermometer in a dish of boiling water and label the resulting point 100°C. Divide the region in between on the thermometer scale into 100 equal parts. Unfortunately, there are all sorts of problems in doing such a simple thing, and even more problems when you are through in deciding whether or not you have a true temperature scale. Consider the following small practical problems:

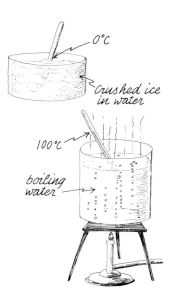

1. The thermometer should be small compared with the system whose temperature is being measured. If you had a small amount of water and a great big air thermometer, you might disturb the temperature of the thing that you are trying to measure or, during calibration, not have the sensitive part of the thermometer at a uniform temperature.
2. There is a subtle problem about when the thermometer should be read. When a thermometer is first moved to a new environment, it takes some time for the reading to settle down. Indeed, when we read the temperature *on* a thermometer, we are reading the temperature *of* the thermometer. We assume that the temperature and the environment have been together long enough so that the two have come to an equilibrium, which we define as being the same temperature.
3. In "Handling the Phenomena" we have already seen that there are problems in trying to measure the temperature of melting ice. An ice cube in water may have a temperature of 0°C at its surface but −10°C inside. Furthermore, the water only a short distance away from the ice cube may be at a temperature of +5°C. To satisfy the definition of a calibration point, the ice-water mixture must consist of finely crushed ice that is kept in equilibrium with the water by continual stirring. Notice, incidentally, that we have repeatedly referred to the melting

point of ice instead of the freezing point of water. Ice melts at 0°C by definition, but water does not necessarily freeze at 0°C. Pure water under clean conditions may go well below 0°C before suddenly crystallizing. This happens frequently in the atmosphere. Ordinary tap water in an ordinary glass is usually sufficiently contaminated so that it will freeze at 0°C or very slightly below.

4. It is relatively easy to keep the thermometer within a degree of the boiling point of water, but there is no guarantee that the temperature is 100°C. The boiling point and, to a much smaller extent, the freezing point depend on the atmospheric pressure. The boiling point is defined to be 100°C at normal atmospheric pressure at sea level. At normal atmospheric pressure in Denver, for instance, about a mile above sea level, the boiling point is about 96°C.

5. There is one other bothersome problem that we must confront when we calibrate a temperature scale. Suppose that you calibrate a mercury thermometer and an alcohol thermometer in the manner that we have described. Clearly, they will both agree at 0°C and 100°C, since they were in identical conditions when they were marked. In each case, then, the range between the two marks was divided into 100 equal parts. If the two thermometers are now placed in a glass of water and the mercury thermometer reads 50°, is there any guarantee that the alcohol thermometer will also read 50°? There is not. We have no reason to expect that the expansion rate of alcohol and mercury are the same with increasing temperature. In fact, the expansion rate is *not* quite the same.

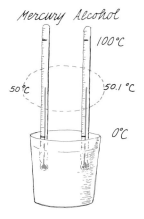

Question 12-6

The question is: Is there a linear relationship between the volume of a material and the temperature? We should ask the same question about other changes in material that are used for thermometers. Is the change of electrical resistance of a platinum wire, for example, proportional to the change of temperature? We ought to be able to determine the answers to these questions by doing experiments. Simply measure the volume or the resistance of the material as a function of temperature. To do this, however, you have to measure temperature. What thermometer would you use?

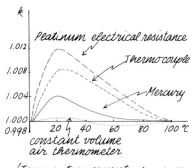

We can't really expect that, as temperature changes, the changes of properties of various materials will all be proportional to each other. For one thing, as temperature increases, many different properties of each material are changing. Perhaps some of them depend on each other. Electrical resistance, for instance, may depend in some way on the density of the metal, which also changes with temperature. Furthermore, we would not expect that the change of volume of a liquid would be the same as the change of volume of a gas of the same material. In fact, it is surprising, though very convenient, to find that certain properties of some materials are sufficiently linear with respect to temperature that we can use them in thermometers. In the diagram, we have plotted the responses of several thermometric materials to various temperatures. It is assumed that all of the thermometers have been calibrated to agree with each other at 0°C and 100°C. They do not, however, agree at any intermediate point, although the differences are not great. Note that 1.01 on the vertical axis corresponds to a 1% difference. Remembering the problem posed in Question 12-6, we must nevertheless assert that there *is* a good way to define temperature, independent of the property of any particular material. That definition of temperature has been used in plotting the graph shown in the diagram. We will see what that definition is in the next section. In the meantime, note that for most everyday purposes the thermometers analyzed in the diagram are almost linear.

For precise work, modern temperature scales are not defined in terms of the temperature interval between the melting point of ice and the boiling point of water. As we will see, there is a so-called absolute scale of temperature, with a zero degree reading that is about −273°C. That zero point, which has both theoretical and practical implications, is used as one of the fixed points of the absolute scale. The other point is the temperature at which ice, liquid water, and water vapor are in equilibrium with each other. (In equilibrium, the ice is not melting, the water is not freezing, and the amount of vapor stays constant.)

The temperature at this equilibrium of water, ice and vapor is arbitrarily defined to be 273.16° on the absolute scale. The melting point of ice, 0°C, at standard pressure is 273.15° absolute. We will see the significance of the triple point and the absolute temperature scale when we study the properties of material in the next section.

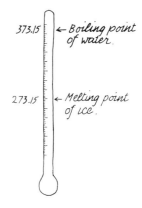

Absolute temperature scale.

THE BEHAVIOR OF GASES AS A FUNCTION OF TEMPERATURE

Gases are much easier to understand than either solids or liquids. Their behavior is particularly simple if their pressure is atmospheric or less and if they are in a temperature region well above the point at which they condense into a liquid. Under these circumstances, the behavior of the gas can be characterized almost completely by any three of the variables of pressure, volume, temperature, and number of molecules in the sample. If two of these variables are held constant, the relationship between the other two is simple. For example, using the experimental setup shown schematically on page 271, the pressure of a gas at constant volume can be measured as a function of temperature. The resulting graphs are shown for several different samples of gas. First, note that the pressure-temperature relationship is linear; the graphs are straight lines. The slope of each graph evidently depends on the kind of gas and the amount of it in the sample. All of the graphs, however, head down toward the same intercept on the temperature axis. That point is at −273.15°C, the temperature that we have defined to be zero on the absolute scale.

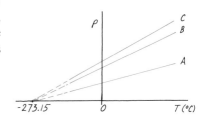

Samples A, B, C, for actual quantities of 3 gases.

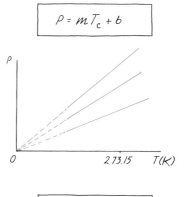

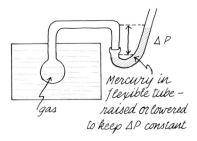

Mercury in flexible tube – raised or lowered to keep ΔP constant

gas

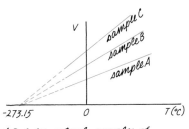

A, B, C for actual samples of different gases

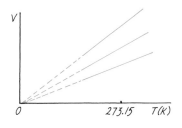

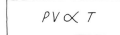

$PV \propto T$

The relationships among P, V, and T for a gas are known by different scientists' names: Boyle's law:
PV = constant, at constant temperature
[Robert Boyle (1627–1691)]
Charles' law:
$P \propto T$, at constant volume
[Jacques Charles (1746–1823)]
Charles' law is also known as Gay-Lussac's law.
[Joseph Gay-Lussac (1778–1850)]

Question 12-7

Does the experimental evidence shown in the graph indicate that at absolute zero the pressure of a gas is zero?

If we change the scale on the temperature axis to absolute temperature, then pressure is proportional to temperature: $P \propto T$. Evidently the absolute scale of temperature is much more convenient to use when dealing with the behavior of gases. This absolute scale of temperature is essentially the same as the Kelvin temperature scale. The S.I. unit for temperature is the kelvin degree. The temperature unit is simply K, not °K. For instance, the boiling point of water at standard sea level pressure is 373.15 K.

With the experimental device shown in this diagram, the *pressure* of a sample of gas can be kept constant while its *volume* is measured as a function of temperature. The resulting graphs for several samples of gas are shown in the accompanying diagram. Once again, note that the slope of these graphs depends on the quantity and type of gas in the sample. The temperature intercept of all the graphs is, once again, at −273.15°C. Accordingly, we can write V proportional to T.

Since $P \propto T$ and $V \propto T$, then PV is proportional to T. The proportionality constant must have something to do with the amount and kind of gas in the sample. In 1811, Avogadro proposed that equal volumes of gas (of any kind) at equal pressure and temperature must contain equal numbers of molecules of gas. In the next chapter, we will show the reasonableness of Avogadro's hypothesis. If we accept it, and call the number of molecules in the sample N, then the gas equation must be $PV = NkT$. The k now becomes the proportionality constant; it is called Boltzmann's constant and plays the role here of making the units agree with each other. When pressure is given in newtons per square meter (also known as pascals), volume is given in cubic meters, N is the actual number of molecules in the sample of gas, and T is given in kelvin, then Boltzmann's constant = 1.38×10^{-23} J/K.

The trouble with this form of the gas law is that it is not easy to actually count the number of molecules in a sample of gas. Instead, we measure the quantity of gas in terms of its mass or its volume. The results of many different types of experiments show that one *gram* of hydrogen atoms (providing we could keep them separate as atoms instead of forming molecules) would contain 6×10^{23} atoms. Two grams of hydrogen *molecules*, the usual form of the gas, contain 6×10^{23} *molecules*. Twelve grams of carbon atoms contain 6×10^{23} carbon atoms. Thirty-two grams of oxygen gas contain 6×10^{23} oxygen molecules. This particular large number, 6×10^{23}, is called a *mole*. A mole of molecules has a mass in grams equal to the molecular "weight" of the molecule. Instead of measuring molecules one by one, we usually measure them in units of the very large number called the mole:

$$N = n \text{ moles} = n(6 \times 10^{23})$$

The gas law can then be written

$$PV = nRT$$

The new proportionality constant, R, which is called the universal gas constant, must be equal to $(6 \times 10^{23})k = 8.314$ J/mole · K

Question 12-8

According to the gas law, if you put one mole of hydrogen molecules in a liter box and one mole of oxygen molecules in another liter box at the same temperature, their pressures would be the same. How can this be? Isn't the mass of the oxygen greater than the mass of the hydrogen? Shouldn't the more massive oxygen molecules produce a greater pressure than the lighter hydrogen molecules?

The number of moles of any gas can be found by dividing the mass of that sample by the molecular mass of that kind of gas:

$$\frac{\text{Mass of sample in grams}}{\text{Molecular mass}} = \text{Number of moles}$$

Another handy thing to know is the volume occupied by one mole of gas at standard temperature and pressure. Let's calculate this value. Standard temperature refers to 0°C, which is equal to 273.15 K. Standard atmospheric pressure is 1.01×10^5 N/m². Inserting these values into the gas equation gives us

$$PV = nRT$$

$$(1.01 \times 10^5 \text{ N/m}^2) V = (1 \text{ mole})(8.3 \text{ J/mole} \cdot \text{K})(273.15 \text{ K})$$

$$V = 22.4 \times 10^{-3} \text{ m}^3$$

The volume of one gram-mole of gas at standard temperature and pressure is equal to 22.4×10^{-3} m³, which is equal to 22.4 liters.

Now let's see how many molecules there are in the air in a standard room of a house at room temperature (20°C). For a bedroom, 10 ft by 12 ft and 8 ft high, the volume is about $(3 \text{ m})(4 \text{ m})(2.5 \text{ m}) \approx 30$ m³. Substituting into the gas equation, we get

$$PV = NkT$$

$$(1.01 \times 10^5 \text{ N/m}^2)(30 \text{ m}^3) = N(1.38 \times 10^{-23} \text{ J/K})(293 \text{ K})$$

$$N = 7.5 \times 10^{26} \text{ molecules}$$

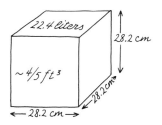

The number of moles is $(7.5 \times 10^{26} \text{ molecules})/(6 \times 10^{23} \text{ molecules/mole}) = 1.25 \times 10^3$. At standard temperature and pressure, 1.25×10^3 moles would occupy a volume of $(1.25 \times 10^3 \text{ moles})(22.4 \times 10^{-3} \text{ m}^3/\text{mole}) = 28$ m³. The volume occupied at room temperature would be $(28 \text{ m}^3)(293 \text{ K})/(273 \text{ K}) = 30$ m³—which is the volume with which we started. Incidentally, a mole of nitrogen has a mass of 28 grams. Therefore, the mass of the air in such a bedroom would be about $(1.25 \times 10^3 \text{ moles})(28 \times 10^{-3} \text{ kg/mole}) = 35$ kg. The air in your bedroom weighs about half as much as you do!

The gas law that we have described is called the *ideal gas law*. It is a good approximation of the behavior of real gases as long as they fulfill our condition of being at low pressures and at temperatures well above their condensation point. In the next chapter, we will see why it is necessary to add corrections for real gases. Note, however, why it is convenient to use the expansion property of gases as a thermometer: there are linear relationships between P and T at constant volume and between V and T at constant pressure. With solids and liquids, the expansion is nowhere near so linear and, furthermore, the size of the effect is different for each substance. In the case of gases, many different kinds satisfy the simple gas law, at least at low pressure and, as we will see in the next chapter, the theory is straightforward and relatively simple.

THE EFFECT OF TEMPERATURE ON SOLIDS AND LIQUIDS

As we have already seen, solids and liquids *usually* expand as the temperature rises. (Gases *always* expand when the temperature rises.) The expansion is not linear with temperature but, for small temperature changes, the fractional increase in volume can be approximated by the following formula:

$$\frac{\Delta V}{V} = \beta \Delta T$$

Thermal expansion joints in the George Washington Bridge over the Hudson River. (The Port Authority of New York and New Jersey)

$$\frac{\Delta L}{L} = \tfrac{1}{3} \beta \Delta T$$

The constant, β, depends on the type of material and, to some extent, on the temperature. The table shows values of β for a number of common materials at several temperatures. Note that mercury has a relatively large coefficient of expansion, at least for a metal, which is a particularly good feature for its use in a thermometer. In general, however, the coefficients of expansion of solids are very small. For iron, a 1° rise in temperature increases the *volume* by only 36 parts per million. The *length* of an iron rod would change by only one-third of that fraction. However, consider the changes in length from summer to winter of a steel bridge that is 1000 feet long. There might be a temperature change of 50 degrees Celsius, which would change the length of the bridge by a factor of $(12 \times 10^{-6}) \times (50°) = 6 \times 10^{-4}$. The 1000 foot bridge would then increase in length by $(1000 \text{ feet}) \times (6 \times 10^{-4}) = 0.6$ feet, approximately equal to 7 inches. The bridge would buckle if provisions were not made in the design for some of the girders and plates to slide over each other. There is the same problem in long railroad rails. On a more homely scale, it is well-known in kitchen circles that the metal covers on glass jars can sometimes be loosened by running very hot water over the metal lid. The expanded metal lid loosens its hold on the glass threads.

Solids and liquids do not always shrink when their temperature decreases. For an exception to the rule, you need go no further than the freezing compartment in your refrigerator. Water expands as it changes from liquid to solid. Furthermore, the expansion begins before the freezing has actually taken place. The diagram shows the volume of water as a function

THE EFFECT OF TEMPERATURE ON SOLIDS AND LIQUIDS

Coefficient of Volume Expansion $\beta = \dfrac{\Delta V}{V} \dfrac{1}{\Delta T}$

Material	Temperature in K	β in K^{-1}
Aluminum	25	1.5×10^{-6}
	293 (room temperature)	69×10^{-6}
	600	85×10^{-6}
Carbon (diamond)	100	0.15×10^{-6}
	293	3.0×10^{-6}
	600	9.0×10^{-6}
Copper	25	1.8×10^{-6}
	293	50×10^{-6}
	600	57×10^{-6}
Iron	25	0.6×10^{-6}
	293	36×10^{-6}
	600	45×10^{-6}
Lead	25	43×10^{-6}
	293	86×10^{-6}
	500	96×10^{-6}
Mercury	100 (solid)	111×10^{-6}
	273–573 (liquid)	$\sim 180 \times 10^{-6}$
Fused quartz	23	-2×10^{-6}
(noncrystalline	293	$+1.2 \times 10^{-6}$
SiO$_2$)	600	$+1.8 \times 10^{-6}$
Water	73 (ice)	18×10^{-6}
	173 (ice)	93×10^{-6}
	273 (ice)	167×10^{-6}
	277–373 (liquid)	430×10^{-6}

of temperature over the small temperature range near 0°C. As you can see, water has its maximum density at a temperature of 4°C. Like many other phenomena concerned with water, this feature seems providential for life on earth. When a pond cools in the wintertime, the cold water settles to the bottom until its temperature is about 4°C. As the temperature continues to fall, the colder water rises to the surface where it freezes, leaving ice on top and water on the bottom, thus sustaining a variety of life forms in the mud of the pond. Furthermore, tall drinks would lose much of their charm if the ice cubes sank to the bottom. Incidentally, the solids of most materials do sink in their liquid. One other exception to this rule is bismuth (a metallic element).

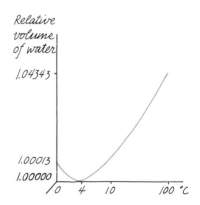

Question 12-9

The percentage change in the volume of a solid or liquid is very small for a 1° change of temperature. What is the approximate percentage change of volume for water when it turns into ice? You can get an answer to this question in two ways:

1. Notice the level of ice in an ice cube tray after the water has frozen.
2. Calculate the relative density of ice and water by observing the relative volume above water of a floating ice cube.

PHASE CHANGES IN MATERIALS AS A FUNCTION OF TEMPERATURE

We usually think of iron as a solid, alcohol as a liquid, and oxygen as a gas. Such a classification is based on our human prejudices about normal temperatures. Most materials can be solids or liquids or gases at particular temperatures and pressures. The most common material that can exist in all three phases within the range of human temperatures and pressures is water. Other examples are benzene and p-dichlorobenzene (mothballs). The boiling and melting points of a number of other materials at *atmospheric pressure* are shown in the table. Note that hydrogen turns into a liquid at 20.4 K and freezes solid at 14 K. Liquid hydrogen is routinely used as a target for high-energy particle beams in research on subatomic particles. Helium can also be liquified at 4.2 K, but cannot be solidified at atmospheric pressure, even at temperatures approaching absolute 0. Instead, helium starts undergoing a transition at 2.18 K, which turns it into a different kind of liquid with fantastic properties. It is a superfluid with zero viscosity.

Boiling and Melting Temperatures of Various Substances at Atmospheric Pressure

Substance	Melting Point °C	Boiling Point °C
Alcohol, ethyl	−114	78
Carbon, graphite	—	3827
Gold	1063	2700
Helium	—	−269
Hydrogen	−259	−253
Lead	327	1737
Mercury	−39	357
Nitrogen	−210	−196
Oxygen	−219	−183
Sulfur	119	445
Sulfuric acid	9	326
Tungsten	3380	5555
Water	0	100

At the opposite extreme, at very high temperatures, note that carbon never melts at atmospheric pressure but instead sublimes—that is, turns directly from a solid to a gas. Tungsten, from which light bulb filaments are made, melts at 3653 K and would boil at 5828 K.

Range of Measured Temperatures in K

Adiabatic demagnetization of nuclei	10^{-6}
Adiabatic demagnetization of paramagnetic salts	10^{-3}
Boiling point of helium (1 atm)	4.2×10^0
Boiling point of hydrogen (1 atm)	2.0×10^1
Boiling point of oxygen (1 atm)	9.0×10^1
Boiling point of water (1 atm)	3.73×10^2
Melting point of lead	6.00×10^2
Melting point of gold	1.336×10^3
Surface of sun	6×10^3
Corona of sun	10^6
Interior of sun	10^7
Hydrogen-helium fusion	10^8

Since both melting and boiling points depend on pressure, this type of information about a material is best shown in the form of a graph of pressure versus temperature or pressure versus volume. Several examples of these are shown in the accompanying figures. The first one shows some of the characteristics of water—the material most familiar to us. First, note the domain of the phases—vapor, liquid, and solid. At one atmosphere of pressure (1×10^5 N/m²), any of the three phases can exist within appropriate temperature ranges. The transition from solid to liquid, for instance, is shown at 273.15 K and the boiling point is at 373.15 K. Below a pressure of 4.58 millimeters of mercury (611 N/m²), however, the liquid phase cannot exist. Below that pressure ice would sublime, turning directly from solid to gas. The triple point of water, where all three phases can exist in equilibrium, is used as the basic fixed point for our modern temperature scale. There is a critical point at a pressure of 218 atmospheres (221×10^5 N/m²) and a temperature of 647 K, above which there is no difference between the liquid and the vapor. Therefore, there would be no heat of vaporization.

At pressures below one atmosphere, the boiling point of water is less than 373.15 K, or 100°C. You can read from the graph the approximate values for the boiling point at various pressures. We have also listed some of the values for various heights above sea level.

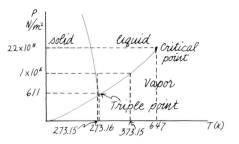

Pressure-temperature phase diagram for water.

Pressures and Boiling Points of Water at Various Heights Above Sea Level

	Pressure Atmospheres	N/m²	Boiling Point of Water in °C
sea level	1	1.0×10^5	100
1 km	0.89	9.0×10^4	96.4
1.5 km (Denver)	0.83	8.5×10^4	94.6
2 km	0.78	8.0×10^4	92.8
4 km	0.61	6.2×10^4	86.4
8 km (Mt. Everest)	0.35	3.6×10^4	72.8
15 km (Concorde jet)	0.12	1.2×10^4	49.6

Another feature to notice on the graph is that the line between solid and liquid slopes upward steeply, but is inclined slightly to the left (the slant has been exaggerated in the diagram). As the pressure increases, the melting point decreases slightly. This behavior is a characteristic of water and bismuth—which both expand upon freezing. If you increase the pressure on ice, you decrease the volume, thus tending to create the condition of a liquid, which can then exist at a slightly lower temperature. A popular legend is that this phenomenon explains why ice skates can glide so easily. Suppose that the weight exerted on the small surface of the skate blade produces such a large pressure that the ice momentarily melts, providing a nearly friction-free water-ice mixture. The melting temperature decreases by only about 1 Celsius degree for an increase of pressure of 1.2×10^7 N/m² (120 atmospheres).

Question 12-10

What pressure is exerted by the blade of a typical skater? Would this additional pressure melt the ice?

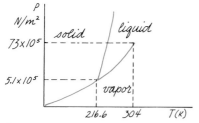

Pressure-temperature phase diagram for carbon dioxide.

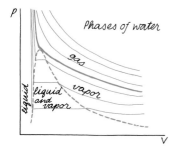

Pressure-volume phase diagram for water.

The pressure-versus-temperature graph for carbon dioxide is different from that of water in several crucial ways. The slope of the line between solid and liquid is positive; this is normal for most materials. The solid form is more dense than the liquid form. However, a major factor in the behavior of CO_2 is evident in the values for the triple point. That point for CO_2 is at $-56.6°C$ and 5.1 atmospheres. At ordinary temperatures and pressures, the liquid phase cannot exist. The solid form of CO_2 is dry ice. If you have handled dry ice, you know that it has a very low temperature and that it does not melt into a liquid. Instead, it sublimes into the gas form.

The *pressure*-versus-*volume* graph for water is shown here in a separate diagram. At temperatures well above 647 K, the *P-V* lines are hyperbolas, corresponding to the equation for an ideal gas: $P \propto 1/V$. The critical temperature of 647 K is the same as that shown on the *P-T* diagram. Above the critical temperature, there is no difference between liquid and vapor. If gas below that temperature is compressed, its pressure will rise until it reaches a crucial pressure and volume. From that point on as the volume is decreased, the pressure stays constant. The gas is condensing into liquid form. Eventually, when all of the gas has turned to liquid, the *P-V* curve rises abruptly, since the liquid is nearly incompressible and a small decrease in volume would create an enormous increase in pressure.

THE DEFINITION OF TEMPERATURE

We have been talking about many different temperature effects and have even described how to make thermometers to measure temperature. So far, however, we have avoided an actual definition of temperature.

THE DEFINITION OF TEMPERATURE

> Question 12-11
>
> If you had two uncalibrated thermometers, how could you tell if one were hotter than the other?

The use of a thermometer depends on the assumption that when a thermometer is placed in contact with an object and the thermometer reading settles down, the thermometer and the object are at the same temperature. That's an unusual property. If two objects with different masses or volumes or densities are connected, they will not end up with the same properties at some equilibrium point. But if two objects are arranged so that they exchange heat with each other but are isolated from everything else, then they will finally come to the same temperature. If you take a thermometer out of your mouth and it reads 100°F, then you assume that the temperature in your mouth was 100°F.

Another peculiar property of temperature is that, if thermometer A and thermometer B are placed in contact and come to the same temperature and then thermometer A is placed in contact with thermometer C and doesn't change its reading, not only are thermometers A and C at the same temperature, but so are thermometers B and C. That fact seems almost too obvious to mention. Remember, however, that thermometer A may be a thermocouple, thermometer B may be mercury in a tube, and thermometer C may be a curved bimetallic strip. When A and B are brought together, the electrical voltage produced by A and the length of the mercury column of B may change until they come to equilibrium. Then if the mercury column is put next to the bimetallic strip and neither one changes, the thermocouple can be put next to the bimetallic strip and not change either. This property of temperature is called the *zeroth law of thermodynamics* (the study of heat).

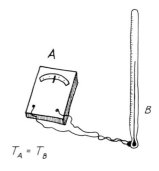

$T_A = T_B$

$T_A = T_C$

$\therefore T_B = T_C$

> Question 12-12
>
> Suppose that a piece of iron, A, is attracted to a magnet, B. If B is attracted to another piece of iron, C, will A also be attracted to C?

All we have done with the zeroth law is to define what it means to be at the *same* temperature. The next question is: How do we know when we make up the temperature scale that there is the same difference of temperature between 10 degrees and 11 degrees as there is between 90 degrees and 91 degrees? As we saw on page 274, the size of a degree on a mercury thermometer is not necessarily the same as the size of a degree on any other kind of thermometer. The expansion of a substance, or electrical resistance, or potential difference is not a linear function of temperature, or at least not a linear function of the corresponding changes in some other thermometer.

For one class of substances, however, there is great uniformity. The graph curves showing pressure versus temperature, or volume versus temperature, for all simple gases are almost identical—so long as the gas pressures are low. For practical purposes, we could make a constant volume thermometer using low-pressure hydrogen gas and assume that the pressure is proportional to the temperature. This assumption would define the intercept point in the gas pressure straight-line graph to be zero temperature. If the triple point of water is then defined to be 273.16 K, the temperature scale is completely defined.

There are two theoretical reasons for thinking that this type of thermometer is a true indicator of temperature. First, as we will see in the next chapter, the simple theory of the microstructure of gases gives a satisfactory explanation of why the pressure of the gas is proportional to the temperature. As a matter of fact, that theory will also provide another interpretation of the meaning of temperature.

There is a second theoretical interpretation of temperature that we will also see in the next chapter. This explanation, which will come from the second law of thermodynamics, is independent of any particular thermometer and thus of the properties of any particular material. This thermodynamic temperature scale agrees with the temperatures measured with a low-pressure constant volume gas thermometer.

There is one additional requirement we must consider in determining the significance of a temperature reading. *We can assign a temperature to a system only if the system is in equilibrium internally and with its surroundings*. If you take an ice cube from the freezer and put it into a glass of water, the temperature of the system is not 0°C. As a matter of actual experimental fact, as well as of theory, the idea of temperature for such a system is meaningless. However, if you put the ice cube with the water into an insulated glass along with a thermometer and stir the mixture until the thermometer reading remains constant for some time, then you can assign a temperature to the system. Similarly, if gas in a cylinder suddenly expands, neither pressure nor temperature of the gas can be defined until the shock waves settle down and equilibrium has been established within the cylinder. Whether a system is in equilibrium is not always easy to decide. The question involves the length of time for thermal mixing within the system compared with the time required for the disruption. For instance, it *is* reasonable to assign temperature to the water in a glass that is slowly rising from 0°C to room temperature. During the short time required for reading the thermometer and stirring the water, the temperature remains approximately constant and uniform throughout the glass.

THE FIRST LAW OF THERMODYNAMICS

We started this chapter by asking where the energy goes that seems to disappear as a result of friction. Our answer has been that the "lost" energy changes the properties of materials, just as the heat from flames can change materials. In many cases (but not all), the changed properties can be characterized in terms of the temperature of the material.

THE FIRST LAW OF THERMODYNAMICS

> Question 12-13
>
> If energy, whether from a flame or from friction, enters material, doesn't the temperature always rise?

The phenomena that we have been studying and the declaration that heat is a *flow* of energy can be summed up in the first law of thermodynamics:

$$Q = W + \Delta U$$

Heat entering = External work + Change of internal
a system   done during the energy of the
            process   system

This law is really just a statement of the conservation of energy along with the claim that energy can be locked up in internal forms of materials. Heat energy, Q, is energy in transit from a hotter to a colder system. It is not a quantity that can be stored in the system into which it passes. Some of the heat energy may be converted immediately into external work. The rest turns into the internal forms measured by ΔU.

The symbol, Q, stands for the amount of heat entering a system. The measuring unit for heat could be the joule, although frequently we use a special unit for heat energy, called the *calorie*. A calorie is approximately the amount of heat it takes to raise the temperature of one gram of water 1 Celsius degree. The present definition is 1 calorie = 4.1840 joule. The kilocalorie, which is equal to 4184 J, is the amount of energy that it takes to raise the temperature of one *kilogram* of water 1 Celsius degree. The diet calorie that many people try to avoid is actually a kilocalorie.

The second term in the first law (W) accounts for the fact that heat entering a system may produce external work whether we want it to or not. For instance, when a gas is heated it may expand, pushing a piston in front of it and thus exerting a force through a distance. If the volume of any material expands by an amount ΔV, then the work done is equal to $P\Delta V$, where P is the average pressure that the material exerts as it expands. For an object being heated in the open air, P would be just the atmospheric pressure. The relationship between $F\Delta x$ and $P\Delta V$ is shown in the diagram. In most cases for a *solid* or a *liquid*, ΔV is so small that the work done in expanding the volume against atmospheric pressure is negligible. However, when calculating the effects of heat on a *gas*, we must take into account the possibility that the external work done by the expanding gas consumes a major share of the incoming heat energy.

The third term in the first law of thermodynamics refers to the change of internal energy, ΔU, of the system. Note how cleverly we account for the disappearance of heat energy. We say that, except for the amount of energy used immediately to do work, the rest is stored in the form of changed characteristics of the material, including any change of temperature. The internal energy of a system may be a function of several variables, including

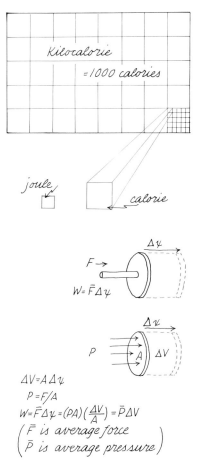

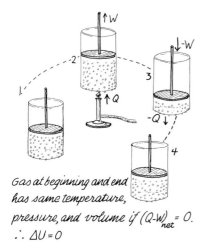

Gas at beginning and end has same temperature, pressure, and volume if $(Q-W)_{net} = 0$. ∴ $\Delta U = 0$

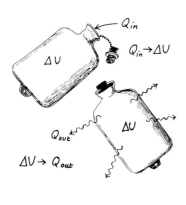

$Q_{in} \rightarrow \Delta U$

$\Delta U \rightarrow Q_{out}$

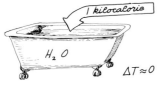

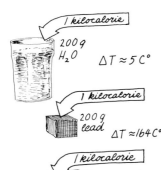

its density, magnetic and electric strains, and also its temperature. For an ideal gas, the internal energy is a function of temperature only.

Regardless of how heat is supplied or how external work is done, if a system starts out from some initial state (a particular pressure, temperature, etc.) and then has heat added and taken away, and work done on it and by it, the system will always end up in its original state as long as $(Q - W)_{net}$ is equal to zero. ΔU does not depend on the details or sequence of the changes, but only on the net value of $(Q - W)$. That makes U a potential energy function. As is the case with any of the other potential energy functions, we are free to choose the zero point in any convenient way. It is only the change of energy, ΔU, that has physical significance.

When we defined potential energy that was the result of springs or gravity, we did it in such a way that kinetic energy could turn into potential energy and back again. Is there the same situation with respect to internal energy? If heat enters a system and no external work is done, is the increase in internal energy, ΔU, stored in such a way that it can be converted into useful work? Whether that's possible depends on how you want to use the energy. If you pour heat into a hot water bottle, then practically all of that heat energy is stored in the internal energy of the water. (Note that the *heat* is not in the water: Q turned into ΔU.) That internal energy can be completely recovered *as released heat energy* to keep your feet warm. However, as we will see in the next chapter, if heat energy goes into the internal energy of a steam boiler, only a fraction of the stored energy can be used later to do external work.

HEAT CAPACITIES—THE TEMPERATURE CHANGE WHEN HEAT IS ADDED

Usually when we talk about "heating up" something, we mean that we are raising its temperature. If you add a kilocalorie to a bathtub of water, the temperature rise will be negligible. If you add a kilocalorie to a glass of water (about 200 grams of water), its temperature will rise about 5 Celsius degrees. If you add a kilocalorie to 200 grams of lead, its temperature will rise 164 Celsius degrees. If you add one kilocalorie to 200 grams of ice at 0°C, the temperature won't rise at all, but some of the ice will melt. We will study these effects in three different realms: first, for exchanges of heat between systems where only small temperature changes are involved; second, for effects produced over large temperature ranges where we will see some remarkable regularities in the behavior of various materials; third, for the heating of gases under conditions where varying amounts of external work may be done.

Mixing Systems with Small Temperature Differences

If we assume that no external work is done when a system is heated, then $Q = \Delta U$. Over small temperature ranges, providing that the phase of the system does not change, the change in internal energy is proportional to the change in temperature: $\Delta U \propto \Delta T$ and therefore $Q \propto \Delta T$. The proportionality constant must have something to do with the type of material and the amount of material in the system. One way to specify the amount of material is in terms of its mass. Then the proportionality becomes $Q = cm\Delta T$. The proportionality constant, c, is called the specific heat of the material. The values remain fairly constant over small temperature ranges, although we will see examples where the "constant" is not constant. For example, the diagram shows the value of the specific heat of water as a function of temperature. From 0 to 100°C, the value changes by less than 1%.

The table shows values of c for various common materials in the range from 0 to 100°C. The units shown are derived from the units commonly used to measure the other quantities in the equation: c = calories per gram per degree Celsius = cal/g·C°.

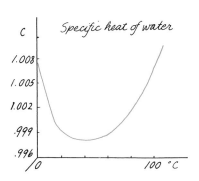

Question 12-14

Consider the table of specific heats. Why is water a useful material to have in a hot water bottle?

Some Specific Heats in Range from 0°C to 100°C

Material	Cal/g·C° = kcal/kg·C°
Aluminum	0.215
Carbon	0.121
Copper	0.0923
Lead	0.0305
Silver	0.0564
Tungsten	0.0321
Water	1.0

Now we can demonstrate why a kilocalorie will produce different temperature changes with different materials. In the case of a bathtub of water, the mass of the water may be 200 kilograms; the temperature change would be

$$\Delta T = \frac{Q}{cm}$$

$$= \frac{(1 \text{ kcal})}{(1 \text{ kcal/kg} \cdot \text{Celsius degrees})(200 \text{ kg})}$$

$$= 1/200 \text{ Celsius degrees}$$

On the other hand, the mass of 200 cc of water is just 200 grams. The temperature change in this case would be

$\Delta T = (1 \text{ kcal})/(1 \text{ kcal/kg} \cdot \text{Celsius degrees})(0.2 \text{ kg}) = 5$ Celsius degrees

For 200 grams of lead, the temperature change would be

$\Delta T = (1 \text{ kcal})/(0.0305 \text{ kcal/kg} \cdot \text{Celsius degrees})(0.2 \text{ kg})$

$= 164$ Celsius degrees

As you saw in "Handling the Phenomena," there is no temperature change produced by heat during a change of phase. While ice is melting, it remains at 0°C; while water is boiling, it remains at 100°C. The amount of energy needed to melt 1 gram of material is called the *heat of fusion*. For water the heat of fusion is 80 calories per gram. The *heat of vaporization* is the number of calories needed to turn 1 gram of a liquid into a gas. For water at atmospheric pressure, the heat of vaporization is 540 calories per gram. Note how much larger the heat of vaporization is than the heat of fusion. Evidently the internal energy of a gram of steam is very large compared to the internal energy of a gram of water. That is why "live" steam can produce such serious burns. To convert one gram of ice at 0°C to water, we need 80 calories; to change that gram of water from 0°C to

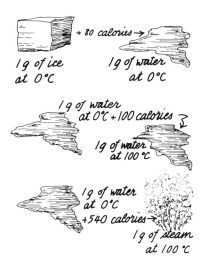

100°C, we need 100 calories; then 540 more calories are needed to turn it into steam. Compare these values with the time taken to turn ice into steam, which you found out in "Handling the Phenomena."

Heats of Fusion and Vaporization at Atmospheric Pressure

Substance	Fusion			Vaporization		
	cal/g	J/g	J/mole	cal/g	J/g	J/mole
Helium	—	—	—	5.0	21	5.0
Hydrogen	14	59	30	108	452	226
Alcohol, ethyl	25	104	2.3	204	854	18.6
Nitrogen	6.2	26	0.9	48	201	7.2
Oxygen	3.4	14	0.4	51	213	6.7
Mercury	2.9	12	0.06	65	272	1.4
Water	80	333	18.5	539	2256	125
Lead	6.0	25	0.12	208	871	4.2
Gold	15.5	65	0.33	378	1580	8.0
Sulfur	9.1	38	1.19	78	326	10.2

The heats of fusion and vaporization for other materials are given in the table. Notice that it always takes more energy to turn the material from liquid to gas than it does from solid to liquid. A small part of the heat of vaporization goes into the work of expanding the material from liquid to gas form. For most materials, the volume of the gas is about 1000 times greater than the volume of the liquid from which it came. If the boiling takes place at atmospheric pressure, the amount of work done in the expansion is equal to $P\Delta V$. In the case of water, the volume occupied by 1 gram of steam at 100°C and atmospheric pressure is equal to 1.7×10^{-3} m^3. The volume of the 1 gram of water is equal to only 1×10^{-6} m^3. Atmospheric pressure is equal to 1.01×10^5 N/m^2. Therefore, the work required for the expansion is equal to

$$P\Delta V = (1.01 \times 10^5 \text{ N/m}^2)(1.7 \times 10^{-3} \text{ m}^3) = 170 \text{ J} = 41 \text{ cal}$$

The other 499 calories per gram of the heat of vaporization must be needed to free the water from the liquid state and convert it into a gas. In the next chapter, we will develop the molecular model to fit these facts.

As everyone knows, if you mix hot water with cold water, you will get warm water. Evidently some of the internal energy of the hot water becomes part of the internal energy of the cold water. The heat lost by the hot water is the heat gained by the cold water. This equality of heat gained to heat lost is generally true of any mixture. Let's try several examples.

A. 100 grams of water at 95°C is mixed with 50 grams of water at 10°C.

$$Q_{\text{lost}} = Q_{\text{gained}}$$

$$(1 \text{ cal/g} \cdot \text{C}°)(100 \text{ g})(95°C - T_f) = (1 \text{ cal/g} \cdot \text{C}°)(50 \text{ g})(T_f - 10°C)$$

$$9500 - 100 T_f = 50 T_f - 500$$

$$150\, T_f = 10{,}000$$
$$T_f = 66.7°C$$

B. 100 grams of aluminum powder at 95°C is mixed with 100 grams of water at 10°C.

$$Q_{\text{lost}} = Q_{\text{gained}}$$
$$(0.217\text{ cal/g}\cdot C°)(100\text{ g})(95°C - T_f) = (1\text{ cal/g}\cdot C°)(100\text{ g})(T_f - 10°C)$$
$$2062 - 21.7\, T_f = 100\, T_f - 1000$$
$$T_f = 25°C$$

Question 12-15

Suppose you performed a heat mixture experiment by putting 100 grams of powdered aluminum in boiling water until you were sure that the aluminum was at 100°C. Then you pour off the boiling water and mix the 100 grams of aluminum with 100 grams of water at 10°C. The final temperature would not at all be what the simple calculation would indicate. Why is this?

C. 100 grams of aluminum at 500°C is mixed with 100 grams of ice at 0°C.

$$Q_{\text{lost}} = Q_{\text{gained}}$$
$$(0.217\text{ cal/g}\cdot C°)(500°C - T_f)(100\text{ g}) = (80\text{ cal/g})(100\text{ g})$$
$$+ (1\text{ cal/g}\cdot C°)(100\text{ g})(T_f - 0°C)$$
$$10{,}850 - 21.7\, T = 8000 + 100\, T$$
$$T = 23°C$$

In all of these heat mixture experiments, it is assumed that no heat is entering or leaving the surroundings. The mixing must take place in a closed calorimeter (an insulated container). In a later section, we will study the flow of heat and the types of insulation. For simple heat mixture experiments, ordinary Styrofoam coffee cups with insulating lids serve very well as calorimeters, providing that the mixing of materials and the temperature equilibrium can happen rapidly.

The Regularities and Irregularities of Specific Heat Values

The specific heats of materials as we have catalogued them so far vary over a large range without any apparent uniformity. Water seems to have a very large value of specific heat compared with any other of the common materials, and most metals appear to have a very low specific heat. In the middle of the nineteenth century, Du Long and Petit noticed that a uniformity does exist for the specific heats of most solid elements. Note from the graph that the metals with small atomic numbers have large specific heats, and the metals with large atomic numbers have small specific heats. If the specific

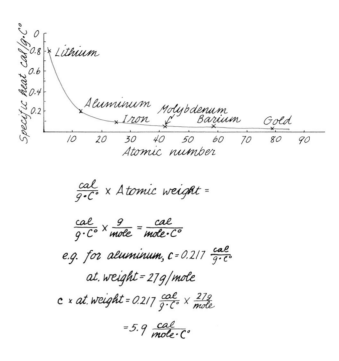

$$\frac{cal}{g \cdot C°} \times \text{Atomic weight} =$$

$$\frac{cal}{g \cdot C°} \times \frac{g}{mole} = \frac{cal}{mole \cdot C°}$$

e.g. for aluminum, $c = 0.217 \frac{cal}{g \cdot C°}$

at. weight = 27 g/mole

$c \times$ at. weight $= 0.217 \frac{cal}{g \cdot C°} \times \frac{27 g}{mole}$

$= 5.9 \frac{cal}{mole \cdot C°}$

heat, in cal/g · C°, is multiplied by its atomic weight, the number turns out to be approximately 6. What we are really doing in multiplying by the atomic weight is finding the *molar* specific heat—the number of calories required to raise the temperature of one mole of the material by 1C°. The table gives the molar specific heats for many materials.

Molar Specific Heats of Some Solids and Liquids at Room Temperature

Substance	Molar Specific Heat in cal/mole · C°	Molar Specific Heat in J/mole · C°
Aluminum	5.82	24.4
Carbon (graphite)	1.46	6.1
Copper	5.85	24.5
Lead	6.32	26.4
Silver	6.09	25.5
Tungsten	5.92	24.8
Uranium	6.7	28.0
Water	18.0	75.3
Bromine	18.1	75.7
Mercury	6.69	28.0

Whenever there is a regularity in nature, we suspect that there is some underlying cause. When we first suggested that $Q \propto \Delta T$, we said that the proportionality constant must be a function of the kind of material and the *amount* of material in the sample. Then we said that the amount of material might be defined by its mass. Another way to describe the amount of material is to give the number of atoms or molecules in it. When we take a mole of material, we are using 6×10^{23} atoms (or molecules). The *molar specific heat* is thus the number of calories that it takes to raise one mole of material 1 Celsius degree. For materials with the same basic structure—metals, for instance—there seems to be a simple relationship between the amount of energy per mole (or per atom) and the temperature.

> Question 12-16
>
> What is the specific heat per individual metal atom? Your units will be in *calories per atom* per Celsius degree. Because those are awkward units for atoms, multiply by 4.17 joules per calorie to turn the answer into *joules per atom* per Celsius degree. Then divide by 1.6×10^{-19} joules per electron volt in order to get a more reasonable value for an atom in terms of *electron volts per atom* per Celsius degree. If the specific heat were constant all the way from 0 K to room temperature at 300 K, about how much internal energy would each atom have?

As a matter of fact, the regularity that we have just observed in specific heats of metals applies only at room temperature, and for the light metals not too well even there. The specific heat of a material is not a constant but depends on temperature. The diagram shows graphs of the specific heats of various materials as a function of temperature. Evidently what is going on in the microstructure of materials is simple in the first approximation, but more complicated when you take a closer look. We will try to explain both regularities and exceptions in the next chapter when we propose the model for the microstructure.

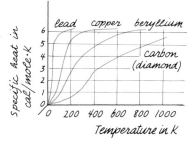

The specific heats of *gases* also have both regularities and differences. The table gives specific heats for several different types of gases; here

Specific Heats of Gases at Room Temperature

Type of Gas	Gas	c_p (cal/mole·K)	c_v (cal/mole·K)	$c_p - c_v$	$\gamma = c_p/c_v$
Monatomic	He	4.97	2.98	1.99	1.67
	A	4.97	2.98	1.99	1.67
Diatomic	H_2	6.87	4.88	1.99	1.41
	O_2	7.03	5.03	2.00	1.40
	N_2	6.95	4.96	1.99	1.40
	Cl_2	8.11	6.15	1.96	1.32
	HCl	6.45	5.00	1.45	1.29
Polyatomic	CO_2	8.83	6.80	2.03	1.30
	SO_2	9.65	7.50	2.15	1.29
	NH_3	8.80	6.65	2.15	1.31
	C_2H_6	12.35	10.30	2.05	1.20

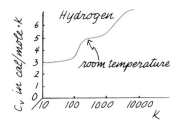

they are grouped according to whether they are monatomic, diatomic, or polyatomic. Notice the uniformities within each group. Also notice that the molar specific heat for monatomic gases is almost exactly half that for the metals. These specific heats are given for room temperature. In the diagram, there is a graph of the specific heat of hydrogen as a function of temperature. Like the metals, hydrogen has a specific heat that is small at low temperature and then increases. However, instead of increasing rapidly or uniformly as in the case of the metals, hydrogen's specific heat seems to rise to certain plateaus and then to remain constant over a brief temperature span. Our atomic model will have to explain these peculiarities.

Special Conditions for Feeding Energy into Gases

The specific heats that we have given so far have been for cases where the volume of the material stays constant; thus no external work is done: $W = P \Delta V = 0$. The specific heat at constant volume should be labeled c_V, as opposed to the specific heat at constant pressure c_P. It is very hard to obtain *experimental* values of c_V for solids or liquids. When heat is fed into a solid or liquid sample, it is hard to keep it from expanding. Usually, c_P is measured. Unfortunately, for *theoretical* calculations the situation is just reversed. It is easier to calculate c_V. For our purposes, the extra work done in the expansion of solids and liquids makes only a negligible difference between c_P and c_V.

In the case of gases, however, there is a large difference between c_P and c_V. The table of specific heats of gases gives both sets of values and also shows the differences between the two as well as their ratios. Notice that the *ratio* c_P/c_V, which is given the special symbol γ (gamma), has special values for each class of molecule. The *difference* $(c_P - c_V)$ has approximately the same value regardless of the type of molecule. The difference between c_P and c_V corresponds to the extra energy that is required to expand the gas at constant pressure as its temperature rises 1 Celsius degree.

Question 12-17

Notice that the quantity $(c_P - c_V)$ is approximately the same for all gases. Compare its value and units with the value and units of the gas constant R in the ideal gas equation.

Of course, energy can be fed into a gas in such a way that the gas expands against *varying* pressure. For purposes of analysis, however, it is usually convenient to use only the limiting cases where either the pressure or the volume is kept constant.

Let's consider the P-V relationship for gases under two limiting conditions: first, the temperature must be kept constant and, second, no heat can flow in or out of the gas. The experimental conditions for doing each measurement are illustrated in the diagrams. In the first case, the gas is trapped in a cylinder with metal walls that can easily conduct heat in or out of the chamber. As the gas expands or contracts slowly to maintain temperature equilibrium with the walls, the piston is shoved out or in. For

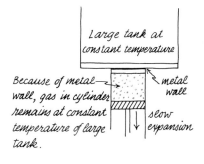

the second case, the cylinder and piston must be made of good insulating material (of negligible heat capacity), and the expansion must be fast so that no heat can flow in or out of the gas as it expands. In both cases, we will allow the gas to expand and we will plot the pressure as a function of volume. When the temperature remains constant, the relationship between P and V is simple; it is given by the ideal gas formula, $PV = nRT$. With T constant, the equation reduces simply to $PV = $ constant, or $P = $ constant$/V$. This is the equation for the hyperbola plotted in the upper left part of the P-V graph.

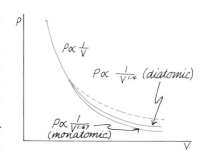

Under most conditions, the temperature of an expanding gas will decrease and the temperature of a gas being compressed will increase. These phenomena are easy to observe with a bike or car tire and a pump. As you pump up the tire, the hose gets hot; if you release the air in the tire, the air coming out and the nozzle will feel cold.

Under the particular and special conditions of the second kind of expansion, with insulating walls, what is the relationship between pressure and volume? As the volume increases, the pressure will drop, but so will the temperature. We cannot therefore set $PV = K$. A calculus derivation is needed to demonstrate that the proper equation in this case is $PV^\gamma = K$. This sort of curve is plotted in the lower part of the P-V graph, for both a monatomic and diatomic gas. Notice that for a given increase of volume the pressure drops more in this case than it does when the temperature remains constant.

■ Optional

Derivation of Pressure-Volume Relationship for Adiabatic Expansion of Ideal Gas

(1)	$Q = dU + W$	First law of thermodynamics
(2)	$Q = 0$	Adiabatic process—no heat in or out
(3)	$W = PdV$	Pressure is changing but has value P over small range dV
(4)	$dU = nc_v dT$	For ideal gas, U is function of T alone
(5)	$\therefore nc_v dT = -PdV$	Linking (1), (2), (3), and (4)
(6)	$PV = nRT$	Ideal gas law
(7)	$PdV + VdP = nR\,dT$	Differentiating (6)
(8)	$PdV + VdP = nR\left(-\dfrac{P}{nc_v}dV\right)$	From (7) and (5)
(9)	$PdV\left[1 + \dfrac{R}{c_v}\right] + VdP = 0$	Rearranging (8)
(10)	$R = c_p - c_v$	For all ideal gases
(11)	$PdV\left[1 + \dfrac{c_p - c_v}{c_v}\right] + VdP = 0$	Combining (9) and (10)
(12)	$PdV\left(\dfrac{c_p}{c_v}\right) + VdP = 0$	Rearranging (11)
(13)	$\dfrac{dV}{V}\gamma + \dfrac{dP}{P} = 0$	Rearranging (12); $\gamma = c_p/c_v$
(14)	$\gamma \ln V + \ln P = 0$	Integrating (13); integral $\int \dfrac{dx}{x} = \ln x$
(15)	$PV^\gamma = K$	Taking antilogs

Surprisingly enough, the value of γ for a gas enters into several formulas concerned with gas behavior. One of these, as we will see later, is the formula for the speed of sound in a gas. The velocity depends on the relationship between the pressure and volume of a gas under conditions where the pressure changes so rapidly that practically no heat flows into or out of the gas.

Any change in pressure, volume, or any other variable that takes place at constant temperature is called *isothermal*. For an ideal gas in which the internal energy is a function of temperature only, an isothermal change implies no change in the internal energy of the system. If $\Delta T = 0$, then $\Delta U = 0$. In this case, the first law of thermodynamics becomes $Q = W$. At the other extreme where the system is insulated so that no heat can enter or leave, any changes that take place are called *adiabatic*. For an adiabatic change, $Q = 0$; therefore, the first law of thermodynamics becomes $0 = \Delta U + W$. In our example of the gas in a cylinder during the isothermal expansion, the incoming heat must have been turned completely into the external work done by the piston. In the adiabatic expansion, the work done by the piston must have come from the internal energy of the gas. The change in internal energy must have been negative, corresponding to a drop in temperature.

Back on page 285, we argued that the work done in expanding an object against atmospheric pressure was equal to $P\Delta V$. In the *P-V* diagram concerned with gas expansion, the pressure does not remain constant. Nevertheless, as you can see in the diagram, the area under a *P-V* curve represents the work done in expanding the gas from an initial volume to a final volume. If the gas is being compressed by an external pressure, then the area under the *P-V* curve shows the external work being done—which either increases the internal energy of the gas or forces out heat energy, or both. Changes that take place in a real gas in the cylinders of a steam engine or a gasoline motor are not completely isothermal or adiabatic. Nevertheless, the area under the *P-V* curve always represents the work that is performed in the expansion or the work that must be provided for the compression.

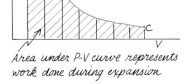

Area under P-V curve represents work done during expansion

> ### Question 12-18
>
> Look at the *P-V* graph showing an isothermal expansion of a gas followed by an adiabatic expansion. The curves for a monatomic gas and a diatomic gas are superimposed, along with a dotted line showing a continuation of the original isothermal expansion to the final volume. Take into account the fact that the work done during the adiabatic expansion must come out of the internal energy of the gas. Give a qualitative explanation for the shape of the adiabatic curves.

HEAT TRANSFER

Heat is energy moving from one place to another because of a temperature difference. The three main ways of transferring heat are *conduction*, *convection*, and *radiation*. There are many familiar examples of each of these in everyday life. Heat is conducted through materials by means of some action of the microstructure of matter. Our atomic model, which we will introduce in the next chapter, will have to explain why some materials like metals are good heat *conductors* and others like most plastics are good heat *insulators*. For conducting heat from the stove into our food, we use metal pots; for preventing the heat from being conducted to our hands, we use plastic or wooden handles on the pots. In heat conduction there must

Home boiler and radiator system

HEAT TRANSFER

be material joining the high-temperature region to the low-temperature region. But there is no transfer of the material itself. With convection, on the other hand, some substance such as water or air is heated in the high-temperature region and physically moved to the low-temperature region where it provides heat taken from its extra internal energy. Most home heating systems rely on convection for transferring heat from the furnace to the house.

A major share of the heat provided to the earth arrives in radiation from the sun. (Less than 1% comes from radioactivity in the earth.) There is no material connecting the sun and the earth, and no transport of material (in the usual meaning of the term) between sun and earth. As we will see when we study electricity, most of this radiant energy is electromagnetic—that is, the same phenomenon that is also responsible for radio, light, and X-rays.

Three centuries ago Newton demonstrated that the amount of heat conducted per second through a material is proportional to the temperature difference across the material. The time rate of transfer is inversely proportional to the thickness, d, of the material separating the two temperatures and is proportional to the area, A. The proportionality constant depends on the type of material.

$$\frac{Q}{\Delta t} = K \frac{(T_{\text{out}} - T_{\text{in}})A}{d}$$

The table gives the thermal conductivity constants for a variety of materials. Note the very large difference between the metals and the insulators. It appears that gases are good insulators but, for practical purposes, they can be used as insulators only if they are trapped in small regions so that they cannot conduct heat by convection. Styrofoam and clothing materials such as wool and fur are effective insulators because they provide a barrier consisting largely of gas trapped in regions too small to allow the transport of heat by convection.

Newton's law explains the way in which hot coffee cools when it is in a closed Styrofoam cup. A typical graph that illustrates this type of cooling is shown in the diagram. At first, the temperature difference between the hot coffee and the outside air is very large. Therefore, the rate of heat loss is large and the temperature falls rapidly. As the temperature difference decreases, however, the rate of heat loss also decreases and, thus, the temperature falls more slowly. This type of behavior is the requirement for exponential decay. If the temperature difference across the wall is 40 Celsius degrees at the beginning of the experiment, and falls to 20 Celsius degrees in 10 minutes, then it will fall to 10 Celsius degrees difference in another 10 minutes, and to 5 Celsius degrees difference in the next 10 minutes. However, it is hard to arrange experimental conditions to demonstrate this law with great precision. For instance, if the Styrofoam coffee cup does not have a lid, then most of the energy will be lost due to convection and evaporation from the top surface. When calculating heat loss through the walls or windows of a house, you cannot assume that the temperature difference across a wall is the difference in temperature between the inside and the outside of the house. As you know, if you put your hand on a window or the inside wall of a house on a cold winter day with the wind blowing

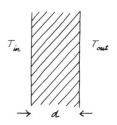

Thermal Conductivities at Room Temperature— kcal/(s·m·°C)

Aluminum	4.9×10^{-2}
Copper	9.2×10^{-2}
Silver	9.9×10^{-2}
Lead	0.8×10^{-2}
Concrete	2×10^{-4}
Glass	2×10^{-4}
Asbestos	0.2×10^{-4}
Ice	4×10^{-4}
Wood	0.2×10^{-4}
Water	1.4×10^{-4}
Air	5.7×10^{-6}
Hydrogen	33×10^{-6}

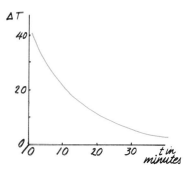

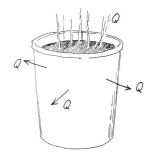

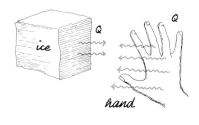

$P = \sigma T^4$

$\sigma = 5.67 \times 10^{-8} \frac{\text{watts}}{m^2 \cdot K^4}$

P is the power radiated per square meter.

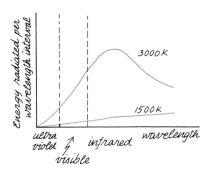

Solar water heater (Grumman Sunstream).

outside, the window or wall will be much colder than the air in the room. If fact, if you do not have storm windows, you may have frost on the inside of the window panes, thus indicating that the inside temperature of the glass must be at least as low as 0°C.

Even when there is no convection and no conduction, heat can be transferred by radiation. Objects don't have to be glowing in order to radiate heat. As a matter of fact, electromagnetic radiation is emitted from all objects at any temperature. You can feel the heat being radiated from an oven door or even a hot pavement. However, it is harder to believe that heat is being radiated from a refrigerator freezing unit or from a very cold bathroom wall. In fact, these cold walls seem to be radiating "cold." However, there is no such thing as negative energy flow. The reason that you feel cold from a cold wall is that you are comfortable when your surroundings are radiating heat back to you at a rate characteristic of materials at about 20°C. A cold wall radiates very little energy back to you compared with the energy that your body radiates toward the wall. Therefore, you sense that the wall is cold. The same phenomenon affects the earth. By day, the earth is warmed by the radiant energy from the sun. On a clear, cloudless night, the earth rapidly cools off as it radiates energy away into outer space.

The total electromagnetic power *radiated* by an object is proportional to the fourth power of its temperature. Compare this very strong temperature dependence with that of heat transfer by thermal conductivity, which is proportional to the first power of the temperature difference. It is the absolute temperature that must be used in the expression for radiation loss. The tungsten filaments in ordinary light bulbs are usually run at a temperature of about 3000 K. If the temperature is lowered to 1500 K, the amount of heat and light radiated is decreased by a factor of $2^4 = 16$. The color of the radiated light is also changed drastically from a bright white to a dull red. We will study this process more when we take up the subject of the emission of light.

The amount of radiant energy emitted or absorbed also depends on the nature and color of the surface of the radiator or absorber. Rough, black surfaces are better absorbers than are smooth, silvery ones. Surprisingly enough, a rough, black surface is also a better *radiator*. The diagram shows a greenhouse-type device for absorbing solar energy for house heating. The surface of the absorber inside must be black and it may get fairly hot in terms of house temperatures. However, even a temperature of 100°C is only 373 K. The amount of radiation from the absorber will be small and in the infrared part of the electromagnetic spectrum (where the wavelengths are longer than those of visible light). Nevertheless, considerable energy could be lost by reradiation, especially at night. To prevent this, the unit is covered with glass. Ordinary glass is transparent to the high-temperature radiation from the sun, but is almost opaque to the low-temperature, long wavelength radiation that is emitted by a black absorber-radiator at 373 K. There has been some controversy in the past as to whether the selective transmission of radiation by the glass has any real effect. Certainly a major role for the glass cover is simply to prevent convection to the outside air. However, modern tests of high-efficiency systems have demonstrated the value of the selective transmission "greenhouse" effect.

A common household item exploits dramatically the various factors concerned with the transmission of heat. The thermos bottle is a remarkably good insulator. If you take one apart (carefully, and with eye protection), you can see that the inner bottle has a double wall of thin glass with a vacuum in between. The nipple from which the air was exhausted is usually at the bottom of the tube. Consider the three ways in which heat might enter or leave the contents of the bottle. *Conduction* can take place only through a small collar of thin-walled glass at the top of the bottle. Since there is a vacuum inside the double wall, *convection* can take place only through the small amount of gas between the contents and the plastic stopper at the top. The inner and outer walls of the bottle are smooth and silvered, thus minimizing *radiation*.

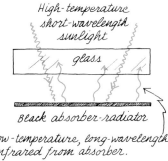

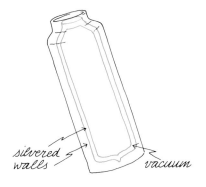

SUMMARY

Mechanical energy is not usually conserved in processes. Friction drains the energy, producing changes in the properties of the materials involved. Many of these changes are characterized by an increase in temperature.

In defining temperature, we started by describing types of thermometers—the mercury thermometer, the thermocouple, the electrical resistance thermometer, and so on. There are both practical and theoretical problems in calibrating thermometers. For everyday use, thermometers can be calibrated using two fixed points: the temperature of melting ice and the temperature of boiling water. On the Fahrenheit scale, these temperatures are 32°F and 212°F. On the Celsius scale, the same points are 0°C and 100°C. A theoretical problem of calibration is caused by the nonlinearity of most thermometric properties. Even if two different kinds of thermometers are calibrated together at 0°C and 100°C, they will not agree exactly at intermediate temperatures. However, the relationships between pressure, volume, and temperature are the same for many gases at low pressure. The response of a constant volume gas thermometer can be used to define a linear, absolute temperature scale where $T = 273.15°$ for the melting point of ice.

For an ideal gas, $PV = NkT = nRT$. N is the number of molecules in the gas sample, and n is the number of moles. A mole is 6×10^{23}.

Solids and liquids usually (but not always) expand as the temperature increases. Over considerable temperature regions the expansion can be approximated with the linear relationship $\Delta V/V = \beta \Delta T$.

Instead of changing temperature, material may change phase as heat is added or work is done on it. The heat of fusion is the energy required to transform a quantity of material from solid to liquid; the heat of vaporization is the energy required to change that much material from liquid to gas. During these phase changes, the temperature of the material does not change. The phase of any substance is a function of temperature, pressure, and volume.

Temperature was finally defined in terms of an equilibrium among the thermometric properties of materials. The zeroth law of thermodynamics says that, if A and B are separately at the same temperature as C, then A and B are at the same temperature. The first law of thermodynamics is Q

$= W + \Delta U$ where U is the internal potential energy of a substance. For ideal gases, U is a function only of the temperature.

When objects at different temperatures are thoroughly mixed in a calorimeter, they come to an intermediate temperature: Q lost by one substance $= Q$ gained by the other. The specific heat of a substance is $c = Q/m\Delta T$. For mixtures, $c_1 m_1 \Delta T_1 = c_2 m_2 \Delta T_2$.

Metals at room temperature have approximately the same molar specific heats: $c_{\text{molar}} = Q/\text{mole} \cdot \Delta T$. Specific heats are, however, a function of temperature. With gases, specific heats depend on the complexity of the molecule. Furthermore, with gases, two specific heats are important: c_P at constant pressure and c_V at constant volume. $(c_P - c_V) = R$ for all ideal gases. The work done during the expansion of a gas in a cylinder is represented by the area under the curve in a graph of pressure versus volume. Two limiting conditions of gas expansion were defined: isothermal, where the gas is in contact with walls kept at constant temperature, and adiabatic, where the walls are insulated so that no heat enters or leaves the gas. For isothermal expansion: $P = K/V$. For adiabatic expansion: $P = K/V^\gamma$, where $\gamma = c_P/c_V$.

Heat can be transferred by conduction, convection, or radiation. For conduction, the time rate of heat transfer across a wall is proportional to the temperature difference across the wall and to the surface area, and is inversely proportional to the wall thickness. The time rate of energy radiated from an object is proportional to T^4.

Answers to Questions

12-1: The time scale taken for your particular experiment will depend on the amount of ice that you used and the type of stove that you have. The ideal graph shown can be obtained only if the ice and water can be stirred steadily. The container should also be insulated, so that heat does not escape out of the walls or through the open top.

12-2: If the glass tube were solid, the whole tube would expand as the temperature increased. If the tube is hollow, the cavity will also expand with increasing temperature. (The expanded hole would be just large enough to accommodate the expanded enclosed glass if the tube were solid.) Some of the expanded volume of the liquid can thus be accommodated by the enlarged tube, but the effect is usually very small and it is easy to make the appropriate correction on the scale.

12-3: If the automatic toaster isn't working, you judge the temperature of the toast by noting its color. If water is frozen, the temperature must be below 0°C. The temperature of the milk in a baby bottle is usually tested by squirting some of it on the inside of your wrist.

12-4: There are (212°F − 32°F) = 180F° between the melting point of ice and the boiling point of water. There are 100 Celsius degrees between those same points. Therefore, the Fahrenheit degree is smaller and is equal to $\frac{100}{180} = \frac{5}{9}$ Celsius degree.

12-5: If the Fahrenheit and Celsius temperatures are the same, then we can use the same symbol, T, for both in the transformation formula.

$$T_F = 32 + \tfrac{9}{5} T_C$$

Since $\qquad\qquad T_F = T_C$

then $\qquad\qquad T = 32 + \tfrac{9}{5} T$

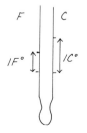

$$\tfrac{4}{5}T = -32$$

$$T = -40°$$

2-6: If we measure the electrical resistance of platinum wire as a function of temperature, then we should use a linear temperature scale. The increase of temperature from 10°C to 11°C must be the same as the increase of temperature from 50°C to 51°C. If we use a mercury thermometer to determine this, we are really assuming that mercury expands as much from 10°C to 11°C as it does from 50°C to 51°C. The only way that we could determine this fact would be to test mercury against some good thermometer such as a platinum resistance thermometer. Of course, this is a circular argument. We need some definition of temperature and some way to measure it that does not depend on the properties of any particular kind of material.

2-7: No experimental evidence is shown on the graph for what happens in the region close to absolute 0. It is always dangerous to extrapolate evidence to a region for which we have no information. In this case, the gases turn to liquids or solids before the temperature gets to absolute 0.

2-8: As we will see in Chapter 13, the pressure of a gas is due to the bombardment of the molecules. The momentum that each molecule delivers to the wall when it strikes it is proportional to its mass and also to its velocity. If there are the same numbers of hydrogen and oxygen molecules producing the same pressure, then the velocities of the hydrogen molecules will be greater than those of the oxygen molecules.

2-9: The volume of an ice cube is about 9% greater than the water from which it came. Consequently about one-tenth of the volume of an ice cube floats above the surface of water.

2-10: One particular ice blade we measured was 27 cm long and 4 mm wide. The surface area therefore was about 11 cm². If a 65 kg skater were balanced on that one blade, the force would be 650 N and the pressure would be $(650 \text{ N})/(11 \times 10^{-4} \text{ m}^2) \approx 6 \times 10^5$ N/m² = 6 atmospheres. This small change of pressure would lower the melting point of the ice by only $\tfrac{1}{20}$ Celsius degree. Even if the blade is properly sharpened and only one-tenth the surface is in contact with the ice, the change in melting point is still less than 1 Celsius degree.

2-11: Wrap the two thermometers together. The one that decreases its reading must have been the hotter one.

2-12: Probably not. The magnetic property is not a situation where an equilibrium is reached between two objects.

2-13: No. As you add heat to ice at 0°C, the ice will melt, but the temperature stays at 0°C. If you add heat to boiling water, the temperature remains at the boiling point.

2-14: The gram specific heat of water is the largest for any common substance. Therefore, it stores a large amount of internal energy and can give it back in the form of heat. Molten lithium near its freezing point has a slightly larger gram specific heat, but would not be so convenient to use in a hot water bottle.

2-15: The crucial flaw in doing the experiment as described is using the hot aluminum powder while it is still wet. Even though the mass of water clinging to the aluminum might be small, the large specific heat of that water leads to a sizable error. On the other hand, you could not wait until the aluminum dried off before pouring it into the water at 10°C. One possible way of doing such an experiment is to put the powdered aluminum into a plastic bag, thus keeping it dry while it is being boiled.

2-16: The specific heat per individual atom doesn't mean much when expressed in terms of calories or joules. However, the number becomes more understandable in terms of electron volts, the natural unit of energy in the atomic realm.

$$6 \frac{\text{cal}}{\text{mole} \cdot \text{Celsius degrees}} \times \left(\frac{1 \text{ mole}}{6 \times 10^{23} \text{atoms}}\right) = 1 \times 10^{-23} \frac{\text{cal}}{\text{atom} \cdot \text{Celsius degrees}}$$

$$1 \times 10^{-23} \frac{\text{cal}}{\text{atom} \cdot \text{Celsius degrees}} \times \left(4.17 \frac{\text{J}}{\text{cal}}\right) = 4 \times 10^{-23} \frac{\text{J}}{\text{atom} \cdot \text{Celsius degrees}}$$

$$4 \times 10^{-23} \frac{\text{J}}{\text{atom} \cdot \text{Celsius degrees}} \times \left(\frac{1 \text{ eV}}{1.6 \times 10^{-19} \text{J}}\right) = 2.6 \times 10^{-4} \frac{\text{eV}}{\text{atom} \cdot \text{Celsius degrees}}$$

It is not a good assumption that the specific heat stays constant all the way from 0 to 300 K. The specific heat actually decreases with decreasing temperature. However, the following calculation is of some interest because it yields an order of magnitude value for the internal thermal energy of each atom at room temperature.

$$\left(2.6 \times 10^{-4} \frac{\text{eV}}{\text{atom} \cdot \text{Celsius degrees}}\right) \times (300 \text{ Celsius degrees})$$
$$= 7.8 \times 10^{-2} \frac{\text{eV}}{\text{atom}} \approx \frac{1}{13} \frac{\text{eV}}{\text{atom}}$$

As we will see in the next chapter, this number is off by only a factor of 2 or 3 from that calculated by more sophisticated methods.

12-17: For an ideal gas, the difference between specific heats is exactly the same in value and units as the gas constant R.

$$c_P - c_V \approx 2 \frac{\text{cal}}{\text{mole} \cdot \text{K}} = 8.3 \frac{\text{J}}{\text{mole} \cdot \text{K}}$$

$$R = Nk = \left(6 \times 10^{23} \frac{\text{atoms}}{\text{mole}}\right)\left(1.4 \times 10^{-23} \frac{\text{J}}{\text{atom} \cdot \text{K}}\right) = 8.3 \frac{\text{J}}{\text{mole} \cdot \text{K}}$$

As a gas expands at constant pressure, the work that is done is equal to $P \Delta V$. Extra heat must be supplied to provide that work. At constant pressure, $P \Delta V = nR \Delta T$. For one mole of gas, the extra work needed to expand at constant pressure during a change of temperature of 1 Celsius degree is just equal to R.

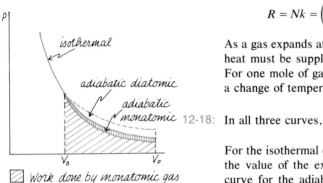

12-18: In all three curves, $P \propto \dfrac{1}{V^x}$

For the isothermal expansion, $x = 1$ and the curve is a simple hyperbola. The larger the value of the exponent x, the more steeply the curve will fall. Therefore, the curve for the adiabatic expansion of the monatomic gas lies below the adiabatic curve for the diatomic gas. Since the work to expand the gas must come out of the internal energy, it appears that the diatomic gas can provide more work for a given expansion than can the same quantity of a monatomic gas. This result is plausible since the diatomic gas with its larger specific heat can evidently store more internal energy than can a monatomic gas.

Problems

1. List five examples where energy lost to friction raises temperatures. Consider common situations with cars or around the home.
2. Suppose that a thermometer bulb is spherical and is connected to a cylindrical capillary tube. If the inside radius of the sphere is 2 mm which is 50 times the inside radius of the tube, what will be the increase in the height of the liquid in the tube if the volume of the liquid increases by 1%?
3. List five temperature indicators that you use in everyday life, besides thermometers and your sense of touch.
4. Estimate the Fahrenheit temperature of the hottest water that you can stand to put your hand in, and the temperature of cold tap water in your house. Convert these temperatures to Celsius.

5. What is the Celsius temperature of 0°F?
6. If you are at a feverish 40°C, what is your temperature in Fahrenheit?
7. A boiler in a modern power plant operates at a temperature of 1100°F. What is this temperature in degrees Celsius and on the absolute scale?
8. A mercury thermometer and a thermocouple are each calibrated at 0°C and 100°C and are marked with linear scales in between. What does each read at 30°C? (See the graph on page 274.)
9. Before a summer trip, pressure in a car tire is set at 26 lb/in^2. (This is *gauge* pressure—above the atmospheric pressure. Total pressure is therefore about 41 lb/in^2.) After an hour of driving on a hot road, the tire temperature has risen from 70°F to 120°F. The tire volume remains essentially constant. What is the new gauge pressure?
10. The small canisters of compressed air for emergency inflation of a tire can raise a flat tire's pressure to 15 lb/in^2 (gauge). If the volume of the can is 5% that of the tire, what is the original pressure in the can?
11. A mole of any (ideal) gas has a volume of 22.4 liters at 0°C and atmospheric pressure. What is the pressure of a mole of gas at -100°C in a volume of 0.224 liters? What is the volume of the same amount of gas at 300°C and 3 atmospheres pressure?
12. A sample of gas at room temperature (20°C) occupies a volume of 100 liters at standard atmospheric pressure. How many moles of gas are there? How many molecules?
13. How many molecules of air are there in 1 cm^3 at room temperature (20°C) and standard atmospheric pressure? How many moles?
14. A copper wire stretches between poles 100 m apart. In the winter at a temperature of -10°C, the sag of the wire in the middle is 0.5 m. What is it in the summertime when the temperature is 35°C?
15. Use your own words and no symbols to write the zeroth and first laws of thermodynamics.
16. If you devour 2000 diet calories every day, how many joules do you eat?
17. The *negative calorie diet* consists of ice water. When you drink it, your body must provide the energy needed to bring the ice water up to body temperature (37°C). How much ice water would you have to drink each day to balance out a food intake of 2000 diet calories? Calculate your answer in liters, which are almost the same as quarts (1 liter = 1.056 quarts). (Instead of drinking the ice water, you can get the same effect by sitting in it.)
18. How much work is done in expanding the volume of one mole of water into steam at atmospheric pressure?
19. Does adding cold milk to coffee lower the temperature appreciably? Assume that you add 10 g of milk at 5°C to one cup of coffee (about 200 g) at 95°C and that the specific heat of milk and coffee are about equal.
20. If 150 g of ice and 200 g of water in thermal equilibrium are heated to 100°C by being mixed with steam at 100°C, how much water will be in the final mixture?
21. An aluminum ball with mass of 50 g is taken out of an oven and dropped into a Styrofoam cup containing 100 g of room temperature water (20°C). If the final equilibrium temperature is 40°C, what was the initial temperature of the ball?
22. Suppose you have a 25 watt heater, and you immerse it in 1 liter of water at 20°C. If all of the heat is transferred to the water, how long will it take to boil?
23. With the temperature at 0°C, a snowball going at 20 m/s strikes a wall. If all the lost kinetic energy goes into the snow, what fraction is melted?
24. Explain in words why $c_P - c_V = R$ for an ideal gas.
25. Explain in words the source of the work done when a piston is moved by the expansion of a gas in an isothermal process and in an adiabatic process.

26. Two samples of gas in cylinders start at the same temperature, pressure, and volume. One gas is monatomic; the other is diatomic. They expand isothermally the same amount and then expand adiabatically the same amount. Which one does more work during the isothermal expansion? Why? Which one does more work during the adiabatic expansion? Why?
27. In most houses, room air temperature in winter is higher than it is in summer for the same feeling of comfort. The difference in relative humidity partly explains the effect. Why? The wall temperatures are also partly responsible. Why?
28. Suppose you must burn 1 gallon of oil per hour to keep your house at 70°F when the outside temperature is 30°F. What rate of oil consumption is necessary if the outside temperature is 50°F? What is the rate in both cases if the house if maintained at 60°F? (Actually, many other factors are involved—wind, activity in house, etc.)
29. The space suits worn by the astronauts were mostly white. On the other hand, some surfaces of space vehicles are black. What determines the choice?

THE MICROSTRUCTURE OF MATTER AND THE SECOND LAW OF THERMODYNAMICS 13

In the previous chapter we expanded the law of conservation of energy by claiming that heat is a flow of energy, and that lost mechanical energy can be accounted for when it apparently disappears into material. As we have seen, when energy flows into materials, many different things happen. Volumes usually increase, electrical properties change, and the materials themselves may change phase or break down into different substances. Usually, but not always, the temperature of an object rises when heat is introduced. The size of the temperature increase apparently depends on the number and kind of atoms present in the sample and how they are combined with each other.

To account for the heating effects, we must design a model of the microstructure of matter. The model will include atoms and molecules, of course, but we must also specify the interactions between atoms and the internal structure and motions of molecules. The thermometer will become our chief instrument for probing the microworld.

HANDLING THE PHENOMENA
1. We will account for the properties of gases by assuming that in the gas phase, molecules are dashing about in chaotic fashion. Qualitatively, such a model would explain how odors spread through air. The vapor of perfume, for instance, must consist of individual molecules of the perfume oil mixed up with the molecules of air. The perfume molecules gradually diffuse or spread to greater distances from their source. We might expect that the speed of diffusion would depend on the mass or size of the molecule. Large massive molecules should move more slowly through the air than small light ones.

You can test this idea qualitatively by using two vapors of very different molecular size. Since your nose will be the detection instrument, the vapors must have very distinctive odors. Try using ammonia and any kind of exotic perfume. (Ammonia is sold in household cleaning solutions.) The molecule (NH_3) is lighter than air molecules (N_2 or O_2). Perfume molecules are usually long chain assemblies of many atoms, mostly carbon and hydrogen. Pour a small amount of ammonia in one covered jar and a small amount of

perfume in another. While you stand in the middle of a closed room, have colleagues open the jars on opposite sides of the room. Measure and compare the times taken for you to smell each odor.

2. In Chapter 12 we gave instructions for cooking ice. Now we will propose a way to cook steam. Put a few cc of water in a pyrex carafe, a coffeemaker, for example. Cover the mouth of the carafe with some aluminum foil that has a small hole pierced in it. Now set the carafe on a stove and boil the water vigorously until it just appears to go dry. Immediately take the carafe off the stove and let it cool.

While the water was boiling, steam filled the carafe and drove out the air. When the liquid water just disappeared, the carafe was still filled with water vapor; that is, H_2O gas. The volume of the water vapor is the volume of the carafe. Watch what happens as the water vapor cools off. It turns back into a liquid condensing on the inside of the glass.

Question 13-1

About what is the volume of the water vapor after it condenses? What is the ratio of volume of gas to liquid for water? (Of course, the gas was at a temperature of 373 K. How important is it to correct for room temperature?).

3. Consider the remarkable fact that at 300 K you are relatively stable, chemically speaking. At twice this temperature your goose could be cooked in an oven and so could you. Even small differences in temperature can have an enormous difference on cooking times.

Find out how long it takes to cook an egg or a potato at 80°C. For the sake of comparison, you should find out how long it takes to cook eggs or potatoes at the normal boiling temperature at sea level of 100°C. You must use some patience and skill to keep the water temperature close to 80°C and some ingenuity to determine when the cooking is done. Of course, you can test the potato with a fork, but if you open the boiled egg too soon you will have raw egg. Perhaps you can cook a number of eggs at once, removing one at a time at fixed intervals. Be prepared to use the 5-minute and 10-minute egg in some other form, such as scrambled.

4. This exercise may be trivial unless you have never noticed the phenomenon before. Feel the radiator or fins in back of your refrigerator (while the refrigerator motor is on).

Question 13-2

Can you cool a room by leaving the refrigerator door open? If you can't cool the room this way, do you at least break even? If a refrigerator with an open door is working in a closed room, will the room temperature remain constant?

ASSUMPTIONS ABOUT OUR ATOMIC MODEL

In Chapter 1 and in Chapter 10 we described some properties of atoms and molecules. All the way from hydrogen to uranium, most atoms are about the same size. Within a factor of 2 their radius is 1×10^{-10} m. (There are a few exceptions, but even these are within a factor of 3. Actually, the "radius" of an atom depends on the method of measurement. For instance, the radius might be half the internuclear distance in a covalent molecule, or in a molecule lightly held by van der Waals' forces.) The mass of an individual atom can be calculated by dividing its gram atomic weight by 6×10^{23}, which is the number called the gram mole. For instance, a hydrogen atom has a mass of $(1 \text{ g/mole})/(6 \times 10^{23} \text{ atoms/mole}) = 1.6 \times 10^{-24}$ g. At the other extreme an atom of the uranium isotope with an atomic mass of 238 has a mass of $(238 \text{ g/mole})/(6 \times 10^{23}/\text{mole}) = 4.0 \times 10^{-22}$ g.

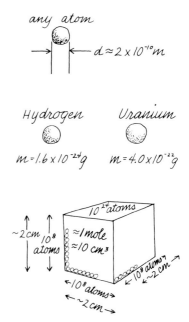

In the solid phase, the atoms are shoulder to shoulder. If a mole of atoms is arranged in a cube, then along one edge of the cube there would be about 10^8 atoms. (The cube root of $6 \times 10^{23} \approx 10^8$). Since the diameter of each atom is about 2×10^{-10} meters, the edge of this cube would be 2×10^{-2} meters or about 2 cm. In other words, a mole of atoms in the solid phase has a volume of about 10 cubic centimeters.

> **Question 13-3**
>
> Can it really be the case that a mole of uranium atoms has the same volume as a mole of carbon atoms? The table gives atomic weights and densities for a number of materials. See for yourself how good an approximation it is that most solids have the same molar volume.

Atomic Masses and Densities

	Atomic Weight Number	Density (g/cm³)
Beryllium	9	1.8
Aluminum	27	2.7
Carbon (graphite)	12	2.3
Sodium	23	0.97
Iron	56	7.9
Lead	207	11.3
Gold	197	19.3
Uranium	238	19.1

There must be attractive forces between atoms which keep them bound together in the solid and liquid phases. In Chapter 10 we proposed that each atom must exist in a potential well created by the attractive forces of all its neighbors. At the bottom of the potential well, the atom is at an equilibrium

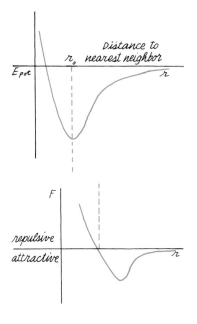

distance from its nearest neighbors. At that point the restoring force on the atom is zero. We would expect, however, that the atom would slosh back and forth in that potential well. In the section on molecular forces in Chapter 10, we made an order-of-magnitude calculation concerning the spring constant, k, responsible for small amplitude oscillations of the atom in the well. Assuming that the atom oscillates in simple harmonic motion about its equilibrium point, we then calculated that the frequency of oscillation must be about 10^{13} Hz. As we will see in Chapter 22, this frequency is in the infrared range. The frequency of visible light is about 50 times greater.

In the solid phase, atoms are bound with a negative total energy. Their positive kinetic energy of oscillation is smaller than their negative potential energy of binding. The situation is analogous to that of our planet moving around the sun with positive kinetic energy, but bound in a negative gravitational potential energy well. If a rocket ship or a comet gets a large enough kinetic energy, its total energy may be positive, enabling it to escape the attractive binding force of the sun. In a similar way, if an atom on the surface of a solid or liquid obtains enough kinetic energy, it may escape the bonds of its near neighbors and become an atom in the vapor or gas phase.

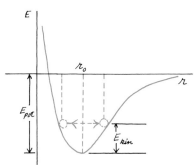

SPECIAL ASSUMPTIONS FOR OUR MODEL OF GASES

It is far simpler to create a kinetic theory of gases than it is to make a quantitative model of solids and especially liquids. In deducing the simple form of the gas model, we will assume some things about the atom that we know are not strictly true. First, let us assume that the volume of the atoms or molecules themselves is negligible compared with the volume in which the molecules are flying around.

> ### Question 13-4
>
> How poor an assumption is this? In ''Handling the Phenomena,'' we compared the volume of water vapor to the volume of the liquid into which it condensed. If the volume of the atoms were really zero, then no liquid water would have been visible on the glass walls of the carafe.

In constructing our model of gases, we will aso assume, *at least at first,* that the collisions of the atoms or molecules with each other and with the walls of the container are elastic. No energy is lost due to friction and, more realistically on the atomic scale, no energy is lost due to the breaking up of molecules or setting them into internal vibration. You might wonder how atoms can collide elastically if there is an attractive force between them characterized by a potential well. As long as the kinetic energy of the bombardment is larger than the depth of the potential well, two atoms will collide by increasing their speeds as they fall into each other's attractive field, come very close together into the region of mutual repulsion, and then go shooting away from each other just as if they were two hard, impenetrable marbles. Indeed, our first model for gases will be completely classical, considering their momentum, energy, and flight times as if they were a swarm of marbles having no interaction with each other unless they collide, possessing no weight, and bouncing off each other according to the laws of classical mechanics.

THE KINETIC THEORY OF GASES

Let's start out with just one atom enclosed in a cubic box of length L. It is bouncing around randomly, striking the walls at various angles. Consider only the component of velocity in the x-direction and what happens when the atom hits the y-z walls opposite that component. For the sake of visualizing the action, we could consider that the atom is simply moving back and forth in the x-direction. In that case, since it is moving to the right, it has momentum, mv_x. When it hits the wall and bounces back, its momentum has become $-mv_x$. The change of momentum, $\Delta(Mom) = 2mv_x$. The wall suffered a recoil impulse to the right and will do so again every time that the atom bounces back. To find the time between bounces, divide the distance traveled across the box and back again by the velocity: $\Delta t = 2L/v_x$. Remember that the general expression for Newton's second law is that $F = \Delta(Mom)/\Delta t$. The impulses on the wall can be equated to an average force; $\overline{F} = 2mv_x/(2L/v_x) = mv_x^2/L$. Of course, one atom won't produce much of an average force but if there are N atoms in the box, all traveling in the same direction, then the force would be $\overline{F} = Nmv_x^2/L$.

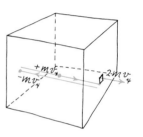

The atoms won't be going in the same direction, however. Their motion is completely random and they are not only bouncing off the walls but off each other. The Pythagorean theorem gives a simple relationship between the square of a vector and the squares of its components in x, y, and z; $v^2 = v_x^2 + v_y^2 + v_z^2$. For any given atom at a particular time, one component may be much larger than either of the other two components. On the average, however, the components in the x-direction will be the same as those in the y-direction or the z-direction. Therefore, for the average $\overline{v^2}$, $\overline{v_x^2} = \overline{v_y^2} = \overline{v_z^2}$. Consequently, $\overline{v_x^2} = \tfrac{1}{3}\overline{v^2}$. The force on the y-z wall of the box averaged over both time and for many different atoms is equal to $\overline{F} = \tfrac{1}{3}Nm\overline{v^2}/L$. Instead of describing the force on the y-z wall, we can make the calculation more general by describing the pressure. Remember that $P = F/A$, where A is the area of the y-z wall.

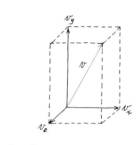

$v^2 = v_x^2 + v_y^2 + v_z^2$

$P = \overline{F}/A = \tfrac{1}{3}Nm\overline{v^2}/LA = \tfrac{1}{3}Nm\overline{v^2}/V$, where V is the volume of the box. ($V = LA$.) The equation now transforms to

$$PV = \tfrac{1}{3}Nm\overline{v^2}$$

There is yet another relationship that involves $m\overline{v^2}$ of the molecules. The average kinetic energy of the atoms or molecules in the box must be

$$\overline{E}_{kin} = \tfrac{1}{2}m\overline{v^2}$$

Therefore, we can rewrite the equation for PV as

$$PV = \tfrac{2}{3}N(\tfrac{1}{2})m\overline{v^2} = \tfrac{2}{3}N\overline{E}_{kin}$$

When we studied the properties of gases in Chapter 12, we found that the *experimental* ideal gas law is

$$PV = NkT$$

If our model and theoretical derivation are correct, these two equations for PV should be identical. If the right-hand sides of the equations are equal to each other, then the temperature of a gas must be related to the average kinetic energy of its molecules:

$$PV = \tfrac{2}{3}N\overline{E}_{\text{kin}} \qquad PV = NkT$$
$$\text{(theoretical)} \qquad \text{(experimental)}$$

Therefore

$$\overline{E}_{\text{kin}} = \tfrac{3}{2}kT$$

Question 13-5

We seem to have pulled a surprising result out of thin air. The impulse delivered by an atom to the wall must be proportional to its velocity. How did the pressure suddenly become proportional to the *square* of the velocity? Furthermore, although we started out describing an atom or molecule with mass m, we have ended up with the result that the temperature must be proportional to the average kinetic energy of the molecules regardless of what kind they are. Can the result really be independent of whether the gas is hydrogen with molecular weight 2 or chlorine with molecular weight 71?

CONSEQUENCES OF OUR MODEL—AVERAGE ENERGY AND AVERAGE SPEED

As a matter of fact, have we really proved that temperature is proportional to the average kinetic energy of the molecules of the gas? We have proposed a theoretical model, deduced a consequence from the model, and *if* the model is correct, the temperature of a gas is a measure of the kinetic energy of its molecules. Such a prediction needs experimental testing. Let's put in some numbers and see what values we get for kinetic energy and speed of gas molecules at room temperature.

As we saw in the previous chapter, k is Boltzmann's constant and has the numerical value of 1.38×10^{-23} J/molecule·K. At room temperature: $E_{\text{kin}} = \tfrac{3}{2}(1.38 \times 10^{-23} \text{ J/mol·K})(293 \text{ K}) = 6.1 \times 10^{-21}$ J/mol. Let's transform the energy units to electron volts, the appropriate unit of energy for chemical interactions of individual molecules.

$$(6.1 \times 10^{-21} \text{ J/mol})/(1.6 \times 10^{-19} \text{ J/eV}) = 3.8 \times 10^{-2} \text{ eV/mol} = \tfrac{1}{26} \text{ eV/mol}$$

The average energy of a gas molecule at room temperature is smaller than 1 electron volt by a factor of 26. At first glance that fact seems to explain why we and other chemically bound objects are stable at room temperature.

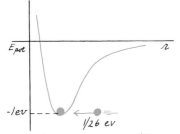

A gas molecule with energy of $\tfrac{1}{26}$ ev cannot bombard and liberate an atom bound by 1 ev.
[But can a gas molecule with energy of $\tfrac{1}{13}$ ev (T=586K) do the trick?]

On the other hand, we know that we are not stable at twice that temperature—at 586 K—where the average energy of a gas molecule is smaller than 1 eV by a factor of 13. Evidently, we have more to learn about the phenomenon.

How fast are the molecules dashing about in a gas at room temperature? The speed of the gas molecules with average kinetic energy can be found as follows:

$$\overline{E}_{kin} = \tfrac{1}{2}m\overline{v^2} = \tfrac{3}{2}kT$$
$$v = \sqrt{3kT/m}$$

For air at room temperature, we can use the mass of the nitrogen molecule and $T = 293$ K.

$$v = \sqrt{3(1.38 \times 10^{-23}\,\text{J/mol}\cdot\text{K})(293\,\text{K})/(4.7 \times 10^{-26}\,\text{kg})} = 5.1 \times 10^2\,\text{m/s}$$

The average speed of air molecules at room temperature is about 500 m/s. That's over 1100 miles per hour!

Question 13-6

When you smelled the perfume or the ammonia, the odor took a number of seconds to cross the room. Doesn't this simple experimental fact disprove our derivation which claims that gas molecules travel faster than speeding bullets?

Although our predicted speed for gas molecules is much larger than the observed speed of the spread of odors, still it is only about $1\tfrac{1}{2}$ times the speed of another influence that propagates through gases. The speed of sound in air at room temperature is about 343 m/s. Since sound in air must be carried by the collision of the molecules with each other, it is encouraging that the speed of sound is the same order-of-magnitude as the theoretical value for the average speed of air molecules. In Chapter 15 we will derive the actual formula for the speed of sound waves in air.

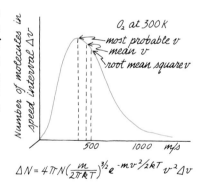

$$\Delta N = 4\pi N\left(\frac{m}{2\pi kT}\right)^{3/2} e^{-mv^2/2kT} v^2 \Delta v$$

■ Optional

The Distribution of Molecular Energy in a Gas
In both gases and human societies, many phenomena are dominated by the nonaverage members of the population. Since the molecules in the gas are continually colliding with each other and with the walls, some will be going very slowly and some very fast. The distribution of energies was worked out by Maxwell and Boltzmann a century ago. The distributions of speeds and energy are shown in the diagrams. Although the formulas look complex at first, there are several simple points that can be seen.

First note that the distributions are not symmetrical about the average speed or the average energy. In fact, with these kinds of distributions, we have to be careful about what we mean by "average." In the speed diagram, we have indicated the most probable speed, the mean speed (the ordinary arithmetic average) and the root mean square speed, which is the average speed we calculated earlier: $v_{rms} = \sqrt{3kT/m}$

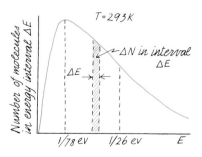

$$\Delta N = \frac{2\pi}{(\pi kT)^{3/2}} N e^{-E/kT} \sqrt{E}\, \Delta E$$

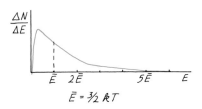

$\bar{E} = \tfrac{3}{2} kT$

For very small values of speed or energy the exponential term is close to unity because the exponent itself is close to zero. The distribution functions are therefore dominated by the v^2 term in one case or the $\sqrt{E}$ term in the other. Apparently, it is very unlikely that a molecule being buffeted about in the gas will end up with zero speed or energy. In the distribution function the rising values of the v^2 or $\sqrt{E}$ terms are eventually overcome by the decreasing exponential terms. Indeed, once past the maximum, the distributions are quite completely governed by the exponential terms. Exponential terms decay very rapidly but they never decay away completely. Therefore it is possible to have some molecules with energies many times those of the average. Notice that the distribution functions are given in terms of the number of molecules per speed or energy interval. Consequently, the total area under each curve is equal to the total number of molecules present.

Question 13-7
The average energy $\bar{E}$ of a molecule is equal to $\tfrac{3}{2}kT$. What is the ratio of the number of molecules with energy $2\bar{E}$ to the number with energy $\bar{E}$? What is the ratio of the number of molecules with energy $10\bar{E}$ to the number with energy $5\bar{E}$? Look out; the answer is surprising and not obvious.

How could we ever measure the speeds or energies of individual molecules in a gas in order to check on the theoretical distribution laws? As we will see, the consequences of the distribution laws lead to predictions about phenomena that can be checked experimentally. Furthermore, the molecular speeds themselves can be measured in a very direct fashion. All you have to do is time the flight of molecules in the same basic way that you might time the throw of a baseball or the flight of a bullet.

Starting back in the 1920s, Otto Stern and his students produced beams of molecules with apparatus shown schematically in this diagram. Many different kinds of atoms and molecules, even those of metals, can be produced in the gas phase. In some cases the source for the beam is an oven that is boiling off the atoms. The source box has a tiny hole through which the gas atoms can pass if they happen to be going in that direction. From that point on they are traveling in a vacuum. Those that can pass through the successive holes in the baffles form a narrow beam. The small group that gets through the opening in the first high-speed rotating shutter starts its flight at a particular time. Only those with a particular speed will arrive at the second shutter in time to get through it and on into the collector. By changing the angle by which the second shutter lags behind the first, we can accept groups of molecules with different speeds. The resulting distribution gives us ΔN versus v experimentally, and the results agree with the theory.

Although we have been talking about the *kinetic* energy of molecules in a gas, the same exponential factor is involved in the distribution of molecules subject to a potential field. For example, the gravitational potential energy of molecules depends on their height, h, above the surface of the earth. Assuming that the temperature is independent of height (a poor assumption over any considerable distance), then the ratio of the number density of molecules at two heights is given by

$$\frac{n_1}{n_2} = \frac{e^{-mgh_1/kT}}{e^{-mgh_2/kT}} = e^{-mg(h_1-h_2)/kT}$$

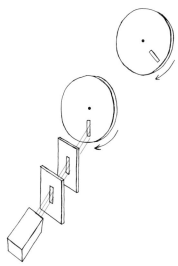

Schematic diagram of apparatus to measure molecular speed.

The number density, *n,* is the number of molecules per unit volume.

So far we have been talking about molecules in the gas form. A similar distribution occurs whenever particles can mix in a chaotic random fashion, subject to the conservation of overall energy and the conservation of the total number of particles. The generalized Maxwell-Boltzmann distribution for such a situation gives the ratio of number of particles in two different energy states, E_1 and E_2, as follows:

$$n_1/n_2 = e^{-(E_1-E_2)/kT}$$

Let's put some numbers into these distribution laws to see consequences concerning some familiar situations.

1. The atmosphere is less dense at high altitudes, and above 50 kilometers is practically nonexistent in terms of what we know as air. Many factors influence the density of the air and also its composition. Since air is mostly transparent to solar energy, heating of the atmosphere is mainly a matter of convection from the earth's surface where the solar energy is absorbed. Consequently, temperature falls rapidly as altitude increases. (At about 50 kilometers there is a layer of ozone created by ultraviolet radiation. This thin layer of chemically active oxygen has a temperature of about 10°C. It acts as a heat source for the atmosphere slightly below and slightly above). In the troposphere, below about 12 kilometers, storms can affect air pressure (the "highs" and "lows" of weather patterns).

In spite of the factors of variable temperatures and storms, the Maxwell-Boltzmann distribution law yields a first approximation to the average density of the atmosphere as a function of height. The ratio of the number of molecules per unit volume at two heights, n_1/n_2, is also the ratio of densities at those two heights, σ_1/σ_2, and therefore is also the ratio of pressures, P_1/P_2. For P_2, use the pressure at sea level, where $h_2 = 0$. Then the pressure at any given height, h, is

$$P(h) = P_0 e^{-mgh/kT}$$

Notice that this formula gives the right value for atmospheric pressure at sea level; namely, if $h = 0$, then the exponential is equal to 1 and $P(0) = P_0$. To find the pressure at any other height, we have to evaluate the argument of the exponential: $mgh/kT = (4.7 \times 10^{-26}$ kg$)(9.8$ N/kg$)(h)/(1.38 \times 10^{-23}$ J/K$)(290$ K$) = 1.15 \times 10^{-4} h \approx 1 \times 10^{-4} h$.

At a height of 1.5 kilometers, about the altitude of Denver, Colorado, the argument of the exponential would become $(1.15 \times 10^{-4})(1.5 \times 10^3$ m$) = 0.17$. For small values of the argument of an exponential, $e^{-x} \approx 1 - x$. Consequently, the normal atmospheric pressure at Denver should be about 0.83 P_0, where P_0 is the atmospheric pressure at sea level. Indeed, the normal atmospheric pressure in Denver is about 630 mm of mercury.

The pressure of the atmosphere is always just sufficient to hold the weight of a column of air of unit cross section going all the way to the top of the atmosphere. Atmospheric pressure at sea level is approximately 1×10^5 N/m². Since most of the atmosphere is in a region where g equals 9.8 N/kg, the mass of a column of air molecules with a cross section of 1 square meter and a height going to the top of the atmosphere must be approximately 1×10^4 kg. These molecules are bounding around.

It is not obvious how they exert a force just equal to what their weight would be if they were all lying quietly on top of each other. Nevertheless, if we use the distribution law to calculate how many air molecules there are in a column with a cross section of one m² reaching from $h = 0$ to $h =$ infinity, the answer is 2.2×10^{29}. This number is obtained by finding the area under the

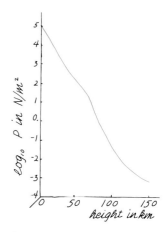

This graph shows pressure readings in the atmosphere as measured in rocket flights. The graph is semi-log. If the pressure followed the formula that we derived, the curve would be a straight line.

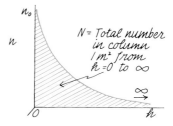

$n = n_0 e^{-mgh/kT}$

$n_0 = \dfrac{(6 \times 10^{23} \text{ mol/mole})}{(22.4 \times 10^{-3} \text{ m}^3/\text{mole})}$

$n_0 = 2.68 \times 10^{25}$ mol/m³ at STP

$N = 2.2 \times 10^{29}$

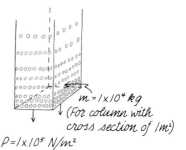

$m = 1 \times 10^4$ kg
(For column with cross section of 1m²)

$P = 1 \times 10^5$ N/m²

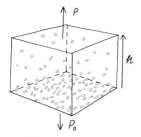

$P = P_0 e^{-mgh/kT}$
$P \approx P_0 [1 - \frac{mg}{kT} h]$
$\Delta P = P_0 - P = P_0 \frac{mg}{kT} h$
For air molecules at
$T = 290 K$, $\frac{mg}{kT} = 1.15 \times 10^{-4} m^{-1}$

For a box with $h = 1 m$:
$\Delta P = (1 \times 10^5 \, N/m^2)(1.15 \times 10^{-4})$
$\Delta P = 11.5 \, N/m^2$

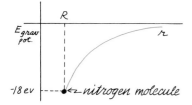

For waves:
(wavelength)(frequency) = velocity
If wavelength is fixed, and velocity increases, frequency increases.

distribution graph as shown on page 311. The mass of that many air molecules (each with a mass of 4.7×10^{-26} kg) is approximately equal to 1×10^4 kg. Their weight is just equal to the atmospheric pressure at sea level.

Question 13-8
According to the energy distribution law for gases, the density of gas at the top of a box must be less than it is at the bottom. If the gas is at uniform temperature, then the pressure at the top must be less than it is at the bottom. The molecular bombardment at the bottom will be greater than it is at the top as calculated in the diagram. How is this extra force related to the weight of the gas in the box?

2. Like everything else on the surface of the earth, an air molecule is in a gravitational potential well. As we saw in Chapter 10, the depth of that potential well is

$$E_{\text{grav pot}} = -GmM/r$$
$$= -(6.7 \times 10^{-11} Nm^2/kg^2)(m)(6 \times 10^{24} kg)/(6.4 \times 10^6 m)$$

For a hydrogen molecule with a mass of 3.3×10^{-27} kg, the gravitational potential binding energy is equal to 2.07×10^{-19} J = 1.3 eV. The molecular thermal energy is equal to $\frac{3}{2}kT$, which, as we have seen, is about $\frac{1}{26}$ eV at room temperature. It appears that hydrogen is tightly bound to the earth since the gravitational potential well is about 34 times deeper than the average kinetic energy resulting from thermal motion. Nevertheless, hydrogen does not stay in the earth's atmosphere. In spite of the mixing of gases within the atmosphere the lighter hydrogen gradually works its way upwards. Above the ionosphere there are regions where the equivalent temperature is very high because of the bombardment of particles coming in from the sun. Hydrogen molecules in the high energy tail of the thermal energy distribution occasionally reach escape velocity in the right direction and are lost to the earth.

The average *thermal* kinetic energy of nitrogen or oxygen is the same as it is for hydrogen. The depth of the gravitational potential well, however, is greater by a factor of 28/2 or 32/2, for nitrogen or oxygen, respectively. A nitrogen molecule, for example, is in a gravitational potential well with a depth of about 18 electron volts. That's why the earth has an atmosphere.

Question 13-9
Why doesn't the moon have an atmosphere?

3. Since the average thermal kinetic energy of different kinds of molecules is the same regardless of the molecular mass, it must be that at a particular temperature, lighter molecules are moving more rapidly than heavy molecules.

$$(\tfrac{1}{2}mv^2)_{H_2} = (\tfrac{1}{2}mv^2)_{O_2}$$
$$v_H/v_O = \sqrt{m_O/m_H}$$

Since the ratio of the mass of oxygen to hydrogen is 16, the ratio of speeds is four.

This dependence of average speed on the mass of the molecule shows up in a variety of ways. Remember that we found that the speed of sound in air is about two-thirds the average speed of air molecules. We should expect to find that the speed of sound in hydrogen or helium is much greater than it is in air. The effect can be demonstrated by flushing helium through an organ pipe (or bugle). The pitch of an organ pipe depends on its size, which determines the wavelength, and on the speed of sound in the pipe. Since the wavelength remains constant, the frequency of the sound rises as helium fills the pipe.

Question 13-10
What is the ratio of the average molecular speed in helium compared with oxygen? If the frequency of the sound in the experiment described is proportional to the speed of sound in the gas in the pipe, what happens to the pitch? (An octave rise corresponds to a doubling of frequency).

Vacuum systems frequently leak. The leak is not caused by gas molecules being sucked into a tiny crack or hole in the vacuum wall. The molecules simply bombard all sections of the wall. If they happen to strike where the crack is, they may get inside. The chance of a molecule striking the wall right in the crack depends on how frequently the molecule hits the wall and therefore on its velocity. Whether the molecule gets through the crack may also depend on the relative size of crack and molecule. Since the lightest gas is molecular hydrogen ($m = 2$), it might seem that hydrogen would leak into or out of a system more readily than any other gas. Indeed, a vacuum chamber or pressure chamber must be very tight to keep hydrogen in or out. It turns out, however, that helium is even harder to hold onto. Helium is monatomic whereas hydrogen is diatomic. The molecular (atomic) mass of helium is twice that of hydrogen and so its average velocity is slower by a factor of 1.4. Still the helium gets through leaks more readily because the single atom of helium is considerably smaller than the double atom of the hydrogen molecule.

The diffusion of gases through air is somewhat similar to the leaking of gases into a chamber. In "Handling the Phenomena," you probably observed that the ammonia diffused faster than the perfume. The ammonia has a molecular mass of only 17. Most perfume molecules consist of long chains of carbon, hydrogen, and other things and so have large molecular mass and size. Consequently, the diffusion speed of the ammonia is faster than that of the perfume.

4. Our simple model of gases ought to apply to any particle that satisfies the assumptions of the model. If an object is suspended in a gas, even though it is large enough to be visible to the naked eye, it ought to take part in the random bombardment and motion. This effect had been seen but misunderstood in 1827 by an English botanist named Robert Browne. He was using a microscope to observe grains of pollen floating on the surface of a liquid. The pollen grains seemed to jiggle and dash around by themselves as if they were alive. This effect, which can be seen with particles suspended in either gas or liquid, is called the Brownian motion. When Maxwell and Boltzmann created the kinetic theory of gases, they assumed that the Browian motion was due to molecular bombardment. The theory was not made quantitative, however, until Einstein did so in 1905. His theory predicted how far a pollen particle would migrate in a given length of time as a function of the temperature and relative masses of particle and molecule.

The Browian motion can be seen with a microscope at a magnification of about 50. Such an observation is about as close as we can come to seeing the result of molecules directly with our naked eye. Einstein's proof of the cause of the phenomenon was historically the capstone of the arguments in favor of atomicity. Until that time, as late as 1905, such influential scientists as the physicist Mach and the chemist Ostwald felt that it was not necessary to hypothesize the existence of atoms.

A grain that is visible in a low-powered microscope must have a diameter of about 10 micrometers (10^{-5} m). The diameter of a gas molecule is about 5×10^{-10} m. The ratio of diameters of grain to molecule is therefore about 2×10^4. The ratio of volumes, and therefore approximately the ratio of masses, is about 10^{13}. The scale of the situation is comparable to that of an ocean liner surrounded and buffeted by a sea of corks. The astonishing feature of the

$\frac{d_{grain}}{d_{molecules}} \approx 2 \times 10^4$

$\frac{m_{grain}}{m_{molecules}} \approx 10^{13}$

phenomenon is not so much that the bombardment of the molecules (or corks) can move the grain (or ship) but that the *fluctuations* in the bombardment produce such motion. After all, at any given instant, the larger object is being bombarded on all sides.

Question 13-11
A pollen grain with a diameter of 10 micrometers has a volume of about 10^{-15} m³. An approximate density for such a grain would be that of water, which is 1×10^3 kg/m³. What is the average speed of such a grain at room temperature? (Remember that the pollen grain, like any other particle in the gas, has an average kinetic energy of $\frac{3}{2}kT$.)

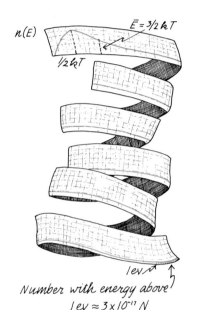

Number with energy above 1 ev ≈ 3×10⁻¹⁷ N

5. In "Handling the Phenomena," you tried cooking eggs and potatoes at 80°C. The potatoes never got cooked. Why should there be such a difference in the result between 373 K and 353 K? The difference in temperature is only 5%. If all the molecules participating in a chemical interaction had exactly the same thermal energy, then we might explain the need for particular temperatures in cooking in terms of a required threshold energy for an interaction. However, the Boltzmann distribution of molecular thermal energy applies approximately to solids and liquids as well as to gases. There are always some atoms or molecules in the high energy tail of the thermal distribution. Is it possible that a relatively small increase in temperature can dramatically increase the population in the high energy tail of the distribution? Let's put in some numbers and see.

The number of atoms or molecules with energies equal to or greater than a certain value, $E_{\text{threshold}}$, is approximately equal to $N\sqrt{E_t/kT}\,e^{-E_t/kT}$, where N is the total number of molecules. Suppose that a certain chemical interaction requires an activation or threshold energy equal to E_t. The order of magnitude of this energy for most chemical interactions is, as we have seen, 1 electron volt. The value of kT, for room temperature, is about $\frac{1}{40}$ electron volt. (Note the calculation on page 308, showing that $\frac{3}{2}kT = \frac{1}{26}$ eV for $T = 293$ K). Substituting these values into the exponential, we find that the number of molecules with energies equal to or greater than the energy required to produce a chemical interaction is equal to

$$N\sqrt{40}\,e^{-(1\text{ eV})/(1/40\text{ eV})} = N\sqrt{40}\,e^{-40} \approx 3 \times 10^{-17}N$$

That is an extremely small factor, but N may be a large number. (Remember that $N = 6 \times 10^{23}$ for one mole.) Notice what happens if the required activation energy is reduced by a factor of 2 to $\frac{1}{2}$eV or equivalently, the temperature is raised by a factor of 2 from 293 K to 586 K. The number of particles larger than the activation energy changes to $\sqrt{20}\,e^{-20}N = 1 \times 10^{-8}N$. The multiplying factor is still a small number but it is larger than the first by a factor of 3×10^8. The time taken for an interaction depends on many factors such as the rate of mixing. However, when a doubling of temperature can increase the number of molecules that can produce the reaction by a factor of 3×10^8, then clearly things can begin to cook. At 293 K you can function comfortably reading this book; at 586 K the book would smolder and you would literally roast.

Question 13-12
Chemists have a rule of thumb that raising the temperature 10 Celsius degrees doubles the rate of reaction. Of course, this is only a crude generalization, which has many exceptions. However, see what happens to the exponential distribution for an activation energy of 1 eV if the temperature rises from 90°C to 100°C (363 K to 373 K).

THERMAL EFFECTS IN REAL GASES, SOLIDS, AND LIQUIDS

1. Few things are ideal in this world including gases. The molecules in our ideal gas model were supposed to have no volume of their own and no interaction until they hit each other. We know that both of these assumptions are wrong. As you saw in "Handling the Phenomena," the volume of the molecules in a gas is about 1/1000 of the volume occupied by the gas at room temperature. We also know that the molecules attract each other at separation distances larger than that where sharp repulsion occurs. In the P-V curves on page 282, the hyperbolic curves called for by the ideal gas equation applied only for temperatures well above the condensation point.

Many "equations of state" have been proposed to describe gas behavior over the complete range of pressure, volume, and temperature. The best known of these is the van der Waals' equation:

$$(P + a/V^2)(V - b) = RT$$

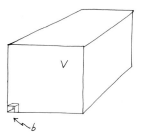

This form of the equation applies only if there is 1 mole of gas. The significance of b is straightforward; it is the effective volume occupied by the molecules of the gas. The actual volume available for the gas is less than V, the volume of the container, because part of the volume is taken up by the molecules themselves. At room temperatures for most gases the correction is only about 0.1%. (In practice, the correction term turns out to be about four times the actual volume of the molecules.)

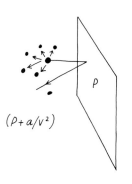

The derivation of the correction to the pressure term is a little more complicated. Evidently, the pressure measured on a wall of the container, P, must be slightly less than the actual pressure in the gas, which is represented by the whole term $(P + a/V^2)$. When a molecule of the gas is surrounded by other gas molecules, the small attractive force that exists between any two molecules cancels out because, on the average, the molecule is being pulled equally in all directions. As a gas molecule starts to strike a wall of the container, however, there is an unbalanced force toward the interior of the gas, thus pulling the punch of the molecule that is striking the wall. The number of molecules exerting this retarding force is proportional to the density of the gas, and the number striking the wall in the first place is proportional to the density. Therefore this effect is proportional to the square of the density. Since we are dealing with a fixed quantity of gas, in this case 1 mole, the density is inversely proportional to the volume. Therefore the correction term is proportional to $1/V^2$. For air at normal temperature and pressure, the correction term for the pressure is about 0.3%.

Question 13-13

The lower the pressure of a gas, the more it is supposed to behave like an ideal gas. However, if the pressure is lowered doesn't the correction term, a/V^2, become more and more important?

2. According to the kinetic theory, the temperature of a gas is related to the average kinetic energy of the molecules. Solids get hot too. Since

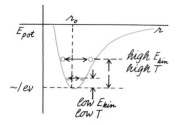

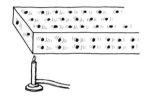

each atom in a solid is locked in place, how does it absorb thermal energy? As we saw in Chapter 10, each atom is locked in a potential well in which it can oscillate. The energy relationships are shown in the diagram. In our simple model we assume that an atom can oscillate around its equilibrium position with any amplitude corresponding to any required energy. The actual situation as required by quantum mechanics is more complicated. Only certain energy states are available to each atom. Nevertheless, the simple classical model gives reasonable explanations of many phenomena. According to this model, as heat enters a solid, the energy spreads out among all of the atoms, causing them to oscillate with larger amplitudes and a distribution of oscillation energy similar to that of the Maxwell-Boltzmann distribution.

What is the relative magnitude of the positive energy of oscillation to the negative potential energy of binding? As we have seen in many different circumstances, the chemical binding energy per atom is about 1 electron volt. The average thermal energy at room temperature is about $\frac{1}{26}$ electron volt. Consequently, the atoms of a solid are deep in potential wells and their oscillations, on the average, do not take them very far up the wall of the well. The temperature of a solid is related to the kinetic energy of *oscillation* of the atoms in the same way that the temperature of a gas is related to the *translational* kinetic energy of the molecules. Heat passes through a metal bar from high temperature to low temperature because the atoms with large amplitude of oscillation at one end share their energy with their neighbors, thus spreading the thermal energy along the bar. There is similar contact between atoms in the surrounding gas and atoms in a solid. The bombarding gas molecules and the oscillating atoms of the solid come to energy—and hence temperature—equilibrium.

Good electrical conductors are also usually good heat conductors. To explain this situation, we must add one particular feature to our model. With insulating materials the electrons are bound in position along with their parent atoms. With metals, however, one or more of the valence electrons are free to move around the whole expanse of the object. These electrical conduction electrons can also spread thermal energy and they do so much faster than when neighboring atoms share oscillation energy.

In Chapter 12 we saw that most solids expand as their temperature increases. At first thought, it might seem that the expansion could be explained in terms of the increasing amplitudes of oscillation of the bound atoms. After all, if each atom oscillates back and forth in a larger region as its average kinetic energy increases, wouldn't the whole object become larger?

Question 13-14

What is the fallacy in that argument? As the amplitude of a vibrating bob on a spring increases, what happens to its average position?

3. Is the thermal behavior of a liquid more like that of a solid or of a gas? The atoms or molecules must be just about as close to each other in

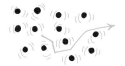

a liquid as they are in a solid. After all, for most materials, the density of the liquid is about the same as the density of its solid. Nevertheless, the binding of individual atoms or molecules in a liquid is sufficiently loose so that they can move around each other. However, the energy associated with this meandering motion is much smaller than the energy of oscillation of each atom about its gradually moving equilibrium point. The period of oscillation of the atom in its temporary cell of neighbors is much shorter than the time it takes to drift past those neighbors.

A quantitative treatment of the liquid phase is still hard to do for most materials. Both the gas phase and the solid phase are much better understood. In liquids there are changing clusters of atoms or molecules, sometimes with sufficient order to their temporary arrangements so that they behave in some ways like crystals. Not only are these clusters continually forming and then breaking up, but there are also fluctuations in the mass movement of micro volumes of the fluid, even when the average velocity of the liquid as a whole is zero.

4. As heat enters a solid, the material can turn into a liquid and then into a gas. (Under certain circumstances, the solid may change directly into a gas.) During the transition between phases, energy is absorbed without any change of temperature. Apparently, the internal energy of the liquid phase is larger than that of the solid phase and the internal energy of the gas phase is greater than that of the liquid phase. However, this does not mean that the kinetic energy of the atoms in the liquid is greater than the kinetic energy of the atoms in the solid, or that the translational kinetic energy of the gases is greater than the vibrational kinetic energy of solids or liquids at their same temperature. Instead, the energy required for phase transition—the heat of fusion or vaporization—must go into freeing atoms from potential wells. Let's compare the depth of these potential wells with those produced by chemical binding.

For water, the heat of fusion is 80 calories per gram and the heat of vaporization is 540 calories per gram. Let's change each of these figures to joules per mole and then to electron volts per molecule.

Heat of fusion: $(80 \text{ cal/g})(4.18 \text{ J/cal})(18 \text{ g/mole}) = 6020 \text{ J/mole}$

$$= 1.4 \text{ kcal/mole}$$

$(6020 \text{ J/mole})(6 \times 10^{18} \text{ eV/J})[1/(6 \times 10^{23} \text{ molecules/mole})] = 0.06 \text{ eV/molecule}$

Heat of vaporization: $(6020 \text{ J/mole})(540/80) = 40{,}700 \text{ J/mole}$

$$= 9.7 \text{ kcal/mole}$$

$(0.06 \text{ eV/molecule})(540/80) = 0.4 \text{ eV/molecule}$

Phase transitions are usually classified as physical changes as opposed to chemical changes. Such distinctions cannot be made solely on the basis of energy effects, although there is usually less energy per molecule involved with so-called physical changes than with chemical changes. In the case of water, the heat of fusion is small compared with the activation energy required by most chemical interactions. But the heat of vaporization is comparable to that of some chemical transitions. The energy required to break up a molecule of water into oxygen and hydrogen is 2.98 eV/molecule.

Energies for Formation of Molecule and for Phase Transitions, in eV/molecule. The Heat of Vaporization Is for Standard Atmospheric Pressure. 1 eV/molecule = 23 kcal/(gram)mole.

	Formation of Molecule from Atoms	Heat of Vaporization	Heat of Fusion
HCl	0.96	0.17	0.021
H_2S	0.23	0.19	0.025
NaCl	1.89	—	0.29
CO_2	4.09	sublimes	0.087
CO	1.15	0.063	0.0087
NH_3	0.48	0.243	0.059
H_2O	2.5	0.42	0.062

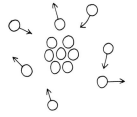

In a saturated gas there is competition between evaporation from the surface of tiny droplets, and condensation. Beyond a critical size, perhaps furnished by a dust particle, condensation wins.

But only 0.23 eV is needed for disintegration of H_2S into hydrogen and sulfur. Note that this "chemical" change takes less energy than the "physical" change of turning water into steam. The table gives phase transition energies and chemical binding energies for a number of materials.

As we saw in Chapter 12, liquids do not necessarily freeze at the freezing point nor do liquids necessarily boil at the boiling point. Similarly, in a dust-free atmosphere, it is possible to have humidity as high as 140% without condensation. Superheating, supercooling, and supersaturation are phenomena that are still poorly understood. The phase transitions proceed most rapidly when there are nuclei or "seeds" for the bubbles or crystals to form on. Each water droplet in saturated air, for instance, formed on a minute dust particle or on an electrical ion left by the passage of an ionizing particle.

SPECIFIC HEATS ACCORDING TO THE ATOMIC MODEL

In Chapter 12, we observed many regularities in the values for the specific heats of materials. At room temperature, monatomic gases have specific heats of about 3 calories per mole·K. Diatomic gases have specific heats of about 5 calories per mole·K. Most metals have specific heats of about 6 calories per mole·K. On the other hand, specific heats change with temperature. A good atomic theory should be able to explain not only the regularities, but also the exceptions.

The specific heat of a material is the ratio between heat added and temperature rise: $c = Q/\Delta T$. The heat added to the material can do two things: it can change the internal energy of the system, and it can do external work. In terms of the symbols for the first law, $Q = \Delta U + W$.

Since we assumed that the atoms or molecules in our model had no size and no internal structure, the total internal energy of the gas is given by the translational kinetic energy of the particles. That kinetic energy is equal to

$$U = N \tfrac{1}{2} m \overline{v^2} = N \tfrac{3}{2} kT = \tfrac{3}{2} RT$$

Therefore, for such a simple, model gas, if there is a change in temperature, ΔT, the change of internal energy is equal to

$$\Delta U = \tfrac{3}{2} R \, \Delta T$$

Any external work done must be produced by the expansion of the gas: $W = P\Delta V$. If heat is fed into a gas at constant volume, then no external work can be done and the specific heat must be labeled c_V. The specific heat at constant volume must then be equal to $c_V = Q/\Delta T = \Delta U/\Delta T + 0 = \frac{3}{2}R\Delta T/\Delta T = \frac{3}{2}R = 3$ calories/mole·K.

> **Question 13-15**
>
> Why did the specific heat come out in terms of moles instead of grams? How was R in terms of calories derived from Boltzmann's constant, k, in terms of joules?

The Universal Gas Constant

$$R = N_0 k$$

Where N_0 is Avogadro's number (the mole), and k is Boltzmann's constant.

$$R = 8.3 \text{ J/mole·K}$$
$$= 2.0 \text{ cal/mole·K}$$

If a gas is allowed to expand while heat is being fed into it, then extra heat will be required to raise the temperature one degree. Under the very special condition that the expansion takes place at constant pressure, the amount of work done is very simple to calculate. Remember that the ideal gas law is $PV = nRT$. If the pressure of the gas is kept constant and yet the temperature rises, it must be that the volume increases: $P\Delta V = nR\Delta T$. The external work done is equal to $P\Delta V$. Therefore the first law becomes

$$Q = \Delta U + W = \tfrac{3}{2}nR\Delta T + nR\Delta T$$

For one mole:

$$Q = \tfrac{3}{2}R\Delta T + R\Delta T$$

The specific heat at constant pressure must then equal

$$c_P = (Q/\Delta T)_P = \tfrac{3}{2}R + R = \tfrac{5}{2}R = 5 \text{ calories/mole·K}$$

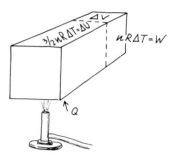

According to our simple model, all gases must have the same values for specific heat and the difference between the specific heat at constant pressure and the specific heat at constant volume is simply equal to 2 calories/mole·K. Look at the table on page 291 which gives the values of specific heats for a number of gases. Let's compare those experimental values with our theory. To start out with success, note first that the difference $(c_P - c_V)$ does indeed have the value of 2 cal/mole·K. Furthermore, argon and helium have exactly the specific heat values that we derived. They apparently act like our simple model. However, the gases that are diatomic or that have even more complicated molecules apparently require a more complicated model.

It must be that energy fed into molecules can be absorbed in some way other than translational kinetic energy. That's not surprising! Two atoms bound together ought to be able to rotate around their center of mass and also ought to vibrate back and forth as if they were held by a spring. Each mode of motion should be able to absorb energy. Perhaps the internal energy, U, should consist of $N\tfrac{1}{2}mv^2$ (translation) + $N\tfrac{1}{2}I\omega^2$ (rotation) + $N\tfrac{1}{2}kx^2$ (vibration). The problem is, how does the energy divide up among the various modes? The translational kinetic energy of the whole molecule can be shared by motion in three perpendicular directions. Surely, in a gas the energy must be *equally* shared among these three directions. A diatomic molecule can rotate like a dumbbell around two axes perpendicular to its

Diatomic molecule vibrating and rotating.

length. There should be as much energy used for one rotation as the other. As the diatomic molecule vibrates, it has both kinetic and potential energy; on the average there is an equal amount of energy in each mode.

In the classical theory of mechanics (before quantum mechanics was known), it was assumed that the principle of equipartition of energy held for all these energy modes, providing that the system could freely interact with all its components for a long enough time. Each such mode is called a "degree of freedom." For instance, translational motion has three degrees of freedom since motion can be broken down into three independent directions, such as the x, y, z-axes. Vibrational motion along a single axis has two degrees of freedom—for kinetic and potential energy. The principle held that in a system of many interacting objects, where all these motions could influence each other and share energy, each degree of freedom would, on a time average, share equally in the total energy. Not only would $\frac{1}{2}m\overline{v_x^2} = \frac{1}{2}m\overline{v_k^2}$ for the same molecule, but also $\frac{1}{2}m\overline{v_A^2} = \frac{1}{2}m\overline{v_B^2}$ for two different molecules in the same gas. Furthermore, $(\frac{1}{2}I_A\overline{\omega_A^2})_x = (\frac{1}{2}I_B\overline{\omega_B^2})_y = (\frac{1}{2}I_B\overline{\omega_B^2})_x$ for rotations of different molecules in the gas about each axis. The principle of equipartition of energy was a cornerstone of classical mechanics in the nineteenth century, even though the truth of the principle is not immediately obvious, nor easy to prove.

If there is really equipartition of energy in a gas, each of the possible modes of motion of molecules should share equally in the heat energy fed into the gas. It is at this point that our simple classical model begins to break down. The next step of explanation suddenly requires quantum mechanics. When we look at the experimental data for the specific heats at constant volume of diatomic molecules, we see that at room temperature H_2, O_2, N_2, and HCl apparently have five degrees of freedom. (Note that each degree of freedom is responsible for a contribution of 1 cal/mole·K to the specific heat value.) From other experiments and from quantum mechanics calculations, we know that the two extra degrees of freedom of these diatomic molecules are related to the possible rotations around two axes. It is reasonable enough that as the molecules run into each other, they set each other spinning.

The specific heat of even the simplest molecule, H_2, cannot be explained on the basis of classical mechanics. As you can see from the graph, it appears that at low temperatures some of the degrees of freedom of the hydrogen molecule are frozen out. Since the molecules are surely free to move about in space, it must be that for some reason their rotational and vibrational modes cannot share in the energy distribution. There appears to be a threshold effect with new degrees of freedom gradually becoming possible as the temperature rises. That is just what happens. Rotations can only occur with certain discrete amounts of energy, because angular momentum can exist only as integral multiples of Planck's constant. The minimum amount of energy needed for *vibrations* is usually even larger. The equipartition of energy theorem assumes that any mode of motion can have any amount of energy, including very small amounts, and that the energy in that mode can change continuously. Since rotational and vibrational energies are quantized, they cannot share in the division of energy until the available energy in a collision is greater than the quantum threshold.

Question 13-16

Why can't the diatomic molecule spin on the axis that goes through the two atoms? For that matter, why can't helium atoms have rotational modes of motion about each of three perpendicular axes? After all, the kinetic energy of a rolling marble is part translational, part rotational.

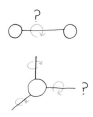

The molar specific heats of most metals have about the same value—6 cal/mole·K. According to our simple model, there must be 6 degrees of freedom for each atom. How can this be? Each atom is locked in place in the metallic crystal. It can oscillate, of course, and in three independent directions. Those modes could provide three degrees of freedom. As we have seen, however, each mode of oscillation absorbs twice the energy allotted to a free motion. One unit of energy goes to the kinetic energy of oscillation, and the other unit goes to the potential energy. Unfortunately, the classical theory has trouble explaining why the specific heats of metals should depend on temperature. As you can see in the diagram, the regularity that we observe occurs at room temperature and even then only if we ignore the elements that are low in the periodic table.

Each bound atom can vibrate in three dimensions. In each dimension there is one degree of freedom for kinetic energy and one degree of freedom for potential energy.

There is an even more embarrassing problem connected with explaining the specific heats of metals on a classical basis. To explain how metals are both good electrical conductors and good thermal conductors, we claimed that the valence electrons from each atom were free to roam the crystal carrying both electric current and also thermal energy. These electrons might be acting just like a gas in the metal crystal. If that is the case, why don't the free electrons absorb heat energy, thus adding three more degrees of freedom to the six possessed by each parent atom? The actual contribution to the specific heat by electrons is very small. They can spread thermal energy but they absorb almost none for themselves. This strange situation is well explained by modern quantum mechanics.

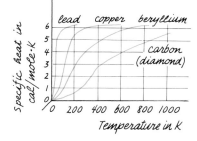

Although the details about how thermal energy is absorbed are evidently very complicated, the value of the specific heat for a material still gives us useful information about the complexity of the microstructure. Take another look at the table on page 291 and notice how the molecules that are very complicated have large specific heats. Some of these values are still surprising, even on a qualitative basis. For instance, notice the large specific heat of water, which has a fairly simple molecule.

Question 13-17

Compare the molar specific heat of water with that of other materials that have the same molecular complexity (two atoms of the same kind linked to a third atom of a different kind). Why should water have such a large specific heat? Where does the extra heat energy go if it doesn't raise the temperature?

THE MICROSTRUCTURE OF MATTER AND THE SECOND LAW OF THERMODYNAMICS

TURNING HEAT INTO WORK

There is no trouble at all in turning mechanical energy into heat or into internal energy. Simply let a heavy block slide down an incline. The gravitational potential energy turns into kinetic mechanical energy. Friction brings the block to rest, raising the temperature of the surfaces and causing heat to flow to the cooler surroundings.

Now let's try to reverse the process. Can we let heat flow into the block and make it slide back uphill? That's not so easily done, of course. The original organized motion of the block as a whole has been turned into the random chaotic motion of the molecules in the block and in the inclined plane.

> **Question 13-18**
>
> What are some devices that produce organized effort out of chaos? A balloon? A water pressure tank? A political leader?

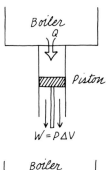
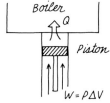
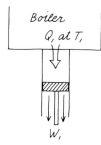
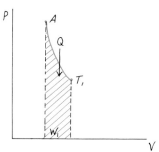

At constant T, $P \propto 1/V$
Stage 1

The Industrial Revolution came about because people learned how to extract work from heat. Heat engines turn most of the generators of our electric power plants and also propel many of our vehicles. Let's analyze a simple version of a heat engine to see how efficient the process can be.

For purposes of analysis, we can study any kind of a system that takes heat from a boiler at a particular temperature and uses it to perform work. The question is: Does all of the heat, Q, turn into useful work, W? If so, the conversion efficiency is 100%. Such an ideal machine would have to be friction-free. It would also have to operate in a way that is technically called "reversible" (in thermodynamics, reversibility of a process corresponds to a friction-free situation in ordinary mechanics). In a reversible process involving heat, everything happens slowly and uniformly so that the temperature and pressure of the material being considered is uniform throughout. For instance, in a cylinder of gas with a piston, a reversible expansion is one where the piston is moving so slowly that there are no shock waves in the gas, and the temperature and the pressure of the gas at any instant are the same throughout the whole volume. Why is a process that satisfies these conditions called reversible? Suppose heat is provided to the gas in the cylinder, forcing it to expand slowly. The piston will be shoved out, performing useful work. Now if the process can be truly reversed, it should be possible to shove the piston back, using the same amount of work, compressing the gas, and driving heat out of the cylinder. The process is indeed possible in principle, providing that there is no friction in the cylinder and that shock waves in the gas don't spread the energy through the walls of the cylinder instead of doing work on the piston. An explosion of gasoline and air in the cylinder of a car engine is not a reversible process since, of course, when the piston moves back in the cylinder, it cannot restore the exhaust gases to their original form of gasoline and air.

The simplest ideal heat engine that we can study consists of an ideal gas in a cylinder with a piston. We could use other materials that expand when heat is provided but for an ideal gas we already know the equations. At first our task seems very simple. Let heat flow into the gas in the cylinder and the gas will expand. The piston will be shoved out, exerting a force through a distance, thus doing a known amount of work. We can make the problem even simpler by sending the heat in at constant temperature. We showed how to do this on page 292 in Chapter 12. A metal wall of the cylinder could be fastened to the boiler. The piston is then allowed to move so slowly that the gas is maintained at the temperature of the boiler. In Chapter 12 this was defined as an isothermal process.

> Question 13-19
>
> If the temperature of the gas in the cylinder is constant, what does the first law of thermodynamics say about the relationship between the heat, Q, and the work done, W?

It might seem that with this process we have extracted order out of chaos with an efficiency of 100%. The heat that left the boiler, Q_1, turned into the work done by the piston, which corresponds to the area under the curve on the P-V graph. No energy was absorbed by the gas itself since its temperature did not change. Unfortunately, we can't repeat the process using the same cylinder because the piston is now shoved partway out. We could get a new piston, but that wouldn't be a practical way to manufacture a heat engine. The thing to do is to shove the piston back to its original position. Then we would have a cyclic process, which could be repeated over and over again. The problem is, how can we shove the piston back without using up the same work that we just gained? The isothermal process that we used was by definition reversible. If the same amount of work is done on the piston, then the gas can be compressed at constant temperature back to its original small volume and high pressure. In the process, heat would be driven out and would be absorbed back by the boiler. We would be right back where we started from with no net heat extracted and no net work done.

Here is a way to produce a cyclic process that will restore the cylinder, the piston, and the gas to their original conditions and yet not retrace the original steps. After Q_1 has been absorbed from the boiler in the isothermal process, insulate the cylinder from the boiler so that no heat can enter or leave. Then let the piston continue to expand, producing more outside work. The energy will be drawn from the internal energy of the gas as it drops in pressure and also in temperature. This second part of the cycle, which takes place without heat entering or leaving, is called an adiabatic process. Incidentally, note that there is nothing sacred about our choice of using first an isothermal process followed by a completely adiabatic process. Each of the steps could have been partially isothermal and partially adiabatic or neither. The only requirement is that the expansions take place slowly enough so that they are reversible. The choice of pure isothermal and adiabatic processes merely makes the computations easier.

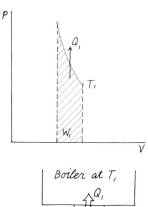

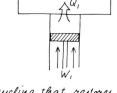

A recycling that restores the system, but yields no work.

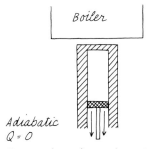

Adiabatic $Q = 0$
Temperature drops from T_1 to T_2.

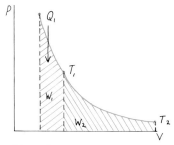

Stage 2

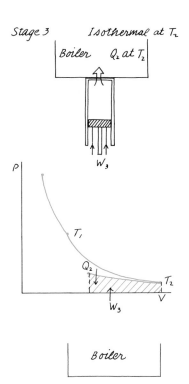

Stage 3 Isothermal at T_2

Boiler Q_2 at T_2

W_3

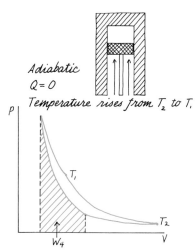

Boiler

Adiabatic $Q=0$

Temperature rises from T_2 to T_1

W_4

Stage 4
W_4 in, raises U of gas from T_2 to T_1

Question 13-20

In the first stage of the heat engine cycle, an amount of heat, Q_1, was extracted from the boiler and an equal amount of work, W_1, was done by the piston. In the second stage, no heat was used but the piston continued to do work, W_2. At this point it appears that the conversion of heat to work has been done with an efficiency of greater than 100%, since $W_1 + W_2$ must be greater than Q_1. Have we really obtained something for nothing?

After the first two stages, our heat engine consists of a piston shoved out as far as it can go and the gas in the cylinder at low pressure and low temperature. If we are going to keep on using the heat machine, we must get the gas and the cylinder back to their original condition. Once again, for the sake of easy computation, the easiest steps to consider are isothermal and then adiabatic. For the third step we must shove the piston back doing work and compressing the gas from pressure P_3 to P_4. However, if this compression is to take place at constant temperature, the heat must be expelled to some reservoir at that temperature. We could do this by connecting one end of the cylinder through a metal wall to a low temperature bath or radiator. The amount of heat expelled, Q_2, must be just equal to the outside work done on the piston. On the P-V graph, that work corresponds to the area under the compression curve.

The fourth stage of the cycle is once again adiabatic with the cylinder walls insulated so that no heat can enter or leave. An amount of work, W_4, is done on the piston and all of this work shows up as the internal energy of the gas measured by its temperature. With this fourth stage of the cycle, we have restored the piston to its original position and the gas to its original pressure, volume and temperature. The energy balance is as follows: an amount of heat, Q_1, was extracted from the boiler at temperature T_1. The work done by the piston was $W_1 + W_2$. In restoring the heat engine to its original condition, an amount of work, $W_3 + W_4$, had to be done. It was also necessary to expel an amount of heat, Q_2, at the lower temperature, T_2. The net work done by the heat engine, W, was $W_1 + W_2 - W_3 - W_4$. As required by the conservation of energy as expressed by the first law of thermodynamics, this net work done must be just equal to the net heat used: $Q_1 - Q_2 = W$. No energy is really lost but, on the other hand, we were not able to convert all the original heat energy to useful work. Furthermore, the expelled heat, Q_2, is only available at a lower temperature. Typically, in an electric power plant, this heat is expelled at the temperature of the local cooling water provided by a river or a lake. In most cases, the heat is wasted, although in some plants it is sent out at temperatures high enough to heat buildings. Nevertheless, as far as being useful for producing mechanical energy, the heat has been degraded below a useful level. The conversion efficiency of our ideal heat machine is equal to the net work done divided by the original heat energy used, Q_1. Since the mechanical work done is equal to $Q_1 - Q_2$, the efficiency is equal to $(Q_1 - Q_2)/Q_1$.

THE CARNOT CYCLE

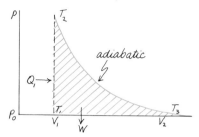

Possible Heat-Work Cycle

$Q_1 = \Delta U = C_v(T_2 - T_1)m$

Question 13-21

Perhaps our heat engine is inefficient because we have chosen a bad sequence of expansions and compressions. Perhaps there is some other way to restore the cylinder and piston to the original position without doing any work. For instance, why not just let the gas expand to atmospheric pressure? Then open a valve and shove the piston back without compressing the gas and thus without doing any work. When the cylinder is at its original volume, the valve could be closed and heat could be admitted to raise the temperature and pressure of the gas to the original condition. Wouldn't this cycle be 100% efficient in changing heat into work?

THE CARNOT CYCLE

There must be an infinite number of ways to turn heat into work. Back in 1824 a young Frenchman named Sadi Carnot solved the general problem of finding the conversion efficiency for any heat engine, using any cycle. Note that he did this about 25 years before the general acceptance by scientists that heat is a flow of energy. The particular cycle of steps analyzed by Carnot is the one that we have just described. It is called the Carnot cycle. What Carnot demonstrated was that any *reversible* cycle has certain universal features that apply to all other *reversible* cycles. Let's take a look at the special features of the Carnot cycle and then see why Carnot argued that these features are characteristic of all ideal heat engines.

In this schematic drawing the details of pressure and volume of gas in a cylinder are stripped away. The basic operation of the heat engine is to extract an amount of heat, Q_1, from a boiler at temperature T_1 and produce an amount of useful work, W. However, to complete the cycle it is necessary to expel a small amount of heat, Q_2, at a lower temperature, T_2. Since each step of the cycle is completely reversible, the whole cycle could be run backwards to create a refrigerator. As shown in the schematic, the refrigerator would require an amount of work, W, to extract a small amount of heat, Q_2, at the low temperature T_2 and expel the sum of the work energy and the low temperature energy as the heat Q_1 at the higher temperature T_1. In our home refrigerators, for instance, W is provided by the electric motor and Q_2 is extracted from the freezing compartment at the low temperature T_2. The total energy, Q_1, is then expelled out the back of the refrigerator at temperature T_1. The room air takes away this heat from the radiator coils.

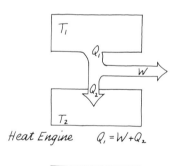

Heat Engine $Q_1 = W + Q_2$

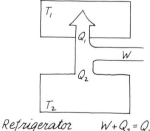

Refrigerator $W + Q_2 = Q_1$

On the P-V graph for the Carnot cycle, the work done corresponds to the area enclosed by the cycle. Graphically, it is obvious why the return path of the cycle must be different from the expansion part. Otherwise there would not be an enclosed area and hence no net work would be done. Regardless of the detailed steps of the conversion cycle, the area inside the graph corresponds to the useful work obtained.

We claimed that one reason for choosing a sequence of isothermal and

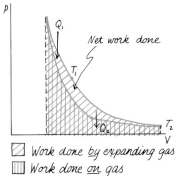

☐ Work done by expanding gas
▨ Work done on gas

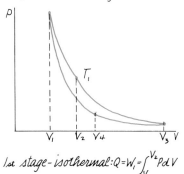

1st stage - isothermal: $Q_1 = W_1 = \int_{V_1}^{V_2} P\,dV$

$P = \dfrac{nRT_1}{V}$ $W_1 = nRT_1 \int_{V_1}^{V_2} \dfrac{dV}{V} = nRT_1 \ln \dfrac{V_2}{V_1}$

2nd stage adiabatic: $W_2 = \Delta U = nc_v(T_1 - T_2)$

3rd stage - isothermal: $-W_3 = -nRT_2 \ln \dfrac{V_3}{V_4}$

4th stage - adiabatic: $W_4 = \Delta U = nc_v(T_1 - T_2)$

$W_2 - W_4 = 0$

$\dfrac{Q_1}{Q_2} = \dfrac{nRT_1 \ln V_2/V_1}{nRT_2 \ln V_3/V_4} = \dfrac{T_1}{T_2} \dfrac{\ln V_2/V_1}{\ln V_3/V_4}$

then adiabatic steps was that it made computations easier. This is indeed the case if we want to calculate the amount of work done in a cycle. The work done for a small change of volume is $W = P\Delta V$. To find the total work done as the volume of the gas expands and its pressure drops, we have to know how the pressure depends on the volume. Those formulas are well known and relatively easy to use for isothermal and adiabatic expansions of ideal gases. The details are shown in the margin. The final result is very simple and very important for an ideal gas used in this particular cycle: $Q_1/Q_2 = T_1/T_2$. The ratio of the heats extracted and expelled is just equal to the ratio of the absolute temperatures of the heat source and the heat sink. We can express the efficiency of the machine in a new way:

$$\text{Efficiency} = \dfrac{W}{Q_1} = \dfrac{Q_1 - Q_2}{Q_1} = 1 - \dfrac{Q_2}{Q_1} = 1 - \dfrac{T_2}{T_1} = \dfrac{T_1 - T_2}{T_1}$$

In the special case of Carnot's cycle it appears that the efficiency of conversion of heat into work depends only on the temperature of the source and of the sink (the reservoirs at T_1 and T_2). Note that although we have assumed that the gas in the cylinder is an ideal gas, apparently the process does not depend on how much gas there is or on the starting pressure or volume. The efficiency is a function only of the two temperatures.

Question 13-22

The boiler of a modern steam plant operates at about 550°C. The exhaust heat might be released into a lake or river at about 20°C. If the heat machine followed an ideal Carnot cycle, what would be the operating efficiency?

Now that we have seen some of the features of the Carnot cycle, we return to the question of whether any other heat machine can be more efficient. First, let's give the competing machine the best possible chance, and assume that it uses only reversible processes. Remember that these

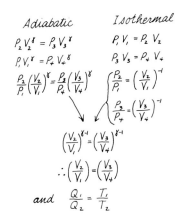

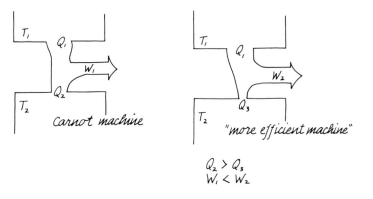

correspond to friction-free situations in mechanics. No energy is wasted in shock waves or lost as heat out the side walls. As long as the processes are reversible, our arguments will apply to any kind of machine using any kind of material. For instance, the machine might use real gas instead of an ideal gas or might even be made up of rubber bands. (Heat machines using the stretch and contraction of rubber bands have indeed been made.) Suppose that we have such a heat machine and that it is more efficient than a Carnot cycle in taking heat from a boiler at T_1, performing work and then emitting heat at a lower temperature T_2. For a given amount of heat extracted, the more efficient machine would produce a larger amount of work and expel a smaller amount of heat at the lower temperature.

Now let's use the more efficient one as a refrigerator. We can operate it with the work produced by the Carnot machine. With that much work, another Carnot machine would just be able to take the expelled heat, Q_2, and shove it up the temperature hill to the boiler at T_1. The original conditions would be completely restored. However, if that same amount of work is used on a more efficient engine acting as a refrigerator, a larger amount of heat from the low temperature, T_2, can be taken up to the boiler temperature of T_1.

The result would be to take more and more energy from the lower temperature and pump it to the higher temperature. No external work would have to be supplied to the pair of machines since the Carnot machine gets its energy from the boiler and supplies it to the more efficient machine.

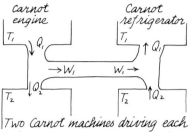

Two Carnot machines driving each other. No net work done or heat transferred.

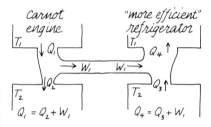

Carnot engine providing work to drive a "more efficient" engine as a refrigerator. Net work done is zero ($W_1 = W_1$) Net heat transferred from T_2 to higher $T_1 = Q_3 - Q_2$

Question 13-23

Wouldn't this combination of machines disobey the law of the conservation of energy? After all, they operate each other with the net result of pumping more and more energy to the boiler.

THE SECOND LAW OF THERMODYNAMICS

The first law of thermodynamics is an assertion that heat is a form of energy and that energy is conserved. Now we have come to the second law of thermodynamics, which can be expressed in many different ways. One of the simplest statements of the second law is that heat cannot by itself flow from a low temperature region to a higher temperature. The second law is not something that can be proved. It is one of the laws of impotence—a statement that something cannot happen. Another such law is that no energy or information or object can be transmitted faster than the speed of light. Although laws of impotence cannot be proved, they do have specific consequences. If all the consequences prove to be valid experimentally, then we have faith in the original hypothesis. In the case of the second law, we will see a number of these consequences. Furthermore, it is a matter of common experience that heat always flows from hotter things to colder things. Notice that there would be no loss of energy if it went the other way. The first law of thermodynamics would be satisfied if we could run our industrial machinery with heat energy taken from the ocean. Such a process would simply cool down the ocean a little. But the heat would have to flow from the lower temperature of the ocean to the higher temperature of a boiler. It just doesn't happen.

Now note one of the consequences of the second law of thermodynamics. If heat cannot flow from cold to hot without work being done, then the combination heat machines that we analyzed will not work. *Therefore, there cannot be any heat machine that is more efficient than the Carnot cycle.* Furthermore, all machines operating with reversible processes between temperatures T_1 and T_2 must have the same efficiency. We do not have to analyze each and every possible cycle. The Carnot cycle or any cycle using reversible processes yields the maximum efficiency possible. A machine using irreversible processes would necessarily be less efficient since it would be wasting energy in friction-like processes.

Since the efficiency of the Carnot cycle depends only on the operating temperatures and not at all on the mechanism of the machine, then we have a new definition of temperature. Remember that in Chapter 12, when we first defined temperature, we worried about the fact that a temperature scale using one material is slightly different from the temperature scale using another material. We defined an absolute temperature scale based on the properties of an ideal gas. Experimentally, an ideal gas is approximately the same as any gas operating at low pressure and at a temperature well above its condensation point. Now it appears that the temperatures defined in that way also define the efficiency of heat engines that do not depend on the properties of the materials at all. In principle, we could set up an absolute temperature scale in terms of some ideal heat engine working between two temperatures. If a heat engine operated between any temperature and 0° absolute, then its efficiency would be 100%.

$$\text{Efficiency} = \frac{T_1 - T_2}{T_1} = \frac{T_1 - 0}{T_1} = 100\%$$

If we then arbitrarily choose a particular temperature as 100°, a temperature of 200° is defined to be such that a heat machine operating between that temperature and the defined 100° has an efficiency of 50%.

$$\text{Efficiency} = \frac{200° - 100°}{200°} = 50\%$$

In practice, temperature scales are not determined experimentally by measuring the efficiencies of heat engines. However, *in principle*, temperature can be defined in a way that is independent of the properties of the thermometer.

The temperatures defined in this way (sometimes called Kelvin temperatures) must be the same as the absolute temperatures that we have already deduced in terms of the properties of ideal gases. Note that we used the ideal gas equations to show on page 326 that $Q_1/Q_2 = T_1/T_2$. Therefore, the efficiency of a reversible heat engine using ideal gas is $(T_1 - T_2)/T_1$, where T is the absolute temperature defined in terms of ideal gases. However, since *all* reversible engines working between T_1 and T_2, regardless of working material, must have the same efficiency, the absolute temperature must be the same as the Kelvin temperature defined by the operation of heat engines.

Since a large share of the energy that we use in this world comes from heat engines, there is obvious practical importance in making the process as efficient as possible. Not only is it wasteful to throw away part of the heat energy at a lower temperature, it is also a serious nuisance. The discarded heat pollutes our rivers and lakes, changing the natural ecology of many places. Carnot attacked this problem long ago because of its practical implications. Real machines are always less efficient than ideal ones. But at least we can use the ideal case as a goal and as a guide in determining the important factors. Large electric power plants use very high temperature, high pressure steam as the operating gas. The expanding steam does not push a piston but instead turns a turbine. The pressure and temperature of the steam decrease as it flows through the turbine until finally the steam is condensed, producing a vacuum at the far end. Some of the steam is led off at earlier stages, reheated and sent back to the original boiler. To keep the final stage cold so that the final pressure will be low, cooling water is circulated through radiators. The expelled heat is then carried out into the stream or lake at the temperature of the source of water, typically around 20°C. In order to maximize the efficiency of such a machine, the operating temperature of the boiler, T_1, should be as high as possible. Why not make it as high as the fires can reach—perhaps 2000°C? The trouble is, steam at such a temperature would have an extremely high pressure and also be very corrosive. No practical boiler could stand it. An engineering compromise is reached so that the steam temperature is as high as can be reached without excessive damage to the boiler or early stages of the turbine. Most large power plants operate at a boiler temperature of about 1100°F, which is about 600°C. With a cooling temperature of 20°C, the ideal efficiency of such a machine is

(Courtesy General Electric)

$$\text{Efficiency} = \frac{873 \text{ K} - 293 \text{ K}}{873 \text{ K}} = 66\%$$

Because of friction of the rotating system and other losses, the actual efficiency of such a power plant is less—usually under 40%. This means that for every joule of electricity produced, $2\frac{1}{2}$ joules of energy has to be supplied by the oil or coal that is burned. Furthermore, the extra $1\frac{1}{2}$ joules of energy is discarded into the cooling water, raising the temperature of the river or lake downstream.

Thermal pollution is a serious concern as we build more and more power plants. Nuclear power plants currently being used are worse offenders than those powered by fossil fuels. The reason is that the fuel elements in nuclear plants cannot be allowed to go to very high temperatures lest they suffer corrosion and spill their radioactivity. Nuclear powered boilers usually operate below 600°F, which corresponds to about 330°C. The ideal efficiency of a machine operating between that temperature and 20°C is about 50%. Because of other losses, the practical efficiency for many nuclear power plants is only about 30%. For every joule of electricity produced, 3 joules have to be supplied and 2 joules have to be drained away in the cooling water.

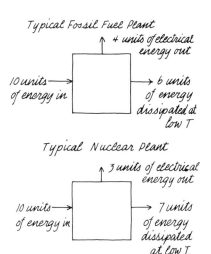

■ Optional

Most of us have heat engines in our homes—refrigerators or air conditioners. By supplying outside energy, usually in the form of electricity, these machines pump the heat from a cold region to a hot region. Let's calculate the efficiency of such a machine, assuming that it is an ideal heat engine being run backward. When converting *heat* into *work*, the efficiency is

$$\eta = \text{efficiency} = \frac{Q_1 - Q_2}{Q_1} = \frac{W}{Q_1} = \frac{T_1 - T_2}{T_1}$$

When we use the machine as a refrigerator, however, we want to pump heat, Q_2, for as little work, W_1, as possible. The *quality factor* for a *refrigerator* is thus

$$\frac{Q_2}{W} = \frac{Q_1 - W}{W} = \frac{Q_1}{W} - 1 = \frac{1}{\eta} - 1 = \frac{T_1}{T_1 - T_2} - 1 = \frac{T_2}{T_1 - T_2}$$

According to this analysis, the quality factor is proportional to the temperature from which we are pumping the heat, and inversely proportional to the temperature difference through which we are pumping it. Reasonably enough, it would be expensive to pump heat from a very low temperature to a very high temperature. It is also slightly cheaper to pump heat from 20°C to 30°C with an air conditioner than it is to pump it from −10°C to 0°C with a freezer.

A typical air conditioner for a bedroom is rated at 5000 BTU's per hour with a consumption of 750 watts. Let's see how the quality factor for this machine compares with that for an ideal heat engine.

$$\frac{Q_2}{W} = \left(\frac{5000 \text{ BTU}}{\text{hr}}\right)\left(\frac{\text{s}}{750 \text{ J}}\right)\left(\frac{1 \text{ hr}}{3600 \text{ s}}\right)\left(\frac{1055 \text{ J}}{1 \text{ BTU}}\right) = 1.95$$

The actual quality factor for this machine is thus equal to about 2. If the room temperature is to be kept at 20°C and the outside temperature is 30°C, then the ideal quality factor for a machine working reversibly is 29:

$$\frac{Q_2}{W} = \frac{293 \text{ K}}{303 \text{ K} - 293 \text{ K}} = 29.3$$

As a practical matter, even if the temperature difference between room and outside is only 10C°, the actual temperature difference between refrigeration coil and radiator exhaust will be more like 30 or 40C°. That would bring the ideal quality factor down to less than 10. Even so, it is clear that an actual machine is using several times more energy than an ideal reversible one. The extra electrical energy is, of course, turning into heat that also has to be blown out into the surroundings, preferably outdoors.

ENTROPY

A complex system of interacting parts, such as a gas of many molecules, has a certain amount of internal energy. We rarely are able to measure that energy directly and, indeed, are usually concerned only with changes in the energy. Even these *changes* in the internal energy are actually measured in terms of such quantities as pressure, volume, and temperature. For a particular number of molecules of a gas, the measurement of any two of these quantities determines the third and also characterizes the internal energy. No matter what changes the gas undergoes, if it ends up with the same pressure, volume, and temperature that it had to begin with, then its internal energy is also the same.

ENTROPY

There is another way to characterize the state of a system. Suppose that the gas in a box were arranged with all the high energy molecules on one side and all of those with lower energy on the other side. The internal energy of the system would be the same as if the molecules were mixed in the regular fashion. There is no conservation law that forbids such a momentary arrangement, but the *probability* of it happening is very small. We ought to have some way to characterize the possibility of having a particular arrangement of the interacting particles of the system. The measure of this probability is called entropy.

We usually cannot measure the entropy of a system directly any more than we can measure internal energy. We are mainly concerned with *changes* in entropy and these can be calculated in terms of such variables as pressure, volume, and temperature. The formal definition of entropy in terms of probability is $S = k \ln W$. The entropy, S, is equal to Boltzmann's constant, k, times the natural log of the probability, W, of the system having that particular arrangement. If the system consisted of 10 coins mixed and cast over and over again, then the most probable arrangement for each throw would be 5 heads and 5 tails. An arrangement of 10 heads or 10 tails would be possible, is not forbidden, but would be unlikely. $[p = (\frac{1}{2})^{10}]$. With the molecules of a gas, each possible combination of the positions and velocities of the molecules is subject to the constraint that the total internal energy must be constant. Each possible arrangement that satisfies that condition has equal probability but there are many more arrangements where the molecules are uniformly distributed in space and have velocities corresponding to the standard distribution than there are arrangements where the molecules are all off on one side or the other. The equilibrium distribution shown in the top diagram is the most likely one because it can be obtained with the greatest number of molecular arrangements. In fact, for a very large number of molecules, the number of ways in which any other appreciably different distribution could be arranged is insignificantly small. Note that if the probability, W, of the equilibrium distribution is larger than for any other distribution, then the entropy as we have defined it is maximum when the system is in its normal equilibrium.

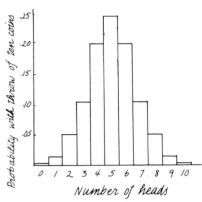

Number of heads

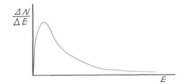

Maxwell-Boltzmann distribution
maximum probability-maximum entropy

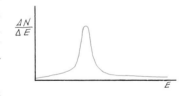

Highly unlikely distribution-low entropy

The distribution graphs show the number of molecules that have a particular kinetic energy, as a function of energy. A detailed graph of the Maxwell-Boltzmann distribution is shown on page 309.

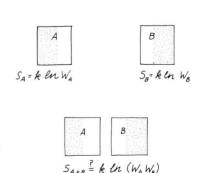

Question 13-24

The entropy is defined to be proportional to the log of the probability. Why not simply define it as proportional to the probability itself? To answer this, suppose that you had two boxes of gas: A and B. The entropy of the gas in the first box is $S_A = k \ln W_A$ and the entropy of the second box is $S_B = k \ln W_B$. If we don't mix the gases, the combined entropies must be simply the sum of the two: $S_{A+B} = S_A + S_B = k \ln W_A + k \ln W_B = k \ln (W_A W_B)$. Is this the way that probabilities should be combined?

Since it is not practical to calculate the detailed probabilities of arrangements of complicated systems, we should find some other way to characterize changes of entropy. We can find a clue about how to do this by reexamining the Carnot cycle. In any reversible heat cycle, the working substance ends up in its original condition, even though during two of the steps heat was added or taken away. As we have seen, if the original conditions in the system are to be restored, the heats and temperatures must be related by

$$\frac{Q_1}{T_1} = \frac{Q_2}{T_2}$$

We now give another definition for the change of entropy and will then have to make it plausible that this definition has the same characteristics as the first one. If a system has heat, Q, added to it at a temperature, T, in a *reversible process*, then its entropy is increased by $\Delta S = Q/T$.

According to this definition, the gas in the cylinder of our reversible heat engine will have an increase of entropy during the first isothermal expansion. That increase will be equal to $\Delta S_1 = Q_1/T_1$. Meanwhile, the boiler will have lost that much entropy. The total system undergoes no entropy change during the adiabatic expansion or compression, since there is no heat loss or gain. During the third stage of isothermal compression, the gas in the cylinder loses entropy equal to $\Delta S_2 = Q_2/T_2$. The cold heat absorber must gain that much entropy. At the end of the complete cycle, the gas in the cylinder has its original entropy since $Q_1/T_1 = Q_2/T_2$. The *total* entropy of the system has also not changed, although the boiler lost entropy, Q_1/T_1, while the cold absorber gained entropy, Q_2/T_2. Entropy has been transferred from the hot source to the cold sink but for the entire system there has been neither gain nor loss. However, for a *nonreversible* heat engine, the amount of heat absorbed from the boiler must be large enough to furnish the energy for frictional and other losses. In that case, Q_1/T_1 is greater than Q_2/T_2. At the end of the cycle, the cooling water ends up with extra heat having been added to it and the overall entropy of the entire system has increased. In fact, a reversible process can be defined by the fact that it creates no change in entropy in an isolated system. An irreversible process always creates an *increase* in entropy. This feature can be used as another statement of the second law of thermodynamics.

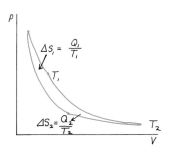

In any process within an isolated system, the total entropy will either increase or stay the same. This statement of the second law is equivalent to the first one, which says that heat cannot by itself flow from a low temperature region to a higher temperature. For instance, if a pair of heat engines working together could pump a quantity of heat Q from a low temperature sink to a high temperature boiler with no other work being done, then the sink would lose entropy equal to Q/T_2. The boiler would gain entropy equal to Q/T_1. But since T_1 is greater than T_2, Q/T_2 is greater than Q/T_1. The loss of entropy by the sink would be greater than the gain of entropy by the boiler. Hence the entropy of the whole system would have been reduced. Since heat cannot flow to higher temperature without work being done, it must be that the entropy of an isolated system cannot be reduced. A corollary to this argument is that in any process involving an irreversible reaction, the overall entropy of an isolated system increases. It is only during reversible processes that the entropy of an isolated system stays the same.

▪ Optional

The equivalence, in the case of an ideal gas, of the two definitions of entropy:

$$P = NkT/V$$

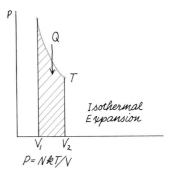

Isothermal Expansion

$P = NkT/V$

Work done during an isothermal expansion

$$= \int_{V_1}^{V_2} P\,dV = NkT \int_{V_1}^{V_2} \frac{dV}{V} = NkT \ln\left(\frac{V_2}{V_1}\right) = Q$$

$$\Delta S_{gas} = \frac{Q}{T} = Nk \ln\left(\frac{V_2}{V_1}\right) \quad \text{(derived from gas laws)}$$

Entropy in terms of probability, W: $S = k \ln W$
The probability of any *one* molecule being confined in V is proportional to V. With N molecules confined in V, $W \propto V^N$ (see answer to Question 13-24).

$$\Delta S = k \ln C V_2^N - k \ln C V_1^N$$
$$= Nk \ln\left(\frac{V_2}{V_1}\right) \quad \text{(derived from probability considerations)}$$

The two expressions for ΔS are the same.

▪

These statements about entropy and the second law have many far-reaching implications. Since most of the natural processes in this world are irreversible, it follows that the entropy of the universe, or at least our local region in it, must be constantly increasing. That means that the energy available to do useful work is continually decreasing because the energy is being degraded into low temperature forms. In terms of the probability definition of entropy, the second law implies that the distribution of matter and energy in our current local universe is very unlikely. As time continues we must be heading toward a more probable equilibrium distribution that, unfortunately, will be at a very low temperature. In effect, the universe, once highly organized (high density in a small region), is running down, with random chaos as the ultimate end. Of course, we have a few more years to go and during that time may find that the universe has other surprises in store for us.

Since entropy is a measure of the probability of the distribution of parts of a system, it is also a measure of the disorder of the system. A random distribution of the parts of a dynamic, interacting system would seem to be most likely. A particular well-ordered distribution should be highly unlikely. Nevertheless, all around us in nature we see very complex systems with an amazing degree of orderliness. Not the least among these is ourselves. A living object is able to take molecules from the earth and air and arrange them in highly organized combinations of cells and larger structures. In effect, living creatures are refrigerators of entropy. They are able to create order out of chaos.

> **Question 13-25**
>
> Can we claim that living creatures defy the second law of thermodynamics?

SUMMARY

In order to understand the thermal phenomena described in Chapter 12, we introduced a model of the microstructure of matter and examined its consequences. We assumed that matter is composed of different types of atoms, all having about the same radius of 1×10^{-10} m. There is an attractive force between most atoms. When they are held together, each can be modeled as oscillating in a potential well formed by its neighbors.

The atomic model for the behavior of ideal gases assumes that the volume of the molecules is negligible and that their collisions are elastic and abrupt. In deriving the kinetic theory of gases, we analyzed the motion of a molecule as it collides with the walls of a box. These collisions produce forces on the wall, resulting in the relationship $PV = \frac{2}{3}NE_{kin}$. We know that for real gases $PV = NkT$. If the model is valid, $E_{kin} = \frac{3}{2}kT$. We link the average kinetic energy of a molecule to the absolute temperature. Putting numbers into this formula, we found that the thermal kinetic energy of a molecule at room temperature is $\frac{1}{26}$ eV, an energy much lower than the usual chemical binding energy. The average molecular speed of an air molecule is slightly greater than the speed of sound. The speed is inversely proportional to the square root of the mass of the molecule. Hence, helium leaks from an enclosure faster than air. Brownian motion can be explained as the movement of a large particle in thermal equilibrium with bombarding molecules.

Molecules have a distribution of energies, extending far beyond the average. The Maxwell-Boltzmann distribution of speeds and energies agrees with experiment and explains many phenomena both with gases and with the oscillations of atoms in solids and liquids. For any system in thermal equilibrium, $n_1/n_2 = e^{-(E_1-E_2)/kT}$. For gases in the atmosphere, $P = P_0 e^{-mgh/kT}$. Because the distribution allows some molecules to have very high energies, gases can escape from gravitational potential wells that are much deeper than the average kinetic energy of the molecules. The exponential nature of the distribution explains why a relatively small temperature rise can produce a large increase in the number of molecules with energy greater than some threshold.

We examined real gases, solids, and liquids to compare them with our atomic model. A good approximation for the behavior of real gases is van der Waals' equation: $(P + a/V^2)(V - b) = RT$. The correction terms account for the fact that the molecules do have volume and affect each other over distances greater than their actual radii. We can account qualitatively for many properties of solids and liquids by assuming that the atoms vibrate in their potential wells, and that the vibration energy is proportional to the temperature. Heats of fusion and evaporation represent the energies needed to free molecules from their neighbors' bonds.

SUMMARY

According to our model, the specific heat of a monatomic ideal gas at constant volume should be $c_v = Q/\Delta T = \tfrac{3}{2}R = 3$ cal/mole·K. At constant pressure: $c_p = \tfrac{3}{2}R + R = 5$ cal/mole·K. The experimental data agree at room temperature. To explain discrepancies, we postulated the Equipartition of Energy and defined degrees of freedom. Diatomic molecules, for instance, can rotate and vibrate as well as move in three spatial directions. Each degree of freedom shares equally in the thermal energy, providing there is enough energy to get above any quantum threshold. Because some thresholds are high enough to lock out some degrees of freedom, specific heats are functions of temperature.

The second law of thermodynamics is concerned with the conversion of heat into work. The law has several equivalent forms: heat cannot by itself go from lower to higher temperature; the ideal efficiency of a heat engine depends only on the temperatures between which it operates, and is equal to $(T_1 - T_2)/T_1$; in any process, the entropy of an isolated system cannot decrease.

We analyzed heat engines and refrigerators, particularly the Carnot sequence that uses only isothermal and adiabatic steps. The efficiencies of heat engines can be used in principle to define an absolute (Kelvin) temperature scale which is independent of the material used in the cyclic process. Real heat engines must expel at a low temperature a large fraction of the heat they take in at high temperature.

Entropy, S, is a measure of the disorder of a system of interacting units. We showed the equivalence of two definitions. In terms of the probability, W, of the distribution of the components of a system, $S = k \ln W$. The change of entropy of a system when heat enters or leaves in a reversible way is $\Delta S = Q/T$. Because most processes are irreversible, the entropy of the universe is steadily increasing.

Answers to Questions

3-1: You will have to judge for yourself the volume of the condensed water droplets on the glass walls of the carafe. The total volume will certainly be no more than 1 or 2 cc³. The volume of the water vapor in the carafe is simply the volume of the carafe. Remember that a liter is about the same as a quart. Compared with room temperature, the water vapor had expanded by a ratio of (373 K)/(293 K). Although this factor is large, it is apt to be no larger than your uncertainty in judging the volume of the condensed water droplets.

3-2: When you open the door of a refrigerator, you may feel cold air pour out onto the floor. When you put your hand near the radiator in back of the refrigerator, however, you will feel hot air rising. The heat taken out of the interior of the refrigerator is expelled into the room. As we will see later in the chapter, you can't even break even in this situation. While the refrigerator is working, there is a net flow of energy into the closed room through the electric cord that carries electricity to make the motor run.

3-3: The molar mass in grams is equal to the "atomic weight number." For instance, a mole of carbon atoms has a mass of 12 grams. The molar volume is equal to the ratio of molar mass and density: molar volume = (molar mass)/(g/cm³). For the elements given in the table on page 305, the smallest molar volumes are for beryllium (5 cm³) and graphite carbon (5.2 cm³). The largest molar volume is for sodium (23.7 cm³). The molar volume of lead is surprisingly large (18.3 cm³). The smaller the molar volume, the smaller the individual atoms, or at least the more tightly they are packed together. Lead is very "undense" considering its place in the periodic table. Note that although the molar volumes vary by almost a factor of 5, most of them are in the range of 10 within a factor of 2. The atomic *diameters* deduced from this calculation have an even smaller variation since the atomic diameter is proportional to the cube root of the molar volume. Thus the beryllium atom has a diameter of 1.7×10^{-10} m while the sodium atom has a diameter of 2.9×10^{-10} m and the uranium atom has a diameter of 2.3×10^{-10} m.

13-4: From what you saw in "Handling the Phenomena," there could have been no more than 1 or 2 cubic centimeters of condensed water on the walls of the carafe. If it had a volume of about 1 quart (or 1 liter), then the volume of the molecules in the liquid phase is only about 1/1000 the volume occupied by the molecules in the gas space. Therefore it seems like a good assumption, at least to a first approximation, that the volume of the molecules themselves is small compared with the volume that they occupy in the gas phase.

13-5: The pressure is proportional to the *square* of the velocity because the velocity affects the momentum transferred to the wall in two ways. The greater the velocity of the molecule, the greater the momentum transfer when it strikes each time. Also the greater the velocity of the molecule, the less time it takes between collisions. The resulting pressure is proportional to the momentum transfer each time divided by the time interval between collisions. According to our model, the average kinetic energy of gas molecules depends only on the temperature. At the same temperature the average *velocity* of hydrogen molecules would be much greater than the average velocity of chlorine molecules. Their average *kinetic energy*, however, would be the same.

13-6: The average speed of a gas molecule is indeed very large. But the distance that it travels before colliding with another molecule is microscopic. Instead of shooting straight across the room at you, a gas molecule slowly moves away from its source along a very zig-zag path. The process is called diffusion.

13-7: At first look, one might think that $(10\ \bar{E})/(5\ \bar{E})$ is the same as $(2\ E)/E$. What is required, however, in the case of $(2\ \bar{E})/(\bar{E})$, is the number of molecules that have an energy of $3kT$ compared with the number that have an energy of $\tfrac{3}{2}kT$. The second comparison is for the number of molecules that have an energy of $15kT$ compared with those that have an energy of $7.5\ kT$. Although both energy ratios are equal to 2, the population ratios are not the same at all. The distribution curve falls exponentially. Here is the calculation:

$$\frac{\Delta N(2\bar{E})}{\Delta N(\bar{E})} = \frac{\sqrt{2\bar{E}}e^{-2\bar{E}/kT}}{\sqrt{\bar{E}}e^{-\bar{E}/kT}} = \sqrt{2}e^{-\bar{E}/kT} = \sqrt{2}e^{-3/2} = 0.32$$

$$\frac{\Delta N(10\bar{E})}{\Delta N(5\bar{E})} = \frac{\sqrt{10\bar{E}}e^{-10\bar{E}/kT}}{\sqrt{5\bar{E}}e^{-5\bar{E}/kT}} = \sqrt{2}e^{-15/2} = 7.8 \times 10^{-4}$$

13-8: As we saw on page 277, the number of molecules per cubic meter at STP is 2.68×10^{25}. At 290 K this number is reduced to 2.52×10^{25}. For air molecules with a mass of 4.7×10^{-26} kg, the mass of 1 cubic meter of air molecules is 1.19 kg. The weight of these molecules is 11.6 N. Thus, the weight of the molecules is just equal to the extra force exerted on the bottom of the box.

13-9: The depth of the gravitational potential well on the surface of the moon is smaller than that on the earth by a factor of 22. (The radius of the moon is 1.74×10^6 m and the mass of the moon is 7.34×10^{22} kg). Therefore, a nitrogen molecule on the surface of the moon is in a gravitational potential well of less than 1 eV. That much binding is still greater than the average thermal kinetic energy but not great enough to prevent the escape of the molecules at the high energy end of the thermal distribution.

-10: The frequency and the speed go up by a factor of
$$v_{\text{He}}/v_{O_2} = \sqrt{32/4} = 2.83 \approx 2\sqrt{2}$$
The pitch rises by $1\frac{1}{2}$ octaves.

-11: The mass of the pollen grain would be 10^{-12} kg.
$$\tfrac{1}{2}mv^2 = \tfrac{3}{2}kT$$
$$(10^{-12}\,kg)v^2 = 3(1.38 \times 10^{-23}\,\text{J/K})(293\,\text{K})$$
$$v = 1.1 \times 10^{-4}\,\text{m/s}$$

Compare this very small average velocity of the pollen grain with the average velocity of an air molecule at room temperature: $v = 5.1 \times 10^2$ m/s

-12: For $T = 363$ K (90°C) and $E_t = 1$ eV, $(E_t/kT) = 31.94$. For $T = 373$ K this quantity equals 31.08. The ratio of molecules above the threshold at 100°C compared with 90°C equals
$$\frac{\sqrt{31.08}\,e^{-31.08}}{\sqrt{31.94}\,e^{-31.94}} = 2.4$$

The 10° rise in temperature increases the fraction of molecules over the threshold temperature by a factor of 2.4. For a factor of 2 increase in this particular case, the threshold energy would have to be 0.8 eV.

-13: Van der Waals' equation, as we have written it, is for a fixed quantity of gas—namely, 1 mole. To lower the pressure of a fixed quantity of gas (at constant temperature) you must increase the volume. Note that in order to lower the pressure by a factor of 2, you must increase the volume by a factor of 2. The correction term would then decrease by a factor of 4.

-14: Regardless of the amplitude of a vibrating bob, its average position remains the same. If the increase of volume of a heated object depended only on the larger amplitude of oscillation of each of the atoms, only the atoms on the surfaces would contribute. The increase in volume would be negligible. Note, however, the shape of the potential well in which each atom is bound. It is asymmetric. The average position within this asymmetric well is shown in the diagram. As the amplitude of oscillation increases, so does the average separation of each atom from its neighbors. As the temperature increases, the average oscillation energy and amplitude of each atom increases and so does the average separation distance. Hence the solid expands.

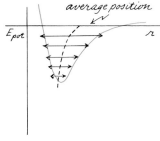

r is separation distance of each atom from its neighbors.

-15: When we wrote the equation for the internal energy of a gas and set $Nk = R$, we were choosing to deal with one mole of gas. If N is 1 mole, then $Nk = R$
$$(6.023 \times 10^{23})(1.38 \times 10^{-23}) = 8.31\,\text{J/mole}\cdot\text{K}$$

There are 4.184 joules in 1 calorie. Therefore $R = (8.31\,\text{J/mole}\cdot\text{K})/(4.184\,\text{J/cal})$ = 1.986 cal/mole·K.

-16: As we have already seen in Chapter 8, angular momentum is quantized. The spin of an atom around an internal axis, or a diatomic molecule about its long axis, must be equal to some integer times $h/2\pi$, where h is Planck's constant. The rotational kinetic energy associated with spin is $\tfrac{1}{2}I\omega^2 = \tfrac{1}{2}(I\omega)^2/I = \tfrac{1}{2}(n^2\hbar^2)/I$. The smaller the moment of inertia, I, of the system, the larger the rotational kinetic energy required to achieve one unit of angular momentum. Atoms have very small moments of inertia about their own axes, since most of their mass is concentrated at a very small radius in the nucleus. Consequently, the quantum requirement prevents spins about internal axes from participating in the small energy exchanges possible in a gas mixture.

-17: Since the molar specific heat of water is greater than that of other materials with the same molecular complexity, water must have some extra way of soaking up heat energy. It is not a case of the atoms in each water molecule having some fancier ways of vibrating or rotating. Water has many unusual properties because the hydrogen atoms are stuck to the outside of the oxygen atoms, almost as raw protons. The oxygen atom has shifted the electrons from the hydrogen atoms, making itself strongly negative in electric charge while the two hydrogen nuclei provide positive electric charge regions. The resulting molecule bonds tightly to other water molecules, forming clusters or supermolecules. As heat is fed into such a system, energy is absorbed in breaking some of the bonds between molecules, thus making the clusters smaller. The heat provides the energy to free some of the molecules from the bonds of their neighbors and thus goes into a form of potential energy.

13-18: In a balloon the gas molecules are moving equally in all directions and each one continually changes velocity as it runs into others. Nevertheless, if the nozzle of the balloon is open, the chaotic motions are turned into an organized surge of the gas out of the nozzle while the balloon goes in the opposite direction. The conversion of chaos into organization is a selection process. Those molecules that happen to have velocity in the direction of the nozzle can escape. The effect of opening a valve in a water pressure tank is somewhat similar. Inside the tank the pressure is the same in all directions. An open faucet provides a selection mechanism for motion in one direction. The way in which political leaders focus public effort is a more complicated business.

13-19: The first law of thermodynamics says that the heat entering a system is equal to the external work done plus the increase in internal energy of the system. With an ideal gas, the internal energy depends only on the temperature. If the temperature is constant, there can be no change in internal energy. Therefore $Q = W$.

First stage:

$Q_1 = W_1 +$ *no change in internal energy of gas*

Second stage:

$0 = W_2 +$ *change in internal energy of gas, characterized by drop in temperature*

13-20: During the second stage, W_2 was extracted from the internal energy of the gas. Note that the temperature of the gas fell from T_1 to T_2. Remember that there are three terms in the first law of thermodynamics: Q, U, and W.

13-21: There are some problems with the proposed cycle. Note that as long as we keep the same quantity of gas in the cylinder, the gas must satisfy the equation $PV = nRT$. After the expansion, the pressure of the gas is back to P_0 but the volume it occupies is V_2, which is larger than the original volume V_1. Therefore, the temperature of the gas must be greater than it was at the beginning: $T_3 > T_1$. In restoring the cylinder to its original condition we have to cool off the hot gas. If we simply open a hole in the cylinder and shove the piston back, we will shove out hot gas and lose some of its internal energy that way. Note also that the work performed, W, must be less than the heat Q_1 injected in the first place. If the internal energy of the gas at the beginning is U_1, then after the heat is admitted in stage 1, the internal energy of the gas must be equal to $(U_1 + Q_1)$, since no external work was done. After the adiabatic expansion (during which no heat is admitted or provided), the internal energy U_3 must be equal to $(U_1 + Q_1 - W)$. However, since $T_3 > T_1$, then $U_3 > U_1$. But to satisfy this requirement, $Q_1 > W$. Once again we have a heat-work cycle in which it is necessary to throw away some heat during the restoration part of the cycle and in which the work done is less than the heat admitted.

13-22: For this steam plant $T_1 = 823$ K and $T_2 = 293$ K. The *ideal* operating efficiency, which the real machine does not attain, is

$$\frac{T_1 - T_2}{T_1} = 64\%$$

13-23: The law of conservation of energy (or the first law of thermodynamics) only forbids the creation or destruction of energy. It says nothing about transferring energy from one place to another. In this case no energy has been created; it has simply been moved from a low temperature region to a high temperature region.

3-24: Yes. The probability of obtaining first one arrangement of a system that is controlled by chance and then obtaining a second particular arrangement is equal to the product of the probabilities of each. For instance, the probability of obtaining any particular number when casting a die is equal to 1/6. The probability of casting a die twice and obtaining first a 1 and then a 2 is equal to (1/6)(1/6) = 1/36.

3-25: Not unless we observe that the living creatures don't eat anything and thus have no source of external energy. As the entropy of living systems decreases, the entropy of their surroundings necessarily increases. For the *whole* system of living objects plus surroundings, the entropy increases.

Problems

1. Find the molar volumes of uranium and gold, using the data on page 305. Compare with lead.
2. List the assumptions made about atoms in our model.
3. List the assumptions about molecules in our model of an ideal gas.
4. Explain why the pressure on a wall due to our model gas is proportional to v^2 instead of v.
5. Write down the steps in the derivation showing that $PV = \frac{2}{3}N\bar{E}_{kin}$, and explain each step in words.
6. Find the total kinetic energy in joules for a mole of molecules at standard temperature and pressure.
7. The pressure in a television picture tube is about 10^{-9} atmospheres. What is the number of molecules/cm^3?
8. Equal *masses* of hydrogen and oxygen are mixed. What is the ratio of numbers of molecules? What is the ratio of average kinetic energy per molecule? What is the ratio of average speeds? What is the ratio of pressures exerted on the walls?
9. What would be the behavior of an ideal gas in a box if the average molecular kinetic energy were less than the *gravitational* binding energy by a factor of 10,000?
10. What is the average *velocity* of hydrogen molecules at 1000°C? What is the average speed?
11. Compare the average kinetic energy of an oxygen molecule at a temperature of 1000 K with its gravitational binding energy on the surface of the moon.
12. If a free electron in a gas at 20°C is in temperature equilibrium with the gas, what is its average speed? (The mass of an electron is 9×10^{-31} kg.)
13. At what height above sea level is the pressure only $\frac{1}{10}$ atmosphere?
14. Write down the steps in the derivation showing that the difference of force between top and bottom of a box is equal to the weight of the molecules producing the force by their bombardment. Explain each step.
15. Find the ratio of molecules with energy above 1 eV for 200°C and 190°C.
16. Explain why the correction term for P in van der Waals' equation becomes less important as P decreases.
17. Explain the temperature dependence of the specific heat of hydrogen.
18. Find the average speed due to Brownian motion of a dust particle 20 microns (20×10^{-6}m) in diameter at room temperature. Do not give more significant figures than are justified.
19. How much energy does it take, per molecule, to raise the temperature of a monatomic gas one degree? Express in joules and electron volts.
20. Why is $c_P > c_V$? Find the difference and explain your reasoning.

21. Describe the steps in a Carnot cycle, using a *PV* diagram. In what step does heat enter the engine? How much? At what temperature? In what step does heat leave the engine? How much? At what temperature? What represents the work done? How is the work related to the heat entering and leaving? Why are adiabatic and isothermal steps used?
22. A feasibility study is being made for installing a heat engine in the ocean off Puerto Rico. There is a deep trench in which the water is always 5°C, while the surface water is always about 25°C. What would be the ideal efficiency for a heat engine operating between these two temperatures?
23. If a power plant can raise its boiler temperature from 1000°F to 1200°F, while keeping the exhaust temperature at 100°F, what will be the increase in the ideal efficiency of the system?
24. Compare the quality factors of a refrigerator operating between −10°C and 35°C and a room air-conditioner operating between 20°C and 35°C.
25. Describe a process that could be explained if heat were really caloric—a substance that could be poured into a system, stored, and then drawn out again. Describe a process that could not be explained by such a theory.
26. Explain why the entropy of a system should depend on the log of the probability of the system being in that particular arrangement. (What is the total entropy of two systems, each with entropy, S_1?)

FLUIDS 14

In everyday life we are familiar with only three phases or states of matter—solids, liquids, and gases. The subject of this chapter is not *liquids,* one of those three phases, but *fluids.* A fluid is something that can flow. Obviously, the category includes liquids, but it also includes gases and even, under some circumstances, solids. Evidently, the definition is not so much concerned with a type of material as it is with a particular response of material to forces. A fluid changes shape *continuously* under the action of a steady shear force, giving rise to the phenomenon of flow.

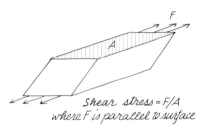

Liquid water is the very prototype of a fluid—"water seeks its own level." Air is also a fluid. To a first approximation, the atmosphere acts like a sea of air. A major difference between the fluid behavior of a gas and a liquid is that a gas is easily compressed, resulting in variable density, while most liquids are almost incompressible. The density of the atmosphere is greater at the surface of the earth than it is at mountaintop level. In contrast, the density of water in the deepest trench of the ocean is not much greater than it is on the surface.

At low velocities, fluid behavior is relatively simple. The flow follows "stream lines" that remain in fixed positions. To a first approximation that has limited usefulness, the *viscosity* of a fluid can be ignored. Viscosity is the friction property of a fluid. Most of the interesting effects of real fluids involve viscosity and *turbulence*. With these added factors, the simple subject of fluids suddenly becomes fearsomely complex. Although fluids have been the subject of scientific study since at least the days of Archimedes 2200 years ago, the analysis of fluid behavior is still one of the most difficult branches of applied science. The equations are known, but the solutions of those equations are usually too complex for calculation or even approximation.

In this chapter we'll take a look at just some of the simple and easy features of fluid behavior. Even an introductory treatment reveals important and fascinating phenomena.

HANDLING THE PHENOMENA

1. Solid objects are buoyed up when submerged in a liquid. You experience this effect every time you jump into the water. If you have access to a lake or a pool or even a bathtub, you ought to demonstrate the phenomenon

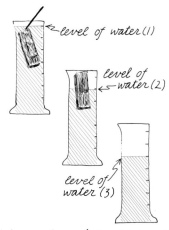

Volume of wood = Level 1 − Level 3
Volume of displaced water needed to float wood = Level 2 − Level 3

again. Get a rock or brick heavy enough so that it exerts a strain on the muscles that you use to lift it. Lower the rock into the water and feel the dramatic decrease in the force that you have to exert to support it. If you want to make the experiment more quantitative, suspend a rock or stone from a rubber band. The stone should be heavy enough so that the rubber band is stretched a considerable amount. Once again, lower the stone into the water and observe what happens to the length of the rubber band.

2. This next observation requires some kind of weighing scales. You should put a container partly filled with water on the scale. Next, suspend an object by a string and lower it into the water so that it is submerged but does not touch the bottom. The size of the container and the weight of the object depend on the size of the scale you use. The object should be heavy enough so that its weight can be easily read on the scale. Compare the weights read by the scale when the object is above the water, when the object is suspended in the water, and when the object is lying on the bottom of the container.

While you have the container and the scale, float a sizable piece of wood on the water. Compare the weight read on the scale when the wood is floating and when it is just lying beside the container.

3. For this next observation you will need a graduated cylinder or a kitchen measuring cup. Float a piece of wood (or anything else that will float) in water in the cylinder or cup. Use a needle or toothpick to hold the floating object under water and then fill the container up to its top notch. Next, let the object float freely and measure the level of the water. Finally, remove the floating object and record the new level of the water. Find the volume change for each of these two cases. Compare these volumes with the measured volume (cross-sectional area times height) of the piece of wood.

4. Anyone who swims has experienced the increased pressure on the ears while swimming or diving to greater depths. If you have access to a pool or lake, take a small rubber balloon filled with air down as deep as you can. There are two effects to be noticed. First, the volume of the balloon will noticeably decrease as you take it down. The second effect is a consequence of the first. The buoyancy of the balloon will decrease the further down it goes.

5. Liquids act as if their surfaces consist of elastic skin. This property is called surface tension. One way to observe it is to blow soap bubbles or to observe bubbles that are clustered together in any kind of foam. First, note that a soap bubble that is free and drifting in the air is spherical. On the other hand, soap bubbles that meet in a cluster have very sharp and specific angles where their walls meet.

You can produce soap bubbles as large as basketballs by using the method shown in the photograph. Almost any good detergent (not the kind for dishwashers) makes a satisfactory solution when mixed with the usual amount of water. Roll a piece of paper into the shape of a cone and soak the large end in the soap water until it is sopping wet. If you draw the paper cone slowly out of the solution and blow gently, you can produce and shake loose a very large soap bubble. Such bubbles will last for 30 seconds or so and you can even hold them in your hand if you wear a woolen mitten.

You can also observe the surface skin of a liquid by floating a needle or razor blade on water. There is an easy way to rest the metal on the liquid surface without penetrating the surface skin. Simply float a piece of tissue paper on the surface and then put the needle or the razor blade on the tissue. With your fingers or a pencil, you can poke the paper gently down into the water and away from the floating metal. Examine closely the surface of the liquid at the edges and underneath the floating metal. Then add a small amount of soap or detergent to the water.

Sprinkle water droplets of various sizes on a piece of wax paper and examine them closely. The large drops will be flat on the bottom and then bulge up, but very fine drops will be almost spherical.

6. If you have never siphoned a liquid from one container to another, you should see for yourself how the process works. It is particularly instructive to use a transparent tube so that you can see what the liquid and any bubbles inside are doing. One easy way to make such a transparent siphon is to use two plastic straws that have corrugated elbows for sipping at an angle. As you can see in the photograph, you can crease the end of one straw, insert it into the other, and end up with an airtight U-tube. Siphon water as shown in the diagram and find the conditions that are necessary to produce a flow. For instance, what happens if the bottom tube is below the surface of the upper liquid but higher than the opening in the upper tube? Find out also if the siphon will work if it starts out with a large bubble in the upper end of the U-tube.

7. The motion of air sometimes produces pressure changes that are surprising. An atomizer (a terrible misnomer) draws the liquid up a tube when you force compressed air across the open end of the tube. The same effect is used in a carburetor. You can produce the effect easily by using two straws and a cup of coffee. Place one straw vertically in the coffee and use the other straw to blow directly across the open end of the first straw. Watch what happens to the level of coffee in the vertical straw.

For another example, hold two small pieces of paper slightly apart with your fingers and blow between them.

8. There's a lot of physics, most of it impossible to solve, involved in the way water comes out of a faucet. Turn a faucet on slowly and watch the way the water flows and twists and even breaks up into droplets if it falls far enough. Notice particularly that the cross section of the stream decreases as it moves away from the immediate region of the outlet.

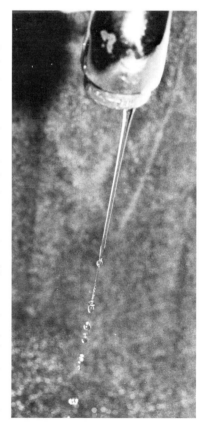

Dangle the convex side of a spoon close to the edge of a smooth stream of water coming from a faucet. You might expect that when the water hits the back of the spoon, it would bounce off in one direction while knocking the spoon in the other. What actually happens is different and surprising.

9. Take a close look at the way fluid moves in a coffee cup. If you use instant coffee, you can observe the phenomena on the surface shortly after you pour in the hot water and before all of the foam disappears. You can also see these effects by adding one drop of milk. Notice particularly how eddies persist and move around. Notice also what happens to their rotational velocity.

10. One way to remove dust from a surface is to blow on it—or so one might think. Why is it, then, that the outside surface of a car remains dusty

even when it has traveled at high speed? For an even more dramatic example of paradoxical situations, look at the blades of a fan after it has been in operation some time. A lot of dust-laden air has blown through the fan and quite a bit of the dust remains on the blade. It is not because the dust is greasy; the dust can easily be wiped off.

Another way to observe this effect is to sprinkle some very fine powder on a smooth tabletop. Use talcum powder or a sugar-cinnamon mixture. Now blow hard but steadily along the surface of the table. The larger clumps or grains will be blown away but a film of fine powder remains. We must conclude that either the fine powder is particularly sticky or else the wind velocity very close to the surface is too small to remove the grains.

11. You can observe the pressure as a function of height in a liquid by letting water run out of holes in the side of a waxed cardboard or plastic milk carton. Punch three holes with a nail, one about one-third of the way from the top, another near the bottom, and the third midway between them. Stagger the holes slightly so that the three streams of water will not run into each other. Now fill the carton with water and observe the trajectories of the three streams. There is a tricky point involved here. Depending on how far the streams fall before they hit the ground (or sink), the top stream may actually have a longer range even though its horizontal velocity is clearly less than that of the bottom stream. The photograph shows this effect.

Question 14-1

How is the horizontal range of the stream related to the horizontal velocity? Doesn't a greater range indicate a greater pressure at the hole where the stream starts?

FLUID PRESSURE

$p_{gas} = 1/3 \, (N/V) \, m \, \overline{v^2}$

Pressure is exerted by gases, liquids, and solids in very different ways. In Chapter 13 we presented a molecular model to explain the behavior of gases. In terms of that model, the pressure of a gas on a surface depends on the density of molecules at the surface, their mass, and their average velocity. The pressure at the surface is created by the transfer of momentum as the molecules strike the surface and bounce off. The primary reason that the atmospheric pressure is greater at sea level than it is at mountaintop is that the density of molecules is greater at sea level.

In the case of liquids and solids, pressure is not transmitted by molecular bombardment. The atoms or molecules vibrate within potential wells created by their neighbors. As molecules press closer together, exerting pressure on each other, the equilibrium position between molecules decreases. It is as if the molecules were being shoved against a hard spring. We know that the equilibrium position does not decrease much because the compressibility of solid and liquids is very small. In Chapter 4 on page 75 we gave a table of values for the bulk moduli of different materials. The bulk modulus β is the ratio of the change in pressure, ΔP, on an object to the fractional change of its volume, $\Delta V/V$: $\beta = -(\Delta P)/(\Delta V/V)$. Remember that the minus sign means that an increase in pressure produces a decrease in volume. On page

FLUID PRESSURE

76 we calculated the change in density of sea water produced by an enormous increase of pressure. At a depth of 10 km where the pressure is 1000 times atmospheric pressure, the density increases by only 5%.

> Question 14-2
>
> If the density increases by 5%, what is the percentage decrease of the intermolecular distance? In other words, by how much has the equilibrium distance between molecules decreased?

The extra pressure exerted by a liquid at the bottom of a tank is not caused by an increase of bombardment from its vibration. Remember that its *temperature* is proportional to the energy of vibration. If the liquid is at constant temperature, the average vibration amplitude of a molecule is the same at the bottom of the tank as it is at the top. Nor is the density of molecules appreciably greater at the bottom than it is at the top. The extra pressure at the bottom is simply from the repulsive forces of the molecules as they support the weight of the molecules above them. As we pointed out in Chapter 4, the resulting pressure in a liquid is perpendicular to any surface placed in the liquid and does not depend on the orientation of the surface. At a given point in the liquid, there is the same force up as there is sideways or as there is down. Otherwise, as shown in the diagram, the net forces on a slab of the liquid would not be zero and so there would be movement of the liquid. For the present, however, we are talking about a static situation. Notice in the diagram that there is a difference of pressure between the top and the bottom of a slab of the liquid. The total difference in the force between top and bottom is just sufficient to support the weight of the slab of liquid.

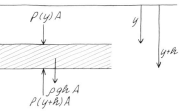

$$P(y+h)A - P(y)A = \rho g h\, A \qquad \rho \text{ (rho) is the density}$$

Since ρ is approximately constant for most liquids, and most depths, $P(h) = P_0 + \rho g h$.

The pressure in a liquid at a depth h is equal to the pressure at the surface, which is usually the atmospheric pressure, plus a term proportional to the depth. Notice how different this is from the case that we looked at in Chapter 13 for atmospheric pressure. In that example the density of the gas is proportional to the pressure. Consequently, the pressure and density of the atmosphere decreases exponentially with increasing height.

Solids transmit pressure more like liquids than like gases. In solids, however, the atoms or molecules are locked in place. In a crystalline structure, the repulsive spring-type constants between atoms may be different in one direction from what they are in another. Consequently, the pressure and pressure response are not the same in all directions. There can be a very complicated relationship between stress and strain. In a liquid, molecules can move around each other, and so static pressure effects are uniform in direction.

The fact that pressure in a liquid depends only on the height gives rise to the so-called hydrostatic paradox illustrated in the diagram. The con-

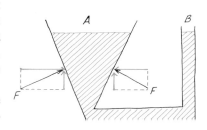

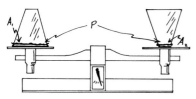

Force exerted by diaphragm 1 = PA_1
Force exerted by diaphragm 2 = PA_2
$PA_1 > PA_2$

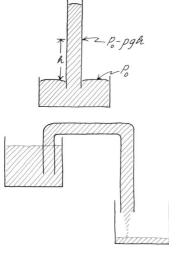

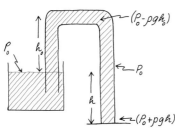

tainers, called Pascal's vases, are named for Blaise Pascal (1623–1662). The liquid stands at the same level in each of the containers. One might naively be surprised that the heavy weight of all the liquid in container A could be balanced by the small weight of the water in the cylinder B. Of course part of the weight of the liquid in A is being supported by the walls of the container as shown in the diagram.

> **Question 14-3**
>
> Two different water towers are shown in the diagram. Will each produce the same pressure at the water main? If so, why build the large and expensive one on the left?

Another form of the liquid paradox is shown in the next diagram. The two conical containers contain equal amounts of water and therefore should have the same weight, as shown by the fact that the weighing pans are balanced. On the other hand, the bottom of each container consists of a rubber diaphragm that transmits the water pressure on the bottom to the pan. Since the height of the water is the same in both containers, the pressure at the bottom must be the same. The container at the left, however, rests on a much larger area and so the total force exerted by the diaphragm must be $F_1 = PA_1$. But $PA_1 > PA_2$, and $PA_2 = F_2$. The weights of the two conical containers are the same but once again the *container* on the right must support a large part of the weight of the liquid in it. That extra force must be supported by the right-hand pan balance acting on the rim of the container.

In Chapter 4 we analyzed the operation of a liquid barometer. The pressures at various points of such a barometer are shown in the diagram. An extension of this analysis explains the operation of a siphon. A siphon is an upside down U-tube that can provide drainage from an upper level of fluid to a lower level as shown in the next diagram. Of course, a hole in the upper container could provide the drainage just as well, but a siphon has the apparently peculiar ability to raise a liquid uphill. The liquid does indeed rise in one leg of the tube but only so long as it falls a longer distance in the other leg. To make a siphon work, you first have to fill the tube with the liquid and keep it filled until the short end is in the liquid.

Suppose that you put your finger over the end of the longer tube, stopping the flow. The pressures in the liquid column are shown in the following diagram. At the surface of the upper container there is atmospheric pressure P_0. At the top of the U the pressure is reduced to $(P_0 - \rho g h_0)$. In the long leg of the tube at the level of the liquid surface in the upper container, the pressure is again atmospheric, P_0. The pressure on your finger is equal to the atmospheric pressure plus the difference in liquid heights in the two legs: $(P_0 + \rho g h)$. If you take your finger off the tube, the liquid that is under pressure will shoot out. The cohesiveness of the liquid keeps the column from separating and so drags water into the short end to keep the flow going. All that is required is that the lower end be below the upper surface of the liquid in the upper container.

Question 14-4

In this case, isn't it necessary for the open leg of the siphon to be below the submerged end?

We emphasized that the siphon must be filled before its action can start. Suppose that you don't fill it or end up with an air bubble in the U section as shown in the diagram. As you can see from the analysis of pressures in the tube, the siphon will not work unless $h_1 = h_2 > h_0$.

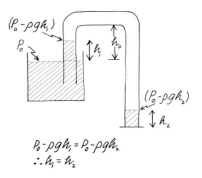

Question 14-5

Is there a limit to the height that a siphon can raise a liquid before sending it back down again? Remember the barometer principle.

Because liquids can transmit static pressures with practically no change in their own density, they can be used in force-amplifying devices. The operation of a pneumatic or hydraulic lever is shown in the diagram. In both legs the pressures at the same level are equal. If a force of 10 N is exerted on a piston with a cross section of 1 cm² on the left-hand side, then the pressure at the top surface is 1×10^5 N/m². That pressure is also exerted on the large piston on the right side. If the piston has a cross-section of 100 cm², then the force on it is $F = PA = (1 \times 10^5 \text{ N/m}^2)(1 \times 10^{-2} \text{ m}^2) = 1 \times 10^3$ N. The force has been amplified by a factor of $A_2/A_1 = 100$.

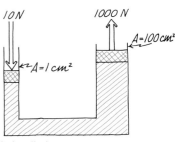

Hydraulic lever.

Question 14-6

In terms of work done, do we get something for nothing?

ARCHIMEDES' PRINCIPLE

Archimedes (287–212 B.C.) was a Greek mathematician and scientist who lived in Syracuse, Sicily. He considered his primary work to be in mathematics, but he was famous in his own day and in subsequent legends because of his inventions. One of these involved the measurement of the density of an object that had complex geometry. According to legend, King Hiero of Syracuse had a crown made of gold. He suspected that silver had been substituted for part of the gold, and asked Archimedes to find out without destroying the crown. Supposedly Archimedes discovered the method while in his bath and ran naked through the streets crying, "Eureka, Eureka" ("I have found it, I have found it"). Succeeding generations have argued endlessly over exactly what it was he found. One possibility is that he noticed the overflow of the water as he got into the tub and realized that he could measure the volume of an irregularly shaped object by means of water displacement. If he could find the volume of the crown in this way, he could weigh it and then calculate the density. He could then compare the density with that of pure gold or with gold-silver alloys of various concentrations. The scheme is suitable in principle but is not capable of yielding great precision. Because of surface tension (which we will discuss later) and the requirement that the water container be at least as large as the object to be submerged, the volume of the displaced water is hard to measure with a precision of more than two significant figures. There is, however, another phenomenon that Archimedes might have noticed when he got into the tub. He got lighter. A fluid exerts a buoyant force on a submerged object. As we will now see, this effect provides another method of measuring the density of an object.

Archimedes' principle is that an object in a fluid is buoyed up by a force equal to the *weight* of the *fluid displaced*. This rule applies to both floating and submerged objects. As we have already seen, the upward force on the bottom is greater than the downward force at the top by an amount equal to the weight of the fluid slab. Consequently, the slab is in equilibrium and from this relationship we derived the formula for the pressure as a function of depth in a fluid with constant density. If the slab of fluid were to be removed and replaced with a slab of any other material of the same size, the pressure at each point of the surrounding fluid would not change. There would still be an excess upward force on the slab equal to the weight of the *fluid* that used to occupy that space. The volume of that fluid is the same as the volume of the object. When the object is submerged, that volume of fluid is shoved up to make room for the object.

In "Handling the Phenomena," you felt the buoyant effect on a rock when you submerged it in water. Suppose that the rock had a mass of 5 kg and therefore had a weight in air of about 50 N. Since the density of most silicon-based rocks is about 2500 kg/m³, the volume of this rock must be

$$V = m/\rho = (5 \text{ kg})/(2500 \text{ kg/m}^3) = 2 \times 10^{-3} \text{ m}^3$$

When you submerged the rock in water, the volume of the water displaced must also have been 2×10^{-3} m³. The mass of that much water is

$$m = \rho V = (1 \times 10^3 \text{ kg/m}^3)(2 \times 10^{-3} \text{ m}^3) = 2 \text{ kg}$$

The weight of 2 kg is about 20 N. While holding the rock in air, you had to support a weight of 50 N; while holding it under water, you need only exert an upward force of 30 N.

Notice that the effect of the buoyant force is to reduce the apparent density of the object by the density of the fluid. Since the mass and the weight of an object are proportional to the density, the ratio of weight in the fluid to the weight in air is

$$\frac{W_{\text{in fluid}}}{W_{\text{in air}}} = \frac{\rho_{\text{object}} - \rho_{\text{fluid}}}{\rho_{\text{object}}} = 1 - \rho_f/\rho_o$$

Question 14-7

Does this formula agree with our previous calculation for the apparent weight of the stone submerged in water?

What happens if the density of the object is equal to or less than that of the fluid? Our formula apparently gives a nonsensical answer. If the densities of object and fluid are the same, then the ratio of weights is 0. If the density of the object is less than that of the fluid, then the weight ratio is negative. In these cases the object is floating. If the densities of object and fluid are the same, the object loses all apparent weight since the upward buoyant force exactly matches its own weight. The object acts just like a slab of the fluid and could be in equilibrium at any point within the fluid. If the density of the object is less than that of the fluid, then the object will float and will not displace its volume. Archimedes' principle holds, however. The object is buoyed up by a force equal to the weight of the volume of fluid that it does displace. That force is just equal to its own weight and so it floats with part of its volume sticking out of the liquid.

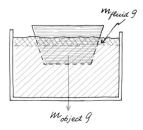

weight of fluid displaced = weight of floating object

Question 14-8

How can a steel boat float if the density of steel is 5500 kg/m³?

Whether or not Archimedes realized the significance of the buoyant effect, we do not know. If he did, he could have determined the density of the crown by weighing it in air and while it was submerged in water. The precision of weighing can be much greater than the precision of determining volume by liquid displacement. Eighteen hundred years after Archimedes considered the problem of the fraudulent crown, another great scientist, Galileo, analyzed the legendary problem. To solve it, he devised a balance that would have been sensitive enough to have made the crucial measurement.

We can use Archimedes' principle to determine the unknown density of a liquid in which an object with known density is floating. Such an instrument, called a hydrometer, is commonly used to determine the charge in storage batteries, or the antifreeze component of radiator coolant. The device is shown in the photograph. If you squeeze the bulb and then release it, you draw up a sample of the liquid to be tested. The internal float is weighted so that it remains upright, but rises or sinks to various depths depending on the liquid density. The level lines on the float are usually calibrated directly in terms of density. For instance, when a lead storage battery is fully charged, the sulfuric acid in it has a density of 1300 kg/m³. As the sulfate combines with the lead during the discharge process, the density of the liquid falls. By the time the density is 1150 kg/m³, the battery is essentially discharged.

The device shown the diagram is called a Cartesian diver. You can make the capsule rise or sink by exerting a small amount of pressure on the diaphragm at the top of the container. The capsule is open at the bottom and contains a small volume of trapped air. At equilibrium, the volume of the capsule, including the air, displaces a weight of water just equal to the weight of the capsule. Consequently, the capsule can float at any level in the water. If you exert a slight extra pressure on the top diaphragm, the pressure throughout the water and capsule increases. The extra pressure decreases the volume of the trapped air slightly. But now the volume of the capsule

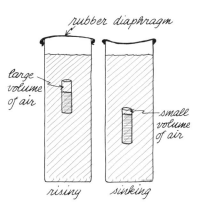

is smaller, although its weight is the same. Since the capsule displaces less water, the buoyant force on it is less and it sinks. If the pressure on the diaphragm is reduced, the volume of the trapped air increases slightly and so the capsule displaces more water. The buoyant force on the capsule increases and the capsule rises.

Buoyant forces are exerted by gases as well as by liquids. That's why dirigibles or toy helium balloons float. In everyday life, we do not feel buoyed up by the surrounding air, but the effect is not negligible in precision weighing with a pan balance. Standard comparison weights are usually made of brass, which has a density of 8600 kg/m³. The density of air is only 1.3 kg/m³. The buoyant force exerted by the air on the standard weights is equal to the product of their volume and the weight density of the air:

$$F_{\text{buoyant}} = Vg\rho_A = (m_o/\rho_o)g\rho_A \qquad \text{where } V_{\text{object}} = m_o/\rho_o$$

The fractional loss of weight of an object submerged in air is

$$\frac{F_{\text{buoyant}}}{W} = \frac{(m_o/\rho_o)g\rho_A}{m_o g} = \rho_A/\rho_o$$

In the case of brass standard weights, the percentage of error due to the buoyancy in air is (1.3 kg/m³)/(8600 kg/m³) 100% = 1.5×10^{-2} %.

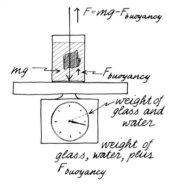

> ### Question 14-9
>
> Such a small effect seems hardly worth bothering about. Besides, wouldn't the same correction apply to the sample being weighed in the other pan?

In "Handling the Phenomena," you submerged an object in a liquid that was being weighed. If you held on to the object with a string, you still had to support its weight minus the buoyant force being exerted by the liquid. This extra buoyant force in turn had to be supplied by the scales. The forces involved are shown in the diagram. Of course, if you let go of the string supporting the object and allow it to sink to the bottom of the container, the scales would have to support the entire weight of the object.

SURFACE TENSION

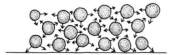

The molecules at the surface of a liquid experience forces that are not symmetrical. Within the body of the liquid a molecule is, on the average, pulled equally in all directions by the attractive forces binding it to its neighbors. This force is called *cohesion* and is simply the short-range force between atoms and molecules that we have described in Chapter 10. It is electromagnetic in nature but is more complicated than the simple attraction of a positive charge to a negative one. The molecules in the *surface* of the liquid experience this cohesive force only in one direction.

SURFACE TENSION

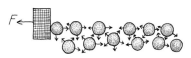

Question 14-10

If there is a downward force on a molecule as shown in the diagram, why doesn't the molecule accelerate downward into the liquid?

If the surface of the liquid expands, perhaps because one boundary is moved as shown in the next diagram, then more molecules must come from the body of the liquid to take their place in the surface. As they do so, they move against a restraining force. The potential energy of the system increases. At the molecular level, this potential energy is stored in the compression between the surface molecules and those underneath that support them. This extra energy comes from the work done in expanding the surface. Since the surface expands along a direction x, a force must be required in that same direction in order to do the work that increases the potential energy. Therefore, the force must lie in the surface and be in the direction of the expansion. This property of a liquid surface is called *surface tension*, γ (gamma). It is measured in newtons/meter since the force needed is proportional to the width of the surface being stretched. Values of surface tension for various liquids in contact with air are given in the table.

In Chapter 10 we used a rough model of atomic sizes and energies to get an approximate value for Young's modulus. Let's see if that same model can give us a reasonable value for the magnitude of surface tension. We must make an assumption about the energy involved when a molecule is pulled into the surface from the body of the liquid. That energy must be considerably less than one electron volt, which is the order-of-magnitude of binding energy of atoms within molecules. We are not removing the atom from a chemical bond when we move it to the surface. We might guess that the energy involved would be about half of the fusion energy per molecule as the substance changes from solid to liquid. The factor of one-half is used because the surface molecule is still held by bonds from below.

For water, the fusion energy is 80 cal/g = 335 J/g = 6024 J/mole = 1 × 10⁻²⁰ J/molecule. Let's assign this value to the depth of the local potential well in which a water molecule is held by its immediate neighbors. To allow a new molecule to come to the surface, we must exert a force along the surface to separate the layer by one full molecular diameter. That force must be equal to the force required to pull a molecule partway out of its potential well to the position where it is held only by bonds from below. The force is $F = \Delta U/\Delta x$, where $\Delta U = \frac{1}{2} \times 10^{-20}$ J/molecule, and $\Delta x = 3 \times 10^{-10}$ m, the molecular diameter of water.

$$F = \Delta U/\Delta x = (\tfrac{1}{2} \times 10^{-20} \text{ J})/(3 \times 10^{-10} \text{ m}) = (\tfrac{1}{6})10^{-10} \text{ N}$$

In one meter along a surface there are $\tfrac{1}{3} \times 10^{10}$ water molecules. The force per meter to separate all of these is

$$(\tfrac{1}{6} \times 10^{-10} \text{ N})(\tfrac{1}{3} \times 10^{10} \text{ molecules}) = 0.06 \text{ N/m}$$

Compare this calculated value for surface tension of water with the table

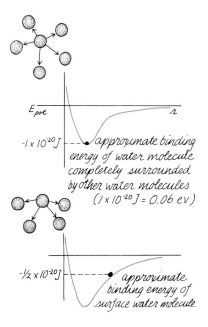

Surface Tension of Liquids at Room Temperature in Contact with Air

	N/m²
Water	0.073
Carbon tetrachloride	0.027
Ethyl alcohol	0.022
Mercury	0.49

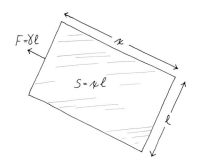

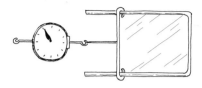

value of 0.073 N/m. The agreement is probably better than we deserve, as you might surmise by looking at the values for other liquids, particularly mercury. Nevertheless, the order-of-magnitude agreement is encouraging; our crude atomic model must be pretty good.

The available energy (sometimes called the "free energy") due to surface tension must be equal in magnitude to the work done in pulling out the surface from zero size to its final area. Since the force required is constant (for a flat surface), the work done is equal to $F \cdot x$. The force required is equal to the product of the surface tension and the length of the surface: $F = \gamma l$. Therefore, the free energy due to the surface tension is

$$U_{\text{surface}} = F \cdot x = (\gamma l) \cdot x = \gamma S \qquad \text{where } S \text{ is the surface area.}$$

The actual energy involved in forming a liquid surface is slightly greater than U. When any liquid surface is enlarged, the temperature of the liquid drops. The extra thermal energy is stored in the surface layer.

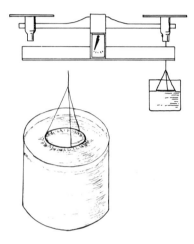

One way to measure surface tension is shown in the diagram. The force required is actually twice the product of the surface tension and the length since two surface films are involved—top and bottom. A more practical way to measure surface tension is shown in the next diagram. The force required to pull the ring out of the liquid is easy to measure with a sensitive pan balance.

> ### Question 14-11
>
> How sensitive a balance must we use? If the wire ring has a diameter of 6 cm, what is the mass of the balancing weights that would be required to pull the ring out of the water?

Because of surface tension, any volume of liquid tends to decrease its surface area, thus decreasing its potential energy. Surface tension is one of the elastic forces responsible for the motion of ripples. The water in the uplifted surface is pulled down by gravity and also by surface tension trying to make the surface smooth again. Tiny water droplets in air are almost spherical since the sphere has a smaller ratio of surface area to volume than any other geometrical shape. Larger raindrops are more tear-shaped as they fall because of the effects of moving through air. The surface tension of mercury is so great that small drops on a surface are almost spherical, as you can see in the photograph. You probably observed the same effect in "Handling the Phenomena" if you sprinkled fine drops of water on waxed paper.

Another example of the tendency of surfaces to decrease their area is shown in the next photograph. A soap bubble film is picked up on a wire loop that has a thread across one diameter. The middle section of the thread is divided into two parts. When the film between these two parts is pierced, the outer film withdraws to its minimum area, which is obtained when the center hole is a circle.

The tendency of a surface to achieve minimum area also explains why a needle can be floated on the surface of water as you saw in "Handling the

Phenomena." If you looked closely, you could see the slight depression of the surface on which the needle rested. The restoring forces trying to make the surface flat again supported the weight of the needle. When you reduced the surface tension by adding soap or a detergent, the needle promptly sank.

> Question 14-12
>
> If soap reduces surface tension of water, why do we blow soap bubbles instead of water bubbles?

Because of surface tension, the surface of a liquid acts as if it were composed of an elastic skin. There is a major difference, however, between the elasticity produced by surface tension and that produced by an elastic material such as rubber. For a flat surface on a liquid, the strength of the restoring force is independent of the amount by which the surface has been stretched. In fact, the "amount of stretch" of such a surface is meaningless. You need exert only a constant force to increase the surface area by pulling more and more molecules into the surface layer. To stretch a sheet of rubber, on the other hand, you must exert a force that is roughly proportional to the stretch.

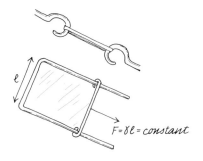

You may have noticed in blowing soap bubbles that if you take your mouth off the pipe, the bubble will contract, blowing the air back out again. Evidently, there is higher pressure inside the bubble than there is outside. The forces caused by air pressure on a hemisphere of a bubble are shown in the diagram. The forces on the spherical surface add vectorially to a resultant force tending to separate this hemisphere from the opposite one. In the diagram we show that the x-component of the force on a segment of the curved surface is just equal to the product of net pressure and projected area. Since the projected area of a hemisphere of radius R is equal to πR^2, the x-component of the force on the hemisphere is $(P - P_{atm})\pi R^2$. The two hemispheres are held together by the force of the surface tension, which has a magnitude of $(2\gamma)(2\pi R)$. The factor of 2 arises again because there is both an inside and an outside surface to the bubble. The force from the excess internal pressure must be equal to the surface tension holding the two hemispheres together:

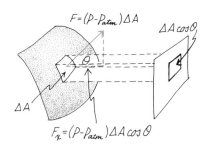

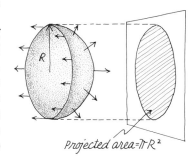

$$(P - P_{atm})\pi R^2 = 4\pi R \gamma$$

$$P - P_{atm} = 4\gamma/R$$

Surprisingly enough, the smaller the radius of the soap bubble, the greater the excess pressure must be. Because of this feature, whenever two bubbles come together, the smaller feeds air into the larger and the larger grows at the expense of the smaller. The effect is shown in the photograph, which shows two bubbles that were about the same size one minute earlier. Because the two pipes were connected by a straw, the bubble that was slightly larger to begin with was able to grow at the expense of the smaller one.

A similar analysis of the pressure relationships and surface tension

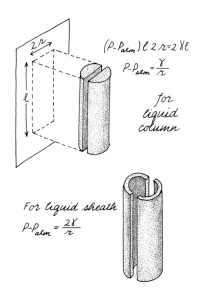

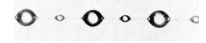

Silk strand of a spider web. The sticky globules formed from a continuous cylindrical sheath. (Courtesy Cornell U. Press, from "The Spider Book" by John Henry Comstock, 1912, 1980.)

forces in a cylindrical column is shown in the next diagram. Once again, we find that the excess internal pressure that balances the surface tension is inversely proportional to the radius. This fact is responsible for the instability of long cylindrical columns of liquid. You must have observed this effect in "Handling the Phenomena." A thin stream of water coming from a faucet breaks up into a spray of droplets. If the column could maintain exactly the same radius all along its path, there would be no reason why a pinch should occur in one place as opposed to another. However, if for any reason, such as fluctuations in the streamlines or variation in outside pressure, one part of the stream becomes momentarily narrower than the surrounding parts, then the internal pressure at that point will become greater. The liquid will flow from that point to the surrounding lower pressure region, which will make the stream even narrower. The pressure will rise still higher at that point, making the stream yet narrower until it pinches off and forms a separate droplet.

In earning their livelihood, web-spinning spiders routinely make use of the instability of fluid cylinders. When the spider first produces the filament, the sticky fluid forms a cylindrical sheath around the strand. Since the long liquid column is unstable, it soon breaks up into tiny globules that are ideal for catching flies.

So far we have been talking about the surface between a liquid and a gas. At the edge of a container the liquid will be in contact with both a solid and a gas. A surface molecule at that boundary is affected not only by the forces of cohesion within the liquid, but also by the force of *adhesion* to the solid. The liquid surface lines itself up perpendicular to the resultant of these two forces. Three cases are shown in the diagram. In the first, the force of adhesion is strong enough to produce an upward curvature of the liquid at the solid surface. To describe this situation, we say that the liquid has "wet" the surface. This is the normal situation with water touching glass. In the second case the force of adhesion is small and the resultant force on the liquid at the junction remains directed back into the liquid. The resulting curvature of the liquid surface is convex, away from the solid container. This is the situation with mercury in a glass container. In the third case, the resultant of the forces of adhesion and cohesion is parallel to the wall and hence the liquid surface is perpendicular to the wall. This is the situation with water in a silver container or with water in some types of plastic.

The curved surface of the liquid in a cylindrical tube is called a *meniscus*.

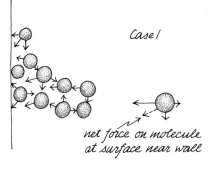

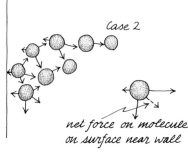

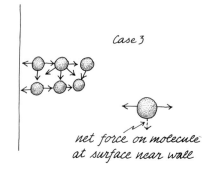

In order to establish a meniscus with a curvature such that the surface is perpendicular to the resultant of adhesion and cohesion forces, the edge of the liquid must rise or fall some distance along the solid surface. This is called *capillary* action. The effect in small cylindrical tubes is shown in the photographs. The surface tension pulls a column of water up a glass tube but pulls a column of mercury down a glass tube.

> Question 14-13
>
> In all cases, the resultant of the forces of cohesion and adhesion is more "down" than "up." How then can water rise in a capillary tube?

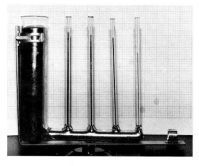

The forces acting on the column of water in a capillary tube are shown in the diagram. Notice that at the upper surface where the liquid touches the glass, the geometry is very much like the hemispherical soap bubble we analyzed before. The surface tension force lies in the liquid surface. Therefore, it is pointed upward at an angle θ from the vertical. The upward component of this force is equal to the product of the surface tension, cos θ, and the circumference of the contact ring.

$$F_{\text{vertical}} = \gamma \cos \theta (2\pi r)$$

This upward force is balanced by the weight of the liquid column, which is

$$W = \rho_L g h (\pi r^2)$$

When the upward capillary force balances the downward force of the weight of the column, we have

$$\gamma \cos \theta (2\pi r) = \rho_L g h (\pi r^2)$$

$$h = \frac{2\gamma \cos \theta}{\rho_L g r}$$

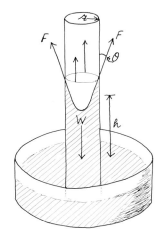

As shown in the next diagram, the depression of the column of mercury in a glass capillary is essentially the same effect and is governed by the same formula. Notice that the smaller the radius of the capillary tube, the higher the liquid can rise (or sink in the case of mercury). Capillary action is responsible for the absorption of liquids in paper towels and is partially responsible, along with osmotic pressure, for the rise of sap in plants and trees.

We have already seen that there is excess pressure inside a spherical surface of liquid. The smaller the radius of curvature, the greater the excess pressure. The same arguments show that if the curvature is negative, the pressure in the liquid is less than on the outside. Both these situations occur in capillary tubes. In the case of water in a tube, the radius of curvature of the meniscus is negative and the pressure of the liquid inside the surface is less than the pressure on the outside. Note that the pressure *must* be less than atmospheric since the column of water is higher than that of the reservoir from which it comes. With the meniscus formed by mercury, however, the radius of curvature is positive. The pressure of the mercury

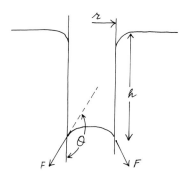

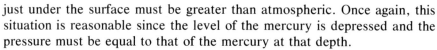

An alternative derivation of capillary rise

For water on glass, θ=0
If surface is spherical:
$\Delta P = P_{atm} - P = \frac{2\gamma}{r}$

[Compare with formula on page ___ . Note that there is only one surface in this case. Also, in this case, $P < P_{atm}$.]

$A\Delta P$ = upward force = weight of column
$\frac{2\gamma}{r}\pi r^2 = \rho g h \pi r^2$

$h = \frac{2\gamma}{\rho g r}$

just under the surface must be greater than atmospheric. Once again, this situation is reasonable since the level of the mercury is depressed and the pressure must be equal to that of the mercury at that depth.

Let's calculate two cases of capillary rise. In the case of mercury, the contact angle between the surface of the mercury and the lower wall is 140°. The surface tension of mercury is 0.47 N/m and the density of mercury is 13.4×10^3 kg/m³.

$$h = \frac{2(0.47 \text{ N/m})(\cos 140°)}{(13.4 \times 10^3 \text{ kg/m}^3)(9.8 \text{ N/kg})r} = \frac{-5.4 \times 10^{-6}}{r}$$

If the capillary tube has an inside radius of 1 mm (10^{-3} m), then the mercury in it will be depressed by 5.4×10^{-3} m = 5.4 mm.

Water completely wets clean glass so that the angle of contact between the surface of the water and the glass wall is 0°. Therefore the height that water will rise in a glass capillary is

$$h = \frac{2(0.073 \text{ N/m})}{(1 \times 10^3 \text{ kg/m}^3)(9.8 \text{ N/kg})r} = \frac{1.5 \times 10^{-5}}{r}$$

For a glass capillary tube with a radius of 1 mm, the water will rise to a height of 1.5 cm.

Some of the cells just inside the bark of trees are long and hollow with interior dimensions less than 1 micrometer. At first thought, one might expect that capillary action could explain how sap rises in trees even to heights of several hundred feet. Simply make the cellular channels thin enough. Note, however, that by the time the water has risen to a height of about 34 feet, the internal absolute pressure is approximately 0. Standard atmospheric pressure can support a column of mercury that is 76 cm high. The comparable height of a free-standing column of water is 10.34 m. The pressure in such a column starts out at atmospheric pressure at the bottom and decreases steadily to 0 at the top.

$$P = P_0 - \rho g h$$

Can liquids sustain negative pressures? If so, the liquid would be under tension! Apparently this situation can and does occur, although the complete explanation of the rise of sap is still a matter of controversy. To be sure, the pressure in the roots, P_0, is greater than atmospheric pressure because of osmosis. The osmotic effect is small, however, in comparison with the large pressure differential needed to raise water to the top of tall trees.

"IDEAL" FLUID FLOW—A POOR APPROXIMATION

The motion of fluids—fluid dynamics—is one of the most complicated fields of physics. Our standard method of dealing with complex phenomena is to make simplifying assumptions so that we can calculate a first approximation. For instance, in the study of mechanics we started out by assuming that there was no friction. The equivalent assumption in the study of fluids is that the *viscosity* is 0. We will define viscosity later but for now keep in mind that the viscosity of ketchup at room temperature is high, the viscosity of water is low, and the viscosity of air is even lower. We cannot arbitrarily

Low viscosity water and high viscosity ketchup.

say, however, that the effects of viscosity can usually be ignored when objects move through air. The motion of dust particles in air is completely dominated by the viscous effects.

In spite of the fact that we must be very careful in applying our first approximation, let's analyze fluid flow under the special conditions of 0 viscosity and incompressibility. Most *liquids* are incompressible, and in regions where the pressure does not change rapidly, the volume changes of gases can be ignored. The fluid motion must be smooth and without turbulence. We will also ignore flow that is rotational with small whirlpools. Does fluid ever flow in such a simple way? Yes, the approximation is good when considering the movement of water in slow-moving streams or large pipes, as long as one stays away from the banks or walls. The approximation also helps to explain some of the phenomena that occur when air moves over surfaces at low speeds. We can characterize this flow in terms of *streamlines* that remain constant in time. Streamlines are tangent to the fluid velocity at each point. If we form a bundle of streamlines as shown in the diagram, we enclose a quantity of fluid that moves within the bundle and never passes in or out of the walls formed by the streamlines. Note that streamlines never cross each other and can never just end.

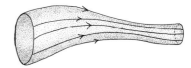

Question 14-14

If the fluid is not compressible, and yet must stay inside the tube formed by the bundle of streamlines, how can the tube have different cross-sectional areas? As the tube gets larger, wouldn't the fluid have to expand?

A chunk of fluid passing within a bundle of streamlines has *kinetic energy* due to its motion. If in following the streamlines the fluid rises or falls, its *gravitational potential energy* will change. If it moves from a region of low pressure to high pressure, some *work* must be done on it. Each of these three forms of energy can change, but the work done on the fluid must equal the change in its kinetic and potential energy.

Let's analyze the three types of energy change. In the diagram we show a slug of fluid (indicated by solid lines) enclosed by streamlines. During a short interval of time the fluid moves up and to the right (shown by dotted lines). The cross-sectional area on the left side is A_1; on the right side it is A_2. The left end moves a distance, Δx_1, while the right end moves a greater distance, Δx_2. Since the fluid is incompressible, the volume change on the left must equal the volume change on the right. Therefore,

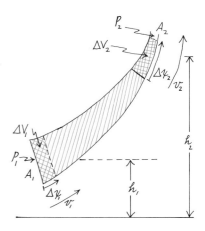

$$\Delta V_1 = \Delta V_2 = \Delta V \qquad A_1 \Delta x_1 = A_2 \Delta x_2$$

The end volumes that are under consideration have a mass, m. Since the density of the fluid is $\rho = m/\Delta V$, the volume of the small end element is $\Delta V = m/\rho$.

The change of kinetic energy of the entire slug of fluid is simply the difference in kinetic energy between the two end elements:

$$\Delta(\tfrac{1}{2}mv^2) = \tfrac{1}{2}m(v_2^2 - v_1^2)$$

The change in gravitational potential energy of the whole slug of fluid is just the change of potential energy of the two end elements:

$$\Delta E_{\text{grav}} = mg(h_2 - h_1)$$

As far as the slug of liquid is concerned, P_1 is pushing it from the left and meanwhile the slug has to exert a pressure of P_2 on the right. The work done *on* the slug is $P_1 A_1 \Delta x_1$. The work done *by* the slug is $P_2 A_2 \Delta x_2$. The net work done is

$$W = (P_1 - P_2)\Delta V = (P_1 - P_2)m/\rho$$

The work done on the slug of fluid must be equal to its change of kinetic energy plus the change in potential energy.

$$W = \Delta(\tfrac{1}{2}mv^2) + \Delta E_{\text{grav}}$$

$$(P_1 - P_2)m/\rho = \tfrac{1}{2}m(v_2^2 - v_1^2) + mg(h_2 - h_1)$$

$$P_1 + \tfrac{1}{2}\rho v_1^2 + \rho g h_1 = P_2 + \tfrac{1}{2}\rho v_2^2 + \rho g h_2$$

This is Bernoulli's equation, named for Daniel Bernoulli (1700–1782), a Swiss mathematician and scientist. Let's examine the consequences that it predicts and compare these with experiments.

First, if the fluid is not moving, we should get the pressure-depth relationship for a static liquid:

$$P_1 + \rho g h_1 = P_2 + \rho g h_2$$

If P_2 is the pressure at the top of a liquid and $(h_2 - h_1)$ is the depth from that top surface h, then we have the same formula that we used previously:

$$P = P_0 + \rho g h$$

If we leave out the potential energy term, we get a relationship between the pressure and velocity along the streamlines of a fluid moving horizontally. Where the velocity is high, the pressure is low. An example of this effect is shown in the diagram. The velocity in the narrow region is necessarily greater than it is in the large region.

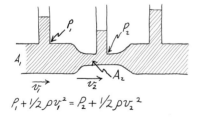

$P_1 + 1/2 \rho v_1^2 = P_2 + 1/2 \rho v_2^2$

Question 14-15

Why?

To find v_1 with Venturi tube:
$P_1 - P_2 = 1/2 \rho (v_2^2 - v_1^2)$
$v_2 = (A_1/A_2) v_1$
$P_1 - P_2 = 1/2 \rho [(A_1/A_2)^2 v_1^2 - v_1^2]$
$P_1 - P_2 = 1/2 \rho v_1^2 [(A_1/A_2)^2 - 1]$

The pressure in the fluid is measured by the height of the static fluid in the standpipes. A tube like this, with a narrow section and standpipes, is called a Venturi tube. When it is inserted into a pipe in which fluid is flowing, it can be used to measure v_1, the velocity of flow in the main pipe. Bernoulli's equation gives one relationship between v_1 and v_2 in terms of the pressures P_1 and P_2. One other relationship is needed to eliminate the variable v_2. That is the equation of continuity of flow:

$$A_1 v_1 = A_2 v_2$$

"IDEAL" FLUID FLOW—A POOR APPROXIMATION 359

A type of air speed indicator called the Prandtl tube or sometimes a Pitot tube is shown in the next diagram. The trick of its operation is that it measures both the pressure in the moving stream and the pressure at 0 velocity and takes the difference automatically. Note first of all that opening #1 is adjacent to the main flow. The pressure will be less than it would be if the fluid were standing still. Opening #2 samples the pressure in a region called the stagnation point. Observe the way the streamlines divide and avoid the leading edge of the probe. At that leading edge the velocity of the fluid relative to the probe is 0. The pressure at the stagnation point is read by the right-hand part of the U-tube, while the pressure in the moving stream is read by the left-hand part of the tube. The difference in levels of the liquid in the two legs measures the difference in pressure at the two points. Bernoulli's equation requires

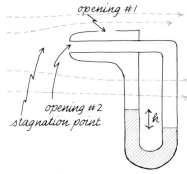

Prandtl or Pitot tube.

$$P_0 + \tfrac{1}{2}\rho v_0^2 = P + \tfrac{1}{2}\rho v^2$$

Since $v_0 = 0$ and the difference in pressures is

$$(P_0 - P) = \rho_m g h$$

$$\tfrac{1}{2}\rho v^2 = \rho_m g h$$

The instrument can be calibrated to read velocity directly in terms of the height difference h.

Suppose that such an air speed indicator uses oil with a density of 1×10^3 kg/m³ for the indicating fluid. What is the air speed if $h = 10$ cm?

$$(1 \times 10^3 \text{ kg/m}^3)(9.8 \text{ N/kg})(1 \times 10^{-1} \text{ m}) = \tfrac{1}{2}(1.2 \text{ kg/m}^3)v^2$$

$$v = 40 \text{ m/s}$$

With a choice of indicating fluid with the proper density, the instrument can be made sensitive in any of several different velocity ranges.

When liquid flows out of a hole near the bottom of a tank, the streamlines converge as shown in the diagram. Note that they are not parallel to each other as they leave the tank. The cross-sectional area of the stream ends up smaller than the area of the hole. Let's apply Bernoulli's equation to find the velocity of the stream as it leaves the hole. The top surface of the liquid is at a height h with respect to the hole.

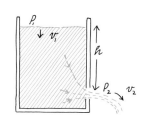

$$P_1 + \tfrac{1}{2}\rho v_1^2 + \rho g h = P_2 + \tfrac{1}{2}\rho v_2^2$$

For a container like the milk carton you used in "Handling the Phenomena," $P_1 = P_2$. Both pressures are atmospheric. (Note that P_2 is the pressure of the stream out in the air.) Furthermore, $v_1 \ll v_2$. The top level of the liquid does not fall fast compared with the velocity in the stream. If we say that v_1 is approximately 0, then we are left with the equation

$$\rho g h = \tfrac{1}{2}\rho v_2^2$$

$$v_2 = \sqrt{2gh}$$

When you observed the three streams coming from the holes in the milk carton, you probably could not compare velocities just by the appearance of the trajectories. If the water level over the bottom hole were four times that over the top hole, the horizontal velocity of the bottom stream would

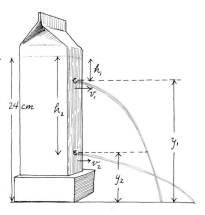

be twice that of the top stream. The horizontal *range* of the two streams, however, may not have been very different since the top stream had further to fall. In order to compare the velocities quantitatively, it would be necessary to express the horizontal range in terms of the horizontal velocity and the height.

> **Question 14-16**
>
> Take another look at the formula for the velocity of the stream coming out of the hole. If the liquid is sent through a tube with a 90° turn and shot straight upward with the same velocity, how high would it rise (assuming zero viscosity and therefore no friction losses in going through the 90° turn)?

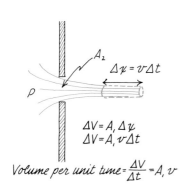

$\left(\dfrac{momentum}{unit\ volume}\right) = \left(\dfrac{mass}{unit\ V}\right) v = \rho v$

$\Delta x = v \Delta t$

$\Delta V = A_1 \Delta x$
$\Delta V = A_1 v \Delta t$

Volume per unit time $= \dfrac{\Delta V}{\Delta t} = A_1 v$

$F = \dfrac{\Delta(momentum)}{\Delta t}$

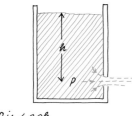

P is $< \rho g h$

In using Bernoulli's theorem to find the velocity of the stream, we were of course applying the law of conservation of energy. The velocity did not depend on the area of the hole and, furthermore, did not depend on the fact that the cross-sectional area of the stream is actually less than the area of the hole. Now let's take into account the fact that the stream is also carrying momentum. The time rate of delivery of momentum by the stream must be equal to a driving force within the tank. Presumably, the driving force results from the pressure of the liquid and ultimately must be provided by the opposite wall of the tank. There is an interesting problem in connection with this calculation, however. The momentum per unit volume in the stream is equal to the density times the velocity: (momentum/volume) = ρv. The volume of the stream that is emitted in unit time is equal to the product of the cross-sectional area A_1 and the velocity v, as shown in the diagram.

$$\text{Volume of stream per unit time} = A_1 v$$

Therefore the momentum carried away in the stream per unit time is equal to (momentum/unit volume)(volume emitted per unit time) = $(\rho v)(A_1 v)$ = $\rho A_1 v^2$. The momentum carried away per unit time must be equal to the force provided by the container: (momentum/Δt) = $F = PA_2$. The area of the hole is A_2, which we know is greater than A_1. In general, however, we do not know the ratio A_2/A_1 nor, surprisingly enough, do we in general know P. At first thought we would expect that $P = \rho g h$ since the pressure in a static liquid depends only on the density and depth of the liquid. The catch is, this liquid is not static. Near the hole inside the container the liquid is beginning to move fairly rapidly and therefore the pressure in the immediate vicinity of the hole must be less than it would be at that depth in a static situation.

Note the trap that we would fall into if we were not aware that $A_1 < A_2$. If we said $A_1 = A_2$, and $P = \rho g h$, then Newton's second law concerning the time rate of change of momentum would give us:

$$\rho v^2 = P = \rho g h \qquad \text{and} \qquad v = \sqrt{gh}$$

Compare this incorrect velocity with that we derived from the conservation of energy using Bernoulli's formula: $v = \sqrt{2gh}$.

Optional

In one special case we can choose a geometry so that we can be sure of the value of P and therefore find the ratio A_1/A_2. Instead of a hole in the container, we will use a re-entrant tube such as shown in the diagram. If the hole is small, the liquid in the container will be moving only in the immediate vicinity of the hole that is not near any of the walls. Therefore the walls will sustain only the static pressure. In particular, the section of the wall directly opposite the hole in the re-entrant tube will have only the static pressure and that must be the source of the force that provides the momentum change of the liquid. Our momentum formula therefore becomes

$$P = \rho g h = \rho v^2 (A_1/A_2)$$

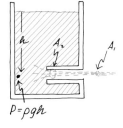

Since we already know that $v = \sqrt{2gh}$, it must be that $(A_1/A_2) = \frac{1}{2}$. In this particular geometry the contraction of the cross-sectional area of the stream can be calculated in a relatively simple manner. For an ordinary hole in the bottom of a tank, the ratio is between $\frac{1}{2}$ and 1 but the calculation is very difficult. For a round hole with a sharp edge, $(A_1/A_2) = 0.62$.

LIFT PRODUCED BY FLUID FLOW

When the streamlines passing around an obstacle in their path are not symmetric, the velocity of the flow on one side of an object may be different from the velocity on the other side. If the flow satisfies Bernoulli's law, then there will be lower pressure on the side with higher velocity. The resulting force on the object will have a major component perpendicular to the flow.

In the diagram we show the streamlines around two air foils. The first is symmetric and so are the streamlines around it. Note that a higher concentration of streamlines through a perpendicular area implies a higher velocity. Near the top and bottom of the air foil, the streamlines are closer together indicating that the fluid velocity in those regions is greater than it is far away from the foil. In the second case, the foil is asymmetric and so are the streamlines. There are two important characteristics of these lines. First, they crowd together above the humped section of the foil, indicating that the velocity on top is greater than it is on the bottom or in the rest of the stream. According to Bernoulli's law, the higher velocity must be accompanied by lower pressure. Consequently, we would expect a net lift force on the foil. This result would provide a very primitive explanation of how airplanes can stay up in the air. Notice another vital fact, however. The streamlines coming from both above and below the wing are deflected downward. The presence of the foil in the airstream generates a downward momentum in the air. This situation is absolutely necessary if the foil is to remain aloft and not fall. The upward force on the foil is equal to the time rate of change of momentum of the air. That change of momentum is necessarily downward. The down-wash from a wing can be felt in the immediate vicinity of the wing and produces particularly dramatic effects in the case of a hovering helicopter. The reason that you do not feel the down-wash or the excess pressure from a high flying plane is simply because the effect is spread over a large area.

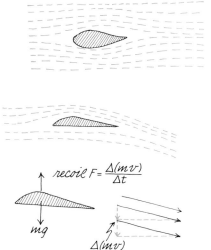

There is a way to create asymmetry of streamlines with a symmetrical

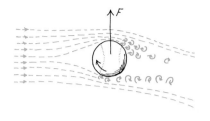

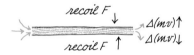

If a ball is thrown without spin (a knuckleball), turbulence builds up pressure at random on one side of the ball or the other, giving the ball an erratic flight path.

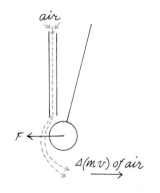

obstacle. The method is shown in the diagram. The streamlines going around a stationary baseball (or a baseball that is moving slowly without spin through still air) are symmetrical. Actually, for any reasonable velocity of throw, the streamlines behind a moving baseball break down into eddy currents or turbulence. This turbulence, representing friction, drastically affects the ball's motion. Nevertheless, the streamlines can persist partway past the ball and to the extent that they remain as streamlines, are subject to Bernoulli's law.

In the diagram, an asymmetry has been introduced by spinning the ball. The effect is legendary. The path of the spinning ball is curved. Baseball pitchers can throw curves, but the effect is even more noticeable with ping-pong balls. As you can see in the diagram, before the streamlines break up into turbulence, they crowd together more on one side of the ball than the other. A spinning surface of the ball increases the velocity on one side and decreases it on the other. Since the pressure is lower in the region with higher velocity, the ball moves sideways in that direction. Necessarily, in order to conserve momentum, air is swirled sideways in the opposite direction.

In "Handling the Phenomena," you were asked to blow between two small pieces of paper. The naive expectation would be that the pieces of paper would be blown apart. Instead, they are pulled together. Once again the explanation is that the region with higher velocity also has lower pressure. Remember, however, that in this case, too, the air must end up changing its momentum in the opposite direction from the thrust exerted on the obstacle. In this case if the two pieces of paper are drawn together, the air escaping out the edges must spread out in the two directions perpendicular to the paper. The recoil action of the stream that exerts a low pressure pull on an obstacle is easy to see when you hold a spoon in a faucet stream as suggested in "Handling the Phenomena." In this case the spoon is snatched into the stream because of surface tension as well as the higher velocity and lower pressure on the convex side of the spoon. Notice, however, what happens to the direction of the stream after it leaves the spoon. A force to the left on the spoon must be balanced by a force to the right on the water. The same demonstration can be performed with a stream of air as shown in the diagram. Again the naive expectation would be that the air would bounce off the ball toward the left, knocking the ball to the right. Instead, the ball is swung to the left, requiring that the stream of air swirl around the bottom of the ball toward the right.

THE REAL WORLD OF VISCOSITY

So far we have assumed that there is no friction in fluids. The great physicist-mathematician, John von Neumann, characterized this model as the study of dry water. Aside from the fact that turbulence is associated with most motion, even in streamline flow there is energy lost because of friction. Consider, for example, the flow of water through a long horizontal garden hose. According to Bernoulli's law, the pressure in the tube is constant since the cross-section and height remain constant. We all know very well, however, that the pressure along such a tube drops steadily as shown in the diagram. Indeed, since there is friction, if there is not a higher pressure at the beginning of one section than there is at the end of that section, why should the fluid in that section flow?

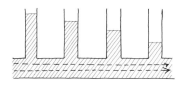

Pressure gradient in a pipe with liquid flow

We have also assumed that all the fluid at a particular point in a pipe is moving with the same velocity. The velocity at the center of the pipe is supposed to be the same as it is at the walls. The real situation, however, is radically different. *The velocity of a fluid at the wall of a container is always zero.* That is why fine dust remains on fan blades and car bodies. A leaf or a stone will be blown off the top of a car, but a fine grain of dust does not stick high enough above the surface to be affected by the moving air.

Suppose that we have a liquid between two flat horizontal plates. Let the top plate move with constant velocity to the right with respect to the bottom stationary one. Then there will be a layer of liquid attached to the bottom plate that has zero velocity and a layer of liquid attached near the surface of the top plate that has the velocity, v, of the top plate. Each layer of water in turn affects the layers next to it so that there is a continual change of velocity from top to bottom as shown in the diagram. This situation is very much like that produced by the shear forces described in Chapter 4, and pictured in the next diagram. A shear stress on a *solid* produces a *finite* strain measured by an angular deformation. For a given material, the strain is proportional to the stress. In a *fluid*, any shear stress can produce an *infinite* shear strain. The fluid does not provide spring-type resistance that could eventually buck out the shear stress. However, not all fluids yield to the shear stress at the same rate. If you apply a shear stress to tar, for instance, the shape of the tar will change continuously but slowly. If you apply the same shear stress to water, the change in shape will be very rapid. The *time rate of change of shear strain* in the case of fluids is equal to v/y where v and y are the variables shown in the diagram. The shear stress in this case is the same as it is for solids:

Velocity distribution in cross section of liquid flowing in circular pipe.

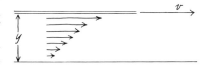

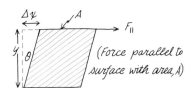

$\dfrac{F_{\parallel}}{A} \propto \theta$ (For solids)

Shear stress ∝ angular strain

$$\text{Shear stress} = F/A$$

In the case of simple fluids, the *time rate of change* of shear strain is proportional to the shear stress:

For fluids:

$$\Delta\theta = \dfrac{\Delta v}{y}$$

$$\dfrac{\Delta\theta}{\Delta t} = \dfrac{\Delta v/y}{\Delta t} = \dfrac{v}{y}$$

$$\text{Shear stress} = F/A = \eta \times (\text{time rate of change of shear strain}) = \eta(v/y)$$

$$\eta = \dfrac{F/A}{v/y}$$

The proportionality constant η (eta) is called the *viscosity* and for simple fluids is independent of the velocity. The table lists the values of viscosity for a number of fluids. The S.I. units for viscosity are determined by its definition.

$$\eta = \dfrac{\text{N/m}^2}{(\text{m/s})/\text{m}} = \text{N} \cdot \text{s} \cdot \text{m}^{-2}$$

This unit is not yet in common use. Instead, we use the old unit based on the centimeter-gram-second system. The old unit is called the *poise*, named after a nineteenth-century French physician and scientist Jean Poiseuille.

$$1 \text{ poise} = \tfrac{1}{10} \text{ N} \cdot \text{s} \cdot \text{m}^{-2}$$

Table of Viscosities—in Centipoise

Temperature °C	Water	Air	Mercury	SAE 10 oil	SAE 30 oil	Glycerin	Honey
0	1.79	0.017	1.68				
20	1.01	0.018	1.55	70	300	500	1500
40	0.66	0.019	1.44				
60	0.47	0.020					
80	0.36	0.021					
100	0.28	0.022	1.21				

To transform viscosity in centipoise to $N \cdot s \cdot m^{-2}$, multiply by 10^{-3}.
e.g. $\eta_{\text{water}} = 1 \times 10^{-3}\, N \cdot s \cdot m^{-2}$ at room temperature

Notice the enormous difference in viscosities between air, water, and viscous liquids. (Also notice the temperature dependence of η for water. As the temperature rises, the hydrogen-bonded clusters break up.) Notice also that the viscosity of air does not change much with temperature but that the change is one of an *increase* in viscosity as the temperature increases. On the other hand, for most liquids viscosity *decreases* with an increase in temperature. The viscosity of an ordinary motor oil is high at low temperature, making it hard to start the engine. When the engine is hot, the viscosity of the oil decreases, thus affording less protection to the engine just when it needs it most. Hence the fabricated oils now available are designed with several components so that the combination effect will be just the opposite.

Newton derived the viscosity relationship that we have shown. If a fluid has a viscosity that is independent of velocity, it is called Newtonian. Not all fluids act so simply. Fluids composed of suspensions such as colloids frequently have different viscosities at different velocities. For instance, as the velocity of blood increases, the cells orient themselves for minimum resistance to the flow and the viscosity decreases.

Kinematic Viscosity

The motion of fluids is frequently characterized by the *kinematic viscosity,* which is equal to the viscosity divided by the density:

$$\text{Kinematic viscosity} = \eta/\rho$$

The viscosity of air is much less than that of water and yet if a whirlpool is set up in a pail of air and a similar whirlpool in a pail of water, the air will come to rest sooner than the water. The viscosity is a measure of the internal friction in the fluid. The internal energy loss as the fluid moves is proportional to the viscosity. On the other hand, the kinetic energy that a moving fluid can lose is proportional to its density. The friction in moving air is very small but moving air has little energy to lose in the first place.

Viscosity of Gases: Temperature and Pressure Dependence

We might expect from our atomic model that the viscosity of liquids would decrease with increasing temperature. The hotter the liquid, the more easily the molecules can move past each other. With gases, however, the molecules are actually moving freely between collisions. As the temperature increases, the average velocity of the molecules and hence the average momentum transfer increases as the square root of the temperature. Since the internal friction, or the viscosity, depends on momentum transfer, we might expect that $\eta \propto \sqrt{T}$. Experimentally, the viscosity increases slightly faster with temperature because of the increasing penetration into each other of the molecules at higher velocities.

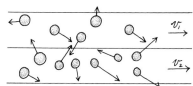

molecules in top layer have slow drift velocity, v_1, molecules in bottom layer have high drift velocity, v_2. The two layers affect each other by momentum transfers of their molecules. Since $v \propto \sqrt{T}$
$\Delta(mv) \propto \sqrt{T}$
and $\eta \propto \sqrt{T}$

Surprisingly enough, viscosity of gases is almost independent of the pressure. You might think that as the pressure decreases, the number of collisions would decrease, which would reduce the number of molecules coming from one region to transfer momentum to the adjacent plane. On the other hand, as the pressure decreases, the mean free path of molecules increases and so molecules transferring momentum from one moving plane can affect a region further away. These two effects just cancel each other so that the viscosity in gases is independent of the pressure over a large range. When motion through a gas is dominated by viscosity effects, such as with drifting dust particles, the terminal velocity of an object is independent of the air pressure. Although it is true that at 1% atmospheric pressure, a feather will fall with about the same acceleration as a coin, the gradual fall of a tiny dust particle would be no faster at this low pressure than it would be at normal atmospheric pressure. In the next section we'll take a closer look at the conditions that determine whether motion through a fluid is dominated by viscosity or by inertial effects.

Fluid Flow Through Tubes

Poiseuille, for whom the poise is named, investigated the flow of liquids through thin tubes. Being a physician, he was primarily interested in the circulation of blood. As we have pointed out, the flow of a real liquid does not follow Bernoulli's law because of the loss of energy due to viscosity. Since the velocity of a real liquid is zero adjacent to the wall of the container, the velocity pattern of a liquid moving through a tube follows the pattern shown in the diagram. The effective size of the tube decreases drastically with radius since, for small radii, the low velocity near the wall affects most of the liquid in the tube. Poiseuille demonstrated that the volume of the liquid passed in unit time through a round tube with radius r and length L is given by the formula

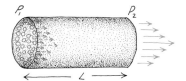

$$\frac{V}{\Delta t} = \frac{\pi r^4 (P_1 - P_2)}{8 \eta L}$$

The effective pressure across the tube is $(P_1 - P_2)$. We might expect that the volume per unit time would be proportional to the net pressure per unit length along the tube and that the volume of flow should be inversely proportional to the viscosity. If the velocity of flow were uniform across the entire cross-section of the tube, then we might expect that the volume of flow would be proportional to the area of the cross-section: πr^2. Instead, because of the wall effect, the volume of flow is proportional to r^4.

Superfluids

There is such a thing as a superfluid. Below 4.2 K, at atmospheric pressure, helium becomes a liquid. As the temperature is reduced below 2.19 K, the viscosity of part of the liquid drops to zero. Without internal friction, the superfluid can penetrate tiny cracks much more rapidly than gaseous helium can. Because the atoms in the superfluid can slide over each

Solid Fluids

Glass flows when it's hot enough.

other without energy loss, they can rapidly transmit disturbances throughout the liquid. Consequently, superfluid helium can conduct heat better than any other substance (about 100 times better than copper).

The difference between a liquid and a solid is determined by their response to shear forces. A crystalline solid will respond to a shear force with an angular distortion, the strain, which is proportional to the stress. But the distortion does not keep on increasing. With a fluid, however, the distortion continues indefinitely even though very slowly, so that the fluid eventually assumes the shape of the container that it is in. Many solids, however, are not crystalline and actually flow. Glass is a good example. If you support a glass rod horizontally at the two ends, it will eventually sag permanently in the middle. When glass is hot enough, it is obviously a fluid. For instance, at the temperature at which it is blown, the viscosity is about 10^7 poise.

Even crystalline solids can flow. If the strain exceeds the elastic properties of the material, the crystalline layers start sliding over each other. This effect is evident when you stretch a thin wire to the breaking point. At the point where the break begins, the wire not only lengthens, but becomes much narrower. Under sufficient pressure, such as those that exist deep in the earth, any material responds like a fluid.

■ **Optional**

The Flow of Real Fluids

The motion of real fluids or the motion of objects through fluids can be strongly influenced by the viscosity of the fluid and the adhesive forces between the molecules of the fluid and the object. The existence of a boundary layer around an object where there is a velocity transition from zero to full velocity may have a major effect on the shape of the surrounding streamlines. Furthermore, in many situations the fluid does not follow streamlines but changes paths in a chaotic way called turbulence. Turbulent flow not only mixes up the mass elements of the fluid but also averages out velocity and momentum throughout the turbulent region. The turbulent region contains extra kinetic energy that must have been provided by the source of energy of the flow.

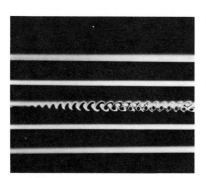

von Karman vortices. (Courtesy F. N. M. Brown, U. of Notre Dame.)

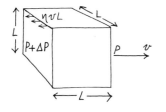

Let's analyze the forces acting on a tiny chunk of fluid that is being swept into a current from an initial speed of zero. The driving force is equal to the product of the cross-sectional area of the chunk and the pressure difference from one end to the other. A friction force reduces the effect of the driving force. On page 363 we showed that the viscous stress on a small chunk of fluid is equal to $\eta(v/y)$. The viscous force is equal to the product of this stress and the cross-sectional area of the fluid chunk. Let's assume for the moment that the chunk is cubic with a side length L. In our formula for viscous stress, y is thus equal to L. The total viscous force acting on the cube of fluid is equal to $\eta(v/L)L^2 = \eta vL$.

The net force acting on the fluid cube will produce an acceleration:

$$L^2 \Delta P - \eta vL = ma$$

The term on the right is the inertial response to the net force. We can express this inertial response, at least approximately, in terms of the density, size, and

THE REAL WORLD OF VISCOSITY

velocity of the cube of fluid. To begin with, note that $m = \rho L^3$. Next, $a = \Delta v/\Delta t$. Since the cube starts from rest, the change of velocity Δv is equal to v. The acceleration cannot occur in less time than it takes the cube to move through its own length: $\Delta t = L/v$. The inertial reaction is then equal to

$$ma \approx (\rho L^3)\frac{v}{L/v} = \rho L^2 v^2$$

The behavior of a moving fluid can be characterized in terms of the relative importance of the inertial reaction and the viscous drag. The ratio of these two terms, as we have defined them, is called the Reynolds number, named for Osborn Reynolds (1842–1912), an Anglo-Irish scientist.

$$R = \frac{\text{inertial reaction}}{\text{viscous drag}} = \frac{\rho L^2 v^2}{\eta v L} = \frac{\rho L v}{\eta}$$

Question 14-17
What are the *dimensions* of the Reynolds number? (To find the dimensions of viscosity, refer to the units given on page 363.)

The density, viscosity, and velocity of a fluid are specific attributes of the flow, but the significance of L is not so obvious. We originally defined L as the side of the tiny cube of fluid whose behavior we were analyzing. It appears that the Reynolds number depends on the size of this sample cube. In practice, we let L equal the major characteristic length of the fluid under consideration. For instance, for a dust particle or a raindrop falling through air, the diameter of the particle or the drop is approximately the size of the disturbed chunk of air under consideration. An airplane will disturb air all around it in a region with a characteristic dimension of the length of the wing (actually the disturbed air behind the plane extends over much greater distances but that is due to more complicated effects). As you can see, the assignment of the proper size for L is inexact by a factor of 2 or more. Nevertheless, the order of magnitude of the Reynolds number is useful in characterizing the type of motion that the fluid undergoes. For instance, if R is between $1/10$ and 10, both the inertial reaction of the fluid and its viscous drag play an important role in the motion. However, if $R < 1/1000$, we can ignore the inertial reaction. Viscous drag dominates the motion and there is essentially no acceleration of the fluid. If $R > 1000$, the viscous drag is negligible although, as we will see, the viscosity may still produce changes in the flow pattern that will dominate the phenomena. Usually with $R > 3000$ the flow is turbulent.

Another nineteenth-century Anglo-Irish scientist, George Stokes (1819–1903), analyzed the laws of motion for small R. In this region objects move with constant velocity with a drag force proportional to $\eta v L$. Note that the friction is proportional only to the first power of the characteristic size. The drag on a spherical droplet, for instance, is proportional to the diameter and not to the cross-sectional area. The friction force is proportional to the first power of the velocity and is also proportional to the viscosity of the fluid. The drag does not depend, however, on the density of the fluid, *since the fluid is not accelerated.*

Note that viscous behavior depends on the combination of factors represented in Reynolds' number and not just on the viscosity of the fluid. The viscosity of air, for instance, is very small and yet tiny dust particles or droplets travel at terminal velocity in air with zero acceleration *of the air,* and hence are in the viscous regime. It is also interesting, though surprising, that the drag force does not depend on the density of the fluid. As we pointed out on page

365, the terminal velocity of dust particles in air is the same at a pressure of 1 atmosphere or 0.01 atmosphere. Although the fluid density is different by a factor of 100 in these two cases, the viscosity of a gas is independent of pressure.

Let's compute the Reynolds number for a dust particle and a raindrop. Assume that the dust particle has a characteristic size L of 0.1 mm, and that for the rain droplet, $L = 5$ mm. The dust particle is usually not solid and therefore has a density less than that of water. Its velocity might be 10^{-3} m/s. The velocity of a large raindrop might be 10^{-1} m/s. The density of air is 1.3 kg/m³. The viscosity of air at room temperature is 180 micropoise = 1.8×10^{-4} poise = 1.8×10^{-5} Nsm^{-2}. For the dust particle:

$$R = \frac{\rho L v}{\eta} = \frac{(1.3 \text{ kg/m}^3)(10^{-4}\text{m})(10^{-3}\text{m/s})}{(1.8 \times 10^{-5}\text{ Nsm}^{-2})} = 10^{-2}$$

According to the assumptions that we made, this particular dust particle is in the viscous regime. The Reynolds number for the rain droplet is larger by a factor of 5×10^3 (L is larger by a factor of 50 and v is larger by a factor of 100). Since, according to our assumptions, R for the raindrop is equal to 50, this means that both viscous drag and inertial reaction are important for the falling rain drop. The drop itself may be falling at terminal velocity but the acceleration of the air as it gets out of the way has an important effect on the resulting motion.

Question 14-18
What is the ratio of density to viscosity for water at room temperature? When you swim or boat are you in the region of large or small R?

Most human actions take place at large Reynolds numbers. However, for microbe-sized creatures in air or water, motion is completely dominated by viscous effects. Surface effects also become very important for such creatures because of surface tension and because the boundary layer for transition between 0 velocity and full velocity becomes as thick as the object itself.

For Reynolds number > 100 or so, the drag force is proportional to $v^2 \rho L^2$. Note that the viscosity does not enter in although, as we will see, it can produce very important effects that act like drag forces. The drag is proportional to the density of the fluid since an object moving through a fluid must shoulder the fluid aside and so accelerate it. The drag is proportional to the cross-sectional area of the disturbed fluid L^2 *and is also proportional to* v^2.

Let's calculate the Reynolds number for a thrown baseball. Let $L = 0.1$ m and $v = 30$ m/s.

$$R = \frac{(1.3 \text{ kg/m}^3)(1 \times 10^{-1}\text{m})(30 \text{ m/s})}{(1.8 \times 10^{-5}\text{ Nsm}^{-2})} = 2 \times 10^5$$

Evidently, the drag on a baseball is proportional to its cross-sectional area and to the square of its velocity. The Reynolds number is also well into the region where turbulence in the flow around the ball must be a major factor in its behavior.

Even at large Reynolds numbers, viscosity can affect the boundary layer where the velocity is very small. The viscous region of the boundary layer itself usually does not add much drag to the motion of an object. It does, however, affect the streamlines in the vicinity of the boundary and thus influences the way turbulence builds up in the wake of the object. In the diagrams we show streamlines around a sphere and a "streamlined" foil. The drag on either shape is caused by two effects. There is resistance to the motion due to the pressure changes in the vicinity of the object and there is resistance caused by the

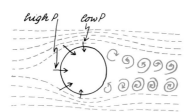

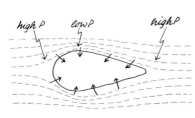

viscous effects as the fluid passes the boundary. In the case of the streamlined foil, a chunk of the fluid passing from in front of the foil up to the region of the shoulder would go from a high pressure region to a low pressure region. In doing so, it exerts a backward shove on the foil as shown in the diagram. However, as the chunk of air moves down the streamline following the long tail, it moves from low pressure back to high pressure and in doing so exerts a thrust on the foil, sending it forward. As long as all the fluid follows streamlines the backward pressure on the foil in front is matched by the forward thrust along the tail. The net pressure resistance would be zero. The only drag would be caused by the viscous friction along the boundary layer. If the viscosity were zero or negligible, a foil would move through a fluid with zero resistance.

Look what happens, however, in the case of fluid moving past a sphere. The transition between zero velocity at the surface and high velocity a short distance away becomes too abrupt and flow becomes unstable. Turbulence sets in. Along the surface where there is turbulence, there is no net thrust on the ball. Consequently, there is resistance to the motion due to the pressure drop from in front of the ball up to the point where turbulence sets in and there is no forward thrust to make up for this loss. The resistance to motion is thus very large. The energy lost to this fluid friction becomes the kinetic energy of the turbulent region.

Although the viscosity drag on a baseball is small compared with the pressure drag, it is the viscosity that creates the conditions leading to turbulence. As a chunk of fluid moves from in front of the ball up the curved surface, it loses velocity and hence energy because of viscosity at the boundary layer. Once the chunk of air gets up to the top of the curve, it does not have enough momentum to work its way down from the region of low pressure to high pressure at the tail and so it stalls. This gives rise to the instability of turbulence. Even a streamlined foil can suffer stall at the wrong angle as shown in the diagram. Once the turbulence begins too high up on the convex surface of the wing, the lift drops abruptly and the plane will drop.

From the diagrams that we have just shown, it is evident that a ball moving through air must meet a lot of wind resistance. Golf balls have a dimpled surface. It might seem that this roughness would produce extra viscous drag and it does. At low speeds, a dimpled golf ball has greater air resistance than a smooth one. At high velocities, however, the pressure drag is more important and the dimpled golf ball has much less pressure drag than a smooth one. The boundary layer of a rough surface is turbulent and clings to the ball far down toward the tail as shown in the diagram. The turbulent *wake* is relatively small and a considerable region behind the ball exists where forward thrust is exerted by the pressure change. Since the boundary layer on a moving object may be very thin, and since the action in the boundary layer determines where turbulence begins, small changes in the smoothness of a surface may produce large effects in the drag forces. Notice that on airplane wings the rivet heads are flush with the rest of the surface.

Flow lines around sphere with rough surface

Although the actual value of the Reynolds number for a particular situation can only indicate flow behavior within an order of magnitude, the comparison of Reynolds numbers for two different situations can be made very precise and useful. It is useful and even necessary to study the possible motions of very large objects by making small models of them. Most air foils and whole airplanes are tested with models in wind tunnels before they are made full size. Many properties of moving ships are also tested in small-scale water tanks. The dynamic effects of the model correspond to those of the real object if and only if the Reynolds numbers are the same for both the model and the real situation.

Consider the consequences of this requirement, however. Suppose that you want to make a small-scale model of a jet plane and test it in a tabletop wind tunnel. For a 1/10 scale model, the characteristic length L of the model is smaller than that of the real object by a factor of 10. If you use air in your wind tunnel, the density and viscosity are about the same as they would be for the jet plane. To make the Reynolds numbers the same, the velocity of the air in the wind tunnel would have to be 10 times that expected for the real airplane. You could not obtain this velocity with a tabletop air tunnel and, even if you could, the analysis would be far more complicated because the velocity would be much greater than the velocity of sound.

Question 14-19
Could you still make a small-scale test arrangement using some other fluid?

SUMMARY

A fluid is any substance that flows continuously under a steady shear force.

Fluids can be liquids or gases or, under extremely high pressures, even solids. While many everyday phenomena can be explained, at least qualitatively, with a simple theory of fluids, the explanation for many other common effects requires very complex calculations.

Fluid pressure is explained at the molecular level by compression of molecules against their neighbors as they are bound in potential wells, rather than by molecular bombardment as with gases. The pressure at any point in a fluid is uniform in all directions. At a depth, h, below a surface: $P = P_0 + \rho g h$. This equation governs the operation of hydraulic levers and siphons.

Archimedes' principle is that an object in a fluid is buoyed up by a force equal to the weight of the *fluid displaced*. Consequently, a submerged object appears to have less weight: $(W_{\text{in fluid}})/(W_{\text{in vacuum}}) = 1 - \rho_f/\rho_0$. A floating object displaces a quantity of fluid with a weight equal to its own.

The cohesion of molecules gives rise to a surface tension at the boundary between fluid and air or solid. The energy needed to expand the surface area is equal to $\gamma \Delta S$, where ΔS is the change in area and γ is surface tension measured in N/m. Surface tension is responsible for many phenomena where the surface to volume ratio of a fluid is large, such as with bubbles, surface ripples, and thin streams. Capillary action is caused by a combination of cohesion and adhesion, and the requirement that a fluid surface be perpendicular to the net force acting on its molecules.

For ideal fluid flow, there must be zero viscosity (no friction losses), no rotation or turbulence, and an incompressible fluid. The flow can be traced by streamlines that never cross. Under these circumstances, conservation of energy requires Bernoulli's law: $P + \frac{1}{2}\rho v^2 + \rho g h =$ constant. According to this law, if height remains constant, the pressure will be low in a region where the velocity is high, and vice versa. Venturi and Pitot tubes work on this principle and provide a means of measuring the velocity of a fluid.

Fluid flowing around an asymmetric object will be deflected asymmetrically. The change in momentum of the fluid in one direction will be matched by a thrust on the object in the other direction. In airplanes this effect provides lift. The velocity of the fluid is faster on the side away from the subsequent deflection, providing lower pressure on the lift side.

In the real world, fluids have viscosity. Furthermore, the velocity of a fluid is zero at a boundary (with respect to the boundary). Viscosity, η, is the proportionality constant between shear stress, F/A, and the time rate of change of shear strain, v/y. $\eta = (F/A)/(v/y)$. For simple fluids, viscosity is independent of velocity. Some phenomena are better characterized by the kinematic viscosity, η/ρ. The atomic model explains why for gases $\eta \propto \sqrt{T}$ and why η is almost independent of the pressure. For liquids, η usually increases as T decreases. Because of friction losses, and because the velocity of flow is zero at the walls, the flow in a tube in terms of volume per second is inversely proportional to the length of the tube and proportional to the fourth power of its radius.

The flow of real fluids past objects can be characterized by a dimensionless quantity, the Reynolds number. $R = $ (inertial reaction)/(viscous drag) $= \rho L v/\eta$. For $R < 1/1000$, the inertial effects of the fluid are negligible. In this region, drag is proportional to $\eta v L$, and is not a function of the density of the fluid. For $R > 1000$, the viscous drag is negligible (although it may affect the turbulence pattern). In this region, which concerns most human motions in air or water, the drag is proportional to $\rho v^2 L^2$. To be valid, scale models of objects moving through a fluid must have the same Reynolds number as their real counterpart.

Turbulence, rather than streamlines, dominates the motion of most objects in a fluid when $R > 3000$. The turbulence absorbs energy and produces drag.

Answers to Questions

4-1: The horizontal range of the stream is proportional to the horizontal velocity. However, the range is also proportional to the time of flight. The flight time is proportional to the square root of the height through which the stream falls. ($y = \frac{1}{2}gt^2$). A stream coming from a hole near the top of the carton where the pressure is low still may have a long range because the flight time is long.

4-2: If the density *increases* by 5%, the volume of material with a particular mass must *decrease* by 5%. In each of the *x, y, z* directions, the intermolecular distance must decrease by $\frac{5}{3}$%.

4-3: Since the heights of the two towers are the same, the pressure heads provided will be the same. However, the purpose of having a water tower is not to provide a pressure head but rather a *constant* pressure head. After all, the pressure had to be supplied by the pump that got the water up there in the first place. If that pump were to feed directly into the mains, the fluctuations in pressure would ruin the system. An ideal water system should supply constant pressure regardless of how much water is being drawn. The larger the reservoir, the more constant the pressure can remain in spite of fluctuating supply from the pump and fluctuating demand from the system.

4-4: No, as long as the lower end of the exhaust tube is below the upper surface of the liquid being drained, the siphon will work. Note the pressure relationships in the diagram on page 346.

4-5: The pressure in the upper leg of the siphon decreases steadily from the atmospheric pressure at the upper surface. $P = P_0 = \rho g h_1$. The siphon will not raise water higher than the point where the pressure is equal to 0. For mercury, this height is 76 cm (at STP). For water the height is 10⅓ m or about 34 feet. Actually under the right circumstances, liquids can withstand negative tension and so can rise to even greater heights in siphons and capillary tubes.

14-6: You know we don't. In order to raise the system on the right by 1 cm we must push the piston on the left 100 cm down. The work done on the left piston is equal to $W = (10 \text{ N})(1 \text{ m}) = 10$ J. The resulting work that can be done by the piston on the right is equal to $(1 \times 10^3 \text{ N})(1 \times 10^{-2} \text{m}) = 10$ J.

14-7: For the stone, $1 - \rho_f/\rho_o = 1 - (1000 \text{ kg/m}^3)/(2500 \text{ kg/m}^3) = 0.6$

$$W_{\text{in fluid}} = (0.6)W_{\text{in air}} = 30 \text{ N}$$

14-8: The density of the *hollow* boat is considerably less than that of water and so the whole object floats, displacing a volume of water considerably less than the volume of the whole boat.

14-9: If you were weighing brass, the corrections would cancel. However, if you are weighing material that has a density close to that of water, 1000 kg/m³, then the correction due to the buoyancy of that material in the air would be about $13 \times 10^{-2}\%$. The net effect would therefore be an error of about 0.1% or 1 part per thousand. Chemical weighings frequently must be done to precision better than that.

14-10: The attractive forces of cohesion pull the surface molecules toward the greater number of molecules in the body of the liquid. But that force is balanced by the repulsion of the molecules directly underneath as the surface molecule presses slightly closer to them than the normal intermolecular spacing.

14-11: Note that the ring picks up a film along both its inner and outer edge. Therefore the force needed is

$$F = 2\gamma(\text{circumference}) = 2(0.073 \text{ N/m})\pi(0.06 \text{ m}) = 2.75 \times 10^{-2} \text{ N}$$

This force is the weight of 2.8 g.

14-12: Air bubbles are formed most easily in a liquid that has low surface tension, low density, and relatively high viscosity. Consider an extreme case. Have you ever seen an air bubble in mercury?

14-13: The surface of the liquid is always perpendicular to the net force acting on the surface molecules. In a horizontal surface, for instance, the net force of cohesion on each molecule is down. (Note that the net force on a molecule is not an unbalanced force that produces acceleration. The attractive net force of cohesion and adhesion is balanced by the molecular compressive force of other molecules). The *surface tension* is always directed along the surface, and it's the surface tension that pulls the liquid up, or down, depending on the curvature of the surface.

14-14: The fluid can move with different velocities at different points within the tube. Where the cross-section is large, the velocity is small and vice versa.

14-15: Since the fluid is incompressible, there must be the same volume per second transported past any point.

$$(\Delta V_1/\Delta t) = (\Delta V_2/\Delta t)$$

$$A_1(\Delta x_1/\Delta t) = A_2(\Delta x_2/\Delta t)$$

$$A_1 v_1 = A_2 v_2$$

$$v_1/v_2 = A_2/A_1$$

14-16: The velocity of the liquid coming out of the hole is just sufficient to shoot it back up to the top level of the surface that provides the pressure head. (With most tubes and 90° bends, friction will take a large toll.)

14-17: The first term in Reynolds number is ρ, the mass density. It has dimensions of M/L^3. The characteristic length L, has the dimensions L. The velocity v has the dimensions

L/T. The units of viscosity η are Nsm^{-2}. Therefore the dimensions of viscosity are $(ML/T^2)(T)(L^{-2}) = M/TL$. When we combine these, we find that they all cancel. Reynolds number is dimensionless.

-18:
$$\frac{\rho}{\eta} = \frac{1 \times 10^3 \text{kg/m}^3}{1 \times 10^{-3} \text{Nsm}^{-2}} = 10^6 \quad \text{(water)}$$

Note that the ratio of density to viscosity for *air* is equal to

$$\frac{\rho}{\eta} = \frac{1.3 \text{kg/m}^3}{1.8 \times 10^{-5} \text{Nsm}^{-2}} \approx 10^5 \quad \text{(air)}$$

Since your characteristic length when you are swimming or are in a small boat is about 1 m and any reasonable motion would have a velocity of the order of 1 m/s, then your Reynolds number would be equal to or greater than 10^6. You are in the regime of large R. Note that the same argument holds for your motion through the air.

-19: Remember that for water the ratio of density to viscosity is over ten times greater than it is for air. A $\frac{1}{10}$ scale model in water has the same Reynolds number (for the same velocity) as a full scale object in air.

Problems

1. What is the fractional increase in pressure at the bottom of a glass of water?
2. What is the total pressure (in N/m² and in atmospheres) at the bottom of a swimming pool that is 3 m deep?
3. In a rectangular backyard swimming pool with sides of 10 m and 5 m, the water is filled to a depth of 1.5 m. What is the total force of water pressure on the bottom and on each side?
4. Give in your own words the operational definition of a fluid.
5. A water container is made in the shape of a thin horizontal disk with the large circular bottom consisting of a rubber diaphragm. Water can be poured into the container through a tall cylindrical tube or chimney rising along the axis of the disk. The water in the tube does not add much to the weight of the water in the disk, but the pressure everywhere on the bottom must be proportional to the height of the water in the tube. Why won't the total force exerted downward by the rubber diaphragm be more than the weight of the water and disk?
6. The cross-sectional area of a piston in a foot brake is 5 cm². The brake cylinder has a cross section of 75 cm². What force must be exerted by the foot to create a force of 1500 N on the brake? If the foot pedal is depressed 8 cm, how far does the brake itself move?
7. Some siphons have an attached squeeze bulb to create a partial vacuum to start the siphoning process. If gasoline must be raised 30 cm to get it over the top of a tank so that it can start its downward flow, what pressure must be created in the tube? The density of gasoline is 700 kg/m³.
8. An ice cube with a density of 920 kg/m³ floats in a glass of water. What fraction of the volume is out of the water? As the ice cube melts, what happens to the level of the water?
9. Styrofoam has a density of about 100 kg/m³. If a piece of it is weighed on a pan balance, using brass weights with a density of 8500 kg/m³, what is the percentage error due to the buoyant effect of the air, which has a density of 1.3 kg/m³?
10. The density of gold is 19.3×10^3 kg/m³ and the density of silver is 10.5×10^3 kg/m³. If King Hiero's crown were 50% silver by weight and had a mass of 4 kg, what would have been its volume? What would have been its volume if it were pure gold? If the crown were submerged in a cylinder 25 cm in diameter

(big enough for the crown), what would be the rise of water in the two cases?
11. Suppose that Archimedes had weighed the crown in air and in water. Find the ratio of weights for the two cases given in Problem 10. What would have been the difference in grams between the gold crown and the one with 50% mixture?
12. One type of formed brick has a density of about 1200 kg/m^3. Calculate the ratio of the brick's weight in water to its weight in air.
13. Draw a diagram of a Cartesian diver and explain in your own words how you can make the diver rise or sink.
14. Is the surface tension of water really sufficient to support a needle? A typical needle has a length of 3.5 cm and a mass of 0.1 g. On the surface of the water it lies in a trough. The tension along the surface must be at least as great as the downward weight of the needle. Calculate the forces to see if this is the case.
15. How high can water rise in a capillary tube with a radius of 2 microns (2×10^{-6} m)? Assume that the water wets the tube, so that $\theta = 0$.
16. In a Venturi tube the main section of the pipe has a radius of 4 cm and the constricted section has a radius of 2 cm. The standpipe in the large section supports a water column of 30 cm; the water column in the narrow region stands only 10 cm high. What is the velocity of the water in the main pipe?
17. In a particular Pitot (or Prandtl) tube, mercury with a density of 13,500 kg/m^3 is used. What is the air speed if the difference in liquid levels is 30 cm?
18. Use your own words and an air flow diagram to explain the lift of an airplane wing.
19. Use your own words and diagram to explain the rise of coffee in the straw as described in "Handling the Phenomena."
20. Use your own words and diagram to explain what happens when you hold the convex side of a spoon under a stream of water. Be sure to indicate the direction of water flow after it leaves the spoon.
21. Explain why a whirlpool persists longer in water than in air.
22. Mercury and water are being piped through separate tubes but with the same pressure head. The mercury tube is 1 m long and has an inside radius of 0.5 cm. The water tube is 1000 m long and has an inside radius of 1.0 cm. What is the ratio of water flow to mercury flow?
23. Explain in your own words how and why motion through a fluid differs in the two extremes of R value: <1000 and >1000.
24. A bead with a diameter of 3 mm is dropped into a heavy oil with a viscosity of 1×10^{-1} Nsm^{-2} and a density of 1800 kg/m^3. It falls with constant velocity of 1 cm/s. What is the Reynolds number?

TRAVELING PULSES AND SIMPLE WAVES 15

We are surrounded by vibrations. The vibrating air brings sounds to our ears. Vibrating electric and magnetic fields bring light to our eyes. Beneath our feet the earth vibrates like a giant bell. In all these cases, a disturbance in one region causes energy and information to be transmitted to another region.

However, no actual material is transported by waves. In the case of sound or water waves or earth waves, the disturbance is propagated through changes in an elastic medium. Air can oscillate between high and low pressure. The surface of water can rise and fall. The earth can vibrate both up and down and back and forth along its surface and throughout its interior. The transmission of light requires no physical medium, but oscillating electric and magnetic fields recreate and sustain each other as the pulse travels.

The remarkable feature about wave motion is that under the right circumstances it can transport energy over vast distances with very little loss. The earth is heated by electromagnetic radiation coming from the sun. We, in turn, can send back radio waves to the planets or beyond. A sudden shift of the earth in one location can produce vibrations that destroy houses many miles away. An earthquake under an ocean can create a *tsunami,* sometimes called a tidal wave. On April 1, 1946, an earthquake near the Aleutian Islands made the Pacific Ocean slosh with such an amplitude that villages on shores 4000 miles away were destroyed.

In this chapter, we will study phenomena concerned with the transmission of single pulses or simple waves. Why does a disturbance in an elastic medium propagate? How does the velocity of propagation depend on the properties of the medium and of the wave itself? What happens when a wave strikes a boundary of the medium, or passes from one medium to another, such as from air to water? What determines the amount of energy transmitted in the wave? Why is a "simple" wave simple? For such a wave, what relationships exist among frequency, wave length, and velocity?

We will derive the few basic principles that govern wave motion by examining waves that are easily produced and seen in long ropes or on the surface of water. We can then apply these principles to explain many phenomena that we see or hear or feel every day.

HANDLING THE PHENOMENA

1. You can see pulses and waves travel and reflect by using a rope or heavy wire, the longer and heavier the better. Fasten one end to something that can't move. Pull the rope taut and send a pulse down the line with a quick sideways snap of your hand. Note what happens to the pulse as it hits the fixed end. In particular, if you send a pulse down the rope on your right side, on what side does the pulse return?

Watch what happens to the speed of the pulse as you hold the rope slack or very taut. If you have several different thicknesses of rope, compare the pulse speeds in light and heavy ropes.

Tie a heavy rope to a very long light cord or string. Tie the other end of the string to a fixed support and then send pulses down the rope. Watch what happens to the pulses as they travel from one medium to the other. What happens to the pulse speed? If there is a reflected pulse at the junction, does it come back on the same side or the opposite side?

An excellent "rope" to use for these studies is a garden hose. You can easily change the density of the hose by a large amount simply by filling it with water (providing that you have a shut-off nozzle at the end). Observe the difference in the velocity of a pulse if the hose is full of water.

2. You can see pulse and wave phenomena of many different types on the surface of a glass of water or by looking at the water in a washbowl or, best of all, a pond. At home, try using both a washbowl and a baking pan. Watch the waves spread out on the water surface in a partially filled washbowl with a slowly dripping faucet. Notice the shape of the advancing wave fronts and see what happens when those wave fronts hit the sides of the bowl.

With a large flat baking tin, you can easily control the depth of the water. Keep the depth below 1 cm. By raising the pan at one end, you can produce a gradually changing depth. See what happens to the velocity of a ripple as the depth goes to zero.

With the pan, you can produce pulses from a point source by dipping your finger in the water, or you can produce long straight waves with a ruler or piece of cardboard. The ripples will reflect from the boundaries of the pan. Look at the reflections from ripples that strike a boundary at an angle of about 45°.

You can produce a small scale tsunami by raising one end of the pan slightly and then letting it back down again quickly. Notice how little energy is absorbed from the advancing wave until it strikes a boundary. Even then there will probably be enough energy left to produce many sloshes back and forth.

3. If you have a pond available, you can produce many of these effects on a grander scale without the interference of close boundaries. In particular, toss a stone into a pond and observe the amplitude of the wave front as it spreads away. Try estimating the velocity of ripples on the pond. If you can reach out with a stick or with your finger and produce an oscillatory motion up and down in the water, you will observe a steady train of waves going away from the source. Try two different frequencies of oscillation that you can measure conveniently—say, once a second and twice a second. Estimate the distances between crests in each case. The relationship among the

distance between the crests, the frequency, and the velocity is: (distance between crests)(frequency) = (velocity). Rough estimates of these quantities are usually good enough to demonstrate that the relationship is at least plausible.

4. If you have access to any kind of a stringed instrument, observe the relationships among pitch, the length of the string, the tension of the string, and the density or thickness of the string. In a violin or guitar, the strings are all approximately the same length. But the active part of the string can be shortened by pressing the string against one of the frets. There are obvious differences in density or thickness between the highest pitch string and the lowest pitch. The pitch of the string can be adjusted by changing the tension. Change these various factors enough so that you know qualitatively how each affects the pitch produced by the string.

Open up a piano and observe the structure of the mallets, the strings, and the sounding board. Note particularly the difference between the high pitched strings and the low pitched strings, not only in length but also in thickness or density. With the piano open, sing or whistle as pure and loud a note as you can at the strings. You should be able to hear at least one of the strings singing the same note back at you.

Stretch a rubber band between your fingers and pluck it to produce a sound. You can easily change the tension to produce higher or lower notes. These sounds will probably be weak and not very musical. Now press the stretched rubber band across a drinking glass. Pluck the band again and you will observe quite a different sound. Note that most stringed instruments have sounding boards or resonant air chambers of some type.

5. You do not need a mountain valley in order to produce and hear echoes. A large flat brick wall will do just as well. However, the wall must be large enough so that even if you back off 30 to 40 meters from it, the wall will still subtend a sizable angle from you. Clap your hands sharply and listen for the echo which will occur almost immediately. When you are sure that you have heard it, you can measure the time it takes for the sound to leave your hands and return again by rhythmically clapping your hands at such a frequency that the echo is between claps. If you are clapping at three times per second, then the flight time of the sound must be $\frac{1}{6}$ of a second. Therefore, the sound must take $\frac{1}{12}$ of a second to get from your hands to the wall (and $\frac{1}{12}$ of a second to return). You can find the speed of sound by pacing off your distance to the wall and dividing by that $\frac{1}{12}$ of a second.

6. If you have a good stereo sound system, you should invest in one of the many hi-fi test records that are available. Most of these have a band that plays several seconds of sound at each frequency range, from very low pitch sounds to very high. If you can use one of these, test your own set and your own hearing to determine your frequency range and also to find the frequencies of familiar notes. For instance, find the frequency of middle C or the A to which an orchestra tunes.

DISTURBANCE OF AN OSCILLATOR

In Chapter 6, we described the special kind of oscillation of a bob on a spring. The action was called simple harmonic motion.

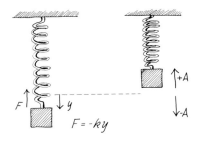
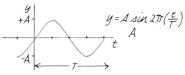
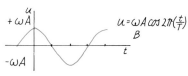
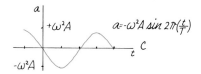

> **Question 15-1**
>
> What is simple about the motion, and why is it called harmonic?

If Hooke's law applies, and it does for most springs (at least for small displacement), the restoring force on the bob is proportional to the displacement from equilibrium. The resulting displacement is given by $y = A \sin \omega t = A \sin(2\pi f t) = A \sin 2\pi(t/T)$. This function is shown in the A part of the diagram. The frequency in cycles per second is symbolized by f. The standard S.I. name for "cycles per second" is hertz, abbreviated Hz. The angular frequency in radians per second is $\omega = 2\pi f$. The period for one oscillation, T, is equal to $1/f$. The amplitude of the oscillation, A, is the maximum displacement. The bob moves between $+A$ and $-A$.

In Chapter 6, we showed that the velocity of the bob is given by $u = \Delta y/\Delta t = \omega A \cos \omega t$. (Note that we are calling the oscillation velocity u. We will reserve the symbol v for the velocity of a pulse or wave traveling through a group of oscillators.) This function is plotted in the B part of the diagram. Note that both displacement and velocity are sinusoidal functions, but are out of phase by 90°. When the bob is at its 0 or equilibrium position, its velocity is maximum, either shooting up positive or shooting down negative. On the other hand, when the displacement is maximum, the velocity is 0.

The acceleration of the bob is given by $a = \Delta u/\Delta t = -\omega^2 A \sin \omega t$. This function is shown in the C part of the diagram. Once again, it is a sinusoidal function that is 90° out of phase with the velocity and 180° out of phase with the displacement. Whenever the displacement is 0 and the velocity is maximum, the acceleration of the bob is 0. When the displacement is maximum positive, the acceleration is maximum negative. This relationship is just what we should expect, since the acceleration is proportional to the force on the bob which, in turn, is proportional to the negative of the displacement. When the displacement is large in one direction, the force and hence the acceleration is large in the opposite direction, always attempting to restore the bob to its equilibrium position.

If the bob is displaced a distance A and then released, it will oscillate in the way that we have described. The period of oscillation is determined by the spring constant, k, and the mass of the bob, m. The spring constant is the constant in Hooke's law: $F = -ky$. The period of this free oscillation is $T = 2\pi\sqrt{m/k}$. The frequency is just the reciprocal: $f = \left(\frac{1}{2\pi}\right)\sqrt{k/m}$. Reasonably enough, a stiff spring with a large constant, k, will tend to produce a fast return of the bob toward its equilibrium position and hence a high frequency of oscillation. If the bob has a large mass, on the other hand, its motion will be sluggish and the oscillation frequency will be low.

However, an oscillating system does not have to vibrate at its natural frequency; you could push a child's swing back and forth at whatever frequency you could physically manage. Nor would your driving motion have to be sinusoidal. But if you do push a swing at its natural frequency, or even give it a brief shove each time at that frequency, the response of the swing is much more satisfactory. The driving force is said to be in resonance with the natural frequency. The corresponding amplitude of oscillation may become very large even for weak pushes. That is just what happens, of course, when you "pump" the swing at the right frequency.

If an oscillator has no friction and is vibrating at its natural frequency without being driven, its energy must remain constant. At times the energy is all kinetic and at other times all potential. At the maximum displacement of the bob, when the velocity is 0, the potential energy is equal to $U = \frac{1}{2}ky_{max}^2 = \frac{1}{2}kA^2$. As the bob passes through the equilibrium point, its potential energy is 0 and its velocity is maximum. Its energy at that point is equal to $E_{kin} = \frac{1}{2}mu_{max}^2 = \frac{1}{2}m\omega^2A^2$.

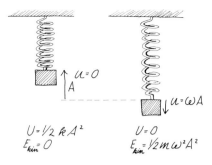

Question 15-2

Do these two expressions agree? Is the maximum potential energy equal to the maximum kinetic energy?

In most physical systems there is some friction. Every time the oscillator swings, some energy is lost and so the amplitude gets smaller and smaller. A graph showing the typical decay of oscillation of a spring is shown in the diagram. If enough energy is supplied to the oscillator at its natural frequency, this frictional decay can be arrested. If more than enough energy is being supplied in each period to make up for the friction loss, then the amplitude of oscillation will steadily increase. There is a limit to such buildup because the friction losses usually increase with amplitude and velocity. When an oscillator is receiving energy in resonance with its natural frequency, a large amplitude is produced such that the energy losses are just matched by the energy gained in each period. In the next diagram, we show the amplitude as a function of driving frequency for several oscillators. In each case, there is a resonant increase in amplitude as the driving frequency approaches the natural frequency of the oscillator. If there is a large frictional loss, the resonance response is broad and does not get very large. However, if there is very little frictional loss, the oscillator does not respond strongly to the driving force until the driving frequency is very close to the natural frequency. Then the resonance response is very large.

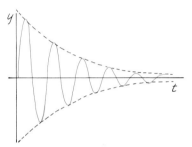

Decay of the amplitude of an oscillating spring with friction.

Resonance phenomena can be observed in many different types of systems. Your radio or TV has a resonance response to electromagnetic vibrations. In Chapter 22, we will study how these radio waves are produced and intercepted. When you tune your radio, you are changing its natural frequency of oscillation to the particular radio frequency that you want to hear. The very small amount of energy extracted from the antenna at that frequency is enough to set the sharply tuned electrical circuit in your radio into large voltage oscillations. Earlier when you sang a note into a piano, your voice provided the driving frequency for the oscillators, which consisted of stretched strings. The string that had the same natural frequency as your voice was driven into resonant vibration. According to folklore, when soldiers march across a footbridge, they should break step lest the frequency of their steps should happen to be close to one of the natural frequencies

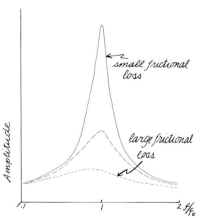

f_o is natural frequency of system. When $f = f_o$, there is resonance. The system responds to the resonant driving force by oscillating with large amplitude.

DISTURBANCE THAT PROPAGATES ALONG A LINE

On an atomic scale, most particles are held in potential wells and, for small disturbances, act as if they were held by springs. If a piece of metal is stretched or compressed, each atom is moved from its equilibrium position by a small amount. The movement of any one atom affects its neighbors, and their movement in turn affects their neighbors. The resulting displacement travels as a pulse through the material. If the displacement of each particle is in line with the traveling pulse, the wave is said to be *longitudinal*. This is the way that sound travels. A series of high-pressure and low-pressure regions, or compressions and rarefactions, travel away from the source. If you hit a nail with a hammer, a longitudinal pulse of high density hurtles down the nail, driving the point further into the wood.

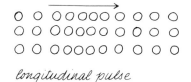

longitudinal pulse

Displacement of a medium can also produce a pulse that travels perpendicular to the displacement. This is the effect that you saw when you snapped a pulse along a rope. Such wave motion is called *transverse*. As we will see in Chapter 22, electromagnetic radiation is transverse. The electric and magnetic field directions are perpendicular to the direction in which they travel. When shaken by earthquakes, the earth produces both longitudinal and transverse waves. Water waves are usually composed of a mixture of the two. Any particular drop of water disturbed by a passing wave travels in an elliptical motion, both up and down, and forward and backward.

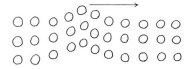

transverse pulse

Regardless of whether the wave motion is longitudinal or transverse, the motion of any individual particle can be described in terms of its displacement as a function of time. The motion of the particle is not necessarily, or even usually, a sine wave. In the diagram, we show a graph of the displacement as a function of time, $y(t)$, of a particle in a rope along which a pulse is traveling. The graph looks just like a snapshot of the rope, with the pulse in it. That's reasonable, since the snapshot of $y(x)$ shows what any particular particle will be doing at an earlier or later time.

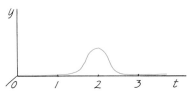

Position, y, of a particular particle in the rope at position x_0, as a function of time during which a transverse pulse sweeps by. At $t=0$, the particle was motionless. At $t=2$, the particle was pulled the maximum distance from its equilibrium position.

The snapshot of the pulse in the rope showing $y(x)$ does not give any clue about the direction of the traveling pulse. How does the pulse itself know which way to go? Note that the rope is under tension, and any particular section is subject to the opposing tensions on either side of it. As long as the rope is straight, the forces on any section cancel out. Where the rope is curved, however, the forces on either side of the section are equal in magnitude but not in line. Therefore, there will be a net transverse force. In part (A) of the next diagram, we show the tension pairs operating on several sections of a rope as a pulse sweeps past. In part (B), we show the resulting net transverse *forces*. At Section A, the pulse has not yet arrived, and the opposing tension forces cancel. At B, there is a net upward force tending to raise that section and give it an upward velocity. At point C, where the shape of the pulse is straight, the opposing tensions cancel, and the resulting force is 0. Section C must still be moving upward, however, because of the previous acceleration. As soon as the pulse has a curvature downward, such as in Section D, there is a transverse force down, acting to stop the upward motion of the rope in that region. In Section E, the rope is traveling downward but the net force is 0. From that point on to section F, there is a net upward force that slows the downward movement of the rope, bringing it to rest. This same sequence is repeated for each section of the rope over which the pulse passes. In part (C), we show how the resulting transverse *velocities* of the rope produce a pulse that travels to the right.

The same sort of diagram could be drawn for a longitudinal pulse. Instead of showing the transverse displacement, the graph could show the displacement in the line of travel of the pulse. The same form of graph would also portray the pressure or density variation along the direction of travel. As a pulse of high pressure passes through a region, a particular section will experience a net force only if the pressure on one side is greater than it is on the other. The resulting forces on a section of the medium as the pulse passes over are shown in the diagram. Once again, as the pulse crosses over, the effect of the forces in the first direction is eventually canceled by the forces in the opposite direction.

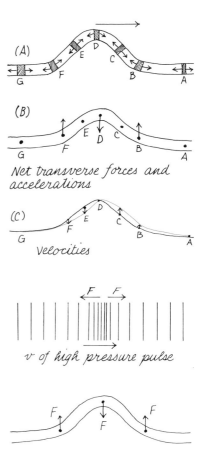

Net transverse forces and accelerations

velocities

v of high pressure pulse

Question 15-3

What would happen if a rope were plucked into the form of a pulse as shown in the diagram and then released? How would the pulse know which way to go?

THE VELOCITY OF A PULSE

The velocity of a traveling pulse depends on the restoring force that brings any disturbed section back toward its equilibrium position. The velocity must also depend on the mass of the disturbed section. As you saw in "Handling the Phenomena," the greater the tension in a rope, the faster the pulse travels. The greater the density of the rope, the slower the pulse travels. For instance, the pulse in a hose filled with water travels much more slowly than when the hose is empty and lighter. The velocity of a pulse may also depend on the shape of the pulse. Perhaps large displacements travel faster or slower than small displacements. It is also possible that very sharp short pulses travel at a different speed than do broad gradual pulses.

Wave velocity is function of: restoring force, density of medium, perhaps shape of pulse.

Let's consider first the simplest situation where the velocity of a disturbance in a medium depends only on the characteristics of the medium and not on the shape or magnitude of the disturbance. In that case, the velocity is a function only of the restoring force, F_r, provided by the medium, and the density of the medium: $v = f(F_r, \mu)$. For a pulse traveling along a line, the mass density, μ (mu), should be the *mass per unit length*. With these assumptions, we can derive a form for the velocity of a pulse without detailed considerations of the mechanism involved.

The next few steps will seem as if we are almost getting something for nothing. The method is called *dimensional analysis*. Consider that we know that the velocity must be some function of a force and a linear mass density:

$v = f(F, \mu)$. The question is, what function? Does the velocity depend on the square of the restoring force or on the cube root of the linear mass density? Whatever this functional dependence is, the combination must end up with the *dimensions* of a velocity. Let's say that the velocity depends on the force to the power a and on the linear mass density to the power b:

$$v = F^a \mu^b$$

Now let's express velocity, force, and linear mass density in terms of their dimensions of mass, length, and time.

$$L^1 T^{-1} = (MLT^{-2})^a (ML^{-1})^b = M^{a+b} L^{a-b} T^{-2a}$$

Since the dimensions on one side must be the same as those on the other (in both cases the dimensions of a velocity), it must be that

$$a + b = 0 \quad a - b = 1 \quad -2a = -1$$

Therefore,

$$a = \tfrac{1}{2} \quad \text{and} \quad b = -\tfrac{1}{2}$$

We now know the functional dependence of the pulse velocity on a restoring force term and the linear mass density of the medium.

$$v \propto F_r^{1/2} \mu^{-1/2}$$

$$v = K \sqrt{F_r / \mu}$$

Dimensional analysis requires that the velocity of a pulse in the medium be proportional to the square root of the restoring force term and inversely proportional to the square root of the linear mass density. This is a very general result applying to all wave motion that meets our original assumptions. The pulses must be small enough and broad enough so that the velocity does not depend on the characteristics of the pulse. Dimensional analysis cannot, however, give us the proportionality constant.

Question 15-4

Does the velocity formula that we have derived agree qualitatively with the result that you observed in "Handling the Phenomena"?

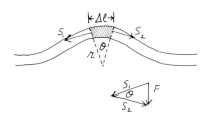

■ Optional

Now let's try to derive the velocity of a pulse in a *stretched rope*, paying attention to the actual dynamics of the process. In the diagram, we show an idealized pulse in a stretched rope that has a tension, S. We assume that the shape of the pulse is smooth enough so that at the top it can be approximated by the arc of a circle. The small section of rope in this arc has length Δl and mass $\mu \Delta l$. The small section is being pulled in both directions by the tension of the rope, but these two forces are not quite in line. Since the tension is always in line with the rope, the force acting at each end of the arc is tangent to the arc and therefore perpendicular to the radius. The net force acting on the small section is radial downward and has the value:

THE VELOCITY OF A PULSE

$$F_r = 2S \sin \frac{\theta}{2} \approx S\theta$$

We have made the approximation that $\sin \theta \approx \theta$ for small angles. The angle θ is equal to the arc it subtends divided by the radius. Therefore,

$$F_r = S \frac{\Delta l}{r}$$

This radial restoring force must be centripetal, acting on the small section of the rope.

Question 15-5
A centripetal force is required to force a moving object to travel in a circular path. The *pulse* is traveling in this case, but isn't any particular section of the rope just moving up and down instead of along an arc?

Now we equate the formula for centripetal force to the radial restoring force: $F_{cent} = mv^2/r$.

$$\frac{(\mu \Delta l) v^2}{r} = S \frac{\Delta l}{r}$$

$$v = \sqrt{\frac{S}{\mu}}$$

Notice how this formula agrees with the one that we derived on the basis of dimensional analysis. In this case, we have replaced the restoring force term, F_r, with the actual tension in the rope. The restoring force is, as we have seen, proportional to the tension in the rope. With this change, the proportionality constant, K, is equal to unity. We have obtained a formula for the velocity of a pulse that is independent of the shape of the pulse—but notice how the restrictions enter. The amplitude of the pulse must be small and the shape must be smooth so that $\sin \theta$ can be replaced by θ and so that we can assume that the tension is constant throughout the rope, including in the region of the pulse.

Now let's find the velocity of a longitudinal pulse, such as might be propagated through a gas or a fluid or even a solid. (A solid can transmit either transverse or longitudinal waves; an ideal gas can support only longitudinal waves of compression; waves in most fluids are usually longitudinal except at the surface between fluids such as between air and water.) To make the derivation easy, we will assume that a square pulse of *uniform* pressure is moving to the right, as shown in the diagram. Once again, we will change reference frames and travel along with the pulse. In our frame, it appears to us that the fluid is moving from right to left with a velocity v. When a region of this fluid strikes the high-pressure zone, its velocity will be reduced because it will be meeting a force directed to the right. Let's choose a particular volume of fluid and set up Newton's second law for it just as it enters the high-pressure zone. We will choose a cylinder of fluid, as shown in the diagram, with a cross section A and a length $v\Delta t$. The short interval Δt is just the time that it takes this cylinder to completely enter the high-pressure zone. The mass of the fluid cylinder is equal to $(\rho v A \Delta t)$. (Now we use ρ, the *volume* mass density in units of kg/m³.) During the time that the fluid cylinder is entering the high-pressure zone, the pressure on its left end is $(P + \Delta P)$, while the pressure on its right end is simply P. Therefore, the net force on the cylinder exerted toward the right is: $A\Delta P$. Because of this net force to the right, the fluid cylinder undergoes a negative acceleration equal to $(-\Delta v/\Delta t)$. Newton's second law for this action is

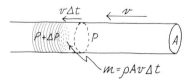

In the reference frame of the high pressure region:

$$F = ma$$
$$A\,\Delta P = (\rho v A\,\Delta t)(-\Delta v/\Delta t)$$

This expression reduces to

$$\rho v^2 = \frac{-\Delta P}{\Delta v/v}$$

The relative change of velocity $\Delta v/v$ is related to the relative change of volume of the fluid cylinder.

$$\frac{\Delta V}{V} = \frac{A\,\Delta v\,\Delta t}{A v\,\Delta t} = \frac{\Delta v}{v}$$

Question 15-6
Why should the volume of the fluid cylinder change as it enters the high-pressure zone?

We can replace the relative change of velocity with the relative change of volume:

$$\rho v^2 = \frac{-\Delta P}{\Delta V/V}$$

Although this expression may at first glance look very complicated, we have seen the right-hand side before in Chapter 4. The ratio of excess pressure to fractional change of volume is called the *bulk modulus, B*. It is negative because a positive increase in pressure produces a decrease in volume.

bulk modulus: $B = -\dfrac{\Delta P}{\Delta V/V}$

In terms of the bulk modulus, we now have a simple formula for the velocity of a longitudinal pulse.

$$v = \sqrt{B/\rho}$$

Notice that once again we have a formula for velocity that is proportional to the square root of a force term divided by a density.

> Question 15-7
>
> According to our dimensional analysis, the velocity should be proportional to the square root of a force term divided by a linear mass density. In the formula that we just derived, we have a pressure term, rather than a force term, and the density is a volume density. Is this all right?

For pulse transmission along a solid rod, the same analysis would lead to a velocity formula with Young's modulus in the numerator and the linear mass density in the denominator. Although we assumed a particularly simple form of high-pressure pulse for this derivation, the formula is good for a pulse of any shape as long as the bulk modulus remains constant. For a pulse that starts at one *point* in a solid and then spreads throughout the volume, a more complicated expression must be used since the solid is not only compressed but also suffers tangential or shearing forces.

$v = \sqrt{\dfrac{Y}{\rho}}$

$\rho = m/\ell$ *linear mass density*
Y *is Young's modulus*

The bulk modulus of a gas (for adiabatic changes) is equal to γP, where γ is the ratio of specific heats, which we defined in Chapter 13, and P is the pressure. The value of γ depends on the structure and number of atoms in the gas molecule. For monatomic gases, $\gamma = \frac{5}{3}$. For diatomic gases, including air, $\gamma = \frac{7}{5}$.

The formula for the velocity of sound in a gas can be transformed so that it is in terms of the more usual gas constants. From the gas law, we have $P = NkT/V$. The density is equal to $\rho = M/V$. The ratio of the number of molecules in a sample to the mass of the molecules is equal to the ratio of the number of molecules in a mole to the mass of one mole:

$$N/M = N_0/A$$

where A is the mass of one mole. Let's substitute these values into our formula for the velocity of sound in a gas.

$$v = \sqrt{\gamma P/\rho} = \sqrt{\frac{\gamma(NkT/V)}{(M/V)}} = \sqrt{\gamma k T(N/M)} = \sqrt{\gamma k T(N_0/A)}$$

Notice that the speed of a longitudinal pulse in a gas, which is the speed of sound in the gas, is proportional to the square root of the absolute temperature and to a number of fixed constants characteristic of that type of gas. Let's substitute the values for air and see what the formula gives us for the velocity of sound in everyday life.

$$\gamma = 1.4$$

$$N_0 = 6 \times 10^{23} \text{ molecules per mole}$$

The Boltzmann constant $k = 1.38 \times 10^{-23}$ J/(molecule K). The mass of one mole of an oxygen-nitrogen mixture of air is equal to 0.0288 kg. Substituting these values into the equation, we calculate that the velocity of sound in air is $v = 20.1\sqrt{T}$ m/s. At room temperature, where $T = 293$ K, the speed of sound is 344 m/s. A convenient approximation of this formula is to take the velocity at 0°C and add a small amount for each degree above 0°C. This approximate formula gives us $v = (331 + 0.6t)$ m/s. In this formula, t is the temperature in degrees Celsius.

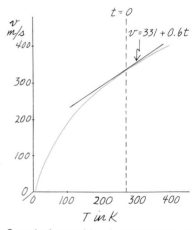

Speed of sound in air as function of temperature.

Question 15-8

How good is the approximate formula at a room temperature of 20°C?

The table shows the speed of sound in various materials. Notice how much faster sound travels in hydrogen and helium than it does in air. The large difference in atomic mass accounts for the difference in speed. Also notice that the speed of sound in metals and water is much greater than it is in air. In this case, the density of metal is much greater than that of air, but the relative bulk modulus is even greater. If you have access to a long pipe, such as a railing or a railroad track (without any nearby trains), you can observe the relative speed of sound in steel and in air. Put your ear on

Speed of Sound in Various Materials

Material	m/s
Air (20°C)	343
Helium (20°C)	981
Hydrogen (20°C)	1328
Carbon Dioxide (20°C)	267
Water	1490
Seawater	1530
Mercury	1450
Iron	~5000
Glass	~4500
Lucite	~1800
Aluminum	~5000

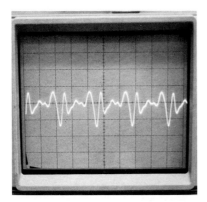

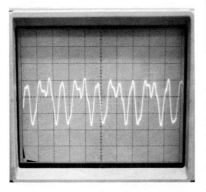

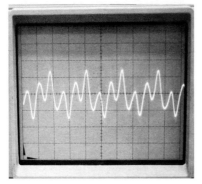

Oscilloscope traces of wave patterns produced by flute (a), voice (b), and trumpet (c). All were sounding middle A at 440 Hz. Scope sweep was 2 ms/cm.

the track and have your friend strike the track at least 30 to 40 meters away. Instantly, as far as you can tell, you will hear the sound through the steel, but there will be a brief delay before you hear the same sound transmitted through the air.

WAVE TRAINS

Most of the signals that we send, including sound and radio and light, consist of long trains of pulses. In most cases, each pulse has approximately the same shape and amplitude as its immediate neighbors. The most common shape is that of the sine wave.

> Question 15-9
>
> Why is the sine wave signal so prevalent?

In the next chapter, we will see how complex wave forms can be considered as superpositions of sine waves of various amplitudes and frequencies. The physical nature of this concept is familiar to many people in musical terms—the difference between one musical sound and another of the same pitch and loudness is the number of harmonics accompanying each tone. In the photographs, we show a number of wave trains produced by various instruments. We are now going to define terms and derive some relationships using a sine wave as our model. In general, the terms and relationships will apply to moving waves regardless of the shape of the individual pulse.

To begin with, we will describe only waves moving along one direction, such as waves in a rope or a parallel beam of light or sound. The y-axis represents either the actual displacement of the rope perpendicular to the direction of the wave, the pressure of a longitudinal wave, or the electric field strength of an electromagnetic wave. At a particular point along the line, any of these quantities oscillates from a maximum positive value to a maximum minus value. The quantity at each point is a sinusoidal function of time: $y = A \sin \omega t$. The maximum displacement from equilibrium is called the amplitude, A. As time progresses, $\sin \omega t$ oscillates between $+1$ and -1, with an angular frequency, in radians per second, of ω. In cycles per second, or hertz, the frequency is $f = \omega/2\pi$.

> Question 15-10
>
> What is the name for this type of response?

In the diagram, we show a plot of $y(t)$ and $y(x)$. The first shows the behavior of some particular point x during the time that the wave is passing by. The second graph provides a snapshot of a section of the line at a particular time. We have labeled several variables in each graph. The period T is the time that it takes for y to go through a complete excursion from $+A$

to −A and back to +A again. If the period of an oscillation is one-tenth of a second, then there must be 10 such oscillations per second. The frequency is 10 Hz ($f = 1/T$). The wave length λ is the distance between crests or between any two points along the line that have the same value of y and the same slope.

If a wave is traveling with a velocity v, there is a basic relationship among the velocity, the wave length, and the frequency. This relationship is shown in the diagram, where we have marked off the number of complete waves that will pass a particular point in 1 second. That number is just equal to the frequency, f. The first of these waves is already at the marker point. If there are 10 waves per second and the wave length is 1 cm, then the last wave that will pass the marker point in 1 second starts 10 cm away. The velocity of the wave must therefore be 10 cm/s. In general,

$$\lambda f = v$$

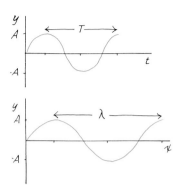

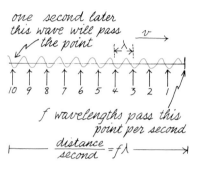

Let's use this relationship to find the wave length or frequency of some familiar waves. As we have seen, the velocity of sound in air at room temperature is 344 m/s. The frequency of the musical note, A, to which the orchestra tunes is 440 Hz. According to our wave equation,

$$\lambda(440 \text{ Hz}) = 344 \text{ m/s}$$

$$\lambda = 0.78 \text{ m} = 78 \text{ cm}$$

As you can see, sound waves have human-size dimensions. In the table we list various frequencies and their corresponding wave lengths in the audio spectrum. As far as humans are concerned, the audio spectrum extends from 20 to 20,000 Hz. Actually those limits are only for very young humans. Men begin losing the upper range of hearing sooner than women do. Most middle-aged men cannot hear notes higher than about 12,000 Hz. Beyond the age of 50, both men and women lose more of the upper range. Notice that our response range is much greater than the small range of frequencies

Frequency and Wavelength Range for Sound

	f in Hz	λ	
Lower limit of audio	16	21.5	m
Lowest piano note	27.5	12.4	m
"60 cycle hum"	60	5.7	m
Lower limit of tabletop radios	100	3.4	m
Middle A for orchestra tuning	440	78	cm
"High C"	1048	33	cm
Highest piano note	4186	8.2	cm
Upper limit for cassette tape	8000	4.3	cm
Upper limit of audio	20,000	1.7	cm
Ultrasonic used by bats	100,000	0.34	cm
Ultrasonic medical scanning	2 MHz	0.17 mm in air	
		0.74 mm in water (or flesh)	

that we use for speaking or singing, which is usually between 100 Hz and 1000 Hz. Cheap sound systems such as small radios or tape recorders usually only reproduce sounds in the range from 100 to 200 Hz to about 6000 Hz. Although speech and tunes can be reproduced with this limited range, the sound seems to lack "quality," "presence," "richness." These qualitative words describe the effect that you feel. In the next chapter, we will learn about overtones or the higher frequency components of the sounds that we produce. When these high frequencies are cut off, the notes lack sharpness or brilliance. When the low frequencies are cut off, our ears may interpret the pitch correctly but the notes will sound muffled or dull.

Other creatures can make and hear sounds far beyond the human range. Bats and porpoises use sound as a form of radar called sonar to locate objects.

> Question 15-11
>
> As we will see later, you can "see" an object by bouncing waves off it, but only if the object is larger than the wave length. What size objects can bats locate with their sonar pitched at 100,000 Hz? Porpoises also use frequencies up to 100,000 Hz. What is the minimum size of the objects that porpoises can locate? (It's not the same as it is for bats! Remember that porpoises use their sonar under water.) Does the minimum size in each case correspond to an object of interest?

On the next page, we show a piano keyboard with frequencies and musical notation for many of the keys. You can also see the approximate range of other instruments and voices. We will refer to this chart again in a later section when we describe more of the characteristics of musical sound.

The other type of waves that we use in daily life is electromagnetic, including radio and visible light. We will describe many other details of this radiation in Chapter 22. While we are talking about waves, however, let's find a few electromagnetic wave lengths and frequencies of particular interest. The center of your AM radio dial, for instance, is at a frequency of 1 megahertz, or 1000 kilohertz (1×10^6 Hz). All electromagnetic radiation, including light and radio, travels at the same speed in a vacuum and, for most purposes, the same speed in air. Since the speed is so special, as we have already seen in the chapter on relativity, it is given the special symbol, c. For all observers, $c = 3 \times 10^8$ m/s. For the center frequency of the AM dial,

$$\lambda(1 \times 10^6 \text{ Hz}) = 3 \times 10^8 \text{ m/s}$$

$$\lambda = 300 \text{ m}$$

The wave length of the middle region of ordinary radio is about $\frac{1}{6}$ mile. The wave length of visible light is a little smaller than the smallest object that we can see with an ordinary microscope (once again, the probe with which you "see" an object must be smaller than the object itself). The limit of microscopic resolution is 1 micrometer = 1×10^{-6} m. Let's find the

WAVE TRAINS

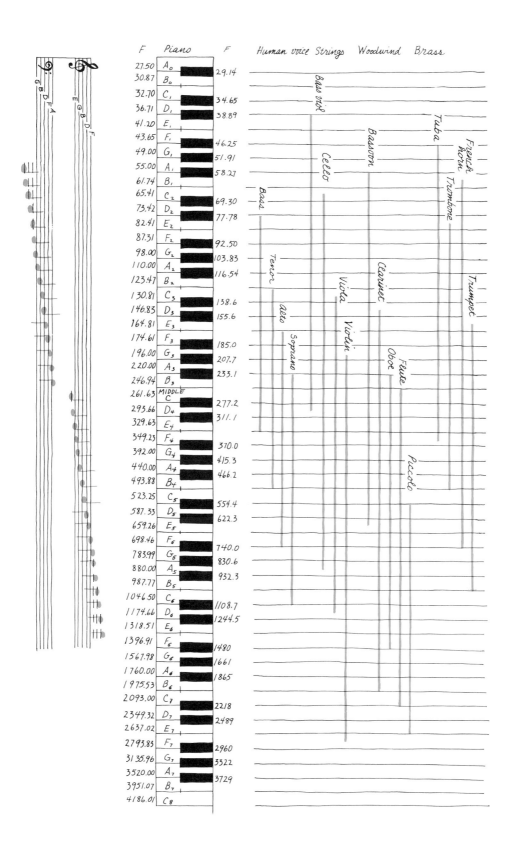

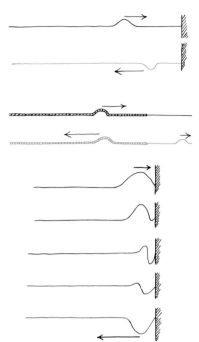

frequency of green light, which has a wave length of 0.5 micrometers = 5×10^{-7} m:

$$(5 \times 10^{-7} \text{ m}) f = 3 \times 10^8 \text{ m/s}$$

$$f = 6 \times 10^{14} \text{ Hz}$$

It is hard to imagine something oscillating at a frequency of 6×10^{14} cycles per second. As we will see later, frequencies of electromagnetic vibrations can go much higher than that.

REFLECTION OF PULSES AND WAVES

In "Handling the Phenomena" you saw what happens to a pulse in a rope when it comes to the end of the line. The pulse bounces back, traveling in the opposite direction. If the rope is tied to a fixed wall, the pulse returns on the opposite side, as shown in the diagram. At the other extreme, if the rope is tied to a long light string, the reflected pulse will return on the same side, as shown in the lower part of the diagram. These are two extreme cases of what happens when a pulse passes from one medium to another. In general, there is partial reflection and partial transmission.

Consider why the reflection from the fixed end produces a pulse on the opposite side from the original one. As the pulse arrives at the fixed boundary, the leading edge, which is accelerating upwards, pulls on the boundary, which does not move but does pull back. In effect, the spring constant of the rope at the leading edge is very great since the next section of the rope cannot move. Consequently, the rope at the leading edge springs downward as if it had been pulled down by a force at the boundary. The higher the advancing pulse tries to pull the material near the boundary, the greater the restoring force on that material to snap it down in the opposite direction. The result is that the same pulse is formed, now traveling in the opposite direction and on the opposite side of the rope.

At the open-ended boundary, the situation is just the opposite. The light string has very low mass density and so, when the leading edge of the pulse reaches it, the upward force produces large acceleration and a high velocity upward. The junction point whips upward to an amplitude almost twice that of the original pulse before being pulled back down again. The effect is the same as if someone had raised and lowered a free end of a rope, sending a pulse back down the rope on the original side.

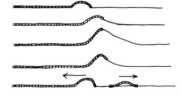

The moment-by-moment sequences of events during these two different kinds of reflections are shown in the diagram.

Plane waves in a tilted baking pan.

We can produce waves traveling in one direction in a medium that has two or three dimensions. For example, when you produce ripples in a bathtub or pan of water by dipping in a ruler or piece of cardboard, a "plane" wave sweeps along the surface. An illustration of such a phenomenon is shown in the photograph. Radio or light waves can advance in a plane wave front in three dimensions, as shown in the diagram. So can sound waves. In Chapter 17, we will describe such radiation in terms of parallel *rays*.

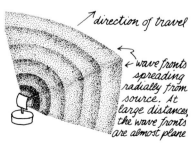

Back in 1678, Christiaan Huygens (1629–1695) in Holland proposed a way to predict the motion of waves in various geometries. His idea is simple

REFLECTION OF PULSES AND WAVES

if you don't look at it too closely, but can become very sophisticated if you examine the implications in detail. For now, let's just use the simple application. Huygens proposed that each point of a disturbed medium (such as water or air) should be the origin of a new wave front coming from that particular point.

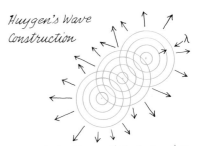

Each of the three disturbed points is a source of new waves spreading from the point.

Question 15-12

What will be the shape of a wave front coming from a *point* disturbance in a two-dimensional wave, such as on the surface of water? What will be the shape of a wave front coming from a point disturbance in three dimensions, such as in open air?

Let us assume that Huygens' principle is correct and consider what would happen to a *line* pulse on the surface of water. If each point on the crest sends out a *circular* disturbance, then one period later we might expect to see advancing wave fronts in *every* direction. The more complicated analysis, part of which we will see in the next chapter, shows that in every direction, except one, the disturbance from any point along the line is canceled out by the corresponding disturbances from all other points. The surface motion at any point ahead of the advancing wave will be determined by contributions from the *entire* wave front. It turns out that the resulting wave front one period later is located along the line, tangent to the circular wave fronts that originate from each of the points in the original crest. The line tangent to all of these circles is parallel to the original crest and represents a front that has moved one wavelength to the right in one period.

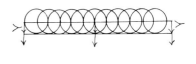

An easy way to observe this effect experimentally is to dip a coarse comb in shallow water. Each tooth of the comb will be the source of circular wave fronts. The resultant disturbance, however, will be a plane wave, the same as if the comb had been a solid ruler.

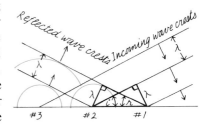

Now let's use Huygens' construction to see what happens when a plane wave is reflected at an angle from a boundary. In the diagram, we show the parallel wave fronts advancing at an angle i toward a boundary. We construct the reflected wave fronts by drawing tangents to the reflected circular wave fronts generated at each point. However, we must connect wave crests of the same phase. One period before the positions shown in the diagram, crest 1 was in the position now occupied by crest 2. Therefore, we must connect point 1 with the crest reflected earlier from point 2.

The resulting reflected waves are also plane and parallel and, furthermore, they maintain a special angular relationship with the boundary and the incoming waves. The two right triangles showing this relationship are emphasized in the diagram. They have a common hypotenuse, and the short side of each has the same length, λ. Therefore, the triangles are identical and $i = r$. The reflected wave fronts leave the boundary at the same angle at which the incident wave fronts strike the boundary. *The angle of incidence is equal to the angle of reflection.*

Reflection of plane waves in ripple tank.

Refraction of plane wave crossing boundary into shallow water.

REFRACTION OF WAVES

When light goes from air into water or glass, it is going from a medium with high velocity into a medium of lower velocity. If water ripples move across a boundary where the depth suddenly becomes very shallow, their velocity decreases at the boundary, as shown in the photograph.

> **Question 15-13**
>
> If the velocity changes as waves go from one medium to another, is there a resulting change in the frequency or in the wavelength? After all, the wave equation is $\lambda f = v$.

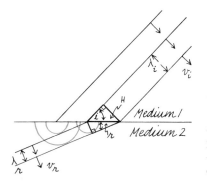

Let's use Huygens' construction to find out what happens to the wave fronts as they pass into a medium with lower velocity. The geometrical construction is shown in the diagram. Because the velocity is less in Medium 2, the spread of the circular wave fronts during each period is less than it is in Medium 1. The tangent lines to these smaller circles are parallel to each other but not to the incident wave fronts. They have been *refracted*. The special angular relationship between the incident angle and the refracted angle can be obtained from the geometry of two adjacent triangles. There is a common hypotenuse to the two triangles. Each of them is a right triangle, with one leg being the wavelength in that particular medium. Therefore,

$$H \sin i = \lambda_i = v_i T_i$$
$$H \sin r = \lambda_r = v_r T_r$$

The periods T_i and T_r are equal in the two media because the frequency is the same. Therefore,

$$\frac{\sin i}{\sin r} = \frac{v_i}{v_r}$$

The amount of refraction or bending of the wave path depends on the ratio of velocities in the two media.

> **Question 15-14**
>
> In the case of light going from air into water, will the angle of refraction be greater or smaller than the angle of incidence? What happens to a beam of light starting out in the water and entering the air?

Sound waves reflect from buildings, as you observed in "Handling the Phenomena," and also reflect within lecture halls. Musicians sometimes test the resonant qualities of a hall by clapping their hands sharply from the middle of the stage. Unless the reverberations can still be heard dying away 1 or 2 seconds later, the hall is too dry or "sec" for music. If the hall is coated with sound absorbent material, reflections will be weak and the sound will also appear weak. On the other hand, the walls and ceiling should not be smooth and solid. In that case, the wave fronts may bounce back and forth in unison, causing actual echoes to be heard. Unfortunately, the reverberations that make music sound richer in a hall tend to blur the sounds of speech. A *lecture* hall should be acoustically dry.

Sound waves also refract. Remember that the velocity of sound in air is proportional to the square root of the absolute temperature. Under normal conditions, the temperature of the air decreases with increasing height above the ground. Therefore, the velocity of sound is faster near the ground.

Question 15-15

If a plane wave starts out moving parallel to the ground, how will it be refracted if the temperature decreases with height?

The refraction of sound has caused many strange and unfortunate occurrences in connection with warning signals, such as from lighthouses. Joseph Henry, an American physicist in the first half of the nineteenth century, investigated and solved several such mysteries. Ships would report that they had heard a warning bell but that as they advanced toward it the sound disappeared, only to reappear too late. The sound had been refracted upward and then, due either to a temperature inversion or air movement, had been refracted downward, leaving a region of silence in the middle. A similar phenomenon occurs if you try to shout upwind. Since the velocity of sound in air is about 340 m/s, you would not expect that a wind of 10 m/s or so would make much difference in the transmission of sound. A wind in the opposite direction would indeed reduce the speed of sound slightly, but that is not why a person upwind has a hard time hearing you. The secret is that the velocity of the wind near the ground is almost always less than it is higher up. Therefore, the opposing velocity of the wind is less near the ground than it is higher up and so the velocity of your sound is faster near the ground. Once again, refraction is produced and the wave front of your voice bends upward, passing over the person who is trying to hear you. If the person upwind is on a balcony or in a tree, he may hear you very well. People living near water frequently hear sounds from boats that are far away. If there is a temperature inversion so that the air is warmer higher up than it is near the surface of the water, then sound will be refracted downward and will travel long distances along the surface.

THE VARIATION OF WAVE VELOCITY WITH FREQUENCY

When we derived formulas for the velocity of longitudinal and transverse waves, we found that the velocity depended only on the restoring forces and the inertial properties of the medium. The velocity did not depend on the shape or magnitude of the pulse or wave, but that is because we made approximations based on the assumption that a disturbance of the medium was small in magnitude and gradual in time. Let's see whether these assumptions are justified by looking at some of the waves we have been studying.

> Question 15-16
>
> Does the velocity of sound depend on the frequency? Does the sound of an orchestra which has both piccolos and bassoons get confused as you move further away from it?

The velocity of light in a vacuum is independent of the frequency. This fact has been experimentally tested for frequencies all the way from radio waves to X-rays. If the velocity of light did depend on the frequency, there would be some very strange astronomical phenomena. For instance, certain binary or twin stars are in such a plane that as we view them they eclipse each other. Each star blanks out the other part of the time. If one of the stars emerged from the shadow of the other one and its blue light traveled faster than the red, we would see it first as a blue star, then white and, when it started into eclipse, as a red star. No such effect has been seen. However, when light passes through transparent materials, the velocity does depend on the frequency. When we study optical instruments in Chapter 17, we will see how refraction can be used with lenses and prisms to control the path of a light beam. Since the amount of refraction depends on the velocity of light in the glass, the amount of bending of the light path will be different for red light than for blue light. This feature causes complications in lens design but is the main virtue of gem stones, such as the diamond. Not only does a diamond produce larger refraction than almost any other material, but, even more important for engagement rings, there is a large *difference* in the refraction between blue light and red light. The effect is best seen by flickering candlelight (though any moving point source of light will do). The incident light is broken up into a spectrum of brilliant colors.

We did not derive the formula for the velocity of water waves because the phenomenon is very complicated. When water is disturbed from its equilibrium position, restoring forces are provided by gravity, water pressure, and surface tension. For ordinary beach waves, the simplifying assumptions that we made do not apply. Even for small ripples on ponds or in kitchen pans, the phenomena are complex. The velocity of ripples on water depends on the depth of the water and also on the frequency of the waves. In the diagram, we show a graph of these complicated relationships.

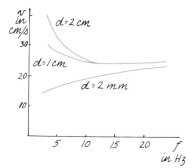

Velocity of water ripples as function of frequency and water depth, d.

> Question 15-17
>
> If you want to produce strong refraction of ripples in a tray, you must have two regions where the wave velocities are very different from each other. Considering the graphs of ripple velocity, how can you obtain such conditions?

POWER TRANSFERRED BY WAVES

Energy, or at least the means to produce energy, can be transported in railroad coal cars or oil tankers or through gas pipelines. You can also deliver energy by throwing a rock or, better yet, an electric battery. In all these cases, some material containing the energy is actually moved from one place to another. You can also transmit energy without transporting any material. Ocean waves grind away the shore, delivering energy that may have been provided by wind or earthquakes far away. The water in between stays roughly where it is. A supersonic jet trails a compression wave that can break windows. But the air in between does not permanently change its position. An ordinary electric light in a house consumes energy from the oscillation of electric charges in the wire. These electrons oscillate because an electric wave with a frequency of 60 Hz is in the wire. No electrons are transported along the wire.

Wave damage (Fredrik D. Bodin/Stock, Boston).

The important consideration about the energy of wave motion is not how much energy is *in* the wave but, instead, how much energy is *delivered* per second by the wave. In other words, we want to know the *power* delivered by the wave. In the case of a wave spreading out in three-dimensional space, such as a sound wave or an electromagnetic wave, we should calculate the power delivered per square meter.

First let's calculate the power delivered by a continuous sine wave in a rope. We can almost see the energy in this case since the rope is actually vibrating back and forth.

■ Optional

Any particular point on the rope oscillates with a displacement perpendicular to the velocity of the wave and with a value:

$$y = A \sin \omega t$$

The velocity of that segment of the rope is

$$u = \Delta y / \Delta t = \omega A \cos \omega t$$

As we have seen before in the case of simple harmonic motion, the velocity of the particle is 90° out of phase with its displacement and is proportional to the frequency. ($\omega = 2\pi f$). The higher the frequency of the oscillating object, the faster it must travel to go between $+A$ and $-A$ during half a period.

Averaged over time, any segment of rope in the wave has the same amount of energy as any other segment of the same length, and the average kinetic energy must be equal to the average potential energy. Both kinetic and potential energy go hurtling down the rope together. When a particular segment has zero displacement, $y = 0$, its velocity, u, is maximum. Therefore, its kinetic energy is maximum. The potential energy of that segment is also maximum because, at that moment, there is the greatest vertical stretch on the segment. The segment has twisted its maximum amount from the equilibrium position. Half a period later the same segment is momentarily at rest at $y = A$ and, thus, has zero kinetic energy. Its potential energy is also zero because the segment itself is not stretched vertically. The net vertical force on the segment is maximum, but this force produces a vertical acceleration, not a strain representing potential energy.

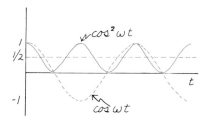

Average value of $\cos^2 \omega t$ over full cycle = 1/2

Let's calculate the kinetic energy of a segment of the rope. The mass of the segment is m.

$$E_K = \tfrac{1}{2}mu^2 = \tfrac{1}{2}m\omega^2 A^2 \cos^2 \omega t$$

The average value of $\cos^2 \omega t$ over one period is exactly $\tfrac{1}{2}$. Therefore, the average kinetic energy of a rope segment is

$$E_K = \tfrac{1}{4}m\omega^2 A^2$$

The average potential energy has the same value. Therefore, the average *total* energy of the segment is:

$$E_T = \tfrac{1}{2}m\omega^2 A^2$$

The total energy in one wavelength is:

$$E_T = \tfrac{1}{2}(\rho\lambda)\omega^2 A^2 \qquad (\rho \text{ is the mass/unit length})$$

The power delivered is the energy contained in one wave length divided by the period of the oscillation.

$$P = (\tfrac{1}{2}\rho\lambda\omega^2 A^2)/T$$

Since $\lambda/T = v$, the velocity of the wave, the power is also:

$$P = \tfrac{1}{2}\rho v \omega^2 A^2$$

The velocity of a wave in a stretched string is equal to $v = \sqrt{S/\rho}$. Therefore, the power is

$$P = \tfrac{1}{2}\rho\sqrt{S/\rho}\,\omega^2 A^2 = \tfrac{1}{2}\sqrt{S\rho}\,\omega^2 A^2$$

Energy/meter = $\tfrac{1}{2}\rho\omega^2 A^2$

$v = \Delta x/\Delta t$

Power = $\dfrac{\text{Energy delivered}}{\text{second}} = \dfrac{\tfrac{1}{2}\rho\Delta x \omega^2 A^2}{\Delta t}$

= $\tfrac{1}{2}\rho v \omega^2 A^2$

Look at the various terms in this expression. The power delivered by a wave in a rope is proportional to the square of the amplitude of the wave and to the square of the frequency.

Question 15-18

Where in the mathematical derivation did ω and A enter? Physically, why should the power depend on the square of these terms?

In our expression for the power delivered by an oscillating rope, the frequency and amplitude terms are characteristics of the waves that we generate. The first term $\sqrt{S\rho}$ is a characteristic of the rope itself. Note that it is made up of the two properties that control the velocity of the waves— the tension S and the mass per unit length ρ. This time, however, we are not dealing with their ratio but with their product. This combination is given a special name. It is the *impedance* of the rope, usually given the symbol Z. The power can thus be expressed in terms of the impedance:

$$P = \tfrac{1}{2}Z\omega^2 A^2$$

As we shall soon see, the impedance of a medium not only is a factor in the power delivered by a wave, but also is the major consideration in determining the extent to which a wave will be reflected from a boundary with another medium or will be transmitted. We will describe impedances for other types of waves in other media. In each case, they will play a similar role, but the impedance is not always simply related to a force term and a mass density term.

> Question 15-19
>
> What are the units and dimensions of the impedance of a rope? Notice the way that the impedance depends on the tension and mass density of the rope. Does it make physical sense that the power transmitted should be proportional to the impedance?

In longitudinal pressure waves such as sound, each particle oscillates back and forth in line with the direction of the wave, as shown in the diagram below. If we let y equal the displacement of each particle from its equilibrium

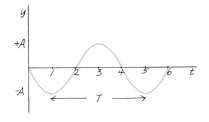

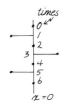

Movement of particle at $x = 0$ back and forth along x-axis, as longitudinal wave passes to right.

At $x = 0$:
$y = -A \sin \omega t$
$\quad = -A \sin (2\pi/T)t$

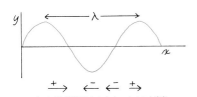

Displacements of particles along x-axis at $t = 0$.

At a point where displacements of particles on the left are positive and on the right are negative, there is high density and excess pressure. Where conditions are reversed, density and pressure are low.

At $t = 0$:
$y = A \sin (2\pi/\lambda)x$

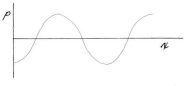

Excess pressure, $p = (P - P_0)$, along x-axis due to density fluctuations caused by particle displacements

At $t = 0$:
$p = -B(2\pi/\lambda)A \cos (2\pi/\lambda)x$
At $x = 0$:
$p = -B(2\pi/\lambda)A \cos \omega t$

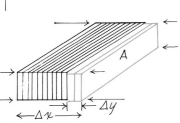

$A \Delta x = V$
$A \Delta y = \Delta V$

Bulk modulus, $B = -\dfrac{\Delta P}{\Delta V/V} = -\dfrac{p}{\Delta V/V}$

$p = -B\dfrac{\Delta V}{V} = -B\dfrac{A \Delta y}{A \Delta x} = -B\dfrac{\Delta y}{\Delta x}$

$p = B\dfrac{\Delta(A \sin \frac{2\pi}{\lambda}x)}{\Delta x} = B(2\pi/\lambda)A \cos (2\pi/\lambda)x$

position (even though the particle is moving back and forth along the x-axis), then for the particle at $x = 0$, $y = -A \sin \omega t$. As shown in the diagram, the excess pressure at that same point is equal to $p = -B(2\pi/\lambda)A \cos \omega t$. This excess pressure p, which is the sound pressure, is very small compared with normal atmospheric pressure. The sound pressure at which humans begin to feel pain is about 30 N/m², whereas atmospheric pressure is 1×10^5 N/m². The pressure amplitude of the faintest sound that we can hear (depending on frequency) is about 3×10^{-5} N/m².

The longitudinal velocity of particles in a sound wave is

$$u_{\text{particles}} = \Delta y/\Delta t = -\omega A \cos \omega t$$

Notice that the longitudinal velocity of the particles is in phase with the pressure, both of which are 90° out of phase with the displacement. Both the velocity and the pressure are proportional to the frequency.

The power transferred across unit area is equal to the product of the pressure differential and the velocity. (Remember that pressure = force per unit area and that the product of force and velocity is work per unit time.)

$$pu = [-B(2\pi/\lambda)A \cos \omega t][-\omega A \cos \omega t] = \omega B(2\pi/\lambda)A^2 \cos^2 \omega t$$

The average value of $\cos^2 \omega t$ over a full cycle is just one-half. Therefore, the average intensity or power per unit area is

$$I = \tfrac{1}{2}\omega B(2\pi/\lambda)A^2$$

■ Optional

This term can be developed into something more familiar by using simple substitutions. Call the velocity of the wave v to distinguish the wave motion from the oscillatory velocity of the individual particles. Here are the relationships that we will use:

$$f\lambda = (\omega/2\pi)\lambda = v \qquad (2\pi/\lambda) = \omega/v$$
$$v = \sqrt{B/\rho} \qquad B = v^2\rho$$

Substitute these values into our expression for the intensity:

$$I = \tfrac{1}{2}\omega B(2\pi/\lambda)A^2 = \tfrac{1}{2}(\omega^2/v)BA^2 = \tfrac{1}{2}\rho v \omega^2 A^2 = \tfrac{1}{2}\sqrt{B\rho}\,\omega^2 A^2 = \tfrac{1}{2}Z\omega^2 A^2$$

where $Z = \sqrt{B\rho}$.

We end up basically with the same formula for the power transmitted in a longitudinal pressure wave that we had for power transmitted in a rope. The power is proportional to the square of the amplitude of the wave and to the square of the frequency. It is also proportional to the impedance of the medium. In this case, the impedance Z is equal to the square root of the product of the bulk modulus and the mass density.

Question 15-20

What is the relationship between this impedance and the velocity of the wave in the medium? What are the units and dimensions of this impedance?

Our formula for the power transmitted per unit area in a longitudinal wave in a gas is derived in terms of the amplitude of oscillation of the particles. Actually it would be very difficult if not impossible to measure that amplitude. Let's calculate the amplitude for the weakest sound that we can hear. The excess pressure amplitude is only 3×10^{-5} N/m². As we saw above, the excess pressure amplitude, p, is related to the displacement amplitude, A, by

$$p = B(2\pi/\lambda)A$$

For a frequency of 1000 Hz, to which the ear is very sensitive, $\lambda = (343$ m/s$)/(1000$ Hz$) \approx 0.3$ m. Therefore, the displacement amplitude is $A = p(\lambda/2\pi)/B = (3 \times 10^{-5}$ N/m²$)(0.3/2\pi$ m$)/(1.4 \times 10^5$ N/m²$) = 1 \times 10^{-11}$ m. (The bulk modulus for air, B, is equal to $\gamma \times P_{atm}$, where γ is the ratio of specific heats, c_P/c_V, and P_{atm} is atmospheric pressure. The ratio of specific heats for air is 1.4.) Compare this amplitude of oscillation with the diameter of an atom which is 1×10^{-10} m. Nevertheless, our ear drums must respond to that amplitude when they detect such a weak signal.

Optional

Let's find the intensity of sound in terms of the maximum pressure variations. Once again, we must make a series of substitution steps:

$$I = \tfrac{1}{2}Z\omega^2 A^2 = \tfrac{1}{2}Z\omega^2 \frac{p^2}{B^2(2\pi/\lambda)^2} = \tfrac{1}{2}Z\frac{f^2\lambda^2}{B^2}p^2 = \tfrac{1}{2}Z\frac{v^2}{B^2}p^2 = \tfrac{1}{2}Z\frac{1}{B\rho}p^2 = \frac{p^2}{2Z}$$

(Remember that $v = \sqrt{B/\rho}$ and $Z = \sqrt{B\rho}$.)

The power transmitted per unit area, the intensity, is proportional to the square of the pressure amplitude and inversely proportional to the impedance. Notice that in this form the intensity, I, does not depend on the frequency.

Question 15-21

Why not? At what point did the frequency dependence disappear?

Let's calculate the intensity in a sound wave that is produced by normal conversation. The excess pressure amplitude is about 1×10^{-3} N/m². The impedance of air is equal to $Z = \sqrt{B\rho} = \rho\sqrt{B/\rho} = \rho v = (1.3$ kg/m³$)(343$ m/s$) = 4.5 \times 10^2$ kg/m² · s. The intensity is $I = p^2/2Z = (1 \times 10^{-3}$ N/m²$)^2/(4.5 \times 10^2$ kg/m² · s$) = 1 \times 10^{-9}$ W/m². If you assume that your voice has this intensity over an area of about 10 m², then the total power that you deliver is about 1×10^{-8} W. If you can deliver this intensity to the back of a lecture hall with an area of about 100 m², then you would be producing about 10^{-7} W. Compare this power with that produced by hi-fi sets that are rated from 10 to 100 W from each speaker.

Like most of our sense organs, our ears respond logarithmically to stimuli. The *intensity level* of sound is defined to be 10 log (I/I_0). The reference intensity, I_0, is usually chosen to be 10^{-12} W/m², which is about the threshold of normal human hearing at 300 Hz. The *unit* of intensity level for sound is the decibel (db). Intensity levels were originally defined in terms of $\log_{10}(I/I_0)$, and the unit was called the *bel* in honor of Alexander Graham Bell. The decibel is more convenient because the human ear can detect a difference in intensity level of about 1 db. Note that if the actual power doubles, the ear hears an increase of 3 db. If the intensity increases by a factor of 10, there is an increase of 10 db.

The intensity levels of various sounds are given in the table. They are only approximate because of the qualitative method of describing them and because the sense of hearing is strongly dependent on the frequency of the sound.

In Chapter 22, we will derive the equations for electromagnetic waves. When an electromagnetic wave passes a point in empty space there is no vibration of any physical object. If an electrical conductor, such as an antenna, is present, then electrons may respond to the passing field. They will oscillate and extract power from the wave. In free space, however, the intensity of the wave is defined in terms of the amplitude of the electric and magnetic fields and an impedance, which is characteristic of the electric and magnetic properties of space. The formula for the intensity is

I Sound Intensity in W/m²	Intensity Level as heard by our ears—in db	Sensation	Conditions
10^{-12}	0		threshold of hearing
		barely audible	
10^{-11}	10		murmuring breeze
10^{-10}	20		
	30	quiet	quiet office
10^{-8}	40		
	50	moderate	average home
10^{-6}	60		
	70		normal conversation
10^{-4}	80	noisy	
	90	very noisy	city street
10^{-2}	100		subway train
	110	intolerable	
			auto horn (1m)
1	120		
		pain	jet takeoff (60 m)

$$I = \tfrac{1}{2}E^2/Z$$

The amplitude of the sinusoidal electric wave is E. The formula for the impedance is the square root of a product of a magnetic characteristic of space that corresponds to an inertial effect and an electric property of space that corresponds to a tension. The units for this impedance turn out to be those of electrical resistance or ohms. The value of Z for free space is 377 ohms. As we will see on page 403, the effective transmission of the TV signal from the antenna to your set depends on matching this value of impedance with the electrical impedance (or resistance) of the lead-in wire and the input of the TV set.

THE SPREAD OF WAVE ENERGY

So far we have been dealing with waves whose wave fronts are plane and parallel to each other and all move in one direction. We have assumed that the amplitudes of these waves stay constant and that no energy is lost. Now we must examine situations where wave energy is dissipated or spread, in either case changing the amplitude, A. Waves can lose energy if they set into oscillation some medium that has friction. For instance, radio waves do not penetrate steel buildings because they set up electrical currents in the metal surface that both reflect the beam and also absorb part of it. The absorbed energy turns into internal energy of the metal. If you stretch a rope along a floor and send a pulse down it, the amplitude and hence the transmitted energy will steadily decrease as the wave progresses. The lost energy is absorbed thermally by the rope and the floor. If sound waves strike a smooth, solid wall, most of the energy will be reflected since the solid mass is barely set into vibration. If sound is directed through low-density material, however, the waves penetrate and may be damped. The sound energy will be reduced. Energy loss of sound in air is proportional to the square of the frequency. Although the absorption is small, the effect can be noticed if you listen to an orchestra at a great distance; the high notes are lost.

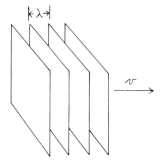
Plane parallel wave fronts.

The intensity of waves (the power per square meter) can decrease even when no wave energy is lost. You can observe this effect by watching circular spreading waves on the surface of a pond. The frictional energy loss for ripples is very small and approximately the same amount of power is spreading out from each circle. The circumference of the circle, however, is increasing. Consequently, the intensity (the power per meter in this two-dimensional case) is decreasing. Since the *power* per meter is inversely proportional to the distance from the center, the *amplitude* of the two-dimensional wave must be inversely proportional to the square root of the distance: $A \propto \sqrt{1/r}$

In the case of waves spreading out uniformly in *three* dimensions, the wave fronts must be spherical. With the same amount of power carried by each spherical front, the intensity steadily decreases with distance from the source. Since the spherical surface of the wave front has an area proportional to r^2, the *intensity* ought to be proportional to $1/r^2$. Consequently, the *amplitude* is proportional to $1/r$.

There are cases where waves seem to travel long distances without a

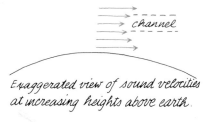

Exaggerated view of sound velocities at increasing heights above earth.

decrease in intensity. In some cases, the intensity even increases at particular points. In Chapter 17, we will study the special geometries of materials that can focus waves either by reflection or by changing selectively the wave velocities in certain regions. We have already pointed out how sound waves can be refracted either upward or downward in the air. At the top of the atmosphere there is a region where the temperature begins to rise again with height. This region provides a natural sound channel, since the temperature and hence the wave velocity is higher at both the upper and lower edges of the channel. Sound entering this region, for instance from large explosions on earth, is refracted back and forth so that the energy spreads out in a thin shell around the earth rather than spreading into three-dimensional space. Because of this channeling, sounds from large explosions can be detected around the world if the detecting instruments are at the right height. There is a similar narrow channel in the deep oceans. A combination of water temperature and salinity creates a thin shell where the wave velocity is less than it is above or below. Any sound energy released in this region propagates essentially in two dimensions rather than in three and hence can be detected at large distances. This channel has been used for signaling across oceans.

THE DEPENDENCE OF WAVE REFLECTION ON IMPEDANCES

Earlier when we described the reflection and refraction of waves, we assumed that the waves were either totally reflected or totally refracted. To be realistic, we should take into account the fact that both usually occur at once. If you shine a flashlight at an angle through a window at night, some of the light will go through the glass and some will be reflected as from a mirror. If you send a pulse into a thick rope that is tied to a thinner one, part of the pulse will be reflected but part will continue on into the thin one. If a sound wave comes to the end of an open pipe in an organ or a flute, some of the energy is reflected and some is transmitted.

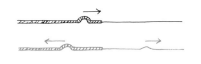

The angle of *refraction* when a wave goes from one medium to another depends on the ratio of wave *velocities* in the two media. The fraction of energy *reflected* depends on the relative *impedances* of the two media. Both reflection and refraction are also affected by the angle of incidence.

For wave fronts parallel to the boundary between two media (i.e. at 0 angle of incidence), the ratio of the reflected wave *amplitude* to the incident *amplitude* is

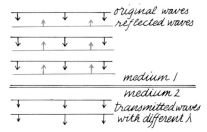

$$R = \frac{Z_1 - Z_2}{Z_1 + Z_2}$$

The impedance Z_1 is that for the medium of the incident wave and Z_2 is the impedance in the second medium. Let's take some extreme cases. Suppose that $Z_2 = 0$. In the case of ropes, this would mean that the second rope has 0 mass density. An approximation to this situation that we have already described involves joining a heavy rope to a long thin string. According to our formula, the ratio of reflected amplitude to incident amplitude is 1. Indeed, when a pulse meets a boundary between a heavy rope and a thin string, it bounces back with the same amplitude and on the same side of the rope. Let's take a look at the other extreme case. If Z_2 is very large compared with Z_1, it corresponds to a rope tied to a solid wall. In that case the ratio of amplitudes is equal to -1. The pulse bounces back with the same amplitude, but this time on the opposite side of the rope. If Z_2 is between these two extremes, then the reflected amplitude will be smaller than the incident amplitude.

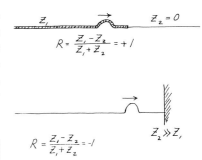

Question 15-22

Suppose that $Z_2 = \frac{1}{2}Z_1$. What fraction of the *energy* will be reflected from the boundary and what fraction of the energy will be transmitted? What fraction of the energy is reflected if $Z_1 = Z_2$?

If we want to send wave energy from one medium to another, we should try to match the impedances of the two media as closely as possible. Ideally, the electrical impedance of the antenna and input circuit of a radio or TV set should be 377 ohms. As a practical matter, this situation is hard to arrange for AM radio but is approximated for TV and FM sets.

In the case of musical instruments, it is necessary to set air into vibration if the sound is going to be heard by a listener. Obviously a vibrating string cannot push much air back and forth. If the string is attached to a sounding board, however, and sets the board into vibration, the board in turn has a much larger surface area with which to push the air. In terms of impedance, it is more closely matched to the impedance of the air.

In the case of wind instruments, it might seem that the vibrating air in the pipe would have the same impedance as the air outside the pipe. According to our simple formula for impedance, that is the case. But that formula was derived for the case of a plane parallel wave. The impedance of an extended medium not only depends on the density but also on the total amount of the medium that must be disturbed. When an organ pipe or a flute transmits energy into the surrounding air, a small area of oscillating pressure must influence an increasingly larger area. The impedance of the open air is greater than that of the air in the open pipe even though the velocity of sound in the two cases is approximately the same.

It's a good thing that the impedance of open air is different from that of a small pipe. If it were not so, there would be no reflection from the end of the pipe back into the instrument. As we will see in the next chapter, it is just this reflected energy that permits and controls the sound that is produced. Meanwhile, of course, we do want some energy to come from the pipe or else we would never hear the internal music. In the case of the flute or most organ pipes, we settle for the small fraction of energy that is released across the barrier. In the case of some wind instruments such as the brasses, there is a flaring bell that provides a gradual transition from a small diameter pipe to the open air. The corresponding impedance changes smoothly and gradually, allowing a larger fraction of the sound energy to leave.

> **Question 15-23**
>
> What is the effect of cupping your hands in front of your mouth or using a megaphone to project your voice over a distance?

SUMMARY

We are surrounded by wave phenomena. You can demonstrate the many effects of waves by sending pulses down stretched ropes, by creating ripples in shallow water, and by playing with stringed musical instruments.

Waves are frequently generated or detected by oscillators. The individual regions through which a wave passes also respond like oscillators, vibrating longitudinally or transversely, but not moving steadily with the wave. For simple harmonic motion, $y = A \sin \omega t$, where y is the displacement of the particle from equilibrium, A is the maximum displacement or the amplitude, and $\omega = 2\pi f$ where f is the frequency in cycles per second or hertz. The natural period is $T = 1/f = 2\pi\sqrt{m/k}$, where m is the mass of the particle and k is the spring constant in Hooke's law: $F = -ky$. The velocity of the vibrating particle is $u = \omega A \cos \omega t$. An oscillator can be subjected to an oscillating force at any frequency, not necessarily its natural frequency. The amplitude of response depends on how close the forcing frequency is to the natural frequency, and how much energy is dissipated in each cycle by the system. Left to itself, an oscillator with an initial disturbance and with internal resistance or damping will vibrate with decreasing amplitude.

We derived the velocity of a traveling transverse pulse in a stretched rope and of a longitudinal pulse traveling along one dimension. If the velocity is a function only of the restoring force, F_r, and the linear mass density, ρ, of the disturbed particle or region, dimensional analysis requires that $v = K\sqrt{F_r/\rho}$. A detailed analysis of the dynamics of small pulses in a stretched rope yields the equation: $v = \sqrt{S/\rho}$, where S is the tension in the rope, and ρ is the linear mass density. For a longitudinal pulse along one dimension, $v = \sqrt{B/\rho}$, where B is the bulk modulus and ρ is the mass density. For a gas, $v = \sqrt{\gamma P/\rho} = \sqrt{\gamma k T (N_0/A)}$. For sound in air, this last formula reduces to $v = 20.1\sqrt{T}$. The velocity of sound in metals is much greater because, although the density is greater than for air, the bulk modulus is even greater.

For continuous waves, $\lambda f = v$. We found the wavelength, λ, and frequency, f, of various pitches of sound and various regions of the electromagnetic spectrum.

Pulses or waves traveling in one medium reflect when they run into a boundary with a different medium. If a pulse on a rope runs into a fixed end, the reflected pulse goes back on the opposite side of the rope; if the pulse runs into an open end, the reflection travels back along the original side. The propagation and reflection of waves in two (or three) dimensions can be described by Huygens' assumption that each disturbed point is the source of a new circular (or spherical) wave. Using Huygens' construction, we showed that for a reflected wave the angle of incidence is equal to the angle of reflection.

When a wave goes from one medium with velocity v_1 to another with velocity v_2, the frequency stays the same but the wavelength changes. If the wave strikes the boundary with an incident angle, i, the angle of refraction in the second medium must satisfy the equation: $\sin i / \sin r = v_i / v_r$. Sound may be gradually refracted in air as it passes through regions where there is a vertical difference in temperature and hence of speed of sound.

For ordinary sound waves and for light in a vacuum, the velocity does not depend on frequency. This is not true, however, for the light in material such as glass or water, and it is not at all true for ripples in shallow water.

Energy is transferred by waves. The power delivered by a one-dimensional wave is $P = \frac{1}{2}\sqrt{S\rho}\,\omega^2 A^2$. The quantity $\sqrt{S\rho}$ is called the impedance, Z. The power transmitted per unit area in a three-dimensional wave is called the intensity, $I = \frac{1}{2} Z \omega^2 A^2$. For a sound wave (or any longitudinal wave), $I = p^2/2Z$, where p is the maximum excess pressure, and $Z = \sqrt{B\rho}$. Since the response of the human ear to sound intensity is logarithmic, the intensity level for sound is measured in decibels, which are log increments above an arbitrary reference level. Intensity level (in db) = $10 \log_{10} (I/I_0)$.

As energy spreads in a wave, the amplitude of the disturbance can decrease because of damping in the medium. It can also decrease because the wave front is increasing as r, in two dimensions, or as r^2, in three dimensions. In the first case, the intensity must be proportional to $1/r$ and the amplitude to $1/\sqrt{r}$; in the second case, the intensity is proportional to $1/r^2$ and the amplitude to $1/r$.

When a wave goes from one medium to another, part of the energy is reflected and part is refracted. The ratio reflected depends on the impedances of the two media and on the angle of incidence. For zero angle of incidence, or in one dimension, the reflection ratio for amplitudes is

$$R = \frac{Z_1 - Z_2}{Z_1 + Z_2}$$

To get a large fraction of wave energy from one medium to another, the impedances must be matched.

Answers to Questions

5-1: The motion is simple because it follows a simple force law and because only one frequency of oscillation is produced. The force law is $F = -kx$, where x is the length of the stretch or the amount of distortion. The motion is "harmonic," as we will see later in detail, because more complicated vibrations can be described in terms of harmonics of the fundamental vibration.

5-2: Does: $\frac{1}{2}kA^2 = \frac{1}{2}m\omega^2 A^2$? The equation is true if $\omega = \sqrt{k/m}$. This condition gives us the angular frequency for a bob vibrating on a spring. Note that it is the same frequency that we obtained in Chapter 6 from a geometric argument based on the relationship between SHM and circular motion.

5-3: If no part of the rope has initial momentum, then the pulse will spread in both directions symmetrically. There will be a pulse traveling to the right and a pulse traveling to the left.

5-4: Without careful measurements you probably could not determine the square root dependence of the velocity. However, it is easy to see qualitatively that the velocity of the pulse increases as you increase the tension in the rope. It is also clear that

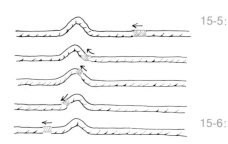

Motion of rope segment as seen from reference frame of pulse.

15-5: pulses travel more slowly in ropes with greater density. For example, the pulse in the hose travels much more slowly when the hose is filled with water.

15-6: By watching the rope from our reference frame, we see any particular section just move up and down. In the reference frame of the pulse, however, the rope is traveling along in the opposite direction. As a particular piece of rope approaches the origin, it swings upward, traveling in an arc (the shape of the pulse), and then swoops down again. We are free to choose any reference frame to analyze a problem. In the reference frame of the pulse, the rope must undergo centripetal acceleration in order to travel in a circular arc.

15-7: The volume of a fluid is always less at higher pressures, even though for liquids the volume change is small. Furthermore, since the cross-section of our sample doesn't change as it enters the high-pressure region but its velocity is reduced, its volume must decrease.

Our original dimensional analysis is still satisfied. Notice that the units and dimensions of B are those of pressure. A force divided by a linear mass density has the same units and dimensions as a pressure divided by a volume mass density:

$$\frac{F}{m/l} = \frac{P}{m/l^3}$$

15-8: The approximation formula gives us $v = 331 + 0.6\ (20°C) = 343$ m/s. This approximation is off by only 1 in 344, or less than $\frac{1}{3}$%.

15-9: When something vibrates according to Hooke's law ($F = -kx$), the resulting motion is sinusoidal. Most small disturbances can be described by Hooke's law.

15-10: Simple harmonic motion.

15-11: For bats:

$$(1 \times 10^5 \text{ Hz})\lambda = 343 \text{ m/s}$$

$$\lambda = 3.4 \times 10^{-3} \text{ m} = 3.4 \text{ mm}$$

That's about the size of an insect which would be of interest to the bat if not to you. For porpoises, the same frequency produces a longer wavelength since the sound is traveling in water. The velocity of sound in water depends on both temperature and salinity. Let's take a typical value of 1500 m/s. Then,

$$\lambda = (1.5 \times 10^3 \text{ m/s})/(1 \times 10^5 \text{ Hz}) = 1.5 \times 10^{-2} \text{ m} = 1.5 \text{ cm}.$$

The porpoise can detect small fish.

15-12: A point source in a two-dimensional surface produces circular wave fronts. A point source in three-dimensional space produces spherical wave fronts.

15-13: The particles in the boundary are being driven with the original frequency of the incoming wave. They drive the particles in the second medium at that same frequency. Therefore, the *wavelength* must change to match the change in velocity.

15-14: Light travels more slowly in water than it does in air. Since $v_r < v_i$, then (sin r) < (sin i). The angle of refraction is less than the angle of incidence. With light coming from the water into the air, the incident angle is in the slower medium. Therefore, the refracted angle in the air will be larger.

15-15: The section of the plane wave close to the ground will move faster than the upper sections. The face of the plane will tilt upwards so that some distance away the sound can be heard better at a height than on the ground itself.

15-16: Some orchestras sound even better further away than they do close up. Apparently both high-frequency and low-frequency sounds travel at the same velocity.

15-17: First, you must go to low frequencies, preferably 5 Hz or less. Second, one of the regions must be very shallow—only a couple of millimeters. It doesn't make much difference whether the deep region is 1 cm or 2 cm.

-18: The total energy of any segment of the rope is part kinetic and part potential, but the total remains constant as the segment executes simple harmonic motion up and down. Since the transverse velocity of a segment of the rope is proportional to ωA, its kinetic energy is proportional to $\omega^2 A^2$. All of the energy in the segments contained in Δx can be delivered in a time Δt where $\Delta x/\Delta t = v$, the velocity of transmission.

-19:
$$N \cdot kg/m = \left(\frac{ML}{T^2}\frac{M}{L}\right) = \frac{M}{T}$$

The impedance of a rope has the units of kilograms per second. It is certainly reasonable that the greater the tension in the rope, the greater the energy contained in a segment of it when it is oscillating and, therefore, the greater the power transmitted. The energy contained in a segment and the power transmitted should also increase as the mass density of the rope increases.

-20: $v = \sqrt{B/\rho}$ and $Z = \sqrt{B\rho}$. The units of B are N/m^2. The units for ρ are kg/m^3. Combining these, we find that the units for this impedance are kilogram per second per square meter: $kg/m^2 \cdot s$. Using these units in the formula for I, we get the proper units of joules per square meter per second.

-21: The frequency disappeared from the intensity expression when it was combined with the wavelength and turned into the velocity of transmission. However, the pressure amplitude is related to the displacement amplitude in terms of the wavelength.

-22: If $Z_2 = \tfrac{1}{2}Z_1$:
$$R = \frac{Z_1 - \tfrac{1}{2}Z_1}{Z_1 + \tfrac{1}{2}Z_1} = \frac{\tfrac{1}{2}Z_1}{\tfrac{3}{2}Z_1} = \frac{1}{3}$$

The reflected amplitude is $\tfrac{1}{3}$ of the incident amplitude. Therefore, $\tfrac{1}{9}$ of the energy is reflected and $\tfrac{8}{9}$ is transmitted. If $Z_1 = Z_2$, then $R = 0$. No energy is reflected.

-23: There are two effects. The megaphone does provide an impedance match between your mouth and the open air. Probably even more important, the cupped hands or megaphone focuses the sound into a relatively narrow beam and thus concentrates the energy along a preferred direction.

Problems

1. Describe the qualitative behavior of ripples that are produced in the deep end of water in a tilted pan as they progress toward the shallow end. What happens to the wavelength? Is there reflection?
2. Investigate the sound produced by a stretched string. If possible, use one on a violin or guitar; otherwise, stretch a string over a box. Describe qualitatively the dependence of the pitch on the length and thickness of the string, and on the tension.
3. When a mass of 1 kg is hung on a particular spring, the spring stretches 10 cm. What is the natural period of oscillation of this system?
4. Use diagrams and your own words to explain why a traveling pulse keeps going in one direction. After all, if you were to take a flash picture of a pulse on a rope, it would look symmetrical and you could not tell which way it was going.
5. The final velocity v_f of an object, starting from rest and subject to constant acceleration, is a function of the acceleration, a, and the distance, x. Use dimensional analysis to find the *functional* dependence: $v = f(a, x)$. Make sure that your formula is in agreement with the one you already know for this situation.
6. A heavy rope has a linear density of 0.5 kg/m. What is the wave velocity in the rope if it is under a tension of 2.50 N?

7. A Slinky (large diameter coiled spring) with a mass of 0.5 kg is stretched to its full length of 8 m with a force of 5 N. What is the transverse wave velocity?
8. The mass of one mole of helium is 4×10^{-3} kg. What is the speed of sound in helium?
9. What is the speed of sound in air at 30°C, the temperature of a hot summer day? What is the percentage change of this speed to the speed of sound on a cold winter's day with the temperature at -10°C?
10. The bulk modulus of steel is 1.6×10^{11} N·m² and the density is 7800 kg/m³. What is the speed of sound in a railroad rail? If a person struck a rail at a distance such that it took 1 s for the sound to reach you through the air, how long would it take through the rail?
11. A flute is brought to a winter concert in the trunk of a car that was at 0°C. The flutist rushes in, tunes her instrument, and starts to play. What happens to the pitch (frequency) as the temperature of the flute and air inside rises to 30°C? (The air temperature rises to midway between room temperature and body temperature.) Assume that the size of the flute does not change appreciably and therefore the wavelengths stay the same. Calculate the effect of the change of temperature on the speed of sound and, hence, on the frequency.
12. The FM band is just above TV channel 6. What is the wavelength of the FM station whose frequency is 100 megahertz?

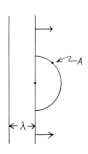

13. Draw a line to represent one crest of an advancing wave. Use one point as the center of a circle, with a radius equal to one wavelength. Draw that circle to represent a Huygens' wavelet coming from a point in the advancing wave. Mark the 90° (forward) point on that circle, B. Mark another point at 45° on that circle, A. Now find another point along the original line that will produce a wavelet that always cancels the first one at A. Does this second wavelet also cancel the motion at B?
14. In a ripple tank, the speed in one region is 20 m/s and in another region is 15 cm/s. What happens to an incident wave crossing the boundary at an angle of 60°? Consider the problem for two cases; crossing the boundary from fast to slow and from slow to fast.
15. A wave of frequency 12 Hz travels at 30 cm/s in the shallow part of a ripple tank. It enters a region of deeper water with an incident angle of 35°. If the wavelength in the deeper water is 3.7 cm, what is the angle of refraction?
16. A ship's foghorn is sounded, and an echo is heard 6.4 s later. If the temperature is 5°C, how far away is the reflecting surface?
17. Suppose that you want to know the speed of a *longitudinal* wave in a spring. Since the speed must be the square root of an elastic quantity divided by an inertial quantity, perhaps the speed is simply the square root of the spring constant divided by the linear density. Use dimensional analysis to test whether such a formula is possible.
18. A transverse wave of amplitude 0.20 m, with a wavelength of 1.50 m, travels at 3.5 m/s along a spring. If a small segment of the spring is moving with simple harmonic motion, find (a) its maximum speed and (b) its maximum acceleration.
19. Draw a diagram that illustrates Huygens' construction for the reflection of plane waves from a boundary for an incident angle of 15°. Use your diagram, labeling all necessary angles and line segments, to show that $i = r$.
20. Draw the diagram showing Huygens' construction for the refraction of plane waves where $i = 30°$ and $v_i/v_r = 2$. Label all necessary angles and line segments.
21. People live on a cliff overlooking a deep lake on which there is boat traffic spring, summer, and fall. Describe the difference in boat sounds heard on the cliff in fall compared with spring.

22. Using a 5 Hz vibrator, what depths could you use in a ripple tank to produce refraction with an incident angle of 30° and a refraction angle of 15°?
23. A railroad whistle produces 100 W of sound. If the energy spreads out spherically, what is the intensity at a distance of 100 m? What is this intensity level in db?
24. How much power can be delivered by sending sine waves down a stretched rope? The rope has a density of 0.3 kg/m and is under a tension of 100 N. The rope is shaken with a frequency of 2 Hz and an amplitude of 15 cm.
25. At an excess sound pressure of 30 N/m² at 1000 Hz, the human ear feels pain. What is the intensity in watts per square meter and the intensity level in decibels?
26. If a ring (perhaps made of rope) is dropped into a pond, what happens to the amplitude of the pulse created outside the ring? What happens to the amplitude of the pulse created inside the ring?

16 COMPLEX WAVES AND INTERFERENCE

Tuning forks execute simple harmonic motion. The sound waves they produce are almost pure sinusoidal tones. How distressing it would be, however, to listen to music produced by an orchestra of tuning forks! Musical sounds not only have pitch or frequency but also timbre. The waves are periodic but are not sinusoidal. The irregular shapes and the jagged edges of the harmonics produce tonal richness. Now we must study these complex waves.

For many types of wave motion, the wave velocity depends on frequency. This phenomenon is called *dispersion*. It gives rise to many dramatic results, some of them familiar in everyday life. For instance, the dispersion of white light in a glass crystal yields a rainbow spectrum of brilliant colors.

The observed frequency of waves depends also on the relative velocity between the source and the observer. If a train blows its whistle as it approaches you, the pitch is higher than normal until the train rushes past you, whereupon the pitch abruptly drops. This is the Doppler effect, named for its Austrian discoverer, Christian Johann Doppler (1803–1853).

In most cases waves can pass right through each other in a medium without disturbing the progress of each other. In the region where they overlap, however, the disturbance of the medium at each point is the sum of the separate disturbances. With the right geometry, this effect gives rise to standing waves where certain points appear to experience no disturbance at all, although both waves are passing through.

The interference of waves is usually dependent on wave length. Interference phenomena can produce spectra similar to the phenomenon produced by dispersion in prisms. The phenomenon of interference also produces *diffraction* where a beam of waves interferes with itself to cancel out propagation in some directions but not others.

All these complex wave effects help produce the sounds of music. After learning how to describe complex waves, we will apply the results to the structure of musical tones and the instruments that produce them.

Standing waves produced by reflection of water ripples.

HANDLING THE PHENOMENA

1. Fill the bathtub or a cookie pan with water again and look for more complex effects of ripples. Start out by dipping two fingers at a time into the

water. Produce a steady beat of several cycles per second so that a continuous progression of waves spreads out from your fingers. Observe the way that the two circular wave fronts interfere with each other some distance from your fingers. See if you can detect the nodal lines where there is no disturbance at any time, just as the photograph shows.

Form a barrier of two pieces of wood or cardboard across one end of a pan of water. Create a gap in the middle and then send plane waves marching toward the barrier and the gap. Do the waves that get through the gap continue in the original direction? Do they continue as a narrow beam or do they spread out? Try increasing the width of the gap and observe what happens in the regions where you might expect shadows.

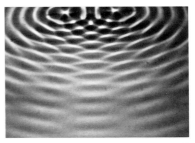

Ripples were created by two small spheres bobbing up and down in ripple tank. The picture shows the shadows of the ripples.

These effects are more dramatic if you have access to a large water surface such as a pond.

2. In a pond or even in a bathtub, you can see the Doppler effect. Tap your finger in the water to produce ripples but move your finger steadily along one direction. The successive series of crests will be bunched up in the direction of your finger motion and spread out in the opposite direction. Consequently, the wavelength will be smaller than normal in the forward direction and larger than normal in the backward direction. While you are at it, simply trail your finger through the water at several different speeds and observe the angles of the bow waves produced.

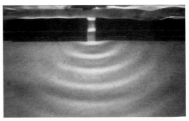

Plane waves advancing from the top penetrate a narrow gap, producing circular waves.

If you live near a railroad crossing that still has trains on it, you have probably heard the Doppler effect. Notice that the drop of pitch is abrupt when the train hurtles by you. The lonesome whistle in the distance may get louder as it comes toward you but the frequency remains roughly the same until the sudden change in velocity occurs. The train doesn't change velocity with respect to the earth, but its relative speed *toward* you abruptly turns into that same speed *away* from you. In case you have no nearby trains, you might be able to arrange a similar demonstration with someone driving a car past you, while the horn sounds continuously.

3. The interference of two sounds with almost identical frequency produces the phenomenon of *beats*. You can hear this for yourself with any two musical instruments. If you have two strings, preferably on the same instrument, producing the same pitch, change the pitch of one of them very slightly and you will hear a rapid rising and falling of intensity. If the pulsation is slow enough, you can actually count the number of beats per second. The number of beats per second is equal to the difference of frequencies of the two notes. Beats can also be heard with two wind instruments that can be played simultaneously while the frequency of one of them is slightly changed. Two toy slide whistles work well for this purpose, as do two trombones.

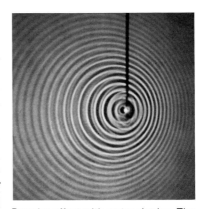

Doppler effect with water ripples. The source at the end of the rod, moved steadily to the right. (From the film "Ripple Tank Phenomena, Part III," Educational Development Center, Newton, MA.)

4. If you have access to any kind of a wind instrument, measure the longest and the shortest length of the instrument that can be produced. The shortest length for the trombone occurs, of course, when the slide is fully in. For instruments with valves, such as the trumpet, each valve that is pressed down adds a length of tubing. Find the lowest note and the highest note that you can play on the instrument. To discover the frequencies of these notes, you can match them against piano notes and then find the frequencies in the chart on page 389.

For any particular length of tubing in the instrument, you may be able to play several notes. After all, a bugle always has the same length and yet it can be used to play at least five different notes.

Find the wavelengths of the high and low notes and compare them with the length of tubing needed to sound them. You will notice first that the wavelength of the lowest note is of the same order of magnitude as the length of pipe. The actual relationship depends on the type of instrument and also on whether the lowest note is the fundamental or the first overtone. We will describe some of these relationships in a later section.

Take a similar measurement of the length of a string in any stringed instrument and compare it with the wavelength of the sound produced (once again, if you don't know the musical name for the note, such as middle C, you can compare the tone with a piano note and then find the frequency from the chart on page 389. You can then find the wavelength, knowing that the velocity of sound in air is approximately 340 m/s). In the case of the stringed instrument, the length of the string is considerably shorter than the wavelength produced. In a later section, we'll see why these length relationships exist.

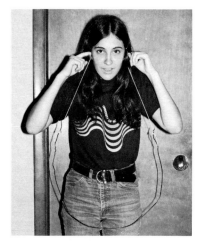

5. There is a spectacular way of demonstrating that low frequency vibrations have a hard time getting out into the air without a sounding board with large surface area. In the process of showing this, you can make a musical instrument that no one will believe can make such beautiful sounds. All you need is a wire coat hanger and two pieces of string or stout thread. Unwind the coat hanger so that it becomes just a long piece of wire, and tie the strings to each end as shown in the diagram. Press the strings into your ears with your index fingers so that the string and wire hang freely without touching anything except your fingers and ears. Now let the coat hanger bump into something hard or have a friend tap the wire with a pencil. Anyone near you will hear only a weak clanking sound from the wire, but you will hear the awesome tones of a great bell.

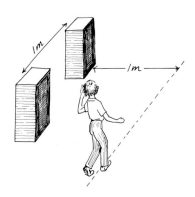

6. If you got a hi-fi test record as suggested in Chapter 15, you can play the same pure sine wave tone into both speakers of a stereo system. Arrange the speakers about 1 meter apart and play a 1000 Hz note. Move parallel to the line connecting the speakers at a distance of about 1 meter. One ear will be toward the speakers. See if you can detect the null regions as you move along the line. If you have the speakers connected correctly, there will be maximum loudness at the center point between the two speakers. You should find null regions where the sound is noticeably fainter at points about 20 cm on either side of the central maximum.

With your hi-fi test record, send a signal into just one speaker. If you have a room large enough so that there are not strong reflections from the walls, place an upholstered chair or a substantial person directly in front of the speaker and a couple of meters away. See if you can detect with your ear any shadow cast by the chair or person, using a frequency first of 100 Hz and then several thousand Hz. You can also experience this same effect if you can project the sound toward an open door leading to a much larger room or to the outdoors. In this case, instead of sending the sound against an obstacle, you are sending it through a slit. Hear how the frequency of sound waves determines the extent to which they can curl around corners.

A FORMAL DESCRIPTION OF TRAVELING WAVES

7. Another way to send waves through narrow slits or openings is to observe point sources of light through pinholes or very narrow slits. The source of light should be bright but as small as possible. A distant street lamp works well, but so does a candle flame some distance away in a darkened room. You can make pinholes or extremely narrow slits in aluminum foil (the slit should be narrower than 1 mm). You can also form such holes or slits in a crude fashion with your fingers. In either case, hold the opening close to your eye and look at the point source of light. The light will appear to spread out either in a circular fashion in the case of the pinhole or with broad side bands in the case of the slit. Notice particularly that what happens is not simply a matter of fuzzy light spreading out. There are definite bands of light and darkness corresponding to what we have called null regions or nodes in the case of sound waves or water waves.

A FORMAL DESCRIPTION OF TRAVELING WAVES

In the last chapter we described pulses and waves traveling in many different kinds of media. To understand these phenomena we must experience them with hands and eyes and ears but we should also be able to describe them in the language of physics and mathematics. This language, like any other, helps us to see things in natural phenomena that are not immediately apparent to the senses. Surely a falling stone must look different to you after you have seen the algebra and the graphs of $x(t)$, $v(t)$, and $a(t)$. Anyone can see that a child's swing goes back and forth. But there is an extra pleasure (and sometimes usefulness) in knowing that the motion is sinusoidal. For the sake of both pleasure and usefulness, let's now formulate an algebraic description of a traveling wave.

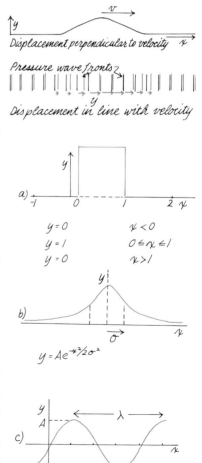

We will describe a wave traveling in the x-direction with velocity, v. If the wave is in a rope, there are displacements, y, in the rope, perpendicular to x, as the wave passes through. If it is a sound wave, there are plane wave fronts traveling in the x-direction and the disturbance can be described in terms of the back and forth displacements, y, of particles, or of the pressure, P. In any case, let us use the variable y to indicate the disturbance of the medium. This disturbance is a function of both time, t, and distance along the axis, x: $y = f(x, t)$.

The disturbance need not be a sine wave. Indeed, in the last chapter we explored the behavior of single pulses. Two of these pulses are shown in the diagram, parts (a) and (b). We show the square pulse and the exponential pulse (technically called a Gaussian) as if they were stationary and simply a function of x. The functional dependence on x is given with each curve. The description of the square pulse requires several statements. The description of the Gaussian is simpler though perhaps less familiar. Note that because the Gaussian depends on x^2, it must be symmetrical about $x = 0$. When $x = 0$, the exponential is unity and $y = A$. The constant σ is a measure of the width of the curve. When $x = \sigma$, $y = 1/\sqrt{e} = 0.61$ of its maximum value.

In the (c) part of the diagram, we show a sine wave once again, frozen in time so that y is a function only of x. Note that we cannot describe this function simply as $y = A \sin x$. We can take a sine only of an angle, but x is a distance. We must describe distance along the x-axis in terms of multiples of angles, measured either in degrees or radians. Since the disturbance is periodic with a wavelength λ, we can define the distance in terms of $(2\pi$ radians$)(x/\lambda)$ or $360°(x/\lambda)$.

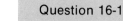

> **Question 16-1**
>
> What is the value of y if $x = 0$ or $\frac{1}{4}\lambda$ or $\frac{1}{2}\lambda$ or $3\frac{1}{4}\lambda$?

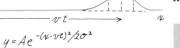

Now we must get all those disturbances moving along the x-axis. Let's assume that they move to the right. After a time t, the whole pattern will have moved a distance vt. Our mathematical description matches this situation if we simply substitute $(x - vt)$ for the original position x. The two pulses and the sine wave are shown once again in the next diagram, as they travel to the right. Notice their new algebraic descriptions.

> **Question 16-2**
>
> How would you describe the Gaussian pulse traveling to the left? Remember that at $t = 0$, the peak of the Gaussian is at $x = 0$. According to your formula for the pulse traveling to the left, where is the peak when $t = 1$ second?

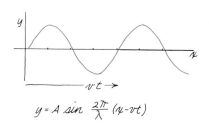

There are several other useful rearrangements of the formula for a traveling sine wave. To make these transformations remember that

$$\lambda f = v \qquad 1/f = T \qquad \lambda = vT$$
$$\omega = 2\pi f \qquad \omega = 2\pi/T$$

Another symbol you may see in advanced work is the spatial analogy to the angular frequency. The *angular wave number*, κ (kappa) is the number of wavelengths in 2π meters: $\kappa = 2\pi/\lambda$. Notice how the angular wave number, κ, corresponds to the angular frequency, ω: $\omega = 2\pi/T$. (In some texts, k is used instead of κ.) In terms of these variables:

$$y = A \sin \frac{2\pi}{\lambda}(x - vt) = A \sin 2\pi \left(\frac{x}{\lambda} - \frac{t}{T}\right) = A \sin(\kappa x - \omega t)$$

In the middle expression you can see the parallel way in which the traveling wave depends on x and t. As the wave progresses along x, the argument of the sine function progresses from 0 radians through fractions (x/λ) of 2π radians and finally through multiples of 2π radians. The same thing happens as time progresses in terms of the period T.

TRAVELING WAVES SEEN FROM A DIFFERENT REFERENCE FRAME—THE DOPPLER SHIFT

In "Handling the Phenomena," we described what happens to the lonesome whistle of a train as it passes you at a crossing. The whistle emits a wave

TRAVELING WAVES SEEN FROM A DIFFERENT REFERENCE FRAME—THE DOPPLER SHIFT

traveling toward you but the source itself is also traveling first toward you and then away from you. In the diagram, you, the receiver, are standing still in the medium, which is the air, and the source is moving. The wave fronts bunch up in front so that the wavelength is shorter. The number of these wave fronts passing your position per second is greater and so the frequency is greater. With sound you hear a higher pitch. As the vibrating source moves away from you, however, the distance between wave fronts stretches out and the frequency of these wave fronts passing your position is less. With sound waves you hear a lower pitch.

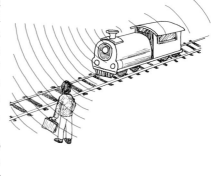

In the diagrams below, we derive the actual frequency changes for two cases. In the first, the train problem, the observer is standing still while the source moves. In the second case, the source is stationary and the observer moves. In either case, if the source and observer are approaching each other, the pitch is higher; if the source and observer are going away from each other, the pitch is lower.

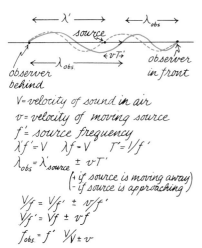

(a) Receiver is stationary in the medium; source is moving.

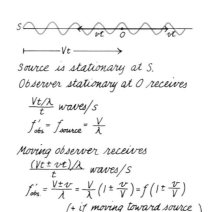

(b) Source is stationary in the medium; receiver is moving.

■ Optional

Since the Doppler shift arises because of a difference between reference frames of the source and the observer, the easiest way to describe the effect is to use the algebra of reference frame changes.* We saw how to do this in Chapter 11. For the sake of simplicity, we will use Galilean relativity where we assume that time is the same in both frames of reference and that therefore velocities add simply.

The rules of the Galilean transformation are

$$t' = t \quad y' = y \quad z' = z \quad x' = x - vt$$

These equations describe the transformation for two systems whose axes

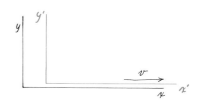

---

* This treatment was described by Eli Maor in the October, 1975, issue of *The Physics Teacher*, p. 399 (13, 399).

coincided at $t = 0$. The primed system moves along the x-axis in the positive direction with a constant velocity v. We will use the symbol V for the velocity of sound in air so as not to confuse that velocity with the velocity of the moving reference frame, v.

Let's start out by assuming that the waves are generated in our (unprimed) reference system. As they spread along the x-axis, we describe them according to the formula

$$y = A \sin 2\pi\left(\frac{x}{\lambda} - \frac{t}{T}\right) = A \sin 2\pi f\left(\frac{x}{f\lambda} - \frac{t}{fT}\right) = A \sin 2\pi f\left(\frac{x}{V} - t\right)$$

The air is stationary with respect to us in the unprimed system. Therefore, the velocity of sound that we measure is the "true" velocity but not the velocity measured by the observer in the primed system. In the Galilean type of relativity, velocities add directly. Therefore the observer in the primed reference frame will find the speed of sound to be equal to $V' = V \pm v$ ($-$ if observer is moving in $+$ x-direction, away from the source; $+$ if observer is moving in $-$ x-direction, toward the source)

We can now transform the formula for the traveling wave into the primed reference system, moving away from the source, by simply substituting the transformation for x.

$$y' = y = A \sin 2\pi f\left(\frac{x' + vt}{V} - t\right)$$

The primed observer hears a sinusoidal wave that she would describe with the same formula we have used in the unprimed system. Of course, the primed observer would use variables from that system including a different frequency and a different velocity for the wave.

$$y' = A \sin 2\pi f'\left(\frac{x'}{V'} - t\right)$$

The two expressions must be identical. Therefore,

$$A \sin 2\pi f\left(\frac{x' + vt}{V} - t\right) = A \sin 2\pi f'\left(\frac{x'}{V'} - t\right)$$

Since V' equals $V - v$, the equality becomes

$$f\left(\frac{x' + vt}{V} - t\right) = f'\left(\frac{x'}{V - v} - t\right)$$

We now have an expression for the frequency observed in the primed system: f'. Let's reduce the algebra:

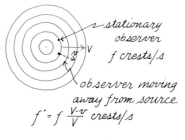

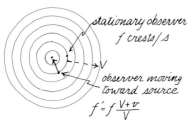

$$f'_{obs} = f\frac{(x' + vt - Vt)/V}{(x' + vt - Vt)/(V - v)} = f\frac{V - v}{V} \quad \text{(observer moving away from source)}$$

Now that we have the algebra taken care of, look at the diagram to see the physical situation that is described. The source of sound (such as a factory whistle) is stationary with respect to us in the unprimed system and with respect to the air. The traveling observer is moving away from the source of sound. Consequently, the wave crests are not moving past her as rapidly or as frequently as they would if she were standing still in the air. Therefore, the frequency that she observes is lower. That is just what our formula tells us.

Question 16-3

What happens if the velocity of the traveling observer is 0? What happens if $v = \frac{1}{2}V$? What happens if $v = V$?

If the observer is traveling toward the source of sound instead of away from it, her velocity would be $-v$. Everything in our derivation would remain the same except that $+v$ should be replaced by $-v$. The final formula would become

$$f'_{obs} = f\frac{V+v}{V} \quad \text{(observer moving toward source)}$$

According to this formula if the observer is moving toward the source that is stationary in the air, the observer will hear a higher frequency. This prediction agrees with the situation shown in the lower part of the diagram. The waves in the air are moving out toward the observer and the observer in turn is moving in toward them. Consequently, the wave crests move past the observer faster and more frequently than they would if she were standing still. She hears a higher frequency.

Now let's take up the case of the stationary observer (stationary with respect to the medium) and a traveling wave source. This is the train problem. In the train's reference frame, it is producing a signal:

$$y' = A \sin 2\pi f'\left(\frac{x'}{V'} + t\right)$$

If the train is rushing away from us, $x' = x - vt$. This situation corresponds to our previous calculation where source and observer were traveling away from each other. Note that the train observes that the speed of sound is V' since the train is moving through the medium. If the waves in the medium are moving with a velocity V, then the train, looking back toward the observer, observes the crests to be moving away from itself with a velocity: $V' = V + v$.

Now let's transform the variables of the wave equation from the train's reference frame to that of the stationary observer.

$$y = y' = A \sin 2\pi f'\left(\frac{x - vt}{V + v} + t\right)$$

The stationary observer hears sine waves passing over him but with a frequency f. Therefore he observes

$$y = A \sin 2\pi f\left(\frac{x}{V} + t\right)$$

Once again we can find the relationship between frequencies in the moving and the stationary reference frame by equating the arguments of the sine functions.

$$f\left(\frac{x}{V} + t\right) = f'\left(\frac{x - vt}{V + v} + t\right)$$

$$\frac{f}{f'} = \frac{Vt + vt + x - vt}{V + v} \cdot \frac{V}{Vt + x} = \frac{V}{V + v}$$

$$f_{obs} = f'\frac{V}{V + v} \quad \text{(source moving away from observer)}$$

This equation agrees with our experience that a train whistle moving away from us will have a lower pitch. If the train were moving toward us, everything in the derivation would remain the same except that we should replace $+v$ by $-v$. Our equation would then become

$$f_{obs} = f'\frac{V}{V - v} \quad \text{(Source moving toward observer)}$$

Question 16-4
For a moving source of sound, what is the frequency ratio f/f' if $v = 0$; if $v = \frac{1}{2}V$; if $v = V$?

The two cases, moving observer or moving source, are not symmetrical. The waves move in a medium and when we say "moving" or "stationary" we mean with respect to the medium. Let's compare the two sets of equations for the situation where the source and observer are moving away from each other. Whether the source or the observer is stationary with respect to the air, the observer will hear a lower frequency:

Source stationary with frequency f; observer hears f':

$$f'_{obs} = f\frac{V-v}{V} = f(1-v/V)$$

Source moving with frequency f'; observer hears f:

$$f_{obs} = f'\frac{V}{V+v} = f'\frac{1}{1+v/V}$$

$(1+x)^{-1} = 1 - x + x^2 - x^3 + \ldots$ for $x < 1$

This last expression can be changed with a binomial expansion. If v/V is much less than 1, then only the first two or three terms of the expansion are needed.

$$f = f'[1 - v/V + v^2/V^2 - \cdots]$$

Now we can see the extent to which there is symmetry between a moving source and a moving observer. If v/V is much less than 1 there is little difference in the frequency changes between moving source and moving observer.

We claimed that a common example of the Doppler shift is the change in pitch when a train rushes past you. Let's use our equations to calculate the frequency shift that might be observed. If the train is traveling at 60 mph, its speed is 27 m/s. Let's take the velocity of sound in air to be 340 m/s and find out what happens if the train's whistle has a basic frequency of 1000 Hz. Since $v/V = 27/340, \ll 1$, we could use either formula. However, using the one for a moving source, $f' = 1000$ and

$$f = (1000\,\text{Hz})\frac{1}{1\pm(27)/(340)}$$

We used the expression ± in the denominator to account for the train moving toward us (minus) or away from us (plus). Since $27/340 = 8\%$, the frequency that we hear as the train approaches is (1000 Hz) (1.08) = 1080 Hz. If we are standing right by the crossing, the shriek of the whistle gets louder as it comes toward us but does not change pitch appreciably. As soon as it passes us, however, it is moving away from us and the frequency drops to 920 Hz. In musical terminology, the basic frequency of 1000 Hz is a little higher than high B (an octave above middle B on the piano). The approaching train sounds like high C sharp and the departing train sounds like high B flat. This abrupt shift of pitch is dramatic and easily heard.

In our derivation of the Doppler shift, we made crucial use of the fact that the waves exist in a medium that we can use as an absolute reference frame. It was furthermore crucial to the derivation that the wave velocity measured by anyone moving through the medium would be more or less, depending on whether they were traveling with the waves or against them: $V' = V \pm v$. As we saw in Chapter 11, this Galilean transformation cannot be used for light signals. There is no medium in which electromagnetic waves propagate and which serves as an absolute reference frame for all other kinds of motion. The measured velocity of light for an observer in any reference frame, regardless of the relative velocity of the source reference frame, is always $c = 3 \times 10^8$ m/s. If we apply the correct Lorentz transformation to our previous analysis, we get just one formula for the Doppler shift. All that matters is the *relative* velocity between source and observer: there are not two special cases depending on who is moving. The Doppler shift for electromagnetic radiation is

$$f_{obs} = f_{source}[1 \pm v/c + \tfrac{1}{2}(v/c)^2 \pm \cdots]$$

Question 16-5
How large would v/V or v/c have to be to produce a 1% difference in frequency, according to each of the three formulas? What speed would this correspond to in the case of sound or in the case of light?

The Doppler shift is a prime tool in determining velocities. Traffic radar depends on it. When a beam of microwaves strikes a moving car and bounces back to the police radar observer, the frequency of the waves has changed because they have been reflected from a moving mirror (the car). The instrument is calibrated to measure the frequency shift in terms of miles per hour of the moving car. When moving atoms or subatomic particles give off light or X-rays, the radiation frequency is affected by the motion of the source.

Perhaps the most dramatic use of the Doppler shift is in astronomy. When light comes to us from a distant star or galaxy, it may contain very little direct geometrical information. With a few exceptions, all stars are truly point sources for telescopes on earth. We cannot see their diameter or structure or internal motions. However, as we will see, we can spread the light out into a spectrum so that different frequencies of the light appear at different places on our photographic plates. Much of the light is concentrated in particular discrete frequencies corresponding to certain atomic processes. Each atom produces its own "fingerprint" of certain frequencies. A particularly simple pattern of these special frequencies is shown in the diagram. If the light from a star or galaxy contains this same pattern, then we know that it contains that particular atom. However, if the star or galaxy is moving away from us, every frequency will be shifted lower and we will see the same pattern but in a different frequency range. Sometimes the spectral pattern from a star will shift from lower frequencies to higher frequencies and back again. We interpret this to mean that the star is revolving around another one so that sometimes it is moving toward us and its frequencies are increasing and later it is moving away from us and its frequencies are decreasing.

The most sensational finding using the Doppler shift has been that the distant galaxies are all moving away from us. Since the spectral patterns move to lower frequencies for a source moving away from an observer, and since in the visible light range this would mean a shift toward redder light, this effect has become known as the red shift. Not only are the distant galaxies moving away from us, but their speed of recession is proportional to their distance from us, as determined by other methods. We have detected systems at the far reaches of the universe whose Doppler shift indicates that they are moving away from us with v/c as large as 0.6. Many attempts have been made to interpret the red shift in terms of some other phenomenon but so far the Doppler shift seems the only reasonable explanation.

Summary of Doppler Shifts

Source stationary in medium in which wave velocity is V
(a) *Observer moving away from source with velocity v:*

$$f'_{obs} = f_{source}\frac{V-v}{V} = f\left(1 - \frac{v}{V}\right)$$

(b) *Observer moving toward source with velocity v:*

$$f'_{obs} = f_{source}\frac{V+v}{V} = f\left(1 + \frac{v}{V}\right)$$

Observer stationary in medium
(a) *Source moving away from observer:*

$$f_{obs} = f'_{source}\frac{V}{V+v} = f'\frac{1}{1+v/V}$$
$$= f'[1 - v/V + (v/V)^2 - \ldots]$$

(b) *Source moving toward observer:*

$$f_{obs} = f'_{source}\frac{V}{V-v} = f'\frac{1}{1-v/V}$$
$$= f'[1 + v/V + (v/V)^2 + \ldots]$$

For electromagnetic waves in vacuum, the speed of light is c in all reference frames. For relative velocity, v, between source and observer:
(a) *Toward each other:*

$$f_{obs} = f_{source}[1 + v/c + \tfrac{1}{2}(v/c)^2 + \ldots]$$

(b) *Away from each other (red shift)*

$$f_{obs} = f_{source}[1 - v/c + \tfrac{1}{2}(v/c)^2 - \ldots]$$

PHASE EFFECTS AND COHERENT WAVES

In describing sinusoidal waves, we have assumed that there is only one wave without beginning or end. The wave is being generated continuously with constant amplitude and is always moving in the same direction. We must begin to deal with situations where wave motion starts and stops and where waves change direction and bounce back upon themselves.

Even when two oscillations have the same frequency and the same amplitude, they may differ in *phase*. An example of phase difference is shown in the diagram. When one pendulum bob is at its maximum displacement, the other is passing through the equilibrium position. They are 90° or $\pi/2$ radians out of phase. We could describe one motion as $y = a \sin \omega t$ and the other as $y = a \cos \omega t$. Alternatively, we could describe both motions in terms of the sine function, but include the phase angle:

$$y_1 = a \sin \omega t \qquad y_2 = A \sin (\omega t + \alpha)$$

In this case $\alpha = 90°$. If there is a single oscillating system, the value of the phase angle simply tells us when we started our clock to measure time. If there are two oscillating systems, however, the phase angle gives a specific relationship between the two motions.

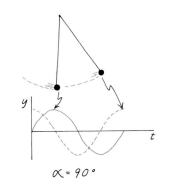

$\alpha = 90°$

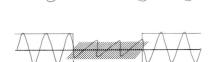

There is clearly a physical difference between sending waves down a rope continuously with the displacement always along the same vertical line, and an alternative way of disturbing the rope by starting and stopping the sine waves at arbitrary times, with the displacement sometimes vertical, sometimes horizontal, and sometimes in between. This alternative way is analogous to the wave motion of light from an ordinary incandescent bulb. The duration of the continuous sine waves at any point, coming from a distant light bulb, is less than 10^{-8} s. Each time a new wave train of light starts, its perpendicular displacement is in a new direction. (In the case of light, this perpendicular displacement is in the direction of the vibrating electric field.) This type of wave motion where the relative phase (with respect to an external clock or to another wave) changes from instant to instant is called incoherent.

Whether we consider waves coherent or incoherent, depends to some extent on the length of time during which we observe the waves. The 60-cycle alternating current produced by one power company is not necessarily in phase with the 60-cycle produced by the power company in the next county (unless they are locked together in a common system). The waves are partially coherent with each other, however, since whatever phase difference exists remains roughly constant over observing times of many minutes. Over periods of hours or days, however, the phases would shift. If you measured the phase difference between light waves coming from two different incandescent bulbs, it would remain constant and the waves would be coherent for 10^{-8} s or so. Over observing times more appropriate to the human eye, however, two ordinary sources of light are incoherent. The sound waves coming from two different loudspeakers driven by the same oscillating electrical signal are coherent. The phase difference between the two sound waves may be different at one point in the room from what it is in another, as we soon will see. However, at any one point the phase difference remains constant.

Until recent decades it was not possible to generate high intensities of coherent infrared or visible light. As we will learn in Chapter 22, radio waves and even microwaves down to a wavelength of 1 mm can be generated coherently by devices such as radio tubes where we can actually picture electrons oscillating back and forth at high frequency. With the invention of the laser (an acronym for *l*ight *a*mplification by the *s*timulated *e*mission of *r*adiation), it has become possible to generate intense beams of coherent infrared and visible light waves. If light is sent back and forth between mirrors in a tube containing excited atoms, the wave trains of light trigger the excited atoms to produce more wave trains in phase with the original.

THE LINEAR SUPERPOSITION OF WAVES

An elastic medium can support many waves at once. In general, the waves pass through each other or march along together without interfering with each other at all. You may have noticed one example while sending pulses down a stretched rope. If you send two pulses in succession, the first one will reflect from the far end and then pass right through the advancing second one without changing its velocity or shape. You can also see this effect if you and a colleague stretch a rope between you and each sends pulses toward the other.

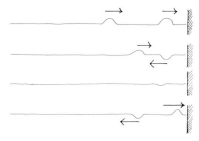

At the point where the pulses are passing each other, the displacement of the medium is the simple sum of the two separate disturbances. If the original displacements were in the same direction, they interfere *constructively*. If the displacements were in opposite direction, they may interfere *destructively* while they are passing through each other. They have not really destroyed each other, however, except momentarily while they are at the same point.

The superposition of waves is linear in a medium as long as the disturbance obeys Hooke's law. As long as the restoring force is proportional to the displacement, the displacements can add algebraically because the restoring forces will add algebraically. If the combination amplitude is too large, perhaps because a spring has been stretched too far, then the superposition of the waves will not be linear. Often energy is lost to other forms under such circumstances and sometimes new frequencies are generated. In fact, if a wave all by itself has sufficient amplitude so that the medium does not obey Hooke's law, then the wave shapes will be distorted. As we will see, this phenomenon can be interpreted in terms of the creation of harmonic frequencies. Intense laser light passing through certain crystals can emerge with some of its energy at twice the original frequency.

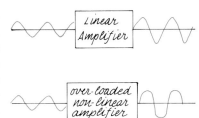

A non-linear amplifier (such as a cheap sound system at too large an amplification) introduces distortion into the pure sine wave

Because of the linear superposition characteristic of most wave motion, we can describe the interference of two waves by adding their displacements graphically or by using the trigonometric addition of two sinusoidal functions. During the rest of this chapter we will see numerous examples of both techniques.

BEATS BETWEEN WAVES

The simplest case of interference between two waves occurs when the waves have equal amplitude and are moving in the same direction with the same velocity but have slightly different frequencies. In "Handling the Phenomena," you produced such a situation with two musical tones that had almost the same pitch. The situation is shown graphically in the diagram. The upper part of the diagram shows the two separate sine waves moving along the same line. In the lower part of the diagram, the waves have been superimposed by adding the displacements of the two waves at every point. Wherever the two separate signals are approximately in phase, the resultant amplitude is large. When the separate signals are approximately 180° out of phase, the interference produces a signal with very small amplitude. In examining the resultant pattern, the first thing to observe (and it is not obvious that it should be true) is that the sum looks like a sine wave with changing amplitude. The second thing to note is that the change in amplitude is itself sinusoidal.

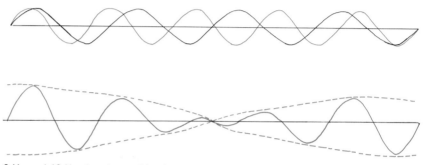

8 Hz and 10 Hz signals combined.

Suppose that the frequencies of the two signals are 8 Hz and 10 Hz. If they start out in phase, they will be back in phase half a second later. In that time one of them will have been through 5 periods and the other will have been through 4 periods. At $\frac{1}{4}$ second, the 10 Hz note will have gone through $2\frac{1}{2}$ periods and the 8 Hz note will have gone through 2 periods. At that time they are 180° out of phase since the 10 Hz note has gone through an extra $\frac{1}{2}$ period. They destructively interfere at that time producing a minimum of the amplitude of the resultant signal. During one second the resultant amplitude goes through 2 minima and 2 maxima. The frequency of these "beats" is 2 Hz, which is the difference in frequency between the two separate signals.

■ Optional

We can describe the general situation algebraically by using one of the trigonometric identities to be found in the Appendices. The displacement of each signal separately at a given point is given by

$$y_1 = A \sin 2\pi f_1 t \qquad y_2 = A \sin 2\pi f_2 t$$

If the superposition is linear, the resultant displacement is

$$y = y_1 + y_2 = A (\sin 2\pi f_1 t + \sin 2\pi f_2 t)$$

The trigonometric identity that we need is

$$\sin\theta + \sin\phi = 2 \sin \tfrac{1}{2}(\theta + \phi) \cos \tfrac{1}{2}(\theta - \phi)$$

In terms of the two frequencies, the resultant displacement is

$$y = \left[2A \cos 2\pi \left(\frac{f_1 - f_2}{2} \right) t \right] \sin 2\pi \left(\frac{f_1 + f_2}{2} \right) t$$

Here we have the algebraic proof that the superposition of the two sine waves produces another sinusoidal wave. The fine structure of the oscillation is described by the sine curve. The frequency is $(f_1 + f_2)/2$, which is just the average of the two separate frequencies. In our example for instance, the resultant average frequency is $(10 + 8)/2 = 9$. The amplitude of this sinusoidal high frequency wave is given by the term in the bracket. It is an amplitude that depends on time with a frequency equal to $(f_1 - f_2)/2$. In our example the frequency of this slowly varying amplitude is $(10 - 8)/2 = 1$.

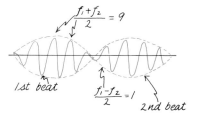

It appears at first that the algebraic derivation for the beat frequency (the frequency of the slowly varying amplitude) is different from the one we derived graphically. However, the frequency of $(f_1 - f_2)/2$ is the frequency for the amplitude to go through a complete cycle including a maximum amplitude, $+A$, and a maximum negative amplitude, $-A$. Each of these, however, is heard as a beat and therefore the frequency of the beats is twice that of the varying amplitude term. The beat frequency is indeed $(f_1 - f_2)$.

Skilled musicians use the beat phenomenon for tuning purposes. When two tones have almost the same frequency, you can hear a definite pulsation. You can even determine the difference in frequencies by counting the number of beats per second, providing the number is under about five. The tuning proceeds by adjusting the frequency of one instrument until the beat disappears.

The presence of beats between two notes is a factor in our pleasure in listening to the notes. Each string in the middle and upper register of a piano is really a triplet of strings, each of which can be tuned separately by adjusting the tension. If the frequencies of the separate strings are slightly different, beats will be produced. The effect is not particularly noticeable unless the beat frequency is more than several per second. If they are out of tune by that much, the strings produce a jangling sound typical of the sound supposedly produced by barroom pianos.

As we will soon see, actual musical tones are not pure sine waves; they have harmonics with frequencies that are integral multiples of the fundamental frequency. If you strike a chord of several notes, some of the harmonics of the individual notes may be very close to each other in frequency (perhaps a second harmonic of one and a third harmonic of another) and these harmonics will produce beats. If the beats are too pronounced, the chord will not sound "harmonic" in the classical sense.

RESONANT LINES

A wave traveling along a straight line can interfere with itself if it is reflected. Under special conditions the interference can provide reinforcement to the original wave so that the oscillations build in amplitude. This phenomenon of *resonance* is crucial to such things as musical instruments and radio receivers. In all resonant systems there must be some way in which the incoming energy pulses can be added *in phase* to the existing oscillations, and the energy in the incoming pulse must be greater than or equal to the energy loss in the system during one cycle.

Let's consider first what would happen to a single pulse in a stretched line fastened to a solid wall at each end. (We will worry later about how the pulse got in there.) The pulse reflects from each wall with a phase change of 180°. The total time taken for a round trip is equal to $2L/v$. At the end of this time the pulse is back to its original position and traveling in the original direction. Now let's introduce a whole train of these pulses both up and down (plus and minus). All of these pulses join the reflection sequence and go back and forth along the path. It might seem as if the displacement of any particular point in the line would get greater and greater with all these pulses superimposed. But in general, the opposite happens. The time taken for a reflected pulse to return to its original condition is not usually related to the period of the incoming train of pulses. Therefore, the result at any point is a random sum of displacements (plus and minus), and quickly tends to zero.

There can be one type of exception to this situation. If the round trip time of the pulses is such that a pulse can join in phase with a new pulse being introduced, then they will reinforce each other and not cancel. The round trip time is $2L/v$. To arrive back in phase with the next pulse, the travel time must be equal to the period of the pulses, T:

$$2L/v = T$$

For a wave train, $vT = \lambda$. Therefore,

$$L = \tfrac{1}{2}\lambda$$

The buildup of such synchronized pulses is shown in the diagram.

There are other privileged frequencies where the reflections will produce reinforcement. If the pulses occur at twice the frequency, or three times, or any integral number times the fundamental frequency, the phase of the reflected pulse after a round trip will match the phase of the starting pulse. The condition is that $2L/v = nT$, where n is an integer. In terms of wave length, the condition is

$$L = (n/2)\lambda$$

The particular relationships that we have derived between the length of the line and the possible wavelengths of the resonant pulses were for the case of a line fixed at both ends. This feature provided a phase change of 180° at each reflection. The same relationships would be derived if the line were open at both ends. In this case there would be no phase change at a reflection but the prime condition would be the same. The travel time of a pulse from one end to the other and back again must be such that the reflected pulse is ready to make the return journey in phase with a new pulse. Once again, the travel time must be equal to the period or an integral multiple thereof:

$$2L/v = nT$$

If one end of the line is open and the other end is closed, the conditions for reinforcement change. Since there is only one reflection at a closed end, providing one phase change of 180°, the travel time must provide a phase difference of another 180°. Therefore, the new condition is $2L/v = T/2$, or $\frac{3}{2}T$, or $\frac{5}{2}T$, and so on. For resonance on a line where one end produces a 180° phase change and the other does not:

$$L = nvT/4 = (n/4)\lambda, \text{ where } n = 1, 3, 5, \text{ etc.}$$

It may be difficult to imagine circumstances where pulse trains can be introduced into lines like those we have described. For instance, if the line is actually a stretched string, such as a violin string, then it is certainly fixed at each end, but how can you get pulsed energy into it without disturbing the existing chain of reflected pulses? Similarly, how can you have a line with one or two free ends and still keep it stretched taut and feed energy into it? Actually, the theoretical line with its resonant chain of pulses is a good model for many different practical systems. We will study the details of some of these in connection with musical instruments in a later section. First, however, we must examine a special case of resonance where the pulse train consists of a sine wave.

Conditions for resonance:

Both ends closed or both ends open $\Big\}$ $T = \frac{2L}{nv}$, $f = \frac{nv}{2L}$, $\lambda = \frac{2L}{n}$

where $n = 1, 2, 3 \ldots$

One end open, one end closed $\Big\}$ $T = \frac{4L}{nv}$, $f = \frac{nv}{4L}$, $\lambda = \frac{4L}{n}$

where $n = 1, 3, 5 \ldots$

STANDING WAVES

We have gone to some trouble to describe the behavior of traveling waves. It may seem like a regression to turn our attention now to waves that stand still. How, indeed, can a wave not move?

The standing wave phenomenon is similar to the phenomenon of beats. Once again, we deal with two sine waves moving along the same line. For the sake of simplicity, we will assume that they have equal amplitude. They must also have the same frequency. Their speed will be the same but their velocities must be opposite in direction. If one wave travels to the right and a duplicate wave travels to the left, there will obviously be constructive and destructive interference along the path. All these conditions will exist if sine wave disturbances are reflected without loss of amplitude from each end of a path. Of course, in a real situation with a vibrating string or an air column, energy losses occur along the path and particularly during the reflections. We will have to take a closer look at how we can pump energy into the system so that the waves will bounce back and forth with constant amplitude.

Let's consider first what would happen to a sine wave in a stretched line fastened to a solid wall at each end. The wave reflects from each wall with a phase change of 180°. If the round trip time of the waves is such that a wave can join in phase with the new wave being introduced, then they will reinforce each other and not cancel. The round trip time is $2L/v$. To arrive back in phase with the next wave, the travel time must be equal to the period of the wave, T.

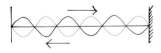

$$2L/v = T$$

For a wave train, $vT = \lambda$. Therefore,

$$L = \tfrac{1}{2}\lambda$$

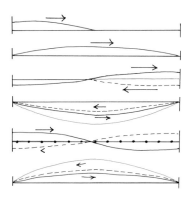

These are the resonant conditions we derived in the previous section. The buildup of such synchronized waves is shown in the diagram.

> **Question 16-6**
>
> Where does the reflected wave traveling to the left reinforce the wave traveling to the right and where does it produce a cancellation? Are these cancellations and reinforcements continual?

There are other privileged frequencies where the reflections will produce reinforcement. If the waves occur at twice the frequency or three times or any integral number times the fundamental frequency, the phase of the reflected wave after a round trip will match the phase of the starting pulse. As we saw in the case of pulses, the condition is that $2L/v = nT$, where n is an integer. In terms of wave length, the condition is

$$L = (n/2)\lambda$$

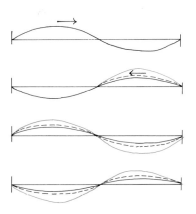

Here is another sequence of diagrams showing the buildup of these synchronized waves for $n = 2$. Note that the envelope of the resultant displacements for the line is that of a sine wave.

The resultant motion of the line is due to the superposition of a wave traveling to the left and a wave traveling to the right (which in this case was furnished by reflection). Although the resultant was built up out of traveling waves, there is no indication in the final appearance that there is any transfer of energy along the line. Instead, the line appears to be oscillating up and down. The amplitude of oscillation varies along the line and there are nodal points where there appears to be no disturbance at all.

> **Question 16-7**
>
> If energy must be fed along the line to keep it in oscillation, how can energy pass a nodal point?

■ **Optional**

The mathematical analysis of a standing wave involves only a simple trigonometric substitution. The wave traveling to the right is described by

$$y_1 = A \sin (2\pi/\lambda)(x - vt)$$

The reflected pulse traveling to the left is given by

$$y_2 = A \sin (2\pi/\lambda)(x + vt)$$

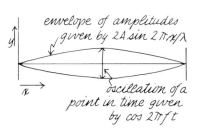

The resultant displacement is equal to

$$y = y_1 + y_2 = A[\sin (2\pi/\lambda)(x - vt) - \sin (2\pi/\lambda)(x + vt)]$$

The trigonometric identity that we need is

$$\sin\theta + \sin\phi = 2\sin\tfrac{1}{2}(\theta+\phi)\cos(\theta-\phi)$$

Therefore,

$$y = [2A\sin(2\pi/\lambda)x]\cos(2\pi/\lambda)vt = [2A\sin 2\pi(x/\lambda)]\cos 2\pi ft$$

Any point along the line is vibrating up and down with a frequency f and an amplitude determined by the quantity in the brackets. The shape of the line at any given time is sinusoidal.

Standing Waves
For a line closed at both ends: $\lambda = 2L/n$
$n = 1, 2, 3\ldots$

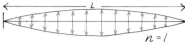

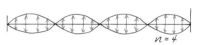

If the resonance conditions are met with other combinations of L and λ, corresponding to various integers, n, the resulting standing waves remain sinusoidal but contain more loops. Examples are shown in the diagram for lines that are closed at both ends (or open at both ends), and for lines that are open at one end and closed at the other.

Standing waves can be established in stretched strings and in air pipes. There must be some way to feed in energy at one of the resonant frequencies. In "Handling the Phenomena" in Chapter 15, you accomplished this feat with piano strings. By singing a loud note at the bare strings, you set up a standing wave in the fundamental mode in the particular string that matched your pitch. You could then hear the string sing back to you as the standing wave died down.

In this photograph we show another way to feed energy into a stretched string. In this case, the mechanical oscillator feeds in the necessary energy to compensate for the friction losses and also acts as an open end of the string. For a standing wave to exist, the length of the string must be an odd-integral number of quarter wavelengths. To produce standing waves with such a device, either the frequency of the oscillator must be adjusted to yield the right wave lengths or else the tension in the string must be adjusted.

For a line open at one end and closed at the other: $\lambda = 4L/n$
$n = 1, 3, 5\ldots$

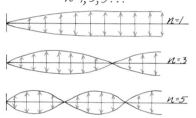

Question 16-8

Why should the tension have anything to do with the conditions for a standing wave?

For a line open at both ends:
$\lambda = 2L/n$

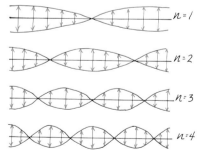

The conditions for compressional standing waves in a pipe are the same as those that we have derived for transverse waves in a line. In describing the situation, however, we must be careful to distinguish between longitudinal displacements of the particles such as gas molecules and the variations in pressure. In general these two variables are 90° out of phase.

Consider the situation in a column of air that contains a standing wave. At a displacement node, the molecules are moving in one direction on one side and in the opposite direction on the other side. At one instant they are rushing toward the node and half a period later they are rushing away from it. These surges produce maximum pressure fluctuations at that point, creating a pressure anti-node. On either side of a displacement anti-node, the molecules always move in the same direction as each other. Consequently,

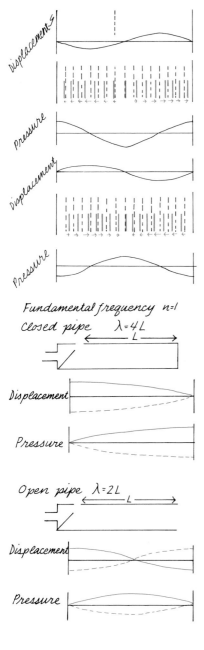

Fundamental frequency n=1
Closed pipe λ=4L

Open pipe λ=2L

First Overtone

there can never be a pressure difference across this region. Since the molecules move together, they do not pile up there. At a closed end of a tube containing gas, such as an organ pipe, there must be a displacement node. The molecules just cannot vibrate back and forth into the solid wall. Reasonably enough, there is a pressure anti-node at a closed end. The molecules pile up there, creating high pressure, and then as that pulse is reflected with a 180° phase change, the region at the wall goes to low pressure. At an open end the situation is just reversed. The molecules are free to move back and forth in unison but the pressure does not change since the molecules never pile up.

In the diagram we show both the displacement and the pressure amplitudes for standing waves in the fundamental mode in a closed and open pipe. (The driven end of either one acts like an open end.) The schematic method shown for driving the pipe is a reasonable representation of the geometry in an actual organ. Compressed air is forced in one end and spills into the tube close to a whistle-type opening in the wall of the tube. The region around this opening is approximately a displacement anti-node, which is why this end acts as if it were "open." Although the compressed air comes from a constant pressure box, it enters the pipe in a turbulent fashion controlled by the pressure waves moving back and forth in the pipe. The simple result of the complicated phenomena is that enough air is added at the right phase to make up the losses of the traveling waves and to maintain a standing wave.

In general the amplitude envelope for the displacement or pressure variations is not part of a simple sine wave as we have shown. Most singing pipes have overtones produced by traveling waves that are not sinusoidal. We will soon see how these overtones arise. A pipe can, however, sing at any of its harmonics if the pressure pulses are fed into it at the proper frequency. In the diagram we show the displacement and pressure patterns for standing waves in open and closed pipes that are singing at the second harmonic. For the open pipe, $n = 2$, but for the closed pipe only the odd harmonics are present. Therefore, for the closed pipe, $n = 3$ for the first overtone above the fundamental.

Note that a pipe sings at a higher pitch only if the source of excitation can feed energy into the system at the higher frequency. To change a singing pipe from its pure fundamental to a pure harmonic requires some change in the driving system such as higher pressure or deflection of the input air to

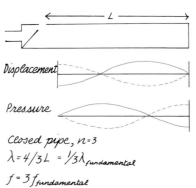

Closed pipe, n=3
$\lambda = 4/3 L = 1/3 \lambda_{fundamental}$
$f = 3 f_{fundamental}$

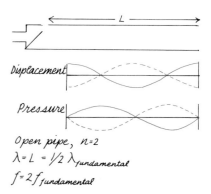

Open pipe, n=2
$\lambda = L = 1/2 \lambda_{fundamental}$
$f = 2 f_{fundamental}$

FOURIER ANALYSIS OF COMPLEX WAVES

a different angle, thus altering the turbulence pattern and so changing the frequency at which the extra pressure is admitted. In organ pipes this is seldom done, although for lip-controlled wind instruments it is the standard method for sounding notes in different frequency ranges.

FOURIER ANALYSIS OF COMPLEX WAVES

When an object hung on a simple spring is disturbed, it vibrates with simple harmonic motion. The displacement, velocity, and acceleration are sinusoidal functions of time. Systems more complex than a simple spring usually respond to disturbances in a variety of ways, most of them complex. Although a tuning fork produces pressure waves in the air that are sinusoidal, a metal platter that is struck will respond with a great jangle of pressure variations. Each muscial instrument produces periodic pulses with characteristic shapes, at least during the initial attack of the tone. We showed several of these on page 386. The notes were picked up by a microphone and turned into electrical signals that were displayed on an oscilloscope.

Back in 1822 Jean Fourier (1768–1830) published a treatise demonstrating how repetitive mathematical functions could be represented by the algebraic sum of a series of sine and cosine waves, each term of which has a different frequency. If the function to be modeled has a frequency f_0 then the first sinusoidal term has a frequency f_0, the second one has a frequency $2f_0$, and so forth. The term with the frequency f_0 is called the fundamental, or first harmonic, and the other terms are called higher harmonics. The amplitude of each term is chosen to produce the required shape when all the terms are added together.

It is not at all obvious that Fourier's claim is true and this book is not the place to prove it. However, we can make the idea plausible by demonstrating what happens when you add several sine or cosine waves. In the adjacent graphs we show the addition of a fundamental and the second harmonic. In the first case the harmonic has the same amplitude as the fundamental. In the second case it has half the amplitude. In the second pair of graphs we show the corresponding additions of a fundamental and the third harmonic ($f = 3f_0$). Notice how the resulting wave shapes differ from the first case.

If a complicated wave shape must be reproduced, many sinusoidal terms must be used to get a good approximation. This means that many harmonics are required, extending to high frequencies. Sharp corners on the basic wave form are particularly hard to reproduce. In the diagram we show the successive addition of the first three terms of a Fourier series for a square pulse. The algebraic formula giving the amplitude of each term is shown below the diagram. Notice the effect of each stage of the approximation. Many more terms would be needed to reproduce well the corners of the square pulse. A similar summation for a sawtooth wave is shown on the next page. Once again, the approximation is not bad for the gradually rising part of the sawtooth but many more terms would be needed to reproduce the abrupt drop of the original.

If a trumpet and oboe both sound middle A, each is producing 440 repetitive pulses per second. The oboe pulses look different from the trumpet pulses, however. Pulse shapes from a number of different instruments were shown on page 386. Each characteristic pattern can be represented by the

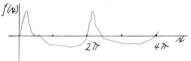

Fourier series
$f(x)$ is a continuous function that is periodic in 2π
For instance:

Then:
$f(x) = \tfrac{1}{2}a_0 + a_1 \cos x + a_2 \cos 2x + \ldots$
$\qquad + b_1 \sin x + b_2 \sin 2x + \ldots$

where $a_0, a_1, a_2 \ldots$
and $b_1, b_2 \ldots$
are constants that depend on the shape of $f(x)$

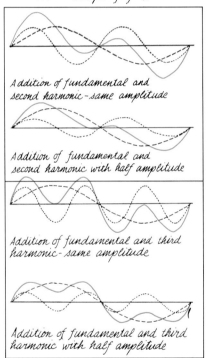

Addition of fundamental and second harmonic - same amplitude

Addition of fundamental and second harmonic with half amplitude

Addition of fundamental and third harmonic - same amplitude

Addition of fundamental and third harmonic with half amplitude

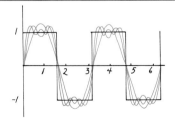

Successive Fourier approximations for a square wave.

$f(x) = \dfrac{4}{\pi}\sin x + \dfrac{4}{3\pi}\sin 3x$
$\qquad + \dfrac{4}{5\pi}\sin 5x + \ldots$

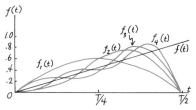

Successive Fourier approximations for a sawtooth.

$$f(t) = \frac{2}{\pi}\sin\frac{2\pi}{T}t - \frac{2}{2\pi}\sin\frac{2\pi}{T}2t + \frac{2}{3\pi}\sin\frac{2\pi}{T}3t - \frac{2}{4\pi}\sin\frac{2\pi}{T}4t + \ldots$$

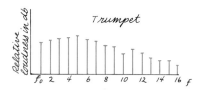

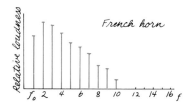

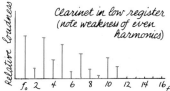

Relative loudness in db of fundamental and harmonics for instrument tones after initial attack.

sum of a fundamental and many harmonics. The difference between one pattern and another is determined by the distribution of amplitudes of the harmonics. Each instrument has its own recipe for these amplitudes. A graphical representation of these recipes for several instruments is shown in the diagram below.

When we were discussing the audio range in Chapter 15, we pointed out that most instruments and the human voice stay below a fundamental frequency of 2000 Hz. Nevertheless, hi-fi audio reproduction systems should be able to produce frequencies up to 20,000 Hz. Now we can understand the reason why this extended range is required. A good rule of thumb for the reproduction of musical notes is that harmonics up to the tenth are required. A 2000 Hz note produced by a voice or an instrument is not a 2000 Hz sine wave. The peculiarities of its shape may require harmonics up to 20,000 Hz for faithful reproduction. Of course, this rule of thumb depends on the nature of the instrument. Any note produced by plucking or striking is apt to have repetitive pulses with sharp edges involving harmonics of high frequency. Reproducing faithfully the sounds from bells or triangles, or even pianos, is one test for a hi-fi system.

THE CHANGE OF PULSE SHAPE DURING TRANSMISSION

As we have seen, odd shaped repetitive pulses traveling along in the same direction can be considered the resultant of many sinusoidal waves with harmonic frequencies. If there is no absorption in the medium or if the absorption does not depend on the frequency, and if the velocity of the waves does not depend on frequency, then the pulses will retain their original odd shape. These are the usual circumstances with sound traveling in air or with electromagnetic radiation traveling in a vacuum. Under many other circumstances, however, the velocity of wave motion does depend on the frequency and so does the absorption or radiation of wave energy.

The dependence of velocity on frequency is called *dispersion*. We can see the reason for this strange term if we observe a crest of a water ripple. An initial sharp crest soon becomes more rounded and spread out as it travels. The reason for this becomes apparent if you notice the tiny short wave length ripples creep out ahead of the main pulse. The high frequency components travel faster and so sneak out ahead. Without these components the main pulse becomes more rounded, eventually assuming the shape of the fundamental sine wave.

The wave velocities that we have used so far have been *phase* velocities. For instance, the equation, $f\lambda = v$, describes the phase velocity. The *group* velocity is the velocity of the center of the pulse, which may be considerably slower than any of the phase velocities. The pulse itself is a cooperative effect of many sinusoidal terms. The individual sinusoidal waves slide into the pulse and out of it again, interfering constructively in one region to produce the pulse. The traveling energy is contained within the pulse and is cancelled out elsewhere through destructive interference of the separate waves. Depending on how the phase velocity varies with frequency, the group velocity can be greater or smaller than the phase velocity. For certain frequencies of electromagnetic radiation, usually in the X-ray region, the phase velocity in most materials is actually greater than c, the velocity of light in a vacuum. The group velocity however is slightly less than c. It is still the case that no energy (or information) can be sent faster than c.

> Question 16-9
>
> If a sinusoidal signal of a particular frequency has a velocity in material faster than c, why can't we use that signal to send information or energy?

The shape of pulses is also affected by what happens while they are being transmitted through a medium and also what happens to them at boundaries. Frequently absorption along the path is frequency-dependent. Similarly, at boundaries, the ratio of reflected to transmitted intensity may depend on frequency (remember that our simple formulas for impedance depended on several assumptions about the waves and the medium). In many musical instruments the high frequency components of pulses are absorbed internally, or radiated from the instrument with higher efficiency. In either case the wave motion loses its high frequency harmonics and settles down to a fundamental and a few weak harmonics. Since the particular pattern of harmonics provides the characteristic tone of the instrument, even skilled musicians have a hard time identifying an instrument if the initial stages or attack of each note are electronically removed. Sustained tones from trumpet, trombone, and even oboe sound very much alike. The sounds from plucked or struck strings rapidly turn into tones that are almost purely sinusoidal.

Optional

Production of Sounds by Musical Instruments

Every musical instrument consists of some source of vibrations and some arrangement for transmitting the energy to the air with reasonable efficiency. The basic vibration is rarely sinusoidal but instead consists of a series of pulses of various shapes. The vibrating system is part of or is connected to a resonating system with its own pattern of preferred frequencies of oscillation. The resonating system responds to the fundamental and overtones of the driving pulse and radiates these into the air with varying efficiency depending on the frequency of each and the shape of the resonator. The initial vibrator may be a stretched string, or vibrating lips, or vocal cords, or turbulent air in a constricted channel. The resonating system may be a carefully shaped wooden box that can vibrate in response to a whole range of frequencies or it may be a column of air enclosed in a pipe that will respond only to certain harmonic frequencies.

There is a large range of relationships between the vibrator and the resonant system to which it is attached. In a mechanical siren, for example, compressed air blows through holes in a rapidly revolving disc. The frequency is strictly determined by the number of holes passing the air blast each second. Even if there is a horn attached in front of the air blast to concentrate the sound in one direction, reflections of pressure from the end of the horn exercise no control over the frequency of the source.

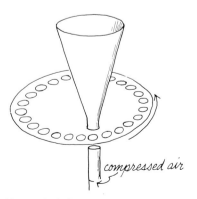

Mechanical siren.

The frequencies of the stringed instruments are also independent of the resonator. The basic frequency and the shape of the individual pulses on the string depend only on the mass, tension, and length of the string and how it is disturbed in the first place. Music that we hear from a stretched string, however, depends on the response of the resonant system to the various harmonics present in the pulse shape and to the efficiency with which the resonator can send these frequencies into the air.

At the other extreme of frequency control, the source vibrations of an organ pipe are caused by the self-triggered release of pulses of compressed air into one end of a pipe containing traveling pulses or pressure waves. The reflected pulses trigger their own replenishment; thus the vibration frequency is determined by the shape of the resonating system. However, even in this case the source has some privileges. If the air pressure, or the direction in which the air is admitted, is arranged correctly, the turbulent air at the opening may admit an extra pulse into the pipe when the first one is up at the other end. There will then be two pulses in the tube at the same time passing through each other, each triggering its replenishment when it gets back to the opening. The pipe will then sound its first harmonic, which in the case of an open pipe produces a note one octave higher.

Organ pipes are designed to play only the pitch corresponding to the fundamental frequency. Consequently, there is only one pressure pulse at a time in the pipe. With many other wind instruments, however, the driving vibrator can be controlled to send in one, or two, or three, or many pulses into the pipe before the first one has returned. The pitch of the horn is determined by how many pulses per second that it contains. Thus, with a flute, for a given setting of the holes that determine a given length of pipe, the player can change the position of his mouth to send from one to seven pulses at a time into the pipe. Seven different pitches can thus be produced (though not necessarily seven different octaves). Similarly, a bugler can sound five (or more) different pitches with an instrument that contains no keys. In the case of the bugle (and most other brass instruments), the lowest *musical* pitch is actually the second harmonic containing two pulses of pressure in the tube at a time. (The fundamental, which is the first harmonic, can sometimes be sounded. It sounds like a growl, and is called the "pedal" note.)

With the reed instruments such as the clarinet or the oboe, the combination of players' lips and reeds merely takes the place of the turbulent air gate of the organ pipe. The frequency at which the reed opens to admit another pulse of air is controlled by the reflection of the previous pulse as it returns to the reed. The player chooses different frequencies by opening or closing holes in the pipe. These actions change the effective length of the pipe and therefore the travel time for the pulse to make the round trip. For a particular setting of the holes, the player can raise the pitch to a harmonic by opening one of a set of holes that throws the pipe into the two-pulse-at-a-time mode.

Production of sound by the human voice and by the brass instruments is intermediate between control of the frequency by the source or by the resonator. The vocal cords in the throat vibrate as air rushes past them. Men usually have more massive muscles in these cords than do women. Consequently, you might expect that male vocal cords would vibrate at a lower frequency. It is indeed true that some small men can have deep bass voices while some large women can be lyric sopranos.

In the case of brass instruments played with the lips pressed against a mouthpiece, the lips vibrate at the particular frequency that they wish to sound. In fact, a bugler can play recognizeable bugle calls with just a mouthpiece and

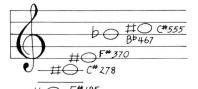

The five bugle notes normally used. (Higher notes can be played.) Note that the frequencies are multiples of 92 Hz, which is f_0, the fundamental

$n = 1 \quad f_0 = 92$
$n = 2 \quad f_1 = 184$
$n = 3 \quad f_2 = 276$
$n = 4 \quad f_3 = 368$
$n = 5 \quad f_4 = 460$
$n = 6 \quad f_5 = 552$

no bugle. On the other hand, the bugler's lips are not completely free agents in determining their frequency of vibration. The motion of the lips depends to some extent on the arrival of reflected pulses to trigger and maintain the proper frequency of vibration. As evidence for this, try playing a trumpet or bugle that is filled with helium. The pitch of any note will be much higher than expected, but then will return to normal as the helium drains out of the instrument. Because of the low atomic mass of helium, the speed of sound in helium is several times that in air. A pressure pulse can travel to the end of the instrument and back in a much shorter time and will trigger the lips into a higher vibration rate.

There is a standard lecture demonstration apparently showing the same effect with the human voice. If a lecturer inhales one good breath of helium, his next few sentences will sound as if they were spoken by Donald Duck. While this effect is a stunt for a lecturer, it is a serious problem for deep-sea divers who breathe a mixture of helium and oxygen instead of nitrogen and oxygen. Their words cannot be understood or easily decoded by surface listeners. The frequency shift is a complicated one because the mechanism is more complicated than it is for wind instruments. Apparently, the vocal cords are not much affected by any feedback. Instead, they produce a large number of overtones that resonate in various cavities in the throat, nose, and mouth. The basic frequency of the vocal cords does not change when the throat is filled with helium, but the resonating frequencies of the cavities are shifted upward. The higher overtones are thus amplified more than the lower ones. (If you ever have occasion to try the helium stunt, inhale no more than one deep breath of helium. Helium is not poisonous but it is not oxygen, either. It takes only about three breaths to fill the lungs. If you deplete your lungs of that much oxygen, you will pass out immediately.)

Any of the instruments that help control their own vibration rates also determine the shape of the pressure pulses that are in them. Instead of a smoothly varying sinusoidal type of pulse, the pulses are usually abrupt, containing many harmonics. The harmonic recipes for several instruments were shown on page 430. Remember, however, that the pulse shapes, and thus the harmonic pattern, change from the beginning to the end of the note. The beginning of each note, depending on how the note is attacked, usually has many more high frequency harmonics with larger amplitude.

Question 16-10
There seem to be two ways of producing harmonics. If you admit just one pulse at a time into the pipe, it may have harmonic frequencies because of the shape of the pressure pulse. On the other hand, in some instruments you can have two or more pulses in the pipe at the same time, also creating harmonics. Is there any difference between these two types of operation?

When a stringed instrument is plucked, the kink hurtles back and forth along the string, gradually reducing in amplitude and also gradually losing the sharp point of the kink and turning into something resembling half of a sine wave. At the beginning, the shape of the kink represents many harmonics. If the string is not attached to a resonating system, the sound transmitted to the air is weak and nonmusical. The higher harmonic frequencies disturb the air more efficiently than the fundamental does, making the sound seem thin. The basic frequency of the note is determined by the time that it takes the kink to make one round trip.

When you bow a violin string, you are merely plucking it in a very special way. The resin on the bow hair helps to grab the violin string and pull it along with the moving bow. When the kink gets too sharp, the violin string is freed from the bow and snaps back. The disturbance travels in both directions along the line—toward the bridge and toward the far end—and reflects at both ends. Meanwhile the bow has grabbed the string again and has pulled it up to form a new kink. The crucial part of the sequence now arises. The part of the previous pulse that started toward the bridge, after a complete round trip, triggers the release of the next kink and the whole sequence repeats itself. The fundamental frequency is therefore determined by the time it takes the kink to make a round trip along the string. There is only one such traveling kink in the string at a time. Under exceptional and unfortunate circumstances, however, two kinks can be introduced into the string as the result of improper bowing. This gives rise to a harmonic screech and happens particularly frequently with the cello or other bass stringed instruments. It is one of the reasons why children should never be allowed to play stringed instruments in public.

With the strings attached to a resonating box, such as a violin body, the kink-shaped pulse drives the wood and the air in the cavity of the instrument. The entire system now resonates to all of the harmonics that make up the kink-shaped pulse. The larger surface area of the instrument helps to transmit the lower harmonics including the fundamental. Consequently, the tone that we hear is mellow and rich.

Notice that we have now answered the question raised in the section on resonant lines: "How can you feed energy into taut strings or pipes at the resonant frequency?" In the case of bowed strings or blown pipes, the reflected pulse triggers its own replenishment, *and so the proper phase is maintained.*

Question 16-11
If the violin case helps to radiate energy into the air, where does the extra energy come from?

So far we have described two apparently different responses of strings and pipes to repetitive disturbances. First, we showed how standing waves can be set up by reflections within a system. These standing waves are sinusoidal with frequencies that are simple multiples of each other, and with wavelengths that are integral fractions of the length of the line. Although we explained standing waves in terms of the interference of two equal waves traveling in opposite directions, a standing wave itself does not appear to be traveling. Second, we claimed that in musical instruments the sound is produced by a pulse of some sort that hurtles the length of the instrument, reflects at the end, and then triggers the next pulse when it gets back to the origin. Which of these two views is correct? They both are.

The traveling pulse, which may be a pluck in a string or a high pressure region in a pipe, can be represented mathematically by the Fourier sum of many sinusoidal waves. These waves include one with a fundamental frequency equal to the frequency with which the traveling pulse completes its round trip. The other sinusoidal waves have harmonic frequencies that form the overtones of the fundamental. Each of these harmonic waves has one of the special wavelengths required to form standing waves in the line or pipe.

The pitch that we hear in a sound usually (but not always) corresponds to the frequency of the fundamental. If a sustained tone in an instrument gradually loses its overtones, it is because the higher harmonics have been absorbed or radiated away more efficiently. What remains is a standing wave of the fundamental. From the viewpoint of the traveling pulse model, the sharp edges of the pressure pulse or the pluck shape have been lost, leaving a sinusoidal pulse interfering with its own reflection.

The size and shape of instruments is obviously associated with the particular sounds they produce. In general, the small instruments produce soprano notes and the large instruments produce bass notes. In all cases the frequency of the fundamental is determined by the time it takes a pulse to make a round trip within the instrument. The longer the instrument, the longer that time, and therefore the lower the frequency.

With wind instruments there is a direct relationship between the length of the pipe and the wavelength of the sound in air. We have already deduced these relationships for the case of organ pipes and the same formulas apply for most other instruments. Since most instruments act like pipes open at both ends (with the driving vibration occurring at one end), the fundamental wavelength is approximately twice the length of the pipe (see page 00). As we have already mentioned, this fundamental or "pedal" tone cannot be sounded easily in most brass instruments and is nonmusical. The frequency of the lowest note that can be used is usually twice that of the pedal note. For that note two different pulses are introduced into the instrument during the time for one round trip. The length of the pipe can be changed by pressing down valves and adding extra lengths in the case of the trumpet, by sliding out a length of pipe in the case of the trombone, or by opening and closing holes in the side of the tube, which effectively changes its length, in the case of the flute or clarinet.

The wavelength in air of the fundamental note produced by a violin string is not twice its length.

Actual length of bugle = 176 cm
$f_0 = 92\ Hz$
$\lambda = \dfrac{343\ m/s}{92\ Hz} = 3.7\ m$

For pipes with both ends open:
$\lambda_{fundamental} = 2L$
$L = 1.85\ m$

Lowest note that bugle can sound:
$f_1 = 184\ Hz \qquad \lambda = 1.85\ m$

Question 16-12
Why not? Incidentally, notice that all four strings of a violin are the same length, yet each produces a different fundamental note.

The shape and size of the instruments are related in yet another way to the frequency range in which they work. Not only is the oboe shorter than the bassoon and the trumpet shorter than the sousaphone, and the violin strings are shorter than those of the bass viol, but the soprano instruments are also thinner or smaller in diameter. This feature is connected with the problem of transferring energy from the internal vibrating system to the external air. The lower the frequency, the larger the area of the surface boundary that is needed for effective transfer. Remember in "Handling the Phenomena" that the pulse in the vibrating coat hanger had a low frequency of oscillation but only the high frequency harmonics of the pulse were transferred directly to the air. You heard the rich tones of the fundamental and the lower frequency harmonics only by coupling the energy directly to your ear drums.

As a rule of thumb, an instrument does not produce or at least send into the air a note that has a wavelength in air very much larger than the size of the instrument itself. In the case of a flute, the lowest note produced has a wavelength about twice that of the length of the flute. In the case of the trumpet, the wavelength of the lowest note that can be sounded is about the same as the length of the instrument (about 5 feet). With the violin the lowest note produced by the G string is about four times the length of the case.

Violin, viola, cello, bass. (Courtesy Scherl and Roth, division of C. G. Conn, Ltd.)

Harmonic Relationships of Music
All God's chillun got rhythm but not all of us hear harmony in the same way. Indian and Chinese music often sound jangling and confused to people brought up in Europe or the United States. Cultural experiences make a large difference in how we respond to combinations of notes. To people accustomed to western classical music, certain notes on the piano seem to go together while certain other combinations sound harsh or even wrong. Some instruments such as bells produce sounds that just cannot go together no matter what the basic pitch is. Sometimes people who play a chime try to produce chords by striking two bells at once. The result is usually a clanging noise instead of a pleasant harmony.

The diagram on the next page shows the musical notations and the

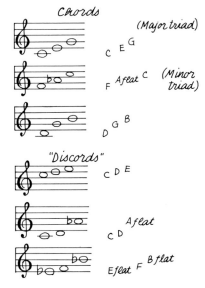

respective frequencies for several different tonal clusters. The first three correspond to classical chords. The next three would have been considered discords in classical music although modern compositions frequently make use of such chords. What is there about certain notes that make them sound good together? Why does this work only with certain instruments? It turns out that the shapes of our familiar music instruments are carefully designed so that they will produce harmony when played with other instruments.

There is something very special about the overtones or harmonics of musical instruments. They do not consist of just any combination of higher frequencies, as we have seen. If you have a fundamental note of 500 Hz, then the musical harmonics are 1000 Hz, 1500 Hz, 2000 Hz, and so on. In other words, if the overtones are musically harmonic, they must be at twice the fundamental frequency or three times the fundamental frequency and so on. As we have seen, vibrating strings automatically produce higher frequency harmonics as do most, but not all, musical pipes. Bells, however, produce lots of overtones but not *harmonic* ones. Besides the fundamental frequency, which depends on how the bell is struck, there may be overtones that are $2\frac{1}{3}$ or $3\frac{1}{4}$ times the frequency of the fundamental. The resultant sound is interesting in itself but when you try to mix sounds coming from two different bells, their overtones don't match at all. Just why this should make a difference to our ears we will soon learn.

So far, all we have said about harmony is that musical notes have overtones with very special combinations of higher frequencies. When a note has *harmonic* overtones, the note sounds rich and pleasant. What happens if you start out with two notes played at the same time? If they are pure sine waves and differ only by a few vibrations per second, then you will hear beats. The combination tone will seem to pulsate in a rather unpleasant way.

If we want to produce a chord of music with two or more notes in it, we must make sure that the notes are far enough apart in pitch so that the beat frequency is higher than 20 Hz or so. Otherwise we will be annoyed by the beats. But we must also face the fact that real musical tones are not sinusoidal but contain many harmonics. In producing a chord, we must make sure that none of the harmonics produce loud beats. In many cases the amplitude of the second or third harmonic is almost as great as that of the fundamental and so might produce loud beats. As an example, suppose we play together three notes with the fundamental frequencies of 311 Hz, 392 Hz, and 523 Hz. Those frequencies correspond to the notes D# (D sharp, three and one-half notes below middle A), G (one note below middle A) and C (two notes above middle A). In the diagram at the top of the next page you can see the frequencies of the various harmonics of these notes. If the notes were played by tuning forks, there would be no discord. But if the notes are played by instruments that produce lots of harmonics, then these frequencies will produce beats with each other.

Since harmony and the quality of music create such subjective experiences, there is considerable discord among musicians over whether or not certain tonal clusters are pleasant to listen to, and there is also controversy about the physical reasons for this situation. Yet another feature of the ear's response to tonal clusters is that most people find it pleasant to hear sounds that are multiple frequencies of each other—in other words, harmonics. That is why the A produced by a musical instrument such as the violin sounds richer and more pleasant than the rather mechanical sound of a pure A from a tuning fork. When two or more notes are sounded together from different instruments, the ear is pleased if their various overtones have a harmonic relationship. The more combinations there are of overtones that are the same as each other, or two or three or four times the frequency of each other, the better the ear likes it.

THE CHANGE OF PULSE SHAPE DURING TRANSMISSION

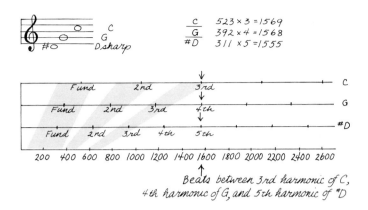

Certain combinations of notes produce the same overtones and so reinforce each other. The most important of these is produced by two notes, one octave apart. The upper note has twice the frequency of the lower note and many of their harmonics are the same. For instance, if one note has a frequency of 400 Hz, then the note an octave higher will have a frequency of 800 Hz. If these notes are produced by musical instruments, each of them will have harmonics at 2, 3, 4, and 5 times the fundamental. Every harmonic of the upper note matches one of the harmonics of the lower note. There are similar matches of harmonic overtones if two notes have frequencies with a ratio of 3:2; for instance, 600 Hz:400 Hz. Other ratios that have some matching overtones are 4:3, 5:3, 5:4, and 6:5.

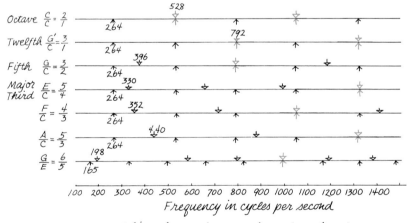

Matching harmonics are shown by colored arrows

One combination of three notes is very important as the basis of chords and also for forming the standard scale of the notes that we know as "do, re, mi, fa, sol, la, ti, do." This basic trio of notes must have frequency ratios of 4:5:6. It is called a major triad. Notice in the diagram at the top of the next page how many of the harmonic overtones of these three notes match each other. A fourth note can always be fit into such a triad if its frequency is just twice that of the first note (the fourth note would thus be the octave of the first note).

The scale we sing with "do, re, mi, fa, sol, la, ti, do," is based on this major triad. In the second diagram on the next page we show the notes and frequencies of the C scale. Note that the middle A to which the orchestra tunes has the standard frequency of 440 Hz. It is "la" in the "do, re, mi" scale where "do" is the middle C. Notice the way that the whole scale is made up out of three major

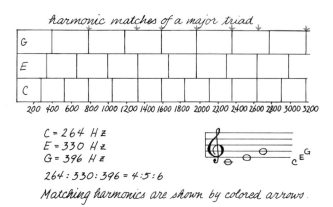

triads. If it were not for the preference of our ears, it might seem logical to divide an octave into eight equally spaced notes. However, this is not the way the standard scale works. As you go up the scale, you sometimes advance a whole tone and sometimes only a half tone. For instance, the ratio of frequencies between D and C (between "re" and "do") is 9:8. The "re" note has a frequency about 12% larger than the "do" note. The next step up to "mi" is almost as large—in this case a ratio of 10:9 or a frequency that is higher by about 11%. As you go on to "fa," however, you go up only half a tone. The frequency ratio is 16:15 or an increase of about 6½%. As you go on to "sol," "la," and "ti" you go up whole tones; but between "ti" and "do" there is only a half tone again. That's the way it sounds good. If you think that it would sound better to go up a whole tone every time, try it on a piano. If you start out with middle C and use only the white keys, you will be playing the standard scale we have just described. To go up a whole note each time, you will have to use some of the black keys in a way that you can figure out by looking at the diagram and the piano scale shown on page 389.

Major scale in C

N	C	D	E	F	G	A	B	C'	D'
Triads	4	5		6					
				4		5		6	
						4		5	6
Name	do	re	me	fa	sol	la	ti	do	re
Frequency	264	297	330	352	396	440	495	528	594
Interval		9/8	10/9	16/15	9/8	10/9	9/8	16/15	9/8

Major scale in D

Note	D	E	F#	G	A	B	C#	D'	E'
Triads	4	5		6					
				4		5		6	
						4		5	6
Name	do	re	me	fa	sol	la	ti	do	re
Frequency	297	333	371.3	396	445.5	495	556.9	594	668.3
Interval		9/8	10/9	16/15	9/8	10/9	9/8	16/15	9/8

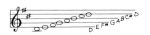

If you play a musical instrument, you may have had to spend time learning how to play scales. The easy one is the scale in C. There are no sharps or flats. With the piano, you need use only the white keys. Try to do that, however, starting one note higher in the key of D. If you go marching up from D using no sharps or flats, you will get a scale that doesn't at all sound like the standard "do, re, mi, fa, sol." That's because you haven't maintained those same frequency ratios that are determined by the major triad frequencies of 4:5:6. In the key of D, for instance, the scale can start with D, go to E, but then must go to F#. This scale and its frequency ratios are shown in the diagram on page 438.

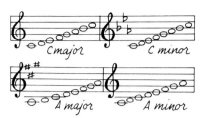

There is another important combination of frequencies called the minor triad. The frequency ratios of this triad are 10:12:15. This combination of notes has matching harmonics that also sound pleasant to the ear but western culture interprets the result as a sad sound.

As a matter of fact, the exact frequency ratio that we have described cannot work for more than one scale without producing all sorts of complications. For instance, suppose that we say we would like to build a scale starting with D, based on the same magic triad: 4, 5, 6. We must start out with a frequency for D that we got from the scale in C. That is, 297 Hz. The scale that we get from doing this is the major scale in D, shown above. Notice that the note that has a frequency 6/4 that of D, should be A at 440 Hz. Instead, 6/4 of 297 is 445.5. Most of the other notes would also be slightly off. If a person is playing an instrument with continuous tuning like a slide trombone or a violin, and that person is very skillful, he or she can make the slight adjustments so that the frequency ratios of the notes are exactly right. With valve instruments, however, that is harder to do and with a piano it is impossible. If we had to have a different set of keys and strings on the piano for every key that might be played, the piano couldn't fit into a living room and no human would be able to play it. This problem was recognized long ago and various piano makers have used various compromises so that a reasonable number of notes will be almost right for a large number of scales. The most common system is called an equal tempered scale. The chart below compares an octave of standard notes of the equally tempered scale of a piano with the frequencies of the notes in the perfect key of C. Using the equal tempered system, only 12 notes are needed to play all the tones and half tones of a full octave. As you can see, none of the notes are off frequency by very much. The progression of frequencies in the equal tempered scale is geometric. The frequency of each of the 12 notes in an octave is higher than the preceding note by $\sqrt[12]{2} = 1.05946$.

Comparison of piano frequencies (equally tempered scale) with frequencies of perfect key of C

Note	C	C#	D	D#	E	F	F#	G	G#	A	A#	B	C
Name	do		re		me	fa		sol		la		ti	do
Frequency of equally tempered scale	264	279.7	296.3	314	332.6	352.4	373.4	395.6	419.1	444	470.4	498.4	528
Frequency of perfect key of C	264		297		330	352		396		440		495	528

Question 16-13
Why can't you make a reasonable musical scale by dividing an octave arithmetically into seven equal intervals? Suppose that the ends of the octave are at 350 Hz and 700 Hz. What would be the piano notes if "do, re, mi . . ." were represented by notes with frequencies of about 350, 400, 450 . . . ?

Receptions of Sound by the Ear

So far we have described various things that can vibrate and set up pressure waves in the air. What happens when those waves hit our ears? We say that we hear sound or music, but how do changes in air pressure turn into music in our brains? The first parts of that process are mechanical and are fairly well understood. To be sure, the mechanical structure of the ear is extremely clever. Consider the problem— air pressure changes of a millionth or even a billionth of an atmosphere have to produce some sort of effect on nerves inside the ear. Remember that it is hard for a vibrating object to set air into vibration unless there is a large surface area pushing back and forth against the air. In the same way it's hard for the air to be very effective producing vibrations in something solid like flesh.

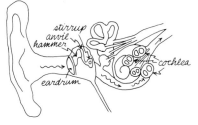

Look at the cross-sectional view of the ear in the diagram. Note first that the outer ear helps direct the air pressure changes down into the ear canal. Without your outer ears you could not hear quite so well, and you would not have any place to hang your glasses. About 2 cm inside the opening the ear canal ends in a diaphragm, or ear drum. This thin membrane of flesh vibrates back and forth as the air pressure changes. The distance that it moves back and forth is extremely small. As we calculated on page 399, we can hear sound when the vibration amplitude is no larger than the diameter of an atom. The motion of the ear drum is not itself felt by any nerves. Instead, the ear drum is fastened to three bones that connect to yet another diaphragm in the inner ear. Inside this third part of the ear the pressure waves travel in a jellylike material. The function of the bones and diaphragms in the middle part of the ear is to turn air pressure changes into liquid pressure changes. The surface area of the ear drum is about 25 times that of the inner diaphragm and the connection between them is made with bones acting like levers. The combination produces pressure changes on the inner liquid about 35 times greater than the changes in air pressure.

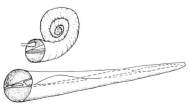

Cochlea of ear shown in normal coiled position and the way it would look if uncoiled.

The vibrations in the liquid pass along the tube called the cochlea, which is shown in the drawing. Note that the cochlea is separated along its length into two main parts, one of which is filled with nerves. Some of the action of this organ is still unknown, but it is certain that different parts of the cochlea are sensitive to different frequencies. Although the pressure waves of all different frequencies travel the whole length of the cochlea, resonance for a particular frequency occurs at a particular region of the long canal. At that particular point, the dividing wall between the two parts of the cochlea expands and contracts more than at any other point along the tube. The nerves that are attached to that region are therefore more easily triggered to send messages on to the brain. High frequencies are detected near the beginning of the cochlea and low frequencies toward the far end. Apparently, therefore, different pitches excite different nerves. A louder sound makes the same set of nerves send more signals to the brain. Just how the brain sorts out all this information is still a mystery.

INTERACTION OF WAVES IN TWO DIMENSIONS

Everyone knows that light travels in straight lines. Actually, it doesn't. Like any other form of wave motion, the light at a given point is the superposition of the light waves reaching there from all other points. This superposition

INTERACTION OF WAVES IN TWO DIMENSIONS

may produce constructive or destructive interference, resulting in regions of light or darkness. The region of darkness may occur even where there is a straight-line path back to a light source! Similarly, a bright region of wave reinforcement may occur in the geometrical shadow of a light source. Light waves can bend around obstacles just as water waves bend around breakwaters.

In the photographs we show two views of a shadow cast by a razor blade. In the first view we see why light has a reputation of traveling in straight lines. In the second illustration, however, we see a different story. The light for this picture came from a laser that produces very parallel light (consisting of plane waves) in a very narrow range of frequency. The light is said to be monochromatic; not that the light has exactly one particular wavelength but the spread of wavelengths is very small.

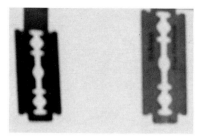

Shadow (at left) of razor blade (at right).

The shadow in the second picture shows one small spot on the edge of the blade. The shadow is not at all a sharp line but is not fuzzy, either. Definite bands of light and darkness extend into the region that we might expect would be evenly illuminated. Furthermore, some light extends into the shadow region.

In the next picture we show the shadow of a round disc, once again illuminated with a laser beam. Around the rim of the shadow are alternate circles of light and darkness. (Note that the first light band is brighter than the general illumination beyond.) The most remarkable feature of this kind of shadow is that there is a bright spot of light right in the center of the shadow. Evidently, the waves of light have curled around the disc in such a way that they reinforce each other in the central region.

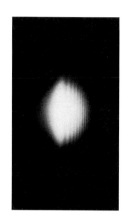

Since waves seem to bend as they go past a barrier, let's see what happens when we attempt to make a narrow beam out of a large plane wave. We could do this with water ripples by introducing two barriers in the water to block the waves except for the part going through a narrow slit. The same thing could be done with sound waves, but the barriers would have to be solid and sound absorbent. One way to produce the effect with sound waves is to arrange a loudspeaker inside a room and point it toward a door leading outside. In this case, the width of the opening cannot be changed but the wavelength of the sound can be changed. As we will see, the bending effect depends on the *ratio of the wavelength to the slit width.* With light waves you can produce the bending effect with an adjustable knife edge slit. The largest opening of these slits should be no more than a millimeter. As we suggested in "Handling the Phenomena," you can see the qualitative effect with light simply by looking at a distant light source through the gap in two fingers pressed nearly together.

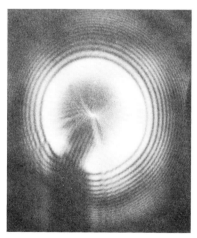

"This is the shadow of a ball bearing illuminated with a He-Ne laser. (The ball was held by a vertical rod.) The bright spot at the center of the shadow is known as Poisson's or Arago's spot."

In all of these cases a narrow opening (but very large compared with the wavelength) produces a narrow beam, as we might expect. However, if you try to make the beam narrower by closing down the gap as shown in the sequence of photos on the next page, you begin to make the beam wider. The angular spread of the beam increases until finally, when the slit width is about one wavelength, the slit acts as if it were a line source emitting cylindrical waves.

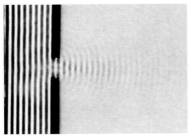 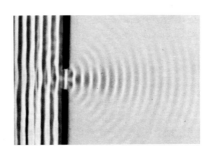 

Plane waves passing through gap in ripple tank. As wavelength increases, angular spread of beam increases. (Educational Development Center)

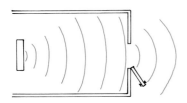

Sound waves going through door

Question 16-14

If you try to observe this beam-spreading with sound waves by sending them through an outside door, what frequency would you use to produce a wavelength about equal to the door width?

Now we must explain why waves bend when they pass by obstacles. The explanation is very closely connected with the reason that plane waves continue to advance in a straight line without going off in all directions. Back on page 391 when we introduced the model of Huygens' wave construction, we proposed that each point on an advancing wave front can be considered a source of a new set of waves. If the wave front is plane to begin with, then it continues on as a plane, and in the same direction. However, according to Huygens' construction, each point is the source of *spherical* wave fronts. We claimed that interference cancels out any wave effects except along the tangent of crests. For complete cancellation in any direction except forward, contributions of waves are needed from *every* point along the original plane wave front. If no contributions arrive from part of the original plane wave, then there may be reinforcement of the waves at one or more angles besides the forward direction. If an advancing plane wave goes through a narrow slit, then the wavelet contributions on either side of the slit are eliminated. What comes through the slit will not completely be a plane wave. In the limit of a slit smaller than a wavelength, for example, the wave getting through the slit will act as if it came from a line source and will spread out radially in two dimensions.

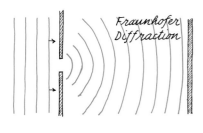

Fraunhofer Diffraction

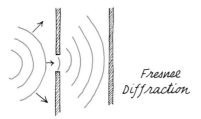

Fresnel Diffraction

For certain simple geometries, it is easy to add up the effects of the Huygens wavelets and so predict the interference patterns of the waves after they pass through the slit. One of our simplifying factors will be to consider that it is indeed a plane wave that advances on the narrow opening. The situation would be more complicated if the waves had originated from a point source only a small distance from the opening. In that case, the original wave would be spreading out radially and consequently the wave that enters the opening would not be all in phase. If we use plane waves to begin with and look at the interference pattern far away from the barrier, the phenomenon is called Fraunhofer diffraction. If the original wave front is not plane and if we study the interference pattern just past the barrier, the phenomenon is called Fresnel diffraction. Frequently, Fraunhofer diffraction, which is easier to calculate, makes a good approximation to what happens in a real system of Fresnel diffraction.

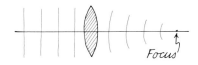

With light waves there is an additional feature that makes the Fraunhofer approximation a good one. Because the wavelength of light is so small and the slits producing diffraction are so narrow, the diffraction pattern is usually observed at a large distance compared with the slit opening. Furthermore, the light from the slit frequently passes through a lens before it interacts to form the diffraction pattern. If we form a slit with our fingers and observe a distant light, we have all the conditions necessary for using the Fraunhofer approximation. Since the source light is distant, the wave fronts striking the slit between our fingers are plane. After the light passes between our fingers, it enters the lens of our eye and is focused on our retina at the back of the eye ball. In Chapter 17 we will study the effect of lenses on wave fronts. The crucial point that we have to know now is that a lens focuses a plane wave front to a point. In the diagram we show several plane wave fronts striking a lens and then converging to their point images. Notice that each plane wave focuses to a point. It is not the same point for all plane waves. The focal point depends on the angle of the plane waves with respect to the lens' axis.

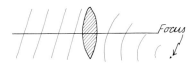

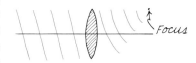

Now let's analyze the interference of the waves coming from a slit opening in a barrier. In two-dimensional geometry, we can divide the opening into an arbitrary number of small points, each of which, according to Huygens, can be the source for circular wave fronts. All of these point sources are being driven in phase because we assume that they are points on the original plane wave front. In the original forward direction, the wave fronts remain in phase and will be brought to a central focus point by the lens. For small angles on either side of the forward direction, the wave crests are almost in phase and these wave fronts will be focused to points on either side of this central image with only a small amount of cancellation. However, at larger angles from the forward, the wave crests are not all in phase with each other and there will be some destructive interference reducing the intensity of the light when the lens brings it to a focus. At certain particular angles, this destructive interference is complete.

In this diagram we show the geometry that produces complete destructive interference of the light coming from the narrow slit. We have divided the narrow slit into two sections: a top and a bottom. Within each half there are numerous point sources. A line has been drawn perpendicular to the direction θ. At that line the light coming from the top point of the top half has had to travel $\lambda/2$ further than the light from the top point of the bottom half. Therefore the light from the upper point is 180° out of phase with the light from the bottom point. This same geometrical relationship holds for each pair of points in the bottom and top half that are separated by $\frac{1}{2}a$. For every point source in the bottom half, there is a matching point source in the top half. At the particular angle θ the contributions from these matching pairs cancel each other. Consequently, there is no light intensity at all in the interference pattern at the point corresponding to the angle θ.

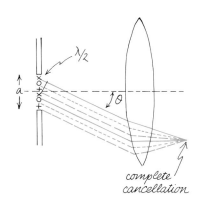

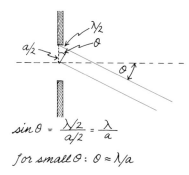

$\sin\theta = \dfrac{\lambda/2}{a/2} = \dfrac{\lambda}{a}$

for small θ: $\theta \approx \lambda/a$

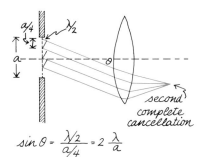

second complete cancellation

$\sin\theta = \dfrac{\lambda/2}{a/4} = 2\dfrac{\lambda}{a}$

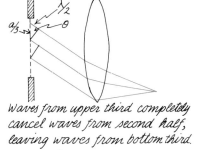

Waves from upper third completely cancel waves from second half, leaving waves from bottom third.

$\sin\theta = \dfrac{\lambda/2}{a/3} = \dfrac{3}{2}\dfrac{\lambda}{a}$

The critical angle θ is determined by a relationship between the wave length of the light and the width of the slit.

$$\sin\theta = (\lambda/2)/(a/2) = \lambda/a$$

As long as θ is small (less than 30° or so), we can use the approximation that

$$\theta \approx \lambda/a$$

The same geometrical argument can be used if the slit is divided into four or six or eight equal parts. In each case there will be an angle such that pairs of points will produce waves that cancel. For a division of the opening into four parts, which is shown in the diagram, the critical angle for cancellation is given by

$$\sin\theta = (\lambda/2)/(a/4) = 2\lambda/a$$

In general, there will be complete cancellation for any angle that satisfies the condition:

$$\sin\theta = n\lambda/a \qquad n = 1, 2, 3 \ldots$$

We would expect to see bands of light between the lines of complete cancellation. It is not quite so easy to calculate exactly the intensity within these bands or even the position of maximum brightness. Notice, for example, that the intensity must fall off on either side of the central maximum, since for any angle different from the forward direction, there will be partial cancellation. One way to estimate the amount of light entering the first bright fringe (between the first and second cancellation bands) is to divide the slit into three sections as shown in the diagram. If we draw a line such that $\sin\theta = (\lambda/2)/(a/3)$, then matching points in two of the sections will produce wave contributions that cancel each other.

Question 16-15

Will all the contributions from the remaining third then arrive without interference?

The crude argument that we have used predicts that the *amplitude* of the resulting wave in the first bright fringe will be only $\tfrac{1}{6}$ of the total amplitude of the original wavelength. That does not mean, however, that the *intensity* of the light in the first bright fringe is $\tfrac{1}{6}$ of the original intensity available. Remember that the intensity of a wave is proportional to the square of its amplitude. Therefore the intensity of the light in the first bright fringe is $\tfrac{1}{36}$ of the original intensity, or about 3%. The actual intensity, which can be derived from a more complete analysis, is 4.7%. To analyze the interference in the second bright band, we would have to divide the slit opening into five equal parts. The contributions from two of these would cancel the contributions from two others and the contributions from the fifth section would cancel about half the remaining amplitude. The resulting amplitude would be only about $\tfrac{1}{10}$ of the original and the resulting intensity only about 1% of the original. The actual intensity in this band is 1.7%. Since there are bright fringes on either side of the central maximum, 9.4% of the energy goes into the first side band pair, and 3.4% into the second side band pair. Altogether, the side bands get about 15% of the energy. The remaining 85% is in the central maximum. Here are some photographs of diffraction patterns produced by several different slit widths. Note once again that as the slit width decreases, the width of the interference pattern spreads.

Other patterns produced by diffraction can be analyzed with the same method we used for the single slit pattern. In each case, the diffraction is caused by the fact that some of the contributions from the original plane wave are blocked. The contributions from the remaining sections interfere constructively or destructively both inside and outside their geometrical shadow region. Along straight edges there will be parallel light and dark fringes. Around circular holes or obstacles there will be circular fringes. One of the most spectacular examples of diffraction is the bright region in the center of a shadow such as the one shown on page 441. At other points of the shadow, the wave contributions being received from the surrounding region cancel out to zero. At the center, however, the contribution from the ring closest to the disc dominates and the contributions with varying phase from the surrounding concentric regions are not enough for complete cancellation.

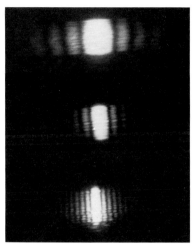

Deflection patterns of laser beams passing through slits. Widest slit is top, narrowest on bottom. Note that fringe separation is largest for narrowest slit.

Any time that waves pass through an opening or over an obstacle, the angular location of the first minimum of the fringe pattern depends on the ratio λ/a, where a is the width of the slit or the diameter of the hole or obstacle. For the circular fringes surrounding a pin hole, or any hole or obstacle with circular geometry, the angle for the first minimum is given by $\sin \theta \approx \theta = 1.2\lambda/a$.

In "Handling the Phenomena" you may have tried to create a sound shadow by placing a stuffed chair or human between the loudspeaker and your ear. The waves bend around the obstacle as shown in the diagram with an angle given approximately by λ/a. What sort of shadow would you expect for a frequency of 2000 Hz?

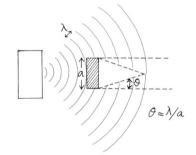

Although we have referred to these diffraction patterns produced by *pin* holes, an opening of any size will produce a diffraction pattern. For example, the largest diameter of your eye pupil is about 8 mm. (The size of the pupil varies, of course, with the amount of light present.) The image on your retina of a point source will not be a point but a circular diffraction pattern. The angular half width has a size: $\theta = 1.2\lambda/a$. For visible light, and for a pupil with a diameter of 8 mm, $\theta = 1.2(5 \times 10^{-7} \text{ m})/(0.8 \times 10^{-2} \text{ m}) = 7.5 \times 10^{-5}$ radians $= 4.3 \times 10^{-3}$ degrees.

It might seem that such a small angle is hardly worth bothering about. Suppose, however, that you are looking at two point light sources very far away, such as the headlights of a distant car. If the headlights are close enough, the pattern on your retina will look like the (a) part of the diagram. If the headlights are far away, the pattern on your retina may look like part (b) or even part (c). In this last case, the maximum of one diffraction pattern falls on a minimum of the diffraction pattern of the other light. If the lights were any further away, the two images would be so completely merged that

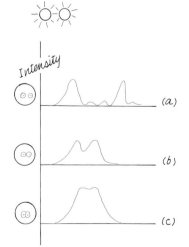

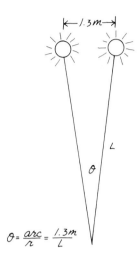

your eye could not detect whether it is seeing two sources or one. It could not *resolve* the two sources. The resolution power of a lens or a circular opening is given as $\theta = 1.2\lambda/a$. As we have calculated, the resolution limit of a human eye is approximately 7.5×10^{-5} rad. How far away would a car have to be so that headlights subtend an angle of 7.5×10^{-5} rad? The headlights of a car are about 1.3 m apart. From the geometry shown in the next diagram:

$$(7.5 \times 10^{-5} \text{ rad}) = (1.3 \text{ m})/L$$

$$L = 1.7 \times 10^4 \text{ m} = 17 \text{ km}$$

(Glare will make the lights appear to fuse at a shorter distance.)

The resolving power of a microscope or a telescope is a crucial measure of its performance. Since the objective lens of a microscope is about the same size as the human eye, its resolving power is approximately the same as that of the eye. If the focal length of the lens is 1 cm, then the separation of two objects that can just be resolved is

$$\Delta x = (7.5 \times 10^{-5} \text{rad})(1 \times 10^{-2} \text{ m}) = 7.5 \times 10^{-7} \text{ m}$$

A microscope with these characteristics could just resolve two point objects if they were 0.75 micrometers apart. The resolving power of the 200-inch reflecting telescope on Mount Palomar is much greater. Since the diameter of the mirror is about 600 times that of the human eye, the minimum angle for resolution must be 600 times smaller than that for the eye. Therefore, the resolving angle for the big mirror is approximately 1×10^{-7} rad, which is equal to about 0.02 seconds of arc.

Diffraction limits the parallelism of a laser beam. For most lasers, this tiny aperture is not the diameter of the opening or of the beam as it leaves but rather a narrower cross-section in the middle of the lasing column. In a typical helium-neon laser, the minimum diameter of the beam within the instrument is only one-quarter millimeter. We might expect the diffraction angle to be

$$\theta = \frac{1.2(5 \times 10^{-7} \text{ m})}{(0.25 \times 10^{-3} \text{ m})} = 2.4 \text{ milliradians}$$

The actual diffraction angle for a laser beam is only about one-half that which we just calculated. The intensity of the laser beam is not uniform across its diameter but is Gaussian. In effect, the limiting aperture is "softer" and produces less diffraction than an aperture with sharp edges.

The spread of a laser beam is limited by diffraction but, as we will see in the next chapter, the spread of most searchlight beams with ordinary incandescent sources is limited by the finite size of the filament or light source. There is a conservation requirement imposed on any kind of beam. If the beam *cross-section* is made smaller, with lenses or mirrors for example, then the *angular divergence* must increase. Conversely, in the case of the narrow laser beam, the beam can be expanded to have a larger cross-section and consequently will have a smaller divergence.

THE INTERFERENCE PRODUCED BY TWO WAVE SOURCES

Historically, the interference of waves coming from the *same* source or through the *same* opening was called diffraction. The interactions between

THE INTERFERENCE PRODUCED BY TWO WAVE SOURCES

the waves coming from *separate* sources or through *separate* channels was called interference. There is no physical difference between these processes. In one sense the wave interference produced by two sources is easier to understand than diffraction. To calculate a diffraction pattern we have to divide the wave source into many small regions and then calculate the phase and intensity arriving from each of these sources. If we have just two sources or two slits through which the waves may come, we can assume that each one acts like a line source and that at the sources the waves are in phase with each other.

This last point is an important requirement. In the diagram we show several possible ways of creating waves from two sources. In the case of the two radio antennas, the phase relationship between the two signals is controlled by the radio station that drives both antennas. In the case of the two light bulbs, the phase relationship in the light they cast at a particular point is completely random. On the human time scale the light is *incoherent*. At a point some distance from the bulbs, the phase between the light coming from the two sources stays constant for no more than 10^{-8} seconds.

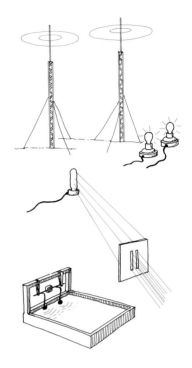

If two slits are illuminated by the light from a single source far enough away so that it is essentially a point source, then the light reaching the slits is in phase. The wave train reaching the slits may start and stop abruptly every 10^{-8} s, but at every instant the phase in one slit is the same as the phase in the other.

The light in a slide projector beam is not coherent (in phase) from one point to another in its cross-section. The light has come from the surface of a filament that is not even approximately a point. The light from one point on a filament is not in phase with that coming from any other point. A laser beam, however, consists of light waves that are in phase throughout a cross-section of the beam.

In the next section we will observe yet other ways in which light from two or more sources can occur in phase. These phenomena, many of which occur in nature, originate when a beam of light is divided into two or more beams by reflection or refraction and then these beams are recombined as if they had come from two or more different sources. Since the divided beams come from just one beam, their phase relationships remain constant.

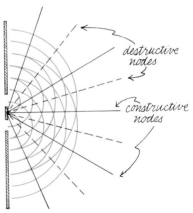

Let's analyze first the classic double slit experiment described in 1800 by the English scientist Thomas Young (1773–1829). The geometry is shown in the diagram. An illustration of the effect with water ripples is given in the first photograph and the pattern produced with light is shown on page 448.

Notice in the geometrical pattern showing the wave fronts, and in the illustration of the water ripples, that there are nodal lines of destructive interference. In the pattern produced by light interference, these nodal lines show up as dark bands.

In producing and analyzing these patterns, we use slits that are narrower than the distance between the slits and assume that the slits act essentially as line sources. With our two-dimensional geometry the wave fronts are cylindrical. We also must use waves with a particular wavelength, which in the case of light means that the source must be as monochromatic as possible.

There is one additional feature about double slit interference that is not obvious. At a large distance from the double source, the *basic* pattern

Interference produced by plane waves going through double slit in ripple tank

Interference produced by laser light going through double slit.

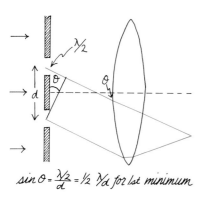

$\sin\theta = \frac{\lambda/2}{d} = \frac{1}{2}\lambda/d$ for 1st minimum

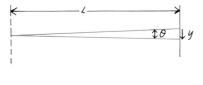

(except for the interference we will examine) is essentially that resulting from a *single* source. You can see this in a pond by throwing two stones into the water close together. The spreading ripples eventually merge into one large circular pattern. In the case of light, the screen is usually far away from the slits in comparison with their separation distance. The central diffraction patterns of the two slits overlap. Frequently, there is a lens that focuses the slit images on a screen (or the retina of our eye). Then the central maxima of the individual diffraction patterns overlap exactly.

Now let's examine the particular geometry shown in the diagram. The light from the upper slit travels an extra half wavelength. Therefore, the light reaching the screen along the upper path is 180° out of phase with that coming along the lower path. The two beams interfere destructively at the screen, producing a dark region or a node. The geometrical relationships are

$$\sin\theta = (\lambda/2)/d$$

There will also be destructive interference between the two beams if the extra path distance for the upper one if $\frac{3}{2}\lambda$ or $\frac{5}{2}\lambda$. In general, there will be a *node* if

$$d\sin\theta = (2n+1)\lambda/2 \qquad n = 0, 1, 2\ldots$$

If the extra distance traveled by the upper beam is an *integral* number of wavelengths, then the waves will reinforce each other and produce bright bands or antinodes. The geometrical condition for *antinodes* is

$$d\sin\theta = m\lambda \qquad m = 0, 1, 2\ldots$$

The light and dark bands produced by the two-slit *interference pattern* exist within the *diffraction envelope* produced by each slit.

Let's calculate the *combination* double slit interference and diffraction pattern for two slits, each of which is 0.1 mm wide and 0.5 mm apart. In our notation, $a = 0.1$ mm and $d = 0.5$ mm. The first minimum for the *diffraction* envelope occurs for

$$\theta = (5 \times 10^{-7} \text{ m})/(1 \times 10^{-4} \text{ m}) = 5 \times 10^{-3} \text{ rad}$$

On a screen 1 meter away from the slits, the first minimum of the diffraction pattern would occur at a distance y from the central maximum where $y = \theta L = (5 \times 10^{-3} \text{ rad})(1 \text{ m}) = 5$ mm. The total width of the central *diffraction* maximum is thus 10 mm. The first minimum of the *interference* pattern occurs for $\theta = (\lambda/2)/d = (2.5 \times 10^{-7} \text{ m})(5 \times 10^{-4} \text{ m}) = 0.5 \times 10^{-3}$ rad. On the screen one meter away this first minimum of the interference pattern occurs where $y = \theta L = (0.5 \times 10^{-3} \text{ rad})(1 \text{ m}) = 0.5$ mm. The width of the central maximum of the *interference* pattern is 1 mm. The resulting pattern is shown in the diagram. The diffraction pattern provides an envelope for the interference maxima. A photograph of this effect is shown above.

All the interference effects depend on the ratio of the wavelength to the opening or separation distance. The same equations control the behavior of light waves, water waves, and sound waves. One of the reasons why it is difficult to see interference phenomena in light waves is the small size of the wavelengths. For the patterns to be visible, the slit openings or separation distances must be very small. Another reason the effects are not obvious with light, under geometrical conditions where you might expect to see them, is that the interference bands will be colored and overlap if white light is used as a source.

> ### Question 16-16
>
> Suppose that in a two-slit interference pattern there is a node for red light. What will be the situation for blue light?

INTERFERENCE PRODUCED BY SPLITTING AND RECOMBINING A SINGLE BEAM

We actually see many interference patterns in everyday life. Since these are usually produced by white sunlight, they appear as colored spectra. One of the most common examples is the colored rings seen on the surface of thin oil films or the colored regions in a soap bubble film. We show the geometry of one typical situation in the diagram. The incoming beam is partly reflected and partly transmitted from the top surface of the thin film. The reflected beam is reversed 180° in phase at the surface since it is a reflection from a low impedance (air) to a high impedance medium. The transmitted beam travels through the thickness of the film. A part of this beam is reflected, within the film, at the second surface. This reflection occurs without change of phase because the beam is in a high impedance medium and reflects from a boundary with low impedance. The internally reflected beam returns to the first surface and is partially transmitted, emerging parallel to the beam first reflected at this surface. If the extra distance traveled by the second part of the beam is half a wavelength, then it will re-emerge in phase with the first part of the beam. The waves will reinforce each other for that particular wavelength, that particular thickness of film, and that particular angle of the light entering the film. For a different thickness of film or a different wavelength, reinforcement will occur for a different angle of incidence. Consequently, different colors are seen in different regions and at different angles. Notice the colors in the soap bubble photograph on the cover of this book.

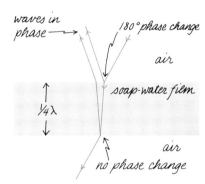

> ### Question 16-17
>
> It is easy to blow soap bubbles and to observe the interference colors produced by the thin film. In any such film the liquid continually drains downward, leaving a thinner film at the top. Shortly before the film gets too thin and breaks, it turns black. Apparently all colors of light undergo destructive interference when the film is thin enough. Why should this be?

A pattern of concentric colored circles is frequently seen if thin films are bound together. The classic demonstration of this was described by Newton and is called Newton's rings. Although Newton studied this phenomenon, he did not ascribe it to wave motion. Here is an illustration

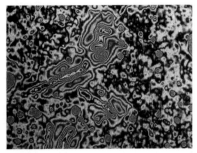

Newton's Rings. (Courtesy Eastman Kodak Company).

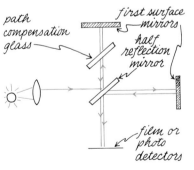

of Newton's rings and a diagram showing the geometry of the effect. The situation is very much like that for the thin film interference. In this case the film is air, sandwiched between two objects with greater impedance. The circular fringe pattern is produced by the circular symmetry of the air film.

Several instruments that have been extremely important for research exploit the interference properties of a beam of light that has been split into two parts and then recombined. The classic instrument of this type is the Michelson interferometer. Albert Michelson (1852–1931) was an American physicist who was proudest of his work in the measurement of the speed of light but who is most famous for performing the Michelson-Morley experiment we described in Chapter 11. In that chapter we also described his interferometer. Modern versions are still used in research and a sketch of the geometry is shown in the diagram. The original light beam is split into two parts by the first mirror. If both beams of light travel for exactly the same length of time, they will arrive at the photographic plate in phase. The effect will be the same as if there had been no splitting of the light at all. In order to provide equal-time paths, a compensating sheet of glass must be added to the second beam path.

Question 16-18

Why?

If the times taken for the two different paths are slightly different, it will appear to the observer that there are two different sources, one slightly behind the other. Since these two sources have a constant phase relationship with respect to each other, they will produce an interference pattern at the observer's location. If the far mirrors are exactly perpendicular to each other, the interference pattern will look like the one at the top of the next page. The concentric circles are caused by the circular symmetry of the extra path lengths between the two sources. In practice, one mirror is frequently tilted very slightly, which produces fringes that are almost parallel to each other at the location of the observer.

The great virtue of the Michelson interferometer is that the beam paths can be separated from one another by distances up to several meters. Each beam can then be handled separately, including introducing transparent materials into the beam path or moving the mirror to extend the beam path. For instance, if part of one of the beam paths is in vacuum and a gas is gradually admitted to the region, the travel time through the column will gradually increase. At the observation point this increase in travel time produces a continuous change of the interference pattern. The alternate dark and light bands move across the field of view, with each alternation corresponding to an extra path length of one wavelength. In this way, the speed of light in the gas can be measured as a function of pressure.

If one of the mirrors is slowly moved, the fringes will march across the observer's field of view with each fringe corresponding to a mirror movement of one-half wavelength. Michelson used this technique in a more complicated way to measure the length of the standard meter in terms of wavelengths of light.

THE GRATING—INTERFERENCE PRODUCED BY MULTIPLE SOURCES

If instead of just two slits, we use many slits to produce interference, the same geometrical considerations will apply for the location of the nodes and antinodes. In the pictures we show the effects of using different numbers of slits. In each case, the slit width and the slit spacing are the same. One obvious effect of using more than two slits is that the intensity of each maximum gets greater. More light is being admitted. With three slits, three amplitudes are being added at each maximum instead of two. The *intensity* therefore increases by 9:4. An even more important effect of increasing the number of slits can be seen by examining the photographs. The maxima not only get brighter; they get narrower. The contributions from many different slits add constructively at exactly the proper angle but for angles slightly smaller or slightly larger, there are many contributions to produce destructive interference.

The angular position of the maxima produced by multiple slits depends on the wavelength. Consequently, we have an instrument that can produce a spectrum, spreading white light into its component colors. During the past 100 years an enormous amount of scientific work has depended on the analysis of light according to its distribution of wavelengths or frequencies. The tool for this research is called the *spectrometer*. A picture of a typical spectrometer is shown in the margin. The analyzing device, which deflects the beam according to wavelength, can be either a refracting prism or a multiple slit grating.

Gratings have been created with as many as 24,000 lines per centimeter. The spacing between the slits is therefore 4×10^{-7} m, the wavelength of blue light. The smaller the spacing between slits, the greater the angle for constructive interference of a particular wavelength. The greater the total number of lines (not just the lines per cm), the narrower the width of an antinode. The latter property is connected with the resolving power of a grating, $\lambda/\Delta\lambda$. The width of the antinode provides the uncertainty, $\Delta\lambda$, in the measurement of the wavelength, λ. If $\Delta\lambda/\lambda$ (the fractional uncertainty) is small, the resolving power is high.

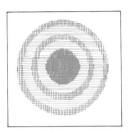

Interferometer picture for ⊥ mirrors

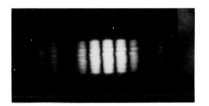

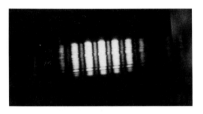

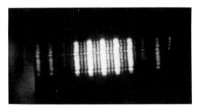

Question 16-19

Suppose that you have a 2-inch and a 1-inch grating, each with the same number of lines/cm. Because there are twice as many slits in the 2-inch grating, the amplitude of the waves at the maxima of the 2-inch grating must be twice that for the 1-inch grating and therefore the intensity of the light in the maxima must be four times as great. How can the intensity in the maxima be four times as great if the total amount of light entering the 2-inch grating is only twice that entering the 1-inch grating?

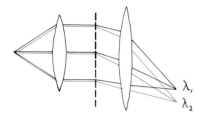

Although white light entering a grating will spread out into a continuous spectrum of colors, light from a glowing gas produces only discrete lines of different colors. Each element, when heated to incandescence in the gas phase, produces its own particular pattern of wavelengths. The "lines" themselves are created by the geometry of the observing system. The light enters through a collimation slit and then passes through (or reflects from) the grating, which consists of ruled lines. The spectrum line itself is an image of the collimation slit, but its angular position is determined by the atomic process that determined its wavelength.

OTHER INTERFERENCE PATTERNS

Long-playing records and the feathers of starlings frequently display glimpses of color, usually blues or purples. These effects are produced by the interference of light glancing off regular arrays of narrowly spaced reflectors.

Another example of wave interference is X-ray diffraction. X-rays that can penetrate some distance into solids have wavelengths very close to the spacing between atoms of a solid. Many materials, such as the metals or the salts, are crystalline or contain other complicated but regular arrays of atoms, such as in the long chain molecules of living material. When X-rays scatter through such regular arrays, their wave motion can be analyzed in terms of the interference from many different regularly spaced sources. The analysis of these interference patterns is very complicated because the scattering takes place in three dimensions. Nevertheless, each crystalline arrangement produces a unique pattern of destructive and constructive interference of the X-rays. The photograph shows an example of this pattern. Working backward from interference patterns like these, crystallographers have managed to identify atomic arrangements in many crystals, including such complicated ones as the heredity-carrying DNA.

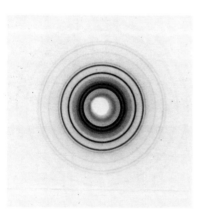

X-ray diffraction of aluminum. The thin film target was rotated in order to produce the circular pattern. (Courtesy Bell Labs).

An ordinary photograph records the amount of light reflected from each point of an object to the camera lens and hence to the photographic film. Another type of photograph, called a *hologram*, records the *interference* between coherent light striking the film directly and reflections of that same coherent light from the object being photographed. The geometry for making such a hologram is shown in the diagram. Notice that no lens is needed in front of the film. Every section of the film is recording information from the entire object. The coherent light is provided by a laser that must be intense enough to expose the film in a time short enough so that vibrations do not move the relative position of light source, object, and film. Since an interference pattern is being recorded, the relative positions must remain stationary to within a small fraction of a wavelength of light.

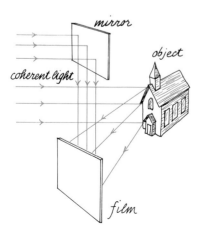

The resulting photograph does not contain a picture that can be seen with ordinary light. What has been recorded is an interference pattern containing information about the phase differences between light reflected from an object and light that reached the film directly. If the photograph is viewed with transmitted coherent light from a laser or even with nearly coherent light from a strong point source, the original scene is reconstructed

OTHER INTERFERENCE PATTERNS

and can be seen as a real image on the near side of the film or a virtual reflected image on the far side. The light reaching the eye has been influenced by the interference pattern, passing through where there was a node and being blocked where there was an antinode. The resulting light forms an interference pattern (of the hologram—which is itself an interference pattern) in your eye—a pattern that is exactly the one that would have been formed had you been looking directly at the object in the first place. Furthermore, this pattern is different in the left eye from what it is in the right eye just as it would have been if you had viewed the object directly. Hence you see the object in three dimensions. By moving your head with respect to the film, you can look around the object to the same extent that you could if you had been in that position looking at the object directly.

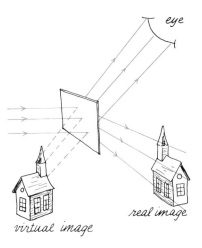

One way to see why an interference pattern allows you to see the original object is to analyze the interference pattern produced by a very simple geometry. Suppose that the object being photographed is a long rod as shown in the diagram. The coherent light reflected from the rod interferes with the light striking the film directly. The resulting interference pattern is a series of dark and light bands covering the entire film. If a positive transparency is made of the film and it is then illuminated with a plane wave of coherent light, the cylindrical wave fronts originating from the transparent bands on the film will produce a reinforcement maximum at the position P. The waves diverging from that point will look as if they had come from the original rod. Another set of wave crests diverge from the film as if they were coming from an illuminated rod on the other side of the film.

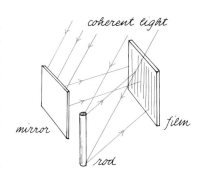

Note that the interference pattern corresponding to the shape of the rod existed in every part of the film. The position of the rod was determined by the spacing of the light and dark fringes. If the rod is close to the film, the spacing is large at first and then decreases rapidly. If the rod is far away from the film, the spacing is almost uniform. If the object consisted of two rods perpendicular to each other forming a cross, the resultant image on the film would consist of perpendicular fringes. If the object were a point, the interference pattern would be a series of concentric circles. Of course, a complicated object produces a complicated interference pattern, but when coherent light shines through that pattern and produces its own interference pattern in our eyes, we see exactly the pattern of light and darkness that we interpret as the image of the original object.

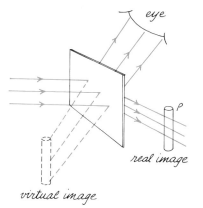

Note that if parallel light is used, a diffraction pattern of <u>a diffraction pattern</u> is the original source pattern:
Diffraction pattern of a pinhole → ◎
Diffraction pattern of ◎ → point image
Diffraction pattern of a slit → ||||||
Diffraction pattern of |||||| → slit image

Similarly, interference pattern produced by coherent light mixing with reflections of coherent light from object → hologram

Diffraction pattern of hologram → image of original object.

SUMMARY

A wave traveling along the x-axis with velocity, v, wavelength, λ, and period, T, can be described as

$$y = A \sin \frac{2\pi}{\lambda}(x - vt) = A \sin 2\pi \left(\frac{x}{\lambda} - \frac{t}{T}\right) = A \sin(\kappa x - \omega t)$$

$\kappa = 2\pi/\lambda$ (angular wave number)

$\omega = 2\pi/T$ (angular frequency)

If there is relative motion between wave source and observer, the observed frequency will differ from the source frequency by the Doppler shift.

$$\text{Observer moving through medium:} \quad f'_{\text{obs}} = f_{\text{source}} \frac{V \pm v}{V}$$

$$\text{Source moving through medium:} \quad f_{\text{obs}} = f'_{\text{source}} \frac{V}{V \pm v}$$

f is frequency in medium, V is velocity of wave in medium, v is velocity of source or observer through medium.

The Doppler shift of spectra from stars and galaxies is a key tool of astronomy. For electromagnetic radiation, where there is no "medium" for the wave motion, the Doppler shift depends only on the relative velocity between source and observer:

$$f_{\text{obs}} = f_{\text{source}}[1 \pm v/c + \tfrac{1}{2}(v/c)^2 \pm \ldots]$$

The interference of waves depends on their relative phases, and on whether the phase relationships stay constant. Most ordinary light sources produce incoherent radiation; the phase changes rapidly and at random.

Waves can pass through each other in many media without disturbing each other. Where they intersect, however, the amplitudes can add constructively or destructively. For most of our examples, we assume that the addition is linear—otherwise new frequencies are produced. If two waves traveling together have frequencies f_1 and f_2, they combine to produce beats—a wave with an envelope frequency of $(f_1 - f_2)/2$ and a fine structure frequency of $(f_1 + f_2)/2$:

$$y = \left[2A \cos 2\pi \left(\frac{f_1 - f_2}{2}\right)\right] \sin 2\pi \left(\frac{f_1 + f_2}{2}\right) t$$

If periodic pulses travel back and forth along a line with length L, reflecting at the ends with little loss, and if the round trip travel time is equal to the period of the pulses, then the line is resonant. The amplitude of the pulses will build up (depending on losses).

$T = 2L/v$ or $L = \tfrac{1}{2}\lambda$ for 0° or 180° phase change at each end. The resonance condition also holds if $L = (n/2)\lambda$ for closed or open lines where $n = 1, 2, 3 \ldots$, or if $L = (n/4)\lambda$ for lines open at one end where $n = 1, 3, 5 \ldots$. If a sine wave satisfies the resonance conditions, standing waves are set up in the line. The *envelope* of the disturbance does not move along the line, although the effect is produced by two waves traveling in opposite directions, maintained by reflections.

Complex, nonsinusoidal waves can be represented as the sum of two or more sine waves of harmonic frequencies. Musical tones consist of a fundamental frequency that usually determines the pitch, plus numerous harmonic overtones.

Musical instruments have some source of vibration attached to a resonating and radiating system. In many instruments, the signal reflected on a resonant line triggers its own replenishment, whether from a bow or lips or air turbulence. Musical scales are determined by the human preference for harmonic ratios of overtones, and the desire to avoid beats in the fundamental or higher harmonics. Reception of sound in the ear depends on matching the impedance of the air to the impedance of liquid in the ear canal, and then triggering nerve cells along a column that has varying resonances.

When coherent waves interact in two (or three) dimensions, complex patterns of nodes and antinodes are established. Parallel wave fronts passing through a narrow slit of width a, spread out into an angular pattern with a central maximum that has a half-angle defined by $\sin \theta = \lambda/a$. Parallel light going through a hole with diameter a produces a set of concentric conical nodes, the first one at an angle of $\theta = 1.2\ \lambda/a$. Two distant point sources separated by θ produce images that can just be resolved as coming from two sources instead of one.

The interference of waves from two coherent line sources, such as light from a distant source going through double slits separated by d, gives rise to a pattern of nodes with geometry:

$$d \sin \theta = (2n + 1)\lambda/2$$

Light from a single source can be split into two beams that then interfere with each other. This effect is responsible for Newton's rings and is used deliberately in Michelson's interferometer.

A grating of many parallel lines produces brighter and narrower antinodes. Gratings are used in spectrometers to produce spectra of light sources. The larger the number of lines in the grating, and the smaller the separation between lines, the greater the resolving power, $\lambda/\Delta\lambda$.

With intense coherent light now available from lasers, the interference of light is easy to observe. Holograms are photographs that record the interference of a choherent light source with a fraction of the same beam that has reflected from an object. By viewing the hologram with partially coherent light, the eye sees an interference pattern of an interference pattern that appears as an image of the original object.

Answers to Questions

16-1:

If $x = 0$, then $y = 0$

If $x = \frac{1}{4}\lambda$, then $y = A \sin \frac{1}{2}\pi = A$

If $x = \frac{1}{2}\lambda$, then $y = A \sin \pi = 0$

If $x = 3\frac{1}{4}\lambda$, then $y = A \sin 6\frac{1}{2}\pi = A \sin \frac{1}{2}\pi = A$

16-2: To describe pulses or waves traveling to the left, substitute $(x + vt)$ for x. In other words, reverse the sign of v. Therefore, the Gaussian traveling to the left is described by

$$y = Ae^{-(x+vt)^2/2\sigma^2}$$

The peak exists where $(x + vt) = 0$. If $t = 1$ s, then $x = -(v$ m/s$)$ $(1$ s$)$.

16-3: If $v = 0$, the observer is stationary and $f' = f$. If $v = \frac{1}{2}V$, then $f' = f\dfrac{V - \frac{1}{2}V}{V} = \frac{1}{2}f$.

The frequency that the moving observer hears is down one octave from the source frequency. If $v = V$, the observer is traveling with the speed of sound and hears nothing from the source.

16-4: If $v = 0$, there is no Doppler effect and $f = f'$. For the other velocities, let us find the effects when the source is moving away from the observer. We can compare these effects with those calculated in Question 16-3 where the observer was moving away from the source. The difference in this case is that the observer is stationary with respect to the air through which the source is moving away. If $v = \frac{1}{2}V$, then $f/f' = V/(V + v) = V/(V + \frac{1}{2}V) = \frac{2}{3}$. Remember that in the comparable situation in 16-3, the ratio of frequencies was $\frac{1}{2}$. If $v = V$, then $f/f' = V/(V + v) = \frac{1}{2}$. In the comparable case in 16-3, the ratio of frequencies was 0. The observer would hear no sound. It clearly makes a difference whether the source or the observer is moving with respect to the air.

16-5: In the case of sound, if $v/V = 0.01$, then there is a 1% difference between f and f', regardless of whether the source or the observer is moving through the air. Note that the two expressions are different by only 1 part in 10^4 since $v^2/V^2 = 1 \times 10^{-4}$. A 1% effect would be produced if $v = 3.4$ m/s. For a 1% Doppler shift for electromagnetic radiation, $v/c = 0.01$. Therefore $v = 3 \times 10^6$ m/s.

16-6: Note that in the third frame the first reflected wave exactly cancels the right-hand side of the advancing wave. A quarter of a cycle later, however, as shown in frame 4, the reflected wave overlaps the incident wave (although they are traveling in opposite directions) and the two reinforce each other. From that time on, there is continual *destructive* interference of the wave near the boundaries and *constructive* interference halfway between. The resultant pattern looks like half a sine wave, anchored at the boundaries, and oscillating between + and − amplitudes at the center.

16-7: But why does energy have to pass a nodal point? In the ideal, frictionless case, the amplitude of the reflected wave is equal to the amplitude of the incident wave. If no energy is lost, no energy has to be fed into the system, and therefore no energy need cross a nodal point. In a real situation, there is energy lost along the line, or in the reflection. To make up for the loss, energy must be fed in, the incident wave is larger than the reflected wave, and there are only approximate nodal points.

16-8: The conditions for a standing wave involve a particular relationship between frequency, velocity, and length of the medium. Since the velocity of the wave depends on the tension in the medium, the privileged frequencies also depend on the tension.

16-9: A *pure* sine wave never starts and never stops. Information can only be sent by pulsing the wave, or at least turning it on or off. A continuous wave would tell the listener nothing. A pulsed wave, however, necessarily consists of some sine waves of other frequencies that combine to cancel the main sine wave before it starts and after it stops. The pulse travels at the group velocity, which can never be greater than the speed of light.

16-10: When the fundamental is sounded, there is one pulse at a time in the pipe. It hurtles back and forth, being reflected from one end and replenished at the other. However, it is not a standing wave of the type that we pictured on page 426 where a traveling *sine* wave had a particular relationship between its wavelength and the length of the pipe. Since the pulse in a musical instrument is usually not a pure sine wave, it can be represented by a Fourier sum of a sine wave with the same fundamental frequency and other sine waves with harmonic frequencies. Although the pitch heard corresponds to the fundamental frequency, the tone is richer because of the harmonics. Mathematically, the single pulse, shuttling back and forth, can be analyzed into a linear addition of standing sine waves of the fundamental plus harmonic frequencies.

When the second harmonic is sounded, *two* equally spaced pulses shuttle back and forth, each with its set of harmonics. If there were 100 pulses per second originally, there will now be 200 pulses per second. In this case, the pitch (usually) corresponds to the frequency of the second harmonic. (Under certain circumstances, the ear may subjectively dub in the pitch corresponding to the fundamental.)

6-11: The violin case allows the string to radiate the same amount of energy more quickly to the air. After the string is plucked, the energy is rapidly drained.

6-12: The wavelength *in air* of the fundamental is determined by the driving frequency of the string and the velocity of sound *in air*. The frequency of the pulse *in the string* is determined by the length of the string and the velocity of pulse transmission *in the string*.

6-13: The piano notes would be, approximately:

F	G	A	B	C#	D	E	F
350	400	450	500	550	600	650	700

sequence of notes that approximately have frequencies: 350, 400, 450...

As for whether or not such a scale sounds good, try it for yourself and listen.

6-14: An ordinary house door is about 70 cm wide.

$$f = v/\lambda = (340 \text{ m/s})/(0.7 \text{ m}) = 490 \text{ Hz}$$

This note is between B and the A to which the orchestra tunes.

6-15: No, because the light from the top part of the remaining section must travel half a wavelength further than the light from the bottom part of that section. So there is partial cancellation of the contributions within this remaining section. As a first approximation we could guess that about half the points cancel each other. If we add the amplitudes of the wavelets that survive, we have only about one-half of the one-third of the total sources in the slit.

6-16: The equation for a node from a double slit is

$$d \sin \theta = \frac{(2n + 1)}{2}\lambda = (n + \tfrac{1}{2})\lambda$$

The equation for an antinode with a double slit is

$$d \sin \theta = m\lambda$$

Consider the case for the first node of the red light: $n = 0$. That node occurs where $d \sin \theta = \tfrac{1}{2}\lambda = \tfrac{1}{2}(7 \times 10^{-7} \text{ m}) = 3.5 \times 10^{-7} \text{ m}$. The first antinode for blue light occurs where $d \sin \theta = \lambda = 4 \times 10^{-7} \text{ m}$. The first antinode for blue light is almost at the same position as the first node for red light. Therefore, that point will appear blue. The overlapping of nodes and antinodes for red and blue light occurs throughout the pattern. The interference pattern looks like a group of rainbows running into each other.

6-17: As the film thickness becomes much smaller than 1 micrometer and less than one-quarter wavelength, only a small phase change takes place because of the extra path length through the film and back again. There still remains a phase change of 180° between reflection at the outer surface and that at the inner surface. Consequently,

16-18: The light that is reflected upward from the half-reflection mirror does not penetrate the mirror. The light that goes through that mirror heading to the right goes through that thickness of glass twice before being reflected downward. By inserting the same thickness of glass at the same angle in the path of the light going up and down, the two paths are made equal in optical length.

16-19: The width of each maximum, which determines $\Delta\lambda$, is inversely proportional to the total number of grating lines, N. Therefore the lines produced by the 2-inch grating are only half the width of those produced by the 1-inch grating. The 2-inch grating accepts twice as much light as the 1-inch grating and puts it into lines that are only half as wide. Therefore, the intensity of the lines produced by the 2-inch grating is four times that produced by the 1-inch grating.

Problems

1. Draw a diagram showing qualitatively what happens when plane waves in a ripple tank pass through a narrow gap.
2. Measure the length of one particular musical instrument (toy or real) and find the frequency and wavelength of the lowest note that it can produce.
3. Define in words and in their other relationships all the symbols in this equation:

$$y = A \sin(\kappa x - \omega t)$$

4. The displacement of a rope segment is given by $y = A \sin(\kappa x - \omega t)$. What is the value of y at $x = 10.0$ m and $t = 1.5$ s if the amplitude is 2.0 cm, the wavelength is 0.5 m, and the period is 0.1 s?
5. What is the value of y in problem 4 if $x = 7.6$ m and $t = 1.5$ s?
6. Is it practical to create a demonstration of the Doppler effect by swinging an alarm clock in a circle? Assume reasonable values for the pitch and frequency of the alarm; choose a reasonable radius and speed of rotation; assess the feasibility of determining the resultant variation of pitch.
7. A car traveling at 60 mph past you sounds its horn at 400 Hz. What is the frequency you hear before and after the car passes you?
8. If a distant galaxy rushing away from us at half the velocity of light emits light that has a wavelength of 4×10^{-7} m in the galaxy, what is the wavelength we receive?
9. Ninety years ago in Ohio, Michelson did a precision experiment with his spectrometer using visible light. He sent two halves of a light beam over different paths and then recombined them. He could tell if they were out of phase by as little as 30°. What flight *time* difference was he thus able to determine? (Assume $\lambda = 6 \times 10^{-7}$ m.)
10. A wave is traveling down a spring. Two points 20 cm apart are 60° out of phase with each other. If the frequency is 2 Hz, what is the wave velocity?
11. If the oboe sounds 440 Hz to tune the orchestra, and the trumpet frequency is 450 Hz, what is the frequency of the combined tone and how many beats are there per second?
12. What is the frequency of each of the first five harmonics of a violin string tuned to 192 Hz?
13. If the length of a violin string is 40 cm and the fundamental frequency is 400 Hz, what is the velocity of the wave in the string? What is the wavelength in the

string? What is the wavelength in air? (At room temperature, the speed of sound in air is 344 m/s.)

14. What is the wavelength of the second harmonic in a pipe that is open at both ends and is 2 m long? What is the wavelength of the first *overtone* if the pipe is open at one end and closed at the other?
15. Make a careful graph of $A \sin \theta$ and $A/2 \sin 2\theta$ on the same graph for $\theta = 0$ to $\theta = 360°$. Add the two curves and show the resulting curve on the same graph in another color.
16. Sketch the interference pattern produced by a double slit opening with the separation distance between slits equal to four times the width of each space. ($\lambda \ll$ width.)
17. Sketch the diffraction patterns produced by a single slit opening ($\lambda \ll$ width) for light consisting of two wavelengths: blue at 4×10^{-7} m and red at 7×10^{-7} m.
18. For what frequencies of sound would you expect to find a sharp shadow cast by an upholstered chair?
19. Two small loudspeakers, located at about head height, are connected to the same signal generator so that they sound a 344 Hz tone in phase. They are separated by a center-to-center distance of 1 m. A person is right in front of one of the speakers and then moves further away until the sound goes to a first minimum. How far is he from the first speaker? How far is he from the second speaker?
20. Calculate the lengths of pipes closed at one end whose fundamentals form the scale starting with middle C. Obtain actual numbers for the end notes, but do not do unnecessary arithmetic for the rest.
21. A violin string 40.0 cm long is tuned to sound C. How much would it have to be shortened to sound G?
22. How far apart are two parallel slits if red fringes ($\lambda = 7 \times 10^{-7}$ m) are 2 mm apart on a screen 1 m from the slits? ($\theta = 2$ mm/1 m.)
23. What is the angular width of the central maximum in a slit diffraction pattern for blue light ($\lambda = 4 \times 10^{-7}$ m) and a slit width of 0.1 mm?
24. The beam from a traffic radar gun must be narrow enough so that it covers only one car at a distance of 100 m or so. About what wavelength must be used to produce such a beam if the opening of the gun has a diameter of 30 cm?
25. When the mirrors of the Michelson interferometer are exactly perpendicular to each other, the interference pattern consists of a set of concentric nodes and antinodes. Explain the reason for this in your own words.

17 GEOMETRICAL OPTICS

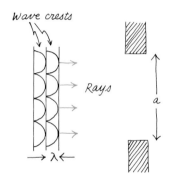

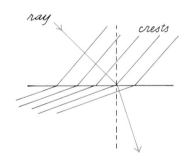

For most purposes, light really does travel in straight lines. For that matter, so do sound waves, or radio signals, or ripples, or any other kind of wave motion where wave length is short compared with any boundary constraints ($\lambda < a$). Under such conditions, we can use the approximation that the wave travels along a path normal to the wave front. These paths are defined as *rays*. While a ray is a mathematical line, we have all seen light beams that approximate rays. For instance, a projector beam, cutting through a smoky auditorium, is outlined by rays. We sometimes see the sun's "rays" coming through an opening in the clouds or through a narrow crack created by a partly open door.

Instead of dealing with wave fronts reflecting, refracting, or diffracting, it is frequently easier to deal with the rays that are perpendicular to the wave fronts. This approximation is called *geometrical* optics, as opposed to the full wave treatment, which is called *physical* optics. The rules and techniques that we will learn can be applied to any kind of wave motion. Most of our examples, however, will come from optics, since the approximation that $\lambda < a$ applies to so many common situations in the case of visible light.

Most of the phenomena to be studied in this chapter depend on simple rules that describe the reflection and refraction of rays. Yet complications can arise because the mirrors and lenses that produce reflection and refraction may have a variety of shapes. The rays can be guided from one place to another to produce images of several types.

Combinations of lenses and mirrors form instruments used in everyday life as well as in research. In studying the behavior of rays, we will also study the human eye, the camera, and the giant telescopes that probe the universe.

HANDLING THE PHENOMENA

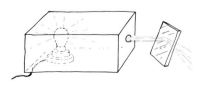

1. The first thing that you should do is to take another look at the beams of light produced by flashlights or by light coming through small openings. You can make a light beam by putting a cardboard collimator over a light bulb, as shown in the diagram. Such a collimator doesn't focus the beam but merely blocks all rays traveling in any direction except the narrow solid angle defined by the opening. Direct your beam of light at various angles to

HANDLING THE PHENOMENA

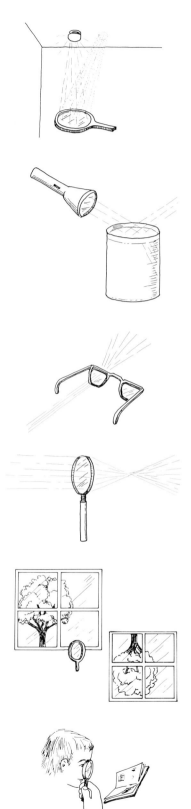

a plane mirror. Observe at least qualitatively that the angle of reflection is equal to the angle of incidence.

2. If you have a hand mirror that is slightly concave or magnifying on one side, and a plane mirror on the other, observe what happens to the light beam when you shine it perpendicular to the curved side and move it over the face of the mirror. The reflected beam does not return to the source. You can see this same effect in another way by reflecting a ceiling light back onto the ceiling. Do this first with the plane mirror and then with the curved side. In the latter case, if you move the mirror up and down, you will be able to focus an image of the light back on the ceiling. If you move the mirror above or below the point which produces a good focus, the image on the ceiling will become blurred.

3. Shine a narrow beam of light into a glass jar filled with water. If the water is slightly dirty, or better yet if you add a pinch of powdered milk, you can actually see the light rays in the water. Observe how the path of the light bends as it enters the water. Try different angles of incidence and see if, qualitatively at least, you confirm Snell's law ($\sin i/\sin r = \frac{4}{3}$, for water).

4. Shine a narrow light beam through eye glasses and also a magnifying glass. If the eye glasses come from someone who is nearsighted, the effect on the beam will be opposite to that produced by the magnifying glass. The lens for a person who is farsighted is basically the same as a magnifying glass.

5. Use a magnifying glass to produce an image of a light source. The light source can be the sun or an ordinary filament lamp. In either case, cast the image on a sheet of paper. If the source is the sun and the magnifying glass large and your hand is steady, you may be able to produce such an intense image that you can set the paper on fire. Be careful, however. An image hot enough to do that is also bright enough to hurt your eyes if you stare at it.

With a magnifying glass, you can also produce an image of any bright scene. Notice the nature of this image. Is it upside down? Has it been reversed right for left?

Another way of producing an image with a magnifying glass is to hold the glass close to your eye and use it to magnify something. Try reading print with it. Where is the image in this case? Is it upside down or reversed left for right? Also notice the position of your eye with respect to the print while you are using the magnifying glass, and compare this with the distance between eye and print when you are reading without the magnifying glass.

6. Several times a day you probably look in an ordinary mirror. Where is your image in the mirror? Use a ruler or tape measure and see if you and your image together can measure your separation distance. If you come closer to the mirror, what happens to your image? Notice that your image is not upside down, but that it is reversed left for right. Why is this? What happens if you turn your head 90° sideways?

7. If you can get hold of two magnifying glasses or any two convex lenses, you can make a telescope or microscope. Arrange the lenses as shown on page 462. Look at something very bright such as a lamp or an outside scene. Let the first lens produce an image on a thin transluscent sheet of paper. Use the second lens as a magnifying glass so that you can get your eye close to the sheet of paper in order to examine the image.

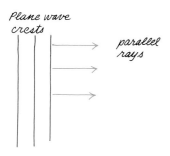

Plane wave crests — parallel rays

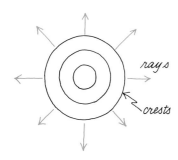

rays — crests

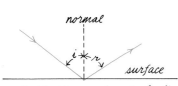

angle of incidence and angle of reflection measured with respect to normal to surface: $i = r$

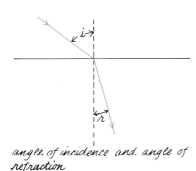

angle of incidence and angle of refraction

When you have the image in good focus, take away the sheet of paper and you will be able to see the image even better. Where does the image appear to be? Does it appear to be larger than the original object? Is it right side up? Or upside down?

THE BEHAVIOR OF RAYS

Ray optics is just a convenient approximation for wave optics. The rays themselves are best thought of as the normals or perpendicular lines to the wave fronts. If the wave fronts are planes, then the rays are perpendicular to those planes, parallel to each other, and evenly spaced. If the wave fronts are circular (in two dimensions) or spherical (in three dimensions), then the rays are radial.

The rays coming from a source travel in straight lines as long as the medium does not change. The rays are continuous if the medium does not absorb the energy of the wave motion. The intensity of the wave can be made quantitative in terms of the number of rays passing through the unit area. When we describe electric and magnetic fields, we will make use of such a quantitative model. In the case of light rays, however, we will merely note in passing that if the light rays are diverging from each other, the intensity must be getting less. If they are converging, the intensity is increasing. If waves spread in two dimensions from a point source, the rays are radial and evenly spread around a circumference. Since the circumference is proportional to the radius, but the number of rays stays constant as the radius increases, the number of rays per unit distance on the circumference must be inversely proportional to the radius. Indeed, the intensity of a wave spreading uniformly from a point source in two dimensions is inversely proportional to the distance from the source. Similarly, the intensity produced by a point source radiating uniformly in three dimensions is inversely proportional to the square of the distance from the source. Once again, the number of rays coming from the source is independent of the distance, but the surface area of the sphere increases as $4\pi r^2$. Therefore, the number of rays per unit area is proportional to $1/r^2$.

As we follow rays through reflection and refraction, we must define their direction with respect to boundary surfaces. The standard convention, which produces the simplest rules, is to measure the angle between the ray and the *normal* to a surface.

> ### Question 17-1
>
> What is the relationship between this angle and the angle between a wave front and the surface of a boundary?

REFLECTION FROM PLANE SURFACES

We have already seen in the previous chapter that when a plane wave reflects from a plane surface, the angle of reflection equals the angle of incidence. The similar construction in terms of rays is shown in the diagram.

If a surface is uniform over a large area, so that all of an incident parallel beam is reflected at the same angle, then the surface and reflection process are said to be *specular*. (The Latin word for mirror is *speculum*.) Of course, a good mirror must do more than just maintain the parallel nature of reflected rays. We expect mirrors to do this without absorbing an appreciable fraction of the incident rays and without discriminating between colors. Most mirrors are only approximations of the ideal. An ordinary mirror consists of a metallic layer deposited on the back of a smooth sheet of glass and then coated with some material to keep the metal from tarnishing. The glass front serves both as a smooth flat surface for the thin metal reflector and also as a protective agent to keep the metal mirror from being scratched or tarnished. Since the front surface of the glass also reflects some light, you can usually see a double image in an ordinary mirror. The weak image is slightly separated from the main image since the glass mirror is slightly closer to the object.

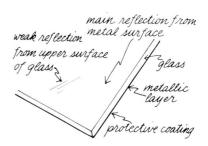

Reflection accounts for most of ordinary visibility. Light from the sun or from a lamp strikes objects around us, and a small fraction of that light is reflected into our eyes. In each case, the reflected ray left the object at an angle equal to that of the incident ray. However, most objects are rough enough so that the rays striking any region of the object reflect in many different directions. Each point on the object acts like a point source of radiation. Such an object is a *diffuse reflector*. Whether an object is diffuse or specular depends on the wave length of the incident radiation. If the reflecting surface changes direction in a random way within a wave length, then the reflection is completely diffuse. The surfaces of many objects are smooth enough to produce some specular reflection along with the diffuse reflection. This effect is seen as glare and shows up at the particular angle such that the incident angle between the light and the normal to the whole surface is equal to the reflected glare angle.

Since diffuse reflection causes every region of an object to act like a point radiator, we can see illuminated objects from any angle. Furthermore, the reflected light provides information about the surface of the object and less information about the geometry of the source. In particular, most objects absorb varying amounts of the incident light, depending on the texture of the material. Since this absorption is a function of frequency, we live in a colorful world. If a sweater absorbs green and blue light at the high-frequency end of the visible spectrum, then the only light reflected into our eyes is yellow, orange, and red. The sweater will probably appear orange.

The efficiency of reflection at a surface, compared with absorption or transmission, depends on the polarization of the light. As we saw in Chapter 16, light waves consist of transverse vibrations of electric and magnetic fields. In free space, these vibrations are always perpendicular to the direction of propagation. But the line of vibration can be at any angle in the transverse plane. With unpolarized light, the vibration line stays constant for no more than 10^{-8} s and then changes randomly to some other angle. Reflection of light occurs with higher efficiency if the vibration line of the electric field is parallel to the reflecting surface. Therefore, reflected light is partially polarized, as you can see in the diagram. At the angle of incidence which produces the 90° angle between the reflected ray and the transmitted

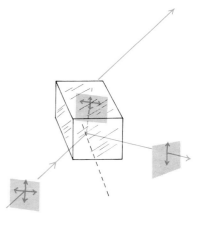

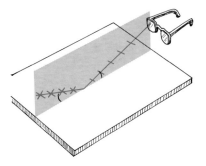

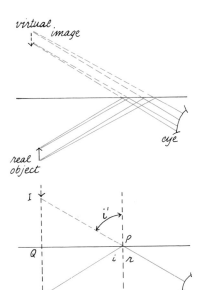

ray, the reflected ray is completely polarized. This incident angle is known as *Brewster's angle*. Since glare light from a surface is usually strongly polarized, you can eliminate much of it by wearing Polaroid glasses, with the preferred direction perpendicular to the reflecting surface. Since usually we see glare from horizontal surfaces, Polaroid sunglasses transmit light with vertical polarization and block light with horizontal polarization.

Through a looking-glass, we see a mirror world just as if we were looking through a window. Although we see images behind the mirror, we cannot touch them or put a photographic plate behind the wall to record them. Such images are called *virtual*.

The image of each object is as far behind the mirror as the object is in front of it. The rays reaching our eyes appear to come directly from the image. The actual paths of the light rays from object to mirror to eye are shown in the diagram. Note that the construction depends on the fact that the reflected angle equals the incident angle.

> ### Question 17-2
>
> Why does this geometry require that the image be as far behind the mirror as the object is in front?

Here are pictures of mirrors for two other kinds of waves. The first, which is an open wire mesh, is a good reflector of microwaves. The second, which is a poured concrete wall, produces specular reflection of sound waves. In both cases, two requirements for specular reflection are met. The first requirement is that the mirrors should not absorb much of the incident radiation. For microwaves, such a mirror must be a good conductor of electricity; for sound waves, the surface must be rigid enough so that it will not vibrate appreciably and absorb energy through friction. The second requirement is that the mirrors be smooth over local regions that are large compared with the wavelength. Both the microwaves and the sound waves have wavelengths many times the distance between wires (or between bumps on the concrete).

A radar reflector. (National Center for Atmospheric Research)

A concrete band shell reflector for sound.

REFRACTION

> **Question 17-3**
>
> Would the wire mesh be a good reflector for sound waves, and would the concrete wall be a good reflector for microwaves?

REFRACTION

In Chapter 16, we saw that waves advancing from one medium into another change their direction of motion. The amount of refraction depends on the relative velocities of the waves in the two media. This relationship, which we derived in Chapter 15 on page 392, is called *Snell's law*. Willebrord Snell (1591–1626) was a Dutch astronomer and mathematician who worked in this era just before Huygens and Newton. In the diagram, we illustrate this law with rays rather than with wave crests. Snell's law in terms of velocities is

$$\frac{\sin \theta_1}{\sin \theta_2} = \frac{v_1}{v_2}$$

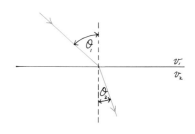

> **Question 17-4**
>
> The velocity of light in water is about three-fourths that of the velocity of light in air. If a light beam goes from air to water with an incident angle of 30°, what is the angle of refraction in the water? For this situation, how close is the ratio ($\sin \theta_1 / \sin \theta_2$) to ($\theta_1 / \theta_2$)?

The ratio of the speed of light in a vacuum, c, to the speed of light in a transparent medium is called the *index of refraction* for that medium:

$$\text{Index of refraction} = n = \frac{c}{v}$$

For most transparent materials and for most wavelengths, $n > 1$. (Since v is the phase velocity, there can be circumstances in which $n < 1$ but, as we saw in Chapter 16, the group velocity cannot be greater than c.)

The speed of light in the medium, and hence the index of refraction, depends on the frequency. As we have seen earlier, this property is called dispersion. In the table on the next page, we list the indices of refraction for a number of transparent materials and for wavelengths at the extremes and in the middle of the visible spectrum.

First, let's take a look at some of the phenomena produced by refraction without regard to wavelength differences. Then we will come back and take a look at dispersion. In terms of the index of refraction, Snell's law can be rewritten as follows:

$$\frac{\sin \theta_1}{\sin \theta_2} = \frac{v_1}{v_2} = \frac{v_1/c}{v_2/c} = \frac{1/n_1}{1/n_2} = \frac{n_2}{n_1}$$

$$n_1 \sin \theta_1 = n_2 \sin \theta_2$$

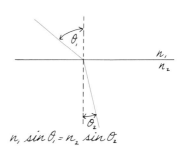

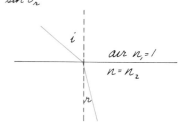

Index of Refraction

	Blue $\lambda = 4.4 \times 10^{-7}$ m	Yellow 5.89×10^{-7} m	Red 6.6×10^{-7} m	Δ(blue-red)
Air	1.000296	1.000293	1.000291	0.000005
Water	1.340	1.333	1.331	0.009
Ethyl alcohol	1.370	1.362	1.360	0.010
Polymethyl methacrylate (Lucite)	1.501	1.491	1.488	0.013
Fused quartz	1.470	1.458	1.455	0.015
Zinc crown glass	1.528	1.517	1.513	0.015
Light flint glass	1.594	1.575	1.570	0.024
Heavy flint glass (leaded crystal)	1.945	1.890	1.875	0.070
Diamond	2.465		2.407	0.058

Δ(blue-red) is a measure of the dispersion of the material. Refraction of white light by diamond or leaded glass produces broad colored spectra. A prism of water or lucite will not yield a broad spectrum.

Snell's law is particularly useful and foolproof in this last form. Frequently, however, we deal with a problem where light comes from air into some transparent medium. In that case, $n_1 \approx 1$ (actually, the index of refraction for air at standard atmospheric pressure is 1.0002957 for blue-violet light and 1.0002914 for red light). For this simple case, $\sin \theta_i / \sin \theta_r = n$, where θ_i is the incident angle, θ_r is the refracted angle, and n is the index of refraction for the transparent medium. Care must be taken in using this simple formula, however, because you have to keep in mind that the incident ray is in air. If the ray starts out in glass or water, it it best to go back to the more complete form of Snell's law.

Although Snell's law and the law of reflection give the observed angles for the refracted and reflected rays, they say nothing about the *fraction* of the light that is transmitted or reflected. That fraction depends strongly on the angle of the incident light. For light shining perpendicularly to clean glass, only about 4% of the light is reflected. At an incident angle of 80°, about 50% is reflected. As we have seen, the reflected light is highly polarized; the direction of vibration of the electric field is parallel to the surface. Indeed, at Brewster's angle, which is the incident angle such that there is a right angle between reflected and refracted rays, the reflected ray is completely polarized.

Question 17-5

From what we have described, is the transmitted ray also completely polarized when the incident ray is at Brewster's angle?

REFRACTION

In the photographs, we show pictures of the relative strengths of reflected and refracted rays for several incident angles. Remember that although refraction depends on the ratios of the *velocities* in the two media, the amount of reflection depends on a relationship between the *impedances* of the two media. As we have seen, waves can reflect from the boundary between the air in a pipe and the air outside, even though the velocity of the wave is the same in the two regions; however, the impedances are different.

  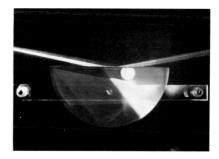

If light starts out from a medium with low velocity such as glass or water, and refracts into a medium of high velocity such as air, then Snell's law applies; but we must be careful about how we label the two media and the angles. In the diagram, we show several such cases. Let's calculate the angles for the case where the incident angle is 30° and the transparent material is glass, with $n = 1.5$.

$$n_1 \sin \theta_1 = n_2 \sin \theta_2$$
$$(1.5) \sin 30° = (1) \sin \theta_2$$
$$\sin \theta_2 = 0.75$$
$$\theta_2 = 49°$$

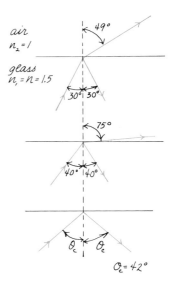

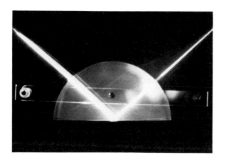 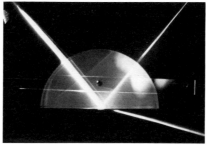 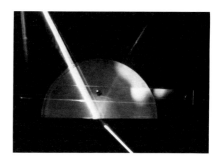

The angle of refraction under these circumstances is larger than the angle of incidence. Notice in the diagrams and pictures what happens as the angle of incidence gets larger. The intensity of the reflected light increases while the fraction of the light that refracts decreases. Eventually the angle of refraction equals 90°, and the intensity of the refracted light goes to zero. This occurs when

$$n_1 \sin \theta_1 = n_2 \sin 90° = n_2(1)$$

$$\sin \theta_1 = n_2/n_1$$

If the second medium is air, then $n_2 = 1$ and $n_1 = n$, the index of refraction of the transparent material in which the ray starts. In this case, the critical incident angle is equal to

$$\sin \theta_c = 1/n$$

Beyond this critical angle, all of the internal incident light is reflected and none is transmitted. This phenomenon is called *total internal reflection*. For water the angle is

$$\sin \theta_c = 1/1.33 = \tfrac{3}{4}$$

$$\theta_c = 49° \quad \text{(for water)}$$

For glass with $n = 1.5$, the critical angle for total internal reflection is

$$\sin \theta_c = 1/1.5 = \tfrac{2}{3}$$

$$\theta_c = 42° \quad \text{(for glass)}$$

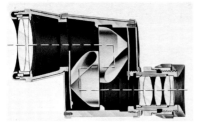

Cutaway view of 7×35 wide angle binocular showing prisms which provide total internal reflection. (Courtesy Bausch and Lomb.)

When light reflects (perpendicularly) from very good plane mirrors, 10% or so of the light is lost. With total internal reflection none is lost. Optical instruments frequently make use of this phenomenon, particularly if several reflections are required. An example is shown in the picture. The light path in binoculars is increased, and the image is presented right side up by means of several reflections. These reflections take place *in* glass prisms instead of from the sufaces of ordinary mirrors, thus reducing the amount of light lost.

The principle of total internal reflection is also used to transmit light through long plastic or glass fibers. The light rays bounce back and forth from the walls, following the twists and turns of the fiber. Once the light is reflected at an angle greater than the critical angle for total internal reflection, it is trapped inside the material and continues to reflect back and forth without loss through the walls. Fibers carrying light in this fashion have been produced with diameters as small as 1 micrometer. These fibers are very flexible and can be bound together in a bundle either to transmit light down curved passageways or to transmit images from hidden or remote locations. (In the latter case, the orientation of the fibers must be maintained, so that the light coming from the upper left-hand corner of the object appears in the upper left-hand corner of the fiber bundle.) These light pipes are now routinely used, particularly in medicine. They can be made small enough to be snaked through blood vessels or through tubes such as the urethra. Here is a picture of such a device.

Fiber optic tube (Courtesy American Optical).

DISPERSION AND REFRACTION

Since the index of refraction depends on frequency, a refracted beam of white light is dispersed into a spectrum of colors. The diagram shows this effect. Notice that the blue or higher frequency light is refracted more than the red. Evidently, the velocity of blue light in glass is less than the velocity of red light.

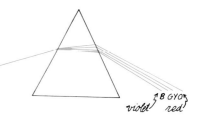

As you can see from the table of index of refraction, a large refractive index and high dispersion do not necessarily go together. Some types of plastic have an index of refraction close to that of glass, but the magnitude of the index is almost independent of frequency in the visible range. Consequently, there is very little dispersion. Take a look in the table at the dispersion produced by a diamond. Not only is the index of refraction large, but there is a sizable difference between the index for blue and that for red. That's the main reason why the diamond sparkles.

Refraction and dispersion can be enhanced by sending light through a prism so that it is refracted twice and suffers a large change in direction. The example shown in the diagram has a particularly easy geometry for analysis. The incident light is at an angle such that the ray inside the prism is parallel to the base. The angle labeled δ is the *angle of deviation*—the difference between the original path direction and the final one. It turns out experimentally, or it can be proven analytically, that δ is at a minimum when the ray inside the prism is parallel to the base. For either larger or smaller angles of incidence, the angle of deviation is larger. However, at the minimum angle, the deviation is relatively insensitive to small changes in the incident angle. On the other hand, it is at the minimum angle condition that refraction in the prism is most sensitive to the frequency dependence of n, providing optimum color resolution.

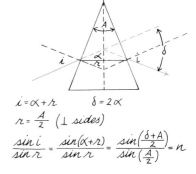

As you can see with the derivation that accompanies the diagram, the angle of deviation is related to the index of refraction and the prism angle. For a prism angle of 60° (or smaller), the ratio of the sines of the half-angles is approximately equal to the ratio of angles. Hence, a reasonable approximation is

$$n = \frac{A + \delta}{A}$$

or

$$\delta = (n - 1)A$$

Question 17-6

For light undergoing minimum deviation from a diamond cut with an angle of 60°, what would be the width of the spectrum viewed from a distance of 1 m? You can see the practical importance of this question—can you detect the spectrum produced by an engagement ring?

Prisms are sometimes used as the dispersive element in spectroscopes. We described the operation of these in Chapter 16 and showed a picture of one on page 451. There are several differences between using a grating or a prism. First, the angular dispersion created by the grating is linked to the structure of the grating by a simple model of interference between line sources. The sine of the angle of deviation is directly proportional to the wavelength. In the case of refraction and dispersion by a transparent prism, the angle of deviation depends in a complicated way on the microstructure of the material. Furthermore, the angle of deviation is not related in a simple way to the wavelength. Second, the sequence of colors in the spectrum produced by a prism is reversed from that produced by the grating. Dispersion produces a greater deviation for blue light than for red.

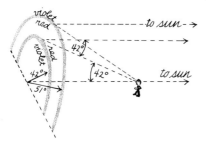

■ Optional

The best known example of the refraction and dispersion of light is first described in the book of Genesis (9:13), which tells how God set his bow in the cloud as a token of a covenant between himself and the creatures on the earth. He would not destroy them again by flood (the fire next time).

The rainbow and the other types of dispersion seen in the sky are caused by light interacting with water or ice droplets in the air. Some of the effects are very complicated. The most common rainbow is illustrated here. It is best seen after a rain storm when the sun is low but is shining brightly through a section of the sky that is clear. To see the rainbow you must turn your back to the sun and look toward a region that still has rain clouds. Under good conditions you can see a colored arc, with blue-violet on the inside and red on the outside. Above the primary bow, there may be a fainter secondary bow with the colors reversed—red on the inside and blue-violet on the outside.

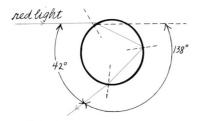

red: 180° − 138° = 42°
blue: 180° − 140° = 40°

Angular relationships for the primary bow.

The light that forms the bows has been reflected and refracted in raindrops, as illustrated in the next diagram. The light for the primary bow enters the top of a droplet and is reflected just once. The light in the secondary bow enters the droplets near the bottom and reflects twice inside the drop. Since the light entering and leaving the droplet undergoes refraction, it also undergoes dispersion, so that each color leaves at a different angle. It might seem at first that this effect is enough to explain the existence of the rainbow. The trouble is that the total angle of deviation depends not only on wavelength but also on the point of entry of the original ray into the droplet. As you can see in the following diagram, a ray near the top of the droplet undergoes a larger deviation; most of the light near the center of the droplet either goes on through or reflects through almost 180°. The light that reaches our eyes from all these regions will be a mixture of colors and hence white, since the sun's rays strike the droplets over their whole surface.

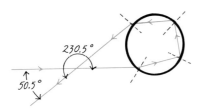

red: 230.5° − 180° = 50.5°
blue: 234° − 180° = 54°

Angular relationships for the secondary bow.

There is a special condition, however, that makes all the difference. There is a particular region near the top of the droplet that produces *minimum* deviation compared with the regions just above or just below. This minimum angle is 138° for red light and 140° for violet. As often happens for such an extremum condition, the rays that strike the droplet just above or just below the critical point are deviated by almost the same angle. Therefore, there is a slight concentration of light deviated at these particular angles.

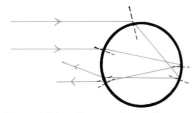

Rays striking above or below the special region of a droplet suffer greater deviation than the rays that form the bow.

The geometry of the sun, the droplets, and the observer is shown in the first diagram. The observer sees the concentration of colors in the rays that have been deviated from droplets in a conical shell. The opening angle of this cone

DISPERSION AND REFRACTION

as seen by the observer is 42° (180° − 138°) for red light and 40° for violet light. For the secondary bow, the angles are 50.5° for red and 54° for violet.

The apex of the light cone is at the eye of the observer. The centerline of the cone extends through the observer to the sun. Consequently, if the sun is on the horizon, the observer will see a full 180° of the rainbow. If the sun is higher in the sky, the light cone is tilted down into the ground, so that only a small section of the upper arc can be seen. If the sun is higher than 42° above the horizon, the rainbow disappears completely. Of course, if you are looking down into a canyon where there is a waterfall or even down at the mist produced by a lawn sprinkler, you can see rainbows even though the sun is high in the sky. Under the right conditions, particularly from an airplane, you may be able to see the whole 360° of the bow.

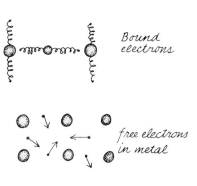

A Model to Explain Reflection, Transparency, and Dispersion

When light enters a sheet of glass, it appears to slow down abruptly. When the beam comes out the other side of the glass, it travels once again at the speed c. Are there forces that slow it down in the glass? If so, how do the forces speed up the light as it leaves?

To explore these problems at least qualitatively, we will present a model of light and materials. We propose that material consists of positively charged ions and negatively charged electrons that are attached to the ions as if by springs. In general, we assume that we are dealing with only one or two electrons per ion. The others are strongly bound to the ion or are shielded by the outer electrons. In the case of metals, the spring is missing; the outer electron is free to wander around the crystal. With other materials the spring constant varies and therefore so does the natural frequency of oscillation of the bound electron.

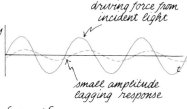

Question 17-7

This model appears to be an approximation to the standard one of atoms. We know, however, that electrons in atoms are not really bound by springs and that their locations and motions must be described by quantum mechanics rather than classical mechanics. Why then use such a crude model for the interaction of light with matter?

The key to our model is that light consists of electromagnetic vibrations. The oscillating electric field in the wave will provide a force on an electron and set it into oscillation. If the electron is bound like a particle on a spring, then the amplitude and phase of the forced oscillation will depend on the relationship between the frequency of the incoming wave and the natural frequency of the bound electron. If the driving frequency is small compared with the natural frequency, then the electron will follow the driving oscillation with small amplitude, but will be slightly lagging in phase. If the driving frequency is close to the natural frequency, a resonance condition will be set up where the electron's amplitude of oscillation is large and the phase difference is 90°. If the driving frequency is larger than the natural frequency, the electron's amplitude of oscillation will be smaller again, but now its phase will be ahead of that of the driving electromagnetic wave.

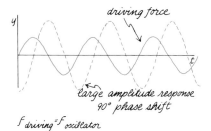

So far with our model we have arranged to produce oscillations of electrons in material when an electromagnetic wave passes through. The next step in the argument is one that we will study in more detail in Chapter 22. An accelerated charge radiates electromagnetic waves. Each electron driven into oscillation by the original electromagnetic wave becomes the source of Huygens' wavelets. The *resultant* wave is produced by the addition of the original wave and the new ones that have been produced. What happens next is very hard to calculate

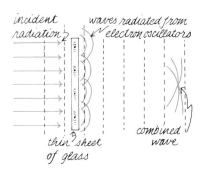

quantitatively, although it has been done and the calculations agree very well with experiments. The problem is that each electron is affected not only by the original radiation, but also by the contributions from all its neighbors. The phase of each contribution depends on the distance of the neighbor—and *its* oscillation in turn depends on all its neighbors.

It turns out that the resultant radiation at a distance from a whole plane of oscillators, being driven in phase, is 90° out of phase with the oscillators themselves. (The radiation from such a group is proportional to the *velocity* of the oscillators, not the acceleration.) The advancing electromagnetic wave is made up of the original wave plus the combined contributions from all of the local oscillators. This contributed wave is 90° out of phase with the original, because of the dependence on velocity of the oscillators, plus the phase lag or lead caused by the mechanical response of the electron oscillators themselves. The resultant wave either lags or leads the wave that would have existed in vacuum. For visible light in glass, the driving frequency is *below* the natural frequencies of the bound electrons. Because the contributed wave from these oscillators is slightly delayed in phase, the advancing wave travels more slowly in the glass than it would in a vacuum. Once the wave leaves the glass, there are no more phase delays from local oscillators, and the wave velocity is c again.

With some materials the oscillation of the bound electrons can produce other motions of the microstructure that dissipate the original energy. This effect may be frequency-dependent. For instance, oscillation at the frequency of red light may convert the energy to random thermal motion. In that case, red light is absorbed, even though the blue light may be transmitted without loss.

The closer the frequency of the light to the resonant frequency of the bound electrons in a material, the greater the phase lag of the driven electrons. In clear transparent materials like glass, the natural frequencies of oscillation of the electrons are greater than the frequencies of visible light. Therefore, blue light with its higher frequency will be closer to the natural frequency of oscillation of the bound electrons and therefore will suffer a larger phase lag than will red light. Consequently, the velocity of blue light in glass is less than that of red light. However, at frequencies higher than the resonant frequency, there is a phase lead for glass. This can occur at certain frequencies in the ultraviolet or soft X-rays. In these regions, the *phase* velocity is actually greater than c. As we have pointed out before, however, no energy or information can be sent at this speed. The group velocity is always less than c.

When light shines on a smooth metal surface, most of the light is reflected. A little bit is absorbed, and none is transmitted far into the metal. The electrical conduction electrons in the metal are not bound to their parent atoms and therefore have no natural frequency of oscillation. At low frequencies, and even up to the frequency of light for some metals, the electrons are free to respond instantly to the driving electromagnetic wave. The field that they radiate in the forward direction is 180° out of phase and completely cancels the original field that is advancing into the metal. This same field radiated in the backward direction becomes the reflected wave. The electrons in the surface of most materials, whether bound or free, radiate some of the incident energy backwards to provide reflected light. The energy from some frequencies is converted into thermal energy. The selected frequencies that are radiated make our world colorful.

Fermat's Principle

Over 2000 years ago, Hero of Alexandria (the same Hero who invented a rotary steam engine) proved mathematically that a light ray reflected from a plane

Wave contributions at a point some distance from glass

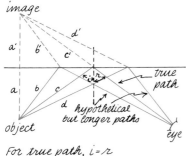

For true path, $i = r$
If $a = a'$
then $b = b'$, $c = c'$, $d = d'$
Shortest distance between *image* and eye is straight line along c'.
Since $c = c'$, shortest distance between *object* and eye is along c.

DISPERSION AND REFRACTION

mirror takes a path shorter than any other path (only paths leading from source to detector *by way of the mirror* are considered). The proof is clever and, of course, classic. The geometry is shown in the diagram. It is essential that the image is as far behind the mirror as the object is in front of it, as was proved in Question 17.2. Of all the rays leaving the object and striking the mirror, the only one to reach the observer obeys the law of reflection. The reflected angle is equal to the incident angle. The diagram shows two hypothetical paths that do not obey this law.

Hero's argument is that the actual path taken by the light is shorter than any hypothetical path. Certainly the shortest distance between the *image* and the observer is along the straight-line path. Because of the symmetrical geometry, however, that distance is equal to the path length from the *object* to the observer. The hypothetical paths from the *image* to the observer are longer than the straight-line path, and each of these equals the corresponding hypothetical path length from the *object* to the observer.

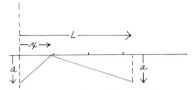

Path length for this special case where both object and eye are at distance a from mirror:

$$\sqrt{a^2+x^2} + \sqrt{a^2+(L-x)^2}$$

The distance from object to mirror and then to observer can be described in terms of the general variable, x, along the mirror, as shown in the top part of the diagram. The path length is plotted as a function of x in the bottom part. When the observer is also at a distance, a, from the mirror, the path is a minimum for $x = L/2$. The variable L is the distance along the mirror between perpendiculars dropped from the object and from the observer.

In the middle of the seventeenth century, Pierre de Fermat (1608–1665) extended Hero's principle to light rays passing through various media, such as glass and air. Fermat transformed the rule from "shortest path" to "least time," raising the possibility of some grand economy of nature. For some reason, light would take a path between two points that would require less time than any nearby path. This principle can serve as the basis for the analysis and design of optical instruments. As we will soon see, however, the path taken by light actually produces an extremum of time—either a maximum or a minimum.

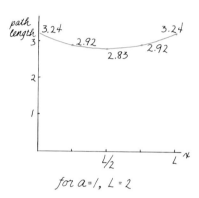

for $a=1$, $L=2$

As an example of minimum path time, consider the path of the refracted ray going from a point in air to a point in water. The actual path is shown in the diagram, along with several hypothetical paths. Richard Feynman in his *Lectures on Physics* points out that the problem is similar to that faced by the lifeguard on a beach who wants to reach a drowning person in the water in the shortest possible time. The lifeguard can run faster on the beach than he can swim in the water. Should he run immediately to the water and then swim in a straight line; should he run in a straight line on the beach to a point opposite the person and then swim the shortest distance out; or should he run and then swim, always moving in a straight line between his original position and the drowning person? None of these paths yield as short a time as the path given by Snell's law: $(\sin \theta_1)/(\sin \theta_2) = v_1/v_2$. The law holds for light and for the lifeguard.

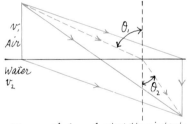

True path for shortest time is broken. This path satisfies Snell's law:
$$\frac{\sin \theta_1}{\sin \theta_2} = \frac{v_1}{v_2}$$

There is no simple purely geometric proof that Snell's law produces minimum time in this case. Once again, it is possible to describe the time in terms of the variable, x, where the path intercepts the boundary. The value of x that gives minimum time also yields the angles corresponding to Snell's law.

The path that light takes in going from one point to another is not necessarily a minimum. Consider the geometry of an ellipse. The distance from one focus to any point on the ellipse and back to the other focus is a constant. If one focus is a point source of light, all of the light will reflect from the ellipse and pass through the second focus (in three dimensions, the ellipse would be an ellipsoid). Since the distance from the source to any point on the ellipse and back to the other focus is a constant, so is the time taken. In this case, therefore, rays of light can and do leave the source and go to any point of the ellipse and

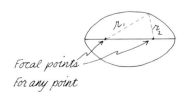

Focal points for any point

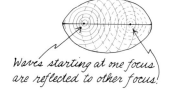

Waves starting at one focus are reflected to other focus!

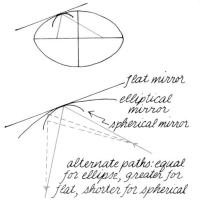

flat mirror
elliptical mirror
spherical mirror

alternate paths: equal for ellipse, greater for flat, shorter for spherical

then reflect back to the other focus, always taking the same time regardless of where they strike the surface of the ellipse.

Question 17-8
Does this mean that there are rays for which the angle of reflection is not equal to the angle of incidence?

There is another important point about the equal time paths for light going from one focus of an ellipse to the other. Since the flight time is the same regardless of the path, the light always arrives in phase at the other focus no matter which way it started out. Therefore, the various waves will all reinforce each other at the second focus. The arrival in phase and the mutual reinforcement is a requirement of having waves come to a focus. We will make use of this requirement in analyzing the focusing properties of mirrors and lenses.

In this diagram, we show one particular ray being reflected inside an ellipse. At the point where the ray reflects, we have drawn a straight line, representing a flat mirror touching the ellipse at that point. We have also drawn an interior curve, which represents a mirror with greater curvature. The particular angle of reflection at that point yields a flight time that is *independent* of position for the ellipse, is a *minimum* for the flat mirror, and evidently is a *maximum* for the mirror of greater curvature. Here we have an example of the generalization of Fermat's principle. The time taken by light in going from one point to another is either a minimum or a maximum or a constant. In all three cases, the time does not change appreciably for small excursions around the actual path.

REFLECTIONS FROM CURVED MIRRORS

As you saw in "Handling the Phenomena," a concave mirror can focus light rays. Usually we think of a concave mirror as being a small segment of a sphere. For practical purposes, most concave mirrors are spherical in shape largely because a sphere is easier to make than a section of a paraboloid or an ellipsoid. It is not obvious, however, that any of these curved surfaces will reflect all the rays incident from one point back to a focal point.

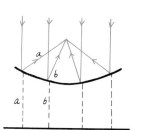

Parallel light rays reflecting to a focus from a parabolic surface.

We have already seen that because of Fermat's principle all rays leaving one focus of an ellipsoid will arrive at the other focus in phase. What about the rays emitted from a point *near* one of the foci? Will all those rays also come to a point, perhaps at the same distance from the other focus? Unfortunately, no, although with certain restrictions the approximation can be fairly good, as we will see.

For astronomical telescopes, the light comes from objects so far away that the rays from a particular star are parallel to each other. Suppose that we want to focus these rays to a point. In the diagram, we show the geometry of parallel rays approaching a curved surface. This particular curve is a parabola; in three dimensions it would be a section of a paraboloid. A *parabola* is the locus of all points such that the perpendicular distance to a line is equal to the distance to a focal point. The construction is shown in the diagram. If all of the rays coming toward the parabola are parallel to each other, they represent a plane wave which would arrive at the line with all points in phase (if the mirror were not there). For each ray striking the mirror, the distance to the line is the same as it is to the focal point.

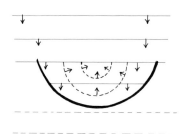

Parallel wave fronts of light reflect from parabolic surface in such a way that they arrive in phase with each other at the focus.

Therefore, all of the rays will arrive at the focal point in phase and will reinforce each other there. This phase relationship is the requirement for a focus. A detailed analytical or graphical examination of the geometry shows that each ray satisfies the law that the angle of reflection must equal the angle of incidence.

A parabolic mirror is ideal for bringing parallel light to a focus *if* the light is parallel to the axis of the parabola. What happens if the light is coming from another star that is not on the axis? The incoming rays are parallel to each other as shown in the next diagram, but not parallel to the axis. Will they also come to a point focus? Unfortunately not; although if the angle between the rays and the axis is small, the focal region will be very small and almost a point.

> Question 17-9
>
> To focus rays from a point object at a finite distance, a curved mirror should have the shape of an *ellipsoid*. To focus parallel rays, a curved mirror should have the shape of a *paraboloid*. Where is the dividing line between these two shapes as the point object gets further and further away from the mirror?

A small segment of a spherical mirror is hardly distinguishable from a small segment of a paraboloid or an ellipsoid. In fact, the standard method of making a parabolic astronomical mirror is to start out with a spherical mirror and then deepen the center very slightly to turn it into a paraboloid.

We are going to derive approximate rules for image formation with spherical mirrors. These rules are good as long as the distance between the rays and the axis is very small compared with the radius of curvature and as long as the angles between the rays and the axis are very small. In the photograph, we show an example of what happens when these conditions are not met. The distortion of the images is called *spherical aberration*.

The first thing that we must find out about spherical mirrors is what happens to light that is approaching parallel to the axis. Since a small section of a sphere is an approximation to a small section of a paraboloid, we might expect that the rays would converge on the axis to a small focal region, if not to a point. The geometry in the diagram shows a ray that is parallel to the axis being reflected back down through the axis. The radius line drawn from the center of the sphere is the normal to the mirror at the incident point. Since the angle of reflection is equal to the angle of incidence, and since interior angles between parallel lines are equal, the angles have the equalities shown on the diagram. Notice that two approximations are made in the derivation. It is assumed that δ is small compared with f or R. It is also assumed that h is small compared with f or R, so that $\tan \alpha \approx \alpha$ and $\tan 2\alpha \approx 2\alpha$. Both of these require that we deal with *paraxial* rays, ones that are close to the axis and make only a small angle with it. When this condition is satisfied, the derivation shows that all the rays parallel to the axis reflect back through a focal point on the axis that is at the location: $f = R/2$.

Now we can ask what happens to the rays coming from an object in

Here is an extreme example of spherical abberation. The original grid had perpendicular and parallel lines. (Grant Heilman)

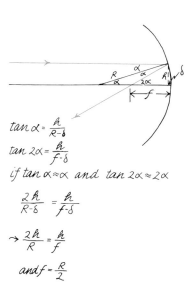

$$\tan \alpha = \frac{h}{R-\delta}$$

$$\tan 2\alpha = \frac{h}{f-\delta}$$

if $\tan \alpha \approx \alpha$ and $\tan 2\alpha \approx 2\alpha$

$$\frac{2h}{R-\delta} = \frac{h}{f-\delta}$$

$$\rightarrow \frac{2h}{R} = \frac{h}{f}$$

and $f = \frac{R}{2}$

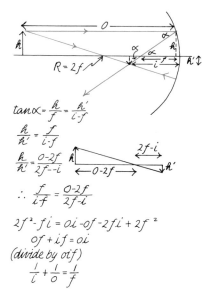

front of the mirror. In forming an image of a real object with a mirror, the rays from different points on the object must converge at corresponding points on the image. Each point on the object emits rays in all directions. Only a small fraction of these are intercepted by the mirror and reflected back to the image region. In the diagrams, we represent the object by a small arrow. We want to know if all of the rays coming from the arrowhead converge at a particular point. If they do, we want to know the location of this point. Rather than draw all the rays coming from the arrowhead, we start out by tracing only two of them whose paths we already know. The first is parallel to the axis. As we have just seen, this ray will intercept the axis at the focal point of the mirror, f. Another ray whose path is easy to determine goes through the center of curvature of the mirror and hence reflects right back on itself. These two rays intercept each other at a distance i from the mirror. Although there is no guarantee yet that all of the other rays from the arrowhead would also pass through this particular point, let us calculate the distance i in terms of the object distance, o, and the focal distance of the mirror, f. The geometrical derivation is shown with the diagram. In the final form, there is a simple and symmetrical relationship among the image distance, the object distance, and the focal distance:

$$\frac{1}{i} + \frac{1}{o} = \frac{1}{f}$$

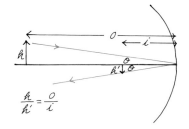

We could now find the direction of other rays that leave the arrowhead and strike the mirror by drawing each one and requiring that the angle of reflection equal the angle of incidence. There is one other ray for which this construction is easy. That is the ray that leaves the arrowhead and strikes the mirror on the axis. If this ray also goes through the same image point formed by the first two rays, as shown in the next diagram, then there is a simple relationship among the height of the object, the height of the image, and the object and image distances. This relationship is

$$\frac{h}{h'} = \frac{o}{i}$$

Conversely, if the relationship between the height of the object and the height of the image agrees with the geometry formed by the first two rays, then the third ray *must* also go through the intersection point of the first two. According to the geometry provided by the first two rays:

$$\frac{h}{h'} = \frac{f}{(i-f)}$$

In terms of the object and image distances:

$$f = \frac{oi}{(o+i)}$$

Therefore,

$$\frac{h}{h'} = \frac{f}{i-f} = \frac{oi/(o+i)}{i-(oi)/(o+i)} = \frac{oi}{oi+i^2-oi} = \frac{o}{i}$$

Since these geometrical relationships match, we have demonstrated that the third ray also goes through the intersection point of the first two rays.

Indeed, all of the rays leaving the arrow and reflected from the mirror pass through an image point of the arrow. To find this point, however, we need use only two of the three principal rays that we have shown. The use of these principal rays makes the construction easy to do graphically (even in a sketch) and also makes the algebraic derivations easier. Note that in the process of demonstrating that all three rays go through the same point, we have also derived a simple expression for the magnification of the image. The height of the image compared with the height of the object is

$$\frac{h'}{h} = \frac{i}{o}$$

We traced the rays from the point of the arrowhead only, but the same procedure could be used for every point on the arrow. The whole image exists in a flat image plane perpendicular to the axis. This result, however, is only approximately true. For a large object with points far removed from the axis, the image would not lie in a flat plane and the image point far from the axis would be fuzzy.

Notice the characteristics of the image that is formed by a spherical mirror with the geometry that we have just described. In the first place, the image is upside down and is smaller than the object. The rays actually go through the image, diverging from it just as they would from a luminous object. If we put a piece of paper in the image plane, the image would be seen on the paper just like the image on a projection screen. Such an image is called *real* as opposed to the *virtual* image which our eyes see on the other side of a plane mirror.

In "Handling the Phenomena," you observed the real image produced by a spherical mirror, by reflecting a light bulb from a curved bathroom mirror. You can also see such an image by looking into the bowl of a shiny spoon. Notice with the spoon that your image is upside down. If you use a pencil point as an object, you can actually locate the image of the pencil point in front of the spoon and touch the image with the object. Notice the distortion of the image at distances away from the axis of the spoon's bowl. Of course, the bowl of a spoon is only approximately spherical, but then a sphere is only approximately the correct shape.

We now have two ways of analyzing the location of images produced by a spherical mirror. We can draw sketches using the principal rays or we can use the simple algebraic formula that we have derived. Let's use both of these techniques together and find the location and size of images produced by objects that are at various distances from the mirror. We will soon see that under certain circumstances i, the image distance, becomes negative. As the sketch will show us, this condition arises when the image becomes virtual and appears to be on the far side of the curved mirror. In dealing with mirrors, we must adopt a sign convention that determines the meaning of positive or negative distances. In all cases, we will call a distance positive if it is in the region where the light is actually traveling. Hence, in the geometry we have used so far, i, o, f, and R are all positive. We will

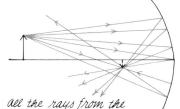

all the rays from the arrowhead, which strike the mirror, reflect to the same intersection point (as long as the approximation conditions are met).

Principal rays:

Parallel to axis → reflects through focus

Through centerpoint (2f) → reflects back on itself

To center axis of mirror → reflection with equal angle from axis

Because of the approximation made in deriving formulas for the spherical mirror, the principal rays can be drawn most conveniently if they reflect as if from a flat mirror at the distance of the curved mirror's axis point

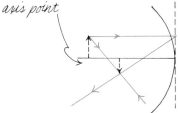

Real image of an upside down light bulb, cast on the wall by the curved side of a bathroom mirror. Plane side reflects ordinary virtual image of the person holding the mirror.

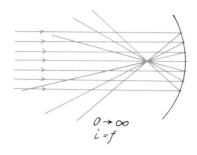

soon take up the case where R and f are negative. In such a case the mirror is convex, with the radius on the far side of the mirror where the light does not travel.

Let us now analyze a series of cases of image formation by paraxial rays for various object distances starting out with the object at infinity.

1. Where is the image if the object is at infinity? If an object is really at infinity, all of the rays coming from it are parallel to each other. In that case, they all converge at a point at a distance f from the mirror.

$$\frac{1}{\infty} + \frac{1}{i} = \frac{1}{f}$$

$$i = f$$

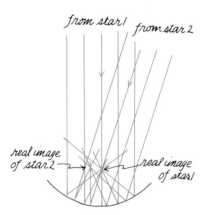

However, the image point is not necessarily on the vertical axis that we have shown in the diagram. A spherical mirror has no unique axis; any diameter is an axis. If the parallel rays are coming at an angle to the vertical axis, then the image point will lie at a distance f on a different diameter. We would have a situation like this if a telescope mirror were observing a field of stars, as shown in the diagram. The rays from each star are parallel to each other, but not parallel to the rays coming from another star. Each set of parallel rays focuses to a different point on a spherical focal surface. For paraxial rays, this surface is almost a plane.

If the distant object is the sun or the moon, then each region of the object is the source of rays that are essentially parallel as far as the size of our mirror is concerned. The rays from the top edge of the moon, however, are not parallel to the rays from the bottom edge of the moon. Therefore, the image of the sun or the moon lies in the focal surface, but has a finite size.

Question 17-10

What is the "magnification" of the moon's image?

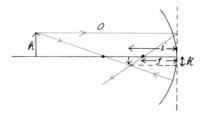

2. Now let us bring the object in closer, but still at a distance larger than $2f$. This is the geometry that we used when we first started talking about curved mirrors. The image is real, upside down, and between f and $2f$. In fact as the object moves from infinity to $2f$, the image moves from f to $2f$.

Suppose, for instance, that $f = 10$ cm and the object distance is 40 cm.

$$\frac{1}{i} + \frac{1}{(40 \text{ cm})} = \frac{1}{(10 \text{ cm})}$$

$$\frac{40 + i}{40i} = \frac{1}{10}$$

$$40i = 400 + 10i$$

$$30i = 400$$

$$i = 13.3 \text{ cm}$$

The image is reduced in size, since

$$\frac{h'}{h} = \frac{i}{o} = \frac{13.3 \text{ cm}}{40 \text{ cm}} = \frac{1}{3}$$

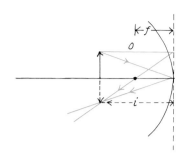

3. When the object is at $2f$ (or approximately at the center of the sphere), $o = 2f$. Consequently,

$$\frac{1}{i} = \frac{1}{2f} = \frac{1}{f}$$

$$i = 2f$$

As you can see both by the algebra and by the diagram, the image distance is equal to the object distance when $o = 2f$. The image is real and upside down, but is the same size as the object. When you form the image of a pencil point with a shiny spoon and then touch the image of the point with the real pencil point, both object and image are at the center of curvature of the bowl.

4. As the object moves from $2f$ to f, the image moves from $2f$ to infinity. The image remains real and upside down, but is larger than the object since $i > o$.

5. If the object is at f, the image is at infinity. This geometry is the standard one used with flashlights, car headlights, and searchlights. The object is a glowing filament centered on the axis in the focal plane and made as small and as bright as possible. As we have seen, the point source at the focus of a paraboloid would produce a parallel beam; hence, the image would be at infinity:

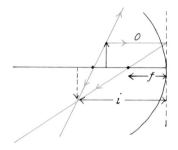

$$\frac{1}{i} + \frac{1}{f} = \frac{1}{f}$$

$$i = \infty$$

As a practical matter, a point source would have no intensity. The filament must have a surface area and, therefore, the focusing condition cannot produce a parallel beam. The larger the curved mirror compared with the filament, the better the approximation. In order to see the difference between a parabolic mirror and a spherical mirror, take a look at an automobile headlight. The reflector is clearly deeper than a spherical surface.

6. Next, we must move the object even closer to the mirror. For the mirror with $f = 10$ cm, let's place the object at $o = 5$ cm. According to our algebraic equation,

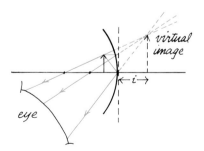

$$\frac{1}{(5 \text{ cm})} + \frac{1}{i} = \frac{1}{(10 \text{ cm})}$$

$$2i + 10 = i$$

$$i = -10 \text{ cm}$$

The image distance is -10 cm. According to the convention that we adopted, positive distances for mirrors are on the side where the light actually travels. A negative distance is on the far side. The image is virtual, like the image in a plane mirror.

Notice how in the diagram we have drawn the principal rays according to our previous rules. These rays diverge from the mirror and do not intersect at any point in space. However, if these diverging rays enter our eyes, we see them just as if they were diverging from an object on the other side of the mirror. In the diagram, we have projected the rays backward to the point from which they appear to be diverging. We see the collection of those points as a virtual image. Notice that it is behind the mirror, right side up, and larger than the object. This is the type of image we see with the ordinary spherical bathroom mirror.

> **Question 17-11**
>
> When we look into the concave side of a shiny spoon, we see an upside-down image of our face. When we look into a curved bathroom mirror, we see an enlarged image of ourselves right side up. Why do we get these two different effects?

7. There is another type of spherical mirror that usually produces only virtual images. You can see an example of this mirror by using the shiny spoon again. If you turn the spoon over so that the side toward you is convex, your image is on the other side of the spoon and is right side up. In this case, according to the convention that we have adopted, the radius of curvature of the mirror, and hence its focal length, is negative. Our basic formula for image and object distance still applies. Furthermore, we can still use principal rays to determine the position of the image.

For our analysis, we take a convex mirror with a focal length of -10 cm. Let's place the object in front of the mirror at a distance of 10 cm.

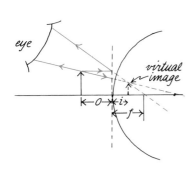

$$\frac{1}{(10 \text{ cm})} + \frac{1}{i} = \frac{1}{(-10 \text{ cm})}$$

$$i + 10 = -i$$

$$i = -5 \text{ cm}$$

The image distance is negative—meaning that the virtual image is located behind the mirror. The image is smaller than the object, in this case only half as high. In fact with a convex mirror, the image must always be smaller than the object. To see this, let us solve for the image distance in the general case.

$$\frac{1}{o} + \frac{1}{i} = -\frac{1}{f}$$

By making the right-hand side negative, we are using f as a positive number.

$$fi + of = -oi$$

$$i(f + o) = -of$$

$$i = -\frac{of}{o + f}$$

$$\frac{i}{o} = -\frac{f}{o + f}$$

The magnification, which is given by the ratio i/o, is always less than one for a convex mirror. The negative sign indicates that the image is virtual and erect.

Notice that when tracing the principal rays with a convex mirror, all of the rays actually travel on the left side of the mirror. The direction of the reflection from the parallel ray is found by projecting that ray backward through the mirror to the virtual focal point. The same thing is true for the ray that is perpendicular to the mirror and thus bounces back along its own path. That ray when projected through the mirror would end up at $2f$.

Convex mirrors are frequently used in side-view mirrors, particularly with trucks. They can capture a large angle of view and present it as a virtual, smaller image.

The focusing of waves from curved surfaces can also be seen and heard with ripples and sound. Fill any circular bowl with water and produce circular ripples by touching your finger in the water. The ripples will bounce off the walls and converge again. If the bowl is circular and you dip your finger in the center, the reflected ripples will focus back at the center. If you start the ripples from a point between the center and one side, the reflected ripples will converge to a rough focus at a symmetrical point on the other side. This effect is an approximation to what would happen if the bowl were elliptical and you dipped your finger in at one focal point.

There are many famous whispering galleries such as the Mormon Tabernacle in Salt Lake City and the Capitol Dome in Washington. These usually exist in a large hall that has a domed ceiling. Some of these approximate half of an ellipsoid. If you are standing at one of the focal points, you can hear whispers spoken at the other focal point, even though it may be far away. The sound waves spreading upward reflect from the ceiling and converge again at the far focal point.

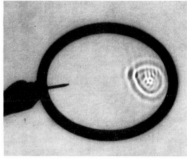

The converging pulse originated from a circular wave that spread out from water dripped from a tube. (From PSSC COLLEGE PHYSICS, 1968, D.C. Heath & Co., with EDC, Newton, Ma.)

IMAGES FORMED BY LENSES

Images can be formed by *refraction* as well as by reflection. With the prisms shown in the diagram parallel rays are brought to a crude focus. If a lens were made of an infinite number of prisms, each cut at the proper angle, it ought to be possible to deviate each parallel ray to the same point on the axis. Whether such a lens would have a simple shape is not immediately obvious. Furthermore, it is not obvious that a lens that would focus parallel light would also produce an image of an object at a finite distance. If a lens shape had to be calculated anew for each position of the source object, lenses would not be very practical or useful. As we will see, there is a practical solution for the design of lenses even as there was for the design of mirrors. Once again, the easiest surface to grind is that of a sphere. Most simple lenses have spherical surfaces, although an actual lens for a camera or an instrument may be a compound lens made up of many simple ones with varying curvatures and optical properties.

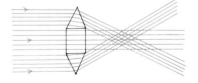

One way to design a lens would be to follow each ray into the lens and out again, using Snell's law to determine the path. Another way, which is actually the basis of some methods of lens design, is to use Fermat's principle. If light comes to a focus having followed various paths, all of the wave segments must arrive in phase so that they will reinforce each other

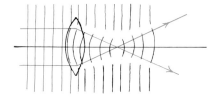

at the focal point. Therefore, all flight paths must take the same time. Consider the simple lens diagram shown here. Since the incoming rays are parallel to each other, they represent plane wave fronts. All of the light is in phase as it enters the lens.

> Question 17-12
>
> Take a closer look at the diagram. The ray coming from the top of the lens clearly has to travel further to get to the focal point than does the ray traveling along the axis. Therefore, how can the rays be in phase when they arrive at the focal point?

A glass lens that will focus parallel visible light must be thicker in the middle than at the edges. In this diagram, we show a lens that brings parallel microwaves to a focal point. The lens consists of metal channels or wave guides. Notice that it is *thinner* in the center than at the edges. The *phase* velocity of microwaves in such a channel is greater than c. Therefore, the microwaves going through the channels at the edge of the lens spend less time in the lens and arrive at the focal point in phase with the rays on the axis, even though they have traveled a longer distance.

■ Optional

Since for practical purposes most lenses are made with spherical surfaces, let's see what such a lens does to parallel rays. In following the geometry of the rays, we will see the nature of the approximation that we make in assuming that all of the rays focus at the same point. The geometry is shown in the diagram. Because the lens is plano-convex (the first side flat; the second side curved so that the lens converges the rays), we do not have to worry about refraction at the first surface. The parallel rays remain parallel as they enter the lens. The refraction at the curved surface follows Snell's law, where the angle of refraction, ϕ, is greater than the angle of incidence, θ, since the ray is going from glass into air.

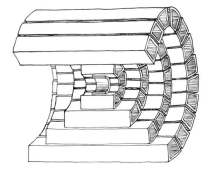

In the analysis accompanying the diagram, we start with exact expressions and then transform to approximations. For these approximations to be valid, the angles must be small enough so that the sine and tangent of each angle are approximately equal to the angle. Furthermore, we assume that the lens is so thin that we can ignore its thickness in comparison with the focal length. The logic of the derivation is to start with Snell's law in terms of θ and ϕ and then eliminate θ and ϕ by substitutions that involve the radius of curvature of the lens surface, r, and the focal distance, f. As you can see, this derivation yields a focal length that is independent of the angle θ (or α), therefore implying that all parallel rays entering the lens focus at the same point on the axis. Remember, however, that this conclusion is good only to the extent that the approximations are good. The spherical aberration gets worse for the rays far removed from the axis.

Note the relationship that exists among the focal length of the lens, its radius of curvature, and its index of refraction. If the index of refraction of the glass is 1.5, then the focal length of the plano-convex lens is twice the radius.

IMAGES FORMED BY LENSES

The smaller the radius, the more the lens bulges and the greater its thickness at the center compared with the edge. As intuition might tell us, the focal length is smaller for such a lens. Also, the greater the index of refraction for a given curvature, the smaller the focal length.

Our next step is to demonstrate that the light from a point *on the axis* of a spherical lens will be deflected to a focal point on the other side of the lens. The argument is in two steps, as shown in the next two diagrams. First, we follow the geometry of the rays leaving a point on the axis in air and entering a long piece of glass with a curved surface. The question is: Does a single spherical surface bring light to a focus within the glass; and if so, where? The sequence of steps is essentially the same as those we used with the plano-convex lens. However, an extra angle is involved here. The individual steps, the reasons for them, and the approximations accompany the diagram. Once again our aim is to eliminate θ and ϕ and express them in terms of the object distance, o, the image distance, i, and the radius of curvature of the surface, r. The approximations required are the same as those we used in the previous derivation.

The rays will come to a focus in the long piece of glass. Suppose the glass is long enough to allow the rays to pass through the focus and to reach the second surface of the lens *at the same height from the axis* at which they entered. The trip through the long glass would merely reverse the up-down position of the rays with respect to the axis. Each ray would make the same angle with the second surface that it would make if the second surface *immediately* followed the first surface. As they leave the second surface, the rays are aimed as if they had come from an object inside the glass at a distance $-i$. For the second stage, we can then use the same formula as the one we derived for the first stage. We need only reverse the indices of refraction to account for the fact that the rays are now going from glass into air, and for the new object distance we will need to use the negative of the image distance that

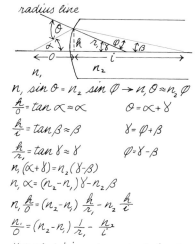

$n_1 \sin\theta = n_2 \sin\phi \rightarrow n_1\theta \approx n_2\phi$

$\dfrac{h}{o} = \tan\alpha \approx \alpha \qquad \theta = \alpha + \gamma$

$\dfrac{h}{i} = \tan\beta \approx \beta \qquad \gamma = \phi + \beta$

$\dfrac{h}{r} = \tan\gamma \approx \gamma \qquad \phi = \gamma - \beta$

$n_1(\alpha + \gamma) = n_2(\gamma - \beta)$

$n_1\alpha = (n_2 - n_1)\gamma - n_2\beta$

$n_1\dfrac{h}{o} = (n_2 - n_1)\dfrac{h}{r} - n_2\dfrac{h}{i}$

$\dfrac{n_1}{o} = (n_2 - n_1)\dfrac{1}{r} - \dfrac{n_2}{i}$

Note that i is independent of α. All rays from the object on the axis (which satisfy the small angle condition) arrive at the same image point.

$\dfrac{n_2}{i} = \dfrac{(n_2 - n_1)}{r} - \dfrac{n_1}{o} = \dfrac{(n_2 - n_1)o - n_1 r}{or}$

$i = \dfrac{or\, n_2}{(n_2 - n_1)o - n_1 r}$

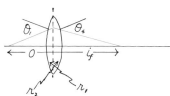

The long lens and the thin lens are equivalent (except for reversing up-down) providing the rays leave the second surface at the same height that they entered the first surface.

Applying the previous formula to the rays coming through the internal focus point, which becomes the new object point:

$\dfrac{n_2}{o_{interior}} = (n_1 - n_2)\dfrac{1}{r_2} - \dfrac{n_1}{i_{final}}$

$o_{interior} = -i_{original} = -\dfrac{or\, n_2}{(n_2 - n_1)o - n_1 r}$

(from previous derivation)

$\dfrac{n_2}{\dfrac{-or\, n_2}{(n_2-n_1)o - n_1 r}} = \dfrac{n_1 - n_2}{r_2} - \dfrac{n_1}{i}$

$\dfrac{[(n_2 - n_1)o - n_1 r]}{or_1} = \dfrac{n_1 - n_2}{r_2} - \dfrac{n_1}{i}$

$\dfrac{n_2 - n_1}{r_1} + \dfrac{n_1}{o} = \dfrac{n_1 - n_2}{r_2} - \dfrac{n_1}{i}$

$\dfrac{1}{i} + \dfrac{1}{o} = \dfrac{n_2 - n_1}{n_1}\left[\dfrac{1}{r_1} - \dfrac{1}{r_2}\right]$

For object at infinity, $i = f$

$\therefore \dfrac{1}{f} = \dfrac{n_2 - n_1}{n_1}\left[\dfrac{1}{r_1} - \dfrac{1}{r_2}\right]$

If $n_1 = 1$ for vacuum, or approximately for air:

$\dfrac{1}{f} = (n - 1)\left[\dfrac{1}{r_1} - \dfrac{1}{r_2}\right]$

This is the lens-makers equation for thin lenses.

we calculated in the first stage. The result of this argument is shown in the derivation accompanying the diagram.

Question 17-13
Does the focal length for this double-convex lens agree with the one that we derived first for a plano-convex lens?

Notice a crucial point about the formula for the focal length of a double-convex lens. The sign of r_1 must be opposite that of r_2 (otherwise if $r_1 = r_2$, which is the case for many lenses, the formula would require that $f = \infty$). The sign convention for lenses requires that object distances and image distances be positive if they are in the regions where the light actually travels. As we will soon see, a negative image distance implies a virtual image. In the double-convex lens, r_1 is positive because the center of curvature of the first side is on the positive side of the axis. The second curved surface, however, has its center of curvature on the negative side of the axis; r_2 is negative. The formula for the relationship among the focal length of a thin lens, the index of refraction, and the radii of curvature is called the *lensmaker's equation*.

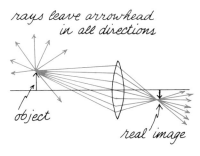

$\frac{1}{i} + \frac{1}{o} = \frac{1}{f}$
for both lenses and mirrors

rays leave arrowhead in all directions

object

real image

The basic formula that we have derived for the relationship among object distance, image distance, and focal distance is the same for a lens as it is for a spherical mirror. There are corresponding relationships between the details of image formation with lenses and with mirrors. One of these we have already seen. If the object is at infinity, the rays coming from it are parallel. With both mirror and lens, the resulting image is at the distance f from the lens or mirror. Note in the lens formula that if $o = \infty$, then $i = f$. This is also the result that we derived for the simple case of a plano-convex lens.

Let's trace the rays from an object through a lens to an image. We assume that each point on the object is emitting rays in all directions, as shown in the diagram. The rays within a small cone have to pass through the lens. They are refracted according to Snell's law at each surface and all pass through an image point on the far side, unless the image is virtual. We will study the situation for virtual images later. As is the case with mirrors, there are principal rays whose behavior is easy to predict. A ray moving parallel to the axis will intersect the axis on the far side at the principal focal point. A ray going through the principal focal point on the near side of the lens will leave the lens parallel to the axis on the far side. A ray going through the center of a thin lens will be undeviated.

The lens at the center is essentially a plane sheet of glass. Rays going through a plane sheet of glass are refracted both entering and leaving but the exit angle is equal to the entrance angle. Although the angular direction of the ray does not change, its position is slightly offset. If the lens is thin and the rays are paraxial, the effect will be negligible. (If the rays are not paraxial, the spherical surfaces do not produce a good focus anyway.)

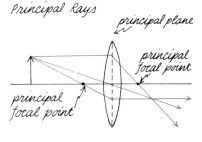

Principal Rays

principal plane

principal focal point

principal focal point

As you can see in the diagram, all three of the principal rays intersect at the same point. Detailed calculations or a careful scale drawing would show that all other rays leaving the same point on the object would intersect at the same point at the image. In drawing these principal rays, we assume that all the refraction takes place in a plane through the center of the lens.

IMAGES FORMED BY LENSES

It is easy to sketch these lines with sufficient precision to find the approximate size and location of the image.

Each point on the object has a corresponding image point in a plane at the distance i from the lens. The image is upside down and is real. The rays actually converge on the image and could be seen on a screen placed at that position. The magnification of the image is easily calculated from the geometry shown in the diagram. Consider the triangles formed by the line going through the center of the lens to the tips of the object in the image, the axis itself, and the object and image. Those two triangles are similar. Therefore,

$$\text{Magnification} = \frac{h'}{h} = \frac{i}{o}$$

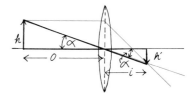

light from star 1

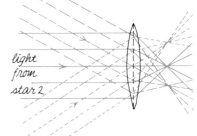

light from star 2

light from star 3

Rays coming from star 1 and star 3 are clearly not paraxial. The angles are exaggerated for clarity in the diagram.

Now let's use both formula and diagrams to analyze different types of images that can be produced with a thin lens.

1. We have already seen what happens when the object is at infinity and parallel rays enter the lens along the axis. The rays focus to a point on the axis at the distance f. If the rays from the distant object are parallel to each other but not parallel to the axis, then they will converge at an image point off axis that is approximately in a plane at the distance f. For paraxial rays, we can assume that all of the focal points are in a plane. The particular point in that plane where these parallel rays converge can be found by tracing the ray that goes through the center point of the lens, as shown in the diagram. The light coming from a group of stars consists of many bundles of rays. The rays from each star are parallel to each other but not necessarily to the axis of the lens. Each bundle of rays converges to a separate point in the focal plane. (Don't forget, however, that each focal point is really a diffraction image with an angular radius of $1.2\lambda/d$, where d is the diameter of the lens.) Stars are so far away that their geometric images are smaller than the diffraction image of a point; therefore, no magnification of the image of a star can be obtained. When we study the construction and use of telescopes, we will have to explain what good it does to use a telescope if you don't obtain magnification.

The lens formula for the case of the object at infinite distance is

$$\frac{1}{\infty} + \frac{1}{i} = \frac{1}{f}$$

Therefore, $i = f$.

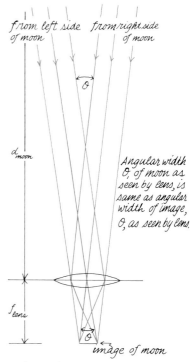

from left side of moon   from right side of moon

Angular width θ, of moon as seen by lens, is same as angular width of image, θ, as seen by lens.

image of moon

width of moon = $\theta \, d_{moon}$
width of image = θf_{lens}

As we pointed out in the case of the spherical mirror, the image of the moon in a converging system is not a point but a disk. The rays from the top of the moon enter the lens at a different angle than do the rays from the bottom of the moon.

The image of the moon produced by the lens is obviously much smaller than the moon itself. The *angular* size of the image, however, is the same as the angular size of the object *as seen by the lens*. This angular size is shown in the diagram. The angular size of the object is equal to the ratio of the diameter of the object and the object distance. The actual diameter of the image is then equal to the product of the angular width and the focal length of the lens. The longer the focal length, the greater the diameter of the image.

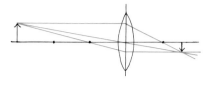

2. If the object distance is greater than $2f$, but less than infinity, then the image distance is less than $2f$ but greater than f. In the diagram, we show the principal rays for an object that is at a distance of 30 cm from a lens with a focal length of 10 cm. The image is between f and $2f$, upside down, and real. According to the lens formula,

$$\frac{1}{(30 \text{ cm})} + \frac{1}{i} = \frac{1}{(10 \text{ cm})}$$

$$i + 30 = 3i$$

$$i = 15 \text{ cm}$$

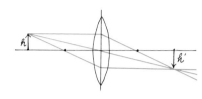

The image is smaller than the object: $M = h'/h = 15/30 = \frac{1}{2}$.

3. If the object is at the distance $2f$, then the image is also at $2f$. The formula for the 10 cm lens is

$$\frac{1}{(20 \text{ cm})} + \frac{1}{i} = \frac{1}{(10 \text{ cm})}$$

$$i = 20 \text{ cm}$$

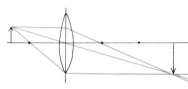

The image is real, upside down, and the same size as the object.

4. When the object is between $2f$ and f, the image is between $2f$ and infinity. The image remains real and upside down, but is larger than the object since $i/o > 1$.

5. A *point* object on an axis at the principal focus would produce parallel rays on the other side of the lens. An object with finite size in the focal plane produces diverging rays on the other side, but these rays do not form an image. The rays from each point in the object form a bundle of parallel rays. The angle of these rays is different for every point on the object.

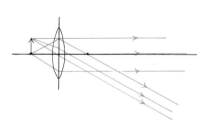

6. If the object is inside the focal distance of a lens, no real image is formed. We can still use the principal rays to trace the geometry, but these rays now diverge on the far side of the lens. One of the three principal rays must be drawn in a way slightly different from that of the previous cases. The ray from the arrowhead going through the focal point on the same side of the lens clearly will not go through the lens in this case. However, a ray leaving the arrowhead as if it had come from the left focal point, will end up going parallel to the axis after it passes through the lens. As you can see in the diagram, all three of these rays, when projected backward to the left side of the lens, meet at a virtual image point. The eye, or a camera lens, over on the right side of the system would intercept these diverging rays and interpret them in terms of rays coming from an object on the left side. The virtual image formed by these diverging rays is right side up and larger than the object.

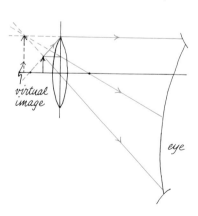

Let's find the position of this virtual image when the object is at a distance of 5 cm from a 10 cm lens.

$$\frac{1}{(5 \text{ cm})} + \frac{1}{i} = \frac{1}{(10 \text{ cm})}$$

$$2i + 10 = i$$

$$i = -10 \text{ cm}$$

The minus sign indicates that the image is virtual and is on the side of the lens opposite to that of the rays which actually form it. The magnification of the image is

$$M = \frac{i}{o} = \frac{10}{5} = 2$$

When a lens is used in this geometry, it can be a reading glass or a magnifying glass. It is a good thing that the image is virtual, since it would be inconvenient to have an upside-down image in a magnifying glass. When a magnifying glass is used for maximum magnification and also maximum field of view, it is held close to the eye rather than close to the object. When you use such a glass, notice that it allows you to get *much closer to the object*. The magnification is really produced by this circumstance. If you want to see a small object, you naturally get as close to it as possible. You can see it better when you are holding it than when it is across the room. The closer your eye to the object, the larger the angular size of the object as seen by your eye.

> Question 17-14
>
> If you really want to see the details of a small object, why not hold it only about a centimeter away from your eye?

Let's calculate the magnification produced by a magnifying glass if the image is to be at a distance of 25 cm.

$$-\frac{1}{(25 \text{ cm})} + \frac{1}{o} = \frac{1}{f}$$

$$-of + 25f = 25o$$

$$25f = o(25 + f)$$

$$o = \frac{25f}{25 + f} \quad (o \text{ and } f \text{ are in cm})$$

Magnification is given as usual by

$$\frac{i}{o} = \frac{25}{25f/(25+f)} = \frac{25 + f}{f} = 1 + \frac{25}{f} \quad (f \text{ is in cm})$$

In focusing a magnifying glass (or the eyepiece of an instrument, which is essentially the same thing), it is not necessary to get the image at exactly 25 cm. That minimum distance for comfort varies from person to person anyway. Sometimes for long-term viewing, it is most comfortable to have the image focused at infinity. Such a change makes very little difference in the apparent size of the image. To focus the image at infinity, get the magnifying glass and your eye close enough to the object so that the object is almost at the focal point of the lens (if $i = \infty$, then $o = f$). Under these conditions, the angular magnification is given by the ratio of the distance between eye and object for normal viewing to the distance using the magnifying glass. The usual viewing distance is 25 cm, and so the magnification is $25/f$.

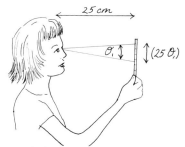

Normal viewing: angular size of object = θ_1

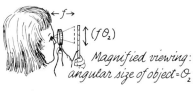

Magnified viewing: angular size of object = θ_2

size of object = $25\theta_1 = f\theta_2$

$$M_{visual} = \frac{\theta_2}{\theta_1} = \frac{25}{f}$$

> **Question 17-15**
>
> If you use a magnifying glass with a focal length of 5 cm, what magnification can you get and what difference does it make whether you focus the image at 25 cm or at infinity?

A person who is very nearsighted has troubles with or without glasses, but in one situation the nearsighted person has an enormous advantage. To examine a small object, the person need only take off his glasses and bring the object up to the comfortable viewing distance. For a viewing distance of 8 cm, for example, there is a built-in magnification of about 3.

7. So far we have dealt only with converging lenses. These need not have two convex surfaces. We have already studied the case of a plano-convex lens where one side is plane. As long as the center of the lens is thicker than the edges, the lens will converge parallel rays to a focal point. If the lens is thinner in the middle than at the edges, then it will *diverge* parallel rays. Two examples of such lenses are shown in the diagram, including the type of diverging lens that is used in eyeglasses for people who are nearsighted. Notice that the latter lens is slightly convex on the outer side, but strongly concave on the inner side. (It is called a diverging *meniscus* lens.)

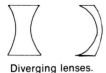
Diverging lenses.

The focal length of a diverging lens is negative. The focal length can be found by sending in parallel rays which then diverge. These diverging rays can then be traced back to their apparent origin on the axis. The diverging lens forms only virtual images of real objects. The location of the image can be found by tracing the principal rays in the usual way or by using our formula. Let's take the case of a symmetrical concave lens with a focal length of -10 cm. We will place the object at a distance of 20 cm. The principal ray that leaves the arrowhead parallel to the axis diverges at such an angle that its projection backwards passes through the first focal point. The principal ray that goes through the center of the lens is undeviated. The third principal ray starts out as if on a straight line to the second focal point. At the lens it is deviated, so that it continues parallel to the axis. Its projection backward to the left side of the lens passes through the intersection point of the projections of the first two rays. The eye or the lens of a camera on the right side of the lens would consider that these diverging rays had come from the image point shown on the diagram.

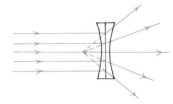

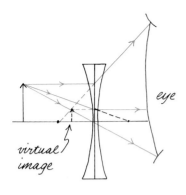
virtual image

The lens formula confirms the qualitative prediction of the diagram.

$$\frac{1}{(20\text{ cm})} + \frac{1}{i} = \frac{1}{(-10\text{ cm})}$$

$$i + 20 = -2i$$

$$3i = -20$$

$$i = -6.7 \text{ cm}$$

As seen by the eye on the right side of the lens, the image is virtual, right side up, and smaller than the object.

Question 17-16

Is it possible that a real image could be obtained from a diverging lens if the object were placed inside the focal length of the lens?

Real lenses do not work so nicely as the simple ones that we have diagramed. There are several aberrations caused by rays that are at too large an angle from the axis or that intercept the lens too close to its edge. The most prominent of these is spherical aberration, which was pictured on page 475. One way to reduce this defect is to "stop down" the lens with an opaque diaphragm, so that only the center part of the lens is used. Of course, doing this reduces the amount of light that can be used.

Another aberration, which a curved *mirror* does not have, is called *chromatic aberration*. Since the index of refraction depends on the wavelength, different colors will form images at different distances. The situation for parallel white light is shown in the diagram. Since the index of refraction is greater for blue light than for red, the blue light is brought to a focus first. In that focal plane, the image will be blue in the center with a red fringe. A short distance further back, the image will be red in the center with a bluish fringe.

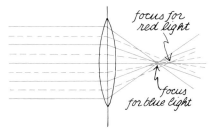

These aberrations can be cured, or at least reduced, by putting together combinations of lenses each of which has spherical surfaces. However, the lenses are made of different types of glass with differing dispersion and refraction properties. In the illustration, we show a cross-section of a compound lens that might be used in a good camera. A lens that reduces chromatic aberration is called an *achromat*.

A doublet achromat is shown schematically in the next diagram. A double-convex lens made with crown glass is in contact with a planoconcave lens made of flint glass. Flint glass contains lead oxide as one of the salts in the melt and has both a high index of refraction and much higher dispersion than the softer crown glass. Flint glass is used for fine crystal. The negative curvature of the flint glass lens only partially reduces the convergence produced by the crown glass lens, but the chromatic dispersion of the first lens is largely compensated by the effects of the second lens.

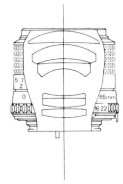

Cutaway view of a compound camera lens.

Double achromat lens.

OPTICAL INSTRUMENTS

The Camera

The simplest camera also has the fewest aberrations. It is literally a "pinhole" camera. All that you need in order to make one is a shoe box, some black masking tape, aluminum foil, and a sheet of film. Cut a hole at one end and tape a piece of aluminum foil over it. Make a pinhole in the foil. While in a dark room, place the film in the other end of the box and then tape the lid. You can put a piece of tape over the pinhole until you have the camera in position to take a picture. Since the exposure of a sunlit scene may take five minutes or more, you do not have to be particularly cautious about how you open or close the pinhole. Extremely sharp and distortion-

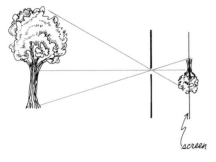

free pictures of stationary scenes can be made with such a device.

The simple ray geometry of this camera is shown in the diagram. Each source of light in the object scene creates a spot on the film, which is just the shape of the pinhole. The image consists of a pattern of many such spots. No focusing is necessary. With straight-line geometry, all objects in front of the pinhole are in focus.

You may think that the way to get sharper pictures with a pinhole camera is to reduce the size of the pinhole. This scheme works up to a point. Clearly if the hole has a diameter of 1 cm, then the image is composed of a montage of 1 cm disks overlapping each other (some semblance of a picture can be seen even under these circumstances). As the hole is reduced beyond a certain size, however, the spot size begins to *increase*.

Question 17-17

Why does this happen?

Reducing the size of the pinhole also reduces the amount of light reaching the film. If you reduce the diameter by a factor of 2, you reduce the amount of light by a factor of 4. For a suitable image, the exposure time would have to increase by a factor of 4. For practical photography, there must be some way to get more light onto the film without increasing the spot size.

The solution to our problem is to replace the pinhole with a lens. Every *distant* light source can be focused on the film in the principal focal plane. Furthermore, all of the light from each distant point that strikes the lens is focused to a particular point. We have greatly increased the size of the "pinhole," but we have maintained a small spot size in the image. Naturally, there are some drawbacks to the scheme. In the first place, only those objects with the same specific distance from the lens will form focused images in the image plane. If the film is placed in the plane of the principal focus, then the geometry is correct for all objects that are essentially at infinite distance. This requirement may not be too strict. For example, in a typical inexpensive camera, the focal length of the lens, and hence the distance between the lens and the film, may be only 4 or 5 cm. Almost any object at a distance greater than a couple of meters will be almost in focus as you can see from the basic lens formula.

Question 17-18

For a lens with a focal length of 5 cm, where is the image if the object is at a distance of 200 cm?

The larger the diameter of a camera lens, the more light it can gather and focus on each point of the film. However, as we have seen, a large-diameter simple lens produces many aberrations, particularly with rays far away from the axis of the lens. A good camera lens like the one shown on page 489 is a complex combination of a number of simple lenses. Even when the aberrations are corrected, however, the larger the diameter of the lens, the more critical must be the focusing. Remember that for a pinhole, objects at any distance form good images on the film. As the lens size increases, the rays from the outer edges arrive at the focal point at a larger angle. Unless the film is right at the proper image plane, the image will be blurred.

For any particular lens diameter and focal length, the "depth of field" is given by the formula shown in the margin. The formula provides only a rule-of-thumb guide, since there is no definite criterion for the required sharpness of an image point. Notice that if you want a large depth of field so that many objects are in focus, you should have a small-diameter lens focused on objects that are far away from the camera. If, on the other hand, you use a lens with a very large diameter and focus on an object very close to the camera, then the depth of field is very short. Only objects within a small range will be in focus.

The amount of light reaching a point on the film is proportional to the area of the lens opening and inversely proportional to the square of the distance from lens to film. The latter factor is just a result of the inverse square law. To a first approximation, the distance between lens and film is just the focal length of the lens. A small correction would be needed if the lens were focused on a nearby object. To this approximation, the amount of light reaching a point on the film is proportional to d^2/f^2, where d is the diameter of the lens and f is the focal length. The f-number of a lens is the square root of the reciprocal of this quantity: f-number $= f/d$. For instance, an $f/2$, 50 mm lens has a focal length of 50 mm and a diameter of 25 mm. A camera lens can be "stopped down" by closing a diaphragm composed of metal leaves that progressively close off the outer sections of the lens. If our lens is "stopped down" to $f/4$, then the diameter has been reduced to 12.5 mm. Reducing the diameter by a factor of 2 increases the f-number by 2 and decreases the amount of light entering the camera by a factor of 4. Consequently, for the same exposure on the film, an increase of the f-number by a factor of 2 requires an increase of exposure time by a factor of 4.

Depth of Field

Lens, with f-number, N, and focal length, f, is focused on object at distance, X. A point object between $\mathcal{N}$ and $\mathcal{N}'$ will produce an image on the film, with diameter smaller than d, if:

$$\frac{1}{\mathcal{N}} - \frac{1}{X} = \frac{1}{X} - \frac{1}{\mathcal{N}'} = \frac{dN}{f^2}$$

For example:
$d = 0.005$ cm (common criterion for 35-mm camera)
$f = 5.0$ cm (the most common lens for 35-mm camera)
$X = 400$ cm

For $N = f/2$ $\quad \frac{400 - \mathcal{N}}{\mathcal{N} \cdot 400} = \frac{\mathcal{N}' - 400}{\mathcal{N}' \cdot 400} = \frac{0.005(2)}{25}$

$\mathcal{N} = 345$ cm $\quad \mathcal{N}' = 476$ cm

For $N = f/8$ $\quad \mathcal{N} = 244$ cm $\quad \mathcal{N}' = 1110$ cm

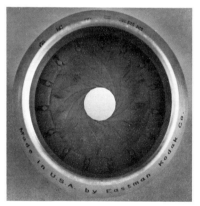

Lens diaphragm (Grant Heilman).

Question 17-19

The labeled f-stop numbers on a particular expensive camera lens are 1.8, 2.8, 4, 5.6, 8, 11, and 16. If you take a sequence of pictures of a stationary scene with one picture at each of these f-numbers, what should you do about exposure times if you want all of the pictures exposed the same amount? If you are successful in producing the same exposure in each picture, how could you tell which picture was taken with $f/1.8$ and which picture with $f/16$?

The Eye

In some respects the eye and the camera are very similar. There is a lens in front and a light-sensitive material in the focal region. In the case of the eye, however, both lens and detector are very sophisticated.

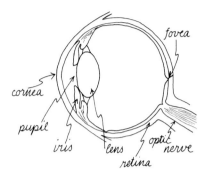

With a camera, focus is achieved by moving the lens forward or back with respect to the film. With the eye, the lens position is fixed, but the power of the lens can be changed by changing its curvature. The muscles holding the lens can stretch it, making it thinner, or compress it, making it bulge more. The lens actually provides only a small part of the convergence of the system. The outer transparent membrane, the cornea, is curved and encloses a liquid called the aqueous humor. Both this liquid and the liquid behind the lens have an index of refraction about the same as that of water: 1.34. Therefore, a considerable amount of the total refraction occurs at the outer surface where the rays first go from air into the cornea. The lens itself has an index of refraction of about 1.44. Changes in its curvature create small adjustments in the overall convergence of the system.

Like the camera lens, the lens of the eye can be stopped down by a diaphragm. In the case of the eye, the diaphragm is called the *iris*. This is the brown, blue, gray, or green disk in the eye. The iris can increase or decrease the circular opening at its center, which is called the *pupil*. In bright light the pupil has a diameter a little less than 2 mm, while in very low illumination the pupil diameter may increase to slightly greater than 8 mm. The feedback system that makes these changes takes several seconds to operate and even longer times with older people. The variable pupil size can compensate for a change in light intensity of only about a factor of 20. However, the entire eye can operate to some extent with light intensities that vary by a factor of one million. Most of this accommodation takes place in the "film"—the nerve cells in the *retina*.

There are two main kinds of cellular detectors in the retina. Because of their shapes, these are called rods and cones. They are not evenly dispersed across the retina. The rods dominate toward the outside, while the cones are concentrated near the center. In one particular small region called the *fovea,* which is less than 1 mm in diameter, there are only cones packed tightly together. When we peer at something to examine it in detail, we are focusing the image on the fovea.

The rods and cones have different wavelength response. The cones are sensitive to red light while the rods are not. The cones require a medium to high intensity of light while the rods remain sensitive to very low levels of intensity. When there is not much light, we lose much of our color vision. Since the rods are more effective in dim light than the cones, we can sometimes see objects in the dark better by looking at them out of the corners of our eyes. Although peripheral vision cannot yield a very detailed image, or one with much color, it is surprisingly sensitive to the motion of objects. If a leaf stirs or a small animal moves, we are more apt to see it if it is off to the side rather than straight ahead.

The rods and cones are interconnected and process some of their information before sending it on to the brain. The messages are transformed to electrical signals that travel along the main nerve trunk, which leaves at

OPTICAL INSTRUMENTS

the back of the retina. What happens in the brain at that point is extremely complex and still not well understood.

Since there are no light receptors in the small region where the nerve trunk leaves the retina, there is a blind spot in each eye. You can detect this blind spot by closing your left eye and focusing your right eye on the cross below. Start with the page at a comfortable viewing distance of 25 or 30 cm from your eye and then move your eye slowly toward the page, still focusing on the cross. First you will see the letter B disappear and then reappear as the letter A disappears.

        X   A   B

The eye, including the analyzing system in the brain, is a marvelous optical instrument and under certain conditions is almost completely corrected for aberrations. However, the eye can be subject to all sorts of illnesses and imperfections. The most common defect is *myopia,* or nearsightedness. The lens is too strong for the size of the eyeball, leaving the image formed in front of the retina. This situation can be resolved by placing diverging lenses in front of the eye. The effect is shown schematically in the diagram. The opposite condition is called *hypermetropia,* or farsightedness. In this case, the image is formed behind the retina. To cure this problem, eyeglasses must be slightly converging, as shown in the bottom part of the diagram. Frequently eyeglasses also have to have a slight cylindrical shape imposed on their spherical surfaces. This shape compensates for astigmatism in the eye, which is usually caused by a nonspherical cornea that makes circles look like ovals. A small prism correction in the eyeglass is also sometimes necessary, to enable the lens in the eye to focus more easily on the fovea.

The focal distance inside the eye is about 1.8 cm. In the terms of camera lenses, its f-number when the pupil is wide open is f-number $= (1.8 \text{ cm})/(0.8 \text{ cm}) \approx f/2$. One other point of interest about the human eye is that it does not work well under water. Since most of the refraction normally takes place at the outer surface of the cornea, the eye loses much of its converging power when the cornea is in contact with water. The index of refraction of water and that of the aqueous humor behind the retina are about the same. Very nearsighted people do better in seeing underwater than do people with normal eye sight, since the loss of convergence at the cornea has the same effect as diverging eyeglasses.

Collimators and Simple Telescopes

In many instruments, light from a source must be "collimated" or made nearly parallel, then sent through some analyzer, and finally refocused onto a film or a screen. In the spectroscope, for example, there should be a line source either formed by the light itself or produced by a thin adjustable slit. As you can see in the diagram, this slit is in the principal focal plane of a converging lens system. The rays from each point of the slit emerge from the lens nearly parallel to each other and then are deflected by the prism or grating through an angle which depends on the wavelength. The rays enter the simplest kind of telescope, which consists of a converging lens. All of the rays parallel to each other come to a focus at a point on an image of the original line source. Since each wavelength is deflected by a different angle, an array of line images is formed in the focal plane.

The Projector

A projector—whether it be for slides, movies, or overhead transparencies—requires an optical system that concentrates light on a small transparency and then spreads it out to come to a focus as an enlarged image on a distant screen. One might think at first that the transparency should simply be placed just outside the focal plane of the projection lens. If the object were self-luminous, such a system would work well. The problem, however, is to get as much light as possible from a filament in a light bulb through the transparency and then through the projection lens. The standard scheme is shown in the diagram below. The filament, which has considerable area, is focused by the condensing lens onto the projection lens. The mirror behind the filament reflects the light going backwards back through the filament, in effect doubling its intensity. The role of the condensing lens is to collect the light coming from as large a solid angle as possible and to focus this light onto the projection lens. Following the condenser lens, the transparency is placed at (or just beyond) the focal length of the projection lens. The light coming through any point of the transparency, even though it comes from many angles from all points of the filament, will be focused by the projection lens onto a single point on the screen.

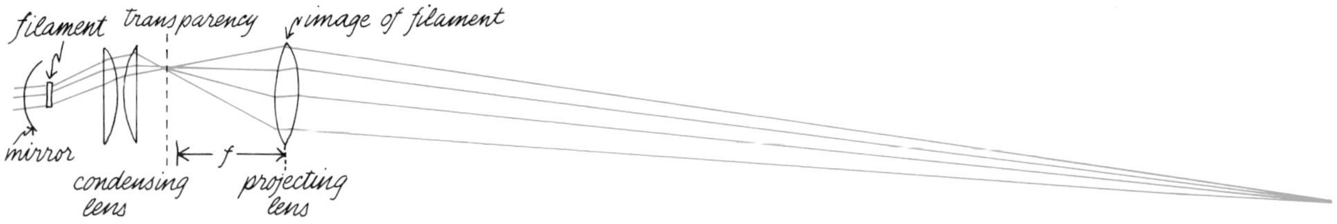

Every part of the filament illuminates each point on the transparency. It is as if that point were luminous with light leaving at many angles. All the rays passing through that point that are intercepted by the projection lens are focused to the same point on the screen.

The Compound Microscope

The schematic of a compound microscope is shown in the diagram. Each of the simple lenses shown in the first diagram becomes a complex lens containing many elements in order to reduce aberrations. Furthermore, the real instrument must have a sophisticated method of illuminating the object with an optical condensing system.

The optical system of a microscope is determined by two main factors that, as we will see, are quite different from those of the telescope. First, a microscope may be placed almost on top of the object. Second, the size

OPTICAL INSTRUMENTS

of the microscope is usually constrained by the requirement that it be convenient to use by a human seated at an ordinary table. Therefore, the eyepiece can only be about 30 cm from the object.

In the compound microscope, the objective or first lens produces a real image in the barrel not far from the eye of the observer. That image, which is already magnified, could be observed by placing a thin translucent screen in the image plane and looking at it with your eye. For comfortable viewing, however, your eye would have to be the standard 25 cm further back. Obviously the thing to do is to examine this real image with a magnifying glass. In an expensive microscope, the magnifying glass turns into an eyepiece, usually made up of several lenses. Its function, however, is to allow the eye to get as close as possible to the real image. Of course, it is not necessary to project the real image onto a translucent screen.

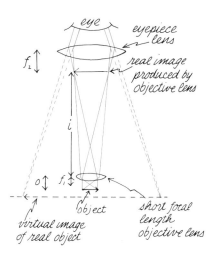

The magnification provided by the objective lens is equal to i/o. The image distance i is limited by our second constraint to a length of about 25 cm. Since the objective lens can be brought very close to the object being studied, the object distance can be just slightly larger than the focal length of the lens. Consequently, the smaller the focal length of the objective lens, the greater the magnification. Since $o \approx f$, the magnification of the first stage of the compound microscope is $m_1 = (25 \text{ cm})/f_1$. The magnification of the eyepiece is just the magnification produced by a magnifying glass; we derived the formula on page 487. If the virtual image is focused at infinity, which produces the most comfortable viewing conditions, the magnification of the eyepiece is $m_2 = (25 \text{ cm})/f_2$.

The overall magnification of the compound microscope is $M = m_1 m_2 = (25 \text{ cm})^2/f_1 f_2$. The magnification, m_2, of a typical eyepiece may be $10\times$ or $20\times$. The focal length of the objective lens may be made very small. Nevertheless, overall magnification for microscopes using visible light cannot get greater than about 1500.

Question 17-20

Why not? Remember that your unaided eye looking into the eyepiece can easily distinguish details in the image as small as 1 mm. With a magnification of 1500, how large would such a detail be in the actual object? Why couldn't you take a picture of such a detail and then magnify it again by a factor of 10 or 100 or 1000?

The Telescope

There are two main factors that determine the construction of a telescope. These factors are just the reverse of those for a microscope. With a telescope the object is far away, practically at infinity. Second, the size of the telescope is not necessarily limited to a length of 25 cm or thereabouts, although in the case of binoculars or opera glasses the physical size must be small. Once again, we show a schematic drawing of a simple system with two converging lenses. A picture of an actual telescope of this type is also shown.

Refractor telescope (Yerkes Observatory).

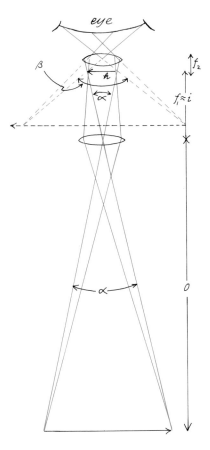

As we saw on page 485, an image of a distant object, produced by a single lens, is located almost in the focal plane of the lens. The angular size of the image is the same as the angular size of the object, as seen from the lens. Therefore, the linear height of the image is proportional to its distance from the lens, which is the focal length. The *linear* magnification is less than 1: $m_1 = i/o \approx f_1/o$. However, as far as the eye is concerned, the real image formed by the objective lens appears to be larger than the object itself. That is because the eye sees the real image from a small distance, but sees the object itself at a remote distance. As you can see in the diagram of the telescope, the *angular* size of the *object* as seen by the telescope or the observer is α; the *angular* size of the *image* seen by the observer is β. The *angular* magnification therefore is $m_1 = \beta/\alpha = (h/f_2)/(h/f_1) = f_1/f_2$. (The linear size of the real image formed by the objective lens is h.)

If the observer looks at the real image with the naked eye, the angular magnification will be $f_1/(25 \text{ cm})$. The obvious way to improve the magnification is to look at the real image with a magnifying glass or, better yet, a good eyepiece. As we have already seen, the magnification of such an eyepiece is $m_2 = (25 \text{ cm})/f_2$. Since f_2 can be made very small (down to 1 or 2 cm), the overall magnification of a compound telescope can be made very large: $M = f_1/f_2$.

Question 17-21

Suppose that you have two lenses with focal lengths of 2 cm, two with focal lengths of 10 cm, and two with focal lengths of 100 cm. If you can use any combination you want, which two lenses would you choose for a microscope and which two for a telescope?

Both the compound microscope and astronomical telescope produce a virtual image of a real image. The resulting image is upside down. Such an inversion makes little difference in most microscope applications or in astronomical observations with a telescope. It would be confusing, however, to have the images upside down in opera glasses. One way to invert the image in a telescope is simply to add another converging lens in the middle. Such an addition makes a telescope unwieldy. Some telescopes are made of nested tubes that can be drawn out in use to provide the necessary length. (The word "telescope" is sometimes used as a verb to describe the action of sliding one component into another.) In binoculars, the image is inverted by an internal reflection, as shown in the drawing on page 468. This arrangement also slightly increases the focal length of the objective lens and thus the magnification. Furthermore, the offset provided by the internal prisms increases the separation distance of the objective lenses and thus enhances the depth perception of the binocular system.

A slightly different optical system is used in opera glasses or in very inexpensive telescopes. The "Galilean" system shown in the diagram uses a converging lens for the objective, but a diverging lens for the eyepiece. Galileo's first telescope was of this type and had an overall magnification factor of only about 3. With later telescopes, he produced a magnification

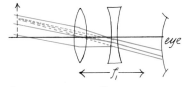

"Galilean telescope."

as large as 30. As you can see from the diagram, the eyepiece intercepts the converging rays and makes them diverge. The eye interprets the diverging rays as coming from the virtual image. Of course, this image is right side up and provides a magnified scene compared with what the naked eye would see. Opera glasses made in this way have a magnification of only 3 or so.

Binoculars or small telescopes are usually described with a double number such as 7×30. The first number refers to the linear magnification (the ratio of height of the image observed with the telescope to the height of the object observed with the naked eye). The second number gives the diameter of the objective lens in millimeters. We will come back to this point in just a moment.

Question 17-22

Most binoculars have magnifications between 5 and 10. Why is this? Consider the practical restrictions on the size of binoculars and then consider the formula for the magnification of a telescope.

From what we have said so far, it might seem that the way to make a better telescope would be to make the focal length of the objective lens as long as possible. Indeed, astronomical telescopes do have long focal lengths, but not enormous ones. There is another factor that is more important than magnification. What is needed is an objective lens or mirror with a large area, in order to gather as much light as possible. The more light, the shorter the exposure time. This is also the reason why binoculars are described in terms of both the magnification and the diameter of objective lens. Remember that the amount of light reaching a point on the focal plane of an objective lens is proportional to the square of the diameter of the lens and inversely proportional to the square of the focal length. The ratio f/d is the f-number of the lens. The smaller the f-number, the more effective it is in putting light on the film. The largest lens ever used for the objective of a telescope is the 40 in. diameter lens of the Yerkes Observatory in Wisconsin. The focal length is 63 ft, yielding an f-number of $(63 \text{ ft})/(3.3 \text{ ft}) = f/18.9$. At Mount Palomar in California, there is one of the largest telescopes ever made for visible light. Its parabolic mirror has a diameter of 200 in., but the focal length is only 55.5 ft, smaller than that of the Yerkes telescope. The f-number of the Mount Palomar reflector is $f/3.3$. Besides providing greater light-gathering power, a large objective lens or mirror provides another vital requirement for astronomy.

The 200 inch reflecting telescope at Mt. Palomar. (Hale Observatories)

Question 17-23

We said that, regardless of magnification, most stars remain point objects and produce point images. But are the images only points? What effect does lens diameter have on the image size?

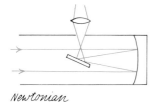
Newtonian

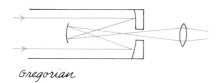
Gregorian

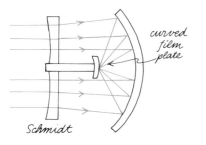
Schmidt

As we have already seen, the objective of a telescope can be a concave mirror instead of a convex lens. It is very hard and expensive to produce very large lenses. The 40 in. objective lens of the Yerkes Observatory is the largest ever made, and that was accomplished in the 1890s. The main problem is to produce a large volume of optical-grade glass that is free of strains and bubbles. Even after a large lens is successfully poured and ground, there are still problems in holding its considerable weight by the edges without producing distortions. It is much easier to pour a large volume of glass out of which a mirror will be made. Only the surface need be free of imperfections, and the mirror may be mechanically supported from underneath its entire surface.

In this diagram, we show schematic diagrams of two different geometries for reflecting telescopes. Another type of telescope, which is a combination of reflector and refractor, is also illustrated. It is called a Schmidt system. The curvature of the front element is exaggerated in this diagram. The extra refraction compensates for the spherical aberration in the mirror, providing a much larger field of view for the telescope. Images at a considerable distance off the axis remain in focus.

SUMMARY

If the wavelength is small compared with any obstructions or openings, then you can approximate the behavior of waves by "rays," perpendicular to the wavefronts. To determine the behavior of rays, measure the angles between the rays and the perpendicular to the boundary where they meet. For reflection, the angle of incidence equals the angle of reflection. Reflection in a plane mirror produces a virtual image behind the mirror. For refraction, $n_1 \sin \theta_1 = n_2 \sin \theta_2$, where $n = c/v$, the ratio of the velocity of light in a vacuum to the velocity in the medium. Since refraction depends on the wave velocity in the medium, and that velocity depends on frequency in most materials, different frequencies refract at different angles in a medium. In all transparent materials, blue light has a larger index of refraction than red light. This *dispersion* gives rise to the rainbow.

We presented a crude atomic model to explain the interaction of light with matter. Light passing through material sets internal electrons into oscillation. They, in turn, radiate electromagnetic waves, which combine with the original wave. In general, the combination wave is continuously delayed inside transparent material, producing a lower velocity.

Another guiding rule for reflection and refraction is Fermat's principle. The time taken for light to travel from one point to another is an extremum, compared with the time taken for any slightly different path. If there exist many paths for light to start out in phase, to travel equal times, and to arrive at a second point in phase, then the second point is in focus.

For a spherical mirror, and for paraxial rays,

$$\frac{1}{o} + \frac{1}{i} = \frac{1}{f} = \frac{2}{r}$$

There are three principal rays that are easy to trace when finding the position of an image in a spherical mirror: the ray parallel to the axis, which goes through the principal focus; the ray through the principal focus, which leaves the mirror parallel to the axis; and the ray striking the center of the mirror, which reflects at an equal angle from the axis. For a concave mirror, if $o > 2f = r$, then $f < i < 2f$. If $o = 2f$, $i = 2f$. If $f < o < 2f$, then $2f < i < \infty$. If $o < f$, then i is virtual. For a convex mirror, the image is virtual.

For lenses and paraxial rays, $\frac{1}{i} + \frac{1}{o} = \frac{1}{f}$. The focal length and radii of a thin lens in air are governed by the lensmaker's formula:

$$\frac{1}{f} = (n-1)\left(\frac{1}{r_1} - \frac{1}{r_2}\right)$$

The geometrical relationships among o, f, and i are the same as for a spherical mirror. To sketch image position, use principal rays (the ray parallel to the axis goes to principal focus; the ray through the center lens is undeflected; the ray through the principal focus leaves the lens parallel to the axis). The rules of light reflection and light refraction govern the construction and operation of many instruments. We examined cameras, the human eye, simple telescopes, projectors, and compound microscopes and telescopes. In many of these instruments, a convex objective lens forms a real image. You can then examine this image at close range through an eyepiece, which acts like a magnifying glass.

Answers to Questions

7-1: The two angles are the same. Note that the ray is perpendicular to the wave front, and the normal is perpendicular to the surface.

7-2: If angle i = angle r, then $i' = r$ and angle QPO = angle QPI. Since these two angles are equal and adjacent to QP, which is common to both right triangles, then the triangles are equal to each other and $QI = QO$.

7-3: The wire mesh would not be a good reflector for sound waves because air vibrations can pass easily through the open spaces. There is not much impedance change in passing through the wire mesh. The concrete wall would reflect microwaves but not well. Concrete is not a good conductor of electricity. Note that portable radios can be heard inside concrete buildings.

7-4:
$$\frac{\sin \theta_1}{\sin \theta_2} = \frac{1}{3/4} = \frac{4}{3}$$

If $\sin \theta_1 = 0.5$, $\sin \theta_2 = \frac{3}{8}$ and $\theta_2 = 22°$. Long before Snell proved that the refraction relationship concerned the ratio of sines, the Greeks knew that refraction approximately followed the law:

$$\frac{\theta_1}{\theta_2} = \text{constant} \qquad \text{(which for air-water is about } \tfrac{4}{3}\text{)}$$

For $\theta_1 = 30°$, this simpler relationship gives

$$\frac{30°}{\theta_2} = \frac{1}{3/4} = \frac{4}{3}$$

$$\theta_2 = 90/4 = 22.5°$$

As you can see, for small-to-medium angles of incidence, it would be hard experimentally to tell the difference between the two relationships.

17-5: No, although at Brewster's angle the *reflected* ray does not contain electric vibrations with a component perpendicular to the surface, the *refracted* or *transmitted* ray is only partially polarized.

17-6: $\delta_{\text{violet}} - \delta_{\text{red}} = (1.4550)60° - (1.4090)60° = 0.76° = 4.84 \times 10^{-2}$ rad. With this angular dispersion, the spectrum at 1 m would have a width: $r\theta = (1 \text{ m})(4.84 \times 10^{-2} \text{ rad}) = 4.8$ cm.

17-7: It turns out that this crude model works very well as long as we do not demand details of the absorption of the electromagnetic radiation.

17-8: No, an ellipse has the property that at any point the radii to the two foci form equal angles to the normal of the curve.

17-9: The circle, the ellipse, and the parabola are all members of the same family of curves—called the conic sections. If the two focal points of an ellipse come together, the ellipse becomes a circle. If one of the focal points moves away to infinity, the curve becomes a parabola. In that case, the light coming from the distant focus would be parallel and would go to the remaining focus after reflecting from the parabola.

17-10: Obviously, the telescopic image of the moon is much smaller than the moon itself, regardless of the size of the telescope. As far as our view of the moon, however, there is useful magnification. We can get our eyes very close to the telescopic image, especially if we use a magnifying glass or eyepiece. The angle of the image subtended by our eyes is then larger than the angle of the moon itself subtended by our eyes. The effective magnification is the ratio of these two angles.

17-11: The bathroom mirror has a large radius of curvature. Usually when we use it, our face is inside the focal length: $o < f$. Therefore, we see an erect virtual image. The bowl of a spoon has a very small radius of curvature. When we look into it, our face is usually outside the focal length. Therefore, we see an inverted real image.

17-12: The ray going through the top of the lens has to travel further in order to get to the focal point, but it travels through only a small section of glass where the velocity is low. The ray along the axis has the shortest distance to travel, but the longest stretch through the glass where the velocity is low. If the lens is the right shape, the *time* of travel for all of the rays to get to the focal point can be the same, even though their path *lengths* are different.

17-13: For a plano-convex lens, $r_1 = \infty$. Therefore, according to our general formula: $1/f = (n - 1)(-1/r_2)$. Except for the minus sign in front of r_2, this formula agrees with the one that we derived for the plano-convex lens. According to the sign convention which is described in the next section, we consider r_2 negative, since its center of curvature is to the left of the lens, in the direction from which the light is coming. With this additional interpretation, the two equations agree.

17-14: Unless you are very nearsighted, you cannot see an object that is only 1 cm away from your eye. Your eye lens cannot focus on such a close object. The reading distance for the normal eye is about 25 cm.

17-15: If you use a magnifying glass with a focal length of 5 cm, hold it close to your eye and bring up the object until the image is at infinity. Your magnification is equal to $(25 \text{ cm})/(5 \text{ cm}) = 5$. If you hold the object so that the image is at the viewing distance of 25 cm, then the magnification is equal to $1 + (25 \text{ cm})/(5 \text{ cm}) = 6$. As a practical matter, you can hardly tell the difference between these two positions.

17-16: Let's solve for the image distance using our standard formula, but remember that the focal length is negative.

$$\frac{1}{o} + \frac{1}{i} = -\frac{1}{|f|}$$

SUMMARY

$$i = -\frac{o|f|}{o+|f|}$$

As long as the object distance, o, is positive, the image distance is necessarily negative and the image is virtual.

-17: The actual spot at each point is not a circle, but rather a diffraction image with an angular radius of $1.2(\lambda/d)$, where d is the diameter of the pinhole. As d decreases, the angular size of the image increases. If the distance between the pinhole and the film is L, then the actual radius of the diffraction image is equal to $1.2(\lambda L/d)$. The size of the diffraction image can become larger than the size of the pinhole itself.

-18:
$$\frac{1}{200} + \frac{1}{i} = \frac{1}{5}$$
$$i = 5.13 \text{ cm}$$

The image of this near object is only 1.3 mm from the main focal plane. Whether or not the image would appear to be out of focus depends on a number of other conditions.

-19: The amount of light reaching the film is approximately proportional to the inverse square of the f-number. The squares of these numbers are given below:

1.8	2.8	4	5.6	8	11	16
3.24	7.84	16	31.4	64	121	256

With the exception of the first pair of numbers, each subsequent number cuts the amount of light arriving at the film by about a factor of $\frac{1}{2}$. To get pictures with the same exposure, as you increase the f-number by one stop, you should increase the exposure time by a factor of two. For instance, if the proper exposure time at $f/4$ is 1/50 s, then the comparable exposure time at $f/5.6$ is 1/25 s. The picture taken at $f/1.8$ would have a very small depth of field. One region would be in focus, and objects closer or further away would be out of focus. At $f/16$, most of the objects would be in focus under ordinary conditions, except those very close.

-20: If an image as seen in the eyepiece has a length of 1 mm and the overall magnification is 1500, then the object itself must have a size of $(1 \times 10^{-3} \text{ m})/1500 = 6 \times 10^{-7}$ m. But that's the wavelength of light! We cannot probe a detail smaller than the size of our probe. The wavelength of our probe provides the ultimate limit to the magnification that we can obtain using it. Even if you enlarge the photograph, the details will be only diffraction patterns.

-21: For maximum magnification in a microscope, both objective lens and eyepiece lens should have small focal lengths. Therefore, for the microscope, the two 2 cm lenses should be used for both objective and eyepiece. For maximum magnification with a telescope, the objective lens should have a long focal length. Therefore, the 100 cm lens should be used. For the greatest magnification with the eyepiece, the smallest focal length lens should once again be used, namely the 2 cm.

-22: The magnification of a compound telescope is: $M = f_1/f_2$. It is hard to make a good eyepiece with a focal length of much less than about an inch. Consequently, if $f_2 = 2.5$ cm and $f_1 = 25$ cm, the magnification of the telescope would be 10. Therefore, the telescope would have to be at least 28 cm long, which is about a foot. With $\times 10$ binoculars, the length is actually shorter because the light path is folded somewhat as the light goes through the prisms. (Another reason for not making hand-held telescopes with magnifications greater than 10 is that there is too much jitter to take advantage of the amplification.)

-23: The image of a point star is actually a set of diffraction rings. The angular radius of the central maximum to the first minimum ring is: $1.2(\lambda/d)$, where d is the diameter of the circular opening, in this case the lens or mirror of the telescope. The larger the mirror or lens, the smaller the diffraction image and, consequently, the greater the resolving power of the instrument.

Problems

1. (a) You are 3 m in front of a plane mirror and want to take a picture of yourself in the mirror. For what distance do you set the focus adjustment on the camera? (b) You throw a ball at a mirror at a speed of 5 m/s. What is the speed of the image of the ball rushing toward you? What is the speed of the image of the ball as seen in the reference frame of the real ball?
2. Traffic radar must use waves that travel and reflect in beams. Ideally, ray optics would govern the operation rather than wave mechanics. What are the limits on wavelength and frequencies?
3. What is Brewster's angle for light going from air into water?
4. A light ray going from air into glass with an index of refraction of 1.5 has an incident angle of 60°. What is the angle of refraction?
5. A light ray going from glass (with an index of refraction of 1.5) into air has an incident angle of 40°. What is the angle of refraction?
6. What is the critical angle for light going from glass ($n = 1.50$) to water ($n = 1.33$)?
7. What are the angles of minimum deviation, δ, for blue light and for red light in a 60° prism made of water?
8. Light passes from air into water at incident angles of 10° and 60°. Compare the angle of refraction with those which you would obtain if Snell's law were simply $i/r = n$.
9. Use ruler and protractor to make a scale drawing that shows that the image in a plane mirror is in line with the object and at an equal distance behind the mirror. The rays must satisfy the reflection law that $i = r$.
10. Describe in words and a diagram the sky conditions necessary to see a rainbow. Show the relationship that exists among the sun, where you are standing, and where you see the rainbow. How high up are the sun and the bow? What is the arrangement of colors?
11. Describe in words why, according to our atomic model, blue light travels slower in glass than red light.
12. Draw a scale diagram of a light ray going from one point to another by way of a reflection in a plane mirror. Make sure that $i = r$. Then draw two other routes from the first point to the second point by way of the mirror. Measure these other routes and show that they are longer than the correct one.
13. Draw a scale diagram of a light ray going from a point in medium 1 to a point in medium 2, ($n_1 = 2n_2$), with an incident angle of 60°. Then draw two other routes from point 1 to point 2. Measure these paths and show that the time that light would take in either of these other routes is longer than the time it takes following Snell's law.
14. For a lens with $f = 20$ cm, find object distances for which the image would be twice as large as the object, the same size, and half as large (in linear dimension). Diagram these cases, using principal rays, and also solve algebraically.
15. Draw the principal rays coming from an object that is 15 cm from a mirror that has a focal length of 10 cm. Calculate the image distance. What is the radius of curvature of the mirror?
16. Draw the principal rays coming from an object that is 8 cm from a mirror that has a radius of curvature of 20 cm. Calculate the image distance. In the ray diagram, show how and where the eye would see the image.

SUMMARY

17. Opticians measure the power of lenses in units called *diopters*. The power in diopters is the inverse of the focal length in meters. Hence, a converging lens with a focal length of 0.50 m is a +2 diopter lens. For a symmetric double-convex lens, made out of glass with $n = 1.5$, what is the radius of curvature if the lens is +5 diopters?
18. A camera can get proper exposure for a particular scene with the lens set at $f/16$ and an exposure time of 2 s. If the lens is opened up to $f/2$, what should the exposure time be? What will be the effect on the depth of field?
19. What is the f-number of the human eye in bright light when the pupil is only 2 mm in diameter?
20. Design and draw to scale the principal rays for a two-lens microscope with magnification of 100.
21. Design and draw to scale the principal rays for a two-lens (both converging) telescope with a magnification of 100.
22. Draw the ray diagram for the objective lens of a telescope that is focused on two distant stars.
23. Draw the ray diagram for a magnifying glass that gives a magnification of 5.
24. A mirror composed of two mirrors at right angles to each other has special properties. Draw a scale diagram of such a mirror, and the rays coming from an object in the shape of an arrow. Trace the rays that go to one mirror and then to the other. Where is the image? Is it reversed left for right, like images in an ordinary mirror?
25. A diverging lens with a virtual focal length of 6 cm is placed 9 cm from an object. Draw the principal rays to scale, and find the image.

18 Electrostatics

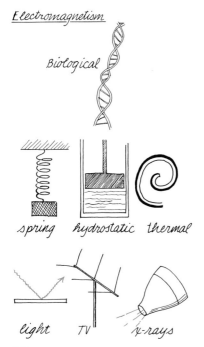

In the first chapter we claimed that in some ways the universe is very simple. On the atomic and nuclear level there are only a few classes of particles, subject to a small number of laws of behavior, and interacting with each other in only four ways. The only physical interactions that have been discovered are: gravitation, electromagnetism, strong nuclear, and weak nuclear. We have already explored some of the properties of the gravitational interaction. In Chapter 23, we will see what is known and not known about strong and weak interactions. Now, and in the next few chapters, let's look at the electromagnetic effects.

Note first that we are combining electricity and magnetism into one phenomenon. That's the way nature is. Although we will start out by studying electric effects alone, we will later find that magnetism can be produced by electric currents. It is not just a case of one phenomenon being produced by another. As we will see eventually, we can transform from magnetic effects to electric effects and vice versa by changing our reference frame. One person's magnetic field is another person's electric field. Electromagnetic forces depend on the relative velocity of source and observer.

The electromagnetic interaction is responsible for most of the phenomena of everyday life. Gravity is important to us here on earth, of course, but the effects of the strong and weak nuclear interactions are so hidden within atomic nuclei that we seldom experience any evidence of them. (When we come in contact with radioactivity, we are dealing with both the Strong and the Weak interactions.) The electromagnetic interaction, on the other hand, is responsible for holding the electrons and nuclei together to form atoms, and is also responsible for all the various forms of molecular bonding. Polar bonding, hydrogen bonding, Van der Waals' forces, and covalent bonding are all electromagnetic. Only the geometry and nature of the atoms produce the apparently different effects. But if molecular forces are electromagnetic, then almost certainly all of the biological phenomena are controlled by electromagnetism. That means you and me. The interactions within biological cells as well as the signals they send to each other are electromagnetic. The forces in everyday life are electromagnetic—whether we classify them as muscular, air pressure, hydrostatic, spring, or thermo-expansion. Furthermore, electromagnetic energy can be radiated. We are familiar with it in the form of light, radiant heat, X-rays, radio, and television. Even sound can be analyzed in terms of electromagnetism since pressure waves in air or solids are caused by density variations between atoms or molecules and those atomic and molecular forces are electromagnetic in nature.

Clearly, we cannot tackle all these phenomena at once. Instead we will divide the subject of electromagnetism into three realms. In the first, the electric charges that are the source of the interaction will be at rest with respect to each other and to us, the observers. In Chapter 19, we will study the second realm and learn how to deal with electric charges that are moving with constant average speed in electrical circuitry. This is the subject that we normally think of as "electricity." In Chapter 20, we will see another aspect of electric charges moving with constant speed. In Chapters 21 and 22, we will learn what happens when the electric charges are moving, but not with constant velocity. In this third realm of accelerated charges, energy is radiated into space.

First, let's learn how to produce these electric charges and see what effect they have on each other when they are standing still. This subject is called *electrostatics*.

HANDLING THE PHENOMENA

In the case of electrostatics, it's hard *not* to handle the phenomena. The situation is shocking, particularly in dry weather. If you pull off a sweater, throw back the bed blankets, or walk across a rug, you become a minor-league Thor. Tiny streaks of lightning flash and the thunder crackles. On a really dry day with a thick rug you can be a hazard to yourself and friends.

Here are some ways to investigate electrostatic effects on a more organized basis.

1. Blow up a small party balloon and you will have an excellent source of negative electric charge. Rub the balloon on wool or fur, or best of all, your own hair, and you will find that the balloon will then stick to the thing that it was rubbed on or to almost anything else, including the nearest wall. In this particular case, as in all others that we will describe, it is not necessary to rub the material vigorously or more than once at a time. It is contact between two different materials that produces the separated electric charge. Rubbing two objects together merely increases the area of contact.

2. Blow up another balloon, preferably the same as the first, and suspend each with a thread to a common point on the ceiling. To begin with, the two balloons will just hang together. But now rub each of them with wool or your own hair and see what happens.

3. Cut several small pieces of paper and aluminum foil. Each should have an area of less than one square centimeter, but make some of them long and narrow. Run a comb through your hair and then pass the comb near the pieces of paper and foil.

4. Wrap a tiny piece of aluminum foil around the end of a nylon or polyester thread. Smush the foil into a little ball and then hang it by the thread so that it is some distance away from any object. Rub a balloon or

your comb and bring it near the aluminum foil ball. What happens next is complicated. Watch the whole sequence. Repeat it, but first touch the aluminum ball with your fingers.

5. To follow some of the arguments we will make later in the chapter, you must know the formulas for the surface area of a cylinder and of a sphere. The curved surface of a cylinder has an area equal to $2\pi rh$ where r is the radius of the cylinder and h is its height. The surface area of a sphere equals $4\pi r^2$. Note that in both cases we have the dimensions of a length squared, a necessary condition for the formula for an area.

These relationships are so important that you ought to experience them with your hands. Take two small cylinders, one having a radius about twice that of the other, and wrap paper around them. Measure the area of the paper corresponding to the area of the curved part of the cylinder. In particular, compare the ratio of areas for the two cylinders with the ratio of their radii. Try the same thing for two small balls such as a golf ball and a tennis ball. Of course, in this case it's harder to make a flat sheet of paper take a spherically curved shape. Try covering only a hemisphere in each case, making folds or cuts in the paper until it is approximately the same as the spherical surface. Once again, see if the area of the paper that you use approximately equals that required by the formula. In particular, compare the ratios of the areas of the two pieces of paper with the ratio of radii of the two balls.

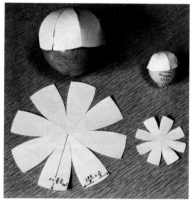

The croquet ball on the left has twice the diameter of the ping-pong ball on the right.

THE ISOLATION OF ELECTRIC CHARGE

In Chapter 9 we noted that in the first edition of the *Encyclopedia Brittanica*, back in 1771, only two sentences are devoted to the topic of energy. Our modern concept of energy did not arise until 80 years later. But the subject of *electricity* occupies many pages, with repeated references to the research of Dr. Benjamin Franklin. According to that first edition, "the word Electricity signifies in general the effects of a very subtle fluid matter different in its properties from every other fluid we are acquainted with. This fluid is capable of uniting with almost every body but unites more readily with some particular bodies than with others; its motion is amazingly quick. It is regulated by peculiar laws and produces a vast variety of singular phenomenon."

Many of our modern words for electricity had already been coined at the time that this article was written. It was known that some materials are good *conductors* for the electrical fluid, and others are *insulators*. Benjamin Franklin had suggested that the many examples of both attraction and repulsion of *charged* objects could be explained in terms of either a surplus (*positive*) or a deficit (*negative*) of the electric fluid. Thus *electrodes* could become either *plus* or *minus*. Franklin described electric charges in terms of plus and minus because two objects that are initially neutral electrically can be rubbed together and each become charged. The charge on one body is apparently different from that on the other because, although the two objects will attract each other, each will repel a similarly charged object. Furthermore, the two objects can be combined so that they are neutral, or zero, again. Calling the charges plus and minus explains the initial and final

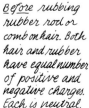

Before rubbing rubber rod or comb on hair. Both hair and rubber have equal number of positive and negative charges. Each is neutral.

After rubbing rubber comb on hair. Hair has net positive charges; rubber has net negative charges.

neutrality and emphasizes how the action of rubbing one material on another merely separates charges that may well have existed together in the first place. Add to this the rule that like charges repel and unlike charges attract and many of the phenomena of electrostatics can be explained—at least qualitatively.

Franklin arbitrarily assigned the name "negative" to the charge that appears on hard rubber when it is rubbed with wool or hair. The wool or hair is therefore left positive. When glass is rubbed with silk, the glass becomes positively charged while the silk is negative.

Question 18-1

Does Franklin's model explain what happened when you brought the comb near the tiny aluminum ball hanging from the nylon thread?

OUR MODERN EXPLANATION OF CHARGE SEPARATION

Nowadays we explain static electricity in terms of the removal of electrons from one material by another. Hard rubber when rubbed with almost any other material will grab electrons; thus, according to the convention started by Franklin over 200 years ago, electrons have a negative electric charge. Since, as we now know, electrons are the charge-carriers in metal wires, there is sometimes confusion as to whether electric current is in the direction of electron movement or in the direction of positive charge motion. The worldwide convention is that *current is in the direction of positive charge flow.* No need to blame Benjamin Franklin, however. In transistors, fluorescent tubes, and electrolytic cells, the charge carriers are as apt to be positive as negative.

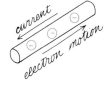

Let's use our modern understanding of the microstructure of materials to explain some common electrostatic phenomena. The outer electrons of atoms are frequently only loosely tied to the parent nucleus. If two atoms come together, the outer electrons may rearrange themselves to bond the two atoms together in a molecule, or one atom may simply steal an electron from the other. Oxygen molecules, for instance, have a tendency to pick up spare electrons forming negatively charged oxygen ions. All materials have this varying power to hang on to electrons whether they are conductors or insulators. However, the effects are most commonly seen with good insulators such as rubber or hair or plastic. The reason for this is that when electrons are added to insulators, they remain fixed in place; if they are taken from a region, it becomes and remains positive. If a piece of metal is rubbed with fur, it too will gain electrons but they spread across the whole surface. If you are holding the metal with your hands, the electrons will spread throughout your body. Unless you are holding the metal with an insulator, any electrical effects will be diluted. However, note that although your body is relatively large and a fairly good conductor of electricity, if you shuffle across a rug you may pick up enough electrons so that your whole body becomes highly charged. When you touch another person or a door knob the charges leap off creating a spark.

THE NATURAL PRODUCTION OF STATIC ELECTRICITY

The separation of electric charges can produce dramatic effects in nature. Almost any material that slides or that is blown past some other object will gain or lose electrons. When this happens with rising or falling droplets of water in clouds, one part of the cloud can become negative with respect to another part, or to the earth. When the concentration of charge in one region is sufficient, part of the charge is driven away, forming a conducting path to ground or to another part of the cloud. Breakdown occurs rapidly, raising the temperature of the path to the point of luminescence, and creating a column of high pressure that radiates outward to be heard as thunder. The main lightning bolt follows a path that has been created by a "leader." The "leader" works its way down to the earth by following the easiest breakdown path, sometimes temporarily ending up in dead ends as you can see in the illustration.

Forked Lightning. The many flashes over Kitt Peak Observatory were photographed with a long time exposure. (Gary Ladd/Kitt Peak Observatory)

Thick wire, without sharp bends, must connect rod to moist earth.

A lightning rod on a building does not prevent lightning from striking, but provides a safe path to earth for any lightning bolt that arrives close to the rod. The wire from rod to earth must be heavy enough so that it does not get hot as the lightning passes through. A ligtning rod provides a cone of protection for everything underneath it out to about 60°. The rod does not discharge the cloud, which is usually far above it. Nor does it make any difference whether the lightning rod ends in a point or a sphere, although a really sharp point might have the effect of creating a small localized cloud of charged air around it, thus increasing its region of protection.

Electrostatic effects can be annoying in industrial processes as well as around the home. Any process that involves moving sheets of material, or moving grains or liquids will produce charge separation. If you throw off wool blankets at night, you may see the whole surface glow. (In some circles, this is known as sheet lightning.) One of the hazards of blowing grain into storage elevators is that the separated charges will produce a spark in an atmosphere filled with the flammable dust. On the other hand, in the making of sandpaper, the electrostatic effect is put to good use. Instead of sifting the abrasive grain onto glued paper, the grain is kept underneath the paper and jumps up into the glue because of an electrostatic attraction that is created in the region. In the process the abrasive grains line themselves up lengthwise producing much sharper sandpaper. Remember the way the scraps of paper jumped up toward your comb.

LOCAL MOVEMENT OR DISTORTION OF CHARGES

When you produce electrostatic effects either in the laboratory or in the games proposed in "Handling the Phenomena," you usually separate the charges with insulators such as a comb or a piece of plastic. If you then touch something else with the insulator you disturb the charges only in the local regions being touched. Actually, at those points small sparks form, clearing small regions of charge around each point. If you want to transfer a lot of charge from the insulator to a metal detector, then you must literally wipe off the charge onto the metal.

While "Handling the Phenomena," you saw evidence of both the attraction of unlike charges and the repulsion of like charges. You also must have observed that either sign of charge attracts a neutral object. For instance, when you charge a rubber balloon you can put it on a neutral wall; the charged comb attracts tiny pieces of conducting foil or pieces of paper, all of which are neutral. In terms of the microstructure of materials, the explanation is that bound electrons in the neutral material shift toward or away from any charged object brought near them. When the negatively charged balloon comes near the wall, it warps electron positions in the molecules of the wall so that they are as far away as they can get from the repulsive negative influence. The electrons do not actually move through the wall unless the wall is a conductor, but the molecular arrangements are warped so that the surface is left with a net charge opposite to that of the outside influence. That situation is of course just what is needed to attract the outside charged object to the wall. The diagrams show how you can use these effects to leave a detector charged with either the same or opposite charge from that of a source.

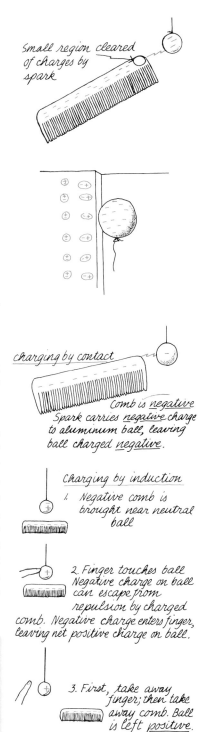

Question 18-2

These electrostatic effects are much more dramatic in the winter than in the summer. Why should this be true? Surely charges separate as well in one season as another.

A RESEARCH APPLICATION OF ELECTROSTATICS

There is an application of electrostatics that combines many of the factors we have been discussing to produce a device that is used both as a toy and as a research instrument. It is the Van de Graaff generator. Here is a schematic sketch of one that can be used as a toy or a school demonstration device. Electric charges are sprayed onto a moving belt and literally carried up to an insulated dome where they accumulate. In the research machines the charges are sprayed onto the belt from needles that are raised to high voltage. Electrons can either be sprayed on or taken off, thus charging the belt and the dome negative or positive. In most of the table-top machines, a rubber belt is driven at the base by an axle that delivers electrons to the inside of the rubber belt as the axle makes rolling contact with the rubber. The inside of the belt is thus made negative but the electrons cannot migrate through the rubber to the outside of the belt. Meanwhile a wire attached to the base of the machine on the outside of the belt, and close to the spinning axle, sees a large *positive* charge on the *axle*. Electrons jump off the wire toward the positive axle, but of course land on the outside of the belt. There they are carried up until they are inside the metal dome; at that point they hop over to a wire connected to the inside of the dome and immediately spread out over the outer surface of the dome. This sequence of actions makes the outside of the dome negative. The electrons are forced into this negative region by the physical motion of the belt.

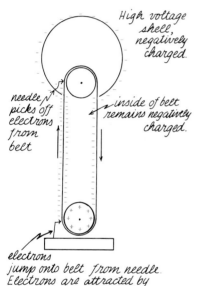

> ### Question 18-3
>
> After the dome has accumulated a lot of electrons, it must be charged highly negative. Why then should the electrons jump off the belt onto the collecting wire inside the dome? If the inside wire is negative why shouldn't the electrons be repelled and stay right on the belt?

The toy Van de Graaffs can raise the voltage of the dome to 20,000 volts or higher, producing sparks several centimeters long. Some of the research Van de Graaffs can produce potentials as high as ten million volts.

QUANTITATIVE ELECTROSTATICS—COULOMB'S LAW

So far we have been talking about the qualitative effects of electrostatics. To probe any deeper into the phenomenon, we must become quantitative. How do we measure the quantity of electric charge? How does the influence between electric charges depend on the distance between them? These questions were answered by several investigators during the last couple of decades of the eighteenth century. The man whose name is associated with the defining law of electrostatics was a Frenchman, Charles Coulomb (1736–1806). He made a very delicate torsion balance, in general form much like the one that Cavendish used to measure the gravitational constant. He had to produce and accurately calibrate a very fine suspension wire.

QUANTITATIVE ELECTROSTATICS—COULOMB'S LAW

Furthermore, the electric charge storage and production system had to be extremely well insulated. Coulomb managed to produce apparatus good enough to allow him to find that the electrostatic force is inversely proportional to the square of the distance between centers of charged spheres. Actually, Cavendish in England had demonstrated the same law to his own satisfaction some years earlier, but never published his results. Most scientists at that time already assumed that the force between charges must depend on the inverse square of the distance between them, since Newton's similar law for gravitation was well established. Coulomb's precision was not great, but was at least sufficient to show that an inverse square force law was plausible.

Coulomb was able to produce equal charges (or known fractions of the original charges) on two different spheres with the following procedure. If one small metal sphere is given an electrostatic charge (of unknown amount), and then is touched to an *identical* metal sphere, the charge must be shared equally between them. If the spheres are then separated, each has half of the original charge. In this way, Coulomb could measure the force between two spheres, each of which had the same charge.

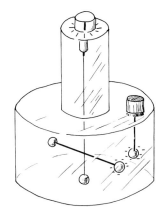

Coulomb's apparatus. The charged ball on the end of the vertical rod remained stationary. The horizontal rod was suspended by a fiber. As the two charged balls repelled each other, the fiber twisted. The knob at the top could twist the rod back. The torque for a given angle was measured separately.

Question 18-4

What procedure could Coulomb then have used to find the force between spheres each of which had half of the test charge he had been using?

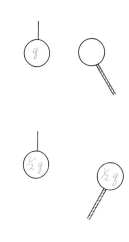

Coulomb found that the force between two charged spheres was proportional to the product of their charges. Calling the proportionality constant k, his law can be written

$$F = k \frac{q_1 q_2}{r^2}$$

The magnitude of the proportionality constant depends on the size of the electric charge chosen as unit charge. Now that we know about electrons and protons and the fact that they have a natural basic electric charge, one negative and the other positive, it might seem reasonable to choose the charge on the electron as our unit charge. That much charge is very small, however, and partly for historical purposes a much larger amount of charge has been chosen as the unit. It is called, appropriately enough, the *coulomb*, with the symbol, C. As we will see in the next chapter, the coulomb is defined in terms of the unit of electric current, since a current is a flow of electric charge. The relation between the two is that when one coulomb of electric charge moves past a point in one second, the electric current is one ampere. A 100-watt bulb, for instance, has an electric current of about 1 ampere. To measure the coulomb you must then know how to measure an ampere. Accept this definition on faith until Chapter 20. With the coulomb so defined, the constant, k, in Coulomb's law is equal to 9×10^9.

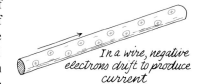

In a wire, negative electrons drift to produce current

1 coulomb/second = 1 ampere

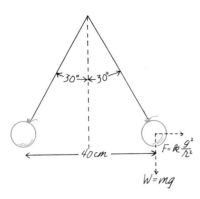

THE MAGNITUDE OF THE CHARGES THAT WE HAVE SEPARATED

Now we can figure out how much charge we separated in some of the experiments described in "Handling the Phenomena." For instance when two balloons are hung by threads from the same point, and then given static charge, they spring apart from each other. For a typical case, how many coulombs are on each balloon? First, we assume that each balloon is the same size and that each has approximately the same amount of negative charge on it, uniformly distributed over the surface. We must also assume that the force between two such spheres depends on the distance between their centers. That assumption will be justified later. The diagram shows a sketch of two balloons hung in this fashion with a separation angle of 60° between them. (The angle that you observed may be smaller or larger but in this calculation we are only after an order-of-magnitude figure for the quantity of electric charge.) For each balloon, its weight is acting down and the electrostatic repulsion is acting horizontally. The two forces form the legs of a 30°–60° right triangle. Therefore, the electrostatic force is about half of the balloon's weight. The mass of a typical small balloon is a few grams. Let us assume therefore that its weight is equal to 0.04 newtons [Weight = $(4 \times 10^{-3}$ kg$) \times (9.8$ N/kg$)$]. The electrostatic repulsion between the two balloons must therefore be 0.02 newtons. With the arrangement shown in the diagram, the center-to-center distance between balloons is about 40 cm. Now let's put all of these numbers into Coulomb's law and find the order-of-magnitude of electric charge on the balloons.

$$F = k \frac{q_1 q_2}{r^2}$$

$$(0.02 \text{N}) = 9 \times 10^9 \frac{q^2}{(0.4 \text{m})^2}$$

$$q^2 = \frac{3.2 \times 10^{-3}}{9 \times 10^9} \approx \frac{1}{3} \times 10^{-12}$$

$$q \approx \frac{1}{2} \times 10^{-6} \text{ C}$$

Notice that this much charge on each balloon is less than a microcoulomb; that is, 1 millionth of a coulomb. It looks as if we chose too large a quantity of charge for our unit. We will see similar situations with units throughout our study of electricity and magnetism. A quantity that is a reasonable unit in one realm of electromagnetism will be inconvenient in another realm. In electrostatics most separated charges are of the order of microcoulombs. With electric currents, however, many coulombs of charge per second can move in a wire.

1 coulomb of isolated charge would be an enormous charge

Question 18-5

Suppose that it were possible to separate a full coulomb of positive charge and put it on one balloon and a full coulomb of negative charge on the other balloon. What would be the force between balloons if their separation distance were one meter?

THE ELECTROSTATIC FIELD

Coulomb's law in its simple form applies only to the force between charge points or uniformly charged spheres. The dependence on the inverse square of the distance is primarily a geometrical property of space, just as it is for gravitational attraction between spheres. If an influence spreads out from a point, or uniformly from the surface of a sphere, without decaying and without being absorbed, then the amount of influence *per unit area* is inversely proportional to the square of the distance from the center of the source. The reason for this is that the surface area of the spreading sphere of influence is equal to $4\pi r^2$. Since the quantity of the influence going out through the surface of each concentric sphere is always the same and since the area of the sphere is proportional to r^2, then the influence *per area* must be proportional to $1/r^2$.

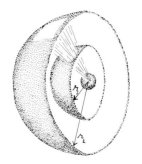

$$\frac{A_2}{A_1} = \frac{4\pi r_2^2}{4\pi r_1^2} = \frac{r_2^2}{r_1^2}$$

Same total influence goes through outer sphere as goes through inner sphere.

$$\frac{\text{Influence/area}_1}{\text{Influence/area}_2} = \frac{\text{area}_2}{\text{area}_1} = \frac{r_2^2}{r_1^2}$$

if $r_2 = 2r_1$,

$$\frac{\text{Influence/area}_1}{\text{Influence/area}_2} = \frac{(2r_1)^2}{r_1^2} = 4$$

We have not given any explanation of how one object influences another some distance away. Is something shot out of one object toward the other? If so, does the influence take some time to travel between objects? A very useful model to explain static effects (where the objects are not moving with respect to each other) is to picture a "field" of influence. This field model was ingeniously developed by Michael Faraday in the early part of the nineteenth century. Instead of asking how one charged particle affects another, Faraday described the effect that one or more charges has at each point in space. He imagined that space is electrically warped by the presence of electric charges. We can find out what this total influence or warping of space is at a particular point by bringing a unit test charge to that point, and measuring the force on it. The field strength at that point is then defined to be the force per unit test charge. $\mathbf{E} = \mathbf{F}/q$. In our units, the field strength is given in terms of newtons per coulomb.

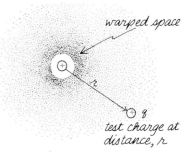

$$E_{\text{at } r} \approx \frac{\text{Force on } q \text{ at } r}{q}$$

Question 18-6

If the electric field is measured in newtons per coulomb then apparently the test charge should be a coulomb. But if you use a whole coulomb as a test charge, wouldn't it upset the position of all of the electric charges whose effect you are trying to measure?

Note that the field model describes the interaction between charged particles in two steps. First, you calculate the effect of one of the charges on the point in space where the other charge exists. The first charge creates a particular field strength at that point. The second charge then experiences that field strength as a force equal to the field strength times its own charge. What advantage does such a system have over Coulomb's simple law? The advantage, as we will see, is that a different mental image can lead to new

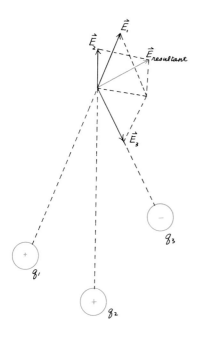

concepts and easier mathematical models. If there are many charges in a region, or if they are spread out over some unusual geometric shape, then we can start out by calculating the combination effect that all the charges have at a particular point in space. Each small segment of charge produces its own particular electric field at that point. These individual contributions must then be added *vectorially* to produce the net field at that point.

Question 18-7

Suppose that we have two equal electric charges at positions A and B. First we calculate the electric field of charge A at position B. Call that field $\mathbf{E}_{A-B}$. The force on q_B must therefore be equal to

$$\mathbf{F}_B = \mathbf{E}_{A-B} q_B$$

However the charge at B must be producing an electric field of its own which influences the charge at A. Is the total force between the two charges equal to the sum of these two forces?

THE ELECTRIC FIELD PRODUCED BY POINT CHARGES

Let's calculate the electric field produced by a point charge, or by two point charges close together.

E at a Distance *r* from a Point Charge

Since the force between the charge and a unit test charge q_U is equal to

$$\mathbf{F} = k \frac{q q_U}{r^2}$$

and since the electric field produced by q is equal to $\mathbf{F}/q_U$, then $\mathbf{E} = kq/r^2$. In such a simple case there is really no need for the electric field concept. While we're at it, however, let's find the magnitude of the electric field strength at a distance of 10 cm from a point charge of 1 microcoulomb.

$$\mathbf{E} = 9 \times 10^9 \frac{(1 \times 10^{-6} \text{ C})}{(1 \times 10^{-1} \text{ m})^2} = 9 \times 10^5 \text{ N/C}$$

The field is almost one million newtons per coulomb. That may seem tremendously strong to be created by such a small source charge. According to the formula, if the test charge were a coulomb, it would experience a force of $\mathbf{F} = q\mathbf{E} = (1 \text{ C})(9 \times 10^5 \text{ N/C}) = 9 \times 10^5$ N—about 90 tons. As we have seen, however, the test charge cannot be as large as one coulomb. If a *micro*coulomb were used as the test charge, then the force that it would experience would be equal to $\mathbf{F} = \mathbf{E}q = (9 \times 10^5 \text{ N/C})(1 \times 10^{-6} \text{ C}) = 0.9$N.

E Produced by Two Point Charges Separated by a Short Distance

The two point charges might both be positive (or both negative), or one positive and the other negative. In the diagrams that follow we find the

THE ELECTRIC FIELD PRODUCED BY POINT CHARGES

electric field at several points in space—on the perpendicular bisector of the line between the charges and along the axis of the line joining the charges. Note that the electric field is a vector and so the fields from the two charges must be added vectorially. In the diagram we have drawn field lines produced by a positive and negative charge of equal magnitude, placed close together. This arrangement is called a *dipole*. The field lines indicate the direction of the force on a positive test charge placed at any point in the vicinity of the dipole. The lines are not, however, the paths that a test charge would follow, since the direction an object moves depends both on the force that it experiences at any given instant and also its previous velocity.

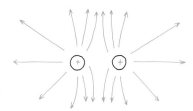

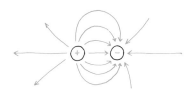

Field lines produced by two charges, equal in magnitude. Upper drawing- same polarity. Bottom drawing- opposite polarity.

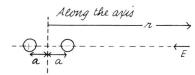

$$\mathbf{E} = kq\left[\frac{1}{(r+a)^2} - \frac{1}{(r-a)^2}\right] = kq\frac{r^2 + a^2 - 2ar - r^2 - a^2 - 2ar}{(r+a)^2(r-a)^2}$$

when $r \gg a$, $\quad (r+a)^2 \to r^2 \quad$ and $\quad (r-a)^2 \to r^2$

$$\mathbf{E} \to -4k\frac{qa}{r^3} = -2k\frac{p}{r^3} \quad \text{where } p = (2aq)$$

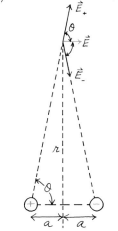

$$\mathbf{E} = 2k\frac{q}{r^2 + a^2}\cos\theta = 2k\frac{q}{r^2 + a^2}\frac{a}{\sqrt{r^2 + a^2}} \to k\frac{p}{r^3} \quad \text{when } r \gg a$$

Note dependence of dipole field on inverse *third* power of distance from center. The dipole field is proportional to the strength of the dipole, defined as $p = 2aq$. An additional factor in the equation varies from a value of 1 to 2, depending on the orientation of the line to the point with respect to the dipole axis.

GAUSS'S LAW RELATING E AND q

If we used the vector addition method to find the net field produced at a point by all the charges spread over a sphere, we would have a very difficult calculation on our hands. (It is, however, possible to do such a calculation using calculus.) An easier method is available for certain geometries. To make use of this method we have to add more details to our model of lines of electric force.

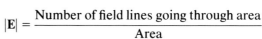

Faraday developed a very useful model out of his mental image of electric field lines. Not only does the direction of the lines indicate the direction of the force that a positive test charge would experience at that point, but the magnitude of the force can be represented by the *density of the lines* near the point. Such a scheme makes qualitative sense. In the diagram the field lines are most crowded close to the source charge. Farther away, where the field is weak, there are fewer field lines per unit area. We can make this idea quantitative by defining the magnitude of the electric field in any region to be the number of field lines per unit area in that region. Of course when we map out the unit area in order to count the number of lines, we must make sure that the surface of the area is perpendicular to the direction of the electric field lines.

$$|\mathbf{E}| = \frac{\text{Number of field lines going through area}}{\text{Area}}$$

Therefore, number of field lines going through area = **E** · **area**

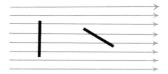

The surface of the measuring area must be perpendicular to the field lines. These two surfaces have the same area. The first is pierced by 4 lines; the second by 2 lines. Nevertheless, the field is the same in both regions.

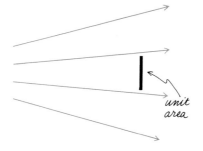

> ### Question 18-8
>
> What if we choose a unit area far away from the source charge as shown in this diagram? In that region of the diagram the field lines are far apart. None of them are going through the unit area that was chosen. Does that mean the electric field in that region is zero?

The reaction of most people to this model of field lines is that it may be a useful qualitative description but that it seems very abstract for quantitative work. Stay with the arguments for just a few more lines, however, and you will see what a powerful calculating device this model can be. Suppose that we say that *one electric field line per square meter* represents a unit field of *one newton per coulomb*. Let's see how many lines are coming from each positive coulomb of charge. If we put the coulomb of charge at the center of a sphere with one meter radius, then the lines of force extend uniformly as radii, piercing the sphere perpendicular to its surface. From Coulomb's law we know that the field produced by one coulomb at a distance of one meter is equal to

$$\mathbf{E} = k\frac{q}{r^2} = 9 \times 10^9 \frac{(1 \text{ C})}{(1 \text{ m})^2} = 9 \times 10^9 \text{ N/C}$$

The surface of a sphere has an area of $4\pi r^2$. Therefore, when r is equal to 1 meter, the number of lines piercing the sphere must be $\mathbf{E} \cdot \mathbf{area} = E(4\pi r^2) = (9 \times 10^9 \text{ N/C}) 4\pi(1 \text{ m})^2 = 4\pi \times 9 \times 10^9$ lines. That may seem like a lot of lines to be coming from a point charge. Of course, a coulomb is a lot of charge and besides this is just a mathematical model and the actual number of lines in the model depends on the units we have chosen. The number of lines, $4\pi \times 9 \times 10^9$, is equal to $4\pi k$. In other systems of units (cm instead of m, for instance), k would have a different value. At any rate, if we use this number, we will end up with the electric field in our standard units.

The next feature of the model was described by Karl Gauss (1777–1855) in a more general form than the one we will use. Gauss pointed out the simple fact that the total number of field lines leaving the surface of a three-dimensional volume must be related to the number of source charges inside the volume. The crucial point is that these field lines can start only on positive charges and can stop only on negative charges. They do not just start or stop in mid-space. Therefore, if there is no electric charge inside a volume, the net number of lines entering and leaving the volume must be zero.

If more lines leave than enter, then there must be positive charge inside; if more lines enter than leave, then there must be negative charge inside. Of course, field lines can enter a volume that has no net electric charge but then they will pass right through and leave again. This simpleminded theorem does not depend on the location of the electric charge inside the volume. The charge can be at the center or near one surface. Furthermore, there can be many charges inside the volume, both positive and negative, located at any point. But only the *net* charge will contribute to the *net* number of lines leaving or entering. Let's apply this theorem to three particular geometries of electric charge.

1. *E produced by a uniformly charged sphere.* If electric charge is uniformly distributed on the *surface* of an isolated sphere (or, for that matter, uniformly distributed throughout the *volume* of the sphere), then symmetry requires that the field lines be uniform and radial. After all, there is no reason why there should be more field lines on one side of the sphere than on the other. If the lines were to leave the surface in some direction other than radial, then the spherical symmetry would not hold.

Under these symmetrical circumstances, the field lines at a distance from the sphere all have the same direction and spacing as if they had come from the center of the sphere itself. You cannot tell whether the lines are coming from uniformly distributed charges on the surface or from that same

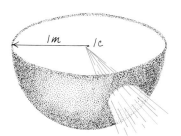

If q at center = 1C,
E at 1m = 9×10^9 N/C

∴ through a spherical surface of area 1m², there must be 9×10^9 field lines. Through the whole spherical surface there must be: $4\pi r^2 \times 9 \times 10^9$ = $4\pi \times 9 \times 10^9$ field lines, if $r = 1m$

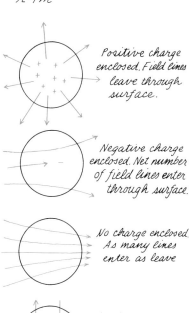

Positive charge enclosed. Field lines leave through surface.

Negative charge enclosed. Net number of field lines enter through surface.

No charge enclosed. As many lines enter as leave

Net charge enclosed is zero. As many lines enter as leave.

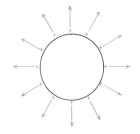

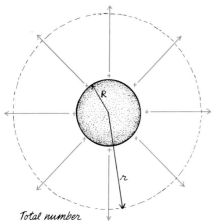

Total number
of lines leaving through surface = $4\pi kq$
Area of spherical shell at radius r = $4\pi r^2$

$$E_r = \frac{\text{total number of lines}}{\text{total area}} = \frac{4\pi kq}{4\pi r^2}$$

$$= k\frac{q}{r^2}$$

amount of charge concentrated at the center. Let's apply Gauss's law to find the electric field at a distance r from the center of the sphere, but outside the surface of the sphere itself which has a radius R. Construct an imaginary sphere with radius r around and concentric to the charged sphere. According to Gauss's law, the total number of field lines passing through the outer sphere is proportional to the net charge inside a volume of that size.

$$\text{Number of field lines} = (4\pi \times 9 \times 10^9)q = 4\pi kq$$

The electric field is radial and must have the same value at every point on the surface of the larger sphere. By definition, the electric field is equal to the number of field lines per unit area: $E =$ number of lines/area. In this case the total number of field lines is equal to $4\pi kq$ and the total area of the sphere is equal to $4\pi r^2$. Therefore, $E = 4\pi kq/4\pi r^2$

$$E = k\frac{q}{r^2}$$

■ Optional

Gauss's law confirms that the electric field produced outside a uniformly charged sphere is just the same as if all of the charge were concentrated at the center of the sphere. The result is not at all obvious. Newton faced the problem long ago in trying to calculate the force of gravitational attraction between the earth and the moon. He assumed that this force was determined by the distance between their centers but he had to invent the calculus to prove it. With Gauss's law, the proof tumbles out easily because of spherical symmetry. Actually, the mass of the earth is *not* uniformly distributed throughout its volume. The core is much more dense than the surface layer. Nevertheless, the spherical symmetry remains because at any given radius the density is the same regardless of the angular latitude or longitude.

In Chapter 10, on page 206, we asked what would happen if you dropped a billiard ball down a hole drilled through the earth. We claimed there that only the mass at a smaller radius from a given point determines the gravitational field at that point. All the mass in the shell of larger radius produces fields that cancel out. With Gauss's law we can now justify that argument. As long as there is spherical symmetry, the field lines must be radial and the gravitational field must be constant at a particular radius. According to Gauss's law, the number of field lines coming out of a closed surface depends only on the number of source charges *inside*. In the case of gravitation, the source charge is mass. The number of field lines coming out of the surface of a sphere with mass m, is equal to $4\pi Gm$. (Notice the parallel treatment of the two inverse square laws for spherical geometries—Newton's and Coulomb's:

$$F_{grav} = G\frac{m_1 m_2}{r^2} \qquad F_{elec} = k\frac{q_1 q_2}{r^2}$$

If $4\pi k$ lines of electric field come from each unit charge, q (a coulomb), then $4\pi G$ lines of gravitational field come from each unit mass, m (a kilogram). For the case of the point in the interior of the earth, at radius r, the number of field lines coming out from the sphere of that radius must be $4\pi G m_2 = 4\pi G(\sigma)(\frac{4}{3}\pi r^3)$. The density is σ, and $(\frac{4}{3}\pi r^3)$ is the volume of the interior mass. Since the field strength at any point is equal to the number of field lines per unit area, and since the field is constant at every point at the distance r:

Number of gravitational field lines penetrating spherical surface of radius, r, must be:

$m_r 4\pi G$, where m_r is mass of inner core with radius, r.

Since $m_r = \frac{4}{3}\pi r^3 \sigma$, where σ is density,
the number of field lines is: $\frac{16}{3}\pi^2 r^3 G\sigma$

Field strength = $g_r = \frac{\text{total number of lines}}{\text{total area}}$

$$= \frac{\frac{16}{3}\pi^2 r^3 G\sigma}{4\pi r^2} = (\frac{4}{3}\pi G\sigma)r$$

Since $M_{sphere} = \frac{4}{3}\pi R^3 \sigma$,

$$g_r = G\frac{M}{R^3}r$$

$$g = 4\pi G(\sigma)(\tfrac{4}{3}\pi r^3)/(4\pi r^2) = \tfrac{4}{3}\pi G \sigma r$$

This is the same expression that we obtained on page 207, although there we derived the force on the mass, *m*, of a billiard ball. That force, of course, is equal to $m_1 g$, and is directed toward the center of the earth.

Incidentally, Gauss's law holds even if the sphere has much higher density at the central core. The only symmetry required is that the field lines be radial. Furthermore, the argument holds whether the sphere is the earth, or a baseball, or ball lightning. (Consider the embarrassment if this were not true; the gravitational field at the center of a baseball might be infinite if $F = (Gm_1m_2)/r^2$ applied all the way to $r = 0$.)

There is yet another corollary to Gauss's law applied to regions enclosed by insulated, closed, *conducting* shells. In such regions, no matter how much charge is placed on the outer shell, there can be no net charge in the interior. For this demonstration, imagine a Gaussian surface the same shape as that of the conducting shell, but lying *in the metal* of the conducting surface, as shown in the diagram. There can be no static electric fields in a conductor that has no power source. If there were, charges would move until static equilibrium were reestablished. Therefore, the electric field is zero everywhere on the Gaussian surface; the number of field lines coming through the Gaussian surface must be zero; which means that the enclosed net charge must be zero. This situation occurs inside the sphere of the Van de Graaff generator, where charges are brought up on a belt into the region enclosed by the metal sphere. Although there is a strong repelling force on the incoming charges as they are hauled up, once inside the sphere the charges are shielded from the charges already on the sphere. Therefore, they can easily hop off the belt onto the collection wire without being repelled.

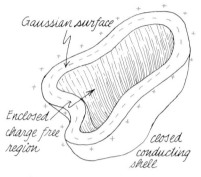

E in conductor = 0 for static conditions

∴ *since* $E = \dfrac{\text{number of lines}}{\text{unit area}} = 0$

no lines cross Gaussian surface

∴ *the net charge in the interior is zero, and any surplus charges must reside on the outer surface of the conductor.*

Question 18-9

But surplus electric charges are not supposed to remain in conductors. Where do the charges go once they get into the metal sphere?

A closed conducting shell is called a "Faraday cage." Even before Faraday's experiments, Franklin had shown experimentally that no external charge penetrates to the inside of a conducting shell. Such a device provides shielding against strong electric fields. A safe place to be in a lightning storm is inside a car—not because the rubber tires insulate you from ground, but simply because even if lightning strikes the car, no charges can transfer to the interior.

Gauss's law holds only if Coulomb's law holds for point or spherical sources. Both depend on the Euclidean nature of three-dimensional space (the fact that the surface of a sphere is $4\pi r^2$), and on the fact that the electrical influence does not decay and is not absorbed in free space. The shielding property of a closed conducting surface depends crucially on

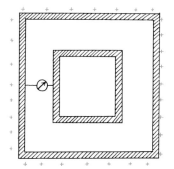

Meter between inner and outer cages determines if there is interior charge movement as outside cage is highly charged. There is none.

∴ Gauss' law holds
∴ $F_{coulomb} \propto 1/r^2$

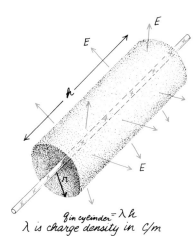

$q_{in\ cylinder} = \lambda h$
λ is charge density in C/m

Number of field lines passing through curved surface at r:

$4\pi k(\lambda h)$

$E = \dfrac{\text{total number of field lines}}{\text{total curved area}}$

$= \dfrac{4\pi k(\lambda h)}{2\pi rh}$

$E = 2k\dfrac{\lambda}{r}$

Gauss's law. Consequently, a shielding experiment is a very sensitive test of the inverse square law. Faraday himself got into a metal cage to see if he could detect any interior charges when the outside of the cage was highly charged. Similar experiments, done in recent years, demonstrate that the exponent, 2, in Coulomb's law is exact to one part in 10^{17}.

2. *E at a distance from a uniformly charged cylinder.* Once again we apply Gauss's law by picturing the field lines and appealing to symmetry. All the field lines from a charged infinite wire or uniformly charged, infinite cylinder must be radial. They are perpendicular to any concentric cylindrical surface. Suppose that there is constant charge per unit length along a wire or cylinder that is very long. This charge density is usually called λ (lambda) and is equal to the total charge divided by the length, L: $\lambda = Q/L$. To find the electric field at a distance r away from the center of the wire or cylinder, apply Gauss's law to a concentric cylinder of radius, r, and height, h. We must assume that the distance r is very small compared to the length of the wire so that we do not have to worry about the asymmetry of the field lines near the ends. The net charge in the cylinder is equal to the charge density λ times the height of the cylinder: net charge = λh. According to Gauss's law, the total number of field lines leaving the outer cylindrical volume of radius r and height h is proportional to the amount of charge in the volume.

$$\text{Number of field lines} = 4\pi k(\lambda h)$$

The field lines pierce only the curved surface of the cylinder. They do not come out through the end caps. Since the electric field is radial and constant at r, then

$$E = \text{number of field lines/area} = \frac{4\pi k(\lambda h)}{2\pi rh} = 2k\frac{\lambda}{r}$$

Notice that h, the height of the Gaussian cylinder, cancels out of the equation.

Once again Gauss's theorem produces something that is not obvious. The electric field outside a uniformly charged wire or cylinder decreases, but only with the inverse first power of the distance, not with the inverse second power of the distance. The electrostatic attraction or repulsion does not necessarily obey an inverse square law. The dependence of the field on distance depends on the geometry of the source. For a spherical source the field is proportional to $1/r^2$; for a cylindrical source, the field is proportional to $1/r$.

Question 18-10

If the charge on a uniformly charged cylinder is made up of a row of point charges, why don't the field lines go off in all directions instead of just radially?

3. *E produced by uniformly charged parallel conducting plates.* The parallel plate geometry shown in the next diagram is the basic form of an

electric capacitor that we will study later. If the bottom plate is uniformly charged positive and the upper plate negative, then the field lines will go from bottom to top, parallel to each other and uniform. Except near the edges, the electric field will be constant and uniform within the volume. If there is a total charge Q on each plate, then the surface charge density will be equal to Q/A where A is the area of the plate. The area charge density is usually called σ (sigma): $\sigma = Q/A$. (Note that the presence of the oppositely charged plate holds the total charge on the inner surface of each conducting plate.) To find the field at any point between the plates, construct an imaginary volume as shown in the diagram. Note that the bottom surface of the cylinder is embedded in the metal of the bottom plate, and the top surface is between the plates at the point where we wish to calculate the electric field. The total electric charge inside this volume is equal to the surface density of charge on the bottom plate times the cross-sectional area of the cylinder. The number of field lines coming from this much charge is equal to $4\pi k(\sigma \times \text{area})$. All those field lines leave the upper flat area of the cylinder. None pass through the bottom because that is buried in the metal and there are no field lines inside the conductor.

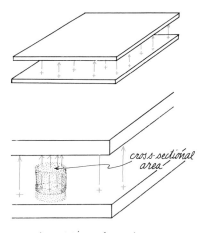

Number of lines through top cap
$= 4\pi k (\sigma \times \text{area})$
σ *is area charge density in* C/m^2
$E = \dfrac{\text{total number of lines out of volume}}{\text{surface area crossed by lines}}$
$= \dfrac{4\pi k (\sigma \times \text{area})}{\text{area}} = 4\pi k \sigma$

Question 18-11

Why not?

No field lines pass through the curved surface of the cylinder because that surface is parallel to the field lines. Therefore, the electric field in the region at the top of the Gaussian cylinder is equal to $E =$ number of field lines/area $= 4\pi k(\sigma \times \text{area})/\text{area} = 4\pi k \sigma$.

In this case, the electric field does not at all depend on distance from the source. As we should expect from the field line model, the field is constant between the plates from top to bottom. (If this were not the case, the electric field lines would not be parallel to each other and uniformly spaced.)

The relationship that we have just developed, between the electric field and the surface density of charge ($E = 4\pi k \sigma$), also applies to the field close to the surface of *any* charged conductor, regardless of its shape and regardless of whether there is another plate nearby. The reason for this generalization is that electric field lines near the surface of a conductor must be perpendicular to the surface. (This is true only in electrostatics. As we will see in the next chapter, when there is a flow of charge in the conductor, there must be field lines in the direction of the flow.)

Question 18-12

Why can't an electric field line leave the conducting surface at an angle other than 90°?

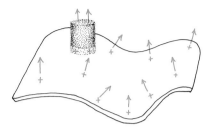

In a very small region near the surface of a conductor we can apply the same Gaussian cylinder that we did between parallel plates. The same arguments apply, and we find that

$$E_{\text{near surface of conductor}} = 4\pi k \sigma$$

Of course, the surface may be curved, and so the field lines are not necessarily parallel to each other. Furthermore, the surface charge density—as we will see—depends on the surface curvature, and therefore the electric field is not constant over the surface.

4. *E produced by an electric dipole.* So far we have applied Gauss's law to find the electric field at a distance from charges in three standard geometries. Another very common charge arrangement is the dipole—equal positive and negative charge separated by a small distance. We have already calculated the field due to a dipole by adding vectorially the fields from the two individual charges.

Question 18-13

Can we use Gauss's law to find the electric field in the region near an electric dipole? Is there a surface surrounding the dipole such that the field lines will be constant on the surface and pierce it in a perpendicular direction?

SUMMARY OF THE DEPENDENCE OF E ON SOURCE GEOMETRY

We have used Gauss's law to find the electric field produced by uniformly charged spheres, cylinders, and planes. What about all the other shapes that may be sources of electric influence? The calculations would be very complicated in general, but for many cases we can approximate unusual shapes with the three that we have studied. Suppose, for example, that you want to know how the gravitational field of the earth depends on distance from its center. True enough, the gravitational field is proportional to the inverse square of the distance from the center of the earth. However, for small distances above the surface of the earth, the gravitational field is almost constant. This is the flat-earth approximation, the one that we use all the time in everyday life. For vertical distances of the order of a few hundred meters, the gravitational field lines are almost parallel to each other and are constant in density—just like the electric field lines in a parallel plate capacitor. As another example, take the case of the light produced by a long fluorescent tube. For distances of a few millimeters away from the surface of the tube, the light intensity must be almost constant. The surface of the lamp looks flat. Farther away, where the distance to the center of the tube is more than about $1\frac{1}{4}$ times the radius of the tube, the geometry should be considered cylindrical. In that case the intensity of the light is inversely proportional to the distance from the center of the tube. That approximation holds only so long as the distance from the tube is small compared with the distance out to the ends of the tube. When you are at a distance from the

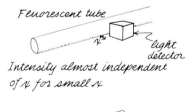

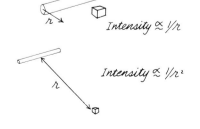

ELECTRIC POTENTIAL

tube many times its own length, the geometry is approximated by that of a point source. The light intensity will be approximately proportional to the inverse square of the distance from the source.

In the table below we summarize the formulas for the geometries we have studied. Notice that they are given in two forms. It is often convenient to express the constant in Coulomb's law in another way. Instead of writing $E = k(q/r^2)$ we can write:

$$E = \frac{1}{4\pi\epsilon_0}\frac{q}{r^2}$$

The second expression may seem more complicated but we will have occasion to use ϵ_0 (epsilon sub zero) later. It is called *permittivity*. The subscript 0 indicates that the constant applies for free space between the charges. If there is water or glass or some other medium between the charges, then the permittivity will have a different value. Since for free space (and air) $k = 9 \times 10^9$, then $\epsilon_0 = 8.5 \times 10^{-12}$. The 4π is not absorbed into the constant because it is a geometrical factor, in this case equal to the number of unit solid angles (steradians) around a point.

Geometry of Source	E ($k = 9 \times 10^9$)	E ($\epsilon_0 = 8.9 \times 10^{-12}$) $k = \frac{1}{4\pi\epsilon_0}$
Point	$E = k\frac{q}{r^2}$	$E = \frac{1}{4\pi\epsilon_0}\frac{q}{r^2}$
Uniformly charged sphere $r \geq R$	$E = k\frac{q}{r^2}$	$E = \frac{1}{4\pi\epsilon_0}\frac{q}{r^2}$
Uniformly charged sphere $r \leq R$	$E = k\frac{Q}{R^3}r$	$E = \frac{1}{4\pi\epsilon_0}\frac{Q}{R^3}r$
Uniformly charged cylinder $r \geq R \quad r \ll L$	$E = 2k\frac{\lambda}{r}$	$E = \frac{1}{2\pi\epsilon_0}\frac{\lambda}{r}$
Uniformly charged plates (region between parallel plates or near surface of individual plate)	$E = 4\pi k\sigma$	$E = \frac{\sigma}{\epsilon_0}$

ELECTRIC POTENTIAL

Separated electric charges attract and repel each other. So far we have been learning how to describe the forces involved. Another feature of electrostatics is even more dramatic than the force fields. When you touch a highly charged object, you draw a spark and are shocked. Energy has been released. In everyday life you expect such effects from objects that are at high *voltage* or high *potential*. Let's define those words and make them quantitative.

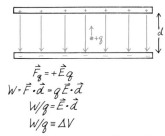

$$\vec{F_g} = +\vec{E}q$$
$$W = \vec{F} \cdot \vec{d} = q\vec{E} \cdot \vec{d}$$
$$W/q = \vec{E} \cdot \vec{d}$$
$$W/q \equiv \Delta V$$

ΔV_{A-B} = work to move unit positive charge from A to B = potential difference between A and B

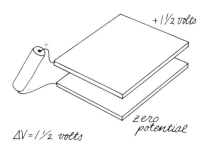

$\Delta V = 1\frac{1}{2}$ volts

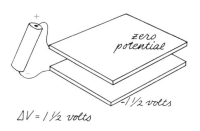

$\Delta V = 1\frac{1}{2}$ volts

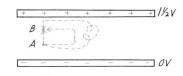

$\Delta V_{A-B} = \frac{1}{2} V$
(A-B is 1/3 distance between plates)

Potential difference, or work done per charge in going from A to B is the same for all three paths shown.

Take a positive test charge and move it from the bottom plate to the top plate of a parallel plate capacitor as shown in the diagram. The electric force on the test charge is constant and opposes the upward motion. That force is equal to $\mathbf{F} = \mathbf{E}q$. The work done in moving the electric charge against this force is equal to $W = \mathbf{F} \cdot \mathbf{d} = q\mathbf{E} \cdot \mathbf{d}$ where $\mathbf{d}$ is the distance between the plates. The amount of work done per charge is equal to $W/q = \mathbf{E} \cdot \mathbf{d}$. The *work done per coulomb* in moving a charge from one point to another in an electric field is called the *electric potential difference*. The unit of potential difference is the joule per coulomb, which is called a *volt*.

The size of the volt is familiar to us from everyday life. A single flashlight cell has a potential difference across it of $1\frac{1}{2}$ volts. Most car batteries these days produce 12 volts. Such ratings do not refer to the force that the batteries exert on charges but instead signify the energy provided to each coulomb of charge as it is pumped from the negative side to the positive side.

Let's connect a $1\frac{1}{2}$ volt cell across the parallel plate capacitor as shown in this next diagram. The potential *difference* across the plate is equal to $1\frac{1}{2}$ volts. But we have not defined the actual electric potential of either plate. The situation is the same as it was with gravitational potential energy. We are free to choose the zero point. In this case we might say that the negative or bottom plate of the capacitor is zero potential, in which case the top plate is at $+1\frac{1}{2}$ volts. Or we could define the top plate to be zero potential, in which case the bottom plate is at $-1\frac{1}{2}$ volts. In either case, it would take $1\frac{1}{2}$ joules to force 1 positive coulomb from the bottom plate to the top plate. (Actually, as we have seen, it would be impossible with this particular geometry to separate a coulomb of charge. However, if 1×10^{-9} coulombs were separated in this case, it would require $1\frac{1}{2} \times 10^{-9}$ joules.)

It is very convenient to be able to assign a potential to each point in space. In describing the *electric field* at any point, we have to give both its magnitude and direction. A *potential*, however, has only magnitude; it is a scalar quantity. That means that we can find the work done in going from one point to another in an electric field simply by finding the potential difference between the two points. The route we take with our test charge makes no difference. In the diagram we show three ways to go from point A to point B in the electric field between the parallel plates. In all three cases the work done is equal to $\frac{1}{2}$ joule per coulomb. (More realistically, if we used 1×10^{-12} coulomb as our test charge, the work done would be $\frac{1}{2} \times 10^{-12}$ joule.) The argument is the same as the one we used in Chapter 10 for gravitational potential energy. Suppose that it takes only two units of energy to go from A to B by the straight path but three units to go by the long path. Then the smart thing to do would be to escort the charge upward on the left-hand path using up two units of energy and then let it slide back down on the right-hand path, gaining three units of energy. We could make money that way if the system worked, but, of course, it doesn't.

It is easy to find the relationship between the electric field in a parallel plate capacitor and the potential difference across the plates. As we have seen, the potential difference, ΔV, is defined as the work done per unit charge, W/q, which is equal to the dot product of force per unit charge and

ELECTRIC POTENTIAL

the distance, $(\mathbf{F}/q) \cdot \mathbf{d}$. The force per unit charge is the electric field **E** and it is constant. Therefore,

$$\Delta V = W/q = (\mathbf{F}/q) \cdot \mathbf{d} = \mathbf{E} \cdot \mathbf{d}$$

At any point in the region between the parallel plates, the electric field is equal to the potential difference divided by the spacing between the plates:

$$|E| = \Delta V/d$$

Note that we now have another way of calculating E and expressing its units. On page 521 we calculated that the electric field between parallel plates is equal to

$$E = (4\pi \times 9 \times 10^9)\sigma = 4\pi k\sigma$$

The units are newtons per coulomb. To calculate the electric field, we would have to know the value of σ which is equal to the total surface charge of the plate divided by the area of the plate: $\sigma = Q/A$. In the new way of determining the electric field for parallel plates, we simply divide the potential difference by the distance between the plates. The units are volts per meter. The two systems are completely equivalent.

$$1 \text{ newton/coulomb} = 1 \text{ volt/meter}$$

Actually, electric fields are most commonly measured in terms of volts per meter. For instance, a radio or TV signal with an electric field strength of 1 microvolt per meter is just at the limit of detectability by a good receiver.

The relationship between E and ΔV in the parallel plate capacitor can be generalized. Even in regions where the electric field is not constant, it is approximately true that $E_x = -\Delta V/\Delta x$ for small displacements, Δx. For instance, suppose that in moving a distance of Δx equal to 1 cm, the potential increases by 1 volt. Then the component of the electric field in that direction, $E_x = -(1 \text{ volt})/(0.01 \text{ m}) = -100$ V/m. The sign is negative because if we take a positive test charge *up* a potential hill of 1 volt, the electric field must be pointing down. Notice that the relationship between E and ΔV allows us to find a vector quantity, **E**, if we know the distribution of a scalar quantity, V. Actually, we find the vector *component* in the direction of the displacement.

The inverse relationship also holds approximately. For small displacements, $\Delta V = E_x \Delta x$. The electric field in the region of a point charge is equal to $E = k(q/r^2)$. For a charge of 1×10^{-9} C, the field at $r = 10$ cm is

$$E = (9 \times 10^9) \frac{(1 \times 10^{-9} \text{ C})}{(0.1 \text{ m})^2} = 900 \text{ V/m}$$

The potential difference between $r = 10$ cm and $r = 11$ cm, is approximately equal to $\Delta V = -(900 \text{ V/m})(0.01 \text{ m}) = -9$ V.

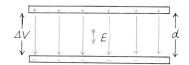

Question 18-14

Why is this expression only approximately true?

ELECTRIC POTENTIAL IN COMMON GEOMETRIES

We found the electric field distribution around spheres, cylinders, parallel plates, and dipoles. Let's find the electric *potential* distribution around those same geometries. In all cases we can find the electric potential at a point in space by measuring the work done in bringing a positive test charge from zero potential to that point. We are free to define the zero point of potential at any convenient place.

In the case of gravitational potential, we found two common definitions of zero potential. For defining gravitational potential close to the surface of the earth, we could define the zero point as being on the surface of the earth. In considering rocket travel or earth-moon relationships, we found it most convenient to choose infinity as the point of zero potential.

Potential near a Charged Sphere

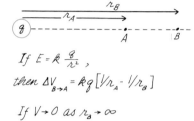

In the case of electrically charged spheres, it is usually most convenient to define zero potential at a point far beyond the influence of the charged sphere. We can call that point infinity. The electric potential of any point in the vicinity of the charged sphere is thus defined as the work done in bringing unit charge from infinity up to that point. We have already solved such a problem in Chapter 10 for an inverse square force (F proportional to $1/r^2$). The potential in this case is inversely proportional to the distance from the center of the sphere:

$$V \propto \frac{1}{r}$$

The complete formula for the potential at a distance r from a uniformly charged sphere (or point charge) is

$$V = k\frac{q}{r}$$

> **Question 18-15**
>
> Suppose that a microcoulomb of positive charge is uniformly spread over the surface of a sphere with 10 cm radius. What is the potential of the sphere itself? How much work would have to be done to take a positive test charge of 10^{-8} coulombs from a distance of 30 cm from the center of the sphere down to its surface?

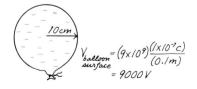

The answer to Question 18-15 has a startling aspect. In "Handling the Phenomena," you electrically charged balloons by rubbing them on your hair, and we calculated that the amount of charge on each balloon must be about $\frac{1}{10}$ of a microcoulomb. To the extent that a balloon is an approximation to a uniformly charged sphere of the same size, it appears that the electric potential of the balloon is about 10,000 volts. Indeed, when you shuffle along a rug and draw a centimeter-long spark by touching someone else, you also are at a potential of between 10,000 and 20,000 volts.

ELECTRIC POTENTIAL IN COMMON GEOMETRICS

> **Question 18-16**
>
> Why aren't you lethal at these voltages? If you have stored up $\frac{1}{10}$ microcoulomb and are at a potential of 10,000 volts, how much energy can you deliver?

Potential near a Charged Cylinder

The potential difference between two points in the region of a charged cylinder is equal to

$$\Delta V_{r_1 \to r_2} = 2k\lambda \ln(r_1/r_2)$$

The derivation of this equation requires calculus. But note how the potential difference depends on the distances. In this case we cannot define the potential at infinity to be 0 because if r_2 goes to infinity, then the ratio goes to 0 and the log of 0 is minus infinity. All we can do in this case is to measure the potential difference between two points that are a finite distance away from the cylinder.

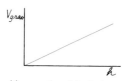

With positive charge on inner cylinder, inner cylinder is at higher potential than outer. Therefore, $\Delta V_{r_1 \to r_2}$ is negative, since there is a drop in potential going from r_1 to r_2. Note that $\ln(r_1/r_2)$ is negative if $r_2 > r_1$. For calculation purposes, remember that $\ln(r_1/r_2) = -\ln(r_2/r_1)$.

Potential Between Parallel Plates

We have already found the relationship between the potential difference across parallel plates and the electric field between them: $E = \Delta V/d$, where d is the distance between plates. If we define the zero of potential to be at the bottom plate, then the potential rises linearly from 0 to V as you go from the bottom plate to the top. The geometry is just like that for the gravitational potential close to the surface of the earth. In that case, the gravitational potential is gh, and the gravitational potential energy of an object with mass m, is mgh. Between parallel plates the electrostatic potential is Ey, and the electrostatic potential energy of an object with charge q is qEy. In both these cases the field lines are uniform and parallel to each other; the force per charge is constant in the region. Therefore, the work done per charge in moving along the field lines is proportional to the distance moved.

The field between charged parallel conducting plates is related to the charge density of the plates: $E = 4\pi k\sigma = \sigma/\epsilon_0$. Therefore, $V = (4\pi k\sigma)y$. The potential difference across the plates is $\Delta V = (4\pi k\sigma)d = (\sigma/\epsilon_0)d$. Since the charge density is $\sigma = Q/A$, where Q is the total charge on each of the plates and A is the area of either plate, we can describe the potential difference across parallel plates as

$$\Delta V = \frac{Qd}{\epsilon_0 A}$$

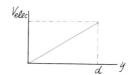

$V_{grav} = gh$ if $V = 0$ on surface, and for flat earth approximation where gravitational field lines are parallel and uniform. g is constant.

$V_{elec} = Ey$ where $V = 0$ at bottom plate where $y = 0$. With parallel plate capacitor, lines of E are parallel to each other and uniform between plates. E is constant.

Potential near a Dipole

The potential due to a dipole depends very much on the angular relationship between the point and the dipole. For instance, at any point on the perpendicular bisector of the dipole, the potential is 0. This must be true because if you started the test charge on its journey far away from the dipole it would start out at 0 potential by definition. As you move the test charge along the perpendicular bisector, you are always moving it perpendicular to the electric field lines. Therefore, no work is done. Since you can move the test charge from infinity right to the center of the dipole without doing any work, all of those points must be at 0 potential. Along the axis line, however, the potential turns out to be proportional to the inverse *square* of the distance from the center of the dipole.

There is an interesting relationship between electric field lines and lines of equipotential. We have just seen one case with the dipole. As long as you move a test charge always in a direction perpendicular to the electric field lines, no work is done and every point on that line must be at the same potential. In the chart we have sketched the field lines and equipotential surfaces around our standard geometries of a charged sphere, cylinder, parallel plates, and dipole. The formulas for the potentials for these cases are summarized.

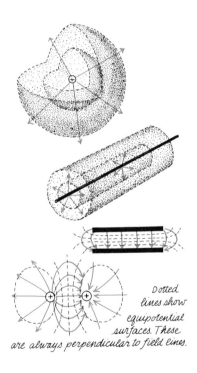

Dotted lines show equipotential surfaces. These are always perpendicular to field lines.

Along perpendicular bisector, $\vec{E} \perp \Delta y$. Since $\Delta V = \vec{E} \cdot \vec{\Delta y}$, $\Delta V = 0$ for any move along bisector.

Source Geometry	Potential
Point or sphere $r \geq R$	$\Delta V = kq[1/r_1 - 1/r_2]$ $V = kq/r$ if $V=0$ as $r_2 \to \infty$
cylinder	$\Delta V = 2k\lambda \ln(r_1/r_2)$
parallel plates	$\Delta V = Ed$ $V = Ey$ if $V=0$ for $y=0$ and $0 \leq y \leq d$ also $\Delta V = \dfrac{Qd}{\varepsilon_0 A}$
Dipole	$V \propto \dfrac{p}{r^2}$ on axis $V = 0$ everywhere on $\perp$ bisector

THE STORAGE OF ELECTRIC CHARGE—ELECTRICAL CAPACITANCE

Every time objects touch, and produce charge separation, the positive charge gets stored on one object and the negative on the other. The stored charge creates an electric field around each object and raises or lowers its potential. In all of the formulas for potential such as in the preceding table, the potential is proportional to the charge on the source. If you double the charge, you double the potential. But some objects require more charge than others to reach the same potential. The *electrical capacitance* of an object is defined as

THE STORAGE OF ELECTRIC CHARGE—ELECTRICAL CAPACITANCE

$$C = \frac{Q}{V}$$

If an object has a large electrical capacitance, then it can hold a large electric charge at a relatively small potential.

The unit of capacitance is called the farad, with the symbol F. It is named after Michael Faraday (1791–1867) who discovered many electrical phenomena by relying on vivid mental models of the mechanism. If a coulomb of charge is placed on an object and the potential increases by one volt, then the capacitance of the object is one farad. That would be a very large capacitance, as we will see. In electrical circuits it is more customary to have capacitances in the range of microfarads. One microfarad equals $1\mu F = 1 \times 10^{-6}$ F. Frequently we must deal with capacitances much smaller than that. One picofarad = 1 pF = 1×10^{-12} F.

Capacitance of a Sphere

Let's see what the electrical capacitance is for a charged sphere. Simply substitute the formula for the potential of a sphere into the formula for capacitance using the radius of the sphere, R, as the particular point for the potential.

$$C = \frac{Q}{V} = \frac{Q}{kQ/R} = \frac{R}{k}$$

The capacitance of a sphere is simply proportional to its radius. Suppose that the sphere has a radius of 10 cm. Then its capacitance is equal to $C = (0.1 \text{ m})/(9 \times 10^9) \approx 1 \times 10^{-11}$ farads. Notice that the capacitance of a sphere in picofarads (10^{-12} F) is about equal to its radius in centimeters. As a matter of fact, it is a good rule of thumb that the approximate capacitance (in picofarads) of any isolated conductor is equal to the radius (in cm) of a sphere that has about the same size as that object.

For an order-of-magnitude approximation of the capacitance of an object, find the radius of an equivalent sphere with about the same surface area.

Question 18-17

What is *your* approximate electrical capacitance?

Let's find the capacitance of the largest sphere that we have available. Try the earth. Its radius is equal to 6.4×10^6 meters. The electrical capacitance of the earth is therefore

$$C = (6.4 \times 10^6 \text{ m})/(9 \times 10^9) = 7 \times 10^{-4} \text{ F} = 700 \ \mu F$$

Even the earth does not have an electrical capacitance of one farad!

Capacitance of Parallel Plates

We can make a device with much more capacitance than a sphere out of a parallel plate arrangement. As we saw on page 527, the potential difference between charged parallel plates is equal to $\Delta V = (Qd)/\epsilon_0 A$. Now we can find the capacitance of the parallel plate arrangement since $C = Q/\Delta V =$

Dielectric Constants—κ

Material	κ
Water	80.4
Ethanol	24.3
Methanol	33.6
Carbon tetrachloride	2.2
Sulfur	4.0
Lucite	2.8
Glass	4–5
Paraffin	2–2.5
Porcelain	6–8
Sodium chloride	5.7
Ordinary mica	3–6
Magnesium mica	275
Lead sulfide	205
Strontium titanate	330
Lead telluride	400
Barium titanates	6000–12000
Air	1.00059

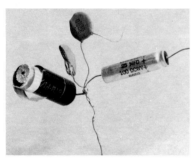

Foil capacitor, partially cut open, is on left. In the center there are two ceramic capacitors, one cut open to show the plates. On the right is an electrolytic capacitor.

$Q/(Qd/\epsilon_0 A) = (\epsilon_0 A)/d$. The capacitance of parallel plates is proportional to the surface area and inversely proportional to the separation distance.

The formula that we have derived applies to a parallel plate capacitor with vacuum (or air) between the plates. The permittivity of other materials is greater than ϵ_0:

$$\epsilon = \kappa \epsilon_0$$

The factor κ is called the dielectric constant of the material. The table gives the dielectric constant of a number of materials that are useful in capacitors. Certain ceramics such as barium titanate can increase the capacitance by a factor of 10,000. The complete formula for the capacitance of a parallel plate arrangement now looks like this:

$$C_{\text{parallel plate}} = \frac{\kappa \epsilon_0 A}{d}$$

PRACTICAL CAPACITORS

Capacitors used in circuits or for power storage usually exploit one or two of the factors controlling the capacitance. For instance, a *paper* and *foil* capacitor consists of a spool of foil and very thin insulator sandwiched together. The total surface area is very large and the spacing between foils is just the thickness of the thin insulating paper or plastic.

In *electrolytic* capacitors the surface area of the foil is not very large but the spacing between foils is produced by a chemical layer that may be only 10 to 100 atoms thick. Since d is so small, the capacitance can be very high. However, in this case, the thin chemical layer is maintained only as long as the electric field is in one direction. An electrolytic capacitor must be polarized with one electrode maintained always positive with respect to the other one.

In the case of *ceramic* capacitors, the area of the plates is not large, nor is the spacing particularly small. The material between the plates, however, may have a dielectric constant of the order of 10^4, thereby producing a large capacitance.

THE MICROSTRUCTURE OF A DIELECTRIC

Why should material between the plates of a capacitor increase the capacitance? The diagram shows a model of how the charges in the material arrange themselves. The original electric field between the plates polarizes the material so that every atomic or molecular cluster is warped. The effect is to produce an inner layer of induced negative electric charges near those on the surface of the positive plate and a similar layer of induced positive charges close to the negative charges on the other plate. The field between the plates in the material is thus greatly reduced. That means, however, that the potential between the plates is reduced. For a given amount of stored charge, if the potential difference between the plates is reduced, then the capacitance is increased since $C = Q/V$.

ENERGY STORAGE IN A CAPACITOR

Capacitors are made for many purposes. In some circuits capacitors allow

ENERGY STORAGE IN A CAPACITOR

rapid changes in potentials to go through easily while blocking slow voltage changes. (In other words, which we will study later, a-c signals can pass through capacitors while d-c cannot.) In other applications capacitors are used to store charge or electric energy for short periods. Here is a picture of a high voltage capacitor made for energy storage. It is rated at 1 microfarad with a maximum potential difference of 2000 volts. It is filled with oil as a dielectric that provides a higher permittivity than air does, and also helps to prevent breakdown sparks between the plates.

To charge a capacitor, we could take positive charge off one plate and force it up to the other plate. If we start out with a capacitor uncharged, there is no potential difference between the plates and so the work that is done in taking the first small amount of charge from the one plate to the other is zero. As soon as that first small charge has been lifted up to the other plate, however, it creates a potential difference between the plates. The work done in taking the next small amount of charge from the bottom plate to the top plate is equal to that small potential difference times the charge being transported:

$$W_2 = V_1 \Delta q_2$$

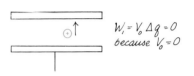

By the time we transport the last of the charge from the bottom to the top plate, the work done is equal to the product of that charge times the full potential difference of the capacitor. The average potential difference through which the charges have been carried is just equal to one-half of the final potential difference. Therefore work done in charging up the capacitor is equal to $\tfrac{1}{2}QV$, where V is the potential difference between the plates, frequently called the "voltage." That is the amount of energy U stored in the capacitor. Since $C = Q/V$, we can express this stored energy in two other ways:

$$U = \tfrac{1}{2}QV = \tfrac{1}{2}CV^2 = \tfrac{1}{2}Q^2/C$$

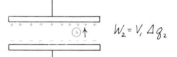

The maximum energy that can be stored in the oil-filled capacitor is equal to

$$U = \tfrac{1}{2}CV^2 = \tfrac{1}{2}(1 \times 10^{-6} \text{ F})(2000 \text{ V})^2 = 2 \text{ joules}$$

Two joules doesn't seem like very much energy. The capacitor has a mass of about 500 grams and thus weighs about 5 newtons. If it were raised only 40 cm in the air, it would have a *gravitational* potential energy increase of 2 joules: (5 N)(0.4 m) = 2 J. The two joules of *electrical* energy are stored in a very convenient way, however. With the right electrical circuit, that energy can be extracted from the capacitor in a time of less than one microsecond. Such a burst would provide momentary power of $(2 \text{ J})/(1 \times 10^{-6} \text{ s}) = 2 \times 10^6$ watts.

When a dielectric is between the plates of a capacitor, the capacitance is greater. Consequently, for a given potential difference across the plates, the stored energy is greater. The extra energy is stored in the distortion of the dielectric molecules. When some types of capacitors are short-circuited, not all the charge drains off the plates immediately. The dielectric remains partially distorted, preventing some of the charge on the plates from leaving. If you touch such a capacitor some time after you have discharged it for the first time, you may get a nasty shock.

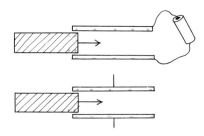

Question 18-18

The diagram shows slabs of a dielectric about to be inserted into two parallel plate capacitors. In the first case, a battery is left connected to the capacitor so that the voltage across the capacitor is maintained as the dielectric enters. In the second case, no battery is connected and the original charge on the capacitor stays constant. What happens to the electrically stored energy in each case? What supplies the work to change the energy in each case?

COMBINATIONS OF CAPACITORS

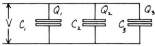

One way to create a large capacitance is to hook up a large number of capacitors in *parallel*. This arrangement is shown in the diagram. The potential difference across each capacitor is the same. The total charge stored is just the sum of the individual charges on the separate capacitors. Therefore, the combination capacitance is equal to

$$C = (Q_1 + Q_2 + Q_3)/V = Q_1/V + Q_2/V + Q_3/V = C_1 + C_2 + C_3.$$

Sometimes, however, capacitors are wired in *series*, as shown in the next diagram. In such an arrangement the *charge*, on each capacitor is the same. (The positive charge on the bottom plate of one capacitor must be just equal to the negative charge on the top plate of the other one—since the two are wired together, and there is no other source of charge for those plates.) On the other hand, the potential across one capacitor is not necessarily the same as the potential across the other one. Instead, the total work done in escorting a test charge through both capacitors is equal to the sum of the individual work done in going through first one and then the second: $V = V_1 + V_2$. Since $C_1 = Q_1/V_1$ and $C_2 = Q_2/V_2$, then $V_1 = Q_1/C_1$ and $V_2 = Q_2/C_2$. Therefore, $V = Q_1/C_1 + Q_2/C_2 = Q/C$. The charge on the series combination is Q, and the combination capacitance is C. Since $Q = Q_1 = Q_2$:

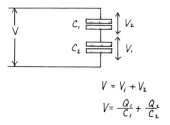

$$\frac{1}{C} = \frac{1}{C_1} + \frac{1}{C_2}$$

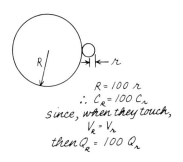

When a charged conducting object touches another conducting object, they share the charge in proportion to their respective capacitances. This must be the case, since while they are touching they are at the same potential. If a small charged sphere with a radius of 1 cm, and thus a capacitance of about 1 pF, is touched by a sphere with a radius 100 times larger, then the little sphere will end up with only about 1% of the original charge and the big sphere will have the other 99%. That's about what happens when *you* touch any of the small classroom electrostatic generating or storage devices with your finger. Your capacitance is probably ten times that of the usual laboratory demonstration device and therefore most of the charge goes over onto the surface of your body. You have "grounded" the device. An even better way to ground something is to fasten it to the largest object around—the earth. If a good conduction path can be arranged to

THE NATURAL UNIT OF ELECTRIC CHARGE 533

moist earth, then most of the charge can be drained off any ordinary object. The "ground" for ordinary house wiring is usually made by making good contact with water pipes that extend out into the surrounding earth.

As we showed on page 521, the electric field close to the surface of a charged conductor is proportional to the density of surface charge on a conductor in that immediate region:

$$E = 4\pi k \sigma = \frac{\sigma}{\epsilon_0}$$

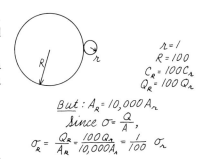

Now we will show that charge density and, consequently, the electric field, is highest in regions where the curvature of the surface is highest. At a sharp point on a conductor, for instance, the charge density is very high and the surrounding field is very strong. Using the idea of capacitance, we can make a simple argument about why this phenomenon should be true. Suppose that a very small sphere with $r = 1$ lies on the surface of a much larger sphere whose radius = 100. They are at the same potential but the big sphere has a capacitance 100 times greater than that of the little sphere and therefore has 100 times the charge of the little sphere. However, let us calculate the respective charge *densities* on the surfaces of the little sphere and the big sphere. The big sphere has a surface 10,000 times that of the little sphere: $(R^2/r^2) = 100^2/1^2 = 10,000$. Therefore the charge *density* of the little sphere is greater by a factor of 100. The electric field in the immediate vicinity of the tiny sphere will be 100 times stronger than that near the surface of the large sphere. In effect, the little sphere is a point on the surface of the big one. Because of the high electric fields produced in the vicinity of points on charged conductors, the air there may become ionized and provide a leakage path for the charge. Any device meant to store charges in air should have a smooth rounded surface.

THE NATURAL UNIT OF ELECTRIC CHARGE

For the unit of electric charge, we chose the coulomb. That turns out to be a very large amount of separated charge. We maintain the unit only because it's a convenient size when charges move past each other to form electric currents. One coulomb per second is an ampere. The charges moving in a wire are on electrons; the positive ions remain locked in their crystalline places. But although the charges move past each other, the wire remains electrostatically neutral. In any region there are just as many negatives as positives.

Since the charge on the electron seems to be nature's unit, why don't we use it? We could, but as we now see, that much charge is very small for most purposes. Let's find out the charge in coulombs on an electron—or, conversely, how many electrons must we gather to form one coulomb?

The charge of the electron was first measured accurately by Robert Millikan in 1909. His apparatus is shown schematically in the diagram. Actually, Millikan measured the electric charge not on electrons but on tiny oil droplets. To duplicate his experiment we spray a mist of oil over the top plate of a parallel plate capacitor. The droplets drift through a hole in the top plate and gradually fall, pulled down by their weight, mg. As seen with

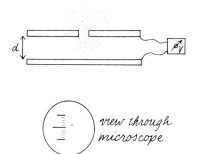

Schematic drawing of Millikan's oil drop apparatus.

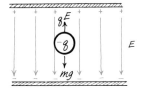

a low powered microscope, they appear to be bright points of light.

When droplets come out of an atomizer, they are usually electrostatically charged. Suppose a droplet has picked up a charge, $-q$. If there is an electric field in the space between the plates, the droplet will be subject to an electrical force, $-Eq$. The electric field, E, is equal to V/d, where V is the potential difference between the plates, and d is the separation distance. We can measure and control the electric field by adjusting a known voltage across the plates. With a small tabletop device and convenient voltages, it turns out to be easy to balance the downward force of gravity with an upward electrical pull. As we watch the pinpoint of light through the microscope, we turn the voltage control and see the point of light slow down, stop, and even start drifting up if we choose. When the droplet is standing still:

$$qE(\uparrow) = q(V/d) = mg(\downarrow)$$

Note that q is the charge on the oil droplet and m is the mass of the oil droplet—not the mass of the extra electrons, which is negligible.

We have no way of knowing how many surplus electrons are on the droplet. However, we can knock some off, or at least change the size of q, by bringing a weak radioactive source nearby. All of a sudden the droplet will start rising or falling again. The voltage source must be readjusted to achieve a new equilibrium. This sequence of operations is done many times with the same droplet, and then with succeeding droplets.

If we knew the mass of the droplet, we could calculate the size of the charge on it. In most school laboratories these days, the "droplets" are actually tiny spheres of plastic of a uniform size manufactured for the calibration of electron microscopes. The diameter of a typical sphere is one micron: 1×10^{-6} m. If you know the diameter and density of a sphere, then you can calculate its volume and mass. Unfortunately, Millikan didn't have spheres of known and uniform size. He found the mass of each droplet by measuring its drift speed as it fell. That speed is related to the diameter and weight of the droplet and the viscosity of air.

Question 18-19

Millikan was watching the droplet through a microscope. Why didn't he measure its diameter just by looking at it through a calibrated eyepiece?

Let's put in some typical numbers for such an experiment. Assume that the plastic spheres have a density of 1×10^3 kg/m³ (the same as that of water). The mass of each one is therefore

$$m = \sigma(\tfrac{1}{6}\pi d^3) = (1 \times 10^3 \text{ kg/m}^3)(\tfrac{1}{6}\pi)(1 \times 10^{-6} \text{ m})^3 = 5.2 \times 10^{-16} \text{ kg}$$

Assume a separation distance between the plates of exactly 1 cm. Then, when the droplet is in equilibrium:

THE NATURAL UNIT OF ELECTRIC CHARGE

$$q = \frac{mg}{(V/d)} = \frac{(5.2 \times 10^{-16} \text{ kg})(9.8 \text{ N/kg})(1.0 \times 10^{-2} \text{ m})}{V}$$

$$q = \frac{(5.1 \times 10^{-17} \text{ N} \cdot \text{m})}{V}$$

If the experiment were done under these conditions, the values for V would be those shown in this table, though they would not be obtained in that particular sequence. Note, first, that only certain values of V produce the condition of balanced force. The mass of the droplet does not change; only the quantity of charge changes as we bombard the droplet from time to time with radioactivity. The data can be explained only if the electric charge is *quantized;* that is, if the charge can exist only in integral multiples of some basic charge. $q = ne$, where $n = 0, 1, 2$, etc.—but never $\frac{3}{7}$ or any other fraction.

V in volts	q in coulombs	n
319	1.6×10^{-19}	1
160	3.2×10^{-19}	2
106	4.8×10^{-19}	3
80	6.4×10^{-19}	4
64	8.0×10^{-19}	5

The minimum charge that can exist on the droplet corresponds to the maximum voltage that is needed. That charge, which is the charge on one electron, is -1.6×10^{-19} coulombs. In one coulomb there are $1/(1.6 \times 10^{-19}) = 6.3 \times 10^{18}$ electron (or proton) charges.

Question 18-20

About how many electrons were on the balloon that you charged in "Handling the Phenomena"?

Now that we know the size of the electron (and proton) charge, let's compare the strengths of the two forces that we have studied so far. What is the ratio of electrostatic attraction and gravitational attraction between proton and electron—for instance, in a hydrogen atom?

$$\frac{F_E}{F_G} = \frac{k(q_e q_p)/r^2}{G(m_e m_p)/r^2}$$

$$= \frac{k q_e q_p}{G m_e m_p}$$

$$= \frac{(9 \times 10^9)(1.6 \times 10^{-19} \text{ C})^2}{(6.7 \times 10^{-11})(9 \times 10^{-31} \text{ kg})(1.7 \times 10^{-27} \text{ kg})} = 2 \times 10^{39}$$

Hydrogen Atom

proton

electron

Both F_{grav} and F_{elec} are attractive forces

The ratio is enormous! Note that the distance between proton and electron cancels out. Both forces are proportional to the inverse square of the distance.

If the electrostatic force is so much stronger than the gravitational force, why do we feel so heavy here on earth? The reason is that although we, and the earth, are filled with electrically charged particles, they are equally distributed between negative and positive. The almost complete neutrality cancels out any large net electrical effect. But there are no negative gravitational charges. Mass is always positive, and the force is always one of attraction. There are about 4×10^{28} protons and neutrons in a human body, and about 3.5×10^{51} protons and neutrons in the earth. The force between each pair of particles, between your body and the earth, is very small—but there are enough pairs to weigh you down.

SUMMARY

There are only four types of physical interactions: gravity, electromagnetism, strong nuclear, and weak nuclear. We have begun the study of electromagnetism, the interaction that is most important for our everyday life. It comprises all the ordinary atomic, molecular, chemical, and biological processes as well as all the phenomena of light, radio, and X-rays. The source of the electromagnetic influence is the electric charge. We divide our study into three realms: the charges are static, the charges are moving with constant velocity, or the charges are accelerating.

In this chapter we link the qualitative and quantitative effects of electrostatics in terms of models. First, we assume that there are negative electrons that can be ripped off some materials and attached to others. If two materials come into contact, one will attract electrons more strongly than the other one and become negatively charged, leaving the other positively charged. Unlike charges attract; like charges repel. A charged object of either sign can polarize the charge distribution in a neutral object, producing an attractive force.

Electrostatics can be made quantitative with the use of Coulomb's law. The force between two point charges separated by a distance r is equal to

$$F = k \frac{q_1 q_2}{r^2}$$

With F expressed in newtons, r in meters, q in coulombs, then $k = 9 \times 10^9$. The coulomb is a unit of charge defined in terms of electric current. There is one ampere of current in a wire if one coulomb of charge passes by a point in one second. Usually in electrostatics we deal with microcoulombs of charge.

The second model that we use directs attention to the way that electric charge distorts or strains the space around it. We picture a force field produced by charges and then calculate the force on a unit test charge that is brought into the field. The field at a particular point in space is defined to be the force per unit test charge brought to that point:

$$\mathbf{E} = \mathbf{F}/q$$

The field is a vector and fields produced at a point by different sources must be added vectorially. We can diagram these fields by drawing at each point an arrow in the direction of the force on a unit charge. These field lines provide a visual representation of the qualitative nature of the electric field around charged objects. With one additional hypothesis they can also provide a quantitative indication of the strength of the field. Where the field lines are crowded, the field strength is large. Using this model, the field strength can be defined to be equal to the number of field lines per unit area, assuming that the unit area is chosen to have its plane perpendicular to the lines: $E = $ (number of field lines)/(unit area).

SUMMARY

Each field line must start on a positive charge and end on a negative charge. Gauss's law, in the form that we have used, asserts that the total number of lines leaving a closed surface must be proportional to the net charge enclosed by the surface. In the S.I. system of units, which we use in this book, each charge provides $4\pi k$ lines of force. We used Gauss's law and Coulomb's law to find the electric field in the region around a charged sphere, a charged cylinder, charged parallel plates, and a charged dipole. The formulas for these fields are summarized in the table on page 523.

Instead of finding the force field at any point near any charged object, we can measure the amount of work needed to take a unit test charge from one point to another in the field. This work per unit charge is called the electric potential difference between those two points. A potential difference of 1 volt is equal to 1 joule per coulomb. The potential difference between two points is independent of the path by which the test charge is moved from one point to another. If some point is defined to be at 0 potential (sometimes this point is chosen to be at infinity) then the potential of any other point is defined as the potential difference between 0 and that point. The potentials as functions of the distance from the standard geometries of charge distributions are given in the table on page 528.

Since potential is a scalar quantity, the potentials at any one point created by different charge distributions can be easily added. Equipotential surfaces are perpendicular to field lines.

The electrical capacitance of a charged object is equal to $C = Q/V$. A capacitance of 1 coulomb per volt is called a farad. Most capacitors that we deal with have capacitances of the order of microfarads or less. The capacitance of a sphere is equal to $C_{\text{sphere}} = R/k$ where R is the radius of the sphere in meters. The capacitance of parallel plates is equal to

$$C_{\text{parallel plates}} = \frac{A}{4\pi k d} = \frac{\epsilon_0 A}{d}$$

If a dielectric is between the plates of the capacitor, then a different value for the permittivity, ϵ, must be used: $\epsilon = \kappa \epsilon_0$. The capacitance of such a system is

$$C = \frac{\kappa \epsilon_0 A}{d}$$

Some capacitors are made with very large surface areas A; others exploit a very small separation distance between plates, d; and yet others have a ceramic with very high dielectric constant κ between the plates.

The energy stored in a capacitor is equal to

$$U = \tfrac{1}{2}QV = \tfrac{1}{2}CV^2 = \tfrac{1}{2}Q^2/C$$

When capacitors are connected in parallel, their capacitances add: $C = C_1 + C_2$. When capacitors are connected in series the equivalent capacity is calculated from $1/C = 1/C_1 + 1/C_2$.

Electric charge is quantized; that is, it can exist only in integral multiples of a basic natural unit. That natural unit is the charge of the electron or proton. (However, subatomic particles called quarks have charges of one-third or two-thirds the natural unit. Many attempts are being made to produce and identify such particles.) The charge of the electron was measured by Millikan to be 1.6×10^{-19} coulombs.

Answers to Questions

18-1: Franklin's model explains part of the action when you bring a charged comb near a tiny aluminum ball. As soon as you touch the comb to the ball some of the extra electrons from the comb jump over to the ball, charging it negative also. Like charges repel; the negatively charged aluminum ball will be forced away from the negatively charged comb. However, with the model that we have developed so far, it is not obvious why the negative comb should have attracted the neutral aluminum ball in the first place.

18-2: There is usually lower humidity in the winter than in the summer, particularly inside heated buildings. In humid conditions there is not only water moisture in the air, but also on the surface of materials. A water film on a surface provides a partially conducting path, so that separated charge can move off an object returning everything to neutral.

18-3: Since the electrons do jump off the belt on to the collecting wire, apparently the *inside* of the dome is not charged negative. Any electrons that get on the collecting wire will try to get as far away from the other electrons as they can. The optimum condition for this is to have all the electrons on the outer surface of the dome. In a later section we will justify this situation from a theoretical point of view.

18-4: If Coulomb had a third test sphere the same size as the other two, he could touch this neutral sphere to the first sphere and remove half its charge. He could then neutralize his third sphere and repeat the procedure by touching it to the second sphere, thus removing half its charge.

18-5:
$$F = (9 \times 10^9) \times \frac{(1\ C)^2}{(1\ m)^2} = 9 \times 10^9\ N$$

That's an enormous force. One kg weighs 9.8 N. 1000 kg weighs 10^4 N. That's a metric ton, which is about the same as our American ton. This electrostatic force of almost 10^{10} N is about equal to 10^6 tons. Such a situation would obviously be impossible.

18-6: As we saw in the answer to Question 18-5, a coulomb of charge would upset everything in its vicinity. Of course the test charge need not be a whole coulomb, just because the units are in newtons/coulombs. The test charge might be only 1×10^{-9} C. If the force on that test charge were 1×10^{-3} N, then the field would be

$$E = (1 \times 10^{-3}\ N)/(1 \times 10^{-9}\ C) = 1 \times 10^6\ N/C$$

18-7: This question is similar to the tug-of-war question. If one team pulls to the left with a force of 1000 N and the other team pulls to the right with a force of 1000 N, is the tension in the rope equal to 2000 N? Of course not. The interaction between two objects produces the same force on each but in opposite directions.

18-8: The "lines" are merely a pictorial model with a quantitative feature concerning the average number of lines per unit area. The actual number of lines depends on how we draw the picture, or on the system of units that we use. Of course, the field strength at all points equidistant from a point source is the same. To find the field at that distance you must take a large enough area so that the number of field lines penetrating that area yields reasonable statistics.

18-9: Once inside the conducting shell the free charges try to get as far away from each other as possible. They do that by going to the outside surface of the conducting shell. If an interior positive charge is held *in* the cavity on an insulator, an equal quantity of negative charge will form on the interior cavity surface. The *net* interior charge (the amount held on the insulator, and the interior surface charge) is still zero.

18-10: The field at any point is made up of the field contributions from all the charges. As you can see in the diagram, in the neighborhood of a uniformly charged cylinder the horizontal component of the field produced by one segment of the cylinder is exactly cancelled by the horizontal component of the field produced by another segment. Only the outward or radial components are not cancelled. They add up to produce the radial field.

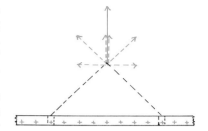

18-11: If there were field lines in a conductor, conduction electrons would experience a force making them move until they gathered and produced a back field to stop the motion. In electrostatics, charges are not moving. The only way this condition can be maintained in the conductor is to have zero electric field in the conductor.

18-12: If an electric field line left a conducting surface at an angle other than 90°, it would have a component along the surface of the metal. But if there is a static situation, there cannot be an electric field in a conductor. If an electric field were momentarily created in a conductor, the electrons would move until they had cancelled out the effect, thus bringing the interior electric field once again to zero.

18-13: Although Gauss's law in its general form always applies, we can use it for practical purposes only if we can choose surfaces that are either perpendicular or parallel to the field lines. Furthermore, every point on the surface must be at the same distance from the source charge. There is no way to surround an electric dipole with such a surface.

18-14: The expression that we used is based on the assumption that E_x remains constant over the displacement Δx. But in this case the field is about 20% smaller at 11 cm than it is at 10 cm. We could have made a better approximation if we had chosen the field strength part way between 10 cm and 11 cm. The actual potential difference in this case is -8.2 volts.

18-15: The potential at any point at a radius equal to or larger than the radius of a uniformly charged sphere is equal to

$$V = k\frac{q}{r}$$

In this case the potential of the sphere itself is equal to $V = (9 \times 10^9)(1 \times 10^{-6}$ C$)(0.1$ m$) = 9 \times 10^4$ V. The potential 30 cm from the center of the sphere is equal to $V = (9 \times 10^9)(1 \times 10^{-6}$ C$)/(0.3$ m$) = 3 \times 10^4$ V. The difference of potential between these two points is thus equal to 6×10^4 V. The work to take the test charge of 10^{-8} coulombs from the low potential to the high potential is

$$W = q\Delta V = (1 \times 10^{-8} \text{ C})(6 \times 10^4 \text{ V}) = 6 \times 10^{-4} \text{ J}$$

18-16: The energy delivered in the spark is equal to $\frac{1}{2}qV$. (The factor of one-half arises because as the charge is drained, the voltage will fall. The average potential difference through which the charge moves is equal to one-half of the original potential difference.) This much energy is equal to $\frac{1}{2}(1 \times 10^{-7}$ C$)(1 \times 10^4$ V$) = 5 \times 10^{-4}$ J. This much energy won't seriously affect anyone.

18-17: Although you are not spherical, you probably have about the same surface area as a sphere 1 m in diameter. Your electrical capacitance therefore should be about 50 picofarads. (5×10^{-11} F).

18-18: In both cases, induced charges on the dielectric provide a force of attraction between the dielectric slab and the plates. The slab is pulled in. To understand what happens when the battery is left connected and the voltage stays constant, use the formula $U = \frac{1}{2} CV^2$. As the dielectric enters the plates, the capacitance, C, increases, and so the electrostatic energy increases. Energy must also be provided to move the slab; the work done is equal to the product of the force and the displacement. This work either provides kinetic energy of the slab or is dissipated in friction. Both the electrostatic and motion energy must be supplied by the battery. The battery maintains constant voltage by sending extra charges to the capacitor as the dielectric enters. This motion of the charges also requires energy, which raises the temperature of the battery and wires.

In the second case, where the charge on the capacitor is constant, the formula to use for the stored energy is $U = \frac{1}{2}(Q^2/C)$. As the dielectric enters the parallel plates, the capacitance increases and the stored energy decreases. Where does the energy go? The lost electrostatic energy turns into the kinetic energy of the dielectric slab. If there were no friction, the dielectric would shoot to the other side and then oscillate back and forth. The action would be very much like that of a pendulum. In this case, electrostatic potential energy would transform into kinetic energy and then back again.

18-19: The droplets, or the calibrated plastic spheres, are only about 1 micron in diameter. The wavelength of light is only about half that size. You cannot accurately measure the size of something if the probe is almost as large. This prohibition is independent of the quality or the magnification of the microscope that you might use.

18-20: On page 512 we calculated that the charge on a typical balloon might be one half microcoulombs = 5×10^{-7} C. The number of electrons on the balloon would therefore be $(5 \times 10^{-7} \text{ C})/(1.6 \times 10^{-19} \text{ C/electron}) = 3 \times 10^{12}$ electrons.

Problems

1. Describe qualitatively how a charged balloon is held to an electrically neutral wall because of induction.
2. List and describe the steps (the actual materials and methods) to charge a small metal sphere negative by contact between the sphere and a charged object.
3. List and describe the steps (the actual materials and methods) to charge a small metal sphere positive without actually touching it with a charged object.
4. Using a diagram, explain why electrons jump onto the outside of the belt at the ground end of a toy Van de Graaff generator.
5. What is the force between a positive point charge and a negative point charge, each 1 microcoulomb, if they are separated by 10 cm?
6. What is the electric field strength 1 meter from a point charge of $+1 \times 10^{-9}$ C? What is the force on an object at that point that has a charge of -1×10^{-11} C?
7. In our (S.I.) system of units, how many electric field lines end up on an electron?
8. If the density of the earth were uniform, the gravitational field would steadily decrease as you went down into a deep mine. Actually, the core is much more dense than the surface mantle. What effect does this fact have on our derivation of the gravitational field as a function of radius inside the earth? Is it possible that the gravitational field is stronger in a deep mine than at the surface?
9. Two positive charges, each 1×10^{-8} C, are 10 cm apart. What is the strength and direction of the electric field along the perpendicular bisector at a distance of 5 cm from the line connecting the charges? What would be the force on a charge of 1×10^{-10} C at that point?
10. The dipole moment of a water molecule is about 6×10^{-30} C/m. What is the electric field strength at a distance of 2×10^{-10} m along the axis of the dipole? (The dipole moment is created because the oxygen atom grabs the two electrons from the hydrogen atoms, making itself doubly negative. The two positive protons, remaining from the hydrogen, cling to the edge of the oxygen atom. This geometry is responsible for the so-called hydrogen bonding.)

11. Write down in sequence the arguments showing that the net electric charge must be zero inside a conducting shell.
12. Write down in sequence the steps in the derivation, using Gauss's law, for the field outside a uniformly charged sphere. Draw the appropriate diagram.
13. Write down in sequence the steps in the derivation, using Gauss's law, for the field outside a uniformly charged cylinder. Draw the appropriate diagram.
14. What is the magnitude and direction of the electric field 10 cm from a long rod that has a uniform charge density of 1×10^{-8} C/m. What is the resulting force on a test charge of $+1 \times 10^{-10}$ C?
15. Write down in sequence the steps in the derivation, using Gauss's law, for the field between the charged plates of a parallel plate capacitor. Draw the appropriate diagram.
16. A square parallel plate capacitor is 10 cm on a side and is given a charge of 1×10^{-9} C. What is the strength of the electric field inside? What is the resulting force on a test charge of 1×10^{-11} C that is between the plates? How does the force depend on the location of the test charge?
17. A parallel plate capacitor has a voltage of 100 volts across its separation of 1 mm. What is the strength of E? What is the magnitude of σ?
18. What would be the potential of a sphere with a radius of 10 cm if you could put 1 C of isolated charge on it?
19. A toy Van de Graaff generator has a metal globe with a radius of 5 cm. How much charge does it take on the globe to produce a potential of 50,000 V?
20. A square parallel plate capacitor is 10 cm on a side with a separation distance of 1 mm. If the charge on one plate is 1×10^{-9} C, what is the potential difference across the plates?
21. List the functional dependence on distance of E and V for sources that are points, uniformly charged spheres, uniformly charged cylinders, infinite planes, dipoles.
22. Sketch the electric field lines (using solid lines) and the equipotential lines (dotted) for the geometries listed in Problem 21.
23. What is the approximate electrical capacitance of an ordinary car? How much charge would it take to give it a potential of 1000 V?
24. What is the capacitance of a flat parallel plate capacitor with a separation distance of 0.1 mm, a length of 1 m, and a width of 10 cm? The dielectric constant, κ, is 3. Would the capacitance be larger if this long strip were made into a four-layer sandwich of dielectric film-conductor-dielectric film-conductor, and then rolled into a cylinder with a diameter of a few centimeters?
25. What is the capacitance of a ceramic capacitor with a cross-sectional area of 1 cm², separated by 1 mm, with a dielectric constant, $\kappa = 10,000$?
26. If you charge yourself up to 15,000 volts by scuffing along a rug, about how much energy have you stored?
27. You have two capacitors: $C_1 = 1 \times 10^{-6}$ F, $C_2 = 2 \times 10^{-6}$ F. What is the combination capacitance if you hook them in series? In parallel?
28. You have two capacitors: $C_1 = 1 \times 10^{-6}$ F, $C_2 = 1 \times 10^{-8}$ F. What is the combination capacitance if you hook them in series? In parallel?
29. Describe the relationships between E and V: given E, how do you find ΔV? Given V, how do you find E?
30. What is the diameter of an oil drop (density = 900 kg/m³) that can just be balanced by one surplus electron in a field of 10,000 V/m?
31. Sketch (qualitatively) the electric field lines and the equipotential lines in the vicinity of a point charge that is a short distance above a conducting plane.
32. Two tiny metal balls, each with a mass of 10 mg and having identical charges, $-q$, are hung from the same point with threads that are 30 cm long. Each thread is at an angle of 15° from the vertical. What is the value of q? How many surplus electrons does this value represent?

19 Electric Current

Now we begin the study of the second realm of electromagnetism. We will see what happens when electric charges move, mostly through wires, and only with constant speed. This type of charge motion is called *direct current,* or dc. In this chapter, we concern ourselves with the motions of the charges and the routes that they take through various arrangements of wires. In the next chapter, we will study another phenomenon that is created by the moving charges.

In *electrostatics,* it is essential that the electric field inside a conductor be zero. Otherwise, the charges would move. To produce current we want the charges to move, and so we must *maintain* an electric field inside the conductor. There must be some sort of power source to maintain this electric field and to pump the charges through potential differences. We will see how electrolytic cells, or combinations of them called batteries, can maintain the needed electric fields.

We have already defined electric current to be equal to $\Delta Q/\Delta t$, a time rate of passage of charge. Now we must find out how the current, I, depends on the electric field, E, and on the materials of the conductor. How much energy must the batteries provide to maintain the electric field, and where does this energy go? To answer these questions, we must build up a model of the microstructure of conductors.

There are also practical questions of circuitry to be answered. Our theories must apply to real wires and resistors and meters. We must learn how to arrange these to produce a variety of practical applications.

HANDLING THE PHENOMENA

1. To study electric circuits, some equipment is needed that is not ordinarily available at home or in dormitory rooms. You need two flashlight batteries, two flashlight bulbs, two bulb sockets, and a couple of meters of thin wire. The first assignment is simply to light one bulb, using only the bulb, one cell, and one piece of wire. By convention, the current direction is defined to be in the direction of positive charge flow. Does the direction of the current in the bulb make any difference in the brightness of the bulb?

Now try wiring various combinations of two cells and two bulbs, as shown in the diagram. These are all "series" connections. Compare the brightness of the light between one cell and one bulb, two cells and one bulb, one cell and two bulbs, and two cells and two bulbs. Next, following the diagram, make "parallel" connections. Compare the brightness of the light with one cell and two parallel bulbs, two cells and one bulb, and two cells and two bulbs. Make note of your results so that when you finish the chapter you can see whether the results have been explained.

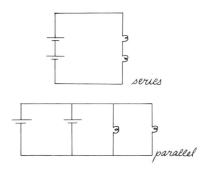

Take a piece of wire just long enough to reach from the positive to the negative electrode of a cell. When you make the contact, you have "short-circuited" the cell. Do this for only a second or two, and note what happens to the wire that you are touching.

2. You can make your own electrolytic cell with a lemon (or any other citrus fruit), a copper penny, and a zinc-coated nail. Roofing nails or ones that appear gray and nubbly are usually coated with zinc. Insert the penny and the nail into the lemon, close together but not touching. This kind of cell cannot provide enough energy to light a bulb but, if you have a good voltmeter, you can determine that there is a potential difference between the copper penny and the zinc nail. If you don't have a voltmeter, use your tongue; tasting this potential difference won't hurt, but it will provide a tingling sensation similar to a sharp sour taste.

3. If you live at home, you should take a look at certain phenomena associated with your house wiring. Of course, household electricity is not direct current; it is alternating current, or ac. Nevertheless, many of the rules that we will develop for d-c circuits also apply to alternating current.

First, take a look at the power meter, located where the power company wiring enters your house. Note that it is not recording power but energy. The name of the commercial unit for quantity of electric energy is the kilowatt-hour. (We will define this unit later.) Have someone turn on a major appliance, such as a stove or a toaster, while you are watching the meter.

The wires from your wall sockets may run some distance to the main fuse box or circuit panel. These wires are supposed to be thick enough so that they do not get hot or cause much of a voltage drop when large amounts of current are drawn from the socket. Most such house circuits are limited to 15 amperes. You can draw almost 10 amperes with a toaster. Plug both a lamp and a toaster into a wall plug and watch what happens to the lamp when you turn on the toaster. If you have an electric shaver, plug that in instead of the lamp and hear the effect of drawing a large current with the toaster. The shaver itself, incidentally, draws very little current compared with the lamp.

CURRENT AND POWER IN FAMILIAR CIRCUITS

We have already talked about current measured in amperes. There is 1 ampere in a wire if 1 coulomb of charge passes in 1 second: 1 A = 1 C/s. The unit was named for André Ampère (1775–1836).

Most of our familiar household appliances are rated in terms of the power they use rather than in terms of the current. The two, of course, are

$$P = \frac{\Delta U}{\Delta t}$$
$$\Delta U = V\Delta Q$$
$$\Delta Q = I\Delta t$$
$$\therefore \Delta U = VI\Delta t$$
$$\text{and } P = \frac{VI\Delta t}{\Delta t} = VI$$

related. If we maintain a potential difference across a wire or a lamp or a motor, then the energy used in shoving a small charge through the device is equal to $\Delta U = V\Delta Q$. The power required is equal to the time rate of using the energy: $P = \Delta U/\Delta t$. But $\Delta Q = I\Delta t$. Therefore, $\Delta U = VI\Delta t$, and $P = VI$. The power is measured in joules per second. The unit of power is a watt: 1 J/s = 1 W. This unit was named for James Watt (1736–1819). (As you can see, many of these early scientists have achieved a certain nominal immortality.)

If we know the power rating of a device and if we know the potential difference across it, then we can find the current in it. A typical flashlight bulb is rated at 1 W. The potential difference is provided by two 1½ volt cells in series. Since $P = VI$, then 1 W = (3 V)I. The current in such a bulb is, therefore, ⅓ A. Consider next the current in a 60 W house light; to be sure, it uses alternating current, but the same relationships hold for average power, current, and voltage. The equivalent average potential difference in this case is between 110 and 120 volts, depending on the locality. If we assume that the value is 120 V, then $I = P/V = (60 \text{ W})/(120 \text{ V}) = $ ½ A.

1 W
3 V
⅓ A

60 W
120 V
½ A

Question 19-1

How can the 60-watt bulb produce so much more light than the flashlight bulb when the current in it is only slightly larger?

Most toasters are rated at about 1000 watts. The current drawn by a toaster would, therefore, be $I = P/V = (1000 \text{ W})/(120 \text{ V}) = 8.3$ A.

A car's starting motor operates on direct current. A typical motor requires about 1 horsepower, the older English unit of power. One horsepower = 746 W. With a 12 V battery, the current drawn by such a motor is $I = P/V = (746 \text{ W})/(12 \text{ V}) = 62$ A. A current that large would blow the fuses in your house wiring. Take a look at the wires used between the battery and the starting motor in a car. They are very thick. In some generators and research magnets, currents of 10,000 A or more are used. The "wires" to carry such large currents are usually copper bars with cross-sections of several square centimeters.

MODEL OF CHARGE MOVEMENT IN A WIRE

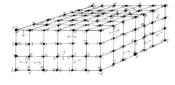

The atoms of a metal are bound in a crystal lattice. The one or two outer electrons, which would be considered valence electrons if the atom were free, are not tied to their parent atom. Instead, they are free to roam throughout the crystal. In the atomic model that we will use, we will assume that the free electrons behave very much like the molecules of a gas. They move about at random, bumping into the massive ions that are vibrating in locked positions.

This classical model of electrons in metal has the same limitations that occur in the Bohr model of the atom. Electrons with the energy of an electron volt or less cannot be localized to regions smaller than an atom. Therefore, we cannot accurately describe their motions in terms of orbits

MODEL OF CHARGE MOVEMENT IN A WIRE

or of particle-particle collisions. Instead, quantum mechanics allows us to predict the probability of finding an electron in a particular region and with a particular energy. The probability functions are in terms of waves that contain all of the familiar wave properties, such as reflection, refraction, and, most important, interference. In this more sophisticated theory, the motion of electrons through a lattice of ions is described in terms of a wave moving through a lattice and being scattered by imperfections in it. Furthermore, there are quantum conditions on the energy levels of the wandering electron. These conditions are similar to the ones that exist for atomic orbits. For instance, no two electrons of the same spin can have the same energy, and only two values of the spin are possible.

Surprisingly enough, however, the crude model of an electron gas in a metal explains most of the ordinary phenomena of electric current. Let's use it as a good first approximation, and borrow additional facts from the more complete theory whenever we need them. Note first the difference between conductors and insulators. If the outer atomic electrons of a substance are being used for molecular bonding, then they are not free to conduct current. It is not the case, however, that materials are necessarily either insulators or conductors. There is an infinite range of possibilities between, including some pairs of materials that permit electrons to move one way across their boundary but not the other. Some materials, such as diamond, are ordinarily very good insulators; but, if a subatomic particle passes through, releasing electrons from their bound positions, the crystal can momentarily become a good conductor.

Assuming that we have a long copper wire filled with free electrons, how can we impose a continuous electric field that will make the electrons move? As a practical matter, we just connect the ends of the wire to the terminals of a battery.

> **Question 19-2**
>
> What does the electric field around a battery look like before you connect the wire?

If the electric field at one point of the wire is stronger than it is at another, then the electrons in the first part will be shoved harder and will accumulate in the second part. But that can't be. The current must be the same at every point in a wire or else the charges would pile up. Therefore, the electric field must be uniform everywhere along the wire (assuming that the wire is uniform and has no divisions). How can the wire have a uniform electric field when it may extend a long way from the battery and may even be tied in knots? The answer must lie in the fact that if at first the electric field is not uniform, charges will pile up and produce forces that tend to slow the motion of charges in one part and speed them up in others. When equilibrium is established, there may be small surpluses of charges along the outer boundaries or at corners. This distribution of surplus charges all along the wire must guide the electric field, causing it to be uniform.

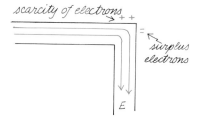

Let's take a particular case with simple geometry to see how much surplus charge is required on the outer wall of the wire in order to lead an electric field around a sharp corner. In the diagram, there is a right angle bend in the wire. In order for the electric field to be uniform in such a geometry, the field lines coming from the left must look as if they were going to terminate in a negative charge at the corner. The field lines starting down must have a pattern as if they had originated in a positive charge on the other side of the corner. For a typical circuit using hookup wire and flashlight batteries, there might be 1 ampere in a wire of 1 mm² cross-section. In order to produce this current, there would have to be a potential difference of 17 millivolts across 1 m of the wire. Therefore, the field in the wire is

$$E = -\frac{(17 \times 10^{-3} \text{ V})}{(1 \text{ m})} = -17 \times 10^{-3} \text{ V/m}$$

Note that in calculating the electric field in the wire, we assumed that it was constant in magnitude throughout the wire and that, therefore, the relationship between field and potential difference was the same as it is in a capacitor where the field is also constant: $E = -\Delta V/\Delta x$.

Now we can find how much charge must be at the corner of the wire in order to terminate this much electric field. Remember that when field lines end on the surface of a conductor the relationship between electric field and charge density is $E = \sigma/\epsilon_0$. In this case,

$$17 \times 10^{-3} \text{ V/m} = \frac{\sigma}{(8.9 \times 10^{-12})}$$

Therefore, the surface charge density is $\sigma = 1.5 \times 10^{-13}$ C/m². Since the total charge is equal to the charge density times the area, $Q = \sigma A = (1.5 \times 10^{-13} \text{ C/m}^2)(1 \times 10^{-6} \text{ m}^2) = 1.5 \times 10^{-19}$ C. This is approximately the charge on one electron! For these particular conditions of electric field and cross-sectional area of the wire, the presence of one extra electron on one side of the corner and the absence of an electron on the other are all that is needed to swing the constant electric field lines through 90°.

When a circuit is first closed, the electric field cannot be uniform all the way along the wire. During the initial surge, however, electrons travel more rapidly where the electric field is strong and pile up in such a way that they reduce the strong field. The surplus charges rapidly establish an equilibrium so that the electric field becomes constant all along the wire, creating a constant current at every point. In Chapter 21, we will calculate how long it takes for the equilibrium to be established in a typical case.

We should emphasize that the current must be constant and the electric field uniform only for the particular case where the circuit consists of just one wire with constant cross section and characteristics. If there are branch points in the wire, the current will divide, some of the charges going one way and some the other. If two different kinds of wire are fastened together to make one long wire, the current will be constant at every point; but in order to produce this condition, the electric fields in the two wires must be different. To understand these complications, we must now derive a relationship between the electric field and the current.

VELOCITY OF CHARGE FLOW IN A WIRE

According to our model, a wire is like a pipe filled with electrons. Although the electrons are dashing about randomly because of their thermal energy, they must also be moving down the pipe because of the imposed electric field. Let's calculate the velocity with which the electrons move along the wire.

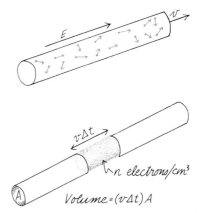

The diagram shows how to compute the number of electrons that will pass a point in unit time Δt. All of the electrons in the cylinder that has length $v\Delta t$ and cross-sectional area A will pass the measuring point in the time Δt. The density of the electrons in the metal is n and, since each electron carries a charge e, the charge density is ne. The amount of charge in the cylinder is, therefore, $\Delta q = neAv\Delta t$. The current is $i = \Delta q/\Delta t = neAv$. If we know the density of the free electrons in a metal, then we can calculate the velocity with which they travel in order to produce a particular current. How many free electrons are in a wire?

Most of the electrons in an atom are bound tightly to the nucleus. In a solid, the outer, or valence, electrons frequently are locked in the molecular bonding. In metals, however, one or two of the outer electrons are free to roam throughout the whole crystal, while the heavy ions remain fixed in the crystalline lattice. Copper provides one free electron per atom. The density of free electrons is therefore the same as the density of atoms. Since we know the number of atoms per mole and the number of moles per gram and the number of grams per cubic centimeter, then we can find the number of atoms per cubic centimeter.

For copper,

$$n = \left(\frac{1 \text{ charge carrier}}{\text{atom}}\right)\left(\frac{\text{atoms}}{\text{mole}}\right)\left(\frac{\text{mole}}{\text{gram}}\right)\left(\frac{\text{grams}}{\text{cm}^3}\right)$$

$$= \left(\frac{1 \text{ charge carrier}}{\text{atom}}\right)\left(6 \times 10^{23} \frac{\text{atoms}}{\text{mole}}\right)\left(\frac{1 \text{ mole}}{64 \text{ g}}\right)\left(9 \frac{\text{g}}{\text{cm}^3}\right)$$

$$= 8.4 \times 10^{22} \text{ charge carriers/cm}^3$$

The velocity of the charge carriers is $v = i/neA$. For a current of 1 ampere in a copper wire with a cross section of 1 mm^2,

$$v = \frac{1 \text{ A}}{(8.4 \times 10^{22} \text{ carriers/cm}^3)(1.6 \times 10^{-19} \text{ C/carrier})(1 \times 10^{-2} \text{ cm}^2)}$$

$$v = 7 \times 10^{-3} \text{ cm/s}$$

Question 19-3

That's an extremely small velocity! Suppose that you have a battery connected to a bulb with two 1 m lengths of wire. How long would it take the bulb to light after you connect the wire to the battery?

The *thermal* velocities of electrons at room temperature are about 8 ×

According to our model, the average energy of electrons should be the same as that of gas molecules at the same temperature. The mass of an electron is smaller than that of an air molecule by a factor of about 60,000. Therefore, the average electron speed must be larger than that of the average molecule speed by a factor of about $\sqrt{60,000} \approx 250$. Since $v_{ave\ mol} \approx v_{sound} = 340$ m/s, then $v_{ave\ elec} \approx 8 \times 10^4$ m/s.

10^4 m/s—much higher than the velocity produced by the electric field. Evidently, the velocity of the electrons created by the electric field is just a small drift velocity imposed on the basic random motion. Nevertheless, there are many electrons drifting at that velocity and, therefore, the current is appreciable. Note, however, that we have been assuming that the current is proportional to the electric field, once the equilibrium is established. But this means that the electron drift velocity is proportional to the electric field.

Actually, our model is a poor approximation in this case. Because of quantum requirements, the electrons in a conductor are not at all in thermal equilibrium with the ions of the lattice. They do not, for example, absorb heat and thus do not contribute appreciably to the specific heat value of the material. The actual random speed of the electron charge carriers in copper is about 2×10^6 m/s, which makes the situation even more dramatic!

Question 19-4

Can the *velocity* of a charge be proportional to the *force* on it? Doesn't Newton's second law apply?

If the current and the drift velocity are proportional to the electric field and if the electric field is constant throughout a wire, then $E = V/L$, where V is the potential difference* across the wire of length L. According to our assumptions, it appears that if the current in a wire is proportional to the electric field in the wire, then it is also proportional to the voltage across the wire: $I \propto E \rightarrow I \propto V$. It is usually easier to measure the potential across a wire or a circuit than to measure the electric fields in it. We will analyze many circuits in terms of the proportionality between current and voltage but, as a matter of fact, it isn't always valid. There are important exceptions. As often happens in science, the exceptions reveal more about the theory than does the rule.

CURRENT AS A FUNCTION OF V

We now have a formula for current as a function of the velocity of the charge carriers, and a relationship between that velocity and the potential difference across the circuit. Let's combine these two expressions.

$I = neAv$ (From now on, we will use capital I for a constant current.)

$$v = KE = K\left(\frac{V}{L}\right)$$

The proportionality constant between the drift velocity and the electric field is K. (It is called the *mobility*.) The constant field in the wire is equal to the potential difference across the wire divided by its length.

---

* From now on in this chapter we will usually use V to indicate the potential difference *across* a circuit. In everyday language, V is referred to as the voltage.

$$I = \left[ne\left(\frac{A}{L}\right) K \right] V \qquad I \propto V$$

The current in a wire is proportional to the voltage across it, although the proportionality factor may change with varying conditions. For instance, a change of temperature may change n, the density of charge carriers; and it may also change K, which relates the electric field to the drift velocity. Furthermore, the proportionality constant between current and voltage depends on the length and the cross-sectional area of the wire.

Let's examine the current-voltage relationship in three cases. It is easy in the laboratory to vary the voltage across a wire or a filament or some other circuit element and to measure the current that is produced. The operation is shown schematically in the diagram, where the circle containing A represents an ammeter, which is used to measure the current, and the circle with V represents a voltmeter. Note that the ammeter is reading the current *in* the circuit, whereas the voltmeter is reading the potential difference *across* the circuit element.

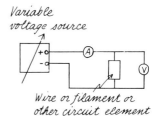
Variable voltage source

Wire or filament or other circuit element

1. *I(V) for a metal wire at constant temperature.* For an ordinary metal wire at constant temperature, the current is proportional to the voltage. The number of charge carriers remains constant and is independent of the direction of the current. (We will soon see a case where n depends on the direction of the current.) The graph of $I(V)$ is shown in the diagram. The positive or negative signs of the current and voltage merely represent the directions. In this case, if you reverse the battery, you reverse the current, and the proportionality constant stays the same.

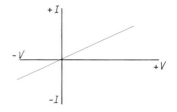

2. *I(V) of a light bulb filament.* The temperature of a tungsten filament in a light bulb rises to about 2000 K when it is fully lighted. The density of charge carriers in the wire does not change appreciably with this much of a temperature change. But the drift velocity does. Remember that $v_{drift} = KE = K(V/L)$. Considering our model of electrons in the metal, we might expect that the drift velocity would be reduced by the increased thermal oscillations of the lattice ions. At a higher temperature and a larger amplitude of oscillation, the ions present larger targets to the drifting electrons, thus reducing their mobility. Notice the effect that a change in K has in our basic equation: $I = ne(A/L)KV$. If K decreases with temperature, then there will be less current for a particular voltage. As the voltage across a filament increases, so does its temperature. The resulting curve of $I(V)$ looks like the one shown in this diagram.

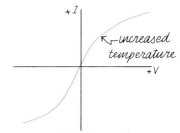

Current as a function of voltage for a tungsten filament. The shape of the curve depends on the temperature of the filament at a particular voltage. Current must be measured after temperature equilibrium is attained.

Question 19-5

Does the region of negative current and voltage represent high temperature or low temperature?

When Thomas Alva Edison started making electric light bulbs in the latter part of the nineteenth century, he used carbon filaments in vacuum tubes. (Our present light bulbs are made with tungsten filaments enclosed

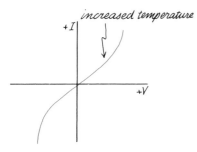

$I = ne(A/L)KV$
With carbon, as temperature increases, K decreases, but n increases.

in a bulb that contains an inert gas at atmospheric pressure.) The carbon filaments were very fragile and also had an electrical characteristic quite different from that of tungsten. The $I(V)$ graph for a carbon filament looks like the one in the diagram. Since an increase of voltage makes the light glow brightly, implying an increase of temperature, it must be that the proportionality constant between I and V *increases* with temperature. Nevertheless, it is still true that the drift velocity constant, K, decreases as the temperature rises. With carbon, however, the density of charge carriers, n, increases strongly with increased temperature. The reason for this is that most of the valence electrons in solid carbon are only loosely tied up in covalent bonds. The number of electrons per atom that are free to roam depends on the temperature (and on the type of carbon crystal). If the voltage increases across a carbon filament, the current increases and the temperature rises. But as the temperature rises, more electrons are freed from the covalent bonds, increasing the current. In turn, the larger current makes the temperature rise still higher. These processes would compound until the carbon filament burned up except for one other factor. The radiation given off by the glowing filament is proportional to the fourth power of the temperature. Before the filament boils away, equilibrium is established between the energy supplied by the voltage source and the energy being radiated. The temperature stops rising and the current stops increasing.

When a voltage is applied across a tungsten filament, the current is large at first because the metal is cold. In about a hundredth of a second, the temperature rises and the current decreases to its equilibrium value. With a carbon filament, the sequence is just the opposite. The initial current is small but then increases to its equilibrium value as the temperature of the filament increases.

3. *$I(V)$ of a diode.* The junction between a copper wire and a silver wire allows electron flow in either direction. This is not true of the junctions of certain materials, such as copper and copper oxide, or specially prepared combinations of germanium or silicon. Such junctions that favor flow of charge in one direction are called diodes. A typical $I(V)$ curve for a diode is shown in the diagram. In one direction, shown as positive in the graph, electrons can move across the junction as if it were an ordinary wire. In the other direction, however, there is great resistance to electron flow. For most diodes, the curve does not sharply change direction at zero voltage. Instead, the forward characteristic slurps gradually into the backward characteristic over a region of several tenths of a volt. Also, in most cases, the changeover takes place on the positive side of the voltage graph. It is necessary to get beyond a threshold voltage of several tenths of a volt before the electrons can pass the junction easily.

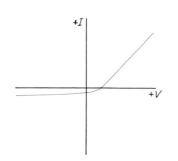

THE SPECIAL—BUT IMPORTANT—CASE OF OHM'S LAW

We have just seen some cases where the current in a circuit element is *not* proportional to the voltage across it. For many circuit arrangements, however, materials are used so that the current *is* proportional to the voltage. The conditions must be such that the number of charge carriers does not depend on the voltage, and such that the proportionality constant, K, between v_{drift} and E must remain at least approximately constant. Usually these restrictions mean that the temperature of the circuit must not change much when the circuit is activated. In that case, we can lump all of the constants relating current and voltage into one new constant:

$$I = ne\left(\frac{A}{L}\right)KV = \frac{V}{R}$$

The new constant, R, is called the *resistance*. The unit of resistance is the *ohm*, defined by a circuit in which there is 1 A if there is a potential difference of 1 V. The symbol for the ohm is the Greek letter Ω (capital omega): $1\ A = 1\ V/1\ \Omega$.

The misanthropic Englishman, Cavendish, measured the proportionality of current and voltage in the 1770s. Characteristically, he did not bother to tell anyone about it at the time. The law was announced for the first time by Georg Ohm (1787-1854) in about 1820, but the law was not generally recognized for another 10 years. Now it is known as Ohm's law: $I = V/R$.

Notice the relationship between the resistance R and the other factors determining current: $1/R = ne(A/L)K$. It may seem unreasonable to you to group those factors together and then define them as being the *reciprocal* of a new constant. A different arrangement might have been chosen with the product of those factors equal to a constant called the conductivity. However, history and custom have ordained otherwise.

The resistance can be expressed in terms of the other factors in another way.

$$R = \left(\frac{1}{neK}\right)\left(\frac{L}{A}\right) = \rho\frac{L}{A}$$

By making this rearrangement, we have separated the geometrical aspects of the conductor from those concerned with its microstructure. The new constant, ρ, is called the *resistivity*. It is a property of the type of material in the conductor and depends on the density of charge carriers and on the relationship between electric field and drift velocity. The geometrical factor (L/A) depends on the thickness of the wire and its length. If we know the size of a wire and the resistivity of the metal from which it is made, we can calculate its resistance. The units of resistivity are ohm-meters (Ω-m). When you multiply resistivity by the length of the wire and divide by its cross-sectional area, the resulting unit of the resistance is the ohm.

$$R(\text{ohms}) = \rho(\text{ohm-meter})\frac{L(\text{meters})}{A(\text{meter}^2)}$$

Question 19-6

The $I(V)$ graph of a conductor that obeys Ohm's law is shown on page 549. How is the resistance of the conductor related to the slope of the graph?

The resistivity of various materials is shown in the table below. Note the enormous range of values. As you can see, there are both good and poor conductors and also good and poor insulators, with a very large difference between the extremes of the two groups.

Resistivities (ρ) and Temperature Coefficients (α)

	ρ in $\Omega \cdot m$	α $(C°)^{-1}$
Aluminum	2.7×10^{-8}	0.0043
Boron	2×10^{4}	
Carbon (amorphous)	3.5×10^{-5}	-0.0005
Copper	1.7×10^{-8}	0.0068
Germanium	0.46	-0.05
Gold	2.4×10^{-8}	0.004
Iron	9.7×10^{-8}	0.0065
Lead	20.7×10^{-8}	0.0034
Mercury	98×10^{-8}	0.0009
Platinum	10.6×10^{-8}	0.0039
Silicon	$100 - 1000$	-0.075
Silver	1.6×10^{-8}	0.0041
Sulfur	2×10^{15}	
Tungsten	5.6×10^{-8}	0.0045
Nichrome (alloy of Fe, Ni, Cr)	100×10^{-8}	0.0004
Fused quartz	10^{18}	
Glass	$10^{10} - 10^{14}$	
Hard rubber	$10^{13} - 10^{16}$	
Hardwood	$10^{8} - 10^{11}$	

Let's calculate the resistance of a copper wire 1 m long with a cross section of 1 mm². This is about the size of thin wire (AWG gauge #18) that we assumed we would use earlier in several of the demonstrations with batteries and light bulbs.

$$R = (1.7 \times 10^{-8} \, \Omega\text{-m}) \frac{(1 \text{ m})}{(1 \times 10^{-6} \text{ m}^2)} = 1.7 \times 10^{-2} \, \Omega$$

Question 19-7

If this 1 m of copper wire is put across a flashlight cell, what current will be produced?

We can calculate the resistance of a flashlight circuit by using Ohm's law. As we have already seen, a typical bulb used with a typical double-cell flashlight has a power of 1 W and uses about $\frac{1}{3}$ A. Therefore,

$$I = V/R \qquad (\tfrac{1}{3} \text{ A}) = (3 \text{ V})/R$$

The resistance of the flashlight circuit must be 9 Ω. This resistance is relatively large compared with that of a copper wire. It is provided by the hot filament, which would have a much lower resistance at room temperature.

The drift speed proportionality constant K, and hence the resistivity ρ, depends not only on the temperature of the metal but also on the crystalline structure. The values of resistivity given in the table for the conductors are for very pure metals. Small amounts of impurities or imperfections in the lattice produced during crystallization can greatly increase the resistivity. The complete explanation of these effects depends on the quantum conditions that we mentioned earlier. In the probability wave model of electron conduction, the electron wave merely diffracts around stationary atoms. Resistivity is caused by the scattering of the wave as it travels down the wire. Any deviation of regularity of the crystal structure, such as would be caused by a gap or an impurity atom, produces an increase in resistivity. Here we have the quantum explanation of how the high-density electrons ($n \approx 10^{23}$ charge carriers/cm³) can act like a low pressure gas (at standard temperature and pressure for a gas, $n \approx 3 \times 10^{19}$ molecules/cm³). Although copper atoms in a solid are only about 3×10^{-8} m apart, electrons in copper at room temperature typically travel 300×10^{-8} m between collisions. In effect, their wave behavior reduces their effective density by a factor of $100^3 = 10^6$.

■ Optional

Frequently, we tend to think of the electrical resistance of a substance as representing some sort of an obstacle course that electrons must pass through. Such a notion leads to apparent paradoxes that are unnecessary. Remember that resistance is just a proportionality constant between current in a circuit and the potential difference across it. Better yet, think of Ohm's law as a relationship between the current and the electric field.

$$I = \frac{V}{R} = \frac{V}{\rho(L/A)} = \frac{(V/L)}{(\rho/A)} = \frac{E}{(\rho/A)}$$

For an example of this altered point of view, consider the two wires in parallel, as shown in the diagram. They are made of the same material and have the same cross section, A, and the same resistivity, ρ. However, $L_2 = 2L_1$. The voltage across the two wires is the same, but their electric fields are different: $E_1 = V/L_1$, and $E_2 = V/L_2 = V/2L_1$. Therfore, $E_1 = 2E_2$. Because the electric field in the shorter wire is twice that in the longer one, the drift velocity of the electrons in the first will be twice that of the second. Hence the current in the first will be twice that in the second. Of course, you can get the same result by calculating that $R_2 = 2R_1$. Since $I = V/R$, it follows that $I_1 = 2I_2$. If you ask how much work you must do to escort a unit charge through the two resistances, you might be tempted to think that you must do more work to shove the charge through the larger resistance, but there is the same potential drop across both of them; the work done per charge must be the same by either route.

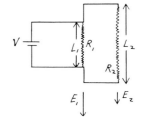

Question 19-8
Why is it that the same work is done in the two cases? After all, the charge going through R_2 must travel twice as far.

Let's look at the meaning of resistance in another case, which is shown in this next diagram. Once again, we have two wires in parallel with the same potential difference across them, but this time they have the same length.

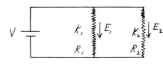

Therefore, there is the same electric field in each of them. They can have the same cross-sectional area, A, and the same charge carrier density, n, but because they are different metals or because they are at different temperatures the drift velocity constant, K, will be different. Let us assume that $K_1 = 2K_2$; then since $v_{drift} = KE$, it must be that $v_1 = 2v_2$. The drift velocity in the first wire is twice the drift velocity in the second. Therefore, the current in the first one will be twice that in the second: $I_1 = 2I_2$. Since $\rho \propto 1/K$, it must be that the resistance of the first one is only half that of the second: $R_1 = \frac{1}{2}R_2$. Once again, we face the question of how much work is done in escorting a charge by each of the two routes. Of course, that work must be the same, since the two routes are in parallel. But route 1 has only half the resistance of route 2. Nevertheless, the fields in the two routes are the same and so are the distances through which the charges must be taken. Since the work per charge is equal to the force per charge times the distance, the work done is the same in both routes in spite of the fact that the resistances are different.

Let's examine one more case where electric fields and resistances are different in two parts of the circuit. The diagram shows a junction between a thin wire and a thick one. Let us assume that the wires are made of the same material at the same temperature but that $A_2 = 2A_1$. Since the wires are in series, there must be the same current in each. Since $I = neAv$, the drift velocity in the larger wire must be just half that in the small wire: $v_2 = \frac{1}{2}v_1$. But the drift velocity is proportional to the electric field: $v_{drift} = KE$. Therefore, the field in the big wire must be only half that in the small wire: $E_2 = \frac{1}{2}E_1$.

Question 19-9
What happens to the electric field lines in the small conductor as they come to the junction with the large one?

If the small wire and the large one in series each have the same length, then the potential difference across the thin wire is twice that across the thick wire because $V = EL$. In this case, the resistance of the thin wire is twice that of the thick wire, and the work done in escorting a charge through the thin wire is twice that of escorting a charge through the same length of the thick one. The reason, however, that twice as much work is done is not because there is more resistance in the wire but because there is a stronger electric field in the wire.

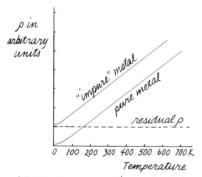

$\rho(T)$ for a "pure" and "impure" metal. The resistivity for a pure metal, without lattice imperfections, is caused by thermal vibrations of the atoms. For real metals, any impurity atoms or breaks in the crystal lattice produce a residual resistivity that does not depend on temperature.

TEMPERATURE DEPENDENCE OF ρ

According to our model, the resistivity should increase with increasing temperature as long as the density of charge carriers stays constant. The resistivity as a function of temperature is shown for several elements in the graph. Over a considerable range, in the region of room temperature, the curve can be described by $\rho = \rho_0 (1 + \alpha t)$. In this formula, t is the number of degrees above room temperature, 20°C. Values of the temperature coefficient, α, are given for several materials in the table on page 552.

On the scale of the graph, it appears that the resistivity of metals goes to zero at absolute zero. In general, this is not true. A residual resistance remains. On the basis of both the electron gas model and the probability wave model, we should expect resistance to decrease as the temperature drops. In both models, the vibrations of the bound ions decrease, presenting smaller collision targets and a more regular lattice. In the electron gas model, the thermal velocities of the electrons decrease with decreasing temperature. During the longer interval between collisions, the electric field can have a greater effect on the electrons, thus increasing their drift velocity and decreasing the resistance. In the probability wave model, a lower velocity for the electrons leads to a longer wavelength, which leads to less scattering and less resistance. However, the lattice vibrations do not stop at absolute zero. Even in a perfect crystal we should expect particle collisions on the one hand or wave scattering on the other, yielding nonzero resistance.

In the regions below 20 K, a number of materials *do* have zero resistance. This phenomenon was discovered in 1911. It is called *superconductivity*. A list of temperatures at which certain materials become superconducting is given in the table below. Note that a superconducting material does not just have small resistance; it has zero resistance. The discoverer of this effect, Kamerlingh Onnes, produced a current in a ring of superconducting mercury and maintained the current for many years without any power supply.

Critical Temperature for Onset of Superconductivity

	$T_c(K)$
Aluminum	1.18
Mercury	4.16
Indium	3.41
Niobium	9.17
Lead	7.23
Tin	3.72
Tantalum	4.39
Vanadium	5.37
Zinc	0.85

Various combinations of molybdenum (Mo), niobium (Nb), tin (Sn), vanadium (V), germanium (Ge), indium (In), gallium (Ga), and aluminum (Al) yield materials that are superconducting at relatively high temperatures. These mixtures are part compound, part solid solutions.

Al_2CMo_3	10
$Al_{0.8}Ge_{0.2}Nb_3$	20.7
$AlNb_3$	18.0
$In_{0.3}Nb_3Sn_{0.7}$	18

Notice that copper and silver do not become superconductors.

> **Question 19-10**
>
> How could Onnes determine that there was still a current in the coil without inserting a meter in the circuit, which would have used up energy?

The theoretical explanation of superconductivity was formulated only within the last two decades. It is strictly a quantum effect without any classical analogy. It turns out that pairs of electrons interact with each other and with the lattice in such a way that the two electrons are loosely held together. Although the binding is weak, it is sufficiently strong so that one electron cannot undergo a collision (its probability wave cannot scatter) from an imperfection or vibration in the lattice without disturbing the other electron. Below a critical temperature, the lattice ions cannot provide sufficient energy in such collisions to break up the electron pair. The electrons cannot exchange any smaller amount of energy, and so they move through the crystalline lattice without any resistance at all.

You might imagine that there would be great commercial advantages in having conductors with zero electrical resistance. Why shouldn't it be possible to transmit power over long distances without any energy loss? There are two problems. First, the superconductor must be kept below its threshold temperature, which in all cases is close to absolute zero. The refrigeration and insulation needed to produce and maintain these conditions is very expensive. Second, the current that can be maintained in a superconductor is limited. Large currents or large magnetic fields can ruin the superconducting property. In spite of these problems, very large superconducting machines, including magnets, have been built and are being used for practical purposes, mostly in physics research laboratories.

POWER LOSS IN CIRCUITS

We have already seen that it takes work to move a charge through a potential difference. This work is equal to $\Delta q \times V$, and the power required is $\Delta W/\Delta t = (\Delta q/\Delta t)V = IV$. The charges move with a drift velocity and so, on the average, do not gain kinetic energy through acceleration. Nevertheless, work is done on them to maintain the drift velocity, and this energy must go someplace. It shows up, of course, in the thermal motion of the ions in their lattice. In other words, a wire with current in it gets hotter.

Since $I = V/R$, we can express the power loss in three different ways:

$$P = IV = I^2R = V^2/R$$

For example, if there is a current of 0.1 A in a resistor of 100 Ω, then the power dissipated is

$$P = (0.1 \text{ A})^2(100 \text{ }\Omega) = 1 \text{ W}$$

One watt is a small amount of power in most circumstances. It is about the power dissipated in a flashlight bulb when it is lighted. However, if the

PRACTICAL RESISTORS

current in the 100 Ω resistor is 10 A, then the power dissipated is much greater:

$$P = (10 \text{ A})^2(100 \text{ Ω}) = 1 \times 10^4 \text{ W}$$

Notice that, while the current in our second example is 100 times greater than it is in the first, the power dissipated is larger by a factor of 10,000.

The fuse in a household circuit consists of a thin ribbon of conductor that has low resistance but which also has a low melting point. If the current in the wire and hence in the fuse exceeds its rating, the fuse wire melts, thus interrupting the circuit and possibly preventing more serious damage. A typical 15 A fuse has a resistance of 0.1Ω. At the rated current, the power being dissipated in the fuse wire is equal to

$$P = (15 \text{ A})^2(0.1 \text{ Ω}) = 22.5 \text{ W}$$

Question 19-11

In one of these expressions, power is proportional to the resistance. In another formula, it is *inversely* proportional to the resistance. If you connect two wires, one with a resistance of 1 Ω and the other with a resistance of 2 Ω, across a battery which wire would have more power dissipated in it?

The unit of power is the watt. A larger unit is the kilowatt, kW, which is about four-thirds of a horsepower. When you buy electricity, you do not purchase power but rather energy. The usual unit is the kilowatt-hour, kWh.

$$1 \text{ kWh} = (1 \times 10^3 \text{ W})(3.6 \times 10^3 \text{ s}) = 3.6 \times 10^6 \text{ J}$$

The cost of energy is rapidly rising but, at a cost of 8 cents per kilowatt-hour for electricity delivered to a house, 1 J costs about 2.2×10^{-8} dollars. Turning the numbers around, you can get about 4.5×10^7 joules of electricity per dollar. Don't consider that price the cost of electricity, however. The number of joules per dollar depends on how the electricity is delivered. A large industry near a power plant may get the electricity at one-third the cost. On the other hand, the total energy in a D cell (a standard flashlight battery) is about 10^4 J. If such a battery costs about 30 cents, the cost of 1 J is about 0.3×10^{-4} dollars, or 3×10^4 joules per dollar.

The electric power used or produced by various appliances or generators is shown in the table on the next page. As you can see, devices around the house that are designed to produce heat from electricity usually consume much more power than devices that perform mechanical work.

PRACTICAL RESISTORS

Many circuit elements such as coils and filaments have built-in resistance. In the case of lamp filaments, the resistance is a desirable feature because

---

P = IV
1 watt = (1 ampere)(1 volt)
  = (1 coulomb/second)(1 volt)
  = 1 joule/second

Household electricity cost
If 3.6×10^6 J costs $0.08
1 J costs 2.2×10^{-8}
or: 4.5×10^7 J/$

Power Production and Consumption for Households

Flashlight (2 D-cell)	3 W	Freezer (15 ft³)	400 W
3 horsepower lawnmower	2¼ kW	Refrigerator (12 ft³)	250 W
Average power (from all sources) used per U.S. citizen	12 kW	Shaver	14 W
		Toaster	1.2 kW
Air conditioner (window—one room)	1.5 kW	Vacuum cleaner	650 W
		Clothes washer	500 W
Electric blanket (average)	175 W	Television	200–300 W
Electric clock	2 W	Stove (4 burners + oven)	12 kW
Clothes dryer	5 kW	Hairdryer	400 W
Dishwasher	1.2 kW	Radio-phonograph	100 W
Small circulating fan	100 W	Water pump	500 W
Attic fan	350 W	Oil burner motor	300 W

it is the I^2R loss that heats the filament to incandescence. In other cases, the heat produced is a nuisance, and provisions must be made to radiate or conduct it away from the circuit. Some electronic devices have metal radiator fins and require forced air cooling.

A basic circuit element is a lumped resistance of a predetermined amount. It is called a resistor. A number of them are shown in the picture.

Question 19-12

Why should resistance be introduced deliberately into a circuit?

22 Ω resistors

Each of the resistors shown in the picture has a resistance of 22 Ω. Yet they are radically different in size. The difference lies in the amount of power that they can safely dissipate. The smallest is ½ W while the largest is rated at 100 W. The amount of current that could be tolerated in each is determined by the I^2R loss. For the 22 Ω, ½ W resistor, the maximum current is 0.15 A; for the 100 W resistor, the maximum current is 2.1 A.

The small resistors shown in the picture are made of a ceramic mixed with carbon. They are commercially available with standardized values of resistance such as 22 Ω (but not 20 Ω) and with wattage ratings of ½ W, 1 W, and 2 W. For power dissipation above 2 W, resistors are usually made of wire coils on ceramic cylinders. Resistors of the compressed carbon type are usually available up to about 10^8 Ω. For higher resistances, the resistor material must be encapsuled in glass and handled in such a way that no grease or dirt gets on the glass envelope. Otherwise, the surface resistance would be less than the rated internal resistance.

Consider what size resistor you should use if you want to put 10 Ω across a power supply. If the power supply is a flashlight cell, then the power that will be dissipated is $P = V^2/R = (1\frac{1}{2} \text{ V})^2/(10 \text{ Ω}) \approx \frac{1}{4}$ W. In this case, you could select a ½ W resistor and have an adequate safety margin. However, if the 10 Ω is to be placed across a house socket, the power needed is $P = (120 \text{ V})^2/(10 \text{ Ω}) = 1440$ W. In this case, your 10 Ω resistor ought to have the heat dissipating characteristics of a toaster.

RESISTORS IN SERIES AND PARALLEL

In electric circuits, there are many combinations of resistors, coils, capacitors, batteries, and other circuit elements. The current wends its way through these, dividing at each junction point. In order to calculate the current in each circuit element and the voltage across it, we must figure out the equivalent resistance of two or more resistors in series or in parallel.

A basic assumption in analyzing a circuit is that there is no potential drop along the connecting wires. This is not always true. In "Handling the Phenomena," you saw or heard what happened when you turned on a toaster next to a lamp or a shaver in a house circuit. The potential at the wall dropped. Evidently, there was a potential drop in the wires that connected the wall socket and the main fuse box. Nevertheless, as a first approximation in the circuits that we analyze, we assume that every point along a connecting wire is at the same potential.

We must also make use of a fact that you have already learned. If two circuit elements are in series, the potential difference across the combination is the sum of the separate potential differences. For instance, if two $1\frac{1}{2}$ V flashlight cells are in series, they provide a 3 V potential difference. The same situation holds true with two resistors in series. If there is a 1 V potential drop across the first and a 2 V potential drop across the second, then the potential difference across the combination is 3 V. In all of these cases, remember that the potential difference is the work done in taking a unit charge through the circuit element. If it takes 1 J/C to go through the first resistor and two more joules per coulomb to go through the second, then of course it takes 3 J/C to go through both of the resistors.

On the other hand, if circuit elements are in parallel, then there is the same potential difference across them. This must be true since there are no potential drops along the connecting wires. When you arranged the two dry cells in parallel in "Handling the Phenomena," you saw that the brightness of the bulb was the same as if you had used only one cell. (If a very large amount of current must be furnished, however, sometimes it is a good idea to have two cells in parallel so that each can furnish half of the current. We will see another reason for such an arrangement in the next section.)

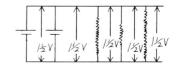

1. *Resistors in series*. The potential difference across a resistor is IR, since $I = V/R$. Such a potential difference is commonly referred to as an "IR drop." The voltage drop across two resistors in series is

$$V = V_1 + V_2 = IR_1 + IR_2 = IR_{\text{equiv}}$$

V_1 and V_2 are the IR drops across resistors R_1 and R_2. The equivalent resistance for the two in series is R_{equiv}. Since the resistors are in series, there is the same current, I, in each. Therefore,

$$R_{\text{equiv}} = R_1 + R_2$$

The equivalent resistance for resistors in series is simply the sum of the individual resistances.

2. *Resistors in Parallel*. In a parallel circuit, the potential drop across each of the circuit elements is the same: $V = V_1 = V_2$. However, the currents in each circuit element may be different. The current in the main part of the circuit divides so that part of the charge goes down one branch and part goes down another:

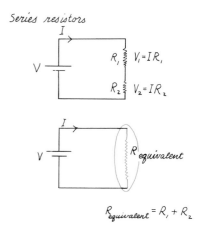

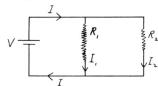

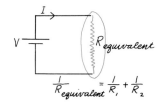

$$I = I_1 + I_2$$

Since in each branch the current and potential drop are related by $I = V/R$, it must be that

$$\frac{V}{R_{equiv}} = \frac{V}{R_1} + \frac{V}{R_2}$$

Since the voltage is common,

$$\frac{1}{R_{equiv}} = \frac{1}{R_1} + \frac{1}{R_2}$$

Compare these formulas for combining *resistors* with the similar formulas for combining *capacitors*. They are just the inverse. Here, if the resistors are in series, their resistances add. If they are in parallel, the equivalent resistance is found from a reciprocal formula.

> **Question 19-13**
>
> What leads to this difference in the combination rules for resistors and capacitors? Weren't similar arguments used in both derivations?

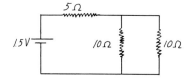

Let's apply these combination rules to a simple, practical example. How much current is provided by the battery in the circuit shown here, and what is the voltage across the parallel network? The general procedure for solving any such problem is to reduce any parallel network to its equivalent resistance, then add that resistance to anything in series with it. In this case, we have two 10 Ω resistors in parallel. The equivalent resistance is easy to find in this case. It is simply equal to one-half of either identical resistor or, in this case, 5 Ω. This solution must be correct for identical resistors, since

$$\frac{1}{R} = \frac{1}{R_1} + \frac{1}{R_1} = \frac{2}{R_1}$$

Therefore, the equivalent resistance, R, equals $\frac{1}{2}R_1$. In our particular example,

$$\frac{1}{R} = \frac{1}{10} + \frac{1}{10} = \frac{1}{5}$$

■ **Optional**

Sometimes, it is useful to solve the formula of reciprocals to find directly the equivalent resistance of two resistors in parallel:

$$R_{equivalent} = \frac{R_1 R_2}{R_1 + R_2}$$

In this form, it is apparent that the equivalent parallel resistance must always be less then either of the separate resistances, since the formula could also be written

$$R_{equivalent} = R_1\left(\frac{R_2}{R_1+R_2}\right) = R_2\left(\frac{R_1}{R_1+R_2}\right)$$

In either case, the individual resistance would be multiplied by a fraction less than one. If one of the parallel resistors is less than one-tenth of the other one, then usually the large one can be ignored.

Question 19-14
If a 10-Ω resistor is in parallel with a 100-Ω resistor, what is the approximate equivalent resistance, and what is the error introduced by ignoring one of the resistances?

Since the equivalent parallel resistance in our example is 5 Ω, the total series resistance in the circuit is 10 Ω. Therefore, the current supplied by the battery is equal to

$$I = \frac{1.5\text{ V}}{10\text{ Ω}} = 0.15\text{ A}$$

The potential drop across the parallel network is the IR drop, where R is the equivalent resistance. In this case, that voltage drop is

$$IR = (0.15\text{ A})(5\text{ Ω}) = 0.75\text{ V}$$

Since the individual 5 Ω resistor is in series with an equivalent 5 Ω, the potential drop across the combination divides equally, so that there is 0.75 V across each.

Let's analyze the current and division of potentials in another simple circuit that is shown here. In this case, we do not have to do a detailed calculation to find the equivalent parallel resistance. It must be slightly less than 1 Ω, differing by only 2%.

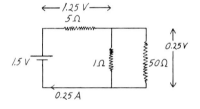

Suppose $R_1 << R_2$. Then the relative difference between R_1 and the true value, R_{equiv}, is

$$\frac{R_1 - R_{equiv}}{R_{equiv}} = \frac{R_1[1 - R_2/(R_1+R_2)]}{R_{equiv}} \approx 1 - R_2/(R_1+R_2)$$
$$= R_1/(R_1+R_2) \approx R_1/R_2$$

In this case, the relative difference is approximately $\frac{1}{50}$, which equals 2%.

In our example, the total series resistance is about 6 Ω. The current provided by the battery is

$$I = \frac{1.5\text{ V}}{6\text{ Ω}} = 0.25\text{ A}$$

The potential drop across the parallel network is
$$IR = (0.25\text{ A})(1\text{ Ω}) = 0.25\text{ V}$$

The other 1.25 V provided by the battery is the potential drop across the 5-Ω resistor.

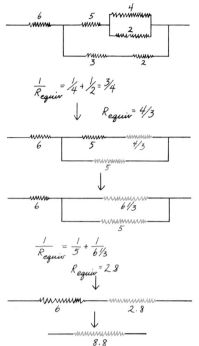

To find the equivalent resistance of the complicated network shown in this diagram, just apply the simple combination rules over and over again. Find the equivalent resistance for each parallel network, starting out with the most interior one. The analysis is shown in a series of steps in the diagram.

INTERNAL RESISTANCE OF THE POWER SUPPLY

So far, we have assumed that a battery in a circuit simply works as an electron pump, producing a continual potential difference between its terminals. Actually, a battery or any other power source must contain an internal resistance. Consider, for instance, what happened when you short-circuited the cell in "Handling the Phenomena." The current in the wire may have been as much as 10 A, certainly sufficient to raise the temperature of the wire so that you could feel it. But the resistance of that short piece of copper wire could not have been more than 0.01 Ω. If there were no internal resistance in the battery, we might expect a current of

$$I = \frac{1.5 \text{ V}}{0.01 \text{ Ω}} = 150 \text{ A}$$

The internal resistance of the battery or any power supply can be represented with the circuit diagram shown in the diagram. The potential difference available is called the terminal voltage, V. For historical reasons, the actual potential difference created by the battery is called the electromotive force, $\mathcal{E}$. In spite of its name, an electromotive force or *emf* is not a force, it is a potential difference—the energy provided per coulomb.

The current in this circuit is

$$I = \frac{\mathcal{E}}{r + R}$$

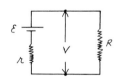

The actual voltage available at the terminal of the battery is

$$V = \mathcal{E} - Ir = \mathcal{E} - \frac{\mathcal{E}r}{r + R} = \frac{\mathcal{E}}{r + R} R$$

If the internal resistance, r, is very small compared to the load resistance, R, then the terminal voltage is almost equal to the *emf*, as you can see from the above expression.

The terminal voltage depends on the amount of current drawn from the source. A new $1\frac{1}{2}$ V D-cell normally has an internal resistance of about 0.1 Ω. Even if the cell provides 1 A, its terminal voltage is not much reduced from its rated *emf*:

$$V = \mathcal{E} - Ir = 1.5 \text{ V} - (1 \text{ A})(0.1 \text{ Ω}) = 1.4 \text{ V}$$

As a battery ages, its internal resistance increases. If you measure the terminal voltage of a dead flashlight cell, your voltmeter might read close to 1.5 V, but that's because the voltmeter draws very little current and, therefore, the Ir drop inside the battery is small. An old dry cell may well have an internal resistance of 5 Ω.

INTERNAL RESISTANCE OF THE POWER SUPPLY

> Question 19-15
>
> What would happen if you tried to use such a cell to provide the $\frac{1}{3}$ A needed by a flashlight bulb?

In "Handling the Phenomena," you noticed the dimming of the light or the slowing down of the shaver when a toaster drew current from the same outlet. In this case, the internal resistance of the power supply may be considered to be the resistance of the wires leading from the wall socket to the main fuse box. Depending on the length of run of the wire in the walls and on the age of the house, the internal resistance may easily be 1 Ω. If the toaster draws about 10 A, the Ir drop in the wires is

$$(10 \text{ A})(1 \text{ } \Omega) = 10 \text{ V}$$

A 10 V drop in line voltage is enough to make itself seen in a lamp or heard in a shaver.

Any power supply has an internal resistance. The size of that resistance in a large city generator is small, but the current provided is large. The resulting internal energy loss is wasteful, though inevitable, and the heat produced must be carried away by cooling water.

■ Optional

There is a surprising condition for extracting maximum power from a power source *for a fixed internal resistance*. Let us combine the expressions that we derived for current and terminal voltage and get the formula for the power supplied to the load resistance, R.

$$P = IV = \left(\frac{\varepsilon}{r+R}\right)\left(\frac{\varepsilon}{r+R} R\right) = \frac{\varepsilon^2}{(r+R)^2} R$$

Let's vary the load resistance to get maximum power. If R is zero, then the current is maximum, but the voltage and external power are zero. On the other hand, if we make R much larger than r, the terminal voltage is nearly equal to the *emf*. However, the power supplied to the load is small:

$$P \to \frac{\varepsilon^2}{R} \qquad \text{for } R >> r$$

The graph of power in the load as a function of R is shown here. Note the two extreme conditions that we have just analyzed. The curve rises to a maximum at the point where $R = r$. With such an arrangement, maximum power is drawn from the supply and consumed in the external load. But the same amount of power also must be dissipated inside the supply. Of course, these are not the conditions under which large power stations operate; they try to minimize the internal resistance of their generators. For maximum power extraction with certain electronic circuits, however, the load is made equal to the fixed internal resistance.

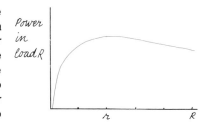

Voltaic Cells and Batteries

When Benjamin Franklin and other eighteenth-century experimenters worked with electricity, they had to use electrostatic generators. It was not until 1800 that Alessandro Volta (1745–1827) created the first practical battery. Some years earlier, Volta's friend, Luigi Galvani, had discovered that when a wire made of iron on one end and copper on the other was touched to various nerve and muscle combinations of dissected animals, the muscles would twitch. Galvani thought that he had demonstrated "animal" electricity. Volta found that as long as two different metals were used, you could substitute a salt solution and do without the animal.

A diagram of a typical "voltaic pile" is shown in the diagram. Alternate layers of silver and zinc are separated by paper soaked in a salt solution. Each layer forms an electrolytic cell producing an *emf* of about 1 V. By arranging the cells in a pile, Volta could produce a source of considerable voltage. With batteries available that could produce direct current for long lengths of time, the nature of experiments in electricity changed drastically. Joseph Henry in America and Michael Faraday in England were able to power large magnets and also to do electrolysis experiments in which various elements and compounds form in conducting liquids.

You formed an electrolytic cell when you made the lemon battery in "Handling the Phenomena." (Technically, an electric battery is a combination of a number of cells but, in everyday language, a voltaic cell is frequently called a battery.) There is a difference of potential across almost any two different kinds of metals that are dipped into an electrically conducting liquid, called an *electrolyte*. In the case of the lemon cell, the electrolyte is the weak citric acid in the lemon. The chemical interactions taking place between the two electrodes and the electrolyte are frequently very complex.

Regardless of the particular details, "oxidation" takes place at one electrode and "reduction" at the other. When an atom is *oxidized* it loses electrons; when it is *reduced* it gains them. If one of the electrodes dissolves in the electrolyte, its atoms go off as positive ions, leaving electrons behind on the electrode. At the other electrode where reduction is taking place, positive ions are gaining electrons. In the reduction in some voltaic cells, metallic ions are plated as solids on the electrode. In many other cells, the reducing electrode turns hydrogen ions into atoms, which then combine and bubble off as a gas.

The oxidation and reduction can continue only if electrons released at one electrode can get over to the other one. An external wire provides such a path. Each electron freed by oxidation circulates through the external circuit (or shoves other electrons ahead of it) and ends up on the other electrode where it reduces a positive ion.

In the lemon cell, zinc dissolves more readily than copper and goes into solution as an ion with a double positive charge. Each zinc ion leaves behind two electrons that can circulate through an external connecting wire. If the external wire is connected to the copper electrode, the electrons are available to reduce hydrogen ions in the acid electrolyte and to turn them into hydrogen atoms.

A voltaic cell turns chemical energy into electrical energy. In the process, some chemical compound breaks down into a more stable arrangement of atoms or ions, thus providing the energy to force an electron into a region of the circuit where there are already surplus electrons. Of course, the chemical energy is really electromagnetic energy in the first place. The amount of energy released depends on the particular chemical reactions (the kind of electrodes and electrolyte), on the concentration of the electrolyte, and on the temperature and pressure. In all cases, the *emf* produced is *of the order of* a volt (in practical cells, from 1 to 2 V). Therefore, the energy provided to each electron emitted by such a cell is of the order of an *electron volt*. As we have seen before, the

"Voltaic pile."

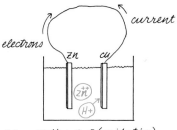

$Zn \rightarrow Zn^{++} + 2e^-$ (oxidation)
$2H^+ + 2e^- \rightarrow H_2$ (reduction)

INTERNAL RESISTANCE OF THE POWER SUPPLY

valence electrons are bound with energies of the order of an electron volt, and that is just the energy associated with the binding of one atom to another.

The standard dry cell that is most commonly used in flashlights was first invented by a man named Leclanché in 1868. The outer casing is zinc, which forms the negative electrode; the top button, which is the positive electrode, is fastened to a carbon rod that goes down the axis of the cylinder. The carbon just serves as a chemically inert connection to a surrounding layer made of a paste of manganese dioxide and powdered graphite. Between the manganese dioxide and the zinc is a wet paste of ammonium chloride and zinc chloride, which serves as the electrolyte. When the zinc dissolves and goes into the electrolyte, it leaves electrons behind on the bottom case. If there is an external circuit connecting the bottom case to the top positive button, electrons can pass down the carbon rod and reduce the manganese dioxide from an oxidation state of 4+ to 3+. This particular reaction produces an *emf* of about 1.6 volts. Commercial batteries may have many variations on this basic scheme, including the use of sandwiched sheets of zinc between the other layers and also small amounts of other chemicals in the electrolyte to control some of the chemical interactions.

Several other kinds of batteries are available on the market. The "alkaline battery" is basically the Leclanché cell, with a high concentration of sodium hydroxide in the electrolyte. It can temporarily withstand a heavier drain of current than the standard dry cell. A "mercury battery" has electrodes made of zinc and mercuric oxide with an alkaline electrolyte. This combination produces an *emf* of 1.34 V, with the advantage that the internal resistance does not change appreciably as the battery ages. A "nickle-cadmium" cell is rechargeable, that is, if an exterior voltage is applied backward across the cell, the chemical interactions will reverse themselves, storing up energy for future use. The nickel-cadmium cell has been known for many years but has become available only recently in a practical sealed container. The problem that had to be solved was how to contain the gases that are generated during the charging process.

The workhorse of batteries is the lead storage battery used in all cars. One electrode is lead and the other is lead dioxide, with the electrolyte being sulfuric acid. The lead would tend to go into solution as a doubly charged positive ion,

Lead storage battery. (*Courtesy of Gould, Inc.*)

releasing two electrons at that electrode. However, the positive lead ion immediately joins with a negative sulfate ion from the sulfuric acid, forming a deposit of lead sulfate on the plate. When the external circuit is completed and electrons arrive back at the lead oxide electrode, the lead oxide is reduced. The oxygen is stripped off and combined with hydrogen ions from the acid to form water, leaving a doubly charged positive lead ion on the plate. This immediately joins with a negative sulfate ion and forms a deposit of lead sulfate. Thus, both electrodes become plated with lead sulfate, and the sulfuric acid slowly turns into water. During the charging process, the chemical reactions are reversed, building up the concentration of sulfuric acid again. The specific gravity of the sulfuric acid is a good indicator of the extent of charge of the battery. When the battery is fully charged, the specific gravity of the sulfuric acid is 1.28. At 50% charge, the specific gravity is 1.18, and at zero charge it is 1.08. The *emf* produced by each cell is about 2 V. Six cells are connected in series to form the standard 12 V car battery.

The amount of charge stored in a typical car battery is 50 ampere-hours. Since 1 ampere-second is a coulomb, such a battery could supply 180,000 C. Let's calculate how many electrons are transferred in a complete discharge of the battery and the mass of the chemicals that undergo transformation. The charge on one mole of electrons is called a faraday.

$$1 \text{ faraday} = (6 \times 10^{23} \text{ electrons/mole})(1.6 \times 10^{-19} \text{ C}) = 96{,}500 \text{ C}$$

As you can see, a 50 ampere-hour battery can supply about 2 faradays. However, the process does not involve two moles of lead, since each lead atom provides two electrons as it transforms into a doubly charged positive ion. Consequently, one mole, or 207 g, of lead atoms are transformed into lead sulfate at the negative electrode. Two moles of sulfuric acid are used up, one at the positive plate and one at the negative plate. Therefore, 196 g of sulfuric acid are transformed during the discharge.

MEASURING CURRENT AND VOLTAGE

An instrument that measures electric current is called an *ammeter,* and one that measures potential difference is called a *voltmeter.* An ammeter must always be connected in series, since it measures the current *in* a circuit. A voltmeter, on the other hand, is connected in parallel with a circuit component, since it measures the potential difference *across* the component.

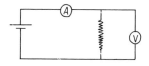

Ammeters and voltmeters of the standard type (as opposed to electronic versions) contain the same basic measuring device, called a galvanometer. It consists of a coil of wire that can rotate in the gap of a permanent magnet. When there is current in the coil, it becomes a magnet with a strength proportional to the current. The angular rotation of the coil, which is opposed by a spring, indicates the amount of current. The sensitivity of the galvanometer depends on its construction, which in turn affects its price. A typical galvanometer movement costing several dollars has a coil that will rotate to a full-scale reading with a current of 1 milliampere. The resistance of the coil winding in this case might be 50 Ω.

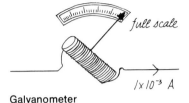
Galvanometer

The galvanometer coil, to which the long needle is fastened, is mounted in the field of a permanent magnet.

A galvanometer is basically a *current* measuring device. How can we make it into a *voltmeter*? Suppose we want a meter to read full scale for a potential difference of 10 V. A resistance can be added in series to the galvanometer coil so that there is only 1 milliampere (abbreviated mA) in the circuit with 10 V across it. The total resistance of the series resistor and the galvanometer coil must be $R = V/I = (10 \text{ V})/(1 \times 10^{-3} \text{ A}) = 1 \times 10^4$ Ω. The 50 Ω of the galvanometer coil itself is only $\frac{1}{2}$% of the total resistance needed. Such meters cannot be read closer than about 2%, and they are seldom calibrated to better than 5%. The 50 Ω could be ignored. With 10,000 Ω in series with this galvanometer, it will read full scale when placed across a potential difference of 10 V.

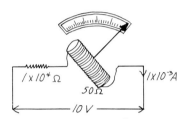
Galvanometer used as voltmeter.

Let's make an *ammeter* out of the galvanometer. By itself, the galvanometer *is* an ammeter, reading full scale with a current of 1 mA. In the diagram we show how to make an ammeter that reads full scale for 5 mA. A "shunt" resistor, R_s, is wired in parallel with the moving coil. If 5 mA is in the parallel network and the galvanometer is reading full scale, then 1 mA must be in the coil and 4 mA in the shunt. The potential drop across both coil and shunt must be the same if they are in parallel.

$$I_C R_C = I_S R_S$$
$$(1 \times 10^{-3} \text{ A})(50 \text{ Ω}) = (4 \times 10^{-3} \text{ A})R_S$$
$$R_{\text{shunt}} = 12.5 \text{ Ω}$$

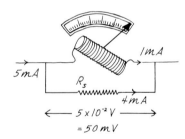
Galvanometer used as ammeter.

We could use the same procedure to find the shunt resistance needed for any other full-scale reading. For large currents, the resistance of the coil can be ignored in comparison with the very small resistance needed for the shunt. For instance, suppose that we want to use this galvanometer for an ammeter that reads full scale with 1 A. The coil requires $\frac{1}{1000}$ of the total current, and the shunt will carry $\frac{999}{1000}$ of the current. Since the parallel shunt carries approximately 1000 times the current of the coil, its resistance must be smaller by a factor of 1000. Therefore,

$$R_{\text{shunt}} = (50 \text{ Ω}) \times 1 \times 10^{-3} = 0.05 \text{ Ω}$$

For ammeters and voltmeters of this type, there are two characteristics

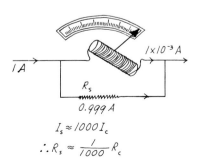

that have to be known when making measurements. First, there is always a potential drop across the coil when the instrument is being used. At full-scale reading, that potential drop is $IR = (1 \times 10^{-3} \text{ A})(50 \text{ }\Omega) = 50 \text{ mV}$. The second feature that must be known if the device is used as a voltmeter is the size of the resistor in series. That value is usually given in small print someplace on the face plate. A typical reading would be 10,000 Ω/V. If the voltmeter has multiple ranges, there is a different series resistor for each range. If one range is 1.5 V full scale, then the series resistor has a value of 15,000 V. If you switch to the 10 V range, you have switched to a series resistor of 100,000 Ω.

> Question 19-16
>
> Suppose that you have switched to the 10 V range with such a voltmeter and then measure 5 V. What is the value of the series resistor?

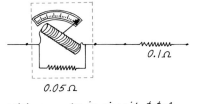

With ammeter in circuit, total resistance is 0.15 Ω

It is hard to make a measurement without interfering with the thing that you are measuring. (You may have experienced this phenomenon while being tested by a professor on a physics exam.) When putting the voltmeters or ammeters into circuits, you may alter the circuit appreciably. Since the inner galvanometer of such a meter has a resistance and requires current, it evidently absorbs energy that has to be supplied by the circuit.

Suppose that you want to measure the current in a 0.1 Ω resistor. Let's use the ammeter whose characteristics we have calculated. In the 1 A range, the internal resistance of the ammeter is 0.50 Ω. If you insert the ammeter into the circuit, you are putting a resistance of 0.50 Ω in series with a resistance of 0.1 Ω. The ammeter will faithfully read the current in this circuit, but the ammeter is now a major part of the circuit since it provides one-third of the total resistance.

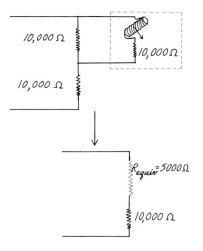

Suppose that you have two 10,000 Ω resistors in series and want to measure a potential drop across one of them. If the expected range is slightly less than 1 V and you use the voltmeter whose characteristics we have calculated, the voltmeter itself will have an internal resistance of only 10,000 Ω. When you place the device in parallel with the 10,000 Ω in the circuit, you are reducing the parallel resistance to 5000 Ω. The voltmeter will give the correct reading for the potential drop of this particular circuit, but the circuit is quite different from what it would be if the voltmeter were removed.

As a rule of thumb, if meters are to be used as test instruments and then removed from the circuit, their internal resistance should not cause more than a 10% change in the circuit. For an ammeter this means that its internal resistance must be less than 10% of the series resistance of the circuit into which it is being inserted. In the case of a voltmeter, the internal resistance should be at least 10 times that of the circuit component across which it is placed.

The internal resistance of a lemon cell of the type that you made in

"Handling the Phenomena" might be anywhere from 1000 to 10,000 Ω, depending on how far apart you place the penny and the nail. The *emf* of the copper-zinc pair is about 1 V. If you tried to measure this potential with a cheap voltmeter having a resistance of only 1000 Ω/V, then you may detect very little terminal voltage. To measure the voltage of lemon cells, you should have a voltmeter with an internal resistance of at least 10,000 Ω/V. In many electronic circuits, particularly if radio tubes are involved instead of transistors, critical control voltages often exist across resistors of 1 megohm (10^6 Ω) or more. If you use a galvanometer-coil voltmeter to measure such voltages, you will completely distort the circuit, and the readings will be meaningless. For such measurements, vacuum tube voltmeters must be used. These usually have input resistance of 10 megohms (million-ohms) or higher.

SPECIAL CIRCUITS

1. *The voltage divider.* It is frequently necessary or convenient to produce continuously variable potential differences. By putting cells in series, all you can get are quantized voltages—$1\frac{1}{2}$ V, 3 V, $4\frac{1}{2}$ V, and so on. The circuit shown in the diagram is a continuously variable voltage divider. Its output could be used in the measurement of $I(V)$ of circuit components as described on page 549. The current in the circuit is

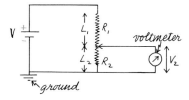

$$I = \frac{V}{R_1 + R_2}$$

The divided resistance ($R_1 + R_2$) is usually a long coil with a tap connection that can slide along the coil. Such a device is called a *rheostat*. Several of them are pictured here. The voltage between the tap and one end of the rheostat is

$$V_2 = IR_2 = \left(\frac{V}{R_1 + R_2}\right)R_2 = V\left(\frac{L_2}{L_1 + L_2}\right)$$

In the circuit diagram, we have used the standard symbol to indicate the "ground" or zero potential point. In most circuits the choice of this point is quite arbitrary. The voltage divider circuit would work just as well if we chose the positive terminal of the battery to be at zero potential. However, the usual convention is to choose the negative side of the battery as zero potential, thus making all the other potentials in the circuit positive.

When various circuits are fastened together, it is usually important to make sure that the ground connections are consistent with each other. In particular, if circuits are used with test instruments powered by alternating current, such as oscilloscopes or vacuum tube voltmeters, then the ground connections may already be determined by the construction of the a-c instrument. The case is usually connected to water pipe grounds through the ground wire of the three-prong, a-c power plug. When wiring a complicated circuit, it is good practice to connect all of the ground lines first. It is also standard practice not to connect the final lead to any power supply until all the rest of the circuit is wired and checked.

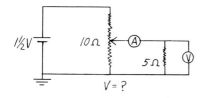

Question 19-17

The circuit shown in the diagram is a voltage divider that you might use to plot $I(V)$ for the 5 Ω resistor. As shown in the diagram, the tap of the rheostat is at the midpoint, so that $L_2/(L_1 + L_2) = \frac{1}{2}$. What is the voltage across the 5 Ω resistor?

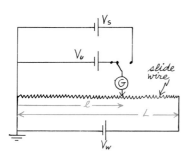

V_u = unknown voltage
V_s = standard reference voltage
V_w = working voltage

2. *The potentiometer.* Precision instruments often involve a null-reading. The chemical balance is an example of such an instrument. The weight to be measured is balanced against a known standard, with the equality of unknown and standard determined by the level of the balance. The circuit of a null-reading instrument for voltage measurements is shown in this diagram. It is called a potentiometer. Basically, it is a voltage divider whose tap voltage is compared with that of the unknown source. The galvanometer, which determines the equality of the tap voltage and the unknown, can be very sensitive because, when the two voltages are nearly equal, there is very little current. You can usually measure zero current more precisely than you can measure the value of some larger current.

In the simple circuit shown in the diagram, the voltage of the unknown is

$$V_U = \left(\frac{\ell}{L}\right) V_W.$$

Laboratory potentiometers are considerably more elaborate than the simple version shown here. For instance, in a good instrument the galvanometer would be provided with several shunts to protect it and make it less sensitive until the zero point region is nearly reached. In our simple circuit, the unknown voltage is found in terms of a working voltage V_W. In the real instruments, another cell is used as a standard. As shown in the diagram, the standard cell can be interchanged with the unknown voltage. In a preliminary procedure, it is used to calibrate V_W, which is then used in the actual measurement. In that way, the standard cell does not have to supply current. A standard cell whose voltage may be known to one part in 100,000 maintains its rated voltage only so long as it does not supply current.

Question 19-18

The accuracy of a potentiometer depends on the accuracy and reliability of the calibration of the standard cell. What is the one other important requirement for the instrument?

3. *The Wheatstone bridge.* There are many "bridge" circuits used in electrical instruments. They are null-reading devices that compare an unknown with a known fraction of a standard value. One of the simplest of these circuits was designed by Charles Wheatstone over a century ago. It

SPECIAL CIRCUITS

is shown in the diagram. The null condition is obtained when there is no current in the galvanometer. Since a very sensitive galvanometer can be used to measure a very small current, the null condition can be determined very precisely. When there is no current in the galvanometer, the circuit consists of just two parallel branches. The current provided by the battery divides into the upper current, I_2, and the lower current, I_1. The voltage drop across the unknown resistor, R_x, must be the same as the voltage drop across R_1; and the voltage drop across the standard resistor, R_S, must be the same as the voltage drop across R_2.

Wheatstone bridge.

Question 19-19

Why is this true?

Let's equate the potential drops, and take the ratio of the two equations.

$$\frac{I_1 R_1 = I_2 R_x}{I_1 R_2 = I_2 R_S}$$

$$R_x = \left(\frac{R_1}{R_2}\right) R_S$$

Note that the value of the unknown resistance depends on the accuracy with which the standard resistance, R_S, is known and on the uniformity of the slide wire or rheostat that determines the ratio R_1/R_2. The voltage rating of the battery and the characteristics of the galvanometer do not appear in the equation. However, those characteristics do affect the sensitivity of determining the null condition.

▪ Optional

Kirchhoff's Rules for Circuit Analysis

It is not always possible to reduce a circuit to equivalent series and parallel resistance. A simple example of the problem is shown here. All circuits, including this one, can be solved at least in principle by an application of two rules that are known by Gustav Kirchhoff's name. The first is that, at any branch point, the current entering must equal the current leaving. This requirement is just common sense but, in applying it, you must be careful to be consistent about the directions of the currents. In the example shown in the diagram, the current entering a junction point is labeled I_1 and the currents leaving are I_2 and I_3. Kirchhoff's law requires that $I_1 - I_2 - I_3 = 0$. The algebraic signs of the currents are determined by the fact that I_1 is entering the junction and I_2 and I_3 are leaving it.

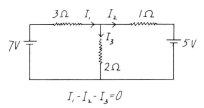

This circuit cannot be solved by using series and parallel equivalent rules.

The second Kirchhoff rule is that if you trace a closed loop path around any part of the circuit, the potential increases and decreases must add up to zero. Again, this requirement is just common sense. If you could go around a closed loop and end up at a higher potential, you could get something for nothing (and make a fortune). The electrostatic potential, like the gravitational potential, is conservative. In Chapter 21, we will study a phenomenon where a charged

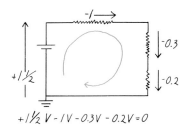

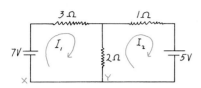

Compare this circuit with the same one on page 571.

particle does gain energy in going around a closed loop. But, in that case, something else furnishes the energy.

Let's apply these laws to the simple circuit shown above and repeated here. Notice that we have labeled the currents in two different ways. In the first diagram, there are three unknowns: I_1, I_2, and I_3. To solve for three unknowns we need three equations. One equation is given by Kirchhoff's first rule: $I_1 - I_2 - I_3 = 0$. The other two equations can be derived by writing down the potential increases and decreases for two different loops. We can simplify the arithmetic by using the notation shown in this second diagram. In this case, there are only two unknown currents: I_1 and I_2. Therefore, only two equations are needed.

Question 19-20

But in this second notation, there appear to be two currents in the parallel 2 Ω resistor. How can this be?

Now let's apply Kirchhoff's second rule to the two loops shown in the diagram. As we go around the left-hand loop starting from the x mark, the potential first rises by the battery potential of 7 V. Then there is an IR drop as the I_1 current goes through the 3 Ω resistor. That potential difference must be listed as $-3I_1$. As our route goes down through the 2 Ω resistor, there is a voltage drop due to the current, I_1 equal to $-2I_1$. However, this route takes us in the opposite direction from the current I_2, which is also going through the 2 Ω resistor. There is a voltage rise due to that current, equal to $+2I_2$. The final equation for the first loop is

$$+7 - 5I_1 + 2I_2 = 0$$

Each term in the equation represents a voltage. If we take the right-hand loop, the arguments are much the same. Starting at the y mark, we travel up through the parallel 2 Ω resistor. We are traveling downstream with I_2 and, therefore, there is a potential drop equal to $-2I_2$. Because we are bucking the current I_1, there is a voltage rise of $+2I_1$. There is a voltage drop going through the 1 Ω resistor, equal to $-1I_2$. Traveling down through the 5 V battery, from positive to negative, there is a voltage fall of -5 V. The complete equation for this loop is

$$-5 + 2I_1 - 3I_2 = 0$$

We have two simultaneous equations that can be solved for two unknown currents. By multiplying the first equation by 2, the second equation by 5, and adding the two resulting equations, I_1 can be eliminated. The solutions are: $I_2 = -1$ A and $I_1 = +1$ A.

Question 19-21

Is something wrong? What does it mean to have I_2 negative? Under these circumstances, what is the current in the 2 Ω resistor?

The Wheatstone bridge circuit was easy to solve because no current existed in the central branch when the bridge was balanced. In this next diagram, we show a circuit with the same configuration, but now there must be current in the central branch. As you can see, there is no way to arrange those branches into equivalent parallel and series resistances. We have shown three possible loop paths along which we can apply Kirchhoff's second rule. Each loop is characterized by a current. In this way, there are only three unknowns, but the net current in three of the resistors is the difference between two loop currents. The three equations are written underneath the diagram, along with the

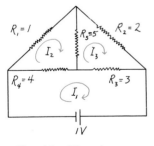

$-7I_1 + 4I_2 + 3I_3 = +1$
$+4I_1 - 10I_2 + 5I_3 = 0$
$+3I_1 + 5I_2 - 10I_3 = 0$

$$I_1 = \frac{\begin{vmatrix} 1 & 4 & 3 \\ 0 & -10 & 5 \\ 0 & 5 & -10 \end{vmatrix}}{\begin{vmatrix} -7 & 4 & 3 \\ 4 & -10 & 5 \\ 3 & 5 & -10 \end{vmatrix}} = \frac{100 - 25}{-7(75) - 4(-55) + 3(50)}$$

$$-\frac{75}{155} \approx -\frac{1}{2} \text{ A}$$

determinant method of solving a triple simultaneous equation. Try deducing the Kirchhoff rule equations for yourself as a test of your understanding of the signs of the potential differences.

Question 19-22
In the two examples that illustrated applications of Kirchhoff's rules, we built equations for just enough loops to solve for the unknowns. In the first case, we had two unknown currents and needed two equations, and so we chose two loops. In the second case, we had three unknowns, and so we wrote equations for three loops. In each case, it is possible to devise other closed loops around the circuit, thus providing more equations. In the first case, for example, we could have followed a path around the outside of the circuit. Would this procedure give us yet another independent equation? Try it and find out.

SUMMARY

Some of the names and concepts used in electricity are familiar to us from everyday life. For instance, a 60 W bulb in a 120 V house circuit must draw $\frac{1}{2}$ A.

In our crude model of charge flow, we pictured electron charge carriers milling about in a metallic lattice. If an electric field is imposed, the electrons will drift slowly in the opposite direction, bumping into the lattice ions and raising the temperature of the solid. If n is the volume density of charge carriers, each with electric charge e, where A is the cross-section of the wire and v is the drift velocity, then $I = neAV$. The drift velocity is proportional to the electric field: $v = KE$. In a long conductor, $E = V/L$. Therefore, $I = [ne(A/L)K]V$. Current in a wire is proportional to the voltage. The inverse of this constant is called the resistance: $R = (1/neK)(L/A) = \rho L/A$, where ρ is the resistivity of the material. Ohm's law is $I = V/R$.

The resistivity depends on temperature, which affects the drift velocity and, in some materials, like carbon, determines the density, n, of conduction electrons. For metallic conductors, over a limited temperature range near 0°C, the temperature dependence of resistance is approximately linear: $\rho = \rho_0 (1 + \alpha t)$.

The work done in moving a charge through a potential difference is $\Delta U = V \Delta Q$. Since $P = \Delta U/\Delta t$ and $\Delta Q = I \Delta t$, then $P = IV = I^2 R = V^2/R$. The unit of power is the watt.

In electric circuits, resistors are arranged in various series and parallel combinations. In series, $R = R_1 + R_2 + \cdots$. In parallel, $1/R = 1/R_1 + 1/R_2 + \cdots$.

Any power supply has an internal resistance which plays a role in the complete circuit. If the *emf*, or no-load voltage, is $\mathcal{E}$ and the terminal voltage under load is V, $I = \mathcal{E}/(r + R)$, and $V = \mathcal{E} - Ir$. Voltaic cells generate current through processes of oxidation and reduction in which some chemical compounds transform to more stable forms.

Ammeters must be wired in series with the device whose current they measure. Therefore, ammeters must have low internal resistance, usually provided by shunt resistances across the internal galvanometer or some other current-sensitive device. Voltmeters must be wired in parallel with the device whose voltage they measure. Therefore, voltmeters must have high internal resistance, usually provided by a resistor in series with the internal galvanometer.

We described three of the most basic circuit measuring arrangements: voltage divider, potentiometer, and Wheatstone bridge. The last two depend on the special detection sensitivity possible in the region of zero current.

Some circuits cannot be analyzed in terms of series and parallel subcircuits. In such cases, Kirchhoff's laws (described in an optional section) can be applied: (1) the current entering a circuit branch equals the current leaving, and (2) the sum of the potential changes around any closed circuit loop is zero.

Answers to Questions

19-1: It is true that the flashlight bulb has about $\tfrac{1}{3}$ C passing through it each second, while the 60 W bulb has only $\tfrac{1}{2}$ C per second. In the 60 W bulb, however, the $\tfrac{1}{2}$ C falls through a potential difference of 120 V, thus losing far more energy than the $\tfrac{1}{3}$ C in the flashlight bulb, which falls through a potential difference of only 3 V.

19-2: Before wires are connected to the poles of a battery, the electric field around the battery must look something like the field of a dipole. To be sure, the conducting parts of the battery casing warp the standard dipole field lines but, in general, the field lines must be concentrated in the region around the battery, going from the positive pole to the negative.

19-3: $\Delta t = (1 \text{ m})/(7 \times 10^{-5} \text{ m/s}) = 1.4 \times 10^{4}$ s $\approx$ 4 hours. Of course, the bulb doesn't take that long to turn on. Regardless of the length of the wire between the battery and the bulb, after you throw the switch, the bulb lights up as fast as the eye can see. (There is a brief delay caused by the thermal lag in the bulb filament; it takes a few thousandths of a second to become hot enough to glow.) The bulb does not have to wait for the electrons to drift all the way from the battery to the bulb. Instead, electrons start drifting all along the wire as soon as the switch is closed. It does take time for the electric field to be established and settled in the circuit, but these times are usually much less than a microsecond, as we will see later on.

19-4: We have seen one other situation where the velocity of an object is proportional to the force on it. When a large object falls through air, the frictional resistance on it is proportional to the square of its velocity. After a while the object reaches terminal velocity, which it then maintains. The friction force at that velocity is just equal to its weight. In such a situation the terminal *velocity* depends on the *weight*. If we find experimentally that the drift velocity of the electrons is proportional to the force on them, then we must conclude that the electrons are suffering from the same sort of continual collisions that produce terminal velocity in a falling object.

19-5: The negative regions for current and voltage merely indicate the direction of current and voltage. The filament gets hot with increased current—regardless of the direction of the current.

19-6: The slope of the linear $I(V)$ graph is $\Delta I/\Delta V$. Since the slope is a constant, and the line goes through the origin, $\Delta I/\Delta V = I/V$. But that is just the reciprocal of the resistance: the slope of the $I(V)$ graph equals $1/R$.

19-7: According to Ohm's law, $I = V/R = (1.5 \text{ V})/(1.7 \times 10^{-2} \text{ }\Omega) = 88$ A. Here's an example of how formulas are models—good only for a *limited* range. An ordinary flashlight cell cannot furnish more than 5 or 6 A. If Ohm's law is to hold, there must be other resistance in the circuit or else the voltage must fall drastically. We will soon learn what happens.

9-8: It is true that the charge going along the second route must travel twice as far, but the electric field is only half as great. Therefore, the work done per charge in the second route is the same as it is in the first route.

9-9: The field lines in the small wire must spread out as they enter the larger wire. The number of lines is the same in the two sections, but the number of lines *per unit area* is reduced by a factor of 2 in the larger wire. Hence, the electric field is less by a factor of 2. Around the surface of the junction, there must be a distribution of static charge, warping the electric field lines into the required pattern.

9-10: As we will see in Chapter 20, a current produces a magnetic field around it. A current loop acts very much like a small bar magnet. The continued existence of the superconducting current could be determined by letting the coil align itself, like a compass needle, in an external magnetic field.

9-11: The answer to this question depends on how the two wires are connected. If they are connected in parallel so that there is the same *voltage* across both, then the power dissipated in the 1 Ω resistor will be twice that in the 2 Ω resistor since $P = V^2/R$. On the other hand, if the resistors are in series, the same *current* is in both of them, and the 2 Ω resistor must dissipate twice the power of the 1 Ω resistor: $P = I^2 R$.

9-12: With a given voltage supply, the current in a branch of a circuit can be controlled by adding a series resistor. A number of resistors in series can also divide the potential difference across them, providing several different voltages at different points with the same supply.

9-13: The arguments in the two derivations are similar, but the basic formulas for capacitance and resistance are reciprocally related as far as the voltage is concerned. Note that $C = Q/V$, but $R = V/I$. When we add voltages in a series circuit containing resistors, we are adding terms proportional to R. With capacitors in series, however, when we add voltages we are adding terms proportional to $1/C$.

9-14: The *approximate* equivalent resistance of 10 Ω in parallel with 100 Ω is 10 Ω. The exact expression is

$$R_{equiv} = R_1 \left(\frac{R_2}{R_1 + R_2} \right) = 10 \left(\frac{100}{100 + 10} \right)$$

The approximation is too high by about 10%. The actual resistance is about 9 Ω. The rated values of most circuit components and the precision of most meters are not guaranteed to better than 5 or 10%. Usually, there is no need for better precision, since many circuit components change with time or with temperature, and circuits can either tolerate such differences or can be adjusted to accept them.

9-15: If a power supply with an internal resistance of 5 Ω provides $\frac{1}{3}$ A, then the internal voltage drop is equal to $(\frac{1}{3} A)(5 \Omega) = \frac{5}{3}$ V. Since a D cell has an *emf* of only 1.5 V, it evidently cannot furnish $\frac{1}{3}$ A when its internal resistance has risen to 5 Ω. Instead, the terminal voltage drops according to the formulas on page 562. Since much less current can be provided to the load, the cell appears to be dead.

9-16: For the 10 V range, the series resistor is always 100,000 Ω for this particular voltmeter, regardless of whether it is reading full scale or only a fraction of the scale.

9-17: Since the rheostat is tapped in the middle, you might expect that the potential at that point would be $\frac{3}{4}$ V—just half the supply voltage. However, the load resistance of 5 Ω is in parallel with the bottom half of the rheostat, which also consists of 5 Ω. Their parallel resistance is $2\frac{1}{2}$ Ω. Therefore, the circuit consists of the 5 Ω of the upper part of the rheostat in series with $2\frac{1}{2}$ Ω. The voltage across that lower parallel combination is just one-third of the supply voltage, or 0.5 V.

9-18: The slide wire that forms the rheostat must be very long and uniform. The measurement actually consists of finding the ratio of lengths along the slide wire. It is assumed that the ratio of lengths is also the ratio of resistances of those fractions of the wire.

19-19: In the null condition, there is no current in the galvanometer and, therefore, the potential difference between A and B is zero. The potential at A can be the same as the potential at B only if $I_1 R_1 = I_2 R_x$.

19-20: The actual current in the 2 Ω resistor is just the algebraic sum of I_1 and I_2.

19-21: The negative I_2 merely means that the current is going in the opposite direction from which we assumed. Since the current in the 2 Ω resistor is $(I_1 - I_2)$, the current is equal to $[+1 - (-1)] = 2$ A.

19-22: As you should find out for yourself, the extra equations are not independent.

Problems

1. What is the current drawn by a household TV set that uses 300 W?
2. Explain with words and a diagram how there can be a uniform electric field throughout a long wire of uniform cross-section, even though the wire is twisted and turns.
3. If a starter motor in a truck draws 100 A through a copper wire 4 mm in diameter, what is the drift velocity of the electrons?
4. Write down and explain or justify the relationships that exist among I, v, E, and V in a wire.
5. Explain what happens, and why, to n and K in a carbon filament as the temperature increases.
6. What is the resistance of a 30 m extension cord, #16 gauge, made of copper wire with a diameter of 1.3 mm? (Don't forget, there are two wires in an extension cord; the current has to go out and come back.) What is the voltage drop if 10 A are drawn?
7. What is the resistance of a tungsten filament in a 60 W bulb *when it is lighted*? Assuming, as a first approximation, that temperature dependence of the resistance of tungsten is linear up to 2000°C, what is the resistance of the filament at room temperature? How much current does it first draw when turned on? (Inexpensive ohmmeters usually contain a $1\frac{1}{2}$ V test battery. If you have one available, test your assumption and calculation by measuring the low voltage resistance of the filament in the 60 W bulb.)
8. You have three identical resistors, each 2 cm long. Connect two in parallel, and these in series with the third. Put 3 V across the whole combination. What is the voltage across each resistor, and what is the field in each? Why don't your answers depend on the resistance values?
9. A long copper wire has a resistance of 0.500 Ω at 20°C. What is the resistance at 80°C?
10. How much energy is stored in a 50 ampere-hour, 12 V battery? What power is furnished at 50 A?
11. Suppose you forget to turn off a 60 W bulb in the attic, and it is on for one whole month. At 8 cents per kilowatt-hour, how much did it cost?
12. For safe operation, what is the maximum voltage that you can put across a $\frac{1}{2}$ W, 22 Ω resistor? A 100 W, 22 Ω resistor?
13. What is the equivalent resistance of a 1 Ω, 2 Ω, and 10 Ω resistor in series? In parallel?
14. A circuit consists of a 2 Ω resistor in series with a parallel combination of a 12 Ω and a 6 Ω resistor. What is the current in the 6 Ω resistor if 4 V is put across the circuit?
15. Standard resistors come in $\frac{1}{2}$ W, 1 W, and 2 W sizes. What are the minimum values that must be used for each resistor in the circuit in Problem 14?

SUMMARY

16. You have *three* D cells with $1\frac{1}{2}$ V each, *three* 1 Ω resistors, an ammeter, and a voltmeter. Use the proper symbols for each, and draw a circuit diagram that contains *all* of them (8 components), arranged so that the ammeter reads 1 A and the voltmeter reads $\frac{1}{2}$ V. (Assume that the ammeter resistance is $\ll$ 1 Ω and that the voltmeter resistance $\gg$ 1 Ω.)
17. The no-load (*emf*) voltage of a battery is 1.55 V. When connected across a load of 3.0 Ω, its terminal voltage is 0.95 V. What is its internal resistance?
18. If a storage battery has to supply 50 A at 12 V for 10 s to start a car, how many grams of sulfuric acid are turned into water?
19. A voltmeter with 20,000 Ω/V input resistance is set on the 100 V scale. By what percent does the voltmeter interfere in the calculated values when inserted into these two circuits?
20. In a potentiometer with a 100 cm slide wire, the galvanometer reads zero when the contact is at 40 cm. The working voltage is 6.0 V. What is the unknown voltage?
21. A lemon cell produces a voltage of 1.0 V with a voltmeter whose internal resistance is listed as 20,000 Ω/V. The voltmeter is set on the 1.5 V scale for this reading. A cheaper voltmeter with 1000 Ω/V, also set on the 1.5 V scale, reads only 0.40 V. What is the internal resistance of the lemon?
22. The internal resistance of an old flashlight battery is 0.50 Ω. A good voltmeter measures its voltage to be 1.5 V at no load. What is the terminal voltage if it is put across a resistance of 1 Ω?
23. Find the unknown resistance in the balanced bridge circuit shown in the diagram. The voltage of the battery is $1\frac{1}{2}$ V. What is the potential at A? What is the potential at B?
24. A galvanometer has a coil resistance of 200 Ω and reads full-scale with a current of 0.1 mA. What shunt resistances must be used to make a milliammeter that can read full-scale for 1 mA, 10 mA, and 100 mA? Draw the circuit.
25. The galvanometer in Problem 24 can be used to make a voltmeter. What resistance must be used for a full-scale reading of 0.1 V, 1.0 V, and 10 V? Draw the circuit.
26. Find the currents in the Wheatstone bridge circuit shown on page 572 if $R_1 = 2$, $R_2 = 1$, $R_3 = 2$, $R_4 = 1$, and $R_5 = 1$.

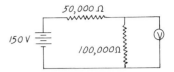

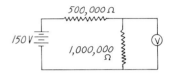

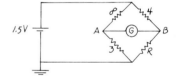

20 MAGNETISM

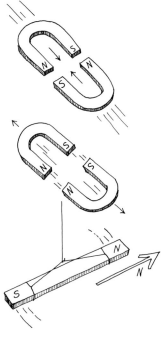

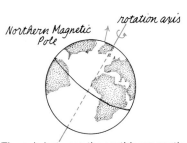

The axis between the earth's magnetic poles is tilted at an angle of $11\frac{1}{2}°$ to the earth's axis of rotation. The northern magnetic pole is currently in Prince of Wales Island and seems to be drifting northwest.

So far in our study of electromagnetism, we can explain everything in terms of the *electrostatic* force. The constant electric field that guides current in a wire is established by the distribution of static charges along the wire. Even the chemical interactions in batteries, which force electrons into regions that are already negative, are based on molecular binding forces that are mainly electrostatic.

We know very well, however, that there are such things as magnets and that they exert forces on each other without the benefit of static electric charges. The study of magnetism developed quite separately from the research on electricity, because the two appeared to have nothing to do with each other. Magnetism seemed to be a property primarily of iron and, mysteriously, of the earth itself.

Magnets can both attract and repel each other, as can electric charges. Therefore, there must be two kinds of magnetic "charges." These are called *poles*; one north and one south. When freely suspended, a bar magnet will rotate, lining itself up in a general north-south direction. The north pole of a magnet is defined to be the one that points north; thus the earth's northern pole is a south magnetic pole.

In 1819, Hans Christian Oersted (1777–1851) discovered (while teaching a class) that an electric current influences a magnetic compass. Shortly afterward, Ampère discovered a force between two current-carrying wires, even though the wires were electrostatically neutral. Evidently electric currents can produce magnetic effects, and in turn magnets can exert forces on currents. Within a decade of these discoveries, Faraday and Henry learned how to make powerful electromagnets and electromagnetic motors and generators.

In this chapter, we define and learn how to describe magnetic fields. We will use the concept of magnetic fields to describe the process by which two electric charges influence each other when they are moving with respect to us, the observers. *The interactions are velocity-dependent.* After describing some practical applications of these principles, we can explain at least qualitatively how magnetism is produced in solid materials. Finally, we take a hard look at the implications of a theory that describes forces as a function of velocity. But whose velocity? Can the physical force between two objects depend on the velocity of an observer?

HANDLING THE PHENOMENA

Many of the phenomena studied in physics are observable in everyday life. Most of the materials that we have needed in "Handling the Phenomena" are common household items. Whether you have the items needed for *this* chapter depends on your household. Even if you don't have them, the items are cheap, are readily available at hardware or toy stores, and are useful things to have around the house. You need a small scout compass, a magnet of some sort, several wires (preferably with alligator clips), and a good $1\frac{1}{2}$ V dry cell. The magnet can be a toy horseshoe, a bar magnet, or the kind used to stick things on steel bulletin boards. The battery can be an ordinary D cell, although a larger one would be better.

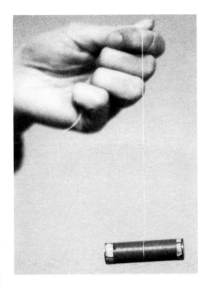

1. A compass needle is just a tiny bar magnet mounted so that it can rotate in a horizontal plane. Explore your immediate surroundings with your compass and make sure that you agree that the direction it indicates is approximately north. Then bring your magnet near the compass and observe the effect that the two different poles of the magnet have on the compass.

2. If you can get two identical bar magnets, find out how to orient them to produce either attraction or repulsion. Use the compass needle to identify the north and south poles of your magnet. How does the force between two magnets depend on the distance between the poles? The difference between a force that depends on the inverse first power of the distance and a force that depends on the inverse square of the distance is easily determined. Compare the force when the poles are separated by 2 cm with the force at a distance of 1 cm. Does the force decrease by about a factor of 2 or a factor of 4?

3. For the following demonstration you need two very fine wires and two thin strips of aluminum foil, about 30 cm long. Fasten the two strips of foil through slits in a cardboard box, as shown in the picture. The strips should be flexible, not taut, facing each other and yet not touching. They should be only 2 or 3 mm apart. Connect the two strips together on each side and, with the fine wires, connect the battery directly across them so that current is in the same direction in each strip. This arrangement will short-circuit the battery and draw between 5 and 10 A, depending on battery size. You can do this for a few seconds at a time without ruining the battery.

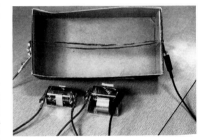

See what the conducting strips do when current is in the same direction and the strips are close together. You can reverse the action by connecting the strips together on one side of the box and connecting the positive and negative leads from the battery to the separated strips on the other side of the box. In this way, you will be sending current in opposite directions in the two strips. The effect that you see in either case is small but easily observable. Note that the effect has nothing to do with the strips being charged. Electrostatically they remain neutral. To see that nothing happens when the strips *are* charged to this low voltage, connect both strips to one side of the battery, or one to one side and one to the other. (But don't close the circuit to allow current in the strips.)

4. Bring a magnet very close to one of the suspended strips. If you have a horseshoe magnet, arrange the geometry so that the aluminum strip passes between the jaws of the magnet. Now send a brief surge of current into the strip and see what happens. Reverse the direction of the current and notice how the effect changes. With most magnets and geometries, this

particular demonstration will be much more dramatic than when the two strips are side by side without any external magnet.

5. Place your compass on a table and drape a wire directly over it. Short-circuit the battery with the wire for a second and see what happens to the compass needle. Try this demonstration with the wire arranged in two different directions: first with the wire parallel to the original direction of the compass needle and second with the wire perpendicular to the compass needle.

THE FORCE BETWEEN PARALLEL CURRENTS

In "Handling the Phenomena," you observed a force that cannot be explained in terms of electrostatics. When the current is in the same direction in two parallel wires, there is an attractive force between the wires. When the current is in the opposite direction, the wires are repelled from each other. The wires are carrying current, but in no region is there a surplus negative or positive charge. The negative electrons are simply drifting past the positive ions.

The actual magnitude of this force between parallel currents, and its dependence on the distance between the wires, can be measured with a simple balance arrangement. Experiments show that the force is inversely proportional to the distance between the centers of the wires: $F \propto 1/r$. Since the force must depend on an influence spreading from one wire to the other, it is not surprising that the *cylindrical* geometry produces a force dependent on the inverse first power of the distance. Remember that the electrostatic field also spreads out from a charged wire with a $1/r$ dependence on distance.

Student laboratory current balance produced by the force between the two parallel wires is balanced by the counter weight in back.

Experiments also show that the force of interaction between the wires depends on the *product* of their currents. From symmetry we would expect that, if the force were proportional to I_1, then it would also be proportional to I_2. That the force is directly proportional to each current is simply an experimental fact, although at the end of this chapter we will see another way of justifying the facts.

Magnetic force between two parallel wires:

$$F \propto \frac{\ell}{r}$$
$$F \propto I_1 I_2$$
$$\therefore F \propto \frac{I_1 I_2}{r} \ell$$
$$F/\ell = \left(\frac{\mu_0}{2\pi}\right) \frac{I_1 I_2}{r}$$

By inserting a proportionality constant, we can now write the formula for the force between two parallel wires. Once again, a proportionality constant will have an irrational factor associated with it, but this time in the form of a 2π, which is not absorbed in the constant. As we have seen before, these extra multiples of π arise because of the geometry of space or because of the sources of the influence. If we did not include them in one formula, they would pop out in another. The force between two parallel wires is given in terms of the force per unit length. The longer the wires, the greater the force:

$$\frac{F}{l} = \frac{\mu_0}{2\pi} \frac{I_1 I_2}{r}$$

The distance between the centers of the wires, r, is measured in meters. The force per meter, F/l, is measured in newtons per meter, and the currents, I_1 and I_2, in amperes. The value of μ_0 (mu sub zero) is then exactly $4\pi \times 10^{-7}$.

> Question 20-1
>
> Suppose that the force exerted by the second wire on the first is 1 N. Then the force exerted by the first wire on the second must also be 1 N. What is the resultant force between the two wires?

Recall that in Chapter 18 we defined a coulomb in terms of the ampere, without defining the ampere, and then took on faith the value of the constant, k, appearing in Coulomb's law. Now we are prepared to look carefully at the definition of the ampere. If we set $\mu_0 = 4\pi \times 10^{-7}$, then the equation for F/l defines the ampere. The constant μ_0 is called the *permeability of free space*. It is analogous to ϵ_0, the electrostatic permittivity of free space. There is an operational difference, however, in assigning values to these two constants. We can choose one or the other to be any value we like, but then the other must be determined experimentally, since the coulomb and ampere are linked. In the SI units, we choose μ_0, then measure ϵ_0.

> Question 20-2
>
> The excuse for a factor containing π has already been given. But why aren't more significant figures given for this constant? Is it $4\pi \times 1.0 \times 10^{-7}$, or $4\pi \times 1.0000 \times 10^{-7}$?

The size of the ampere defined by this formula can also be expressed in words: If the force per meter between two long parallel wires, one meter apart, is 2×10^{-7} N, then the current in each wire is 1 A.

In "Handling the Phenomena," you saw what happens if the currents are antiparallel instead of parallel. There is a force of repulsion instead of attraction, but it has the same magnitude. If the wires are *perpendicular* to each other, there is of course only a very small region of influence where the wires are close together, and so we might expect that the force between the wires would be small. As a matter of fact, the force is zero. Since the force could be described as positive if the currents are parallel, and negative if the currents are antiparallel, it is plausible that, when the wires are perpendicular, the force should be zero, which is in the middle—between positive and negative.

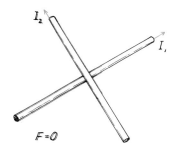

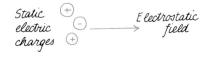

Electric charge, q, in electrostatic field, E, experiences a force, $F = qE$

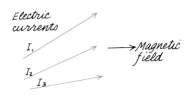

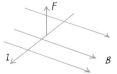

Electric test current, I, in magnetic field, B, experiences a force, F.

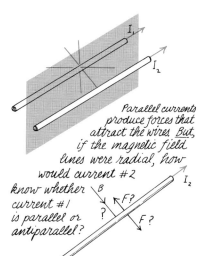

Parallel currents produce forces that attract the wires. But, if the magnetic field lines were radial, how would current #2 know whether current #1 is parallel or antiparallel?

THE MAGNETIC FIELD

In studying electrostatics, we started with Coulomb's law for the force between two point charges. In order to describe the effects produced by more complicated charge distributions, we divided the problem into two parts. We said that electrostatic charge seems to distort or warp the space around it. We described this warping in terms of an electric field, defined at any point as the force per positive unit charge at that point. An actual charge at that point then experiences a force equal to the product of its charge and the electric field strength. Thus, the field is an intermediate stage of calculation, but it also takes on a reality of its own.

We would now like to calculate the interactions of various distributions of electric current. We propose that *moving* charges produce a *magnetic* field; this is analogous to our approach in creating the electric field. The strength and direction of the magnetic field can be calculated in terms of its source of currents. The field can be detected and measured in terms of the force on a unit test current.

In the parallel wire interaction, one of the wires could serve as the testing device to measure the magnetic field produced by the other wire. The magnetic field at a point, which is given the symbol B, would then be measured as the *force per meter on a unit current* at that point. The units of B are $N/A \cdot m$.

Now we have a problem. If we describe magnetic field in terms of field lines, what is the *direction* of the magnetic field produced by a wire? (Remember that the field is a human-made model, which we invent for our convenience.) If we were to make the field lines radial, like those from an electrostatically charged wire, then they would be perpendicular to the parallel unit current. Remember that the force on that test current can be positive or negative or zero, depending on its orientation with respect to the source wire. In all cases, however, the direction of the force on the test wire must be along the radial line that joins the two wires. There would be no way, by just using the orientation of field lines and test current, to predict the direction or size of the force. Such a situation is no good. We must devise a way of portraying the magnetic field so that its interaction with a test current will yield a force in the proper, *unique* direction.

Although it is possible to use a test current to map out magnetic fields, there is an easier way to picture the effects of a field line. We will use small permanent *dipoles* or magnets as our test instrument. Historically, magnetic fields were first defined in terms of the forces on magnetic dipoles.

■ Optional

Why can't we define the direction of a magnetic field in terms of the force on a unit test *pole*? A unit north pole would then be analogous to a unit positive charge in electrostatics. Using such a unit pole, we could map out the fields produced by magnets or by electric current distributions. The trouble is that there are no isolated north or south poles. In the diagram on the next page, you see a proposed solution to this problem. Take a bar magnet that has a north pole on one end and a south pole on the other and cut it in half. According to popular folklore, if you do this you will end up with two separate magnets, each of which will have a north pole on one end and a south pole on the other. Actually, the experiment is much talked about, but seldom done. Good permanent magnets are usually very hard and brittle. If you saw or break one in half, you would probably shatter it or so disturb it in the cutting process that you would end up with no magnets or very weak ones.

It is the case, however, that no one has ever discovered an isolated magnetic pole. It is not for lack of trying; the search for the monopole is as old as the history of electricity and has been pursued in various ways by every generation of physicists. In 1928 Paul Dirac, one of the founders of our modern quantum theory and a man who successfully predicted the existence of antiparticles, devised a simple but rigorous argument concerning the existence of monopoles. If even one magnetic monopole exists anywhere in the universe, then electric charge must be quantized. The inverse of the theorem does not necessarily hold, but it is certainly provocative to note that electric charge is indeed quantized. Recent searches for the monopole have been in the field of cosmic rays and very high-energy subatomic particles. Even if such a particle is found, it will have nothing to do with our everyday problem of defining and measuring magnetic fields.

As a practical matter, an approximation of a test pole can be made by using one end of a magnetized knitting needle. The north pole can sample the field while the south pole is still far away. Or as an alternate method, magnetic fields can be mapped with small dipoles. The north pole of a compass needle, for instance, experiences a force in the direction of the field, while the south pole is forced in the opposite direction. The resulting torque makes the compass line up in the direction of the field.

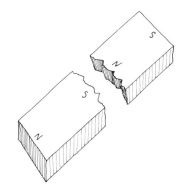

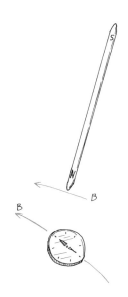

One way to make magnetic dipoles is to scatter iron filings on a smooth surface in a magnetic field. In the field, each little needle of iron is polarized so that it lines up parallel to the magnetic field. A general pattern of the field lines, at least in two dimensions, becomes visible.

Question 20-3

We started out looking for a description of magnetic fields that would allow us to describe the observed forces on test currents. Now we are proposing to map magnetic fields with tiny permanent magnets. How do we know that the fields mapped this way have anything to do with test currents?

In the next section of this chapter, we will look at a very qualitative description of the magnetic field model and the rules for using the model to make qualitative predictions about the interactions of permanent magnets and electric currents. Then we will take a closer look at the quantitative features of the model and develop the several formulas that link forces, fields, and currents.

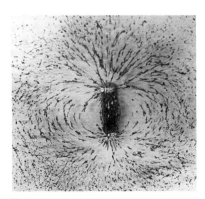

Magnetic field lines, defined by iron filings, around a bar magnet.

FIELD PATTERNS FROM PERMANENT MAGNETS AND ELECTRIC CURRENT CONFIGURATIONS

1. *The bar magnet—a magnetic dipole.* The dipole is probably the most familiar type of magnet. The magnetic field pattern around it looks just like the electric field around an electric dipole. Using the north pole of a long magnetized knitting needle as our test pole, we find that the field lines leave the north pole of the magnet and enter the south pole. What happens inside the magnet is usually not visible. With an electric dipole the lines of force go directly from positive to negative in the region between the charges. In the photograph, we show a picture of a bar magnet with a hole in the middle. A compass in the hole indicates that the field lines inside the magnet are going from south to north, quite different from the case with the electric dipole. Since there are no isolated north and south poles, the field lines can never start or terminate. They must exist in closed loops.

The "poles" of a permanent magnet exist over fairly large regions rather than at points. Although the pole regions are usually at the ends of the magnet, they can be most anywhere, depending on how the magnet was made and how it has been kept. Sometimes there are several pole regions, but no one has ever found a magnet with just a single pole.

The compasses and iron filings portray the direction of the field lines, but do not give a good picture of the strength of the field. Later we will see that the strength of the magnetic field can be represented by the number of field lines per unit area, similar to the situation with electric field strength and electric field lines. In our present qualitative treatment, it is at least plausible that the magnetic fields are strong if the lines are very dense, and weak if the lines are sparse.

2. *Magnetic field due to a C-shaped magnet.* A C-shaped or horseshoe magnet brings the poles close to each other. The continuous field lines are in the iron for much of their length and come out into the air only in a small region. Such magnets keep their strength much better than bar magnets and can be made to have very strong fields in the open region between the poles. The magnet shown in the first illustration has parallel pole faces, and the field lines are uniform in the central region. The pattern is very much like that of the field lines in a parallel plate capacitor.

The field lines will tend to take the shortest magnetic path to complete their loops. The second illustration shows what happens when a nonmagnetized piece of iron is placed between the gap of a C-magnet. The lines concentrate in the iron, producing a stronger field in the iron and a weaker field outside. In the third illustration, we show how to make use of this principle to shield magnetic fields. The region inside the iron cylinder is relatively free from magnetic fields.

3. *Magnetic field between like poles.* Two poles of the same kind repel each other, and the field-line pattern shown in the fourth illustration almost seems to dramatize the situation. Two important points about the field-line model can be seen in this picture. First, field lines cannot cross each other. If they did, an intersection would be a point where the force on a test pole would be in two different directions—a physical impossibility. The second point is one that Faraday stressed, giving a sense of reality to this mental construct of force lines. It appears as if the lines repel each other. We will

FIELD PATTERNS FROM PERMANENT MAGNETS AND ELECTRIC CURRENT CONFIGURATIONS

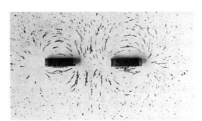

appeal to this aspect of the model when we picture the forces acting on currents in magnetic fields.

4. *Magnetic field due to a current in a long straight wire.* When we first proposed building a magnetic field model, we asked what would happen if the field lines around the current in a long straight wire were radial, pointing out from the wire. That model didn't work.

The vertical wire is carrying 5 amperes. The response of the compasses shows that there are circular lines of magnetic force around the wire. (Before the current was turned on, all the compass needles pointed in the direction shown by the black arrows on the compasses).

Question 20-4

Why didn't this model work?

In this picture, we see that the field lines around a current-carrying wire can be considered to lie on concentric circles. If there were an isolated north pole, it would experience a force tending to take it around the wire. (Its orbit would not be circular because it would be gaining energy constantly and so would spiral outward.) The field is more intense close to the wire; therefore, the density of field lines is greater there.

As you can see, a compass needle placed near the wire lines up tangent to the field lines. The force on the north pole pulls it in the counterclockwise direction, while the equal force on the south pole is in the clockwise direction. Since the forces are not in line, the resultant force on the needle is not zero.

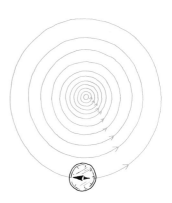

Question 20-5

What is the direction of the resultant force?

The field lines around a current are not only circular but have a specific direction: clockwise or counterclockwise, depending on the direction of the current. In a later section that deals with this material quantitatively, we will see how to determine the direction of the field lines in terms of the vector product of current and radius. In the meantime, here is a simple rule, which is all you will need to figure out the directions of forces and fields in any of the complicated geometries to follow. *Place your right thumb in the direction*

of the current; the fingers of your right hand will curl in the direction of the magnetic field lines.

5. *The magnetic field near two parallel currents.* We have already seen that there is an attractive force between parallel currents. The field-line patterns shown in the first diagram below indicate that the field lines *around* the two parallel currents are strengthened while the field *between* the wires is weakened. If we use Faraday's model in which the field lines are elastic, tending to shorten themselves and to repel each other, then we would argue that the field lines are trying to pull the two wires together into the central region where their mutual fields cancel.

In the second diagram below, we see the opposite situation. The wires are parallel but the currents are antiparallel. Now the fields between the wires add constructively, while there is partial cancellation of the fields in the outer regions. The field lines repel each other and hence try to shove the wires apart.

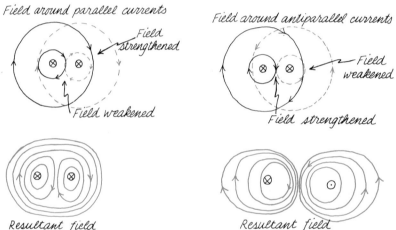

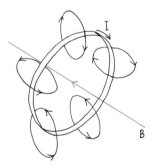

6. *Magnetic field on the axis of a current loop.* The field close to a wire must consist of circular lines in the direction given by the right-hand rule. If we add the field contributions of each little segment of a *loop*, then we get the field pattern shown in this diagram. There is an axial magnetic field spreading out in the surrounding space, coming back again on the other side. In fact, the external magnetic field looks something like that from a dipole.

7. *Magnetic field due to a solenoid.* A *solenoid* is a large number of current loops wound continuously on a cylinder. The field pattern is particularly simple if the length of the cylinder is much greater than its diameter. In that case, the field pattern outside the solenoid looks just like that from a permanent bar magnet. There is a north pole and a south pole, with the field lines coming out of one end and going back into the other. In this case, the behavior of the field lines inside the solenoid is easy to determine. There are no isolated poles, and so the field lines are continuous loops. In the interior of the solenoid the field lines are parallel to each other

in the axial direction and constant in density. Therefore, the magnetic field is uniform throughout the interior except near the ends where the field lines begin to diverge and thus where the field is slightly weaker. The lines leaving the north end spread throughout all space, gradually curving back to enter the south end. The density of lines, and hence the field, close to the outside of the solenoid is nearly zero.

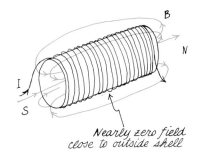

8. *Magnetic field near a current in an otherwise uniform magnetic field.* If a current is started in a wire that has been in a uniform magnetic field, the field due to the current will alter the field lines around it, as shown in the diagram. On one side, the magnetic field is increased and, on the other side, it is reduced by cancellation. According to our model, the crowded field lines to the right will force the wire toward the weaker field on the left. This action is the basis for electric motors and also for the galvanometers used in ammeters and voltmeters.

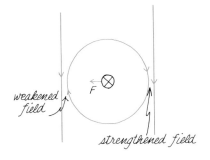
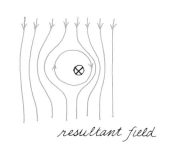

QUANTITATIVE TREATMENT OF MAGNETIC FIELD—AMPÈRE'S LAW

There are two laws relating magnetic field strength to current, even as there are two different forms of laws for the electrostatic field. (Coulomb's law gives the force between two point charges; Gauss's law relates the number of electric field lines coming out of a surface to the enclosed electric charge.) Let's start out with the magnetic equivalent of Gauss's law and relate the magnetic field lines to the current that produces them.

Ampère's law in the simple form in which we will use it is

$$\mathbf{B} \cdot \mathbf{L}_{\text{loop}} = \mu_0 I$$

The dot product of the magnetic field strength, **B**, and the path distance, $\mathbf{L}_{\text{loop}}$, along a *closed loop* is equal to the product of the permeability and the current encircled by the loop. The equation is true for any loop surrounding a current, regardless of where the current is located within the loop. If the magnetic field is not constant around the loop or changes direction with respect to the path, then the dot product must be taken segment by segment and the result added up for the entire loop. (This is called a line integral in the calculus.)

Let's apply Ampère's law to the field around a long straight wire. To find the magnetic field at a radius r from the center of the wire, we multiply the field at that radius times the path length around a closed circular loop of that radius. The field line is also circular, so the magnetic field is always in the same direction as the path. Therefore, the dot product becomes a simple product.

$$B(2\pi r) = \mu_0 I$$
$$B = \frac{\mu_0 I}{2\pi r}$$
{ Ampère's law applied to a long straight wire

Ampère's law specifies that the magnetic field strength for a long straight wire is inversely proportional to the distance from the center of the wire. We should expect just this functional relationship, since a force between two parallel currents is inversely proportional to the distance between the wires. It is reasonable to expect that the force on a wire is proportional to the magnetic field it is in.

Question 20-6

Would Ampère's law also have applied if the route we took around the wire were a square path?

Units for Magnetic Field

$1 \, N/A \cdot m = 1 \, tesla = 1 \, T$
$1 \, T = 10,000 \, gauss = 1 \times 10^4 \, G$

Large research magnets have magnetic fields in the tesla range. The earth's magnetic field is in the gauss range.

The units for magnetic field in Ampère's law are those that we have already deduced from the interaction between two parallel currents.

$$B = \text{force}/(\text{current}) \cdot (\text{length}) = N/A \cdot m$$

The newton per ampere-meter has a proper name: it is called a *tesla* in honor of Nikola Tesla (1856–1943). A tesla (T) is a large unit of magnetic field. A smaller unit is the gauss (G).

$$1 \, T = 10,000 \, G$$

Let's calculate the magnetic field strength 1 cm away from a wire carrying 10 A.

$$B = \frac{\mu_0 I}{2\pi r} = \frac{4\pi \times 10^{-7}}{2\pi} \frac{(10 \, A)}{(1 \times 10^{-2} \, m)} = 2 \times 10^{-4} \, T$$
$$= 2 \, G$$

In "Handling the Phenomena," you short-circuited a dry cell momentarily with a wire that was draped on top of a compass needle. The current in the wire must have been between 5 and 10 A and a little less than a centimeter away from the compass needle. Thus, the conditions correspond to the calculation that we have just made. As you noticed, the current in the wire produced a magnetic field that overcame the earth's magnetic field and made the compass needle swing through almost 90°. The magnetic field of 2 G is of the same order of magnitude as the earth's magnetic field. The strength of the earth's field varies, depending on the location on earth, but in most of the United States it is about $\frac{1}{2}$ G. The further north you go, the more steeply the earth's field points downward. The *horizontal* component of the earth's field in the northern United States is only about $\frac{1}{3}$ G. The table gives values of magnetic field for various natural and technological conditions.

Range of Measurable Magnetic Fields

Source	Strength
Brain's alpha rhythm currents	10^{-15} T
Currents triggering heart beat	10^{-14} T
Typical TV signal	10^{-11} T
Light from 100 W bulb at 3 m	10^{-8} T
1 m from long wire carrying 1 A	2×10^{-7} T
Earth's surface	10^{-4} T
Between jaws of toy permanent magnet	10^{-2} T
Beam-focusing research magnet or electric motor	1 T
Superconducting research magnets	10^1 T
At atomic nucleus from valence electron	10^2 T
Laboratory implosion of trapped fields	10^3 T
Surface of neutron star	10^8 T

QUANTITATIVE CALCULATIONS OF MAGNETIC FIELD—THE BIOT-SAVART LAW

Ampère's law is concerned with the magnetic field along a closed loop. The other law for the magnetic field describes the field at just one point. It is the counterpart of Coulomb's law for electrostatics. It is called the Biot-Savart law, named after Jean Biot (1774–1862) and Felix Savart (1791–1841). There are two slightly different forms of the law, one for the field produced by a small segment of current in a wire and the other for the field caused by an isolated moving charge.

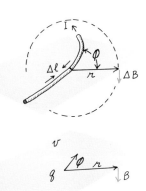

$$\Delta \mathbf{B} = \frac{\mu_0}{4\pi} \frac{I\,\Delta l\,\sin\phi}{r^2} \qquad \text{Field from current segment, } \Delta l$$

$$\mathbf{B} = \left(\frac{\mu_0}{4\pi}\right) q \left(\frac{v\sin\phi}{r^2}\right) \qquad \text{Field from moving charge, } q$$

In the first form of the law, the current in the small segment, Δl, provides a contribution to the field, ΔB. Each segment of the wire yields its own contribution, and the net field at the point would be the vector sum of all those contributions. In the second form of the law, there is only one charge, q, and so it provides the only field there is at the distant point. The geometry on which these laws are based is shown in the diagram. First, note that the field depends on the inverse *square* of the distance between the point and the source.

Question 20-7

But we calculated that the magnetic field near a wire is inversely proportional to the *first* power of the distance from the wire. What's different about this new situation?

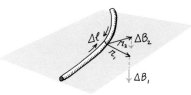

$r_1 = r_2$, but $r_1 \perp \Delta \ell$
$\Delta B_1 > \Delta B_2$
where ΔB_1 and ΔB_2 are contributions of field from $\Delta \ell$. The total field at r_1 or r_2 is the sum of contributions from all the line segments.

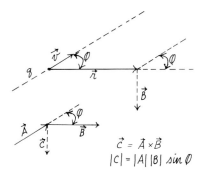

$\vec{C} = \vec{A} \times \vec{B}$
$|C| = |A||B| \sin \phi$

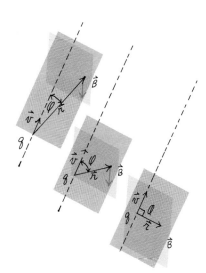

In both forms of the law, the strength of the field is proportional to the current, since a moving charge is a current. The field also depends on the sine of the angle, ϕ, between the velocity of the charge and the line between the point and the charge. The field produced at the point is maximum when the line from point to current is perpendicular to the direction of the current or the charge velocity. The magnetic field produced at the point is a vector and is perpendicular to the plane defined by the current direction and the line from point to moving charge.

Another way to represent the Biot-Savart law is in terms of a vector product. In the chapter on torques (page 81), we defined a vector product. The resultant product is a vector that is perpendicular to the first two and has its maximum value when the first two vectors are perpendicular to each other. These mathematical conditions satisfy the physical condition of the magnetic field produced by a moving charge. In the diagram, we show the geometry and definition of a vector product. Using this terminology, we can write the Biot-Savart law for a moving charge as follows:

$$\mathbf{B} = \left(\frac{\mu_0}{4\pi}\right) q \left(\frac{\mathbf{v} \times \mathbf{r}}{r^3}\right)$$

At first you might think that we have turned the Biot-Savart law into one that depends on the inverse *cube* of the distance between point and charge. However, this is not the case. There is an extra r in the numerator, representing the radius vector from the measuring point to the moving charge. The r is there to represent the vector direction but, since it also has magnitude, we must cancel it out by putting one more power of r in the denominator.

Suppose that you could measure the magnetic field from a single proton as it went past you. Consider how its strength would change as a function of its distance from you. As shown in the next diagram, there are two reasons why the magnetic field would be extremely weak when the proton is far away. First, the inverse square of a large distance makes the field small and, second, the sine of the angle between the velocity of the proton and the line toward you would also be very small. In fact, if the proton headed straight for you, that angle would be 0. You would detect no magnetic field from the proton, even though it went straight through you. If the proton is passing by, however, there will be an instant when the radius vector from you to the proton is perpendicular to its velocity. At that moment, $\sin \phi$ has its maximum value, and the separation distance has its minimum value. Hence, the field would be a maximum. But whether strong or weak, the *direction* of the magnetic field at your position does not change. It is perpendicular to the horizontal plane that contains the radius vector to the proton and its velocity vector. In the case shown, the magnetic field at your position would be directed downward.

Let's see whether it is possible to detect the magnetic field that is produced by a single proton as it goes past you at a very high velocity. We will make the measurement when ϕ equals 90° and r equals 1×10^{-3} m. The proton might be coming down a very small pipe, for instance, and we might try to sense its magnetic presence with a small loop around the pipe. If the proton is traveling with one-third the velocity of light, then the magnitude of the magnetic field that it produces is

$$B = \left(\frac{\mu_0}{4\pi}\right) q \left(\frac{v \sin \phi}{r^2}\right) = \frac{(4\pi \times 10^{-7})}{4\pi} (1.6 \times 10^{-19} \text{ C}) \frac{(1 \times 10^8 \text{ m/s})(1)}{(1 \times 10^{-3} \text{ m})^2}$$

$$= 1.6 \times 10^{-15} \text{ T} = 1.6 \times 10^{-11} \text{ G}$$

The magnetic field produced by a single proton under these conditions evidently is very small. It is on the borderline of detectability. But magnetic detection methods have been used for *beams* of protons, where there may be as many as 10^6 protons passing through the pickup ring at any one time.

We have two laws—Ampère's and the Biot-Savart law—describing the same phenomena. Do they always predict the same results? Can we, for instance, use the Biot-Savart law to calculate the magnetic field near a current in a long straight wire? We already know the result given by Ampère's law. In the diagram, we show how such a calculation can be made by adding up the magnetic field contributions of each little segment of the long wire. While the method is certainly possible in principle, it might seem as if it would be very difficult in practice. The distance from the point to each segment varies, and so does the angle ϕ. It turns out that this calculation is fairly easy to do by using calculus. The result is exactly the same as that given by Ampère's law: the magnetic field strength is inversely proportional to the first power of the distance between the point and the wire.

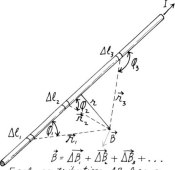

$\vec{B} = \Delta\vec{B_1} + \Delta\vec{B_2} + \Delta\vec{B_3} + \ldots$
Each contribution, ΔB, has a different magnitude, but all are in the same direction.

THE MAGNETIC FIELD ON THE AXIS OF A CURRENT LOOP

From the qualitative treatment, we know that the magnetic field produced by a current loop looks something like the field produced by a dipole. Field lines go through the loop. Let's calculate the strength of the field at the center of the loop on the axis.

Since we have a current loop it might seem that Ampère's law would be the one to use.

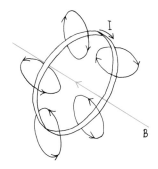

> Question 20-8
>
> Then why don't we use it?

The law of Biot-Savart is easy to use with the geometry of a loop in order to find the resulting field *on its axis*. Each segment of current in the loop yields the same contribution to the total field at the center of the loop. At every point along the circumference, the current is perpendicular to the radius vector. The field contributions, ΔB, are all in the same direction along the axis, and so they simply add up. The contribution to the field by any one segment is

$$\Delta B = \left(\frac{\mu_0}{4\pi}\right) I \left(\frac{\Delta l \sin \phi}{r^2}\right) = \frac{\mu_0}{4\pi} \frac{I \Delta l}{r^2} \quad \text{since } \phi = 90°$$

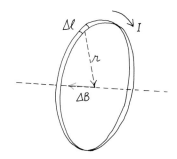

The sum of all the Δl segments is equal to the circumference, which is equal to $2\pi r$. Therefore, the resultant field is

$$B = \frac{\mu_0}{4\pi} \frac{I\, 2\pi r}{r^2} = \frac{\mu_0}{2} \frac{I}{r}$$

Let's see how strong a magnetic field we can get at the center of a small loop of wire. If we have a single turn with a radius of 10 cm carrying a current of 10 A, then the field at the center is equal to

$$B = \frac{(4\pi \times 10^{-7})}{2} \frac{(10\text{ A})}{(1 \times 10^{-1}\text{ m})} = 2\pi \times 10^{-5}\text{ T}$$

$$\approx 0.6\text{ G}$$

The magnetic field has about the same strength as the earth's field. You could make a stronger field using the same current by making a coil of many loops.

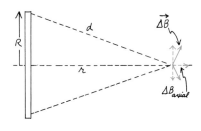

Using the law of Biot-Savart, it is also easy to calculate the magnetic field on the axis of a current loop, but at a considerable distance from the loop. The simplicity of the calculation arises if the distance to the center of the loop r is much larger than the radius of the loop R. The geometry of the calculation is shown in the diagram. The contribution to the field of any one segment is

$$\Delta B = \frac{\mu_0}{4\pi} \frac{I\Delta l}{d^2} \approx \frac{\mu_0}{4\pi} \frac{I\Delta l}{r^2} \qquad \text{for } r \gg R$$

The magnetic field of each segment is perpendicular to the line d. Only a small component of that field is axial. The large component perpendicular to the axis is exactly canceled out by the field contribution from the small segment on the opposite side of the loop. This cancellation of the perpendicular components occurs for all segments of the loop. Only the *axial* components reinforce each other. The axial component of each field contribution, ΔB, is

$$\Delta B_{\text{axial}} = \Delta B \frac{R}{d} \approx \Delta B \frac{R}{r} = \frac{\mu_0 R}{4\pi} \frac{I\, \Delta l}{r^3}$$

The total field due to all of the individual contributions is

$$B = \Sigma \Delta B_{\text{axial}}$$

$$B = \frac{\mu_0}{4\pi} \frac{IR}{r^3} 2\pi R = \frac{\mu_0}{4\pi r^3} 2 (\pi R^2 I)$$

Compare this formula with the expression for the electric field on the axis of an electrostatic dipole, which we derived on page 515 of Chapter 18.

$$E = k\frac{2p}{r^3} \qquad \text{where } p = (ql) \text{ is the electric dipole moment}$$

The field due to a dipole is inversely proportional to the *cube* of the distance from the dipole. As you can see, the magnetic field on the axis of a current loop has this same fucntional dependence on distance. The similarity of the formulas is complete if the magnetic dipole moment of the loop has the value ($\pi R^2 I$). A current loop acts like a magnetic dipole with a magnetic moment (or dipole "charge" strength) equal to the product of the area of the loop (πR^2) and the current in it. We will make use of this relationship when we study the cause of magnetism in solid materials.

MAGNETIC FIELD IN A SOLENOID

We have already described the magnetic field pattern in a long solenoid. The quantitative calculation of the field in a solenoid might seem to be very complicated. If we use the law of Biot-Savart, we would have to add field contributions from current elements with a very complicated geometry. On the other hand, there would seem to be no circular loop along which we could apply Ampère's law.

> Question 20-9
>
> Why not apply Ampère's law to a circular loop concentric with the solenoid, either inside the windings or outside?

There *is* a closed path along which we can apply Ampère's law. This path is shown in the diagram. It consists of four straight segments, along each of which we must compute the dot product of magnetic field and path length. For an easy calculation and a good enough approximation, we assume that the magnetic field close to the outside of the middle of the coil is zero. Therefore, the product of the field and path length along the first segment of the closed path is zero. In segments 2 and 4, the field is either zero or perpendicular to the path length. In either case, the dot product of field and path length is zero. Along path 3, the magnetic field is constant and in the same direction as the path. Therefore, the dot product of field and path length is simply equal to $B\Delta l_3$. According to Ampère's law this product is equal to the product of the permeability and the total current enclosed by the loop. The current going through that loop is equal to the number of turns enclosed times the current in each turn. The number of turns in the loop must be equal to the number of turns per meter, n, times the length of the loop, Δl_3.

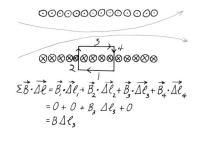

Number of turns of wire in length $\Delta \ell$ is $n \Delta \ell$, where n = turns/meter. Total current carried by these turns is $(n\Delta \ell)I$, where I is the current in each wire.

$$B \Delta l_3 = \mu_0 n \Delta l_3 I$$

$$B = \mu_0 n I$$

The formula for the field in a solenoid is very simple. The magnetic field is proportional to the current and to the linear density of the windings. Note that the field does not depend on the total number of turns in the solenoid but only on the number of turns per meter. The field is uniform in the interior of the solenoid and does not depend on the cross-sectional area so long as the diameter of the solenoid is small compared with its length.

A typical laboratory solenoid, such as that shown in the picture, has about 10^3 turns per meter. If the current in such a solenoid is 10 A, then the interior field is

$$B = \mu_0 nI = (4\pi \times 10^{-7})(10^3 \text{ turns/m})(10 \text{ A}) \approx$$
$$1 \times 10^{-2} \text{ T} = 100 \text{ G}$$

This is a fairly strong magnetic field. It is 200 times the strength of the earth's field and begins to rival the field produced by a good permanent magnet. We will see later how we can make a much stronger magnet out of such a solenoid without expending any more power.

QUANTITATIVE CALCULATION OF THE FORCE ON A CURRENT IN A MAGNETIC FIELD

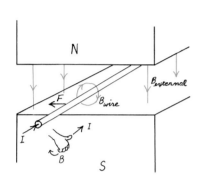

We have already seen what the field patterns look like when a current is in an external magnetic field. The circular lines of the magnetic field produced by the current reinforce the external field lines on one side of the current and diminish them on the other. According to our model of the elastic nature of the field lines, the wire will be shoved toward the region with lower field. With the directions of magnetic field and electric current shown in this diagram. the wire will be shoved to the left.

When we first started analyzing the magnetic force between two parallel currents, we claimed that the equation deduced from experiment is

$$\frac{F}{l} = \frac{\mu_0}{2\pi}\frac{I_1 I_2}{r}$$

At that point in our development we proposed that we create a model of a magnetic field produced by one current with which the second one could interact. Now we have a formula for the magnetic field produced by a long, straight wire carrying current, I. It is

$$B = \frac{\mu_0}{2\pi}\frac{I}{r}$$

The field formula is evidently just part of our original formula for the force between two wires. We can now write that formula as follows:

$$\frac{F}{l} = I_2 B$$

The force experienced by a current that is *perpendicular* to a magnetic field is

$$F = IlB$$

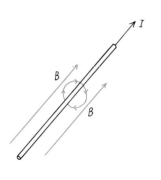

If the current is not perpendicular to the field lines, the force is less. In fact, the force is zero if the current is parallel to the field. We could deduce this conclusion qualitatively by using the right-hand rule and our model of interacting fields. In the diagram, we show the field lines produced by a current that is in the direction of the external field. The combined fields are not stronger on any one side than they are on the other, and so we would not expect that any force would be exerted on the wire.

The quantitative way to describe this geometrical dependence is to use a vector product. The force on a current is a vector, and it is proportional to the product of two other vectors, **B** and **l**. The complete formula for the force on a current in a magnetic field is

$$\mathbf{F} = I(\mathbf{l} \times \mathbf{B})$$

The geometry of this vector product is shown in the diagram. The force must be perpendicular to both the magnetic field and the wire. The direction of the force can be found in terms of the right-hand screw rule for vector products or by appealing to the magnetic field-line model. The magnitude of the force is: $|F| = I\,|l|\,|B|\sin\phi$, where ϕ is the angle between the field lines and the wire. If ϕ is 90°, the force is maximum and in a direction defined to be positive according to the right-hand rule. If ϕ is 0, the force on the wire is zero. If ϕ is 270°, the current in the wire is in the opposite direction from the first case; the force is maximum but now in the direction defined to be negative.

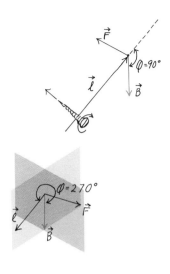

Let's see what magnitude of fields and forces you produced with the two parallel strips of aluminum foil in "Handling the Phenomena." Assume that the short-circuited dry cell provided 5 A and that the aluminum strips were 20 cm long, separated by only 2 mm. The magnetic field produced by one strip at the distance of the other one is

$$B = \frac{\mu_0 I}{2\pi r} = \frac{(4\pi \times 10^{-7})(5\text{ A})}{2\pi(2 \times 10^{-3}\text{ m})} = 5 \times 10^{-4}\text{ T} = 5\text{ G}$$

The force on the second wire in this magnetic field is equal to

$$|F| = I\,|l|\,|B|\sin\phi = (5\text{ A})(0.2\text{ m})(5 \times 10^{-4}\text{ T})(\sin 90°)$$
$$= 5 \times 10^{-4}\text{ N}$$

The force is very small; remember that 1 g has a weight of only 1×10^{-2} N. In order to detect such a small force, we had to use the low-mass and flexible aluminum foil strips.

The aluminum strips displayed a much stronger reaction when they carried current close to the permanent magnet. The field close to an ordinary toy magnet might be between 100 and 1000 G. Therefore, with the same current of 5 A in the aluminum strip, the force would be from 20 to 200 times stronger than the force that we just calculated between the parallel strips.

Electric motors exploit the force on currents in a magnetic field. The fields themselves are usually created by electric coils wound on iron. The resulting fields are usually from 1 to 1.5 T. A cutaway drawing of a typical electric motor is shown here. The fields and coils are arranged to produce fairly constant torque on the center rotor. Suppose that each wire in the rotor carries a current of 6 A. Such a motor would be rated at 1 horsepower for a 120 V circuit.

(Courtesy Reliance Electric Company.)

Question 20-10

How do you calculate horsepower from current and voltage?

Let's calculate the force on a single wire that is 20 cm long and is carrying 6 A in a field of 1.5 T. The force is

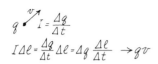

$$F = IlB = (6\text{ A})(0.2\text{ m})(1.5\text{ T}) = 1.8\text{ N}$$

That force by itself is not large, but a motor contains many such windings.

The force on a moving charge in a magnetic field can be deduced from the formula for the force on a current since the current I is the time rate of passage of charge, $\Delta q / \Delta t$. The magnetic force on a moving charge is equal to

$$\mathbf{F} = q(\mathbf{v} \times \mathbf{B})$$

Here is a bubble chamber picture of a positron and an electron traveling in a magnetic field. A positron is an antielectron. It has the same mass and other characteristics of an electron, but it has a positive charge. The two electrons—particle and antiparticle—were created by a high-energy X-ray coming in from the left in the picture. We learn more about the nature of these subatomic particles in Chapter 23. When electrically charged particles dash through a superheated liquid, they cause local boiling that forms bubble tracks. It is indeed the tracks that we are seeing. Notice that the negative electron traveled in a spiral path in a direction indicating that there was a force on it to the right. The positron in the same magnetic field experienced a force that was always to its left. Consequently, the electron traveled in a clockwise path, and the positron went counterclockwise.

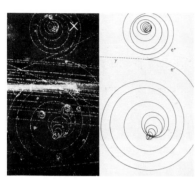

Electron-positron production in a bubble chamber.

Since the magnetic force on the moving charge is always perpendicular to its velocity, the force must be centripetal, tending to make the charged particle travel in a circle. Let's equate the magnetic force to the centripetal force that it provides:

$$qvB = \frac{mv^2}{r}$$

This equation is the basis for many different phenomena and instruments in subatomic physics. For instance, if we cancel a velocity term on each side of the equation, we find that the *momentum* of the particle is equal to

$$mv = qrB$$

If we can see or measure the track of a charged particle in a magnetic field, we can determine its momentum by measuring the radius of its track. In the bubble chamber picture of the electron pair, for instance, it is apparent that the negative electron was produced with greater momentum than the positron because the radius of its track is larger.

Question 20-11

However, the particles are not going in circles. Their tracks are spirals. What is the implication about the momentum of the particles?

Let's rearrange the terms in the centripetal force equation to find the angular velocity, ω, which is equal to v/r.

$$\omega = \frac{v}{r} = \frac{q}{m}B$$

A charged particle going in a circle in a magnetic field has an angular frequency that depends only on its charge-to-mass ratio (q/m) and the strength of the magnetic field. The frequency is independent of the energy of the particle or of the radius of its orbit. The orbital frequency does not depend on the velocity of the particle or its radius of curvature because the velocity and radius are proportional to each other. If a particle has a large velocity, the radius of curvature of its path will also be large, and so the time that it takes to go once around its circular path is the same as if its velocity and radius were small.

The cyclotron is a particle accelerator, or atom smasher, that depends on the principle that the angular frequency of a charged particle in a magnetic field is independent of its velocity or radius. The "cyclotron frequency" is

$$f = \frac{\omega}{2\pi} = \left(\frac{1}{2\pi}\right)\left(\frac{q}{m}\right)B$$

Let's calculate the frequency with which a proton moves in a circle when it is in a magnetic field of 1.5 T:

$$f = \left(\frac{1}{2\pi}\right)\left[\frac{(1.6 \times 10^{-19}\,\text{C})}{(1.7 \times 10^{-27}\,\text{kg})}\right](1.5\,\text{T}) = 2.3 \times 10^7\,\text{Hz}$$

This frequency is midway between that used by AM radio (1 megacycle per second or 1×10^6 Hz) and the FM range (90×10^6 Hz).

A diagram of the working parts of a cyclotron is shown in the diagram. There is a vacuum chamber in the gap of the magnet. A very small amount of hydrogen gas is continuously leaked into the chamber at the center. The gas is bombarded by electrons and ionized so that raw protons as well as molecular ions are available. Meanwhile, the two "dee's" are being driven at the cyclotron frequency by what amounts to a radio station just outside the machine. When a positive proton is formed, it will be attracted toward the negative dee. In the magnetic field, its path will be a circle. While it is inside a dee, it is electrostatically shielded from any electric field lines (but not the magnetic field lines). The proton simply coasts around a semicircle until it emerges across the gap between the two dees. If, at that instant, the other dee is negative, the proton will be accelerated across the gap, gaining energy. The radius of its orbit will now be larger. But once inside a dee, the proton is unaffected by the changing electric field and will coast in a semicircle until it comes to the gap again. If the potential across the gap is changing at the cyclotron frequency, the proton will always arrive at the gap in time to see negative regions facing it, and so it will be accelerated. Its energy increases at each passage by an amount equal to the product of its charge and the potential difference across the gap. The radius of its orbit keeps increasing, along with its energy, until finally it reaches the outside edge of the vacuum chamber. There it either hits the target or is led toward external targets or detectors.

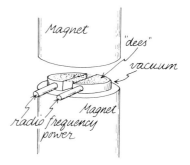

In a typical cyclotron each proton picks up about 10,000 eV during each revolution. After a thousand revolutions, the proton energy is 10 MeV. Cyclotrons of this size were common research instruments during the late

1930s. There is a problem about making machines of this type for yet higher energies. As the proton speed increases, its effective mass increases.

> **Question 20-12**
>
> Why should an increase of the proton mass influence the performance of the cyclotron?

■ Optional

Although ordinary cyclotrons are no longer being built as research tools, one design problem is common to all particle accelerators. The route of the charged particles must contain magnetic or electric fields that keep the particles from leaving the prescribed orbit. For an example of the problem, which is dramatic but turns out to be trivial, consider that a proton beam takes about 1 s to be accelerated in some of the large machines. During the first second, a proton, like anything else in the earth's gravitational field, would fall 5 m. Clearly there must be some focusing force to prevent the protons from falling. Actually, there are far larger forces tending to bump the protons up, down, or sideways. The focusing forces must be strong enough to keep the beam in a narrow path. The diagram illustrates how these focusing forces can be produced. It is essential that the magnetic field *not* be uniformly vertical and constant in strength. Instead, the magnetic field must be slightly weaker toward the larger radius, so that the field lines are slightly curved as shown. If a proton gets above the equilibrium plane, it will then interact with a small horizontal component of the magnetic field that will produce a downward force on the proton. If the proton is below the equilibrium plane, this horizontal component of the field is in the opposite direction and so will produce an upward force on the proton. In any kind of particle accelerator, including the electron accelerators in your dentist's *X*-ray machine and the electron beam in your television tube, electric and magnetic fields are designed to focus the beam in much the same manner that glass lenses focus beams of light.

■

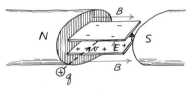

Particle path is straight if:
$qE(\uparrow) = qvB(\downarrow)$

One useful combination of electric and magnetic fields provides a way to select particles with a specific velocity. Send a beam of particles through crossed electric and magnetic fields, as shown in the diagram. With the orientation of the fields as shown, the electrostatic force is directed upward and the magnetic force, downward. If the two forces are just equal in magnitude, the particle will continue moving in a straight line. The balanced forces on the particle are

$$qE = qvB$$

Only particles with a specific velocity can satisfy these conditions. Regardless of its charge or mass, a particle will continue in a straight line if $v = E/B$.

If electric and magnetic fields are crossed in a *conductor*, the charge carriers will drift in the general direction determined by the electric field, but also will experience a weak force due to the magnetic field, tending to make them concentrate along one face of the conductor. The geometry for negative charge carriers is shown in the diagram. The extra concentration of charges along the top face produces a slight difference of potential across the thickness of the conductor. The sign of this potential difference depends on whether the charge carriers are negative or positive. If negative charge carriers drifting to the right experience a magnetic force that makes them turn up so that they concentrate along the top, then positive charge carriers drifting to the *left* in the same electric field would experience a magnetic force tending to make them turn upward also. Therefore, if the charge carriers are negative as they are in metals, the top face of the bar would be *negative* with respect to the bottom face. If the charge carriers are positive as they are in some semiconductors, then the top face of the conductor would be *positive* with respect to the bottom face. The magnitude of the potential difference is also a function of the density of charge carriers and hence of the number of charge carriers contributed per atom. Experiments show that each atom of copper or lithium or sodium contributes exactly one electron, but aluminum contributes three electrons per atom. This phenomenon and experimental technique is called the *Hall effect*, after its discoverer.

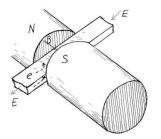

Earlier we said that the path of a charged particle moving with constant speed in a uniform magnetic field is a circle. Actually, that is the case only if the original velocity of the particle is in the plane perpendicular to the magnetic field. If there is also a component of velocity along the magnetic field lines, then the particle will spiral in a helix, as shown in the first diagram. A surprising situation arises if the magnetic field lines are curved. As you can see in the second diagram, a change in the direction of the field line implies that there is a component of the magnetic field perpendicular to the original direction. The charged particle will react to that component in the way shown in the diagram. The resulting forces change the direction of the spiraling orbit so that it follows the magnetic field lines. It is this effect that makes protons and electrons spiral along the magnetic field lines of the earth far beyond the atmosphere. These regions are called the *Van Allen radiation belts*. Except for one other effect, the spiraling particles would come into the atmosphere along with the magnetic field lines near the regions of the north and south magnetic poles. As a matter of fact, many particles do come in along those lines, causing the auroras seen in northern and southern latitudes. But many other particles lie within momentum ranges so that they are reflected from the regions of higher magnetic fields near the poles before they enter the atmosphere. When a particle trapped in this way enters the region of a stronger field near the poles, the pitch of its helical path is reduced and then reversed, so that it starts spiraling back. Particles in such orbits remain trapped in the magnetic fields for many traversals. Eventually they get scattered out through collisions with other particles. But their numbers are continually replenished by protons and electrons shot out from the sun.

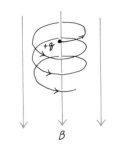

Particle with charge +q with initial component of velocity downward.

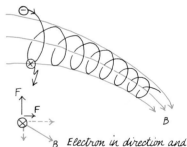

Electron in direction and position shown will experience upward centripetal force from main horizontal component of B, and also a horizontal force to the right from the small vertical component of B.

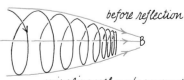

before reflection

spiraling path can be reversed

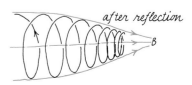

after reflection

THE TORQUE ON A WIRE LOOP

In a uniform magnetic field, a magnetic dipole such as the compass needle does not experience any *net force*. The attraction in one direction on the north pole is balanced by the attraction in the other direction on the south

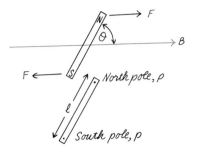

pole. There is, however, a *torque* on the dipole as shown in the diagram. In analogy with the formula for the strength of an electric dipole, the *magnetic moment* of a compass needle could be defined as $m = lp$—if we knew the "pole strength," p, and could measure the distance between poles, l. Actually, this equation can *define* the pole strength. We can assign a meaning and a value to the term by measuring the torque on a bar magnet in a known magnetic field. As shown in the diagram, the magnitude of that torque is equal to $\tau = l\,pB \sin \theta$. Note that the torque is maximum when the needle is perpendicular to the field lines and is zero when the needle is aligned.

Although in the case of the compass needle we have defined the dipole moment in terms of the product of pole strength and length of the needle, the experimentally measured quantity is actually just the magnetic moment itself. The poles cannot be isolated and, as we have seen, exist in a spread out region so that their separation distance is not a definite quantity. The general and most accurate expression for the torque on a magnetic dipole is:

$$\tau = m \times \mathbf{B}$$

The torque, which is a vector, is the vector product of the magnetic moment and the magnetic field. Because of the nature of the vector product, the magnitude of the torque is proportional to the sine of the angle between the magnetic moment and the field. Furthermore, the direction of the torque is along the axis of the dipole, in the sense given by the right-hand rule as shown in the diagram.

Since we know that magnetic *poles* do not exist, how do we explain the existence of magnetic *dipoles*? We have already seen that the magnetic field produced on its axis by a current loop depends on the inverse third power of the distance from the loop. The field on the axis of a dipole has that same functional dependence on the distance. In fact, the formulas for the loop and the dipole are the same if we assign the value of IA to the magnetic moment of the loop. Perhaps, then, all magnetic dipoles are created by current loops. To make sure, however, let's study the other side of the interaction. Does a current loop in a uniform magnetic field behave like a magnetic dipole?

The easiest way to calculate the effect of a magnetic field on a current loop is to make the loop rectangular, as shown in the diagram. When the magnetic field lines lie in the plane of the rectangle, there is no force on the top current or the bottom current, since these are parallel or antiparallel to the field line. The force on the left wire is out of the paper and the force on the right wire is into the paper, thus creating a torque:

$$\tau = 2(\text{force on each vertical wire})(\text{lever arm})$$
$$= 2(BIL)(d/2) = BIA$$

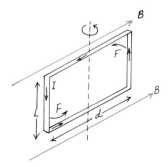

As the rectangular coil turns, there are nonzero forces on the top and bottom wires. These forces are in opposite directions and tend to distort the coil but not to turn it. Meanwhile, the force on the side wires is always perpendicular to the current and to the field. The component of the force that is perpendicular to the lever arm is proportional to $\sin \theta$, as shown in the diagram. Therefore, the torque goes to zero when θ goes to zero, and the coil is lined up with its face perpendicular to the field lines. If we assign a vector direction to area as we have done before, then the torque on the rectangular current loop can be expressed as a vector product:

$$\tau = I(\mathbf{A} \times \mathbf{B})$$

This expression for the torque on the special geometry of a rectangular current loop can be generalized if we define the magnetic moment of *any* plane loop to be equal to the product of the current and its area. This definition agrees with the one that we have already used for the magnetic field *produced* by a current loop. In this diagram, we show some justification for making the generalization from a rectangular loop to any current loop. The torque on any such loop is equal to

$$\tau = m \times \mathbf{B} \qquad \text{where } m = I\mathbf{A}$$

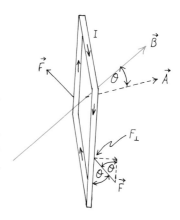

A current loop produces a magnetic field that behaves like the magnetic field of a dipole. A current loop also experiences a torque in a magnetic field as if it were a dipole. Let's consider one other feature of the behavior of dipoles in fields. When a compass needle or a current loop lines up in a magnetic field, its potential energy with respect to the field is a minimum. How much energy does it take to twist the dipole 90° to where the torque is maximum, or to 180° where the dipole is antiparallel? Let's do an approximate calculation.

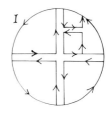

Work is equal to the dot product of force and distance. In the case of the rectangular loop turning on its axis, the force on each of the side wires is constant and equal in magnitude to $I\,|l|\,|B|$. If that were the force in the direction of motion of the wire, the work that would have to be done to move each wire through a $\frac{1}{4}$ circumference (90°) would be (force) · (distance) $= (IlB)[(1/4)(2\pi)(d/2)]$. The work done in twisting the whole rectangle would be $2(IlB)[(\pi/4)(d)] = (\pi/2)IBA = (\pi/2)mB$.

The actual work required to twist the rectangular loop through 90° is less than this amount because the force in the direction of motion is equal to $F\sin\theta$. The effective force in the direction of motion is small at the beginning of the motion and reaches the maximum amount as θ reaches 90°. The *actual* amount of work required is

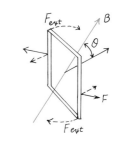

Work to twist dipole through 90° = mB

(*compare with our approximation:* $\frac{\pi}{2} mB$)

The potential energy of a dipole in a magnetic field is defined as

$$U = -m \cdot \mathbf{B}$$

According to this definition, the energy of the dipole is zero when the axis of the loop is perpendicular to the magnetic field. (We can, of course, define the zero of potential energy at any point we wish.) In the perpendicular direction, the torque on the dipole is maximum. When the dipole is lined up with the field, $\theta = 0°$, and the potential energy is negative.

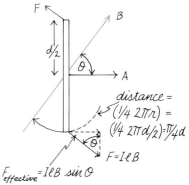

The force shown is the force on the wire caused by the current in the magnetic field. To twist the loop, work must be done by exerting counter forces equal to F.

Question 20-13

What is the physical significance of a negative potential energy?

Let's find the torque and the potential energy of an actual coil in a magnetic field. Can we, for example, send current from a D cell into a coil and see it act like a compass in the earth's magnetic field? If the coil has a diameter of 4 cm, it has an area of the order of 10 cm². Make a coil of 10 turns and connect it across a D cell, obtaining about 5 A. Each turn has a magnetic moment equal to IA. The magnetic moment of a coil with N turns is therefore $m = NIA = $ (10 turns)(5 A)(10^{-3} m²) = 5×10^{-2} A·m². In the magnetic field of the earth, the maximum torque on this coil would be $\tau = mB \sin 90° = (5 \times 10^{-2}$ A·m²$)(\frac{1}{5} \times 10^{-4}$ T$) = 1 \times 10^{-6}$ N·m. That's a very small torque.

> ### Question 20-14
>
> What would be the order of magnitude of the force on each side of the coil to produce such a torque?

The deflection of such a coil in the earth's field could be detected only if the coil were suspended by a very thin and flexible wire. The potential energy difference between the in-line direction and the 90° orientation is $\Delta U = m \cdot \mathbf{B} = (5 \times 10^{-2}$ A·m²$)(\frac{1}{5} \times 10^{-4}$ T$) = 1 \times 10^{-6}$ J. That's the amount of energy required to lift $\frac{1}{10}$ g through 1 mm in the earth's gravitational field. There is a way to increase the magnetic effectiveness of such a coil by a very large factor. In the next section, we see how to make strong electromagnets by winding coils on iron cores.

MAGNETISM IN MATERIALS

Throughout this chapter, we have said that there are no such things as magnetic poles. Current loops or solenoids can duplicate the effects of permanent bar magnets, both in the field they produce and in the torques that they experience. Must we therefore conclude that there are perpetual electric currents in iron magnets that circulate in such a way that we have coils or solenoids without power sources? If this is the case, why are these currents established in iron and a few other elements but not in most materials? Once again, we must appeal to a model of the microstructure in order to explain a natural phenomenon.

Perpetual currents do exist in material. The electrons in each atom are in motion around the central nucleus. The old model of electrons that pictures them orbiting the nucleus, like planets around the sun, has a limited usefulness and is completely wrong in a number of ways. Nevertheless, whenever an electron motion can be characterized as having angular momentum, it seems reasonable to expect that a magnetic moment should be produced. Let's calculate the relationship between the angular momentum of a charged particle and the magnetic moment that we might expect.

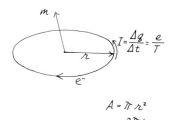

The magnetic moment of a current loop is $m = IA$. If an electron with charge e is going in a circular path with a period T, then the current produced is equal to e/T. The period of rotation is equal to the circumference divided

MAGNETISM IN MATERIALS

by the velocity: $T = 2\pi r/v$. The magnetic moment produced must therefore be equal to

$$IA = \frac{evA}{2\pi r} = \frac{ev\pi r^2}{2\pi r} = \frac{evr}{2}$$

The angular momentum of the electron about the axis is equal to $L = mvr$. If we multiply both numerator and denominator of the expression for the magnetic moment by the mass of the electron, we end up with a formula for the magnetic moment, containing the angular momentum:

$$m = \frac{emvr}{2m} = \frac{e}{2m}L \quad \text{where } L = mvr$$

Atomic electrons have angular momentum due to two different motions: one corresponds to the orbital motion around the central nucleus and the other is due to the apparent spin of the electron on its own axis. Both types of angular momentum are quantized: that is, the angular momentum can change only in integral multiples of a basic unit of angular momentum. We first described this situation on page 169 in Chapter 8. The basic unit of angular momentum is Planck's constant, which has the value (in terms of radians/s) of $\hbar = 1.054 \times 10^{-34}$ kg·m²/s. Although some orbital motion of an electron might produce several units of angular momentum, the *order of magnitude* of magnetic moment produced by any electron due either to its spin or its orbital motion will be

$$m = \frac{e}{2m}L = \frac{(1.6 \times 10^{-19}\text{ C})}{2(9 \times 10^{-31}\text{ kg})}(1 \times 10^{-34}\text{ kg m}^2/\text{s}) \approx 10^{-23}\text{ A}\cdot\text{m}^2$$

Each electron evidently has only a very small magnetic moment, even if it has two or three units of angular momentum plus its spin angular momentum. Still, if all of those magnetic moments lined up for all electrons, every piece of material would be a very strong magnet, as we will see. Actually, most of the magnetic moments cancel out. The structure of atoms is dominated by the *Pauli exclusion principle* which, among other things, requires most of the electrons to exist in pairs with opposing attributes. If two electrons are in the same general region, then the spin of one will be opposed to the spin of the other and, therefore, the spin angular momenta will cancel. Indeed, we might at first expect that any atom with an even number of electrons would have no magnetic moment at all. This is not the case, however; iron has 26 electrons. You might also expect that the magnetic properties of an atom would depend on whether the atoms were free in a gas or linked with other atoms in a solid. A free atom might have a valence electron, with the spin angular momentum not canceled by a paired electron. In a solid, the valence electron of such an atom would usually go into a bonding state, pairing with another electron and thus canceling the magnetic moments of each. The magnetic properites do indeed depend strongly on how atoms are combined with others.

In spite of the cancellation of most atomic angular momenta, many materials do end up with a magnetic moment per atom that is equal to the basic unit that we calculated: approximately 10^{-23} A·m². Suppose that the

The magnetic moments of most electron pairs cancel.

magnetic moments of one mole of these atoms were all lined up. A mole of most atoms in the solid state has a volume about the size of a pencil.

> Question 20-15
>
> A mole (6×10^{23}) of aluminum atoms has a mass of 27 g. The density of aluminum is 2.7×10^3 kg/m³. Is the volume of a mole of aluminum atoms about equal to that of a thick pencil?

The total magnetic moment of a mole of atoms, each with a magnetic moment of 10^{-23} A·m², is equal to $m_{\text{total}} = (6 \times 10^{23} \text{ atoms})(10^{-23} \text{ A·m}^2) = 6$ A·m². Compare that magnetic moment with the one that we calculated on page 602 for a 10-turn coil having an area of 10 cm² and carrying 6 A. The solid bar would have a magnetic moment over 100 times greater. Indeed, 6 A·m² is approximately the magnetic moment of a magnetized *iron* bar of this size. But only iron, cobalt, nickel, and some of their alloys behave in this manner. Why don't the magnetic moments of other elements line up?

The atoms in a solid are continually subject to thermal agitation. Let's compare the potential energy that an atom could gain by lining up in a magnetic field with the average thermal energy that is knocking it around. If the atom had one unit of magnetic moment and swung 180° from the antiparallel to the parallel direction with a magnetic field of 1 T, then its potential energy would decrease by $\Delta U_{\text{mag}} = -2mB = -2(10^{-23} \text{ A·m}^2)(1 \text{ T}) = -2 \times 10^{-23}$ J. The average thermal energy of such an atom at room temperature is $U_{\text{thermal}} = \frac{3}{2}kT = \frac{3}{2}(1.38 \times 10^{-23} \text{ J/K})(300 \text{ K}) = 6 \times 10^{-21}$ J. Both of these energies are small compared with the binding energy of an atomic valence electron, which is of the order of 1 eV (1.6×10^{-19} J). But the average thermal energy of an atom is 300 times the maximum energy that can be gained by an atom lining itself up with a magnetic field as strong as 1 T. Even if materials are composed of atoms that have individual magnetic moments, a random thermal agitation would destroy most of the alignment. In fact, now the question should be phrased the other way. Instead of asking why aluminum is not strongly magnetic, we must figure out why iron is. First, however, let's characterize the two types of very weak magnetism that appear in most materials.

In a uniform magnetic field, a permanent magnet will rotate until it is aligned with the field. So will a rod of unmagnetized iron, and many other materials for that matter, although the torques are very weak—except for iron, cobalt, nickel, and some of their alloys. An even more interesting magnetic effect is produced in a *nonuniform* magnetic field such as that shown in the diagram. If a magnetic dipole is put in such a field, it will not only rotate but will also move along the field lines.

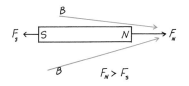

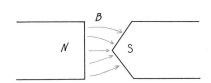

> Question 20-16
>
> Why should a dipole move along a magnetic field line? Isn't the

attractive force on the north pole in one direction balanced by the force on the south pole in the opposite direction?

If a piece of iron is suspended in such a field, whether the iron is a permanent magnet or not, it will swing strongly toward the direction of the strong magnetic field. This effect is a characteristic of *ferromagnetism*. Many other materials, such as chromium, manganese, and palladium, will also move toward the stronger field, but the force on these materials is smaller by a factor of 1000 or so than it is for iron. Materials showing this very weak magnetic effect are called *paramagnetic*.

Surprisingly enough, some materials such as arsenic, mercury, and silver will actually move toward a *weaker* field. They act as if the atomic dipoles preferred to be out of line with the external field. This effect is called *diamagnetism*. Diamagnetism is an *induced* effect, but in one crucial aspect it is different from electrostatic induction. Remember the way that a charged balloon clings to a wall because of electrostatic induction. The electrostatic field induces dipoles in the molecular structure of the wall. The electric charge arrangement in the molecules is warped so that the positive and negative charge centers are separated. The amount of the warping, and thus the strength of the dipoles, is proportional to the strength of the external electric field. In a similar way, when a magnetic field is imposed on any material, the electron *paths* are affected. As we have seen, there is a force exerted on a charged particle that is moving in a magnetic field. The resulting rearrangement of the electron paths produces a magnetic dipole moment proportional to the strength of the external field. However, the induced dipoles are in a direction to *oppose* the field. This diamagnetic effect occurs in *all* materials in whatever state they are in—gas, liquid, or solid. The effect is very small, however, and is washed out if the atom has a permanent magnetic moment due to the orbital or spin motion of one of its electrons. In Chapter 21, we will learn more about how electromagnetic induction produces dipoles that oppose the field that produces them.

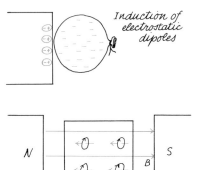

Induction of electrostatic dipoles

Induced electronic currents producing magnetic dipoles that oppose the external field.

In the table on the next page, we list a parameter known as the *magnetic susceptibility* χ (small chi), for a number of elements at room temperature. The magnetic susceptibility times the strength of the external field (divided by μ_0) gives the magnetic dipole moment per unit volume for the material. For our purposes, this parameter serves as a comparison of the relative magnetic responses of various materials. Note that the susceptibility of iron is in the range from 1000 to 10,000. Compared to ferromagnetism, the diamagnetic and paramagnetic effects are extremely weak.

Diamagnetism is not much affected by temperature, since the dipoles are induced by the external field and are not permanently attached to the individual atoms. Their strength is independent of the way the atoms are jostled around by the thermal motion. The electron motions automatically adjust to oppose the external field.

In the case of both diamagnetism and paramagnetism, the resulting magnetic moment of the material is approximately proportional to the strength of the external field. This is true in diamagnetism since the strength of the individual dipoles is proportional to the inducing field. In the case of

Susceptibilities of Various Materials—χ

Paramagnetic		
oxygen (liquid at −183°C)	1.5	× 10⁻³
oxygen gas (20°C)	2	× 10⁻⁶
aluminum	2	× 10⁻⁵
sodium	7	× 10⁻⁶
iron ammonium alum (used in magnetic cooling near absolute zero)	4.8	× 10⁻²
manganese	9	× 10⁻⁴
chromium	3	× 10⁻⁴
palladium	8	× 10⁻⁴

Diamagnetic		
bismuth	−1.7	× 10⁻⁴
mercury (liquid)	−3	× 10⁻⁵
silver	−2.6	× 10⁻⁵
lead	−1.8	× 10⁻⁵
copper	−1.0	× 10⁻⁵
arsenic	−2	× 10⁻⁵

Ferromagnetic	
iron	up to 10,000
cobalt	depending on initial
nickel	magnetization

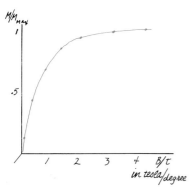

Curie plot for a paramagnetic salt. Note that in a strong field of 1T, at room temperature, B/T ≈ 1/300. At that point, the ratio M/M$_{max}$ is very small. Only a small fraction of the magnetic dipoles are aligned.

paramagnetism, the magnetic moments are intrinsic parts of each atom, independent of the external field strength. The extent to which the individual magnetic moments line up, however, depends on the competition between the external field and the thermal motions. A typical graph of the dependence of magnetization on temperature is shown here. The graph is known as a Curie plot, in honor of Pierre Curie who established both the experimental work on magnetism and the classical theory concerning it in the 1880s and 1890s before helping his wife to discover radium. Note that the units of the horizontal axis are teslas per degree. At room temperature in the strong field of 1 T, the horizontal coordinate of a data point would be at $\frac{1}{300}$. To get to the point 1 on the horizontal axis, we would have to have a magnetic field of 1 T at a temperature of 1 K, or 300 T at 300 K. The vertical axis is in terms of the magnetization produced in the material, divided by the maximum magnetization possible. The maximum magnetization would occur if all of the dipole moments were aligned; thus the curve shows the degree of alignment of the individual atomic dipoles.

FERROMAGNETISM

So far we have given arguments why materials cannot be strongly magnetic. Most of the intrinsic magnetic dipoles within an atom that are due to the orbital or spin motion of electrons cancel out. Even when each atom is left with a small net magnetic moment, the amount of energy that it might gain by lining up with an external magnetic field is much smaller than the thermal energy that makes the directions random. The mutual energy to be gained by lining up with immediate neighbors is even smaller. Nevertheless, iron, cobalt, nickel, and some of their alloys do respond strongly to magnetic fields and even retain the atomic alignment to become permanent magnets. Let's see what is so special about iron, cobalt, and nickel—and particularly iron.

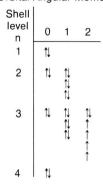

ELECTRON DISTRIBUTION IN IRON ATOM

In the diagram, we portray the positions of the electrons in the iron atom. There are 26 electrons, most of which go into shell-like positions in pairs. The inner shell, which is spherical and is labeled $n = 1$, contains two electrons. These electrons have no orbital angular momentum, and their spins are in opposite direction so that their magnetic moments cancel. In the next region, labeled $n = 2$, there can be eight electrons. Two are in a spherical shell at a larger radius, but with 0 angular momentum, like the electrons in $n = 1$. There are also three other nonspherical regions, at about the same radius and energy, each containing two electrons. Although each electron in these nonspherical regions has 1 unit of angular momentum, all of the angular momenta, including that due to the spins, cancel out because of the geometry and electron pairing. In the next major region, labeled $n = 3$, 18 electrons can be accommodated: 2 with 0 angular momentum, 6 with 1 unit of angular momentum, and 10 with 2 units of angular momentum. Since iron has only 26 electrons, and the three complete shells can hold 28, it might seem that iron would have two empty places in the $n = 3$ region, thus having a valence of -2. Instead, two of the electrons from iron establish themselves in the $n = 4$ region, thus providing a positive valence number. Meanwhile, back in the outer subregions of $n = 3$, there are six electrons that are shielded from the chemical interactions that are performed by the outermost electrons. We might expect that these six electrons would be paired so that their magnetic effects would cancel out. Instead, five of the electrons spin in one direction and one in the opposite direction. The atom is left with 4 units of magnetic moment, all in line.

The detailed calculations about why the electrons arrange themselves in this way are very complicated. The atom actually has a lower energy state if two of the electrons go out to the $n = 4$ shell, away from the densely populated $n = 3$ region, and if the six shielded electrons in the outer subshells of $n = 3$ have their spins mostly aligned, requiring them to stay far away from each other. The situations with cobalt and nickel are similar. Cobalt has 27 electrons and nickel has 28. With cobalt there are seven electrons in the outer subshells of $n = 3$, and in nickel there are eight.

In spite of these almost accidental geometrical arrangements that provide atoms with permanent magnetic moments that are not disturbed by bonding with other atoms, there is still no reason to expect that the individual atoms will line up with each other. The argument is still valid that the magnetic potential energy is small compared with the thermal energy. With the ferromagnetic materials, however, there is another influence that helps align the atoms. It is the Pauli exclusion principle—the same requirement that prevents two electrons from being in the same atomic region without having opposing spins. When an iron atom is surrounded by other iron atoms, the four unpaired electrons in the outer subshell of $n = 3$ constrain nearby electrons wandering between atoms to have the opposite spin; but these spins must also oppose those in the outer subshell of $n = 3$ in the next

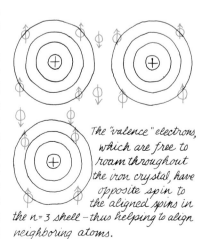

The "valence" electrons, which are free to roam throughout the iron crystal, have opposite spin to the aligned spins in the $n = 3$ shell—thus helping to align neighboring atoms.

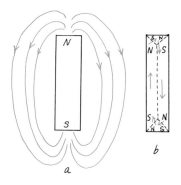

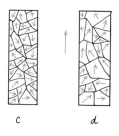

atom. Hence, the magnetic moments of neighboring atoms are aligned, not because of their magnetic influence on each other, but because of a complex application of the simple requirement that electrons in the same region must have opposite spin.

Magnetism is a subject where the obvious consequences of each new addition to the theory are frustrated by yet another condition or circumstance. If the iron atoms can go to a lower energy state by aligning with each other, why isn't *every* piece of iron a strong permanent magnet? In the a part of this diagram, we show the field lines produced by a completely magnetized iron bar. The pattern is the familiar one expected with a bar magnet but, when an iron bar freezes out of a liquid state, the magnetic moments don't line up that way. The field, as pictured, represents a lot of energy stored in space. A lower energy state would be produced as the iron froze if the atoms lined themselves up as shown in part b.. Now the field lines are concentrated within the volume of the iron, reducing the amount of energy. However, it turns out that it takes extra energy to produce a wall between two atomic regions. Furthermore, as the iron liquid freezes into a solid, there are geometrical constraints on the atomic locations other than just the direction of magnetic moments. Crystals of various size form with boundaries that are affected by impurities and imperfections. Within a crystal, magnetic moments are more easily aligned in one direction than in another. The end result of all these complications is shown in part c. The solid iron consists of domains that are microscopic in size.

Question 20-17

If a domain is a cube, 10 micrometers on a side, how many atoms does it contain?

The magnetic moments of all of the atoms within a domain are aligned; therefore, the tiny region is a strong permanent magnet. However, in an ordinary piece of iron just frozen out of the melt, the magnetic moments of the domains are completely random in direction. An iron bar ordinarily has no net magnetic moment. If such a bar is put into a strong external magnetic field, the domains that have magnetic moments in line are in a lower energy state than those with magnetic moments opposed. The boundaries between domains start moving, with the aligned domains gaining volume at the expense of those opposed. The d part of the diagram shows such a transformation.

The growth of the favored domains can be detected experimentally in a dramatic way. Wind a coil of several thousand turns of fine wire around a bar of iron. Connect the coil to an audio amplifier, with the output fed into a loudspeaker. Bring a permanent magnet slowly toward the bar of iron. As the iron is magnetized by the slowly increasing external field, the loudspeaker will make a sound like static. The walls of the domains are not moving smoothly in response to the external field, but adhere to impurities or imperfections in the crystals. As the external field increases, the energy difference between one domain and its neighbor becomes sufficient to shove

FERROMAGNETISM

the wall abruptly past the imperfection, thus creating an abrupt increase in the magnetic moment of the bar as a whole. This phenomenon is called the *Barkhausen effect*. In the next chapter, we will see why the sudden change in magnetic field induces a voltage in the pickup coil, which in turn makes the loudspeaker crackle.

In this diagram, we show a graph of the magnetic field in an iron bar that is being magnetized by an external field, B_0. As the external field increases, the magnetic field inside the iron also increases as the walls of the domains move to produce a larger and larger net magnetic moment of the bar. The growth is not smooth, however, as the magnified portion of the graph indicates.

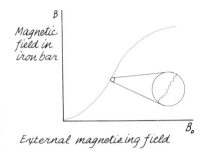

Magnetic field in iron bar

External magnetizing field

The magnetic field needed to align the magnetic domains may be small compared with the resulting magnetic field in the iron. This provides us with a way to amplify greatly the magnetic fields produced by currents in wires. As we have seen, with a tabletop solenoid and currents of a few amperes, we could produce magnetic fields of only 10 to 100 G. With an iron core in the solenoid, the same current could produce a field 100 to 1000 times greater.

The amplification factor of a magnetic material is determined by measuring the magnetic field in a ring solenoid containing a core of the material, as shown in the diagram. Without the core, the field inside the solenoid would be just that which we calculated for an ideal solenoid: $B_0 = \mu_0 n I$. To be sure, this formula was derived for the case of an infinitely long, *straight* solenoid but, if the ring is large compared with its cross-section, the formula applies well. Furthermore, with the ring geometry, we do not have to worry about end effects.

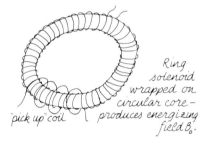

Ring solenoid wrapped on circular core — produces energizing field B_0.

"pick up" coil

Signal produced is proportional to B.

In principle, you could measure the actual field, B, inside the core by cutting a small gap in the material and inserting a small current-carrying wire. The force on the wire would be proportional to B. In practice, another "pickup" coil is wound around the primary coil, as shown in the diagram. The actual measurement of B involves turning on the current in the primary to create the energizing field, B_0. As we will see in the next chapter, a signal is thereby induced in the pickup coil, proportional to B.

In this diagram, we show a B versus B_0 curve for a typical piece of iron. Consider what happens to an unmagnetized iron core as B_0 increases from zero strength. The magnetic domains in the material that have magnetic moments in the direction of the energizing field grow at the expense of their neighbors. The resulting magnetic moment of the material produces a strong internal field. Eventually, however, the domains have grown as large as they can, and most of the atomic magnetic moments of the whole bar are aligned. The material is saturated; an increase of the energizing field can produce no more amplification. If the energizing field is then reduced in strength and finally reversed in direction, the field in the material will decrease, but not along the same curve by which it grew. Depending on the material and the way that it was fabricated, there will be a large or small remanent field left in the material—even when the energizing field is zero. If this remanent field is very large, the material may be a good permanent magnet.

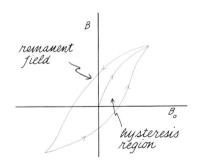

remanent field

hysteresis region

The plot of B versus B_0 is called a *hysteresis curve*. When an energizing field goes through a complete cycle, energy is lost in the material if the domain walls do not move smoothly. A material with large hysteresis and remanent fields loses considerable energy during a complete cycle of the energizing field. The area within the hysteresis curve is proportional to this energy loss. Consequently, only iron with low hysteresis loss should be used in a transformer where the energizing field may be changing at 60 Hz. So-called "soft" iron is used in transformers. Such iron should have few impurities and consist of small crystals, so that the domains can grow or shrink with minimum energy loss.

Another symbol, μ, is frequently given to the ratio, B/B_0. This μ is not the same as the permeability constant, μ_0, but is approximately equal to the relative permeability, μ/μ_0, of the material. The confusion arises from the use of earlier systems of units. The older names and symbols are still embedded in industrial use and even tradenames (such as μ-metal).

The ratio of B to B_0 is approximately equal to the susceptibility, χ. Different kinds of iron or iron alloys have values of χ ranging from 100 to 10,000. Most kinds of iron saturate with internal fields from 1 to 2 T.

There are many special kinds of ferromagnets. One alloy containing 51% of iron, and the rest a mixture of aluminum, nickel and cobalt, is called Alnico. It is a brittle metal with a large remanent field that is not easily destroyed, thus making it a very powerful and long-lasting permanent magnet. Another iron alloy called μ-metal has a value of χ of about 10,000. It is used for magnetic shielding in external fields of under 1 G.

Question 20-18

Why wouldn't μ-metal be effective in fields much greater than 1 G?

The growth of the favored magnetic domains is not instantaneous as the energizing field increases. If the external field is cyclical and changes directions faster than about 20,000 Hz, the growth of the domains cannot follow, and the value of χ will fall. There are materials called ferrites that get around this problem. Ferrites are mixtures of magnetic materials and ceramics. The embedded ferromagnetic materials are small and dispersed. Ferrites are electrically nonconductors and retain high values of χ up to frequencies in the television range of 10^8 Hz.

It is easy to shield a region from *electrostatic* fields just by enclosing it with metal walls. The external field lines will terminate on surface charges on the walls. As we have seen, there cannot be a static electric field inside a conductor. However, it is harder to shield a region from magnetic fields. Since there are no magnetic poles, external field lines cannot terminate on the walls of a shield. All we can do to shield a region from an external magnetic field is to surround it with a ferromagnetic wall that has a large value of χ. Most of the external field lines then concentrate in the shield walls and continue out the other side, bypassing the interior region, as shown in the diagram. For the ultimate shielding, a region could be surrounded by a superconductor. A superconductor sets up opposing currents to block any external field. We will come back to this topic in the next chapter.

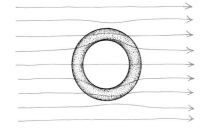

■ **Optional**

The Relative Nature of Electric and Magnetic Fields

In exploring the second realm of electromagnetism, we have described the actions of electric charges moving past us with constant velocity. As you saw in Chapter 11, if we have observers in two different reference frames, measuring the same phenomenon, they must be able to use the same laws—providing they make the right transformations of distances and times. Let's set up a thought-experiment with electric charges and describe the results from two different reference frames moving past each other with constant velocity.

The diagram shows two infinite lines separated by the distance y, each having a uniform, linear, static charge density of λ: $\lambda = \Delta q/\Delta x$. On page 520, we calculated the electrostatic field produced by a line charge. It is:

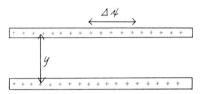

$$E = \frac{\lambda}{2\pi\epsilon_0 y}$$

The force on the other wire is just equal to its total charge times the electric field produced by the first wire. Therefore, the force between the two wires, per unit length, is:

$$\frac{F_y}{\Delta x} = \frac{\lambda^2}{2\pi\epsilon_0 y}$$

So far we have assumed that the charges are standing still, not only with respect to each other but also with respect to us, the observers. Now, however, let us see what happens if the same charged rods are moving steadily to the right along the x-axis with velocities v. As we have seen in Chapter 11, we observe that lengths in the moving system are contracted. A meter stick attached to the moving wires might appear to us to be only 50 cm long. However, there is no relativistic decrease in the actual number of charges. The charge consists of a certain number of electrons, and that number is independent of the velocity of the observer. The linear charge density *does* change, however, because $\lambda = \Delta q/\Delta x$. If we observe that the unit length has shrunk by a factor of two, then we also observe that the charge density has *increased* by a factor of two, since we would count the same number of charges in the contracted meter that we did when the charges were standing still.

In our reference frame, the charge density on the moving rods is higher.

From now on we will label all of the quantities in the moving system, *as measured in the moving system,* with a subzero. These are the quantities that would be measured by an observer riding on the rods along with the moving charges. The symbols without subzero indicate the magnitudes that *we* measure, as we watch the rods hurtle past us. The length relationship, for instance, is given by

$$\Delta x = \Delta x_0 \sqrt{1 - v^2/c^2}$$

The charge density that *we* measure is

$$\lambda = \frac{\Delta q}{\Delta x} = \frac{\Delta q}{\Delta x_0 \sqrt{1 - v^2/c^2}} = \frac{\lambda_0}{\sqrt{1 - v^2/c^2}}$$

Since we measure a larger charge density than does the observer riding on the rods, let's compare the repulsive force between the wires as measured by the two observers. The force per unit length that we measure is $F_y/\Delta x$. The force per unit length as measured by the observer riding on the wire is

$$\frac{(F_y)_0}{\Delta x_0} = \frac{\lambda_0^2}{2\pi\epsilon_0 y}$$

$$\frac{F_y}{\Delta x} = \frac{1}{2\pi\epsilon_0} \frac{\lambda^2}{y} = \frac{1}{2\pi\epsilon_0 y} \frac{\lambda_0^2}{1 - v^2/c^2} = \frac{(F_y)_0}{\Delta x_0} \frac{1}{1 - v^2/c^2}$$

Evidently, we and the moving observer disagree about the magnitude of the repulsive force between the wires. This is a serious problem because it would lead to different predictions about the same event. Let's calculate the magnitude of this difference between the two sets of observers.

$$\frac{F_y}{\Delta x} - \frac{(F_y)_0}{\Delta x_0} = \frac{1}{2\pi\epsilon_0}\left(\frac{\lambda^2}{y} - \frac{\lambda_0^2}{y}\right) = \frac{1}{2\pi\epsilon_0}\frac{\lambda^2}{y}\left[1-\left(1-\frac{v^2}{c^2}\right)\right]$$
$$= \frac{1}{2\pi\epsilon_0}\frac{\lambda^2}{y}\frac{v^2}{c^2}$$

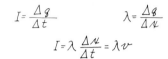

At ordinary velocities the ratio v/c is very small, and so it might seem that the two observers do not disagree by much, although there is a discrepancy. Let us carry this discussion a little further. Note that although the charges are standing still with respect to the observer on the wire, they are moving with respect to us, and so they form an electric current. That current is $I = \lambda v$. Therefore, we can rewrite the size of the discrepancy in observations as

$$\text{Extra } \frac{F}{\Delta x} \text{ observed by us} = \frac{1}{2\pi\epsilon_0}\frac{I^2}{y}\frac{1}{c^2}$$

Now we are faced with another discrepancy. If we observe two parallel *currents*, we know that there is an attractive force between them. That magnetic force between two wires is

$$\left(\frac{F}{\Delta x}\right)_{mag} = \frac{\mu_0}{2\pi}\frac{I^2}{y}$$

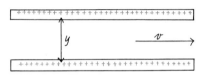

We see:
increased charge density,
increased electrostatic repulsion,
electric currents, and
consequently magnetic attraction

There are now two discrepancies between our measurements and those of the observer on the moving wire. She sees only a static repulsive force. We see a larger, static repulsive force because the charge density that we measure is larger. But we also see an *attractive* force because of the magnetic effect. As far as the traveling observer is concerned, there is no magnetic field.

If the force per unit length is not the same as measured by the two observers, then they would predict different events for the same phenomenon. Such a contradiction would be the downfall of our scientific methods. For instance, what if the magnetic field that we measure were sufficiently strong to cause the wires to come together? (Actually, the repulsive force would always be stronger than the attractive force for the geometry we have assumed. However, there is a situation, called the "pinch" effect, where high-velocity beams do come together. For the pinch effect to occur, there must also be neutralizing charges present between the currents.)

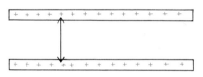

Observer riding with rods sees:
original charge density,
same electrostatic repulsion,
no currents, and no magnetic field

We can escape the controversy between the two observers if we say that the *extra repulsive* force that we observe must be just equal to the *extra attractive* magnetic force. This condition is satisfied if:

$$\left(\frac{F}{\Delta x}\right)_{mag} = \left(\frac{F}{\Delta x}\right)_{\text{extra electrostatic}}$$

$$\frac{\mu_0}{2\pi}\frac{I^2}{y} = \frac{1}{2\pi\epsilon_0}\frac{I^2}{y}\frac{1}{c^2}$$

These two expressions are equal if

$$\mu_0 = \frac{1}{\epsilon_0}\frac{1}{c^2}$$

or

$$\mu_0\epsilon_0 = \frac{1}{c^2}$$

The permittivity, ϵ_0, and permeability, μ_0, are constants measured in static situations. The permittivity is measured in terms of the fields and geometry of a capacitor; the value of the permeability is chosen in an experiment that defines the ampere. Let's substitute those values into our equation.

$$c = \sqrt{\frac{1}{\mu_0 \epsilon_0}} = \sqrt{\frac{1}{(4\pi \times 10^{-7})(8.85 \times 10^{-12})}}$$

$$= \sqrt{\frac{1}{11.1 \times 10^{-18}}} = \frac{1}{3.33} \times 10^9$$

$$= 3 \times 10^8 \, \text{m/s}$$

The equation is satisfied. Astonishingly, the properties of static electric and magnetic fields are connected with the velocity of light!

Whether we observe an electric field or a magnetic field apparently depends on our velocity relative to the sources. One person's electric field may be another person's magnetic field. It appears that magnetism is just a relativistic effect. The magnitude of the effect depends on the square of the ratio of the relative velocity to the velocity of light. We might expect that the phenomenon could be observed only with subatomic particles that have been accelerated close to the speed of light; but we have already observed that the magnetic phenomenon can be observed in ordinary wires fastened across a D cell. The electrons in that current are drifting with a velocity of a fraction of a millimeter per second. Why is it that we can make powerful electric motors that exploit such a weak effect?

The magnetic field produced by electron drift in a wire is indeed extremely small compared with the electric field produced by that same charge. The electron is not alone in the wire, however. All of the negative and positive charges are paired, so that there is very complete electrostatic neutrality. The magnetic field due to the drift velocity of each individual electron, small though it is, is not canceled out. For every ampere in the wire, 6×10^{18} electrons drift past the given point each second. The individual field is small, but the number of contributions is enormous. The result is an influence that can be detected in our human scale of things. Every time that you feel a magnetic force, you are experiencing the consequences of the special theory of relativity.

SUMMARY

There is a force between electric currents that is different from the coulomb interaction. The force per length between currents in long parallel wires is

$$F/l = \frac{\mu_0}{2\pi} \frac{I_1 I_2}{r}$$

If the currents are in the same direction, the force is attractive; if the currents are antiparallel, the force is one of repulsion.

We explained and described this interaction between currents by a model of magnetic fields produced by currents. The lines of magnetic field in space lie in the direction of the force that would be experienced by a unit magnetic test pole. We showed diagrams of the magnetic field lines produced by various shapes of permanent magnets and current configurations.

Magnetic fields can be calculated from two laws with different forms. In Ampère's law, if the magnetic field, **B**, is in line with a path, **L**, then for a complete path loop around a current, I: $\mathbf{B} \cdot \mathbf{L}_{\text{loop}} = \mu_0 I$. According to the law of Biot-Savart, the contribution to the magnetic field, $\Delta \mathbf{B}$, at a point is

$$\Delta \mathbf{B} = \frac{\mu_0}{4\pi} \frac{I \Delta l \sin \phi}{r^2}$$

The field due to an electric charge moving with velocity v, is

$$\mathbf{B} = \left(\frac{\mu_0}{4\pi}\right) q \left(\frac{v \sin \phi}{r^2}\right)$$

Using one or the other of these laws, we derived formulas for the magnetic fields in several current geometries:

Near a long straight wire: $B = \left(\dfrac{\mu_0}{2\pi}\right) \left(\dfrac{I}{r}\right)$

On the axis of a current loop of radius R:

At center: $B = \left(\dfrac{\mu_0}{2}\right) \left(\dfrac{I}{R}\right)$

For $r \gg R$: $B = \left(\dfrac{\mu_0}{4\pi}\right) \left(\dfrac{2}{r^3}\right) (\pi R^2 I)$

In a long solenoid: $B = \mu_0 n I$

The force on a current in a magnetic field is $\mathbf{F} = I\,(\mathbf{l} \times \mathbf{B})$. The force on a moving charge is $\mathbf{F} = q(\mathbf{v} \times \mathbf{B})$. Such a force drives electric motors or provides the centripetal force needed to keep charged particles in circular paths. A current loop in a magnetic field experiences a torque: $\boldsymbol{\tau} = \boldsymbol{m} \times \mathbf{B}$, where $\boldsymbol{m} = I\mathbf{A}$. The potential energy of a current loop is $U = -\boldsymbol{m} \cdot \mathbf{B}$.

Magnetism in materials is caused by the magnetic moments produced by electron spin. For most materials that contain an unpaired electron, the cumulative effect is called *paramagnetism* and is weak. Thermal agitation prevents major alignment of the individual atomic magnets. If the magnetic moments of electrons are paired and cancel, an even weaker effect called *diamagnetism* can be seen. Internal electron motions set up induced fields that oppose the external fields. With a few elements and alloys, alignment of the individual atomic moments takes place because of the exclusion principle. These materials are ferromagnetic.

Magnetic and electric fields are different aspects of the same phenomenon. We showed their relationship by applying the special theory of relativity to a particular geometry of currents. As a result of this analysis, we showed that the speed of light is related to the electric permittivity, ϵ_0, and the magnetic permeability, μ_0:

$$c = \frac{1}{\sqrt{\mu_0 \epsilon_0}}$$

Answers to Questions

20-1: The resultant force is 1 N. It is the same situation in a tug-of-war. If one team pulls with a force of 1000 N, and the other team also pulls with 1000 N, the rope does not move and has a tension in it of 1000 N.

20-2: Since we are free to choose the value of μ_0, we choose it to be *exactly* $4\pi \times 10^{-7}$. In assigning that arbitrary value to μ_0, we effectively decide the size of the coulomb. The size of the coulomb in turn affects the size of ϵ_0. As we will see at the end of this chapter, the constants μ_0 and ϵ_0 are related in another and astonishing way.

SUMMARY

20-3: Indeed, we still face the problem of defining a unique interaction of an electric current with a magnetic field. Once we plot magnetic fields by using a permanent magnetic dipole, we must show how an electric current in that field experiences a force with the magnitude and direction corresponding to any experiment.

20-4: If the magnetic field lines from a long straight wire were radial, then they would intersect a parallel wire at right angles. The model would be unable to predict whether the resulting force between the two wires should be attractive or repulsive.

20-5: The small resultant force on the compass needle is radial toward the wire.

20-6: If we had taken a square route around the wire, the magnetic field would not have been parallel to the path except in four places. Furthermore, we would have no reason to expect that the magnetic field should be constant in magnitude, since the corners of the square are further from the wire than the midpoints of the square. Actually, Ampère's law *always* holds if the dot product of magnetic field and displacement is taken for each infinitesimal segment along the path, and these products are added for the whole path. As a practical matter, we usually use Ampère's law only for cases where symmetry assures us that the magnetic field has constant value along the path. In the case of the infinite wire, the magnetic field at a particular radius must be the same regardless of the angular position around the wire.

20-7: The magnetic field near a long wire is indeed inversely proportional to the first power of the distance from the wire. The Biot-Savart law deals with a point source, however. In the first form of the law, the source is a small segment of the current, Δl. In the second form, the magnetic field is produced by the moving *point* charge.

20-8: To use Ampère's law, we must calculate the magnetic field along a loop where the field is constant. In this case, the *current* is in the form of a loop, and we have no reason from symmetry to expect that there is some other loop encircling the wire, along which there is constant magnetic field.

20-9: Remember that the only practical way to apply Ampère's law is to arrange a closed path that lies along a line of constant magnetic field. A circular loop concentric with the solenoid does not coincide with the magnetic field lines, which are mainly axial.

20-10: First you calculate the power in watts: $P = IV$. If the motor draws 6 A at 120 V, then it uses 720 W. There are 746 W in 1 hp. (In chapter 21, we learn a correction to this first approximation. With alternating current, $P = IV \cos \phi$, where ϕ is the phase angle between the sinusoidal current and voltage.)

20-11: Evidently, the momentum of each particle must have decreased continually as it went through the liquid of the bubble chamber. The particles lost energy and hence momentum by ionizing the material through which they passed.

20-12: Remember that the frequency of revolution of the proton in a cyclotron is proportional to the ratio of its charge to its mass. If the mass increases, the resonant frequency must decrease. The operation of the cyclotron depends, however, on having the same frequency for all the particles, regardless of whether they are at small radius, just starting their journey, or at large radius near the target.

20-13: If an object has negative potential energy, then it is bound or trapped. It would take positive energy to move it from this position. In the case of the magnetic dipole as shown in the diagram it would take positive energy to rotate it from its in-line orientation.

20-14: Since the lever arm of the torque is about equal to the radius of the coil, the approximate force on each side is equal to

$$F = \frac{(\tau/2)}{l} = \frac{(1 \times 10^{-6} \, \text{N} \cdot \text{m})}{(4 \times 10^{-2} \, \text{m})} = \frac{1}{4} \times 10^{-4} \, \text{N}$$

This is a very small force. Remember that the weight of a gram is 1×10^{-2} N.

20-15: A density of 2.7×10^3 kg/m^3 is equal to 2.7 g/cm^3. If a mole of aluminum atoms has a mass of 27 g, then its volume is V = mass/density = (27 g)/(2.7 g/cm^3) = 10 cm^3.

A cylindrical rod with a diameter of 1 cm and a length of about 13 cm has a volume of about 10 cm³.

20-16: A dipole will not move along a field line in a *uniform* field. In a nonuniform field, the field is stronger at the north pole of the dipole than it is at the south pole, and so there is a net force along the field line toward the region of stronger field.

20-17: The atomic diameter is about 1×10^{-10} m. Along one side of a cube that is 10 μ long, there can be 10^5 atoms. Therefore, the domain would contain 10^{15} atoms.

20-18: With a ratio of $B/B_0 = 10{,}000$, the μ-metal becomes saturated with $B_0 \approx 1$ G.

Problems

1. In one of the large particle accelerators, the copper bars in the magnet winding are 25 m long, 10 cm apart (center to center), and carry a pulsed current that reaches 7000 A. What is the force between two of these conductor bars?
2. Calculate the feasibility of demonstrating the magnetic force on a current-carrying wire, using the earth's magnetic field. (Find out what current is needed to produce a force big enough to move a wire large enough to carry the current.)
3. In "Handling the Phenomena" you short-circuited a battery across two strips of aluminum foil. If the current in each strip was 5 A, and the strips were 3 mm apart and 10 cm long, what is the force between strips?
4. Explain with words and a diagram why a magnetic field model would not work if the lines of force were radial.
5. Sketch lines of B for the region around a dipole, for a region near two like poles close together, and for a region around a straight current-carrying wire.
6. Explain with words and diagrams of magnetic field lines why parallel currents attract and antiparallel currents repel each other.
7. Sketch the magnetic field lines in the region around a current-carrying wire that is perpendicular to an external magnetic field. Show why the field lines indicate the direction of the force on the wire.
8. If a lead to the large accelerator mentioned in Problem 1 carries 7000 A, what is the strength of the magnetic field 10 cm from the center of the conductor? Give the answer in both teslas and gauss.
9. If a coulomb of charge could be concentrated in a cubic centimeter, what would be the strength of the magnetic field 10 cm away as the charge traveled past with $v = 1 \times 10^8$ m/s? Express in both teslas and gauss. Compare this B with the static electric field at that distance from such a charge.
10. What is the magnetic field strength at the center of a 100-turn coil with a radius of 5 cm, carrying 10 A? Give the answer in both teslas and gauss.
11. For the coil in Problem 10, what is the strength of the magnetic field on the axis, 1 m from the coil?
12. A solenoid with multiple layers of windings has 1×10^4 turns per meter. What is the magnetic field strength inside when there is a current of 1 A? Express in both teslas and gauss.
13. Use Ampère's law to prove that B on the axis of a long solenoid is equal to B along the inside surface of the coils, at the inside radius.
14. What is the force on a conducting bar 10 m long, carrying a current of 7000 A, in a field of 1.8 T?
15. What is the cyclotron resonant frequency for electrons in a magnetic field of 0.01 T?
16. What is the radius of curvature of the path of a proton traveling at one-tenth the speed of light in a magnetic field of 1.5 T?

17. What is the maximum torque on a rectangular motor coil of 100 turns of wire, 4 cm by 6 cm, carrying 10 A in a field of 1.2 T?
18. What is the maximum potential energy change for the coil in Problem 17?
19. As protons are accelerated in a cyclotron, their mass increases, thus upsetting the cyclotron resonance condition. One way to compensate for this effect would be to increase the strength of the magnetic field as the radius increases. Sketch the shape of the magnetic field lines as a function of radius for such a machine. Use a diagram to analyze and explain the consequences on the vertical focusing of the beam.
20. *About* how much energy would be released if a mole of atoms (each having a magnetic moment, $m = 1 \times 10^{-23}$ A·m²) were to line up with a magnetic field of 1 T? Where does this energy go?
21. For the Curie plot shown on page 606, at what temperature would half of the atoms be aligned in a magnetic field of 20 T?
22. Describe and explain the response of thin needles of para-, dia-, and ferromagnetic materials freely suspended in a uniform magnetic field, and also in a strongly nonuniform magnetic field.
23. The drift velocity of the electrons in the aluminum strips used in "Handling the Phenomena" is only about 10^{-12} that of the velocity of light. Explain why, on the basis of the relativistic increase of charge density, such a small effect can be observed even though no electrostatic effect is detected.
24. In order to sample and measure a magnetic field, a 10 cm conductor carrying 10 A is turned until the force on it is maximum. If the force is then 0.1 N, what is B?
25. Could a homemade coil of wire and a flashlight battery be used as a magnetic compass? Suppose that you could make a 100-turn coil with a radius of 5 cm and send 5 A through it. If the horizontal component of the earth's field is 0.3 G, what is the maximum torque on the coil? About what force on each side of the coil is produced by this torque? What is the maximum potential energy difference?

21 CURRENTS AND FIELDS THAT CHANGE WITH TIME

In our study of the electromagnetic interactions, we have now covered two realms. In the first, the electric charges are stationary with respect to each other and to us, the observers. In the second realm, the charges move with constant velocity. Now our study must enter the third realm where currents are not constant and where electric and magnetic fields change with time.

In this chapter we are concerned only with relatively slow changes in current. The resulting phenomena affect only the circuits themselves or their immediate neighborhood. In Chapter 22 we will see what happens when charges accelerate more rapidly, and where the electric and magnetic fields that are produced travel away from the source charges, carrying energy and momentum which do not come back.

After Oersted discovered that electric currents produce magnetic fields, many attempts were made to find the inverse effect. Can strong magnetic fields somehow produce a current? People wound wires in various ways in magnetic fields and tried with sensitive galvanometers to detect the weak currents that might be produced. It was Faraday who discovered the secret. In order to produce current, there must be motion between the wires and the magnetic fields. Within a year of this discovery in 1831, Faraday and others were making crude generators that turned mechanical energy into electrical energy.

In this chapter we study, as usual, both phenomena and practical techniques. The phenomena involve fundamental and profound questions about the interactions of nature and the way that we observe them. The practical techniques have become fundamental to our industrial civilization and to the conveniences of our everyday life.

HANDLING THE PHENOMENA

We are surrounded by devices that make use of the phenomena to be studied in this chapter. Motors, generators, and transformers depend on currents and fields that increase and decrease sinusoidally. The electricity available in the wall sockets of our homes is alternating current. Even though such devices are common, it is not easy to study their operation in the home. Household a-c voltage is potentially lethal. Skill, knowledge, and caution

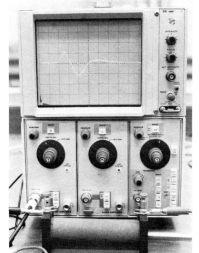

Here we see the insides of two different kinds of small household motors. In both cases the current in the coils creates a strong magnetic field in the laminated iron cores. In the motor on the left the axial armature (which rotates) receives its current through slip rings, seen at the bottom. In the motor on the right, the armature has been removed and rests on top of the magnet. The rotating armature of this kind of motor derives its current from induction without any connecting wires.

are necessary when doing anything more complicated than turning on a wall switch. Nor is it easy or wise to take apart household motors or transformers that are still working. Most of these machines depend on quite sophisticated engineering details that tend to mask the basic principles involved. Of course if you have any old shavers, mixers, or hair dryers with motors that can be sacrificed, get a 12-year-old child to help you and start taking one apart.

There are two experiments you should do qualitatively with your own hands if your school can provide the experimental apparatus. In the first of these, shove a strong cylindrical magnet in and out of a coil containing many turns. The two leads from the coil should be fastened to a sensitive galvanometer or, better still, to the input of an oscilloscope. The picture shows the simple arrangement and also shows an oscilloscope trace as the magnet is shoved first in and then out. Observe the polarity of the current that is produced as the north pole enters and leaves the coil, and then as the south pole enters and leaves. If you are using an oscilloscope, also note how the voltage produced depends on the speed with which you shove the magnet into the coil.

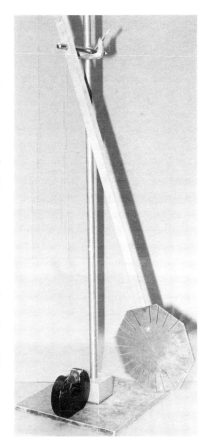

The other qualitative experiment to try requires a device often used for lecture demonstrations. A typical version is shown in the next picture. A pendulum with a bob consisting of a flat sheet of copper is allowed to swing through the gap of a powerful magnet. The pendulum will swing to a stop. If, however, the sheet of copper has slots in it cut as shown in the illustration, the pendulum will swing on through the magnet.

THE DECAY OF CHARGE ON A CAPACITOR

When we described capacitors before, we did not worry about the charging process. Now we must face the fact that it takes time for the charges to build up on a capacitor or to leave during a discharge. No new phenomena

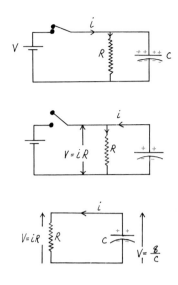

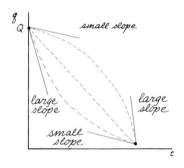

are involved in this process, but some characteristics of this type of changing current apply to some of the more complicated cases that we must study later.

To simplify the analysis, let's see what happens when a capacitor discharges as shown in the circuit diagram. With the switch closed, the capacitor is fully charged with a potential difference equal to that of the battery. Current exists in the resistor, producing a voltage drop of iR equal to the voltage of the dry cell. When the switch is opened, the charges on the capacitor will drain off through the resistor. The resulting current in the resistor will continue to provide a "back" voltage that opposes the field produced by the charges stored on the capacitor. This opposing field weakens, however, as the charges drain out of the capacitor.

We can analyze this current with Kirchhoff's law: the potential rise and drop around a complete loop must equal zero. At any given time the potential difference across the capacitor is q/C. The potential drop across the resistor is iR. (We use the lower-case q and i to indicate that they are variables.)

$$\frac{q}{C} - iR = 0$$

The charge, q, and the time rate of passage of charge, i, are related, but we have to be careful in describing the relationship. In this next diagram we show a graph of the charge on the capacitor as a function of time. When $t = 0$, $q = Q$, the original charge on the capacitor. We know very well that at some later time there will be no charge left on the capacitor. Some function $q(t)$ joins the two data points on the graph. We have sketched three possible graphs that might describe the function. Note that all of them have negative slope. The slope of a $q(t)$ graph represents the current, i. However, in our circuit, we know that the current consists of charges flowing out of the capacitor and down through the resistor in the direction shown by the arrow in the circuit diagram. In order to have a positive current when $\Delta q/\Delta t$ is intrinsically negative, we must define the current to be

$$i = -\frac{\Delta q}{\Delta t}$$

When we substitute this value for i into the Kirchhoff equation, we get

$$\frac{q}{C} - \left(-\frac{\Delta q}{\Delta t}\right)R = 0$$

$$\frac{\Delta q}{\Delta t} = -\frac{1}{RC}q$$

Now we are in a position to choose between the possible functions that describe the decay of the charge in the capacitor. According to our formula, the slope of the $q(t)$ curve, $\Delta q/\Delta t$, is proportional to the magnitude of the charge itself and is negative. Sure enough, the slopes of all three curves are negative but the upper one starts out with a small slope when q is large and ends up with a large slope when q is small. That behavior is not what the equation calls for. The slope of the middle curve is constant, which does not agree with our formula. The bottom curve, however, starts out with a large

THE DECAY OF CHARGE ON A CAPACITOR

slope for large q and then the slope continuously decreases as q decreases. The mathematical function describing the decay of the capacitor charge must look something like the bottom curve in our graph. The mathematical function that satisfies these conditions is the exponential. The complete formula is

$$q = Qe^{-(1/RC)t}$$

Question 21-1

Does this formula agree with the experimental facts when $t = 0$ and when t is very large?

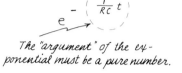

The "argument" of the exponential must be a pure number.

Since the argument of an exponential must be a pure number without any dimensions, it must be that RC has the dimensions of time. That's the case, as you can work out for yourself. Remember that $R = V/I$, $C = Q/V$, and that $I = \Delta Q / \Delta T$. When R is given in ohms, and C is given in farads, the time RC is in seconds.

When we open the switch in our circuit, $t = 0$. At a later time when $t = RC$

$$q = Qe^{-(1/RC)RC} = Qe^{-1} = \frac{Q}{e} \approx \frac{1}{2.7}Q$$

The characteristic time, RC, is called the time constant of the circuit, τ. This is the time that it takes the charge in such a circuit to decay to $1/e \approx 1/2.7$ of its original value. The time that it takes for the charge to decay to half of its original value is called the half-life; $t_{1/2} = 0.69\tau = 0.69RC$.

Question 21-2

If the capacitor in our circuit is 1 microfarad and the resistor is 1 megohm, how long does it take the charge on the capacitor to decay to half its original value? Does the stored charge ever completely disappear?

As the stored charge decays, the capacitor loses energy. When a small amount of charge leaves, the lost energy is equal to the product of that charge and the potential difference across the capacitor at that moment:

$$\Delta U_C = \Delta q (q/C)$$

The power loss at that moment is equal to

$$P = \frac{\Delta U}{\Delta t} = \frac{\Delta q}{\Delta t} \frac{q}{C} = i\frac{q}{C}$$

The Kirchhoff law for the circuit is

$$\frac{q}{C} - iR = 0$$

Therefore the power lost by the capacitor is equal to

$$P = i\frac{q}{C} = i(iR) = i^2 R$$

As you can see, the power lost by the capacitor as it discharges is just the power being dissipated in the resistor and turned into heat.

In Chapter 19 we said that when a d-c circuit is closed it takes a short time for the charges to arrange themselves to produce a constant current. Now we can calculate the order of magnitude of time for a typical circuit. The time constant depends on the capacitance and resistance of the circuit. Wires and other electrical components have capacitance, even if they don't look like capacitors. The capacitance depends on the size of the object and on its proximity to ground or other components. A typical circuit of battery, wire, bulb, and switch might have a capacitance of 10^{-10} F. If the resistance of the circuit is 10 Ω, the RC time constant is 10^{-9} s. Consequently, in a nanosecond or so the charges would be distributed along the circuit in a way that would establish constant current.

MOVING A CONDUCTOR IN A MAGNETIC FIELD

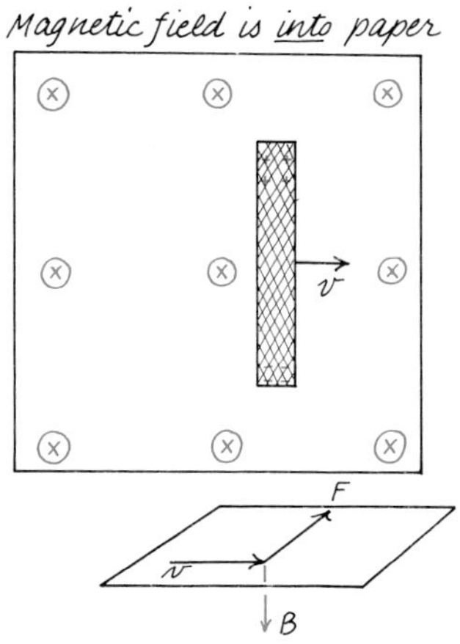

Now we are going to propose a thought-experiment that starts out as a simple application of previous results. But look out! Before we are through we will extrapolate the argument to a conclusion that is not at all contained in the original rules and formulas. Starting simply, however, consider the results of moving a metal bar in a magnetic field as shown in the diagram. The bar moves perpendicular to the magnetic field and also perpendicular to its own length. The neutral conducting bar is filled with an equal number of positive and negative charges. These are being forced to move in a magnetic field and so there is a force on each given by

$$\mathbf{F} = q(\mathbf{v} \times \mathbf{B})$$

In our diagram the resultant force on the positive charges is up and on the negative charges is down. Since some of the electrons are free to move, they will pile up at the bottom of the bar, leaving the top charged positive. An electrostatic field will be created by these concentrations of charges which will oppose any further motion of electrons. When the electrostatic force balances the force produced by the motion through the magnetic field, the two fields must be related in the following way:

$$qE = qvB$$
$$E = vB$$

An observer riding on the bar could explain the motion of the charges by saying that in his or her frame of reference there is an electric field pointed up.

Question 21-3

After the initial surge of charge, a test charge in the moving bar experiences no net force. How could the observer on the bar tell that there were charge concentrations at the two ends of the bar?

So far we have not arranged to produce any continuous current with our thought-experiment. After the initial surge, the piled up charges repel any further motion. We have a similar situation with electric batteries. To produce a current, you must complete a circuit between the positive and negative poles. In the diagram we show how to make a return path for the charges by forming a closed loop. Unfortunately, with this arrangement there is a pileup of charges on *both* of the vertical legs. Once again, no current is produced. We can get around the difficulty in a simple way by keeping the return path out of the magnetic field as shown in the lower part of the diagram. Only the leading side of the loop generates the electric field, which then drives charges around the closed loop. The potential difference from top to bottom of that leading side is equal to the product of the electric field and the length of the side. Such a source of potential difference is what we have called previously an electromotive force or emf, symbolized by $\mathcal{E}$.

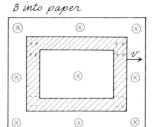

$$-\mathcal{E} = El = (vB)l$$

The negative sign appears in front of the $\mathcal{E}$ because the field is pointed up in the direction of positive charge motion but there is a potential *drop* in that direction. Notice that we have substituted for the electric field, E, its magnitude in terms of the magnetic field and the velocity of the loop.

There is another way to describe the velocity of the loop through the magnetic field. In this particular case the other way comes simply from a geometric substitution, but we will then claim that the result is generally true for any geometry. Note that as the leading side of the loop enters the magnetic field, more and more lines of magnetic force thread through the loop. We can relate the strength of the field, and the size and the velocity of the loop, to the number of lines of magnetic field *entering the loop*:

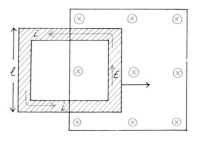

$$-\mathcal{E} = lvB = l\frac{\Delta x}{\Delta t}B = \frac{\Delta A}{\Delta t}B = \frac{\Delta \Phi}{\Delta t}$$

where Φ (capital phi) is the number of lines of magnetic force.

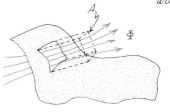

Question 21-4

Why does lv turn into $\Delta A/\Delta t$? The loop doesn't expand. How can there be a time rate of change of its area?

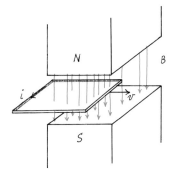

The number of lines of force is called the *flux*, Φ. The magnetic flux in a region is equal to the product of the magnetic field, B, and the area, A, providing that the field lines are perpendicular to the surface of the area. According to our derivation, the emf produced by the motion of this loop into the magnetic field is equal to $\Delta\Phi/\Delta t$, the time rate of change of the magnetic flux through the loop. (Remember that flux can be pictured as being the total number of field lines. The emf does not depend on the *number* of field lines in the loop but rather on how *rapidly* the lines are leaving or entering the loop.)

Let's give some physical reality to our thought-experiment by assigning reasonable values to the magnetic field, the velocity, and the dimensions of the loop. If the length of the leading side of the loop is 1 m, and it is moving with a velocity of 1 m/s into a magnetic field produced by a large research magnet with $B = 1$ T, then the emf produced is

$$\varepsilon = -\frac{\Delta\Phi}{\Delta t} = -lvB = (1\text{ m})(1\text{ m/s})(1\text{ T}) = -1\text{ V}$$

In this case an emf of 1 V is available to produce a circulating current. The direction of the current is counterclockwise, which according to our usual convention agrees with the negative sign for the emf.

EFFECT OF INDUCED CURRENT—LENZ'S LAW

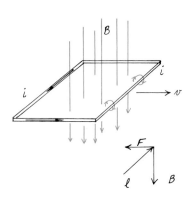

The motion of a *conducting* loop in a magnetic field produces current in the loop. The current is said to be "induced." Consider now the fact that the induced current must interact with the magnetic field that gave rise to it in the first place. A current-carrying wire perpendicular to a magnetic field experiences a force: $\mathbf{F} = i(\mathbf{l} \times \mathbf{B})$. Let's figure out the direction of that force in the case of the loop that we used in the thought-experiment.

The right-hand rule of vector multiplication, as shown in the diagram, predicts that the resulting force will be to the left in the direction opposite to the velocity. We get the same result if we use the argument about combining magnetic fields. As shown in the diagram, the induced current produces field lines that reinforce the field in front of the leading edge of the loop and reduce the field in back. According to our model, the resulting force on the conductor is therefore to the left, in the direction opposite its velocity.

There is another way to predict the direction of the induced current. Note that the field produced by the induced current tends to cancel the field lines that are sweeping into the loop. A generalization concerning the direction of induced current is called Lenz's law. Heinrich Lenz (1804–1865) proposed that induced current is always established in a direction to oppose the change of magnetic field that produced it in the first place. A conducting loop moving into a magnetic field experiences a drag force. Furthermore, the induced current produces a magnetic field that tends to maintain the status quo of magnetic flux in the loop.

Consider the alternative! If the induced current produced fields so that the force on the loop was in the direction of motion, then the loop would move faster in that direction. The resulting higher velocity would induce a

EFFECT OF INDUCED CURRENT—LENZ'S LAW

larger current. This current in turn would produce a larger force which would provide an even larger velocity. We would have perpetual motion, but, of course, it doesn't work that way. (Later in the chapter we will see how to make induction motors that produce an effect something like the one we have just described. The energy, however, is provided by continuously changing magnetic fields.)

If Lenz's law requires that the induced current create a drag force on the moving loop, then we will have to provide an equal force in the forward direction just to keep the loop moving with the same velocity v. That force is

$$F_{\text{needed}} = IlB$$

The induced current I satisfies Ohm's law for the loop circuit:

$$I = \frac{\mathcal{E}}{R} = \frac{Blv}{R}$$

Therefore, the force that we must provide is

$$F_{\text{needed}} = \left(\frac{Blv}{R}\right) lB = \frac{B^2 l^2 v}{R}$$

The power that we must provide to exert this constant force on an object moving with velocity v is

$$P_{\text{needed}} = Fv = \frac{B^2 l^2 v^2}{R}$$

Question 21-5

In order to keep the loop moving at constant velocity, we must supply energy. That energy does not go into the kinetic energy of the loop, however, since the velocity of the loop remains constant. Where does the energy go?

On page 626 we show how three different quantities vary as a function of time while the loop is passing through the magnetic field. We have graphed Φ, the flux through the loop; $\mathcal{E}$, the emf produced; and P, the power required. Up until now we have considered only what happens as the loop enters the magnetic field. Once the entire loop is in the magnetic field, the induced field on the back side of the loop is in the same direction as that on the front side. The total emf is 0 and there is no current. As the leading side of the loop leaves the magnetic field, the flux in the loop decreases and the emf is generated by the back side of the loop with the front side serving as the return circuit. The emf and the current are now in the clockwise direction. Note how the direction of emf and changing flux agree with the formula: $\mathcal{E} = -\Delta\Phi/\Delta t$. The emf is the negative of the slope of the graph of the magnetic flux as a function of time. When the slope of $\Phi(t)$ is positive, the emf is negative and vice versa. The power required to maintain the

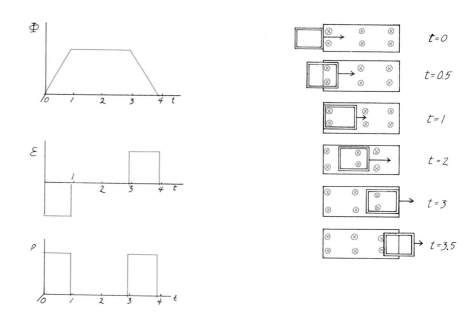

velocity of the loop, which, of course, is also the power dissipated in the loop, is positive, independent of the polarity of the emf or the direction of the current.

FARADAY'S LAW—THE NONOBVIOUS GENERALIZATION

For the particular geometry of the rectangular loop entering a uniform magnetic field, we saw that we could express the emf induced as follows:

$$\mathcal{E} = -\frac{\Delta \Phi}{\Delta t}$$

In that case the emf was just equal to the product of the induced electric field and the length of the region in which the field existed: $\mathcal{E} = -El$. If electric fields are induced in other parts of the loop, the products of those fields and distances add up to produce a total emf around the loop. We now claim that regardless of the geometry of the closed loop, the total emf around it is equal to the time rate of change of the magnetic flux through the loop. This generalization is called Faraday's law.

$$\mathcal{E} = \Sigma\, E \cdot \Delta l = -\frac{\Delta \Phi}{\Delta t}$$

Let's apply Faraday's law to a number of special situations; each will be only a small departure from the geometry or the operation of the preceding example.

1. In this diagram we show a round loop entering a uniform magnetic field. The only difference between this situation and that of the rectangular loop is that in this case the induced electric field is not generally in the direction of the conducting path. The emf produced along a short distance is equal to the dot product of electric field and line segment: $\Delta \mathcal{E} = -\mathbf{E} \cdot \Delta \mathbf{l}$. To add up the emfs produced in each segment, we would have to make use of calculus, but the procedures are straightforward. Actually, there is no

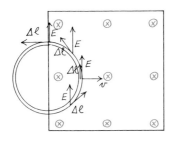

need to add up the products of electric field and line segments. The total emf around the loop is just proportional to the time rate of change of magnetic flux threading through the loop. That change of flux is proportional to the change of area *in* the magnetic field.

2. In the next diagram we show a loop about to pass over the end of a bar magnet that is lying on the axis of the loop. The closer the loop gets to the end of the magnet, the more field lines pass through the loop. As the flux through the loop changes, there will be an emf around the loop as required by Faraday's law. The direction of the induced current in the loop will be such as to produce a magnetic field that opposes the motion. Since a current loop acts like a magnetic dipole, the induced current will make this loop act like a small bar magnet with its north pole in the direction of the north pole to which it is heading. The induced current thus produces repulsion.

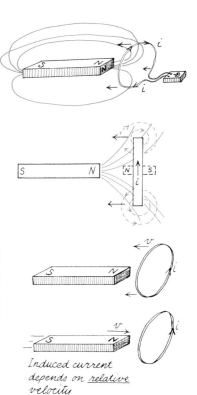

> Question 21-6
>
> What happens to the induced current as the loop passes over the bar magnet and away from it on the opposite side?

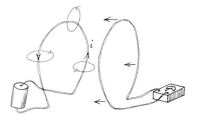

Induced current depends on relative velocity

Note that so far we have been talking about loops moving into and through magnetic fields. The next step of generalization is obviously true in the case of the loop passing over a bar magnet: it makes no difference whether the loop moves over the magnet or the magnet moves through the loop in the opposite direction. Only the *relative* velocity is important.

3. The next case to consider is a simple extension of the one in number 2. Instead of a bar magnet, we have an equivalent magnetic dipole produced by a current loop. It is reasonable to expect the same kind of induced currents in the primary coil as we had when there was a permanent magnet. As before, only the relative velocity between the two coils is important.

4. Once again we have two coils placed face to face, but this time we will not require relative motion between the two coils. Instead, the primary coil is equipped with a switch so that the current can be turned on or off. When the current is off there is no magnetic field produced by the primary and thus threading the secondary. As soon as the switch is thrown, however, the primary produces a magnetic field, which links the secondary coil. This means that the flux through the secondary coil has increased with time and therefore there is an emf around the secondary coil, producing an induced current.

5. Each of these cases of induction has been only slightly different from the preceding one and yet notice how far removed we are from the original situation of a rectangular coil moving in a magnetic field. In that original situation we could explain the induced emf in terms of the magnetic force on a charge moving in the magnetic field. In number 4, there was no physical motion yet, nevertheless, an induced current was produced. At least we can justify the effect, however, by noting that the free electrons in the secondary coil are in a region of changing magnetic field which somehow exerts a force on them.

In the next diagram we show two geometries of a coil in a magnetic

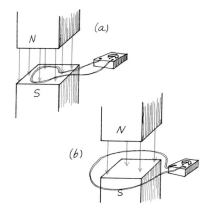

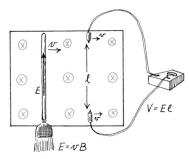

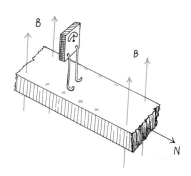

field. In both cases the magnetic field can be turned on, rising from zero strength to a maximum. The situation shown at the top is essentially the same as we saw in the earlier cases, and we would expect to find an induced emf. There is a peculiar feature of the case shown at the bottom, however. There is a changing magnetic flux through the center of the coil, but *nowhere is the conductor itself in a magnetic field.* Nevertheless, according to Faraday's law, there *is* an induced emf in the loop equal in magnitude to the time rate of change of the magnetic flux through the loop.

Now we can see why Faraday's law is not an obvious extension of our earlier formulas concerning the force on a charge moving in a magnetic field. There are really two different laws at work in these cases, both of which can be used to predict the same result in some geometries but each applicable by itself in certain other cases. Consider, for instance, our original geometry of a conducting bar moving perpendicular to a magnetic field. We claimed that the magnetic force on the moving charges in the bar made them move as if they were subject to an electric field: $E = vB$. At that point of the argument, however, we said nothing about the conductivity of the bar. As a matter of fact, the same electric field would be produced in a broomstick shoved across the magnetic field. Even the broomstick is not necessary. Depending on how you arrange the voltmeter and the leads, the same emf produced by the same electric field operating over the length l could be detected by moving the two probes of the voltmeter without any rod between them at all. As we have seen before, a magnetic field in one reference frame can turn into an electric field in a reference frame moving with constant velocity relative to the first.

An example of the way in which electric and magnetic fields depend on the velocity of the observer is shown in the diagram. A long conducting bar moves along the direction of its length and perpendicular to a magnetic field. If an ammeter is arranged with skid contacts, as shown in the diagram, then it will register a continuous current as the bar moves. In a later section we will see how Faraday made a crude generator out of a similar geometry. The effect produced in the case shown here might be explained in terms of the Hall effect, which we described on page 599. The free electrons in the bar experience a force perpendicular to the velocity of the bar and to the magnetic field. Hence they experience an electric field, directed across the bar, that will keep them moving up through the ammeter and down the other side, forming a continuous current. However, if the ammeter and the whole detection system moves along with the bar through the magnetic field, then there will be no current. In such a case the top part of the loop near the ammeter will also have an induced emf, which will buck the effect produced in the bar.

Question 21-7

Suppose that an airplane is flying north through the earth's magnetic field, which has a strong vertical component. Is there a potential difference from one wing tip to the other?

In the case of the moving bar, there is no closed loop in which there is a changing magnetic flux. Nevertheless, we get an induced electric field because of the relative motion. It is also possible to have a case where the flux through a closed loop changes and yet there is no induced emf. We show such a situation in this diagram. The magnetic field through the region remains constant but the area of the conducting region changes drastically, as the tiny contact point slides along the gap. If Faraday's law applied to such a situation, we could get a large emf and induced current even though very little work would have to be done to move the small contact. Evidently Faraday's law cannot apply to a circuit in which the material in the path changes.

Let us summarize the two quite different laws that give rise to induced emfs. The first, which is named after H. A. Lorentz (1853–1928) summarizes the effects of electric and magnetic fields on an electric charge. It also implies that motion through a magnetic field is equivalent to an electric field in a stationary frame.

$$\mathbf{F} = q(\mathbf{E} + \mathbf{v} \times \mathbf{B})$$

The other law of induction is Faraday's:

$$\mathcal{E} = -\frac{\Delta \Phi}{\Delta t}$$

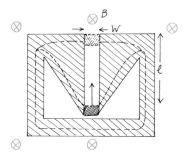

Copper sheet is perpendicular to magnetic field. As zipper is pulled, possible current path changes, increasing area of loop in B. Since there is ΔA, there is apparently $\Delta \Phi$. However, if there is $\mathcal{E} = -\left(\frac{\Delta \Phi}{\Delta t}\right)$, current would be produced and there would be ohmic heating. The work done in pulling the zipper is $W = IBiw$, where l is the path length, i is the current, B the magnetic field, and w the width of the gap. Since w can be made arbitrarily small, the work done must go to 0. Therefore, there can be no $\mathcal{E}$.

AN EMF IS NOT A CONSERVATIVE POTENTIAL

When we first defined electric potential, we were dealing with a property of electric fields around static charges. If you take a test charge around any closed path in such a field, the net work done is zero. Otherwise, as we pointed out in Chapter 18, energy would not be conserved. If you could take a test charge around a closed loop and have a gain of energy, you could provide perpetual motion.

Now we have found how to produce emfs around closed loops through which the magnetic field is changing. In the system shown in the diagram there is a constant electric field around the circular path surrounding the constantly increasing magnetic field in the center. (Of course, the magnetic field can only increase for a limited time.) If an electric charge is allowed to travel along such a path, it will experience a continual force and gain energy. The energy in joules gained per coulomb in one tour around the loop is just equal to the value of the emf in volts. It is not necessary for the charge to travel in a circular path in order to pick up this extra energy. According to Faraday's law, there is the same total emf around any closed path that surrounds the changing flux.

Does this phenomenon violate the law of conservation of energy? Not at all. In the case of static fields, there is no source to provide energy to a charged particle going around a closed path and returning to the original point. In the case of the changing magnetic flux, the energy given to the test charge comes from the source that is providing the change of the magnetic field. When we describe the action of transformers, we will see the mechanism by which the source must provide this energy.

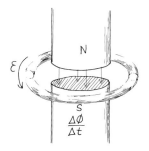

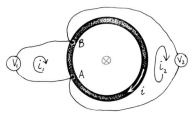

$E = 1 V \quad R = 1 \Omega$
$V_1 = 1/6 V \quad V_2 = 5/6 V \quad i = 1 A$
$R_1 = R \text{ of voltmeter 1}$
$R_2 = R \text{ of voltmeter 2}$
$i_1 \ll i$
$i_2 \ll i$

Although induced emf is measured in volts, it is frequently confusing to think of a potential difference between two points in a changing magnetic field. For instance, suppose that the circular ring in this diagram has a resistance of 1 Ω and is threaded by a changing magnetic flux that produces an emf around the loop of 1 V. While the field is changing at that rate, there will be a current of 1 A in the ring. What then is the potential difference between the points A and B? Since these are the end points of an arc that is $\frac{1}{6}$ the circumference of the ring and since there is an emf of 1 V around the whole ring, you might argue that there would be a potential difference of $\frac{1}{6}$ V between points A and B. On the other hand, A and B are also the end points of the arc that is $\frac{5}{6}$ of the whole circumference. Therefore, the potential difference across A and B should be $\frac{5}{6}$ V. As a matter of fact, if two voltmeters are connected to the ring as shown in the diagram, V_1 will read $\frac{1}{6}$ V, and V_2 will read $\frac{5}{6}$ V. Yet, the two voltmeters are connected to the same two points.

Question 21-8

Do Kirchhoff's laws still apply in such a situation? Try adding up the emfs and the *IR* drops around several closed loops as indicated in the diagram.

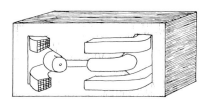

Optional

An electron accelerator invented by Donald Kerst in 1941 made use of the circular electric field lines around a changing flux. It was called a Betatron. A cutaway schematic of the machine is shown here. A doughnut-shaped vacuum chamber surrounds a central core of magnetic fields. The flux in the core rises and falls sinusoidally, usually at 60 Hz. If an electron can be forced to stay on a circular electric field line, it will pick up energy each time it goes around. The energy gained per revolution is $\Delta U = q\mathcal{E} = q(\Delta\Phi/\Delta t)$. In some of the machines that were built to accelerate electrons to energies in the range of 100 MeV, the changing flux was designed so that an electron would pick up about 10,000 eV each time around.

An electron experiencing a constant force won't stay on a circular path unless there is a centripetal force to keep it there. Consequently, the vacuum guide path must be in a magnetic field whose strength is also changing to keep up with the changing momentum of the electrons in the beam. The trick in making this machine work is to synchronize the magnetic fields in the core and over the guide path. It turns out that there is a simple requirement concerning the relationship between these two fields. Let's call the average magnetic field in the core B_{ave}. The average is taken from $r = 0$ out to the radius of the electron orbit r. The emf around the circular path is just equal to the constant electric field times the path length, which is equal to the circumference:

$$\mathcal{E} = \Sigma E \cdot \Delta l = -\frac{\Delta \phi_{\text{ave}}}{\Delta t}$$

$$2\pi r E = \pi r^2 (\Delta B_{\text{ave}}/\Delta t)$$

$$E = (r/2)(\Delta B_{\text{ave}}/\Delta t)$$

The time rate of change of momentum of an electron is just equal to the force on the electron:

$$\Delta p/\Delta t = F = qE = q(r/2)(\Delta B_{ave}/\Delta t)$$

The constant change in momentum provided by the constant change of the core magnetic field is thus equal to

$$\Delta p = (qr/2)\Delta B_{ave}$$

Meanwhile the electron experiences a centripetal force provided by its motion perpendicular to the magnetic field at the orbit, B_{orbit}.

$$qvB_{orbit} = mv^2/r$$

$$B_{orbit} = mv/qr$$

This relationship holds even at relativistic speed as long as we identify (mv) as the momentum p.

$$B_{orbit} = p/qr \qquad p = qrB_{orbit}$$

As the momentum of the electrons increases, the strength of the magnetic field at the orbit must increase proportionately. The relationship is

$$\Delta p = qr\,\Delta B_{orbit}$$

We now have two requirements to satisfy as the momentum of an electron increases. The change of momentum produced by the induced emf is proportional to the change in the average field in the core. In order to maintain a circular orbit, that change of momentum must be matched by a change in the magnetic field at the orbit. Therefore,

$$(qr/2)\Delta B_{ave} = qr\,\Delta B_{orbit}$$

The simple relationship between the average magnetic field in the core and the magnetic field at the orbit radius is

$$\Delta B_{ave} = 2\,\Delta B_{orbit}$$

Betatrons with maximum energy of several hundred MeV were made for high-energy subatomic particle research and also for X-ray generation, both in industry and in medicine. There is a practical limit to the energy that can be attained in such a machine, however, because there is an extra energy loss that we have not yet described. As we will see in Chapter 22, when an electrically charged particle accelerates, it radiates electromagnetic energy in the form of radio waves or light or X-rays. The electrons circling in a betatron are subject to strong centripetal acceleration and therefore radiate energy that must continually be made up by the acceleration process. At very high energies the radiation loss becomes very large and the special betatron field conditions cannot be met. To accelerate electrons to energies above several hundred MeV, we can use either linear accelerators or another type of circular machine, called a synchrotron, which can provide the extra energy needed to make up for the radiation loss.

EDDY CURRENTS

A magnet has almost no effect on a piece of copper that is standing still. However, if the magnet and copper move with respect to each other, or if a magnetic field through the copper increases or decreases with time, there is an interaction. The changing magnetic field or the motion of the conductor through the field sets up emfs that produce current. The induced currents interact with the magnetic field producing a force on the conductor. These currents frequently swirl through the conductor along the leading and trailing edges of the moving magnetic field region. Because of the similarity of their patterns to those seen around an object moving in water, these induced currents are called eddy currents. In some machines they are a nuisance; in others they provide the main mechanism for operation.

In the section "Handling the Phenomena," we showed an illustration of an apparatus used to demonstrate the effect of eddy currents. A flat sheet of copper swings on a pendulum into the gap of a strong magnet. The pendulum is dragged to rest by the electromagnetic interaction. The diagram shows the directions of the eddying currents. These induced currents interact with the magnetic field to produce a braking effect. The effect can be diminished by cutting gaps in the copper as shown in the diagram. The same emf is produced by the motion of the copper through the magnetic field but there no longer exists a continuous path for the current loops.

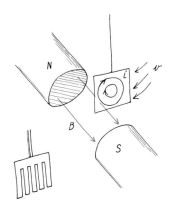

The damping force produced by eddy currents is proportional to the relative velocity of conductor and magnetic field. This effect is deliberately used as a brake in certain machines. Radial-arm saws, for example, are usually equipped with a horseshoe magnet that swings down over the saw blade when the power is turned off. The resulting eddy currents bring the blade rapidly to rest.

The frame of this transformer consists of laminated iron sheets.

Question 21-9

The rotating blade has kinetic energy. Where does this energy go when the eddy current stops the blade?

Eddy currents must be minimized in machinery that rotates in magnetic fields, or in the iron cores of transformers where magnetic fields continually change magnitude and direction. Eddy currents in the rotors of motors or generators would produce drag. In the cores of transformers the eddy currents would create I^2R loss. To avoid the eddy currents, the rotors or cores are made of thin sheets of iron separated by even thinner layers of insulating material such as enamel or mica. Such laminated pieces of iron provide good paths for the magnetic fields but do not provide electric conduction paths in the perpendicular direction. The humming sound heard around some high power transformers is caused by the cyclic compression of the laminations.

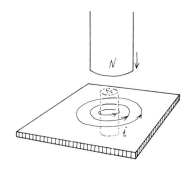

If a bar magnet is moved toward a copper sheet as shown in this diagram the eddy currents in the sheet will oppose the penetration into the sheet of the incoming magnetic field. The induced current loops set up a magnetic dipole with magnetic fields opposing the incoming dipole. Because of the resistance of the copper sheet, the electric currents die out and the upward force on the magnet stops. If the copper sheet were a superconductor, however, such as a lead bowl at a temperature close to absolute zero, then the eddy currents would not die out. These induced currents are sufficiently strong to prevent completely the penetration of any magnetic field into the body of the superconductor. The force opposing the incoming magnet is maintained and is sufficient to keep the magnet floating in air.

Eddy currents can be continuously generated with an alternating magnetic field driven by alternating current. The diagram shows such an arrangement which is frequently used in lecture demonstrations. The magnetic field in the central core is changing in strength and polarity at a rate of 60 Hz. The conducting ring on the core has an emf generated around its circumference that produces circumferential currents. These induced eddy currents also change direction at a rate of 60 Hz, but during more than half of each cycle the current is in the direction to produce a magnetic field that opposes the original field. The result is dramatic. The ring is subject to forces that repel it from the central core and shoot it into the air.

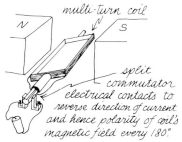

Schematic drawing of simple direct current motor.

ELECTRIC MOTORS

Electric motors are so commonplace in our civilization that it is hard to realize how new they are. Although Faraday produced an electric motor 150 years ago, very few American homes contained anything with an electric motor before the 1920s. Before that time, for instance, cars were started with a crank; rugs were cleaned by sweeping; and alarm clocks were powered by springs.

There are many different kinds of electric motors, most of them designed with very sophisticated engineering techniques. The efficiency of energy conversion, for instance, must be very high. If an appreciable fraction of the electrical energy supplied stayed in the motor, the high temperatures developed would ruin the motor. The diagram shows an elementary sort of motor that illustrates the basic electromagnetic principles but gives little indication of the engineering features required.

The basic principle is simple enough. A current-carrying wire in a magnetic field experiences a force; $\mathbf{F} = I(\mathbf{l} \times \mathbf{B})$. If the wires are wound on a rotor in a stationary magnetic field, then the current in the coil must be changed in direction each half cycle so that the rotor will always experience a force in the same direction. With stationary north and south poles the rotor coil would experience variable torque. The torque could be smoothed out by winding several coils on the rotor and having the stationary magnetic fields divided into several segments.

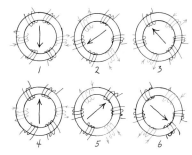

As the coils are energized in sequence, as shown, the return flux in the center of the doughnut iron ring rotates. Three phase A·C, each phase connected to an appropriate pair of coils, would smoothly energize the coils in the required sequence.

Many other variations are possible. The current supplied to the rotor coils need not change direction if the external magnetic fields change their direction. Most large electric motors are fed by a type of alternating current called three-phase. Without any moving mechanical parts, this type of power supply creates magnetic fields in the stationary part of the motor that rotate continuously. The rotor coils just follow along, rotating at the same frequency.

It is not even necessary to supply current to the rotor coils, at least with wires. If the surrounding magnetic field rotates, then current is induced in the coils of the rotor. The effect is basically the same as the drag produced by eddy currents when a conductor moves in a magnetic field. In this case the rotor coils are dragged along following the rotating magnetic field. They cannot rotate at the same frequency, however. If they did, there would be no relative motion and hence no induced currents. Therefore there must be a constant slip. If the stationary coils produce a magnetic field that rotates at 60 Hz, the rotor may follow along at 58 Hz, providing that the motor does not deliver much power. As the external load on the motor increases, the induced current in the rotor coils must increase. This can happen if the relative velocity between rotor coils and rotating magnetic field increases. Therefore the rotor slips a little more, perhaps decreasing in frequency to 56 Hz.

Some motors, such as those used in clocks, must be locked to a specific frequency determined by the alternating current frequency provided by the power station. Here is a photograph of such a motor. The rotor is a steel plate that becomes a semipermanent magnet as the result of eddy currents. A primitive rotating magnetic field is produced with ordinary alternating current by "shading" alternate legs of the stationary magnet. A conducting ring around one leg of an a-c magnet (or sometimes an aluminum plate over the end) experiences induced currents that slow down the change of the magnetic field in that region. The result is that the magnetic field in that leg lags in phase behind the field in the other leg. As the rotor magnet sweeps around, it feels a pull from the unshaded leg and as that force dies down, the rotor is moving into the delayed pull of the shaded leg.

Shaded pole clock motor. Note the coils placed asymmetrically on the magnet poles surrounding the armature.

ELECTRIC GENERATORS

When Franklin studied electricity, he had to use electrostatic devices to separate charges and store them. Experimenters of that era were able to produce high voltage effects but currents were either small or brief. The discovery of electrochemical cells in the early 1800s led to rapid development of electromagnets and electrochemistry, which require large currents. Meanwhile Faraday had discovered how to turn mechanical energy into electrical energy through the induction effect. It took about another 50 years for the widespread development and installation of large electrical generators.

One of Faraday's first versions of a generator was the so-called homopolar generator shown schematically in the diagram. The device uses the same induction principle that we have seen before. A radius of the moving disc sweeps through the magnetic field and hence has an emf generated across its (radial) length. Electric charge of one polarity builds up on the perimeter, and charge of the other polarity on the axis. If the circuit is completed by sliding contacts, then there will be a steady current down the axis, out the radius, through the external circuit, and back again. Such a generator works well but is not particularly efficient in its use of space or materials. There are also problems in extracting high currents from a sliding contact.

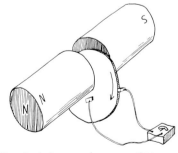

Faraday's homopolar generator.

Generators and motors can have the same basic design. If the rotor windings are provided with current, then the rotor will turn and provide mechanical energy. On the other hand, if some external source of mechanical energy turns the rotor, the induced electric current in the windings can be extracted and used elsewhere. Some motor-generators are designed to be used either way. For instance, electric motors for subways have been designed so that when the trains are braked, the wheels drive the motors as generators sending electrical power back into the system. The resulting drag furnishes a major share of the braking power needed.

In very large generators, such as in a central power station, a small current is supplied to the rotor, which then creates a rotating magnetic field. The large induced current is taken from the stationary surrounding coils, thus eliminating some of the problems caused when large currents have to pass through sliding contacts.

> Question 21-10
>
> Every household has many motors and usually at least one generator. Where is your electric generator?

INDUCTANCE

A current in a coil produces an axial magnetic field. If the current changes, the magnetic field changes. On the other hand, if the magnetic field through the coil changes, an emf will be set up in the coil. The current thus induced will set up a magnetic field in a direction to resist the changing magnetic field that produced the effect in the first place. Consider the consequences of this complicated circular interaction. If you try to change the current in a coil and thus change the magnetic field through it, an extra current will be induced in the coil to prevent what you are trying to do. If you throw a switch to impose a potential difference across a coil, there will be an induced back-emf to slow down the rise of the current and the magnetic field. On the other hand, if a current and magnetic field exist in the coil and you try to turn them off, an emf will be generated that will try to keep the current going.

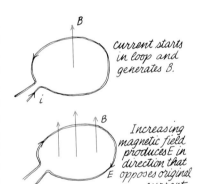

Regardless of whether the magnetic field is generated by the coil itself or is imposed from the outside, the emf of the coil is equal to

$$\mathcal{E} = -\frac{\Delta(N\Phi)}{\Delta t}$$

The number of turns in the coil is N. Since these turns are wound in series, the total emf is simply the product of N and the emf induced in a single turn, $\Delta\Phi/\Delta t$.

If the magnetic field in the coil is generated by the coil itself, then that magnetic field is proportional to the current in the coil: $B \propto I$. The total flux threading the coil is equal to the product of the area of the coil and the average magnetic field through the plane of the coil: $\Phi = AB_{ave} \propto AI$. In all cases, the magnetic flux depends on the geometrical factors involved and is proportional to the current. We can separate out the geometrical factors and describe the self-induced emf in the following way:

$$\mathcal{E} = -\frac{\Delta(N\Phi)}{\Delta t} = -L\frac{\Delta I}{\Delta t}$$

The new constant L, which describes the geometrical factors of the coil, is called the *inductance*. The unit of inductance is the henry (H), named for Joseph Henry (1797–1878). Henry was the American contemporary of Faraday. He discovered many of the same elctromagnetic effects that Faraday did and was particularly knowledgeable about the construction of electromagnets.

Let's calculate the inductance of a solenoid. Since by definition $N\Phi = LI$, the inductance itself is equal to

$$L = N\Phi/I = N(BA)/I$$

In equating Φ to BA, we are assuming that the magnetic field B in a solenoid is uniform throughout its cross-section A. This is approximately the case for a solenoid that is much longer than the diameter of its core. The field of such a solenoid is $B = \mu_0 nI$. Therefore, the inductance is equal to

$$L = (NA/I)(\mu_0 nI) = N\mu_0 nA = \mu_0(N/l)n(Al) = \mu_0 n^2 V$$

In this derivation we first found that the inductance of a solenoid is proportional to the total number of turns, N, and also to the number of turns per meter n. To avoid having two kinds of n in the same formula, we then divided the total number of turns by the length l and compensated by multiplying the cross-sectional area by l. The final result is that the inductance of a solenoid is proportional to the square of the number of turns per meter and to the volume enclosed by the solenoid.

Question 21-11

The inductance is proportional to the magnetic flux of a system. Why, therefore, is the inductance of a solenoid proportional to the *square* of the number of turns per meter?

Here is a picture of a standard laboratory solenoid. Its length is 10 cm and the cross-sectional area of its core is about 75 cm². It has 3400 turns piled into several layers. Let's use the formula that we have developed to calculate the inductance of this solenoid, even though our formula will provide only a rough approximation.

$$L = \mu_0 n^2 V = (4\pi \times 10^{-7})(34{,}000 \text{ turns/m})^2 (0.1 \times 75 \times 10^{-4} \text{ m}^3)$$

$$\approx 1.1 \text{ H}$$

Question 21-12

Why is this result only an approximation?

A coil designed to introduce inductance into a circuit is called an *inductor*. The inductance of an inductor can be greatly increased if the core is filled with iron instead of air. In that case the magnetic field is increased by the χ factor for the iron and the resulting inductance may be larger by a factor of 100–1000. In electronic circuitry these devices are called chokes. We will soon see why.

CHARGING AN INDUCTOR

It takes time for current to rise or fall in an inductor. The circuit shown here is a way to connect a battery across a resistance and inductor in series and then to bypass the battery and let the current in the circuit decay. In the graph we show the current as a function of time for the charging and discharging. When the battery is connected, the current rises to a final value given by V/R. When the switch is thrown to ground, the current eventually dies to zero.

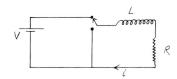

Consider what happens when the switch is first connected to the battery. Current suddenly starts to rise in the inductor but that means that $\Delta i/\Delta t$ is large. The back emf produced by the coil has the value $L(\Delta i/\Delta t)$. The potential difference across the resistor is given by the difference between the battery voltage and the back emf across the inductor; $i = [V - L(\Delta i/\Delta t)]/R$. The current is not blocked completely and as it increases, the *slope* of $i(t)$ decreases. Therefore the back emf decreases allowing yet more current in the circuit. During discharge the same phenomenon takes place. Sudddenly, $V = 0$. The current starts to decrease rapidly but now the slope of $i(t)$ is negative. The emf is opposite in sign to what it was before and tends to keep the current going *in the same direction* that it had been.

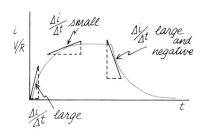

Let's analyze the circuit during the decay process when the battery is not connected. According to Kirchhoff's law, the potential changes are

$$-L\frac{\Delta i}{\Delta t} - iR = 0$$

$$\frac{\Delta i}{\Delta t} = -\frac{R}{L}i$$

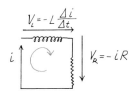

According to our formula the slope of the $i(t)$ curve is proportional to the current at each instant and is negative. We had the same equation and the same graph to describe the decay of charge on a capacitor. The curve is that of an exponential function:

$$i = Ie^{-(R/L)t} \qquad \text{where } I = \frac{V}{R}$$

When $t = 0$ (when we throw the switch to the ground connection) the equation says that

$$i = Ie^0 = I$$

For very long times, as $t \to \infty$, the current goes to 0:

$$i = Ie^{-(R/L)t} = \frac{I}{e^{(R/L)t}} \to \frac{I}{\infty} \to 0$$

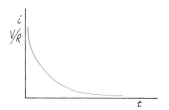

At the particular time when $t = L/R$, the current is equal to

$$i = Ie^{-(R/L)(L/R)} = Ie^{-1} = \frac{I}{e}$$

> **Question 21-13**
>
> The time constant for a circuit with inductance and resistance is evidently equal to L/R. In this time the current falls to $1/e$ of its original value but does the quantity L/R have the dimensions of time?

Let's find the time constant for the solenoid described on page 636. The inductance is $L = 1.1$ H and the resistance is $R = 50\ \Omega$. Therefore the time constant is $\tau = L/R = (1.1\ \text{H})/(50\ \Omega) = 0.022$ s. The time that it would take the current to fall to half of its original value is slightly less than this: $t_{1/2} = 0.69\ \tau = 15$ ms.

The back emf produced by trying to stop the current in a large inductor can be large and dangerous. The inductance of a large research magnet, for instance, might be 10 H. The current in the coil might be 100 A. If you try to interrupt the current by throwing a large switch or if there is an accidental break in the circuit, then the back emf that would appear across the switch or the break would be equal to $\mathcal{E} = -L(\Delta i/\Delta t) = -(10\ \text{H})(100\ \text{A})/\Delta t$. Even if Δt were 1 s, the emf would still be 1000 V. An actual break in the line or the operation of a switch would take much less time than 1 s and so the back emf would be much greater. This large potential difference would appear across the switch or the break and create an arc that would melt the leads. Special provisions must be made for handling large currents that exist in large inductors, even though the original driving voltages may be low.

THE ENERGY IN AN INDUCTOR

If current is suddenly interrupted in a circuit containing inductance, an arc may form across the open leads and melt or evaporate the contacts. Where does the energy come from? An inductor can store a magnetic field. It takes energy to create such a field and the energy becomes available again as the magnetic field decreases.

The power used during the creation of the magnetic field in an inductor is equal to the average value of the product of current and potential difference across the coil.

$$P = \text{average of } i[L(\Delta i/\Delta t)] = \Delta U/\Delta t$$

The stored enery is U. The final stored energy when the equilibrium current I has been reached is

$$U_B = \tfrac{1}{2}LI^2$$

The factor of $\tfrac{1}{2}$ is an averaging factor since the back emf is large to begin with and zero when the equilibrium current has been reached. Compare this

THE ENERGY IN AN INDUCTOR

expression for stored magnetic energy with the similar one for stored electrostatic energy in a capacitor:

$$U_E = \tfrac{1}{2}CV^2$$

Let's find out how much energy can be conveniently stored in a solenoid like the one shown on page 636. The inductance for that solenoid is 1.1 H and its resistance is 50 Ω. If we use a 6-V battery the equilibrium current is equal to $I = (6 \text{ V})/(50 \text{ Ω}) = 0.12$ A. The stored magnetic energy is equal to

$$U = \tfrac{1}{2}LI^2 = \tfrac{1}{2}(1.1 \text{ H})(0.12 \text{ A})^2 = 7.9 \times 10^{-3} \text{ J}$$

Question 21-14

This solenoid has a mass of about 2 kg. How high would you have to lift it to give it gravitational potential energy equal to the magnetic potential energy that we have just calculated?

There would seem to be no danger in handling the amount of energy that we have just calculated. Note, however, that the stored energy depends on the square of the current. Let's find a way to calculate the energy *density* of a magnetic field and then we will have a better way of finding the stored energy in a large research magnet.

In a solenoid the energy density is equal to the stored energy divided by the volume of the solenoid:

$$u_B = \frac{U_B}{V} = \frac{\tfrac{1}{2}LI^2}{V} = \frac{\tfrac{1}{2}\mu_0 n^2 V I^2}{V} = \tfrac{1}{2}\mu_0 n^2 I^2$$

The magnetic field in the solenoid is $B = \mu_0 n I$. Therefore the energy density can be expressed in terms of the magnetic field:

$$u_B = \tfrac{1}{2}\mu_0 n^2 I^2 = \tfrac{1}{2}B^2/\mu_0$$

Question 21-15

We have been talking about the magnetic energy stored in a solenoid and have just calculated the energy density. Whereabouts is this stored energy?

Now let's calculate the energy in a research magnet such as the one shown in the photograph. At full current the magnetic field in the magnet, and hence in the gap, is about 1.5 T. The iron part of the magnet has a much larger volume than the gap alone does but the energy density is inversely proportional to the magnetic permeability, μ. In iron, the permeability is increased approximately by the factor, χ: $\mu \approx \chi \mu_0$. Hence, the value of μ for iron is larger than μ_0 by a factor of from 100 to 1000. Therefore, most of the energy in the magnet is located in the gap between the pole pieces.

Ion beam in vacuum pipe changes direction as it passes through this large research magnet.

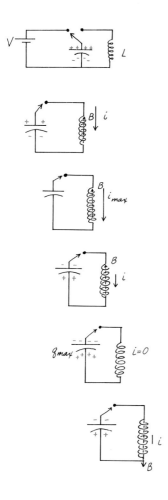

If the volume of the gap is about $\frac{1}{5}$ m³, then the stored energy is

$$U = \tfrac{1}{2} B^2 \, V/\mu_0 = \tfrac{1}{2}(1.5 \text{ T})^2 (0.2 \text{ m}^3)/(4\pi \times 10^{-7}) = 1.8 \times 10^5 \text{ J}$$

If you try to get rid of that much energy in a fraction of a second without special precautions, something will melt. If $\Delta t = 0.1$ s, for instance, the momentary power is $P = 1.8$ megawatts.

INDUCTANCE AND CAPACITANCE IN SERIES

We have seen that it takes time for charge to drain off a capacitor through a resistance. It also takes time for the current in an inductor to rise to its full value after a potential difference has been imposed across the inductor and the resistance in series. In the first case, electrostatic energy that had been stored in the capacitor must dissipate in the resistance. In the second case, the magnetic field energy in the inductor must build up as the current rises.

Let's combine the two actions. In the diagram we show a circuit that allows you to charge the capacitor from a battery and then discharge it through an inductor. Although any real wires and real inductors must contain resistance (unless they are superconducting), let's ignore that resistance to begin with. As soon as the switch is thrown to the right, the charge on the capacitor will start to drain through the inductor. The resulting change of current in the inductor will create a back emf that will slow the decay of charge from the capacitor. Meanwhile the stored electrostatic energy in the capacitor is decreasing and the magnetic energy in the inductor is beginning to increase. Eventually all the electrostatic energy will be turned into magnetic energy. At that time there will be no charge and hence no potential difference across the capacitor. At that same time, however, the current in the inductor is a maximum. Without a potential difference to maintain the current in the inductor, the current and field will start to decrease but the current will continue in the same direction, piling positive charges on the bottom plate of the capacitor. The potential difference across the capacitor rises again but with opposite polarity. Soon the capacitor is fully charged again but in the opposite sense and the current in the inductor is zero. The energy of the system is once again in the electrostatic form. Now the action repeats but this time the positive charges flow in the opposite direction causing the current and the magnetic field in the inductor to increase but in the opposite sense from the original one. The energy surges back and forth between capacitor and inductor. If there is no dissipation in a resistor, the action will continue indefinitely.

In the next diagram we have plotted graphs of the current, the charge on the capacitor, the magnetic field energy, the electrostatic field energy, and the total energy of the system. The current and the charge vary sinusoidally but 90° out of phase with each other. The potential difference across the system is in phase with the charge on the capacitor since $V = q/C$.

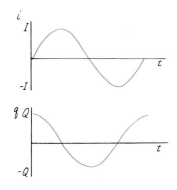

Question 21-16

In the graphs, the charge follows a cosine curve. By definition: $i =$

INDUCTANCE AND CAPACITANCE IN SERIES

$\Delta q/\Delta t$. The slope of the $q(t)$ curve is also equal to $\Delta q/\Delta t$. Should the current therefore be represented by a sine wave?

The equation for the curve representing the charge is

$$q = Q \cos \omega t = Q \cos 2\pi f t$$

The original charge on the capacitor is Q. The frequency of oscillation in radians per second is ω; the frequency of oscillation in cycles per second, or hertz, is f. The formula for the frequency of oscillation is derived in the optional section on page 646. It is:

$$f = \frac{1}{2\pi}\sqrt{\frac{1}{LC}}$$

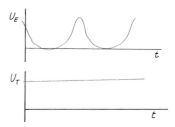

If an inductor of 1.1 H and a capacitor of 1 μF are used, the oscillation frequency will be

$$f = \frac{1}{2\pi}\sqrt{\frac{1}{(1.1\text{ H})(1 \times 10^{-6}\text{ F})}} = \frac{1}{2\pi}\sqrt{0.9 \times 10^6}$$

$$= \frac{950}{2\pi} = 151 \text{ Hz}$$

The period of oscillation is

$$T = \frac{1}{f} = \frac{1}{151} = 6.6 \text{ ms}$$

If you actually wire the circuit shown in the diagram and observe the potential difference across the inductor with an oscilloscope, you will not see a continual sine wave with constant amplitude. Instead you will see a display like the one shown in the photograph.

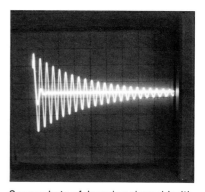

Scope photo of decaying sinusoid with L = 1.1 H, C = 1 μF, and R = 50 Ω. (The heavy horizontal center line was caused by the switching circuit.)

Question 21-17

What does this photograph imply about the electrostatic and magnetic energy in the circuit?

Most real circuits have resistance. For instance, the inductor that we have been using in our example has a d-c resistance of 50 Ω because of the very long length of wire used to wind the many turns. As the current oscillates back and forth through the resistance, there is an I^2R loss. The envelope of the decaying sine wave has a time constant: $\tau = 2L/R$. In that time the maximum charge on the capacitor, or current in the inductor, has decreased to $1/e$ of the original value. The equation for the charge becomes

$$q = Qe^{-(R/2L)t} \cos\left(\sqrt{\frac{1}{LC}}\right) t$$

Variable amplitude of cosine term

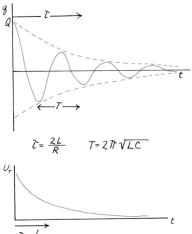

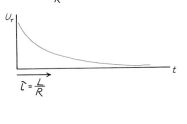

$\tau = \frac{2L}{R}$ $T = 2\pi\sqrt{LC}$

$\tau = \frac{L}{R}$

The cosine term has a variable amplitude, one that decays exponentially.

The energy in the whole system is proportional to the square of the maximum charge in any given oscillation or to the square of the maximum current: $U = \frac{1}{2}q^2/C$. Since the maximum charge is equal to $q_{max} = Qe^{-(R/2L)t}$, then $q_{max}^2 = Q^2 e^{-(R/L)t}$. The energy in the system is dissipated with a time constant $\tau = L/R$. Here is a graph of the electromagnetic energy as a function of time.

For the circuit we have already analyzed, how long does it take the maximum charge to fall to $1/e$ of its original value?

$$\tau_q = \frac{2L}{R} = 2\,\frac{1.1\text{ H}}{50\,\Omega} = 44\text{ ms}$$

Check the scope photograph to see if there is the right number of oscillations within the time constant.

Question 21-18

How much electromagnetic energy is left in the system by that time?

ALTERNATING CURRENT

Thomas Edison thought that electricity provided to homes should be direct current (d-c). The electricity coming out of a rotating generator will always be in the same direction if the current is taken off commutator rings as shown in the illustration. Depending on the number of groups of windings, the resulting current fluctuates in magnitude but is, at least, always in the same direction. George Westinghouse and others championed the cause of alternating current (a-c). In the standard form of alternating current, the voltage and current are sinusoidal. Nowadays the standard frequency for ordinary alternating current is 60 cycles per second, or 60 Hz. In the early years of this century, many public utilities produced alternating current at 25 Hz. As Edison had feared, this low frequency produced a slight fluctuation in the intensity of light from ordinary filament bulbs, which was annoying to some people.

The commutator rings are on the shaft at the right. Each coil is connected to a commutator segment. As the armature rotates, the coils and commutator segments rotate, but a sliding contact touching the commutator always makes electrical contact with the coil that is facing it.

Question 21-19

Motion pictures and television consist of a series of stationary frames changing fast enough so that the eye interprets the action as continuous. What is the frequency of frame changes? Why wouldn't the light produced by 25 Hz filament bulbs be annoying to everyone?

One of Edison's worries about alternating current concerned the higher peak voltage required in order to deliver power equivalent to that produced with a given d-c voltage. The power in a d-c circuit is given by IV or I^2R. These formulas still hold for alternating current but the power delivered changes from instant to instant as the current and voltage change. In these graphs we show current, voltage, and power for alternating current in a resistor. Even though the current and voltage go both positive and negative, the power production is always positive. However, it is pulsating, going from zero to a maximum at twice the frequency of the alternating current. The average power delivered is clearly less than the maximum; as a matter of fact, it is exactly one-half the maximum. (The average of $\sin^2 \theta$ or $\cos^2 \theta$ over a complete cycle is just one-half. This must be the case, since for any angle, $\sin^2 \theta + \cos^2 \theta = 1$, and over a whole cycle the average of $\sin^2 \theta$ must be just equal to the average of $\cos^2 \theta$).

$P_{ave} = \frac{1}{2} I_{max} V_{max}$ (for sinusoidal alternating current in a resistor)

Let's define *effective* current and *effective* voltage for alternating current to be equal to

$$I_{eff} = I_{max}/\sqrt{2}$$
$$V_{eff} = V_{max}/\sqrt{2}$$

The average power produced in a resistor in an a-c circuit is then equal to

$$P_{ave} = \tfrac{1}{2} I_{max} V_{max} = (I_{max}/\sqrt{2})(V_{max}/\sqrt{2}) = I_{eff} V_{eff} = I^2_{eff} R$$

Now let's calculate the peak voltage that Edison was worried about. The standard potential difference across the wall outlets in our homes is 115 *effective* volts. As you can see in the graphs, however, the maximum voltage plus and minus from the wall socket is larger than 115 V by a factor of $\sqrt{2}$, making the power source more dangerous. The maximum voltage is equal to 163 V.

In Chapter 19 we calculated the drift velocity for electrons in a typical wire and circuit. That velocity is a fractional millimeter per second. With alternating current, the charges never get a chance to drift anywhere. They are pulled in one direction for $\frac{1}{120}$ s, and then are shoved in the other direction for $\frac{1}{120}$ s. During that time, however, they are delivering thermal energy to the conductor just like direct current and also are producing and responding to magnetic fields.

Alternating current is usually generated and transmitted with a four-wire, three-phase system. The voltage between any two wires varies sinusoidally, but the phases of the sine waves in the three "hot" wires differ as shown in the diagram. Each phase is driven by a particular pair of coils in the generator. Induction motors are most efficiently driven by three-phase alternating current which can produce a rotating magnetic field in the motor. Although three-phase alternating current is transmitted throughout neighborhoods, usually only single phase from two wires and the neutral is brought into each house, providing either 115 V or 230 V.

With alternating current, capacitors and inductors can become important circuit components.

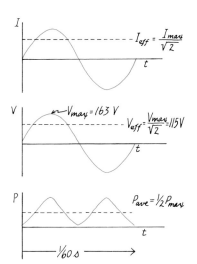

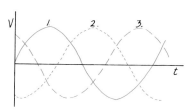

Voltage of each wire with respect to neutral wire as function of time. At neighborhood transformers, the houses are divided into 3 groups - each line of houses fed from one phase. At the cylindrical transformers on local poles, the voltage is reduced, usually from 400-450 V to 115 V. Each house, however, can be fed with three lines. One is neutral, and the other two are 115 V with respect to the neutral, but 180° out of phase with each other. Thus 230 V is available for stoves but it is only single phase.

Question 21-20

How do capacitors and inductors act in d-c circuits?

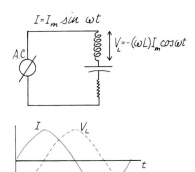

V lags I by 90°

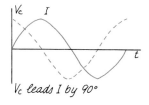

Vc leads I by 90°

Let's see what happens to the current and voltage in an inductor and capacitor in a series a-c circuit. The current in each component will be the same at any given instant. Any charge that is piling up in the capacitor must be coming out of the inductor. The alternating current is described by

$$I = I_m \sin \omega t = I_m \sin 2\pi f t$$

The standard frequency, f, is 60 Hz. The angular frequency, ω is equal to $2\pi f = 377$ rad/s.

The voltage across the inductor is $V_L = -L(\Delta i/\Delta t)$. The slope of the $i(t)$ curve is equal to $(\Delta i/\Delta t) = \omega I_m \cos \omega t$. Note that the values for the slope of a sine curve form a cosine curve. The slope is also proportional to ω. The higher the frequency of the sine function, the greater the slope. The result of this argument is that the voltage across the inductor is equal to

$$V_L = -L(\Delta i/\Delta t) = -(\omega L)I_m \cos \omega t$$

The voltage across the inductor is sinusoidal, just like the current, but *lags* the current by 90° as shown in the diagram. The magnitude of the voltage can be related to the magnitude of the current by an expression similar to that of Ohm's law:

$$(V_L)_{max} = (\omega L)\, I_{max}$$

The expression (ωL) is called *inductive reactance*, and is given the symbol, X_L.

$$X_L = \omega L$$

The unit of reactance is the ohm; it is the equivalent of a resistance in a dc circuit. Note, however, that the reactance is frequency dependent. For a given current in the inductor, the back voltage is proportional to the frequency because the higher the frequency, the faster the current is changing and the greater the slope $(\Delta i/\Delta t)$.

Now let's calculate the potential difference across the series capacitor as the current changes sinusoidally. At all times the voltage across the capacitor is equal to q/C. As we have seen before on page 640, when the sinusoidal current in a capacitor is maximum, the charge on the capacitor is zero. When the charge is maximum, the current is zero. If the current is described by a sine function, the charge and hence the voltage must be proportional to the cosine function as shown in the graph. In this case, the voltage across the capacitor *leads* the current by 90°. The faster the current changes, the less time the charge has to pile up on the capacitor and so the smaller the accumulated charge:

$$q = (1/\omega)I_{max} \cos \omega t$$

The voltage across the capacitor is equal to

$$V_C = \frac{q}{C} = \frac{(1/\omega)I_m \cos \omega t}{C} = (1/\omega C)\, I_m \cos \omega t$$

The maximum voltage across the capacitor is related to the maximum current by an Ohm's law type of equation: $(V_C)_m = (1/\omega C)I_m$. The term

$(1/\omega C)$ plays a role similar to that of a resistance. This term is called *capacitive reactance*, X_C:

$$X_C = 1/\omega C$$

Note that its frequency dependence is just opposite that of inductive reactance. As the frequency ω becomes large, the capacitive reactance becomes small. On the other hand, as the frequency ω goes to 0, the capacitive reactance goes to infinity.

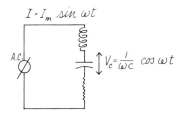

Question 21-21

What is the implication of having alternating current with 0 frequency and an infinite reactance?

Ohm's law as applied to individual components in alternating current is summarized in the box. Let's calculate some typical values of reactances. For instance, what is the inductive reactance of a 1-H coil at 60 Hz?

$$X_L = 2\pi(60 \text{ Hz})(1 \text{ H}) = 377 \text{ }\Omega$$

At the AM radio frequency of 600 kilocycles per second (6×10^5 Hz), the reactance of 1 H would be equal to 3.77×10^6 Ω—greater by a factor of 10^4.

A 1-μF capacitor at a frequency of 60 Hz has a capacitive reactance of

$$X_C = \frac{1}{\omega C} = \frac{1}{2\pi f C} = \frac{1}{2\pi(60 \text{ Hz})(1 \times 10^{-6} \text{F})} = 2.65 \times 10^3 \text{ }\Omega$$

At 600 kHz, the reactance of 1 μF would be only 0.265 Ω.

When resistors are in series in a d-c circuit, the effective resistance is simply the sum of the individual resistances. The derivation of that formula depends, however, on the fact that the total potential difference across the series circuit is equal to the sum of the individual potential differences of the component resistors. With alternating current we have a more complicated situation. The *current* in a series circuit consisting of an inductor, a capacitor, and a resistor is the same in each component at any given moment. The current is all in phase, but the voltages are not. The voltage across the capacitor is 90° ahead of the current and hence 90° ahead of the voltage across the series resistor. Meanwhile, the voltage across the inductance lags the current by 90° and hence is 180° out of phase with the voltage across the capacitor.

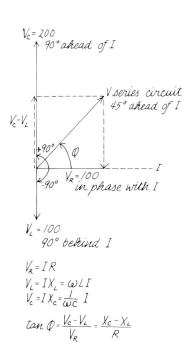

■ Optional

In the diagram we show a way to represent voltage and phase relationships in alternating current. Each voltage is represented by a "phase vector" on a type of graph used to show the real and imaginary components of complex numbers. Any voltage in phase with the current lies on the real (horizontal) axis. A voltage leading the current in phase by 90° lies on the positive imaginary (vertical) axis. A voltage lagging by 90° lies on the negative side of the imaginary axis.

In the case shown, $X_L = 100\ \Omega$, $X_C = 200\ \Omega$, and $R = 100\ \Omega$. If there is a current of 1 A maximum in the series circuit, then the maximum voltage across the inductor is 100 V, the maximum voltage across the capacitor is 200 V, and the maximum voltage across the resistor is 100 V. Since the potential differences across the inductor and capacitor are 180° out of phase, the potential difference across both of them is simply the difference between their two voltages. In this case that difference is 100 V, and it is leading the voltage across the resistor by 90°. The total potential across the whole circuit is the sum of these two sinusoidally varying voltages. That voltage is also sinusoidal with a magnitude equal to the vector sum of the magnitudes of the reactance potential and the resistance potential. In this case that vector sum is $V_{max} = \sqrt{(100\ V)^2 + (100\ V)^2} = 141$ V. Compare this net voltage across the series circuit with the individual voltages across the components.

Here is an easy way to add the effect of reactances and resistances in series. The combination is called an impedance, Z.

$$Z = \sqrt{(X_C - X_L)^2 + R^2}$$

Note the vector nature of this equation. The inductive reactance is subtracted from the capacitive reactance because their voltages are 180° out of phase. The net reactance is then combined with the resistance as though they were two vectors 90° out of phase. The resulting maximum potential difference across the series circuit is given by $V_{max} = I_{max}Z$. In general, the voltage is not in phase with the current. In our example, the current in the resistor is in phase with the potential across the resistor but the potential across the whole circuit is 45° ahead of the voltage across the resistor and hence 45° ahead of the current. As you can see in the diagram, the phase angle between voltage and current is given by

$$\tan \phi = \frac{X_C - X_L}{R}$$

Ohm's law for alternating current can be written

$$I = \frac{V}{Z}$$

In the case of a series circuit, Ohm's law takes the form:

$$I = \frac{V}{\sqrt{(X_C - X_L)^2 + R^2}} = \frac{V}{\sqrt{[(1/\omega C) - \omega L]^2 + R^2}}$$

If the inductive reactance is equal in magnitude to the capacitive reactance, then we have the resonance condition (for oscillations) that we studied in the last section. The resonance condition is:

$$\omega L = \frac{1}{\omega C}$$

$$\omega = \sqrt{\frac{1}{LC}}$$

$$f = \frac{1}{2\pi}\sqrt{\frac{1}{LC}}$$

This is the formula for resonant frequency that we used on page 000. At resonance, there can be a large voltage across the inductance and across the capacitor but they are equal in magnitude and 180° out of phase. Hence they cancel, leaving only the voltage across the resistor.

Question 21-22
Suppose that $X_L = 100\ \Omega$, $X_C = 100\ \Omega$ and $R = 10\ \Omega$. If $I_{max} = 1$ A, what is the maximum voltage across the series circuit? What are the maximum voltages across the inductor and capacitor?

The power consumed in an a-c circuit is equal to the product of the effective current and effective voltage only if the voltage and current are in phase. Otherwise, the power consumed is equal to the product of the effective current and the component of the effective voltage that is in phase:

$$P = I_{\text{eff}} V_{\text{eff}} \cos \phi \text{ (where } \phi \text{ is the phase angle between the current and the voltage)}$$

In an inductor with zero resistance or a capacitor, the voltage is 90° out of phase with the current and therefore no power is consumed. (Of course, an inductor always has resistance because it is composed of many turns of wire. Power is dissipated in the resistance.) In most household a-c circuits the net reactance is small compared with the resistance and so the current supplied to the house is usually almost in phase with the voltage. Motors usually have some inductive reactance and fluorescent lights usually have some capacitive reactance. In large office buildings or in certain industries, the load may be strongly inductive or capacitive. In that case the power company might have to furnish a large current at the standard voltage and yet the product of $I_{\text{eff}} V_{\text{eff}} \cos \phi$ might be considerably smaller than just the product $I_{\text{eff}} V_{\text{eff}}$. Even though the building is using less power because the current and voltage are out of phase, the power company still has to supply a large current which costs money because of the losses along the transmission lines. In such a case, the power company charges extra, depending on the magnitude of the phase angle ϕ.

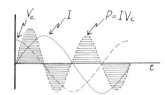

In a pure capacitive reactance, the voltage leads the current by 90°. The instantaneous power fluctuates between positive and negative, absorbing power and then sending it back to the generator. Over one complete cycle, the net power is zero.

TRANSFORMERS

The main advantage of alternating current over direct current is the ease and efficiency with which a-c voltages can be raised or lowered. When electrical power is produced at a central station and then sent over long distances, there is $I^2 R$ loss in the transmission lines. For the same amount of power transmitted, this loss can be reduced by increasing the voltage and thus lowering the current. Transmission lines between cities usually operate at potentials over 100,000 V. Within cities the standard potential is 2300 V. Within neighborhoods, this potential is reduced to 230 V or 115 V for the lines going to individual homes. The voltages are changed up or down by passive devices called transformers, which usually operate with an efficiency of close to 99%.

A basic transformer consists of two coils, a primary and a secondary,

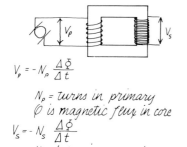

wound on the same iron core, as shown in the diagram. The sinusoidal current in the primary produces a sinusoidally changing magnetic field in the core. The core is usually made of iron laminations separated by thin insulating sheets in order to reduce power loss due to eddy currents. If no current is being drawn by the secondary coil, then the primary coil acts very much like an inductor. In an ideal transformer, without losses, I_P and V_P are 90° out of phase with each other, and so no power is consumed. For that matter, the current drawn under these conditions is usually very small. The inductance of a typical small household transformer is about 1 H. At 60 Hz, its reactance is $X_L = 2\pi\,(60\text{ Hz})\,(1\text{ H}) = 377\ \Omega$. For $V_{\text{eff}} = 115$ V, $I_{\text{eff}} \approx \tfrac{1}{3}$ A.

The primary coil is driven by a generator with a sinusoidal emf. Kirchhoff's law requires that $\mathcal{E} + V_P = 0$, where $V_P = -N_P\,(\Delta\phi/\Delta t)$. V_P is the induced voltage in the primary.

Question 21-23

According to this equation, the induced voltage in the primary completely bucks out the emf of the generator. Then how can there be any current in the transformer to produce the flux?

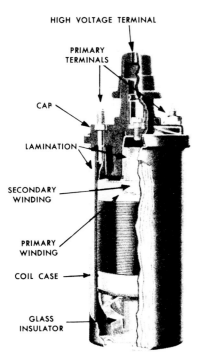

Since the same magnetic flux exists in both the primary and the secondary coil, the induced voltage in the secondary is equal to

$$V_S = -N_S\,(\Delta\phi/\Delta t)$$

The ratio of secondary voltage to primary voltage is therefore:

$$V_S/V_P = N_S/N_P \quad \text{or} \quad V_S = (N_S/N_P)\,V_P$$

In a step-up transformer, $N_S/N_P > 1$. An extreme example of such a transformer is the common spark coil needed in every car. It is used to transform the 12 V from the battery to the 20,000 V needed to produce the ignition spark in the cylinders. To produce the necessary alternating current, the direct current from the battery is interrupted at the crucial moment by the distributor breakers.

If current is drawn from the secondary coil, the secondary current tends to produce a change in the magnetic flux in the core. However, the magnitude of that flux is controlled by the driving emf of the generator, and must remain essentially unchanged. To counteract the effect of the secondary current, the primary must draw current from the source, 180° out of phase with the secondary current. Each current produces an additional flux proportional to the number of turns in its coil and the magnitude of the current, but these fluxes must cancel. Hence, the primary and secondary currents are related by

$$N_P I_P = -N_S I_S \quad \text{or} \quad I_P = -(N_S/N_P)\,I_S$$

The negative sign shows that the primary and secondary currents are 180° out of phase with each other.

> **Question 21-24**
>
> For a step-up transformer, there are more turns in the secondary coil than in the primary: $N_S/N_P > 1$. For such a transformer, what is the relationship between the primary and secondary currents?

If the current in the secondary is in a resistive load, it is in phase with the secondary emf. The extra current drawn by the primary is also in phase with the primary emf, and hence power is drawn from the generator. For the ideal transformer, without core losses: primary power drawn = secondary power consumed.

$$I_P V_P = I_S V_S$$

A transformer not only transforms current and voltage magnitudes, but also the effective resistance or impedance of a circuit. For a resistive load, R_S, in the secondary, the primary coil furnishes current, I_P, at a voltage, V_P. Therefore, the effective resistance in the primary is

$$R_P = V_P/I_P = \frac{(N_P/N_S)V_S}{(N_S/N_P)I_S} = \frac{1}{(N_S/N_P)^2} R_S$$

As you can see, a step-up transformer changes a load resistance in the secondary into a smaller effective resistance in the primary.

If the secondary load is inductive or capacitive, we must analyze more carefully the phase relationships of currents and voltages in primary and secondary. Since the primary and secondary currents must be 180° out of phase with each other, the transformed effective impedances are also changed in sign. A capacitor in the secondary circuit is seen as an inductor by the primary; an inductor is seen as a capacitor. This transformation of impedances, both in sign and magnitude, is a vital function of transformers in circuits transmitting signals. As we saw when studying wave motion, a signal is transmitted without reflection when passing from one medium to another if the two media have the same impedance. For efficient transfer of power from one circuit to another, the "output impedance" must be equal to the "input impedance." For instance, the output resistance of tubes driving audio loudspeakers is usually higher than the resistance of the loudspeaker coils. The final stage of the audio amplifer, therefore, is not the tube output, but instead a step-down transformer.

SUMMARY

This chapter deals with currents that change with time. When the charge on a capacitor builds or decays, it does so exponentially: $q = Qe^{-(1/RC)t}$. The time that it takes the charge to rise $1/e$ of its final value is $\tau = RC$. The half-time is $t_{\frac{1}{2}} = 0.69\,\tau$.

When a conductor moves in a magnetic field, there is a Lorentz force on the electric charges in the material: $\mathbf{F} = q(\mathbf{v} \times \mathbf{B})$. If a return path outside the magnetic field is available, a loop current will be generated. The emf around the loop is equal to the time rate of change of the magnetic flux through the loop: $-\varepsilon = \Sigma \mathbf{E} \cdot \mathbf{l} = \Delta\Phi/\Delta t$. The negative sign indicates the direction of the emf and consequent current. According to Lenz's law, which is required to conserve energy, an induced current has a direction such that it sets up a magnetic field to oppose any change in the original field. Hence, it requires external work to move a loop through a field and generate current; the induced current produces a magnetic field that opposes the motion.

The two laws of induction often can be applied interchangeably or can be derived from each other, but not always:

$$\text{Lorentz: } \mathbf{F} = q(\mathbf{E} + \mathbf{v} \times \mathbf{B}) \qquad \text{Faraday: } \varepsilon = -\frac{\Delta\Phi}{\Delta t}$$

An induced emf is not a conservative potential. An electric charge, traveling a closed path that encloses changing magnetic flux, gains energy.

Electric motors exploit the magnetic force on a current-carrying wire: $\mathbf{F} = I(\mathbf{l} \times \mathbf{B})$. The current can be introduced into the rotating conductor through slip rings (commutators), or can be induced by rotating magnetic fields. Electric generators can be made from motors, by providing a force to move the rotating conductor, and then extracting the induced current.

It takes time to build up the magnetic field in a coil, since the induced emf bucks the charging voltage. The induced emf is equal to the time rate of change of current: $\varepsilon = -L(\Delta i/\Delta t)$, where L, the inductance, depends on the geometry of the coil. For a solenoid with n turns per meter, and volume, V: $L = \mu_0 n^2 V$.

The current builds or decays exponentially in an inductor: $i = I e^{-(R/L)t}$. The time constant for change of $1/e$ is $\tau = L/R$.

It takes energy to build the magnetic field in an inductor. $U = \frac{1}{2}LI^2$. The energy *density* (J/m³) in a long solenoid is $u_B = \frac{1}{2}\mu_0 n^2 I^2 = \frac{1}{2}B^2/\mu_0$. This is the magnetic energy density for free space.

In a circuit containing an inductor and capacitor in series, electric charge will oscillate back and forth through the circuit with the frequency $f = (1/2\pi)\sqrt{(1/LC)}$. Electrostatic energy of the capacitor turns into magnetic energy of the inductor, and then back again. If there is resistance in the circuit, the energy will gradually drain into thermal energy. The i^2R loss has a time constant, $\tau = L/R$.

Most electricity is used in the form of alternating current. The d-c rules for circuits and power are transformed to $I_{\text{eff}} = I_{\max}/\sqrt{2}$. $V_{\text{eff}} = V_{\max}/\sqrt{2}$. Inductive reactance is $X_L = \omega L$. Capacitive reactance is $X_C = 1/\omega C$. The impedance of a series combination of L, R, and C is $Z = \sqrt{(X_C - X_L)^2 + R^2}$. The phase angle between current and voltage is given by $\tan \phi = (X_C - X_L)/R$. When current and voltage are not in phase in a circuit, the power used is $P = I_{\text{eff}} V_{\text{eff}} \cos \phi$.

Transformers change voltage of alternating current with high efficiency. If the number of turns in the primary is N_P and in the secondary is N_S, then $V_S/V_P = N_S/N_P$, $N_P I_P = -N_S I_S$, and $R_P = [1/(N_S/N_P)^2] R_S$.

Answers to Questions

21-1: When $t = 0$, the exponential is equal to unity since $e^0 = 1$. Therefore, when $t = 0$,

SUMMARY

$q = Q$, which corresponds to the original situation in the capacitor circuit. Remember that:

$$q = Qe^{-(1/RC)t} = \frac{Q}{e^{(1/RC)t}}$$

When t is very large, the exponential becomes very large. Since the exponential is in the denominator, q is very small. As t goes to infinity, q goes to 0. This mathematical behavior agrees with our expectation for the physical situation. If we wait long enough, the charge on the capacitor will drain off and reach zero value.

21-2: The time constant of this circuit is $\tau = RC = (1 \times 10^6 \, \Omega)(1 \times 10^{-6} \, F) = 1$ s. The half-life is equal to $t_{1/2} = 0.69 \, \tau = 0.69$ s.

Mathematically, the exponential goes to 0 only as its negative argument goes to infinity. The mathematics, however, is only a model for the physical situation. The electric charges are not infinite in number nor infinitely divisible. When the number of surplus electrons on the capacitor is very small, the mathematical function is no longer a good model of the process. As a practical matter, after 5 half-lives, the charge will have decreased to 1/32 of its original value. After 10 half-lives, only about 1/1000 of the original charge is left.

21-3: A gold leaf electroscope fastened at the end of the moving bar would show a deflection of the leaf indicating a charge concentration.

21-4: The area of the loop does not change but there is a change in the area of the loop *through which there is a magnetic field*.

21-5: The motion of the loop in the magnetic field produces a current in the loop. Therefore there will be an I^2R loss. The thermal power generated is

$$P_{\text{thermal}} = I^2 R = \frac{B^2 l^2 v^2}{R^2} R = \frac{B^2 l^2 v^2}{R}$$

Notice that this dissipated power is exactly equal to the power required to shove the loop through the magnetic field with constant velocity.

21-6: When the loop is around the center region of the magnet, the number of magnetic field lines passing through the loop is maximum. There is very little change in the flux, however, as the loop passes from one end of the magnet to the other. Therefore, the induced current will be small. As the loop passes beyond the center of the magnet, the number of field lines through the loop decreases, slowly at first, then rapidly, inducing a large current in the opposite direction.

21-7: Whether or not there is a potential difference depends on how the measurement is made. If the pilot simply places a voltmeter in the center of the airplane and extends the leads to the opposite tips of the wing, then the voltmeter will read 0. If trailing leads could be dropped from the wing tips to the earth (an unlikely situation), a current in the wing could exist and a potential difference would be measured.

21-8: For loop 1: $-i_1 R_1 + i(\frac{1}{6}\Omega) = 0 \quad i_1 R_1 = V_1 = \frac{1}{6}$V $\quad (i = 1$ A$)$.

For loop 2: $-i_1 R_1 - i(\frac{5}{6}\Omega) + 1$ V $= 0$

$i_1 R_1 = V_1 = \frac{1}{6}$V $\quad$ (which agrees with the result using loop 1)

For loop 3: $+i(\frac{5}{6}\Omega) - i_2 R_2 = 0 \quad i_2 R_2 = V_2 = \frac{5}{6}$V

For loop 4: $-i(\frac{1}{6}\Omega) - i_2 R_2 + 1$ V $= 0$

$i_2 R_2 = V_2 = \frac{5}{6}$V $\quad$ (which agrees with the result using loop 3)

For loop 5: $-i_1 R_1 - i_2 R_2 + 1 = 0 \quad -\frac{1}{6}$V $-\frac{5}{6}$V $+ 1$ V $= 0 \quad$ (check)

21-9: The eddy currents produce an I^2R loss. The blade gets hotter.

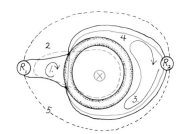

21-10: There is a generator (sometimes called an alternator) in your car. It keeps your battery charged.

21-11: The inductance of a solenoid is proportional to the number of turns in the solenoid and also to the magnetic flux in the solenoid. But the magnetic flux of a solenoid is proportional to the number of turns per meter. Therefore the inductance of the solenoid is proportional to n^2.

21-12: In deriving the formula for the inductance of a solenoid, we assumed that the field was constant throughout the solenoid. This assumption leads to a good approximation if the length of the solenoid is much larger than the diameter of its core. As you can see, the solenoid shown on page 636 does not meet this requirement.

21-13: Since $L(\Delta i/\Delta t) = -V$, $L = -V(\Delta t/\Delta i)$.
Therefore:

$$L/R = -\left(\frac{V}{R}\right)\left(\frac{\Delta t}{\Delta i}\right) = -i\left(\frac{\Delta t}{\Delta i}\right)$$

Therefore L/R must have the dimension of time.

21-14: E_{grav} = (weight)(height) = (2 kg)(9.8 N/kg)(height) = 7.9×10^{-3} J

Height = 0.4 mm

21-15: Since the stored energy of a solenoid is a function of the whole system, it is not obvious that the energy is stored in any particular place. However, we will find it convenient and consistent to assign the stored magnetic energy to the space in which the magnetic field lines exist. According to this model the energy of the solenoid is mostly stored in the core.

21-16: The slopes of a sine curve do indeed generate a cosine curve and the slopes of a cosine curve generate a sine curve. The plausibility of this fact was shown on page 378.

21-17: The electrostatic and magnetic energy in the circuit must be decaying if the amplitude of the oscillating potential decays. As long as the electric and magnetic energy transform back and forth into each other, the amplitude of the voltage oscillation will remain constant. In the case shown on page 641, the energy must be turning into some other form.

21-18: Since the stored energy is proportional to the square of the amplitude of the charge, when the amplitude of the charge has fallen to $1/e$ of its original value, the stored energy has fallen to $(1/e)^2$, or 14% of its original value. Notice that the time constant for energy dissipation is half that for charge decay.

21-19: The standard frequency for motion pictures is 24 frames per second. For television in the United States (where the standard a-c frequency is 60 Hz) the frequency is 30 frames per second. The eye is actually sensitive to flickering *intensity* of light up to frequencies of about 50 Hz, if the light goes from complete darkness to medium brightness. To avoid the flicker problem with motion pictures, each frame is actually blanked out by the shutter in the middle of its presentation, thus raising the flicker rate to 48 Hz. A similar but neater trick is used in television. Each frame is composed of two sets of interlaced horizontal lines. One set is made in $\frac{1}{60}$ s and then during the next $\frac{1}{60}$ s the other half of the lines are created.

The flicker of lights powered by 25 Hz alternating current could be noticed by some people. The effect was not dramatic because it takes the hot filament of a light bulb some time to increase or decrease its temperature.

21-20: In a d-c circuit, a capacitor is simply an open connection producing very large resistance. An inductor simply acts like a resistance with a value determined by the wire in its coil.

21-21: Alternating current with zero frequency is simply direct current. An infinite reactance simply means an open connection.

-22: Since the capacitive and inductive reactances are in series and have the same magnitudes, their reactances cancel, leaving a net impedance equal to the resistance: 10 Ω. Therefore, the maximum voltage across the series circuit is $V = IZ = (1\ A)(10\ Ω) = 10$ V. The maximum voltage across the inductor or the capacitor, however, is given by $V = IX = (1\ A)(100\ Ω) = 100$ V.

In a resonant circuit like this, the large swings of voltage across the inductor and capacitor can sometimes cause problems such as a breakdown of insulation. On the other hand, as we will see in Chapter 22, resonant circuits are vital for radio reception. In effect, they produce amplification of the resonant signal.

-23: The current and the voltage in the transformer are 90° out of phase. The induced voltage does indeed prevent the emf of the generator from providing more current. But the current in the primary that was built up during the turn-on time remains. In a real transformer or inductor, there is some loss of energy due to the resistance and therefore the generator must continually supply some energy. It can do this if the primary current is not quite 90° out of phase with the emf.

-24: You can't get something for nothing. In a step-up transformer the secondary voltage is greater than the primary by a factor of N_S/N_P. In exchange for this amplification of voltage, the primary must provide more current. The ratio of secondary current to primary current is N_P/N_S. Since power is equal to IV, the primary power supplied is equal to the secondary power demanded.

Problems

1. Demonstrate that the dimensions of RC are those of time.
2. When two small spheres (such as balloons) are suspended on threads from a common point and given the same electric charge, they will repel and remain separated from each other for many minutes in dry weather. If the capacitance of each sphere is 10^{-11} F and the time constant is 10^2 s, what is the order of magnitude of the resistance of the leakage paths (through air and threads)?
3. A 0.1-μF capacitor is discharged through a 1×10^6 Ω resistor. If the original voltage across the capacitor is 10 V, *about* how long does it take to fall to 1 V?
4. For thin wire draped across a lab bench from flashlight battery to a 10 Ω resistor, the capacitance is about 10^{-10} F/m. If you use about a meter of wire to connect battery, switch, and resistor, how long does it take the current to establish itself after you throw the switch?
5. Suppose you have a 10-turn loop, 10 cm in diameter, lying flat in the gap of a research magnet with a field of 1.5 T. What is the emf in the loop if you pull the loop out of the gap in 0.1 s? What is the emf if you flip the loop over in the gap in 0.1 s?
6. Can you detect the magnetic field of the earth by flipping a large coil? What emf would be generated by flipping a 10-turn coil, 1 m by 1 m, in 0.5 s? Assume that $B \approx 0.5$ G.
7. Suppose you have a loop flat in the gap of a large magnet. The north pole of the magnet is on top; the south pole beneath. If you suddenly draw out the loop, is the current clockwise or counterclockwise, looking down on the loop? Sketch the geometry and field lines to justify your answer.
8. Faraday's first generator consisted of a copper disk rotating in a magnetic field. Brushes connected to the center and the edge of the disk served as the power source electrodes. If a disk 1 m in diameter rotates in a magnetic field of 1.5 T, with a frequency of 10 Hz, what emf is developed across the electrodes?
9. A torus (doughnut-shaped) evacuated tube lies along the outer circumference of a magnet 1 m in diameter. If the magnetic field increases in 0.01 s from 0 to 1 T, what is the magnitude of the electric field, E, in the tube? If an electron is in the tube, how much energy would it gain each time around?

10. One way of measuring the strength of a magnetic field is to pull a coil out of the field, allowing the induced current to go through a charge-measuring device. If a coil of 50 turns with a cross-section of 1 cm² is in a circuit with total resistance of 10 Ω, and is drawn out of a magnetic field of 1 T, how many coulombs flow? (Note that the problem does not depend on the time!)

11. A 50-turn coil with a radius of 10 cm is 1 m away from an identical coil. They have a common axis. If the primary coil carries 10 A, which is turned off in 0.01 s, what is the induced emf in the secondary?

12. Describe and sketch a case where the Lorentz law gives the induced emf but the Faraday law does not apply. Describe and sketch a case where the situation is reversed.

13. A research magnet with a field of 1.5 T and a cross-sectional area of 0.1 m² has a coil of 100 turns. If a switch is opened and the circuit broken in 0.01 s, what is the average voltage across the terminals?

14. If the field in a betatron rises to maximum in 0.01 s, what must be B_{max} at $r = 1$ m if the electrons gain 10,000 eV each revolution?

15. Sketch a simple generator consisting of a single rectangular loop rotating between two pole pieces. Show the commutator ring connections.

16. What is the inductance of a torus solenoid (like a doughnut) with 2×10^3 turns/m, a cross-sectional radius of 2 cm, and a diameter of the torus (center to center) of 20 cm?

17. What is the time constant, τ, and the half-life $t_{\frac{1}{2}}$ for a circuit with $L = 1 \times 10^{-3}$ H and $R = 500$ Ω?

18. For an inductor with $L = 1$ H, carrying 10 A, what is the voltage across the break if the circuit is opened in 1×10^{-3} s?

19. How much energy is stored in a torus solenoid with 2×10^3 turns/m, a cross-sectional radius of 2 cm, a torus diameter (center to center) of 20 cm, and carrying 10 A?

20. What is the energy density of the earth's magnetic field, assuming that $B \approx 0.5$ G? What is the magnetic energy stored in a region on the earth's surface with area of 1 km² and reaching 10 km high? Compare this energy with that in a gallon of gasoline.

21. Suppose you want a circuit to be resonant with the frequency in the middle of the AM range—1×10^6 Hz. What capacity do you need if you use an inductor with $L = 1 \times 10^{-4}$ H?

22. What is the time constant, τ, for i to fall to $1/e$ of its original value in an inductor with $L = 1 \times 10^{-4}$ H and $R = 1$ Ω? What is τ for the energy, U?

23. In many amplifier circuits the stages are connected with capacitors, with a resistor going to ground from the input. The capacitor blocks direct current but allows alternating current to pass. If the input resistance is 1×10^6 Ω, what value of C is necessary so that the input of 60 Hz to the second stage is only attenuated by $\frac{1}{2}$?

24. What are X_L and X_C for $L = 1 \times 10^{-4}$ H and $C = 1 \times 10^{-8}$ F at 60 Hz and 1×10^6 Hz?

25. An LRC series circuit is resonant at 500 Hz. $L = 0.1$ H and $R = 50$ Ω. With an alternating current of 500 Hz and 100 V amplitude, what is the amplitude of the voltage across the resistor, the capacitor, and the inductor?

26. A transformer for toy trains turns 120 V into 6 V. If the resistance in the train circuit is 10 Ω, what is the effective resistance seen by the primary of the transformer?

ELECTROMAGNETIC RADIATION 22

In Chapter 21 we saw what happened *in circuits* when currents change. A changing current produces a changing magnetic field, which, in turn, sets up an emf. The effects that we studied were all localized within the circuit or the immediate neighborhood. Now we must take a look at a whole new phenomenon that arises when electric charges accelerate. Electric and magnetic fields are produced which radiate from the source, carrying away energy, momentum, and angular momentum. This electromagnetic radiation comprises all the effects of radio, TV, visible light, and X-rays.

We start our exploration of electromagnetic radiation by reviewing the basic laws of electricity that we have seen so far. As they stand, these laws are incomplete. When we add one more term, we will see that the equations predict a radiation phenomenon. The prediction even extends to the velocity of the radiation, which turns out to be the velocity of light.

Once we have seen how electromagnetic radiation arises out of the basic laws, we will explore the various ranges of electromagnetic radiation. Although radio waves and X-rays are basically the same phenomenon, their manifestations are very different.

HANDLING THE PHENOMENA

1. You can transmit and detect radio and TV signals without a license. All you need is a source of current that can be turned on and off rapidly. Electric motors that have commutators (not induction motors) are usually good generators of radio signals. Such motors are usually found in electric shavers or kitchen mixers or sewing machines. A television set or an AM radio is a good detector for locally produced static. TV *sound* is FM, which is not much influenced by static. The video part of TV, however, is AM and subject to local interference from currents that turn on or off abruptly.

Turn a motor on at various distances from the radio or TV. The closer the interrupted current to the detector, the greater the effect. Another way to produce static is to short-circuit a battery momentarily with a loop of wire. To see or hear this effect you must usually have the battery and wire close to the receiver.

Notice that the effect on the TV picture depends to some extent on the orientation of the motor and also on small sideways displacement from the

TV set. The motor is both generator and antenna and so the transmission of the electromagnetic radiation depends on its position and orientation with respect to the receiver.

2. Take a close look at the antennas used for radio, TV, or radar. A radio station for ordinary AM radio must have a tall vertical tower for its antenna. We will see later how the height of the antenna depends on the radio frequency or wavelength. Many AM radios do not have a visible antenna, but if they do, it will also be a vertical rod, such as the one used for car radios. Of course, for practical reasons, the receiver antenna is not as high as the radio station transmission antenna. However, the taller the receiver antenna, the better the reception, at least up to the same height as the transmission antenna. If you have a telescoping car antenna, you can observe this effect by tuning to a weak station and raising or lowering the antenna.

Portable AM radios usually have antennas that are sensitive to the magnetic part of the radiation. If you have such a radio, tune it to a station some distance away, whose compass direction you know, and observe the reception as you turn the radio first around a vertical axis and then around the horizontal axis in the direction of the station. Usually, good reception depends on whether the front or the side of the radio is facing the distant transmitter. If your radio has a telescoping antenna, which is responsive to the electric part of the signal, observe what happens if the antenna is horizontal instead of vertical.

(Charles E. Rotkin/PFI)

FM and TV antennas are very different from those used for AM. Both transmitting and receiving antennas for FM and TV are about the same size and may even be the same shape. The shapes of TV antennas on rooftops can vary but note that they are all about the same size. Furthermore, in a given neighborhood, they are probably all pointed in the same direction. Note that the long cross bars are perpendicular to the direction toward the station.

Radar antennas can be considerably smaller than those used for TV. Consider, for instance, that police radar can be generated and received by a telescopelike device that may have a diameter of only 10 cm or so. The shape of radar antennas is something like that of reflectors for optical telescopes or searchlights.

3. If you have tested the sensitivity of a portable AM radio to various orientations with respect to the station, you have already seen the effect of *polarization* of electromagnetic radiation. You can see some of the same effects with visible light if you have a pair of polarizing sunglasses. Look at the blue sky away from the sun through the polarizer and observe the effect of rotating the glass through 90°. Now look at glare or light reflected from a road or a tabletop and see how the appearance changes as you rotate the polarizer.

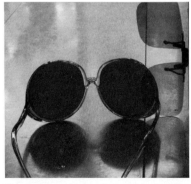
You are looking at table-top glare through two pairs of polarizing sunglasses. The one on the left, which reduces the glare, is in the normal orientation. Note that no light came through the overlap region.

Evidently, electromagnetic radiation, whether radio or TV or visible light, is transmitted by some mechanism that depends on the orientation with respect to the direction of transmission. As we develop the theory of electromagnetic radiation, we will see explanations for all of these effects.

4. A stream of water from a hose makes a very precise analogy of

electric field lines. If you move a hose very slowly (compared to the stream velocity) in a direction perpendicular to the stream, you will have a model of the field lines coming from a slowly moving electric charge. (Ignore the fact that a stream of water also follows a parabolic path in the vertical plane because of gravity. Observe only the horizontal component of the stream.) The stream from the hose will form a straight line leading back to the nozzle, perpendicular to the direction of motion. If you accelerate the nozzle, however, you will form a kink in the stream. This kink has a pronounced component *perpendicular* to the main direction of the stream.

A particularly dramatic effect can be produced by moving the nozzle back and forth perpendicular to the direction of the stream in a simple harmonic oscillation. Loops of wave motion appear in the stream. Notice particularly that the magnitude of the loops *increases* with the distance from the nozzle. An illustration of this effect is shown in the photograph. To understand the radiation from an accelerated charge you should really perform this simple demonstration for yourself.

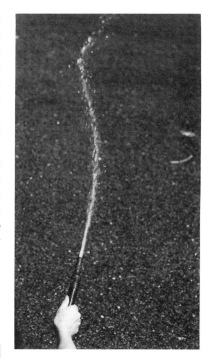

REVIEW OF THE BASIC EQUATIONS OF ELECTROMAGNETISM

> **Question 22-1**
> Are Coulomb's law and Ohm's law the basic equations of electricity?

In the diagram we show the four basic laws of electromagnetism that we have studied so far. The first two are concerned with static situations. In terms of field lines, they link electric and magnetic fields to static sources. Electric field lines start on positive charges and end on negative charges. Since there are no magnetic monopoles, magnetic field lines must be continuous.

The third law, due to Faraday, describes a dynamic way to produce an electric field. A time-varying magnetic flux produces closed lines of electric field. In the fourth equation, due to Ampère, closed lines of magnetic field are produced by the motion of electric charges.

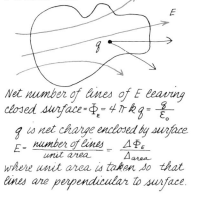
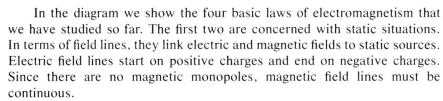
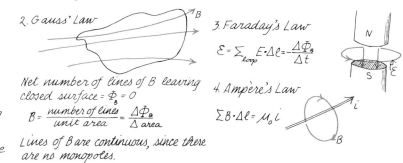

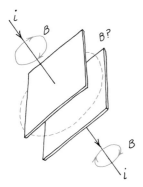

These four equations were gathered together by James Clerk Maxwell (1831–1879). He was impressed by the elements of symmetry between electric and magnetic fields in the equations, but on the other hand he was distressed by some of the obvious lacks of such symmetry. First, as we have already pointed out, there is no magnetic charge and hence no magnetic current. But there is another gap in the equations. Although Equation 3 describes what happens when magnetic fields vary with time, there is no corresponding term to describe what happens if *electric* fields change. In the diagram we show one of the problems caused by this lack. When current is charging a capacitor, there must be magnetic field lines around the lead-in wires. However, a closed loop on the circumference of a circle cutting between the plates of the capacitor would not contain any current. According to Ampère's law, as it stands, there should be no magnetic field lines around the gap in the capacitor. Such an abrupt change in magnetic field would be an unnatural phenomenon. Since the electric field in the capacitor is changing during the charging process, Maxwell proposed that magnetic field lines must be produced around the time-varying electric field. With the additional term, the complete fourth equation looks like this:

$$\Sigma_{\text{loop}} B \cdot \Delta l = \mu_0 i + \mu_0 \left(\epsilon_0 \frac{\Delta \Phi_E}{\Delta t} \right)$$

Question 22-2

In this revised equation, the term $\epsilon_0(\Delta\Phi_E/\Delta t)$ must have the units and dimensions of an electric current. Is this the case?

Let's compare the current and changing electric field in a capacitor that is discharging. In this diagram we show such a capacitor. It has been charged to 100 V and now is discharging through a 1-Ω resistor to ground. As you can see from the calculations that accompany the diagram, the electric field at the beginning is 10^5 V/m and the electric flux is 10^3 lines. When the discharge begins, the charge on the capacitor decays exponentially with a time constant equal to RC. With the dimensions given, that time constant in this case is equal to $\tau = 8.85 \times 10^{-11}$ s. The voltage, the electric field, and the flux also decay with this same time constant. Meanwhile, the charge flowing out of the capacitor creates a current in the lead-in wires. For an approximate calculation, let us assume that the discharge is linear during a short period, instead of exponential. Then at the beginning of the discharge, $I = V/R = (100 \text{ V})/(1 \, \Omega) = 100$ A. This current of 100 A in the wires will produce circular lines of magnetic field around the wires. The equivalent current caused by the change of electric flux in the capacitor is equal to

$$\epsilon_0 \frac{\Delta \Phi_E}{\Delta t} = (8.85 \times 10^{-12}) \frac{10^3 \text{ lines}}{8.85 \times 10^{-11} \text{ s}} = 10^2 \text{ A}$$

It appears that the changing electric field in the capacitor produces a result equal to that of the current in the wire. The magnetic field lines around the gap in the capacitor will be equal to those around the lead-in wires.

THE IMPLICATION OF MAXWELL'S EQUATIONS

The four equations that describe electric and magnetic fields and their sources are called Maxwell's equations, in spite of the fact that some of the individual equations and phenomena were discovered by other people. Thus, Maxwell's equations include those named for Gauss, Ampère, and Faraday. It was Maxwell, however, who gathered the separate equations together and proposed the crucial extra term that describes the effect of time-varying electric flux. Maxwell then proceeded to combine the four equations and predict a whole new range of phenomena.

Note that with the electromagnetic effects that we have seen so far, we would have a hard time transmitting signals through space. We could, of course, start current in a wire or coil to produce a magnetic field, or we could separate electric charge to produce an electric field. In most such geometries, we would be producing a magnetic or electric dipole whose fields fall off as $1/r^3$. At best we would have a point source whose field would fall off as $1/r^2$. We could send signals this way, of course. The detector could be a compass needle for magnetic fields or a small charged ball on a thread for electric fields. You might at first think that the response of these detectors to the new fields created by the distant sources would be proportional to $1/r^3$ or $1/r^2$. The sensitivity of the detector response, however, depends not on the force exerted, which is proportional to the field strength, but on the energy that can be extracted from the field. The energy density of an electric field is proportional to E^2 as we saw in Chapter 18. The energy density of a magnetic field is proportional to B^2 as we saw in Chapter 20. Consequently if we turn on electric or magnetic fields and hope to detect these effects at a distance, we will find that our detection efficiency falls off as $1/r^4$ or $1/r^6$, depending on whether the source is an isolated charge or a dipole. Evidently, we would have a hard time transmitting electric or magnetic signals if detectability faded so rapidly with distance.

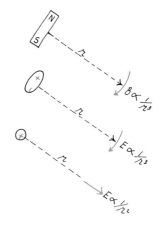

Information cannot be transmitted with a constant field; information is contained in a change of field.

Question 22-3

Why should a signal detector depend on the energy density of a signal rather than on the force available?

Maxwell's analysis of the four equations for electromagnetism showed that there is another way for electric and magnetic fields to propagate through space. In this new mode, the fields fall off only as $1/r$. Let's see how this can happen. Here are the last two Maxwell equations written for free space where there are no electric charges or currents.

$$\Sigma \mathbf{E} \cdot \Delta \mathbf{l} = -\frac{\Delta \Phi_B}{\Delta t}$$

$$\Sigma \mathbf{B} \cdot \Delta \mathbf{l} = \mu_0 \epsilon_0 \frac{\Delta \Phi_E}{\Delta t}$$

A new and remarkable effect is predicted by these two equations. If a magnetic field somehow enters a region of free space there will be a time-varying magnetic flux there. That will produce an electric field. Since the electric field did not exist in that region previously, its arrival must have produced a time-varying electric flux. But a change of electric flux produces a magnetic field. This circular sequence of events generates electric fields from changing magnetic fields and magnetic fields from the subsequent change of electric fields. It is a self-generating system that, as we will now see, is also self-propagating.

Suppose that a wall of magnetic field oriented in the z-direction advances along the x-direction as shown in the diagram. (Later we will ask how such a wall of magnetic field could have been generated in the first place). We now have a problem and a geometry very much like the one with which we introduced the subject of electric induction. If we take a rectangular loop in the x-y plane, the magnetic field will sweep across it. The change of magnetic field inside the loop provides an emf around the loop. The only contribution to the emf comes from the electric field along the side of the loop that is in the advancing magnetic field. That emf or voltage is equal to $-Ew$. The width of the loop, w, is arbitrary and will cancel out. The electric field in that leg of the loop, E, is constant because the magnetic field is sweeping over it at a constant rate. The induced electric field is in the positive y-direction, which makes the emf negative as we go around the loop in the proper x-y-direction. The time-rate of change of magnetic flux in the loop is equal to

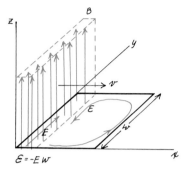

$$\frac{\Delta \Phi_B}{\Delta t} = \frac{\Delta (BA)}{\Delta t} = B\frac{\Delta A}{\Delta t} = Bw\frac{\Delta x}{\Delta t} = Bwv$$

The change of magnetic flux in the loop is proportional to the velocity, v, with which the wall of magnetic field is advancing.

Question 22-4

How would we go about controlling the velocity, v, of the advancing wall of magnetic field?

According to the third Maxwell equation for free space:

$$\Sigma \mathbf{E} \cdot \Delta \mathbf{l} = -\frac{\Delta \Phi_B}{\Delta t}$$

Therefore:

$$-Ew = -Bwv$$
$$E = Bv$$

There turns out to be a simple relationship between the strength of the advancing magnetic field and the electric field that is induced. Note that qualitatively the electric field lines exist only in the region of the advancing wall of magnetic field. There is no electric field ahead of or behind the advancing wall. But this means that we also have a wall of electric field sweeping along the x-direction, with the same velocity, v. We can now use Maxwell's addition to the fourth equation to see how this moving electric

THE IMPLICATION OF MAXWELL'S EQUATIONS

field generates a magnetic field. The same geometry is shown again in the next diagram but this time the wall of electric field is shown with the field lines beginning to penetrate a loop in the *x-z* plane. The induced magnetic field is in the positive *z*-direction but exists only within the region of the electric field wall. Therefore, only the front leg of the loop contributes to the sum of the magnetic field and distance products around the loop. The time-rate of change of electric flux in the loop is equal to

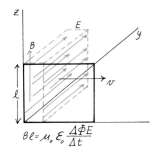

$$\frac{\Delta \Phi_E}{\Delta t} = E \frac{\Delta A}{\Delta t} = E l \frac{\Delta x}{\Delta t} = E l v$$

Substituting into the fourth Maxwell equation:

$$\Sigma \mathbf{B} \cdot \Delta \mathbf{l} = \mu_0 \epsilon_0 \frac{\Delta \Phi_E}{\Delta t}$$

$$Bl = \mu_0 \epsilon_0 E l v$$

$$B = \mu_0 \epsilon_0 v E$$

We have now derived another relationship between the magnetic and electric fields that are generating each other. Once again the relationship involves the velocity with which the fields are moving.

When *E* is generated by a changing *B*:

$$E = Bv$$

When *B* is generated by a changing *E*:

$$E = B \frac{1}{\mu_0 \epsilon_0 v}$$

These two equations cannot both be true at the same time unless there is a special relationship between the velocity, *v*, and the constants of permeability and permittivity. Let's solve for that relationship:

$$v = \frac{1}{\mu_0 \epsilon_0 v}$$

$$v^2 = \frac{1}{\mu_0 \epsilon_0}$$

$$v = \sqrt{\frac{1}{\mu_0 \epsilon_0}}$$

The permittivity ϵ_0 is determined from measurements with electrostatics, which depend on the assigned value for permeability, μ_0. We are not free to choose other values for ϵ_0 and μ_0. Now it appears that the velocity with which the electric and magnetic fields sweep along the *x*-axis cannot be arbitrary either. Only one value is possible and that is determined by the electrical and magnetic properties of free space. Let's put the numbers in and find the magnitude of this special velocity.

$$v = \sqrt{\frac{1}{(4\pi \times 10^{-7})(8.85 \times 10^{-12})}}$$

$$v = 3.0 \times 10^8 \text{ m/s}$$

Note first that this unique speed of propagation of the electromagnetic waves is very fast. Second, as you probably know, it is the speed of light. It must be that light itself is an electromagnetic wave. Consider what a triumph of theory is represented by these simple arguments of the last couple of pages. Maxwell's equations contain only a few terms and these require relatively simple relationships between electric and magnetic fields and their rates of change. When the equations are combined they not only suggest that time-varying electric and magnetic fields will continually generate each other, but also predict that the varying fields will propagate through space with a unique speed. That speed is determined by constants that are measured in static experiments. The permittivity, ϵ_0, is measured with a capacitor experiment. The permeability, μ_0, has an assigned value, which determines the size of the ampere and coulomb. Yet when these values are combined, they lead to the speed of light. We will soon see how Maxwell's predictions were tested and confirmed. First, however, we must explain how these electric and magnetic fields can be generated in the first place.

■ Optional

Generation of Electromagnetic Fields from Accelerated Charges

Maxwell himself did not solve his equations to demonstrate that accelerated charges shake off electromagnetic energy. The detailed calculations are complicated but we can understand the approximate situation through a field-line model. The fields experienced at a distance from a moving electric charge were produced by that charge at an earlier time. If a charge is moving with constant velocity, but with speed considerably less than that of light, the electric field at a point far away will gradually change in strength and direction as shown in the diagram. The magnetic field at that point will remain in the same direction as the charge moves, but will gradually change strength. It takes no energy to keep the charge moving at constant velocity nor does the charge lose energy as it trails the electric and magnetic field lines with it.

To *accelerate* the charge along its direction of motion, a force must be applied to it and hence work must be done. Part of this energy is radiated away in the form of electric and magnetic fields that carry the message of the abrupt change in velocity of the source.

Let's analyze the field produced at a distant point by an electric charge that has been at rest and then is suddenly accelerated upward for a short time δt. Its velocity at the end of that acceleration period is $v = a\delta t$. It continues to drift upward with this velocity v. Consider the electric field line that extends outward horizontally from the charge. Even after the charge has started moving, the electric field at a distant point remains constant and horizontal. No information has reached this point that the charge has started moving. At the time T, which is equal to the separation distance divided by the speed of light, the remote field starts changing its direction. At the time $T + \delta t$, the electric field has been changed in direction so that it is in line with the position of the charge at $T + \delta t$.

Question 22-5
Why shouldn't the field be in line with the position of the charge at δt?

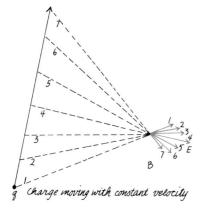

charge moving with constant velocity

The field at a distant point is radial to the position of the charge <u>at that time</u> (not the earlier time when the information was sent).

THE IMPLICATION OF MAXWELL'S EQUATIONS

During the time interval from T to $T + \delta t$, there is an abrupt change in the direction of the electric field at the remote measuring point. Let's look down on the scene and picture the lines of force as shown in the diagram. Beyond a circle with radius $c(T + \delta t)$, the lines of force all point radially back to the charge's original location at $t = 0$. No information about the charge's movement has arrived beyond this circle. Within a smaller circle of radius cT, the lines of force point radially to the position of the charge at $t = T + \delta t$. Both circles sweep outward radially at the speed of light. Between them there is a transition shell where the field lines can have a large component perpendicular to the radial direction. (Note that the field lines must be continuous, and so those in the inner sphere must connect with the appropriate ones in the outer sphere.) The "kink" in the field line is not in the same direction that the field line will assume after the pulse has passed by. That direction will change very little, since the source charge moves only a short distance, vT, while the light signal has traveled a distance of cT. Therefore, the field line will end up changing direction abruptly only by the small angle, v/c.

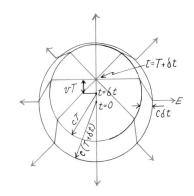

$$\frac{E_\perp}{E_r} = \frac{vT}{c\,\delta t} = \frac{a\,\delta t\, T}{c\,\delta t} = \frac{aT}{c}$$

The radial field is just that due to a point charge, and is given by Coulomb's law:

$$E_r = k\frac{q}{r^2}$$

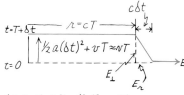

(Note that $\tfrac{1}{2}a(\delta t)^2 \ll vT$)

Consequently, the perpendicular component of the electric field is equal to

$$E_\perp = \frac{kqaT}{r^2 c}$$

The distance between the moving charge and the remote measuring point is equal to r, which is equal to cT. Substituting this value for T yields the following value for the perpendicular component of electric field:

$$E_\perp = \frac{kqa}{r^2 c}\frac{r}{c} = \frac{kqa}{c^2 r} \qquad (r = cT)$$

Remember that this perpendicular component of electric field carries the information that the charge has been accelerated. Its magnitude is proportional to the acceleration but, more remarkably, is proportional only to the inverse *first* power of the distance. Why should a part of an electric field that depends on the inverse square of the distance depend on the inverse first power of the distance? If you watched closely the behavior of the water stream from the hose as you moved the nozzle back and forth sideways, you must have seen that the perpendicular component increased with distance from the nozzle. A picture of the effect is shown on page 657. From an analytic point of view, note that the perpendicular component is proportional to the time taken for the message to go from source to observing point. This is caused directly by the fact that the abruptly shifted field at the observing point lines up with the position of the drifting charge *at that time*. The greater the transmission time, T, the further the charge has drifted upward and therefore the greater the shift in the electric field and the greater the perpendicular component.

The perpendicular component of the electric field—the kink in the field line—also depends on the angle between the field line and the direction of acceleration. We examined the case where the field line is perpendicular to the acceleration, where the effect is maximum. If we had observed the behavior of the field line in the same direction as the acceleration, we would have found no effect. Regardless of the acceleration, the electric field line would maintain its same direction and therefore there would be no perpendicular component. The size of the perpendicular component depends on the sine of the angle between the line of acceleration and the line of the observer, as shown in this diagram. The complete formula for the strength of the perpendicular component is

$$E_\perp = \frac{kqa \sin \theta}{c^2 r}$$

Since the ordinary electric and magnetic fields caused by isolated charges or dipoles or electric currents fall off as the square or cube of the distance, the component of the field created by the acceleration, which depends on the inverse *first* power of the distance, will be the only field detectable at large distances from the source. Note first that this part of the field is perpendicular to the line back to the source. Furthermore, this perpendicular kink is hurtling outward at the velocity of light. As we have already seen, such a pulse of electric field must be accompanied by a magnetic field pulse that is perpendicular to the electric field and also perpendicular to the direction of propagation. The energy density in such a pulse is proportional to the square of the electric or magnetic fields. Therefore, the energy radiated from an accelerated charge must be proportional to the square of the charge, the square of the acceleration, and, inversely, to the square of the distance from the source. The light reaching us from a distant flashlight or a distant star is proportional to $1/r^2$ because by "light" we mean the energy that we can detect from that source. Because the energy produced from a charge accelerated in a straight line is proportional to $\sin^2 \theta$, the energy is radiated in a disk perpendicular to the acceleration.

In a radio transmitting antenna, electrons are sent up and down a vertical line, forming a resonant circuit. The line has length, L, on either side of the power input. The electric charge undergoes simple harmonic motion, with the position of the center of charge given by $y = L \sin \omega t$.
The velocity of the center of charge is equal to

$$v_q = \frac{\Delta y}{\Delta t} = \omega L \cos \omega t$$

The acceleration of the charge is equal to

$$a_q = \frac{\Delta v}{\Delta t} = -\omega^2 L \sin \omega t$$

Question 22-6
How does the magnitude of radiated *energy* depend on the frequency of the signal?

THE ENERGY RADIATED IN ELECTROMAGNETIC FIELDS

When we calculated energy density of static electric and magnetic fields, we made assumptions about the location of the fields. In the case of the capacitor, we said that the electric field exists in the volume between the plates. The magnetic energy calculation was made in terms of a solenoid in

THE ENERGY RADIATED IN ELECTROMAGNETIC FIELDS

the form of a torus or a hollow doughnut. In this case we assumed that the energy was located in the interior volume. In both cases the particular geometries were chosen because the fields are constant in magnitude and confined within those volumes. While it is plausible and useful to picture these interior spaces as filled with energy, we might alternatively have assigned the energy to the stored charges on the capacitor plates, or to the currents in the solenoid.

In the case of radiated energy, the concept of energy actually existing and traveling in space becomes much more physical. A receiver antenna can extract power from empty space. Let's compare the electric and magnetic energy in radiation and see another way of describing its properties.

The static electric and magnetic energy densities are given by

$$u_E = \tfrac{1}{2}\epsilon_0 E^2 \quad (\text{J/m}^3)$$

$$u_B = \frac{1}{2}\frac{1}{\mu_0} B^2 \quad (\text{J/m}^3)$$

For a plane wave of electromagnetic radiation, there is, as we saw, a simple relationship between the strength of the electric and magnetic fields: $B = E/c$. Substituting this value for B into the formula for the magnetic field energy density:

$$u_B = \frac{1}{2}\frac{1}{\mu_0} B^2 = \frac{1}{2}\frac{1}{\mu_0}\frac{E^2}{c^2} = \frac{\epsilon_0 \mu_0}{2\mu_0} E^2 = \tfrac{1}{2}\epsilon_0 E^2$$

Evidently, the magnetic field energy density is equal to the electric field energy density in a plane wave of radiation. The total energy density in the wave is equal to

$$u_{\text{total}} = u_E + u_B = \epsilon_0 E^2$$

The *intensity* of radiation, or *irradiance*, is defined to be the amount of energy passing through unit area in unit time.

$$I = \frac{\text{energy/time}}{\text{area}} = \frac{\text{Power}}{A}$$

In the diagram we show how to calculate the amount of energy passing through unit area in unit time. The long box has unit area cross-section and a length equal to the product of unit time and the velocity of light. In our standard units, the cross-sectional area of the box would be 1 m² and its length would be 3×10^8 m. During one second all the energy in this box would pass through the 1 m² of unit area at the end. The total energy in the box is equal to the product of the energy density and the volume of the box. Consequently, the intensity is equal to

$$I = \frac{u\,(\text{volume})}{(\text{area})(\text{time})} = \frac{u\,(1\text{ m}^2)(c\Delta t)}{(1\text{ m}^2)(\Delta t)} = uc = \epsilon_0 E^2 c$$

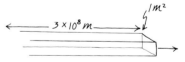

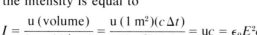

Energy in this box will pass through end during one second

Question 22-7

Are the units right? Is the intensity given in terms of watts per square meter?

There is another way of describing the intensity of electromagnetic radiation. What we have derived so far is the *magnitude* of the intensity. Since the radiation has direction, it would be convenient to represent the intensity as a vector. Let's transform the expression that we have, so that it contains both electric and magnetic fields.

$$I = \epsilon_0 E^2 c = \epsilon_0 E(Bc)c = \epsilon_0 EBc^2 = \frac{\epsilon_0 EB}{\epsilon_0 \mu_0} = \frac{1}{\mu_0} EB$$

We now assert a new definition that was first derived by John Poynting (1852–1914). The *Poynting vector*, a peculiarly descriptive term, is defined as

$$\mathbf{S} = \frac{1}{\mu_0} \mathbf{E} \times \mathbf{B}$$

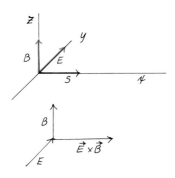

In the diagram we show the cross product relationship between E, B, and the Poynting vector. Note that the magnitude of S is just the intensity that we have already derived and the units are in watts per square meter. The additional feature of the cross product gives us the direction of the power flow. Note also that if the electric and magnetic fields are not perpendicular to each other, perhaps because the electric or magnetic field is twisted in passing through boundaries, the power passing through that region will be less.

Let's calculate the electric and magnetic fields produced by the radiation coming from a 100-watt bulb. In the first place, most of the 100 watts consumed by the bulb is not converted into visible radiation. The efficiency is only about 2.5%. Let's assume that the bulb is a point source and is emitting uniformly in all directions. At a distance of 3 m from the bulb, the surface area of the surrounding sphere is $A = 4\pi r^2 = 4\pi(3 \text{ m}^2) = 113 \text{ m}^2$. The watts per square meter for the visible radiation at the distance of 3 m is, therefore, $I = (2.5 \text{ W})/(113 \text{ m}^2) = 0.022 \text{ W/m}^2$. Half of this power is provided by the electric field and half by the magnetic field. For the electric field:

$$\tfrac{1}{2}\epsilon_0 E^2 c = \tfrac{1}{2}I = \tfrac{1}{2}(0.022 \text{ W/m}^2)$$

$$E = \sqrt{\frac{0.022}{(8.85 \times 10^{-12})(3 \times 10^8)}} = 2.9 \text{ V/m}$$

The value of E that we have calculated is for a steady state field. Since the electric field in a light beam is sinusoidal, the peak electric field is $\sqrt{2}\, E = 4.07$ V/m. As you can see, the electric field strength of the light that you use for reading is fairly large. (The electric field strength of TV or FM for fringe area reception is only a few microvolts per meter). Now let's calculate the strength of the *magnetic* field in the visible radiation from a 100-W bulb 3 m away.

$$\frac{1}{2}\frac{1}{\mu_0} B^2 c = \tfrac{1}{2}I = \tfrac{1}{2}(0.022 \text{ W/m}^2)$$

$$B = \sqrt{\frac{(0.022)(4\pi \times 10^{-7})}{3 \times 10^8}} = 9.6 \times 10^{-9} \text{ T} = 9.6 \times 10^{-5} \text{ G}$$

Once again, because the field in the light beam is sinusoidal, the peak magnetic field is $\sqrt{2}\,B = 1.4 \times 10^{-8}$ T. Although the power in the magnetic field is equal to the power in the electric field, the magnetic field strength is evidently very weak.

The question of the localization of the electromagnetic energy has some strange but revealing consequences if we ask for the direction of energy flow. Let's take the familiar case of a constant current in a wire or resistor. Clearly there is energy flow; the wire or resistor gets hotter. Our intuition would tell us that the energy must in some way flow along the wire. Nevertheless, let's calculate the Poynting vector and find its direction. The electric field in the wire and at its surface is $E = V/l$. The electric field is in the direction along the wire. The magnetic field at the surface of the wire is circumferential. Its magnitude is $B = \mu_0 I / 2\pi r$. Notice that the electric and magnetic fields are perpendicular to each other. The magnitude of the Poynting vector is

$$|\mathbf{S}| = \frac{EB}{\mu_0} = \frac{1}{\mu_0} \frac{V}{l} \frac{\mu_0 I}{2\pi r} = \frac{IV}{(2\pi r)l} = \frac{IV}{\text{(surface area of wire)}}$$

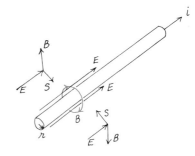

l is length of wire
r is radius
IV is thermal energy provided

The total rate of flow of electromagnetic energy into the wire must be equal to the product of the Poynting vector (watts per square meter) and the surface area of the wire. That product is equal to the thermal power, IV, produced in the wire. Furthermore, as you can see in the diagram, the direction of the energy flow as given by the Poynting vector is radially in toward the wire everywhere on the surface. According to our model, the thermal energy dissipated in the wire is continuously supplied by the electric and magnetic fields at the surface of the wire.

> **Question 22-8**
>
> Aren't these quasistatic fields? How can they supply a continuous flow of energy?

MOMENTUM CARRIED BY ELECTROMAGNETIC RADIATION

If electromagnetic radiation transports energy, does it also carry momentum? Can it exert a force on an object? Certainly we don't experience such a force in everyday life. Furthermore, it is not immediately obvious how a plane electromagnetic wave could produce a force on a charged particle in the direction of the wave propagation. If the electric and magnetic fields are varying sinusoidally in the wave, then the principal reaction of a charged particle would be to oscillate up and down in response to the electric field. This motion is perpendicular to the direction of propagation, and the time average motion is zero. However, as the electric charge is driven by the electric field, it moves in the presence of the accompanying magnetic field. Therefore the charge experiences a $\mathbf{v} \times \mathbf{B}$ force. In the diagram we show the relationship among the directions of E, B, and the resultant force on the

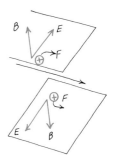

moving electric charge. Half a cycle later when the charge is moving in the opposite direction in response to the electric field in the opposite direction, the magnetic field is also reversed. Once again, the force on the electric charge is in the direction of propagation of the wave.

The radiation force on the charged particle is equal to

$$F_r = evB = ev(E/c)$$

The electric force on the charge is $F_{\text{electric}} = eE$. Therefore the radiation force on the charge is

$$F_r = \frac{(vF_{\text{elec}})}{c} = \frac{power}{c}$$

The power absorbed by the charge is equal to Fv (since the work on the charge is $\Delta W = F\Delta x$ and $power = \Delta W/\Delta t$). Since $\Delta(momentum)/\Delta t = F$ and $\Delta(energy)/\Delta t = power$, the relationship between momentum and energy carried by the electromagnetic wave must be

$$\Delta(momentum)/\Delta t = (1/c)\,\Delta(energy)/\Delta t$$

$$momentum = (1/c)\,energy$$

Question 22-9

Is it dimensionally correct that the momentum delivered to an object by a light wave is equal to the energy absorbed, divided by the speed of light?

As we will see in Chapter 23, the energy of electromagnetic radiation is not contained in a continuous smooth wave, but, at least during interactions, appears to be quantized. Energy is emitted and absorbed in chunks. The corresponding model for light is that it travels in a bundle called a photon whose energy is proportional to the frequency of the light. For a photon: energy = hf. The momentum carried by such a photon must therefore be

$$(1/c)hf = h/\lambda \quad (\text{since } \lambda = c/f)$$

The constant of proportionality h is Planck's constant. Its value is 6.6×10^{-34} J·s. The energy and momentum carried by an individual photon of visible light is very small and yet sufficient to be detected, as we will see later.

Let's do an order-of-magnitude calculation of the pressure of sunlight on the earth. The radiation force is equal to the power delivered divided by the velocity of light:

$$F_r = power/c$$

The power per unit area delivered to the earth in sunlight is approximately one kilowatt per square meter (kW/m²). Therefore, pressure of sunlight on earth = $F_r/A = (power/c)/A = (power/A)/c = (1 \times 10^3$ W/m²$)/(3 \times 10^8$ m/s$) = \frac{1}{3} \times 10^{-5}$ N/m². Evidently sunlight exerts only negligible pressure on the earth.

POLARIZATION OF ELECTROMAGNETIC RADIATION 669

> **Question 22-10**
>
> Assume some reasonable flat area corresponding to the target presented by the spherical earth and calculate the approximate total force on the earth from sunlight.

Comets' tails are blown away from the sun partly by electromagnetic radiation and partly by other particles that are shot out from the sun. Radiation pressure of visible light was demonstrated in the laboratory by E. F. Nichols and G. F. Hull in 1903. The standard technique for such a demonstration is to bombard a small mirror that is part of a torsion suspension as shown in the illustration. The momentum delivered by light to a mirror is twice that which occurs if the light is absorbed. The recoil of the mirror not only accounts for stopping the momentum of the light in one direction but in sending it off again in the opposite direction. The intensity of light that can be focused on a tiny mirror in such a suspension is too small to make a sizable deflection of the system. A standard technique is to turn the light on and off with a frequency equal to the natural frequency of oscillation of the mirror suspension. In this way, like pumping up a swing, the incident light sets the mirror into oscillation. The magnitude of the momentum delivered agrees well with the theoretical prediction.

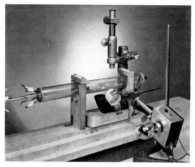

(From PSSC COLLEGE PHYSICS, 1968, D.C. Heath & Co., with EDC, Newton, MA.)

POLARIZATION OF ELECTROMAGNETIC RADIATION

When we first considered how electric and magnetic fields would generate each other and so propagate through space, we assumed that we could create a moving wall of electric field with the field lines all in the same direction perpendicular to the direction of propagation. A radio signal coming from the antenna of an AM radio station is a practical example of such electromagnetic radiation. Because the source electrons accelerate vertically up and down the tall antenna, the electric field of the radiated signal is also vertical. The corresponding magnetic field is horizontal.

The ferrite rod antenna is at the top of this transistor radio.

In "Handling the Phenomena," it was suggested that you turn a portable AM radio in various directions with respect to the radio station. If your radio has a vertical antenna like a radio in a car, then it is designed to respond to the electric field. If you turn the antenna horizontal, the signal will be greatly diminished. Some portable radios have horizontal antennas. They consist of a coil of many turns of wire wound around a rod of ferrite. This ceramic material has a high magnetic χ value but, unlike iron, it can change its magnetization many millions of times per second. A radio that contains such an antenna must not only remain horizontal in order to receive the signal, but is sensitive to the orientation of the radio with respect to the station. The magnetic antenna must not only be horizontal to be affected by the horizontal magnetic field, but must also be lined up perpendicular to the direction to the radio station. You may find strange exceptions to the preceding rules. Radio waves may be reflected from hills or buildings so that the optimum signal may not come from the straight line direction to the station. Furthermore, in these reflections the vertical and horizontal orientation of the field may be twisted.

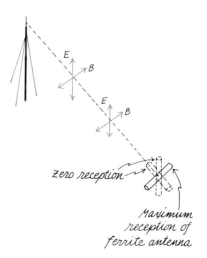

Visible light is not generated by the organized motion of electrons up and down an antenna. The frequency of visible light is so high and the wavelength so short that the generators are usually atomic systems. As we will see in Chapter 23, visible light is usually generated by the transition of an atom or molecule from an excited state to a lower energy level. Each atom or molecule produces one photon at a time with a phase and polarization that is usually unrelated to those of any of the other photons being produced. (There are certain exceptions such as with lasers or with synchrotron radiation, which we describe later.) With ordinary visible light, the electric field vector at a given point at a given instant may be vertical, and a billionth of a second later may be horizontal or at any angle in between. The electric and magnetic field vectors remain perpendicular to each other and to the direction of propagation. But the fields are oriented at random, with continually changing directions.

The efficiency of scattering or reflection of light depends on the orientation of the electric field vector with respect to the interacting surface. Most light that is reflected from surfaces with a large angle of incidence and reflection appears as a glare and contains little information about the surface.

Question 22-11

How can this be? Don't we normally see objects by reflected light?

Since the glare part of reflected light is highly polarized in the horizontal direction, we can reduce the glare by accepting only the light polarized in a vertical direction. That is what polarizing sunglasses do. If you can use such a pair of glasses, notice how they eliminate the glare from any reflecting surface but how the glare returns if you turn the polarizers 90°.

Blue sky is caused by the preferential scattering of blue light from the upper atmosphere. The scattered light is partially polarized as you can tell by examining a section of the sky through polarizing glasses. A polarizing filter on a camera is sometimes used to darken the background sky and enhance the whiteness of clouds.

The long chain molecules embedded in plastic sheet polarizers are oriented during manufacture so that they are all in one direction. If the electric vector of the light is perpendicular to the preferred direction, then the polarizer appears almost black (actually, the effect depends to some extent on wavelength, and blue light can still get through). If light passes through two polarizers, the light will be plane polarized by the first and then "analyzed" by the second. If the orientations of the two pieces are in the same direction, then light will pass through. But if the orientations are at 90°, most of the light will be blocked. If the electric field of the incident light is at an angle θ to the preferred orientation, then only the component $E \cos\theta$ will get through. Since the intensity of the light is proportional to E^2, the intensity of plane polarized light that is passed by a polarizer is $I = I_m \cos^2\theta$.

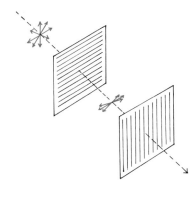

THE EXPERIMENTAL CONFIRMATION OF MAXWELL'S PREDICTION

Maxwell developed his equations and their consequences on the basis of Faraday's model of electric and magnetic fields. The mental models that his mathematics described were more complicated than those we use today. Maxwell and others at that time thought that the fields and the wave motions were physical properties of an actual fluid permeating all matter and called the aether. Nevertheless, in 1862 he proposed that "light consists in the transverse undulations of the same medium which is the cause of electric and magnetic phenomena." By that time he had calculated the speed of electromagnetic waves from his equations and knew that the speed was approximately the same as that which had recently been measured for light.

The experiments that confirmed Maxwell's predictions were first made by Heinrich Hertz (1857–1894), between 1885 and 1889, about 25 years after Maxwell's prediction and only shortly after Maxwell's death. In a brilliant series of experiments, Hertz first learned how to make oscillating circuits with frequencies in the range from 1×10^8 to 1×10^9 Hz. He did this by interrupting currents in induction coils to impose sudden high voltages across spark gaps. The spark that jumps the gap consists of electrons and ions from the air, which oscillate back and forth. The frequency is determined by the inductance and capacitance of the coils or rods that form the gap.

A schematic drawing of the apparatus that Hertz used is shown in this diagram. The receiver consisted of a "split dipole," a rod with a length of half a wavelength, separated in the middle by a small spark gap. The *transmitter* spark jumped between small polished spheres; the resulting spark was fat and carried a large current. However, the radiation energy available at the receiver is relatively weak. Even with the receiving antenna tuned to the radiated frequency, the voltage buildup across the receiver gap was small. The *receiver* gap was formed by sharp points separated by only a fraction of a millimeter. Hertz detected the radiation by looking for the tiny sparks jumping between the points.

At a frequency of 1×10^9 Hz, the wavelength is only 30 cm. Hertz measured this wavelength by reflecting the radiation from metal plates and looking for the nodes in the standing waves. (At a node, there would be no spark in his receiver.) He also demonstrated that the radiation was polarized and that it was subject to refraction (in a prism of pitch) and diffraction.

One of the ironic features of the experiment was Hertz's observation that the gap in the receiver broke down to produce a spark only if the receiver was in a position to be illuminated by the ordinary light from the spark. We now know that the ultraviolet radiation from the original spark helped to release electrons from the receiver electrodes, enabling the detection spark to form. This phenomenon is called the photoelectric effect and is frequently cited as a decisive experiment to demonstrate the photon or particle nature of light. Hertz's experiment, which demonstrated and seems to depend on the wave nature of electromagnetic radiation, also depended crucially on this other aspect of radiation. Hertz was not able to measure directly the speed of the radiation produced by his apparatus but that experiment was done a few years later. The value was indeed equal to the speed of light.

THE ELECTROMAGNETIC SPECTRUM

We claimed that electromagnetic radiation encompasses all of the effects from radio to X-rays. In one sense, the only difference between these very different phenomena is the frequency of the wave motion. That factor makes such a large difference, however, that we should divide the whole range of electromagnetic radiation into various realms. We will spread out the various effects into a spectrum, much as visible light can be analyzed into a spectrum of colors. The organizing variable for this anlaysis could be either the frequency or the wavelength. These two variables are related by the basic equation for wave motion: $f\lambda = c$

If we start with radio waves, we have low frequency and long wavelengths. As we march toward X-rays, we deal with much higher frequencies and much shorter wavelengths.

There are other factors that characterize regions of the electromagnetic spectrum. We have already mentioned that visible light seems to interact with matter in a bunched or quantized way. Each photon of light delivers an amount of energy equal to hf, where h is Planck's constant. It is convenient to characterize radiation in terms of the photon energy at that frequency. Indeed, for visible light frequencies and higher, the photon energy is often the most important parameter.

While we are organizing the electromagnetic spectrum, we should also point out particular features of the transmitters or receivers in each region. The antenna size is linked to the wavelength. Another allied factor is the absorption of a particular frequency. Some materials are transparent for certain wavelengths and opaque to others.

Here is a chart of the electromagnetic spectrum, indicating the ranges that have been given particular names. The scale is necessarily logarithmic since the frequencies and wavelengths extend over such a large range. Notice in particular the very small region for visible light.

	d-c	growlers a-c		AM radio	FM TV	radar μ waves	infrared	visible red blue	ultra-violet	dentist x-ray	γ rays	
f	10^0	10^2	10^4	10^6	10^8	10^{10}	10^{12}	10^{14}	10^{16}	10^{18}	10^{20}	hertz
λ		3×10^6	3×10^4	300	3	0.03	3×10^{-4}	3×10^{-6}	3×10^{-8}	3×10^{-10}	3×10^{-12}	meters
hf		4×10^{-13}	4×10^{-11}	4×10^{-9}	4×10^{-7}	4×10^{-5}	4×10^{-3}	0.4	41	4×10^3	4×10^5	ev

Growlers and Whistlers.

From the earliest days of radio, certain types of receivers have been plagued by a type of static called "growlers" or "whistlers." These are short-lived signals with frequencies in the audio range—a few tens to a few hundreds of cycles per second. The pitch varies during each brief signal. The resulting wavelengths are of the order of 1000 km, almost the width of continents. The waves cling to the surface of the earth, circling it many times before being dissipated. Since, as we have seen, radiated power is proportional to the fourth power of the frequency, it must take an enormous amount of current in the source to produce sizable signals at these low frequencies.

Apparently, these signals are caused by lightning, which provides sufficient power so that signals can travel around the earth.

Man-made radio signals have been used with frequencies in the audio range. In the 1920s and 1930s a 25,000-Hz system was occasionally used for overseas transmission. The wavelength of this signal was 12 km. The antenna consisted of a horizontal wire stretching for 3 km on high towers. The system was designed before high-powered radio tubes were manufactured. A motor generator set with many commutators was built in order to provide sufficient radiated power. The advantage of a transmitter using such long wavelengths was that the signals could creep around the surface of the circular earth rather than flying tangentially off into space. Ordinary radio waves for overseas transmission are directed up to the layer of ionization called the Heaviside layer, high above the atmosphere. The short wavelength signals are then reflected back down to earth again and may make several reflections or skips before arriving at a place where they can be detected. Solar magnetic storms can destroy the Heaviside layers temporarily, but communication between continents could be maintained with the very long wavelength system.

A worldwide navigation system called Omega generates radio signals in the range of 10,000 Hz. A number of transmitters located on various continents send out carefully synchronized signals during alternate periods of several seconds. A ship can compare the relative phase of signals from two or three stations and determine its location on the earth to within 2 miles.

Ultralong wavelengths of electromagnetic radiation can penetrate ocean depths or even the earth itself. For military purposes in contacting submarines, transmitters have been proposed for frequencies in the range of 100 Hz. The transmitting antennas for such systems require formidable land space.

AM Radio

AM stands for amplitude modulation. The transmitter uses one particular frequency for the carrier signal but changes the amplitude of the signal at the much lower frequencies of the audio range. A diagram of such amplitude modulation is shown here. The standard AM band in the United States extends from 535 kHz to 1605 kHz. The center of this range is at 1 MHz or 1×10^6 cycles per second. The wavelength corresponding to this frequency is 300 m. Radiation of this wavelength can penetrate ordinary nonmetallic buildings and can also creep around most man-made structures.

> **Question 22-12**
>
> What does the wavelength of radiation have to do with the question of whether or not an object presents a barrier to a wave?

We seldom have occasion to consider the photon nature of radio waves but let's calculate the energy involved in such a photon.

$$hf = (6.6 \times 10^{-34} \text{ J} \cdot \text{s})(1 \times 10^6 \text{ Hz}) = 6.6 \times 10^{-28} \text{ J}$$

$$= (6.6 \times 10^{-28} \text{ J})(1/1.6 \times 10^{-19} \text{ J/eV}) \approx 4 \times 10^{-9} \text{ eV}$$

The energy involved in a photon of electromagnetic radiation of AM radio frequency is small even by atomic standards. It takes about one electron volt (1 eV) to produce a chemical interaction.

The transmitting antennas for AM radio are vertical towers. Sometimes these are placed on the top of a very high building in the center of a city, although sometimes an array of several towers built up from the ground will be placed on the outskirts of a populated region. By controlling the phase of the signals in the several towers, the direction of the radiation can be controlled so that most of the energy is delivered to the populated region. The towers are usually about $\frac{1}{4}$ wavelength high with the signals being sent in at the ground level. There are several techniques for making efficient radiators out of antennas that are less than $\frac{1}{4}$ wavelength high. Note that for 1-Mhz signal, the $\frac{1}{4}$ wavelength antenna would have to be 75 m high. The complete antenna is actually half a wavelength long. The other $\frac{1}{4}$ wavelength is composed of the ground reflection of the actual antenna. To provide a satisfactory signal, the electric field strength at the receiver must be anywhere from 100 $\mu V/m$ in electrically quiet rural areas to 10 mV in the heart of an electrically noisy city. Most commercial radio stations have a range of from 25 to 100 miles. Reception within this region is due to the so-called ground wave that comes directly from the antenna and is refracted in such a way that it clings to the curved surface of the earth. Some of the radiated energy, however, shoots up to the Heaviside layer and then bounces back down again many hundreds of miles away. The bounce effect is usually small for AM radio but becomes increasingly important for shorter wavelengths, and at night.

The generation and reception of radio waves is crucially dependent on the subject of resonance which we discussed in the chapter on waves and also in Chapter 21 when we described LC circuits. The final stage of the transmitter circuit and the antenna form an LC circuit tuned to the particular frequency of the station. Because of the resonance, large currents and high voltages can be produced in the transmitting antenna.

The receiving antenna and the first stage of the receiving circuit must also be sharply tuned to the frequency of the desired signal. The main reason for this requirement is obvious. The space around the antenna is filled with electromagnetic radiation of many frequencies coming from many sources. If you tune the antenna and receiver to one particular frequency, the signal at that frequency will set the circuit into resonant oscillation. Radiation of other frequencies will not be able to build up such an oscillation.

The magnitude of the voltage buildup in a receiver is surprising. If the electric field strength of the signal is only about 1 millivolt per meter, and the antenna is only a few meters long, then the input voltage to the receiver is only a few millivolts. Nevertheless, crystal radios, without any tubes or transistors, can produce audible signals even though they need inputs of the order of a volt.

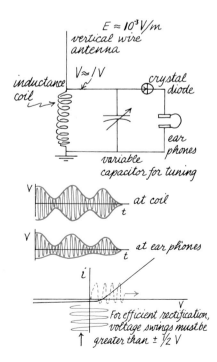

Let's analyze a typical circuit of a simple crystal radio, as shown in the diagram. The purpose of the crystal diode is to rectify the high frequency radio signal; otherwise the loudspeaker or earphones would not respond to the high frequency signal which exerts alternate positive and negative forces.

By rectifying the signal, the circuit produces short pulses of varying magnitude and all in the same direction. The audio part of the circuit can then respond to the slowly varying unidirectional current.

The trouble with such a scheme is that a crystal rectifier changes its forward to backward resistance only over a range of several tenths of a volt. If a signal of a few millivolts were impressed across such a rectifier, the resulting signal would scarcely be rectified at all. The signal across the crystal rectifier is not just a few millivolts, however. The resonant circuit of antenna and input coil is driven into oscillation by the weak radio signal. The resonant oscillation builds up a potential hundreds of times greater than the driving voltage. Consequently, a high frequency signal with an amplitude of at least a volt is imposed across the crystal rectifier so that the resulting rectified signal looks like the one in the diagram. The amplification factor caused by the resonance in such a circuit is referred to as the Q value of the circuit. A high Q circuit has very low resistance. Not only is the resonant amplification great for a high Q circuit, but such a circuit also discriminates sharply against any other frequency.

Television and FM Radio

Each of the television channels is allocated a width of 6 megahertz starting with channel 2, which goes from 54 to 60 MHz. Channel 6 extends from 82 to 88 MHz. The frequency range from 88 MHz to 108 MHz is reserved for FM. Each FM station can use a channel that is 150 kHz wide. Television channels 7 through 13 occupy the range from 174 to 216 MHz, while the UHF (ultrahigh frequency) channels from 14 through 83 use the range from 470 to 890 MHz.

There is a frequency change of about a factor of 15 between channel 2 and channel 83. Let's calculate the wavelengths corresponding to the extremes of this region. The wave length for channel 2 is

$$\lambda = (3 \times 10^8 \text{ m/s})/(54 \times 10^6 \text{ Hz}) \approx 6 \text{ m}$$

The middle of the FM band has a wavelength of about 3 m and channel 83 in the UHF has a wavelength of about $\frac{1}{3}$ m.

Antennas for TV and FM look very different from the simple vertical wires for AM radio. If the transmitter is located in the center of the city, it must be designed to radiate equally in all horizontal directions. The receivers, however, should be directional and aimed at the transmitting tower. Notice the TV antennas on the roofs of houses in a suburb. They will all be pointed in the same general direction, usually toward the center of a metropolis. Furthermore, the receiving antennas are cut to match the approximate size of the wavelength. The picture shows a "yagi" antenna. Its cross bars have a length of $\frac{1}{2}$ wavelength and are horizontal. For FM and TV the electric field is generated to be horizontal, or at 45° to the vertical. The receiving antenna is called a "folded dipole." The geometry is shown in the diagram. The signal is picked up from just one of the bars of a "Yagi" antenna. The others reflect and reinforce the signal. Since a "yagi" antenna has its maximum resonance for a wavelength equal to twice the length of its arms, sensitivity to other wavelengths is low. The more general type of TV antenna will resonate to a broad range of frequencies but has lower gain for any one in

Yagi antenna.

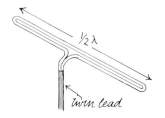

particular. Many different configurations of wires and shapes can be made to resonate at a particular frequency. In all cases, however, the size of the antenna is approximately $\frac{1}{2}$ wavelength. As we have seen, resonance is the key to extracting useful energy out of the electromagnetic wave. The complete resonant circuit consists of the antenna and the tuning stage of the television or FM set.

The energy of a photon in the FM or TV range is very low, but since the frequency is about 100 times greater than that of AM radio, the energy of each photon is about 100 times greater. That means that the photon energy is in the range of 1 millionth of an electron volt.

Although the video part of television is amplitude modulated (notice how a lightning stroke affects the picture), the audio part of the signal is frequency modulated (FM). The diagram shows schematically the appearance of an FM signal. The audio signal controls the *frequency* of the radio signal, which can vary by 75 kilohertz on either side of the central frequency. If there is no audio signal, the FM signal remains at its central frequency and the receiver remains quiet. The louder the original audio signal, the further the radio signal departs from its central frequency, and the louder the resulting tone in the receiver. The frequency of the original audio signal determines the frequency with which the radio signal departs from its central frequency, and therefore the frequency of the rectified signal from the receiver. Ordinary static, whether man-made or from lightning, is not detected as a signal by such a system.

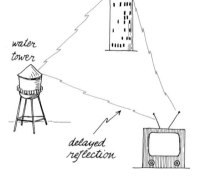
FM signal.

The wavelength of FM and TV is smaller than the size of man-made structures including buildings and airplanes. Consequently, the TV or FM signals frequently reflect from such objects and arrive at your antenna by two different routes. The path difference between two such routes creates a phase difference, which appears on your TV set as "ghosts." Sometimes these ghosts are caused by signals coming from the same channel in a city several hundred miles distant. The high frequency signal did not refract around the curve of the earth but instead shot up to the ionized layer and then bounced back down.

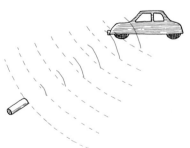

Radar and Microwaves

In a radar system the transmitter is also used as the receiver. A short burst of electromagnetic radiation is sent out in the form of a beam from a reflecting-type transmitter. The same system then listens for any part of the pulse that may be reflected from some object. The time interval between sending the pulse and receiving the reflection is proportional to the distance between the radar set and the object.

In order for the transmitter to produce a beam, the wavelength of the radiation must be small compared with the size of the reflecting transmitter.

Question 22-13

The angular spread of the beam cannot be less than the diffraction pattern of the circular transmitter. What is the relationship among the angular spread, the wavelength, and the diameter of the transmitter?

Various frequencies are used for radar depending on the size of the object being tracked. Police traffic radar uses the smallest wavelength. You can estimate what this wavelength must be by observing the small portable systems that police can aim from their car. The diameter of the radar gun may be as small as 10 cm. If the resulting beam is to be narrow enough to have any discrimination between cars, then the wavelength used must be only about 1 cm. Wavelengths much less than this are absorbed by water vapor in the air.

The frequency of 1 cm radiation is 3×10^{10} Hz or 3×10^4 MHz. The energy of a photon of this frequency is

$$E = hf = (6.6 \times 10^{-34} \text{ J} \cdot \text{s})(3 \times 10^{10} \text{ Hz}) \approx 20 \times 10^{-24} \text{ J} = 1.2 \times 10^{-4} \text{ eV}$$

The photon energy of 1 cm radiation is still well below the energy of molecular binding or the average vibrational or translational kinetic energy of molecules at room temperature. However, microwave photons are in the energy range of molecular *rotations*. The atoms in a molecule can not only vibrate but can also rotate around each other. As we pointed out in Chapter 9, angular momentum is quantized. A rotating system can change its angular momentum only in integral multiples of a basic unit. This basic unit is Planck's constant. When the angular frequency is given in radians per second, the magnitude of this unit of angular momentum is

$$\hbar = h/2\pi = 1 \times 10^{-34} \text{ kg m}^2/\text{s}$$

Let's calculate the energy involved with the simple hydrogen molecule when the two revolving protons are given one extra unit of angular momentum.

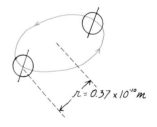

$$\Delta \text{ angular momentum} = \Delta (I\omega) = \hbar$$

Moment of inertia of hydrogen molecule $= I = 2m_p r^2$

Rotational kinetic energy $= E_r = \tfrac{1}{2} I\omega^2 = \tfrac{1}{2}(I\omega)^2/I = \tfrac{1}{2}\hbar^2/I$

$$= \frac{1}{2} \frac{(1 \times 10^{-34})^2}{2(1.7 \times 10^{-27} \text{ kg})(0.37 \times 10^{-10} \text{ m})^2}$$

$$\approx 1 \times 10^{-21} \text{ J}$$

$$= 1 \times 10^{-21} \text{ J} \left(\frac{1 \text{ eV}}{1.6 \times 10^{-19} \text{ J}} \right) \approx 1 \times 10^{-2} \text{ eV}$$

(The actual measured value is 1.5×10^{-2} eV).

As you can see, when a hydrogen molecule changes its angular momentum by one unit, the change of rotational energy is very small, though

Magnetron microwave generator (*Courtesy Pnasonic*).

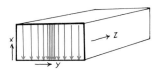

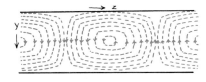

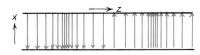

Electric and magnetic fields in a pipe carrying microwaves.

still larger by a factor of about 100 than the energy in a photon with a wavelength of 1 cm. For larger molecules, the change in rotational energy can be much smaller than that for the hydrogen molecule. In the expression for the rotational energy in terms of angular momentum, notice that the moment of inertia term is in the denominator. For a molecule with larger mass and larger radius, the moment of inertia is larger and the rotational kinetic energy for a given angular momentum is less. Consequently, microwave radiation can be used to study molecular structure. Radiation with these wavelengths can be generated with very precisely known frequencies. If a beam of such radiation is absorbed by a material at a particular frequency, then the molecules in the material must be changing their motions using amounts of energy corresponding to the photon energy.

Microwaves can be generated by vacuum tubes that use bunches of electrons to excite standing waves of electromagnetic radiation in small cavities. One type of tube is shown here. Instead of sending this high frequency radiation on wires, we can literally pipe it from one place to another with very low loss. The electric and magnetic fields inside a pipe carrying microwaves are shown in the diagram. For practical reasons this technique can be used only for radiation of wavelengths in the centimeter region since the dimensions of the pipe must be about the same as the wavelength of the radiation.

A radar beam raises the temperature of anything that absorbs it. As a matter of fact, many kitchens have devices that depend on this effect. The radiation in the standard microwave oven has a wavelength of 10 cm. A serious proposal for harnessing solar energy envisions an enormous field of silicon photocells in a geosynchronous orbit 22,000 miles above the surface of the earth. The photocells would absorb sunlight and provide d-c voltage. This energy would then be converted by magnetron tubes, like those in the microwave oven, to 10 cm electromagnetic radiation. The microwaves would be beamed to a rectifying field on the earth close to the city that would use it.

Question 22-14

Wouldn't this radiation cook any birds or airplanes that flew through it? Suppose that the beam supplying energy for a major city has a total power of 10,000 megawatts (10^{10} W). If the average power density in the beam is 100 W/m^2, how large a rectifier field would be required? (The current safety level for microwave oven radiation leakage in this country is 10 mW/cm^2.)

The Infrared Region

The low frequency end of this region has no sharp boundary as it merges with the microwave region. Wavelength, rather than frequency, determines this boundary since for wavelengths less than 1 mm, tubes and hardware of ordinary electronics become too small to be fabricated or used. For a wavelength of 1 mm, the corresponding frequency is 3×10^{11} Hz. The upper frequency limit of the infrared is determined by the beginning of the visible range. That frequency is about 4×10^{14} Hz with a wavelength of 7×10^{-7} m. Notice that this wavelength is a little smaller than one micrometer (0.7×10^{-6} m = 0.7 µm).

The photon energy at the low frequency range is

$$hf = \frac{(3 \times 10^{11} \text{ Hz})(6.6 \times 10^{-34} \text{ J} \cdot \text{s})}{(1.6 \times 10^{-19} \text{ J/eV})}$$

$$\approx 1 \times 10^{-3} \text{ eV}$$

That energy, as we have seen, corresponds to the changes of rotational energy of molecules. At the high frequency end of the infrared, the photons have enough energy to break apart some types of molecules.

$$hf = \frac{hc}{\lambda} = \frac{(6.6 \times 10^{-34} \text{ J} \cdot \text{s})(3 \times 10^{8} \text{ m/s})}{(7 \times 10^{-7} \text{ m})(1.6 \times 10^{-19} \text{ J/eV})} = 1.8 \text{ eV}$$

Because the photons in the infrared have enough energy to disturb molecules, infrared radiation is a prime tool for chemists. Infrared beams are sent through materials and the absorption is measured as a function of frequency. Each frequency where there is absorption corresponds to a particular energy that was absorbed by the material. That energy changed the rotational or vibrational motion of the molecules. From such evidence the actual structure of the molecule can be figured out.

One way to produce infrared radiation is simply to raise the temperature of some solid material. As the temperature of the solid rises, two things happen. The total amount of radiation given off increases rapidly (proportional to T^4); also the radiation shifts to higher frequencies. This latter observation agrees with everyday experience. As the temperature of a coal or a glowing wire increases, its color changes from dull red to bright white.

At any temperature above absolute zero, atomic and molecular motion gives off electromagnetic radiation. We notice the effect with objects hotter than our skin temperature. Glowing electric coils or fireplace coals can warm or even cook us. The effect does not depend on the visible radiation; you can easily feel the radiation from a hot flat iron even though it is not hot enough to glow.

Many types of lasers can produce radiation in the infrared. Radiation from these sources can be generated in a narrow frequency range and also is coherent (locked in phase).

Visible Light

The visible region of the electromagnetic spectrum is very small and has fairly sharp boundaries, determined by the characteristics of human eyes. The high frequency in the blue-violet has a wavelength of 4×10^{-7} m. The low frequency limit in the red has a wavelength of 7×10^{-7} m. The wavelengths and frequencies of various colors are shown in the diagram.

We have already figured out the energy corresponding to the red-infrared border—1.8 eV. The energy of a blue-violet photon is

$$hf = \frac{hc}{\lambda} = \frac{(6.6 \times 10^{-34} \text{ J} \cdot \text{s})(3 \times 10^{8} \text{ m/s})}{(4 \times 10^{-7} \text{ m})(1.6 \times 10^{-19} \text{ J/eV})} = 3 \text{ eV}$$

Notice how many magnitudes of different phenomena are necessarily linked together. Since atoms are bound together with energies of about 1 eV, voltaic cells produce voltages of about 1 V (the standard flashlight cell is $1\frac{1}{2}$ V; the lead storage cell is 2 V). Each time that a molecular combination changes in such a cell, the energy involved is of the order of an electron volt. That much energy is sufficient to shove an electron up a potential hill of about 1 V. Since visible light can clearly produce chemical changes (photosynthesis, photography, bleaching), the energy of visible photons must be of the order of an electron volt.

Visible light can be generated by many different processes. Most of these involve bound atomic or molecular systems. An electron in such a system is pulled away from its ground state by thermal agitation or by electron bombardment. When the disturbed electron falls back to its original condition, the energy is released in the form of a photon. As we have seen, a hot solid emits a continuous spectrum of photons of different frequency. If the atoms are isolated, as they are in a low pressure gas, then the photons that are emitted have only certain energies and hence wavelengths that are characteristic of that particular type of atom. In Chapter 23 we will explore atomic and molecular structure and learn more details about the production and absorption of light.

Ultraviolet and X-Rays

Our main experience with ultraviolet rays comes in the summer when we spend too much time in the sun. Our hair is bleached and our skin is tanned or burned. The ultraviolet region starts at the blue-violet end of the visible spectrum and merges without distinction into the X-ray spectrum. For these photons, energy is a more important parameter than frequency. For a wavelength of 1×10^{-10} m, the atomic radius, the photon energy is about 1×10^4 eV. X-rays with this energy and above "see" individual atoms since their wavelengths are smaller than the atom. X-rays for medical diagnosis usually have energies between 20,000 and 100,000 eV. For radiological destruction of tissue, photons with energies of 1×10^6 eV (1 MeV) or more are used.

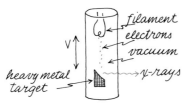

A dental X-ray machine is a classic example of the phenomenon of electromagnetic radiation from an accelerated charge. A schematic of an X-ray tube is shown here. Electrons are accelerated across a voltage between the filament and a high density target. The acceleration through the vacuum is relatively small and produces little radiation. However, when the electrons strike the high density target, they are stopped abruptly, with very large acceleration. Radiated photons stream off the target with energies up to the maximum energy of the bombarding electrons.

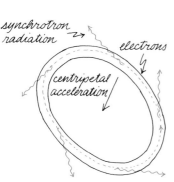

Since X-rays can be used to probe atomic structure, it is important to have high intensity sources of them. A recent addition to the standard X-ray machine is the electron synchrotron. When high energy electrons (with energies of several 10^9 eV) are forced by a magnetic field to travel in an orbit of small radius, they must experience large centripetal acceleration. Consequently, they radiate photons in a narrow beam mostly in the direction of their motion. This synchrotron radiation can be very intense and extends from the infrared to X-rays with wavelengths of atomic dimensions. Several such machines have been built at national laboratories.

X-Rays and Gamma Rays

There is no upper limit to photon energies. Even as there is no real boundary between the ultraviolet and soft X-rays, there is no physical difference between X-rays and gamma rays. Historically, photons coming from energy changes in *atoms* were called X-rays, while those coming from disturbances in atomic *nuclei* were called gamma rays. When the photon energies get above 1 MeV (1×10^6 eV), they can under certain circumstances produce a pair of electrons, one negative and the other positive. In electron pair production, the photon energy provides the rest mass energies of the electron and positron as well as their kinetic energies. It is not a case, however, of "energy turning into matter." A high energy photon is best considered as a subatomic particle. As we will see in Chapter 23, these particles can transform into each other if certain conservation laws are met.

SUMMARY

The separate equations due to Gauss, Ampère, and Faraday were analyzed together by Maxwell. He added one extra term to describe the magnetic field produced by a changing electric flux. The complete equations require that a changing magnetic field produces a changing electric field, which in turn produces a changing magnetic field. The two fields propagate each other, moving with a velocity determined by the static properties of space:

$$v = \sqrt{\frac{1}{\mu_0 \epsilon_0}} = 3.0 \times 10^8 \text{ m/s}$$

An accelerating electric charge radiates an electromagnetic field. This field is perpendicular to the radial line joining charge to receiver. Its strength is $E_\perp = (kqa \sin \theta)/(c^2 r)$. If charges in an antenna are in simple harmonic motion, their acceleration, and consequent radiation field, is proportional to the square of the frequency.

The *electric* energy density in a wave is equal to the *magnetic* energy density: $u_E = u_B$ and $u_{\text{total}} = \epsilon_0 E^2$. The intensity of radiation is given in watts per square meter: $I = (power)/A = uc = \epsilon_0 E^2 c = (1/\mu_0)EB$. This intensity can be described by Poynting's vector: $\mathbf{S} = (1/\mu_0) \mathbf{E} \times \mathbf{B}$. The energy flow is perpendicular to the electric and magnetic fields.

Electromagnetic radiation carries momentum in the direction of $\mathbf{S}$: momentum $= (1/c)$ energy. Since the energy of a photon is hf, the momentum of the photon is $p = hf/c = h/\lambda$.

Radiation coming from an antenna is polarized. With AM radio, the electric vector is vertical. The magnetic vector is horizontal.

Hertz used spark oscillators to produce the electromagnetic waves predicted by Maxwell, and he showed that in many ways the radiation behaved like light. The spectrum of electromagnetic waves can be characterized by f, λ, or the photon energy, hf. The spectrum extends from lightning-induced waves at only a few hertz to gamma rays with photon energies of billions of electron volts.

Answers to Questions

22-1: Coulomb's law applies only to point charges at rest. It can be derived from Gauss's law (and Maxwell's first equation). Furthermore, Gauss's law applies to any group of charges, not just point charges. Ohm's law applies only to the special case of metals where the density of charge carriers remains constant and the charge velocity is proportional to the electric field.

22-2: Gauss's law relates the number of electric field lines (the net flux leaving a closed region) to the enclosed charge.

$$\Phi_E = \frac{q}{\epsilon_0}$$

Therefore the quantity $\epsilon_0 \Phi_E$ has the units and dimensions of an electric charge. Dividing this quantity by Δt produces a quantity that is an electric current.

22-3: Any kind of a detector must have some indicator, such as a pointer on a dial, which will return to its original position if there is no signal. The indicator may be fastened to a spring or might be in a gravitational potential well. In either case, a force must move the indicator through a distance if the signal is to be detected. Work is required and the energy must be provided by the signal.

22-4: We could move a magnet past the observing plane, even as in Chapter 21 we described an experiment where we moved a loop through a magnetic field with a velocity v. In this case, however, we have launched a magnetic pulse by some method of turning on a magnet far away. The magnetic pulse is traveling on its own with an arbitrary velocity v; at least it is arbitrary as far as we know at this stage. In the next step we will see that the arbitrary velocity of the magnetic field is linked with the velocity of the induced electric field.

22-5: One way of considering this problem is to think of the field lines as being generated by photons thrown off from the electric charge. For such a model the mathematical analogy with the motion of a water stream from a hose becomes exact. At time δt, the charged particle has attained a velocity of v. For a photon to be thrown off in a direction perpendicular to the particle velocity (in order to reach the observer), it must start with a *backward* component equal to v (since it is coming from a source with *forward* velocity, v). When the photon reaches the observer, it will appear to have arrived from the line extending to the source charge at time $T + \delta t$.

22-6: Since the acceleration of an electric charge undergoing simple harmonic motion is proportional to the square of the frequency and since the magnitude of radiated energy is proportional to the square of the acceleration of the charge, the radiated energy must be proportional to the fourth power of the frequency. Consequently, for equivalent radiated power, low frequency radio sources require much larger currents circulating in the transmitter than high frequency sources.

22-7: Since the quantity $\epsilon_0 E^2$ is an energy density with the units J/m^3, the product of this quantity and a velocity must yield the units $(J/m^3)(m/s) = (J/s)(1/m^2)$. These are the units of watts per square meter.

22-8: The fields do indeed remain constant but only because there is a steady supply of energy from a battery or a generator. One way to interpret the localization of the energy is that the chemical or mechanical energy of the source continuously supplies energy for the arrangement of the field lines in order to maintain the constant flow of charges and the constant feed of thermal energy to the wires.

22-9: The dimensions of energy divided by velocity are indeed those of momentum. Consider, for example, that kinetic energy has the dimensions of mass times velocity squared. If you divide that by a velocity, you end up with mass times velocity, or momentum.

2-10: Assume that the target surface of the earth corresponds to that of a flat disc with the same radius. The area is $\pi r^2 = \pi(6.4 \times 10^6 \text{ m})^2 = 1.3 \times 10^{14} \text{ m}^2$. The total force of sunlight on the earth is, therefore, $(1.3 \times 10^{54} \text{ m}^2)(\frac{1}{3} \times 10^{-5} \text{ N/m}^2) = 4.3 \times 10^8 \text{ N}$. Since 10^4 N is about a ton, the force of sunlight on the earth is about 40,000 tons.

2-11: The glare part is light that has been reflected as from a mirror. Such "specular" reflection only contains information about the source. In effect, the surface of the mirror has not been probed. When we observe glare mirages from a road ahead of us, we are actually seeing the reflection of the sky.

2-12: As we saw in Chapter 16, waves that have a long wavelength compared to the size of a barrier curl around the barrier and do not create a shadow. If λ/a is less than 1, then the barrier will cast a shadow.

2-13: As we saw in Chapter 16, the angular spread of waves from a circular source is given by $\theta = 1.2 \lambda/a$. This is the half-angular width from the center of the diffraction pattern to the first minimum. The intensity of the central spot of the diffraction pattern is down by a factor of one-half at about half this angle. We can take the full angular width of the diffraction pattern to be approximately equal to $1.2 \lambda/a$. If λ/a equals $\frac{1}{10}$, then $\theta = 1.2/10 = 0.12 \text{ rad} \approx 7°$. At a distance of 100 m, the width of the beam would be $(100 \text{ m})(0.12 \text{ rad}) = 12 \text{ m}$. Such a beam could detect objects at a distance of 100 m providing they were not closer together than 12 m.

2-14: With a power density of 100 W/m² (which meets the current safety levels for microwaves) the rectifier field would have to have an area of 10^8 m². The diameter of the field would therefore have to be a little larger than 10 km. The land underneath the rectifying arrays could be used for many purposes, including farming or grazing. Birds roosting on the antenna might get warm. The metal skin of airplanes would reflect the microwaves without absorption.

Problems

1. A 1.0-μF capacitor, charged to 1000 V, is discharged with a time constant of 1×10^{-6} s. What was the original charge on the capacitor? What was the original electric flux in the capacitor? (Remember the relationship from Chapter 18 between charge and field in a capacitor.) Assuming linear decay of two-thirds of the charge during the decay time equal to the time constant, what is the current?

2. Using your own words and diagrams, describe Maxwell's equations.

3. When we studied waves in Chapter 15, we claimed that any wave velocity must be proportional to the square root of a term representing a restoring force, and inversely proportional to the square root of an inertia term. What corresponds to these terms in the equation for the velocity of electromagnetic waves?

4. A radio antenna for 1 MHz is a vertical tower 75 m high. The length is a quarter wavelength; a reflection from the ground provides another quarter wavelength. If 1 μC of electrons is oscillating up and down the antenna with a frequency of 1 MHz, what is the amplitude of the electric field at a horizontal distance of 1 km?

5. What is the maximum energy density, u, of the space through which a radio wave is passing with the amplitude of the electric field equal to 0.1 V/m?

6. The maximum intensity of sunlight on the earth is about 1.4 kW/m². What are the values of E and B?.

7. A typical classroom laser emits 1 mW in a beam that is about 2 mm in diameter near the exit port. What is the irradiance, or the intensity of radiation, in watts per square meter? Compare this intensity with that of sunlight on the earth.

8. If a radio signal has an effective ($E_{max}/\sqrt{2}$) electric field of 1×10^{-3} V/m, what is the effective magnetic field strength? What is the value of S? Give units.

9. If a radio signal has an effective ($E_{max}/\sqrt{2}$) electric field of 1×10^{-3} V/m, what is the effective momentum density of the beam at that point? What is the pressure produced by the radiation?
10. A spaceship in orbit around the sun, and at the same distance from the sun as the earth, spreads gossamer wings of silvered plastic to use as a sail. If the sail is 10 km by 10 km, what force is exerted by the sun's electromagnetic radiation? (Don't forget that the light is *reflected* by the sail.) If the entire ship has a mass of 1×10^5 kg (100 tons), how long would it take for the radiation force to give the ship a change of speed of 1×10^4 m/s?
11. If the capacitance of the split dipole that Hertz used was 3×10^{-11} F, what was the inductance if the resonant frequency was 1×10^9 Hz? (Note that straight rods do have inductance.)
12. What is the wavelength of a 535-kHz signal, which is at the low end of the AM range?
13. What should be the length of a folded dipole antenna (approximately one-half wavelength long) for an FM signal at 100 MHz?
14. When you make sodium hot enough (perhaps by dropping a pinch of salt in a gas flame), it emits a brilliant yellow light. The color is due to radiation at two wavelengths very close together—5.890×10^{-7} m and 5.896×10^{-7} m. What is the frequency of each of these colors? What is the *difference* in energy (in electron volts) between them?
15. A dentist's X-ray machine may typically use 50,000 V, yielding X-rays with a maximum energy of 50,000 eV. What is the corresponding frequency? What is the corresponding wavelength? (Compare with the size of an atom.)
16. The "solar constant" (the solar electromagnetic energy incident on earth) is 1.4 kW/m². About how many photons/s·m² provide this energy, assuming an average wavelength in the middle of the visible spectrum?
17. One of the tools for exploring our galaxy is a strong hydrogen line with a 21-cm wavelength. What is the energy change in a hydrogen atom that produces this wavelength?
18. Gamma rays with energies of 1×10^9 eV (1 GeV) can be produced with particle accelerators on earth, and are detected coming in from outer space. What are the corresponding f and λ?

THE MICROSTRUCTURE OF THE WORLD 23

This chapter is different from the others in the book. We have already studied many of the important phenomena of the world and, to explain why things happen, we have frequently appealed to an atomic model for matter. Now we should take a closer look at the world's microstructure. However, without special apparatus, there is no easy way to handle the phenomena in this realm. To explore the microworld we will have to rely on descriptions of experiments that others have done or are now doing.

This chapter is different, too, in that we have already dealt with some of the topics that we will now look at again. This time we must tie together various phenomena and explain them in terms of atomic and nuclear behavior. The final chapter thus becomes a summary and reminder of earlier work, seen from a different point of view.

To begin our description of the microworld, it might seem reasonable to list the various subatomic particles and then describe how they form nuclei, atoms, molecules, and us. It isn't that simple. First, there are some new rules or laws that we have had no need to learn so far. But these rules dominate the microworld. Second, the particles cannot be described except in terms of their interactions, and the interactions must be explained in terms of the properties of the particles. Finally, the phenomena of the microworld require description in terms that challenge our everyday view of the universe and our interaction with it. To describe these phenomena, we must raise primitive questions about the nature of reality. We are in the position of trying to describe a play in which the roles of the actors cannot be understood without knowing the plot, which cannot be explained without describing the roles of the actors. Furthermore, some of the stage rules are different from those of the everyday world, which turns out to be an illusion. The play is the real world.

Let us proceed, warily, by taking a closer look at a familiar particle. In describing electricity, light, heat, and other phenomena, we have frequently referred to electrons. Now we describe the parameters, or attributes, of the electron. In the process, we will learn about many of the strange features of the microworld. With this background as a guide, we can then describe the other subatomic particles and how they combine.

THE ELECTRON

The first particle to be actually identified as such was the electron. In 1897, J. J. Thomson measured the charge-to-mass ratio (e/m) of the particles coming from a negative terminal inside a vacuum system. The existence of "cathode rays" had been known for some years and it had been shown that they carried negative charge. It was conceivable, however, that the rays were simply a stream of electric current. Thomson's experiment of deflecting the path of the rays with electric and magnetic fields was most easily explained in terms of the motion of individual particles.

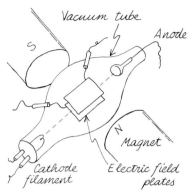

Thomson's device to measure e/m of cathode rays with crossed electric and magnetic fields.

Production

The electron is the easiest particle to isolate. Experimenters ninety years ago were intrigued with the way electrons seemed to leap out of any substance with only the mildest encouragement. Certain minimum conditions are necessary, of course. Electrical instruments are needed to detect most of the effects, and these effects mostly take place in high vacuum. To drive an electron out of a material, it is necessary to give it sufficient energy to escape from its atomic or molecular bonds. With some metals this can be done by heating the whole system and boiling off the electrons. They can also be knocked out by light, as in the photoelectric effect. As we saw earlier, the amount of extra escape energy needed is a few electron volts. Ordinary blue light can provide about 3 eV per encounter, which is sufficient for electron emission from the alkali metals. Ultraviolet, which can deliver energy of 6 to 10 eV to each electron, could knock electrons out of chicken fat, were that desirable (and assuming that the ultraviolet could get through the air without being absorbed by knocking electrons out of the air molecules).

The average vibrational thermal energy of a bound atom at room temperature is about $\frac{1}{25}$ eV (0.04 eV). A tungsten filament in a vacuum tube may be heated to 3000 K, ten times as hot as the 300 K room temperature.

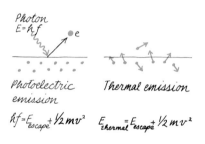

Question 23-1

Even at a temperature of 3000 K, the average kinetic energy of a bound atom is only 0.4 eV. How can the tungsten emit electrons if their binding energy is several electron volts?

Charge

The diagram of the diode shows a way to demonstrate that something negative is coming from the filament to the receiving plate. In spite of the success of Thomson's particle model, however, it is still conceivable that this negative charge is flowing as some kind of fluid instead of traveling in individual packages. A single electron would not make even the most sensitive electrical detector in this circuit respond with a signal greater than the natural fluctuations of the instrument. The individual charge was first measured accurately by Millikan in 1903. In Chapter 18 we described his method of measuring the electric charge on tiny oil droplets. The amount

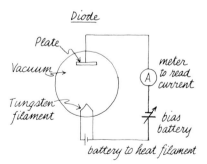

Current is produced only when filament is hot. Therefore, charges must be coming from filament. The current can be diminished or stopped by imposing a negative bucking field with the bias battery. Therefore, charges must be negative.

of electric charge that can cling to a droplet is quantized (exists in multiples of unit quantities). Millikan's experiment yielded a quantitative value for the magnitude of the electron charge. It is negative and equal to 1.6×10^{-19} C.

Mass

The most obvious attribute of the electron is its electric charge. The next question is, does it have any mass? To measure the mass of a baseball we have two methods. First, we can measure the weight or, in other words, the attractive force between baseball and earth. Because the weight is proportional to the mass, $F = G(mM/r^2)$, we can compare the weights (and so the masses) of baseball and a standard mass on a balance. The second method depends on the inertial properties of mass, its reluctance to have its state of motion changed. When a force is applied to an object, the object accelerates: $F \propto a$. The constant of proportionality is called mass: $F = ma$. The mass of the baseball could be found by applying a known force to it and measuring the acceleration produced.

The first method is completely impractical for finding the mass of the electron. No experiment has been performed where the gravitational effect on the electron could be observed. The force—the weight—is just too small. But because the electron has charge, electromagnetic forces can be exerted on it with much larger effects than the gravitational. The bubble chamber tracks in the picture clearly show the result of electron acceleration. The electron velocity was continually being changed by a force perpendicular to the path. Such a perpendicular or centripetal force does not change the speed or energy of the electron but does make it move in a circular path. In the bubble chamber picture, the tracks are spirals because the electrons were constantly losing energy in their passage through the gas. The perpendicular force caused by the movement of the electron in a magnetic field is equal to Bev, where B is the strength of the magnetic field, e is the electron charge, and v is its velocity. A centripetal acceleration must thus be produced with magnitude v^2/r, where r is the radius of the circular path. Since $F = ma$, it follows that

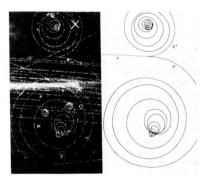

$$eBv = m\frac{v^2}{r}$$

Solving for m, we have

$$m_{\text{electron}} = \frac{Ber}{v}$$

If the velocity of the electron is known, its mass can be determined. There are many ways to combine magnetic and electric forces acting on electrons or other particles so that the velocity can be measured and the mass determined. For the electron, the value turns out to be 9×10^{-31} kg.

Question 23-2

How could you find the velocity of an electron by measuring the electric and magnetic fields that influence it?

There is yet another way of measuring the mass of the electron, depending on an effect that appears quite different from the inertial ones, although really not independent of inertia. Instead of forcing an existing electron to change its motion, create a new electron! This takes energy, of course, in an amount equal to mc^2, where c is the velocity of light and m is the mass of the created electron. The bubble chamber picture shows such an event. A photon came from the left and in the vicinity of a particular atom, turned all its energy into the mass and kinetic energy of two electrons. In the magnetic field the two electrons circled in opposite senses; one was negative, our standard electron, and one was positive, the so-called positron. If only one electron were created, a unit of electric charge would have come from nowhere. This never happens; the total charge remains constant. There was none to begin with and the net amount (+ plus −) afterward is also zero. It takes an *X*-ray of at least 1 MeV to produce such a pair creation. Each electron uses half of this for its mass. If any is left over, the electrons can share it to provide forward motion. So in this other way of measuring the mass of an electron we find that the mass is 0.51 MeV. The arithmetic of course works out:

$$mc^2 = 9 \times 10^{-31} \text{ kg} \times (3 \times 10^8 \text{ m/s})^2$$
$$= 81 \times 10^{-15} \text{ J}$$

Since
$$1.6 \times 10^{-19} \text{ J} = 1 \text{ eV},$$

$$mc^2 = 0.51 \times 10^6 \text{ eV} \approx \tfrac{1}{2} \text{ MeV}$$

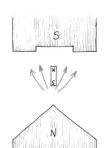

Alignment of magnets in a magnetic field.

A magnetic dipole experiences a translational force in an inhomogeneous magnetic field. The force is stronger on the bottom than on the top of the dipole.

Spin

The electron has yet another attribute that is vital to its interactions. Besides having mass and electric charge, it acts as if it were spinning on its own axis. Such a spin can make itself known through three main effects.

1. *Magnetic Moment.* (a) First of all, a spinning electric charge becomes a small magnet. In a uniform magnetic field, the electrons should align themselves like compass needles all pointing together. If for any reason some of them are forced out of alignment, they will be in a different energy state from the stable ones. When they swing back into line, their energy will be released in some form, since energy had to be provided to get them out of line in the first place. Furthermore, in a nonuniform magnetic field, the electron magnets should not only rotate but should move toward or away from the region of strong field. These effects can be detected with a bit of trickery.

The diagram shows magnet pole pieces that provide a very nonuniform magnetic field. As you can see, a magnet with its north end up will move down in the diagram; the force down on the south end is greater than the force up on the north. If a beam of electrons is shot through such a magnetic field, we might expect the emerging beam to spread out in a vertical line. The deviation of each electron would depend on how its spin was oriented when it entered. We assume that the spin of each would not flip into line while going through the magnet, but instead would precess like a top, still maintaining its original angle to the vertical. But, unfortunately, the experiment cannot be done quite so simply.

> **Question 23-3**
>
> Why not? Aside from the vertical motion of the electrons due to the nonuniform magnetic field, what else would happen to electrons traveling in such a field?

Instead of using bare charged electrons, the actual experiment is done by having the electrons ride along on an atom. Silver atoms were used by Otto Stern and Walther Gerlach in 1924. Silver has just one valence electron, existing in a state such that the only angular momentum or spin must be that associated with the electron itself. All the other electrons in the closed shells produce effects that cancel. The whole atom is electrically neutral so that it will not be influenced by a magnetic field unless the atom is also behaving like a small magnet, in which case it will be deviated up or down. The diagram shows the experimental method and results. Some atoms were indeed forced up and others down, but only into two groups! Apparently, the electron spins were either in line with the field or 180° out of line. The effect is *quantized*. If the whole system were rotated on its axis 90°, we would find that the deviations were then horizontal, but still divided into two definite groups with nothing in between. Thus the directional deviation is not a function of the way the atomic beam was produced. The silver atoms are boiled out of the oven with completely random orientation. We are forced by this phenomenon (and many others) to conclude that the electron spin can exist only parallel or antiparallel with the magnetic field with which we detect it. One is tempted to insist that in the original beam the spins must have had any orientation and somehow arranged themselves in this peculiar way just when they entered the magnetic field. Perhaps so, but of course that is metaphysics and meaningless. If an effect can never be detected, directly or indirectly, then it is rather useless to ponder it. In any experimental arrangement where electron spin is determined, some standard direction or axis is established. The electrons are always found to be lined up either in that particular direction or opposite to it, but never in between.

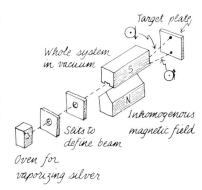

Stern-Gerlach experiment using neutral silver atoms to demonstrate discrete alignment of magnetic moments.

The convenient unit of angular momentum for particles is Planck's constant (divided by 2π). In the literature of physics this is written as the symbol $\hbar = h/2\pi$. It has the value of 1×10^{-34} J·s and the proper dimensions to be angular momentum (Mass · Length^2/Time).

Chapter 8 dealt with the subject of angular momentum. As we pointed out there, it is rather startling that there is a natural unit of angular momentum in this world. We know of very few other quantities that have basic units which exist naturally. There are none for mass, length, or time–as far as we know. Electrical charge has a natural unit, and now we find one for angular momentum. In terms of this unit, the electron has spin $\frac{1}{2}$. As we will see, this is a very significant value. Particles cannot have just any value of spin, but only certain ones. Like electric charge, spin is quantized and can exist in multiples of $\frac{1}{2}\hbar$, but never with values in between, such as $0.65\hbar$. Furthermore, the measured values for a given system, corresponding to the different directions of the spin, must always be different from each other by one whole unit, $\hbar$. Therefore the valence electron of silver is found in only two orientations, with the direction-giving field ($+\frac{1}{2}$) or against ($-\frac{1}{2}$).

(b) The magnetic-moment effect of the electron spin can also be determined in *uniform* magnetic fields. For this it is necessary to use atoms or molecules with unpaired electrons; the goal is to deal with electrons that are relatively free and yet bound sufficiently so that a large number can be held in one place. The conduction electrons in a metal are plentiful enough, but they are by no means independent agents. A conduction electron could not orient itself in a magnetic field without interacting with all the others. Most of the electrons in material are paired off with electrons of opposite spin existing in the same energy level. In paramagnetic substances, however, there are unpaired electrons that do not have to interact with the neighboring atoms to form molecular bonds.

If one of these substances is placed in a magnetic field, the electron spins will point either with or against the field. It would seem at first that all of them would flip around so as to align themselves and thus be in the lowest possible energy state. But actually, energy is being fed into the electron spin system by the random thermal motion of the molecules. It is as though a group of compass needles were trying to line up in a magnetic field but were constantly being jiggled into random directions.

The amount of energy that it takes to flip one electron in a particular magnetic field is a very definite amount:

$$2\mu_e B \approx (2 \times 10^{-23}\, B)\,\text{joules} \approx (10^{-4}\, B)\, eV$$
$$= 10^{-5}\, eV \text{ for } B = 0.1\, \text{T, or } 1000\, \text{G}$$

where μ_e is the magnetic moment of the electron. (Magnetic moment is the measure of strength of a magnetic dipole, and is best defined and measured in terms of this equation.) It would take a chunk of electromagnetic energy of just this amount to flip the electron, and, of course, if the electron flips back, a photon of that energy is emitted. Photon energy is directly related to frequency. Thus,

$$E = hf$$

where h is Planck's constant and f is frequency in cycles per second. In this particular example, a magnetic field of 0.1 T would require photons or radio signals with a frequency of about 3000 MHz.

Question 23-4

What is the wavelength of a 3000 MHz signal?

Normally, such a signal is not heard from the sample. There are as many electrons flipping one way as the other to maintain the particular balance determined by the temperature. If, however, a radio coil driven by a transmitter is placed around the sample, as shown in the diagram, and the frequency is gradually raised, in one particular narrow frequency range the coil will appear to absorb extra energy from the transmitter. This energy has produced the flipping of many electrons out of the magnetic field direction. How long it takes for them to jiggle back depends on the temperature and how easily the substance transmits its thermal agitation to the electrons.

The next diagram shows a typical plot obtained in such an experiment. In this case, the frequency remained constant and the magnetic field was changed.

(c) The different energy states of the electron as it assumes the two orientations (either parallel or antiparallel to the magnetic field) give rise to spectroscopic effects that were observed over a hundred years ago. However, we had to wait until 1924 for the interpretation, which was made by Samuel Goudsmit and George Uhlenbeck. Strong magnetic fields of the order of 10 T can exist naturally inside atoms. The electronic circulation (in the older picture, the orbital motion of electrons) produces fields that interact with the magnetic effects of the intrinsic spin of each electron. Since light is produced by the jump of an outside electron into a vacant position that has been momentarily created in the orbits, the energy of the light pulse will be equal to the difference in energy levels of the jumping electron.

Each energy level really has two possible values, corresponding to the two orientations the electron can assume with respect to the local fields. As can be seen in the diagram, this will give fine structure to most transitions. Instead of producing a photon of light with one particular energy, and so a particular wavelength, each transition produces a double or triple wavelength in the spectrum.

2. *Angular momentum conservation—Another role of electron spin.* Electron spin must also be taken into account in spectroscopy and in particle decays where electrons are emitted, simply because spin means angular momentum, and angular momentum is one of the few things in this world that must be rigorously conserved. If an atomic system emits a photon of light, the photon carries off one whole unit of angular momentum. (The fact that photons possess angular momentum has been experimentally demonstrated.) The remaining system must then be one unit different from what it was to begin with. If the photon carries off clockwise angular momentum, the remaining system must recoil counterclockwise. Within the system various parts are producing angular momentum—some associated with certain electron configurations (as though the electrons were actually orbiting around the nucleus). Also, of course, we must take into account the spin, or intrinsic angular momentum, of each of the electrons. The nucleus has a spin, too. All of these must be added together, but the addition is vectorial; that is, many cancel. The spectral lines can be understood only when the description of total angular momentum includes the spins.

In particle decays, total angular momentum must also be conserved. When a radioactive nucleus decays to emit an electron, a β ray, there is an apparent breakdown of this rule. The parent nucleus recoils, but it can do so with no change in its own spin. Yet the electron goes tearing off, spinning as it goes. It is as surprising as if a person on a friction-free table were able to get a bicycle wheel spinning without recoiling in the opposite sense himself. The very same problem arises when the muon (one of the main parts of cosmic rays at sea level) decays. A single electron comes spinning out.

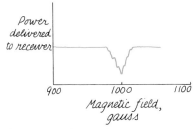

Absorption of radar-frequency energy due to electron magnetic resonance. The jagged sub-peaks along the sides of the major dip correspond to absorption at various local magnetic fields in the structure.

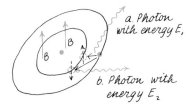

Electron spin splitting of spectral lines. $E_1 > E_2$ since in (b) part of the energy is still retained by the electron when it is oriented against the magnetic field.

Vector combination of angular momenta

Single electron in p-orbit of angular momentum $1\hbar$. Orbit spin Total $3/2$ Total $1/2$

Two electrons, one in s-orbit of angular momentum zero, and the other in d-orbit of angular momentum $2\hbar$. Total electron spin is $1\hbar$.

Orbit spin Total=3 Total=2 Total=1

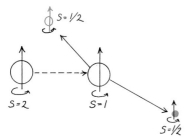

Nucleus with spin 2 ℏ decays to nucleus with spin 1 ℏ, plus electron and neutrino with spin ½ ℏ each.

There is a way out of this apparent breakdown of angular momentum conservation. In all such cases, at least one other particle, the neutrino, ν (nu), is also emitted. The neutrino has no electric charge and so is not easily detected, but it does have spin. The person on the friction-free table would have no trouble getting two bicycle wheels spinning, as long as one went clockwise and the other counterclockwise. The neutrino and its problems will be dealt with later when we define all of the subatomic particles.

3. *The exclusion law—The third role of spin.* The electron has intrinsic angular momentum, and so it acts like a magnet, but its spin of $\frac{1}{2}\hbar$ has an even more profound effect. All particles with half odd-integral spin ($\frac{1}{2}, \frac{3}{2}, \frac{5}{2}$, ...), and this includes protons and neutrons as well as electrons, must obey a special exclusion law. The great physicist Wolfgang Pauli first framed this principle in 1925. Two other great names in physics are associated with the particular rules resulting from Pauli's principle, Enrico Fermi and Paul Dirac. In scientific literature, particles with spin $\frac{1}{2}\hbar$, $\frac{3}{2}\hbar$, and so forth, are

The Pauli exclusion principle determines the periodicity of electron arrangements in atoms. In these diagrams, *n* identifies the main energy level of an electron. The orbital angular momentum is given by *l*. Its orientation (up, down, sideways with respect to a magnetic field) is quantized. The electron spin of ½ ℏ can have only two orientations.

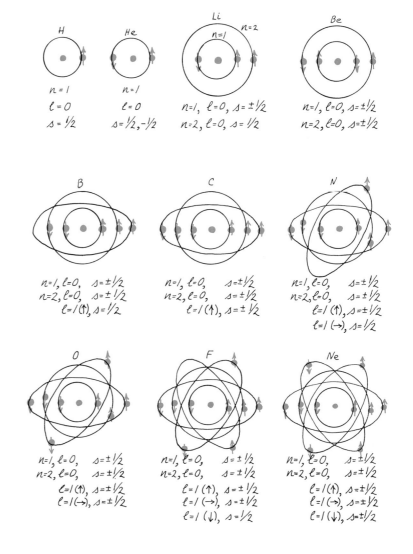

referred to as *fermions*. No two fermions can possess identical properties while inhabiting the same region. It is this saturation restriction that is responsible for the particular arrangement of electrons in the atoms and so determines the nature of the periodic table.

Hydrogen has one electron existing in a region close to the proton nucleus. Helium has two electrons in the same region with all the same properties except that the spins of the two electrons are in opposite directions. Lithium has three electrons, but only two can exist in the lowest allowed energy region. The third electron in this region could have its spin aligned in only one of two ways and, in either case, this would duplicate all of the properties of an electron already there. If there were two kinds of electrons, red and blue, or if the electron had some other kind of variable property—for example, lopsidedness, then more than two could fit into the same energy region. But only two do so. Thus, it must be that the electron has a limited number of attributes, and with spin we have exhausted the list. The third electron of lithium must exist in a higher energy level, farther from the nucleus. The fourth electron of beryllium can, with opposing spin, associate with number three. The fifth electron of boron is almost in the same second energy level, but not quite because it can have one unit of angular momentum (besides its intrinsic spin). The existence of this extra unit of angular momentum slightly changes the energy of the fifth electron and thus makes it distinguishable from electron number four. There can be three directions of angular momentum of magnitude $1\hbar$. In any experimental situation that sets up a preferred direction (e.g., a magnetic field), the axis of a system with one unit of angular momentum can either point with the field, against it, or perpendicular to it.

Question 23-5

Why can a system with an angular momentum of $1\hbar$ assume three different orientations with respect to a magnetic field? For a spin of $\frac{1}{2}\hbar$, there are only two possibilities.

The second main level of an atom's electron system thus can contain four possible configurations: orbital angular momentum zero and three orientations of orbital angular momentum one. Each of these can hold two electrons, with spin "up" and "down." Altogether, eight electrons can fit into the second level. Even without an external field, the internal magnetic fields in an atom cause each of the eight electrons to have slightly different energies. These differences are represented in spectra by photons of slightly different wavelengths, producing "fine structure" or "splitting" of the spectral lines.

Size of the Electron—The Wave-Particle Problem

So far we have assigned only three attributes to the familiar electron: mass, charge, and spin. If we were describing a marble, we would already have inquired about its "looks": color, size, and so on. (Actually, we have been describing the "looks" of the electron; i.e., we have described the experimental responses to probes that can detect electrons.) Of course, the electron has no color, simply because the shortest wavelength of visible light (4×10^{-7} m for blue) is certainly larger than the size of an electron. One cannot determine the structure of an object with a probe larger than the object. There is a special problem concerned with the definition of particle size. In some cases, the problem is as simple as the one we would face in measuring the "size" of a small bar magnet. With a meter stick, and using light and our eyes, we would get one result. Using a magnetic probe, we would apparently have a larger object with rather fuzzy edges. If a particle is responsive to several forces, it is possible that its boundaries may be different for each one.

But there is another size problem more subtle than this. Under certain conditions, the behavior of electrons must be described in terms of the behavior of waves. The wavelength is related to the momentum by the following formula: $\lambda = h/p$. Here λ is the wavelength in meters; h is the ubiquitous Planck's constant; and p is the standard symbol used in the literature to represent momentum. At low velocities, $p = mv$ (rest mass times velocity), the usual definition. At velocities close to the speed of light, the momentum is a more complicated function, as we saw in Chapter 11:

$$p = \frac{mv}{\sqrt{1 - (v^2/c^2)}}$$

Thus experimentally momentum assumes a more important role than either the mass or velocity separately.

Observe now how the size of the electron changes. Surely it is as large as one wavelength and presumably its influence would extend over several wavelengths. For an electron accelerated through 10 V, perhaps in a radio tube, the velocity $\approx 1.6 \times 10^6$ m/s. Then,

$$\lambda = \frac{6.6 \times 10^{-34} \text{ J} \cdot \text{s}}{9 \times 10^{-31} \text{ kg} \times 1.6 \times 10^6 \text{ m/s}} \approx 5 \times 10^{-10} \text{ m}$$

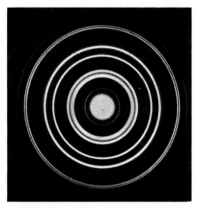

Interference pattern produced by electron beam going through thin foil. (Courtesy, Dr. Lester Germer).

(The outer electrons in atoms have kinetic energy in the range from 1 to 10 eV. Their wavelengths, therefore, are about the size that is indicated by the equation.) Since this wavelength is larger than the size of an atom, the point is emphasized that an electron cannot really be considered to be orbiting about a nucleus. In this sense, the electron is as large as the atom, and indeed there is no other sense. By any experiment that can be performed, the atomic electrons are found to exist over regions comparable in size to the whole atom. It is meaningless to ask how large the electrons are or where they are going during the time when we do not measure them.

There is, of course, good experimental evidence for the wave *behavior* of electrons. For our present arguments, we deliberately avoid the use of the words "wave nature." The similar wave behavior of light is most clearly seen in interference or diffraction patterns. The photograph shows a diffraction pattern produced by a beam of electrons passing through holes produced by the spacing of atoms in a crystal. The pattern is very similar to that produced by light passing through a small hole. The spacing of the rings depends in both cases on the ratio of wavelengths to hole diameter.

With red light, the rings are spaced farther apart than with blue light. With slow electrons, the spacing is larger than with fast electrons. The analysis of the optical phenomenon is in terms of waves originating in different regions of the hole, traveling different distances to the screen, and interfering with each other there, sometimes canceling and sometimes reinforcing. The ring geometry arises solely because with a round hole there must be circular symmetry. Since exactly the same phenomenon is observed with electrons, we use the same language and the mathematics of wave motion to describe the situation.

We need not concern ourselves as to whether an electron *is* a wave. It *is* an electron. In situations where the electron interacts with objects larger than the wavelength given by $\lambda = h/p$, it acts like a hard spinning sphere. When the interaction is with objects comparable in size to λ, or smaller, the effect has to be described in terms of the interactions of waves. There is a similar situation with sound. The behavior of high-frequency sound in a room is very much like that of "rays" bouncing about. The low-frequency sounds, however, curl around corners like waves.

In mathematical descriptions of processes, this dual state of affairs is very common. We may have a particular formula or equation that is complicated and difficult to compute or solve, but that is completely accurate at all times. This is exactly the case for our description of photons and any electromagnetic interactions. To obtain a solution for a particular case, the usual technique is to make certain approximations. An exact solution to most physical problems is never possible, and part of the skill of the mathematician or scientist lies in choosing the right approximation, and knowing to what accuracy the solution should be carried.

For example, consider the equation describing a marble falling in air:

$$ma = mg - Kv^2$$

The product of mass and acceleration of the marble is equal to its weight minus the air friction force. (We assume that the marble is large enough and falling fast enough so that the air resistance is proportional to the square of the velocity.) A graph of the solution to this equation was shown in Chapter 1. The exact solution to the equation is complicated and requires the use of calculus. For two extreme cases, however, approximations can be made that yield simple solutions. First, at the beginning of the fall of the marble, when the velocity is small, we get the simple equation of free-fall in a vacuum: $a = g$, and $x = \frac{1}{2}at^2$. In the second extreme case, the friction is very large: $mg = Kv^2$, and $x = vt$.

Needless to say, we have not produced a paradox about the nature of marbles just because the same equation yields two different solutions. For the same reason, there is no duality paradox about the nature of subatomic particles. Under certain experimental conditions, the best approximation to the equations describing particles makes the equations similar to wave equations. In other experimental situations, the proper approximation produces an equation similar to the kind describing the action of solid balls. We do not claim that the subatomic particle *is* a wave or *is* a hard ball; it merely behaves in this or that fashion under particular conditions. Nor should it be imagined that a subatomic particle (or a photon) moves with a wiggly, wavelike motion. It is the solution to the equation that has the wave character. This solution is a probability function that gives the chances of finding particular values of position or velocity of a given particle at a particular time.

> ### Question 23-6
>
> The equation for the position of a falling marble gives the exact position of the marble as a function of time: $x(t)$. Doesn't the quantum theory yield the same information about subatomic particles?

HEISENBERG'S UNCERTAINTY PRINCIPLE

There is yet another important fact of life concerned with particle measurements. It is not always possible to measure— and so know—certain pairs of related properties of a particle, both with infinite precision. This is not just a matter of failing to have a sensitive enough meter of some kind. There is in principle, as well as in practice, a limit to the precision with which we can measure the position of a particle at the same time that we measure the momentum. Of course, every time any measurement is made of the position of some object a certain amount of error is involved. For example, if you judge by eye the position of a scribe mark on the floor of a room, you might guess the distance between it and a wall to within half a foot: $x = 3$ ft; $\Delta x = \pm\frac{1}{2}$ ft, where Δx describes not any mistake made but simply your assigned limits of accuracy. With a meter stick, the precision could be much greater: $x = 1$ m; $\Delta x = \pm 2$ mm. If there were some reason to attain greater precision it could be done. The distance between two scribe marks on the standard meter stick in Paris was measured by Michelson in terms of the number of wavelengths of a particular spectral line of visible light. His precision with this technique was better than one part in a million: $x = 2 \times 10^6 \lambda$; $\Delta x = \pm 1 \lambda$.

Even though practical considerations might rule out the extension of such efforts, it would seem that there should be no limit in principle to attaining greater precision. The measuring probe, of course, must be smaller than the Δx required. To do better than Michelson, it would be necessary to use light (or particles!) with a smaller wavelength than visible light. Blue light with $\lambda = 4 \times 10^{-7}$ m would be a little better than red with $\lambda = 7 \times 10^{-7}$ m. In some microscopes, this color effect is taken advantage of by using blue light to increase the resolution. Ultraviolet microscopes have been constructed, but the most widely used type is the electron microscope. The wavelength of the electrons in such a machine is about 2×10^{-9} m, 100 times better than can be obtained with visible light. Actually, resolution of even the best electron microscope is not so good as this, because of practical considerations of construction. To get even finer resolution, we need particles or light of yet smaller wavelength. But to get particles with smaller wavelength, we must increase their momentum: $\lambda = h/p$. The same problem faces us when we use light. The amount of energy in each photon of light is $E = hf = hc/\lambda$, where E is the energy in joules, h is Planck's constant, f is the frequency of light in cycles per second, and λ is the wavelength in meters. As we go from radio to radar to infrared to visible to ultraviolet to X-rays to γ rays, we deal with photons with greater and greater energy and with more of the behavior we normally associate with particles. Light does, of course, carry momentum. (Comet tails are blown away from the sun by photons and other particles emitted from the sun.) The amount of momentum is

$$p = \frac{E}{c} = \frac{hf}{c} = \frac{h}{\lambda}$$

But this is just the expression we find for any other kind of particle!

If the measuring probe has higher momentum as we decrease the wavelength, we can still measure the position of a tiny object as accurately as we choose, but in doing so we knock it away with the probe, as shown in the diagram. In measuring position, we add an unknown amount of momentum to the object. The *unknown* momentum is *perpendicular* to the original direction of the wave probe. Instead of casting a sharp shadow of the object, which would determine the position of the object exactly, the wave probe necessarily falls somewhere in a diffraction pattern. Any particular photon (or other probe particle) can fall anywhere within the diffraction pattern in a completely indeterminate way. The opening angle for the maximum of the diffraction pattern is $\theta \simeq \lambda/\Delta x$. The wavelength of the probe is λ, and the indeterminate size of the object (either the vagueness of its edge or, more usually, the whole diameter) is Δx. Of course, the probe particle can also be deflected at an angle two or three times this large and land in one of the outer fringes of the diffraction pattern. Most, however, fall in the central maximum. A small wavelength of the probe produces a small diffraction pattern. But, as a result of the small wavelength, the probe carries higher momentum. This compensates for the smaller angle of deflection, producing a sideways impulse independent of the wavelength.

The dilemma is summarized by Heisenberg's uncertainty relationship. A more careful analysis of the uncertainties leads to a slightly lower limit for the product than the one derived in the diagram.

$$\Delta x \cdot \Delta p_x \geq \frac{h}{2\pi} = \hbar \approx 1 \times 10^{-34} \, \text{J} \cdot \text{s}$$

In words, the product of the uncertainty in position and the uncertainty in momentum must always be equal to or greater than Planck's constant divided by 2π. This rule actually follows from our formula for the momentum of a particle, $p = h/\lambda$. In a glancing collision of a probe particle with an object, an appreciable fraction of its momentum, p, may be given to the object in a direction perpendicular to the original probe direction. Then, that fraction, $\Delta p \approx p$, is the uncertainty in momentum of the object after the measurement. The position uncertainty Δx cannot be much less than the measuring wavelength λ. Since for the particle $p = h/\lambda$, for the object after the collision, $\Delta p_x \approx h/\Delta x$. If angular momentum were not quantized, that is, if Planck's constant were zero, both position and momentum could be measured simultaneously with infinite precision. This world would not be this world at all in that case.

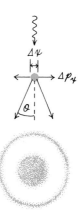

Heisenberg uncertainty principle: the diffraction pattern produced by a light beam with wavelength λ has an opening angle of $\theta \approx \lambda/\Delta x$. The indeterminate momentum in the x-direction, delivered by a probe glancing off at an angle θ, is $\Delta p_x = p_{\text{probe}} \sin \theta \approx p\theta = (h/\lambda)(\lambda/\Delta x) = (h/\Delta x)$. Then, $\Delta p_x \Delta x \approx h$.

Heisenberg's principle also applies to other pairs of quantities that in classical science can, in principle, be known together with infinite precision. Instead of linear position, x, choose angular position, θ. Then $\Delta\theta \times \Delta L \geq \hbar$, where $\Delta\theta$ is the uncertainty in angle of a rotating system and ΔL is the uncertainty in *angular* momentum.

Another pair of variables is of importance in particle physics: $\Delta E \times \Delta t \geq \hbar$, where ΔE is the uncertainty in the energy of a system *at the time we measure it* and Δt is the uncertainty in the time duration of the measurement. One is tempted to say that ΔE is the uncertainty with which we *know* the energy E, implying that actually the system has a definite energy. This is a deluding thought. Once again, if in principle something cannot be measured, then the question as to whether it exists is meaningless.

How can we measure the energy of an excited state of a system? The most general method is to let the system return to normal and see how much energy comes out. A collision, perhaps, throws an atom into an excited state, with one of the electrons in a nonstable position. The uncertainty of the energy of the system presents no problem, however, until the time when we receive the information. As the atom returns to normal, as the electron hops down into place, a photon is emitted. It carries energy: $E = hf$, and this implies that it has a definite wavelength, $\lambda = c/f$. However, the photon is emitted in a time of about 10^{-8} s for an ordinary atomic transition. Before then there is no photon, and after that the electromagnetic vibrations stop coming from the atom. The diagram shows how a meter might record this situation. But a train of sine waves has a definite frequency (or is a pure tone), only if it has been emitted forever and will continue indefinitely. A short burst of oscillations must have overtones associated with it. These are higher frequencies that, when mixed with the main note, produce a beginning and end to the pulse. The shorter the duration, the more important the overtones. Instead of a single sharp wavelength, a narrow band of wavelengths is produced. The sharp spectral line has width.

In the case of visible light where $\Delta t \approx 10^{-8}$ s, we have:

$$\Delta E \times \Delta t \geq \hbar \approx 10^{-34} \text{ J} \cdot \text{s}$$

$$\Delta E \approx 10^{-26} \text{ J} \approx 10^{-7} \text{ eV}$$

This seems like a small uncertainty, indeed. Since the energy of a visible photon is about 2 eV, it appears that the energy is definite to 1 part in 10 million. But spectroscopy can attain precision such that this is observed as the natural line width. The uncertainty in frequency is $\Delta f \approx 10^8$ Hz, and the irreducible width of the spectral line is $\Delta\lambda \approx 10^{-13}$ m, or about 0.3 millionth of the length of the whole spectrum from red to blue.

We have occasion to deal with particles or excited states with lifetimes as small as 10^{-22} s. In this case, $\Delta E \approx 10^{-12}$ J ≈ 10 MeV. Uncertainties of this size can play dominant roles in certain particle reactions.

ATOMIC STRUCTURE

Our preliminary model for the atom is basically the one worked out by Niels Bohr (1885–1962) and Ernest Rutherford (1871–1937) during the second decade of this century. A central nucleus containing most of the mass of the

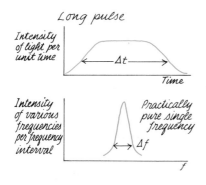
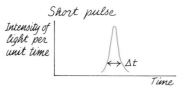
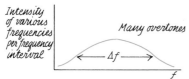

Relationship between duration of light pulse and spread of frequencies

atom has a positive electric charge. The amount of charge ranges from 1 unit for hydrogen to 92 for uranium. Each atom contains a corresponding number of negatively charged electrons, thus making the atom electrically neutral. Each chemical element is defined by the number of electrons in its atom, the "atomic number," or "Z," as shown in the periodic chart of the elements on the front inside cover. Note that for hydrogen the atomic number is 1, for carbon it is 6, for oxygen it is 8, for copper it is 29, for gold it is 79, and for lead it is 82. The numbers are integers, since electric charge is quantized.

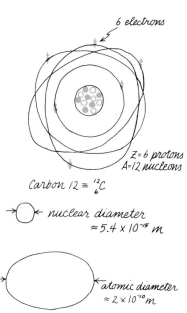

Within the nucleus, there are neutrons and protons. Neutrons have about the same mass as protons, about 1800 times greater than the electron mass. Each proton has one unit of positive charge; the neutrons are neutral. Protons and neutrons are not point elementary particles. Among other internal features, they contain circulating electric currents, which produce magnetic moments. The atomic number of an element is also the number of protons in each nucleus. To a first approximation, each nucleus contains about as many neutrons as protons. In the next section, we will see how the actual ratio of neutrons to protons affects nuclear stability.

The radius of a nucleus is quite accurately given by the formula: $r = r_0 A^{1/3}$, where $r_0 = 1.3 \times 10^{-15}$ m, and A is the total number of protons and neutrons in the nucleus. This is just the relationship expected if the nucleus consisted of tiny marbles packed together, each with radius, r_0.

> Question 23-7
>
> Why should we expect this relationship if protons and neutrons pack like marbles?

For the uranium nucleus, this formula gives $r = (1.3 \times 10^{-15} \text{ m})(238)^{1/3} = 8 \times 10^{-15}$ m. The radius of the atom itself is about 1×10^{-10} m, about 10,000 times greater. Although the electrons in an atom contain only about 1/4000 the atomic mass, they evidently occupy most of the volume.

The Bohr-Rutherford model of the atom pictured the electrons circulating in planetary orbits around the nucleus. Electrostatic attraction provided the centripetal force. For the simplest of atoms, hydrogen, a single electron orbits a single proton.

$$\frac{mv^2}{r} = k \frac{q_1 q_2}{r^2}$$

The undetermined variables are v and r. So far it would appear that an electron could have an orbit at any radius, and thus have the corresponding velocity. Bohr proposed, however, that the angular momentum of the electron must be quantized.

For the electron in a circular orbit:

$$mvr = n\hbar \quad \text{where } n \text{ is an integer}$$

Let's combine the angular momentum requirement with the centripetal force formula:

Centripetal force requirement: $\dfrac{mv^2}{r} = k\dfrac{q_1 q_2}{r^2} \rightarrow m^2 v^2 r^2 = mk q_1 q_2 r$

Angular momentum quantization: $mvr \rightarrow m^2 v^2 r^2 = n^2 \hbar^2$

$$\therefore \quad r = \dfrac{n^2 \hbar^2}{mk q_1 q_2}$$

Now we have a formula for only one variable, r. It appears that only certain radii are allowed, and that their size increases as squares of the integers: 1, 4, 9, 16, and so forth.

> **Question 23-8**
>
> What is the size of the hydrogen atom in its unexcited state when $n = 1$? Since the model is crude, use only order-of-magnitude numbers: $q_1 = q_2 = 10^{-19}$ C; $\hbar = h/2\pi = 10^{-34}$ J·s; $k = 10^{10}$ N·m²/C²; $m = 10^{-30}$ kg.

As you can calculate, the theory yields the same approximate size for the hydrogen atom that we have been using throughout the book. More important for the usefulness of the theory is its prediction concerning the energy of the system. Since the electron and proton are bound together, their electrostatic potential energy must be negative.

$$E_{\text{pot}} = -k \dfrac{q_1 q_2}{r}$$

The electron has positive kinetic energy equal to

$$E_{\text{kin}} = \tfrac{1}{2} m v^2$$

Since the centripetal force requirement is:

$$\dfrac{mv^2}{r} = \dfrac{k q_1 q_2}{r^2}$$

the kinetic energy must be:

$$\tfrac{1}{2} m v^2 = \tfrac{1}{2} k \dfrac{q_1 q_2}{r}$$

The total energy of the system is

$$E_{\text{tot}} = E_{\text{kin}} + E_{\text{pot}} = \tfrac{1}{2} k \dfrac{q_1 q_2}{r} - k \dfrac{q_1 q_2}{r} = -\tfrac{1}{2} k \dfrac{q_1 q_2}{r}$$

Notice that the total energy is necessarily negative, because the electron is bound. Since only certain radii are allowed, only certain energies can exist.

$$E_{\text{tot}} = -\tfrac{1}{2} k \dfrac{q_1 q_2}{r} = -\tfrac{1}{2} k^2 \dfrac{m q_1^2 q_2^2}{n^2 \hbar^2}$$

ATOMIC STRUCTURE

> **Question 23-9**
>
> Use the order-of-magnitude numbers from Question 23-2 to calculate the binding energy of the lowest hydrogen orbit, where $n = 1$.

The actual energies for the various radii are shown in the energy level diagram.

Consider the experimental consequences of such a theory. It would take at least 13.6 eV to "ionize" a hydrogen atom; that is, to rip the electron away from the proton. Indeed, that is the measured ionization energy. If a hydrogen atom is bombarded by other atoms in a hot gas, or by photons, or by subatomic particles passing by, it can accept amounts of energy smaller than 13.6 eV only if the energy is exactly the right amount to raise it to one of the quantized levels. The hydrogen atom cannot exist at energy levels in between. Furthermore, once the atom is excited to a higher energy level, it can lose only the exact energy required to change to one of the lower quantized levels.

In a very hot gas, such as in a fluorescent or spectrum tube, collisions provide the energy needed for an atom to go from its ground state to one of the excited states. Within 10^{-8} s, for most atoms and most transitions, the atom emits a photon and sinks into a lower energy level. Since the atomic energy changes can occur only between quantized levels, the photons emitted must have only certain, discrete energies. Each energy corresponds to a particular frequency and wavelength.

For hydrogen, the allowed energies of photons are shown in the diagram. Note that there are families or series of frequencies, corresponding to the lower energy level of the transition. The series are known by the names of nineteenth-century physicists. In the diagram below, we see a frequency spectrum of three of these hydrogen series. Only the Balmer series is in the visible range. The experimental spectra agree very closely with the predictions of this crude model.

The series of frequencies produced by these energy transitions in the free atoms of gases, as opposed to bound atoms in molecules or solids, are called *line spectra*. When light from a hot gas is analyzed by a prism or a diffraction grating in a spectroscope, the light enters through a narrow slit. The image of that slit is a line. Its position depends on the frequency or wavelength of the light. Thus, the spectrum consists of a number of individual "lines."

The energy levels of atoms can also be observed in the *absorption* of light. A spectrum produced by sunlight, for example, appears at first to be a continuous spread of colors. On close examination, you can see the "Fraunhofer" or absorption lines. Light coming from the "surface" of the sun contains essentially all frequencies. As this radiation passes through the cooler gases in the solar atmosphere, particular frequencies are absorbed, corresponding to the quantized energy levels of the cooler gases. The gases

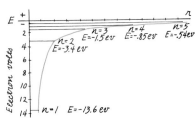

Quantized energy levels of Bohr orbits for hydrogen.

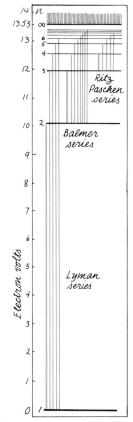

Spectra produced by the quantized energy levels of hydrogen.

Hydrogen spectra in the infrared, visible, and ultraviolet.

in the solar atmosphere have been identified by analysis of these spectra of missing frequencies.

So far we have derived quantitatively the energy levels only for hydrogen. What happens when there are many electrons in each atom? Even when there are two, as in helium, there are complications. To calculate the energy levels, we would have to take into account the repulsion between the two electrons as well as the attraction of both to the nucleus. Furthermore, Pauli's exclusion principle dominates the calculation. The two electrons cannot exist in the same level unless their spins are opposed. Also, transitions between energy levels can take place only if angular momentum is conserved. The emitted photon itself has one unit of angular momentum. The rearranged atomic system, including the electron spins, must account for the rest of the original angular momentum. Only certain transitions can accommodate these requirements. With helium, for instance, the complete spectrum consists of two separate arrays of lines, corresponding to two different sets of final levels. In the lower level, the two electrons are essentially in the same energy state, but have opposite spins. In the other series, one electron is in a higher state, but the electron spins are parallel.

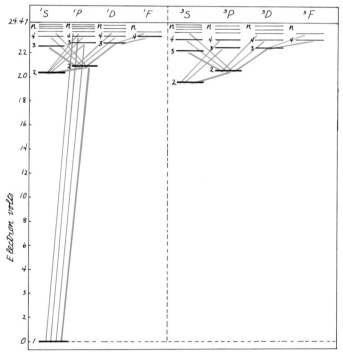

The two systems of spectra produced by the energy levels of helium and the prohibition of Pauli's exclusion principle.

Bohr's simple model agreed astonishingly well with the experimental data for hydrogen, gave qualitative agreement for certain hydrogenlike atoms, such as *ionized* helium or the alkali atoms, but failed completely with more complicated systems, including helium. Consider all the facts about electrons that the simple model ignores. If electrons really traveled in tiny orbits, the calculated speeds would be large enough to require relativistic corrections. By 1920, a number of people had calculated the relativistic results, some of which agreed better with experiments. A more serious objection to the Bohr model was its assumption that electrons actually travel in orbits. As we have already seen, there is a relationship between the uncertainty of the momentum of a particle and the uncertainty of its position. To assume a certain orbit for a particle implies that both position and momentum can be measured precisely enough to determine the orbit.

> ### Question 23-10
>
> One form of Heisenberg's uncertainty principle relates uncertainty of *angular* momentum and uncertainty of *angular* position: $\Delta L \cdot \Delta \theta \geq \hbar$. According to the Bohr model, what is the angular momentum of an electron in the ground state of hydrogen? (n = 1) What is the uncertainty in that angular momentum? (Can the uncertainty be $\frac{1}{2}$ or $\frac{1}{10}$ or $\frac{1}{100}$ of the angular momentum?) Therefore, what is the uncertainty in the angular position?

As we saw in the last section, any particle has a wavelength associated with it, and wavelength is related to particle momentum:

$$\lambda = \frac{h}{p}$$

Let's see how long this wavelength is for an electron in the ground state ($n = 1$) of the Bohr model for hydrogen.

$$\text{Since } mvr = n\hbar \qquad mv = p = \frac{\hbar}{r}$$

Therefore,

$$\lambda = \frac{h}{\hbar/r} = 2\pi r$$

The wavelength is as large as the circumference of the orbit! Evidently the model has been carried too far. The electron cannot be considered to be a point particle orbiting the nucleus.

New laws for the mechanics of the microworld were discovered in the 1920s. The subject is called *quantum mechanics,* or wave mechanics. Systems are described in terms of equations that have wavelike solutions. The solutions can be combined to give the probability of finding a particle in a particular location or the probability of a particle changing from one energy state to another. The quantum mechanics has been completely successful in predicting the results of experiments with atoms or subatomic particles—*if the calculations can be done.* Even the new quantum mechanics faces the old problems of complexity when dealing with more than two particles.

With atoms containing many electrons, approximations have to be made. For example, lithium has three electrons. Two of them can reside in the smallest region with $n = 1$. The Pauli exclusion principle is satisfied because the electrons can have opposite spin. The third electron cannot be accommodated with $n = 1$ or it would be identical with one of the electrons already there, which is forbidden. It has $n = 2$ and acts almost as if the inner electrons formed a closed shell around the nucleus. Since this shell partially masks the positive nuclear charge of 3, the outer electron sees only an effective positive charge of 1. Since it is farther away from the center of that charge than is the electron in hydrogen, it is more loosely held. The ionization energy is only 5.4 eV. In its lowest energy state, the outer electron in lithium is not completely shielded from the nucleus by the two inner electrons. To a first approximation, however, the spectrum from lithium is similar to that from a hydrogen atom.

The analysis of more complicated atoms requires better approximations and lengthy computer calculations. The energy levels and the corresponding spectra of most atoms are very complex. For example, one of the best known spectral lines, and the easiest to produce without fancy apparatus, is the sodium D line. Sprinkle some salt in a gas flame and you can see the bright yellow color produced by the outer electron, jumping from an excited state to the ground state of $n = 3$. When this light is analyzed in a spectrometer, it turns out that there are two lines close together, one with $\lambda = 5.896 \times 10^{-7}$ m and the other with $\lambda = 5.890 \times 10^{-7}$ m. Many of the spectral lines turn out to be doublets, or even higher multiplets.

Not only do the atomic energy levels determine the spectra, but they also control the chemistry of the elements. The periodic table was originally arranged by Dmitri Mendeleev (1834–1907) on the basis of chemical activity and similarities of the elements. The quantum theory now gives a complete explanation of chemistry (at least in principle—the detailed calculations are frequently too complicated). The alkali metals, for instance, such as lithium, sodium, and potassium, have one extra electron outside a filled shell of inner electrons. The spectra are hydrogenlike, and the chemical valence is $+1$. The halogens, such as fluorine, chlorine, bromine, and iodine, lack only one electron to fill a shell. They have a chemical valence of -1. The heavier the alkali atom, the further its valence electron from the nucleus and, hence, the more loosely it is held. Consequently, the ionization energy of potassium is less than that of lithium, and potassium is more active chemically. With the halogens it is just the opposite. Fluorine, which is the lightest, lacks an electron in the second shell, with $n = 2$. The attraction of the nucleus is strong, and fluorine is a strong oxidizing agent (it grabs electrons). With iodine, there is the same net positive charge of the central region, attracting an electron to fill a shell, but in this case $n = 5$, and there is a greater distance to the nucleus. The attraction is weaker, and iodine is a weaker oxidizing agent.

MOLECULES

The simplest molecule is hydrogen, consisting of just two protons and two electrons. It is misleading to think of this molecule as a simple combination of two hydrogen atoms. The energy levels and regions of electron concentration are not related to those of the individual atoms. To calculate the structural details, we start out with twin positive centers and then ask how two electrons can establish themselves in the region. In the lowest energy level, there is a high probability of finding the two electrons (with spins opposed) between the two protons, thus producing binding. When electrons are interchanged like this and neither atom maintains its original outer structure, the binding is called *covalent*. At the other extreme, such as with sodium chloride, the valence electron of sodium transfers quite completely to the outer electron shell of chlorine. The resulting system consists of two spherical ions, one positive and one negative, attracting each other. This bonding is called *ionic*. Most molecules are held together by bonds inbetween pure covalent and pure ionic. In any case, those are merely names for first approximations of the actual energy states and electron configurations.

> Question 23-11
>
> With a hydrogen molecule, there are four particles interacting, all influencing each other's motions. With sodium (atomic number 11) and chlorine (atomic number 17), there are 30 particles interacting. How is it possible to calculate the motions of 30 particles all at once?

Even when there can be no exchange or sharing of electrons between molecules, there may still be forces between them. The van der Waals' force between molecules, which requires a correction in the gas laws and explains physical adsorption, is caused by fluctuations in the electric charge centers of atoms or molecules. Although atoms and molecules are electrically neutral, they are composed of many moving electrically charged particles. There are momentary displacements of the symmetry of the components, producing fluctuating electric dipoles. These, in turn, induce similar displacements in adjacent atoms or molecules. The resulting forces are weak compared to chemical bonding and depend on the inverse sixth power of the distance between molecules.

The spectra of molecules are characterized by bands of colors that can be separated into individual lines only with high resolution. Besides having energy levels due to electron arrangements, molecules have energy levels due to vibration and rotation. These energies are also subject to the quantum conditions. Usually, however, the separation between vibration or rotation energy levels is small, and so there is only a small difference in the energy of photons emitted in transitions between neighboring levels. A typical molecular spectrum in the visible range consists not of lines, but of bands. Each transition between different electron energy levels has many slightly different energies possible, because of the various atomic vibration and rotation states associated with each electron state.

THE SOLID STATE

The complications that arise when a vast number of molecules are bound together are so great that much of the research in this field has been done only since World War II. Out of such studies has come a partial understanding of the art of metallurgy and such practical devices as transistors and lasers.

The most impressive feature of solids is the regularity of the atomic arrangements. Most solids are crystalline with the atoms bound in a lattice. In some cases, the same pattern repeats itself for only a few thousand atoms before running into a similar pattern that is oriented in a different direction, or perhaps the pattern is broken by an impurity atom. On the other hand, if the atoms are assembled very slowly, a single uniform crystal containing many moles of atoms may be formed. A few solid materials, such as glass, do not have crystalline structure and are most easily described as congealed liquids.

Imagine now what happens to the electron arrangements in a crystal in which the atomic centers are uniformly arrayed in all directions and are only about an atomic diameter apart. The inner electron shells will still remain attached to the parent atom and will be relatively undisturbed. The outer electrons, however, will find themselves in completely different energy levels. We saw that in the covalent bond, the electron was as likely to be with one atom as the other. When the bound atoms stretch out in all directions, some of the outer electrons may be shared with the whole crystal.

With an *isolated* atom, the energy levels and probability distributions of outer electrons are calculated in terms of a *central* attractive force, partially shielded by inner electron shells. For the outer electrons of each atom in a crystal, however, the problem is more like calculating the motions and positions of golf balls on a green containing numerous shallow holes. To make such a model a little more like the real situation, we would have to add a rim to the green so that the golf balls would be in a negative potential state compared to the outside and so would not escape. Furthermore, the entire green must be pictured as vibrating with thermal energy. This keeps the golf balls in constant motion and does not allow them to remain trapped for long in any one hole.

After what we have seen about the behavior of electrons, the golf green analogy can scarcely be satisfactory. The wavelength of the probability function describing an electron is larger than the distance between atoms. Two quantum effects dominate the real calculations. First, the Pauli exclusion principle demands that no more than two electrons within the same crystal have exactly the same energy. If we could take out of the crystal all those electrons not bound to their parent atoms and then start feeding them back in one at a time, the first one would drop to the lowest energy level possible. Occasionally the electron would be raised to higher energy positions by thermal collisions.

The second quantum effect results from the interactions of the atoms with each other. In an isolated atom, there are typically several electron volts separating energy levels. But if two atoms are joined, their electrons cannot have the same energy, according to the exclusion principle. What happens is a splitting of energy levels. Each level becomes two, separated by a small amount of energy. Add a third atom, and each level becomes three. In any crystal, the millions of atoms are all joined, and the result is that there is an enormous number of energy levels, barely separated from each other; so the electron energies seem to form a continuum.

Note that in the case diagramed, the energy levels do not form a completely continuous set. The levels are formed by the splitting of atomic levels that may be far apart. Each atomic level splits to form a more or less wide band of levels, but there are gaps between the bands. Within any band, an electron can easily gain or lose a small amount of energy (provided that it can find an unoccupied level to move to), but it must receive a substantial jolt to jump across the gap from a lower band to a higher one.

In some materials, all the energy levels in the lower band are filled right up to the top energy of several electron volts. At room temperature, the average kinetic energy of a free molecule, or an atom bound in a solid, is about $\frac{1}{25}$ eV. Most of the electrons in a solid cannot receive such small amounts of energy, because they would then be lifted to energy states that are already filled with two electrons. But this means that most of the electrons cannot give or receive any thermal energy!

This quantum model explains a number of experimental observations that could not be understood in terms of the classical theories. The older explanations had assumed that in a metal at least one electron per atom must be free to roam. That partially explained electrical conductivity. However, the free electrons should act like a gas enclosed in the crystal box. They would then have a major effect on the specific heat. Here is another example of how gross measurements of something as old-fashioned as the thermal characteristics of a substance give us a clue about its microstructure.

The specific heat of a metal ought to be $3kT$, since atoms can vibrate in three dimensions, and the energy for each dimension is kT—$\frac{1}{2}kT$ potential and $\frac{1}{2}kT$ kinetic. This agrees very well with the observed value of 6 cal/mole·°C. But what about all those electrons? Each of them ought to have $\frac{3}{2}kT$ of energy, just like the molecules of a gas, since they can bounce around freely in all directions. Every time heat is added to a metal with one conduction electron, the electron gas should absorb one-third of the energy.

The electron gas, then, ought to raise the specific heat of a metal to $\frac{9}{2}kT$, or 9 cal/mole·°C. It does not. The quantum model explains why. The electrons cannot absorb energy because most of them are in the lower energy band, which is completely filled. The only way energy could be absorbed would be by giving a lot of energy to the electrons in the upper part of the band, so that they could jump into the higher energy band; but such large amounts of energy are not available. The electrons simply do not have the freedom of gas molecules, because they are bound in quantum states within the crystal.

ELECTRICAL CONDUCTIVITY IN SOLIDS

The quantum model also explains why some materials are electrical conductors and others are insulators. In this diagram we show three different

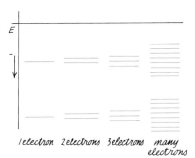

Energy of electrons trapped in a crystal. Each main energy level is subdivided into many others.

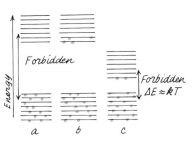

(a) insulator (b) conductor (c) semiconductor

possibilities for the arrangement of energy level bands. In part a, there are just twice as many electrons as there are energy levels, and so every level is filled.

> ### Question 23-12
> How can there be twice as many electrons as energy levels? How can a certain number of levels hold twice the number of electrons?

If an electric field is imposed on material of the type shown in part a, an electron will feel a force, but cannot move and accept energy. To do so, an electron in a lower energy state would have to move into an energy level that is already filled. The electrons at the highest levels could accept energy only if they could be lifted abruptly to energy levels in the next higher band. In an ordinary electric field, this cannot happen because the electron as it is slowly accelerated slowly gains energy. It would have to pass through all the intermediate energy states, which in this type of material do not exist.

The energy level diagram of an electrical conductor is shown in part b of the diagram. Some of the energy levels in one band are not occupied. An electron can accept energy from an imposed accelerating field. It might seem as if it would keep right on accelerating until it had reached the maximum energy available in that band. The electron, however, suffers frequent collisions that keep its speed low and, on the average, constant. Impurities in the crystal structure and the thermal vibrations of the atoms in the lattice are responsible for the collisions. They soak up some of the electron energy that is provided by the imposed electric field and turn it into thermal energy. In general, the hotter the lattice atoms, the more they absorb energy from the conduction electrons. The resistance of a conductor usually rises with increasing temperature.

The semiconductor situation is shown in part c of the diagram. All of the levels in one band are occupied, but only a small energy gap separates this band from a higher one with empty levels. Thermal energy or electromagnetic radiation can provide the small quantum of energy needed for a few electrons to make the jump into the open conduction band. The electrical resistance of such a material will depend not only on the terminal velocity of the charge carriers, which is determined by the nature of the collisions they suffer, but also on how many electrons are available to carry the current. In the case of carbon, in the form of graphite, the electrical resistance *decreases* with increasing temperature. The higher temperature sends more electrons into the upper conduction band.

If two solids are touching each other, the electrons at the junction may be able to move in ways not possible in either solid by itself. The energy levels and forbidden regions in the two materials may be different. An electron at the junction may be able to receive energy from an outside electric field if it moves into one region that contains open energy levels, but not the other region. This electron movement will then leave an open level in the region previously filled, so that another electron can move there. However, this current can only take place in the direction of the material with the open levels. Thus, we have a rectifier that can pass current in one direction, but not the other.

> Question 23-13
>
> An electric field at the boundary between two materials exerts a force on the electrons. Why is it that the electrons can sometimes respond if the field is in one direction but not the other?

The theory of the solid state is now adding to our technology at an enormous rate. The junction diode, described above, was made according to specifications worked out first on paper, based on the quantum theory of crystals. From this has come the transistor, which works on a similar principle, and a great many other devices. The theory is being used to make alloys that are better conductors, that can become superconducting at higher temperatures, that have improved tensile strength, that respond to light and heat in various ways. In this field of the complex arrangement of simple particles, there are no boundary lines between theory and technology.

SPECTRA PRODUCED BY HOT DENSE SOURCES

When isolated atoms are raised to high temperature in a gas, they yield line spectra. If atoms are combined into molecules, they produce band spectra, each band made up of many individual lines. If the atoms are closely packed into solids, or materials at high density, they give off light at essentially all wavelengths. A red hot piece of iron produces such a continuous spectrum. So does the surface of the sun, which is not a solid but a plasma of ions and electrons with a density equal to that of a solid.

Since dense material is still made up of atoms that have discrete energy levels, how can photons of all energies be produced? Isolated iron atoms yield a line spectrum. When the iron atoms are fastened together, the separate atomic energy levels must disappear. Indeed, in a solid metal the outer electrons are no longer part of the structure of individual atoms, but are part of a whole crystal. If any one atom is disturbed, the disturbance is shared with the whole crystal. Although the energy levels of the system are still quantized, there is now an enormous number of such levels, closely spaced.

The spectrum produced by a dense source is a function primarily of temperature and, to a lesser extent, of the surface color and conditions of the hot object. There is a way to eliminate completely the dependence on the surface characteristics. If a closed hollow box or tube has a small hole in its wall, the radiation coming from the hole will be independent of the material out of which the box is made. Furthermore, the radiation coming from the hole will be more intense than that coming from the wall. The hole is brighter! Because a hole in a cavity absorbs all of the light striking it, the arrangement is called a "black body." The radiation coming from a black

An approximation of a "black body." The glowing metal tube has a hole in the side. Almost all radiation entering the hole is trapped so that the hole is a total absorber.

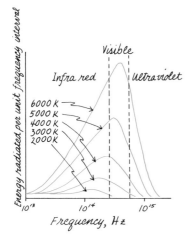

Rate of radiation of a black body as a function of frequency, for several temperatures.

body depends only on temperature. Explaining this dependence became a crucial stumbling block for classical physics and led to the beginnings of quantum mechanics.

The graph shows the intensity as a function of frequency for black body radiation at several temperatures. There are two regularities in the spectra. First, there is a simple relationship between the temperature and f_m, the frequency of maximum radiation; that is, $f_m =$ (constant) T. Second, the total energy radiated is proportional to the fourth power of the temperature. Both of these laws agree qualitatively with everyday observations. As the temperature of a light bulb filament rises, its color changes from red to white hot. Furthermore, the amount of light produced by a hot filament increases dramatically as its temperature rises—recall the bright flash as an incandescent bulb burns out.

Before 1900, classical physics could explain many electromagnetic phenomena. Maxwell's equations successfully predicted radiation through space. Those equations and the basic principles of thermodynamics ought to have been able to explain a type of radiation whose characteristics depended only on the temperature of the source and not on the material from which it was made. The model for black body radiation pictured the solid walls of a hollow box as containing an enormous number of electron oscillators. These could oscillate at any frequency. Because of the equipartition of energy, each mode or frequency of the oscillators should share equally in the thermal energy available. Each mode should have energy equal to $3kT$, since each oscillator has $\frac{3}{2}kT$ for average potential energy and $\frac{3}{2}kT$ for average kinetic energy. However, not all modes or frequencies are possible. The electron oscillators radiate their energy into the hollow region of the box, but they also absorb that radiation. At any given temperature, there is an equilibrium between the electromagnetic radiation in the enclosed space and the oscillations of the electrons in the walls. However, the radiation can exist only at frequencies corresponding to standing waves in the cavity. The situation is analogous to that of the resonance conditions for sound waves in an organ pipe. The actual allowed frequencies depend on the dimensions of the cavity. At low frequencies (long wavelengths), there is the fundamental, and then the next harmonic, and so on, each frequency separated by a considerable interval from the next. At higher frequencies, more and more waves are resonant. The higher the frequency, the more modes exist per frequency interval. If each mode has an energy $3kT$, we would expect black body radiation to be concentrated in the ultraviolet—or even X-rays! Indeed, since the number of modes (overtones) goes to infinity as the frequency goes to infinity, the energy required would be infinite if each mode had to be given $3kT$. Appropriately enough, this paradox of classical physics was called "the ultraviolet catastrophe."

In 1900, Max Planck (1858–1947) resolved the paradox by proposing that the energy exchange between oscillators and electromagnetic field was quantized. The oscillators could exchange energy only in amounts proportional to their frequency:

$$\Delta E = hf$$

The proportionality constant is our old friend, Planck's constant, which has the very small value of 6.6×10^{-34} J·s.

SPECTRA PRODUCED BY HOT DENSE SOURCES

Question 23-14

According to Planck's proposal, is radiant energy quantized like angular momentum or electric charge? Must radiant energy always exist in multiples of some basic unit of energy?

Planck's quantum requirement upsets the equipartition of energy in black body radiation. The equipartition law assumes that energy is transferred from one mode to another by processes that can exchange infinitesimal amounts of energy during each interaction. However, according to Planck's condition, a high-frequency oscillator can change energy only by a large amount, an unlikely event. Hence, the high-frequency modes are seldom excited. The detailed theory yielded a distribution of radiant energy versus frequency that matched exactly the curves shown in the graph. Note the way the distribution falls off at low frequencies—there are not many resonant modes. Note the way the distribution falls off at high frequencies—many modes are possible, but few are populated because of the large jump in energy needed to excite a high-frequency mode.

Planck was uneasy about the implications of his theory. For a long time he preferred to think that the quantization was an artificial peculiarity of the oscillators. Electromagnetic radiation must surely consist of waves, whose energy could assume any value. Albert Einstein (1879–1955), however, made a complete break with the classical theory of radiation. In one of his three great papers of 1905, he explained the photoelectric effect in terms of quanta of radiation. He assumed that electromagnetic radiation acts like chunks of energy when it interacts with matter. Each "photon" carries energy hf and, when it interacts with matter, it does so at a point—not spread out in a wave front. Consequently, a photon's energy can be delivered to a single electron bound in an atom. If the energy is sufficient, the atom will be ionized and the electron freed.

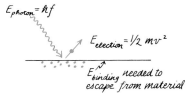

Photoelectric effect

Question 23-15

Blue light has a frequency of 7.5×10^{14} Hz. [$f = c/\lambda = (3 \times 10^8$ m/s$)/(4 \times 10^{-7}$ m).] How much energy can such a photon deliver to an atom? Is the energy sufficient to ionize an atom? (1 eV = 1.6×10^{-19} J.)

Einstein's equation for the photoelectric effect is

$$(\tfrac{1}{2}mv^2)_{\text{emitted electron}} = hf - E_{\text{escape}}$$

In most photoelectric effects, the emitted electron does not escape from the material, and so its kinetic energy cannot be measured. After 1905, several physicists demonstrated the qualitative agreement between Einstein's simple law and experiments. However, it was not until 1916 that Robert Millikan (1868–1953) performed the delicate measurements that confirmed Einstein's photoelectric law and thus confirmed the reality of photons. The measure-

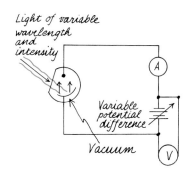

A schematic diagram of the apparatus for investigating the photoelectric effect.

ments had to be done in high vacuum, with emitting surfaces freshly exposed in the vacuum. The kinetic energy of the electrons was determined by measuring the retarding voltage that they could overcome. The binding energy of a metal corresponds to a threshold frequency of the light. As the frequency of the light increased above the threshold, the kinetic energy of the electrons increased.

THE DUALITY PROBLEM

Is light a wave or a particle? All of the evidence of diffraction, interference, and polarization calls for a wave model of light. Furthermore, we know that light is simply a short-wavelength version of electromagnetic radiation, such as radio waves, where the wave nature is observed both in production and detection. On the other hand, the *formation* of light involves features of quantized energy values. The spectra of isolated atoms contain only certain frequencies. The complicated radiation from hot solids is explained if, and only if, it is assumed that the micro-oscillators can exist only in quantized energy states and thus emit only quantized packages of radiant energy. When light interacts with matter, it acts as if it were a particle delivering packages of energy to specific points.

Notice, however, that the two models of light are not completely independent. The package of light, the photon, has an energy proportional to its *frequency!*

$$E_{\text{photon}} = hf$$

But the frequency is a wave concept. How can a particle have a frequency?

When the implications of this situation were realized in the early decades of this century, various philosophical escapes were sought. For instance, it was suggested that the advancing wave fronts might contain photons, much as a breaking water wave could carry a group of surf boards. No quantitative predictions of such a model were successful.

Experiments were performed in attempts to force light to reveal its true nature as either a particle or a wave. The most revealing type of experiment involves the creation of an interference pattern of light at intensities so low that only one photon at a time is likely to exist in the interference region. Such experiments were tried in the 1920s and again, with greatly improved accuracy, in the early 1950s.

The diagram illustrates one such experiment, which was performed in Budapest in 1957. The apparatus is similar to the Michelson interferometer, which was used in the 1880s to calibrate wavelengths in terms of the standard meter and to demonstrate that the speed of light is independent of the motion of the source or receiver. Its operation for that purpose was described in Chapter 16. Light of a single wavelength comes from the left through filters that control the intensity. The half-silvered surface of mirror A transmits 50% of the light and reflects the other 50%. Each beam takes the path shown, and they recombine at the screen. The geometry of the interference pattern depends on the original slit geometry. In this case, alternate bands of reinforcement and cancellation are formed on the screen. In effect, looking

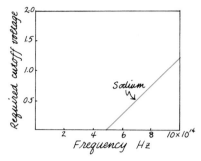

Maximum electron energy in the photoelectric effect as a function of the frequency of the incident light.

THE DUALITY PROBLEM

back from the screen, one sees the single source as two sources at slightly different distances, assuming that the two path lengths are almost but not quite the same.

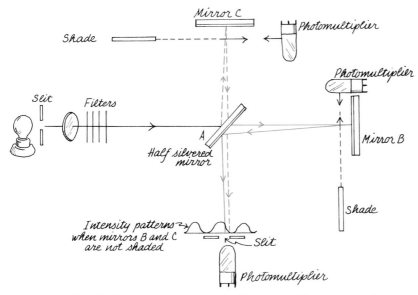

Interference pattern built up by one photon at a time in the Michelson interferometer.

If the slit opening of a sensitive photomultiplier tube is positioned at a dark band of the interference pattern, no signal will be produced except for the intrinsic electrical noise of the tube. (A photomultiplier tube contains a light-sensitive surface that emits electrons by the photoelectric effect. These electrons are then accelerated from electrode to electrode within the tube, generating more electrons at each stage. Millionfold amplification can be obtained.) The light intensity is now reduced by the filters to the level where only a million photons per second are being transmitted. [This number could be determined by measuring the original energy of the beam, dividing by the energy of a single photon (hf), and then multiplying by the filter reduction factor which can be separately calibrated.] To check the reduction factor, other photomultipliers can be used at positions B and C, and the number of photons actually counted. The geometry is such that the travel time for a photon through the whole system is about

$$t_{\text{transit}} = \frac{1 \text{ m}}{3 \times 10^8 \text{ m/s}} = 3 \times 10^{-9} \text{ s}$$

Since about 10^6 photons per second are coming through at random, the total time *per second* that there will be a photon in the apparatus is

$$t_{\text{occupied}} = 10^6 \times 3 \times 10^{-9} = 3 \times 10^{-3} \text{ s}$$

> **Question 23-16**
>
> What is the probability that two photons will be in the apparatus at the same time?

Perhaps, though, a photon can divide at mirror A, half of it going one way and half the other. Eventually both halves might arrive at the photomultiplier and interfere with each other. This possibility can be checked experimentally. When the mirrors B and C are replaced by the photomultipliers, we would expect coincidence of pulses if the photons divide. Modern electronic circuits linked with photomultipliers can detect whether or not pulses are coincident to within 10^{-9} s. These pulses from B and C are usually not in coincidence. Each photomultiplier records half a million counts per second, but only a very few arrive in coincidence, and that number agrees with the probability prediction for pulses arriving at random.

The observation is that, although there is seldom more than one photon at a time in the apparatus, the photomultiplier scanning the destructive interference region at the screen still sees nothing. *However*, if mirror B or mirror C is blocked by a shade, the photomultiplier immediately starts registering counts. With one mirror covered, a photographic plate placed at the screen would not record a pattern, but only the continuous exposure expected from illumination by the central maximum of a single slit pattern. As soon as both mirrors are uncovered, the image of an interference pattern starts building up on the photographic plate, and the photomultiplier records no counts from a cancellation band, *in spite of the fact that only one photon at a time can be in the apparatus.*

Such a stark paradox probably means that we have not yet seen the light and must somehow change our whole viewpoint. Although we usually think that the solutions to Maxwell's equations describe traveling waves of electric and magnetic fields, we actually use the solutions to predict the probability of a meter response or photographic grain development at a particular location. Photons do not travel in or on the waves; the mathematical functions (which are those of waves) yield the probability of a photon interaction.

THE ATOMIC NUCLEUS

Each element is characterized by a certain number of electrons per atom. There is an equal number of protons in the nucleus, leaving the atom electrically neutral. There are also neutrons in the nucleus, usually about as many as there are protons. For a given element, however, the number of neutrons can vary, giving rise to isotopes. Isotopes of the same element have almost identical chemical characteristics, but slightly different masses.

Hydrogen has three isotopes, two of which have special names. *Deuterium*, sometimes called heavy hydrogen, has a nucleus consisting of one proton and one neutron. It is stable but exists naturally only to the extent of one part per 7000. The nucleus of *tritium* contains two neutrons and one proton. It is radioactive, with a half-life of 12.5 years, and exists naturally only with an abundance of 10^{-18} of that of ordinary hydrogen. Because of the large mass ratios of these hydrogen isotopes, their chemical behavior differs sufficiently so that they are easily separated from each other.

Probably the most famous isotope is uranium-235. Its nucleus contains 92 protons (the atomic number of uranium) and 143 neutrons. Its abundance is only $\frac{1}{144}$ that of uranium-238. Elaborate production facilities have been created to separate ^{235}U because it is a fuel for atomic reactors.

Most elements exist naturally as a mixture of several isotopes, usually with one or two predominating. The diagram shows all of the stable isotopes, and many of the unstable ones, up to $Z = 82$ (lead). Note that for the light elements, nuclear stability is achieved when the number of neutrons is about the same as the number of protons (along the 45° line of the graph). For nuclei with larger atomic number, however, extra neutrons are needed for stability. This effect is caused by several factors. Evidently there must be a nuclear binding force between proton and proton, neutron and neutron, and proton and neutron. This force is a "contact" force, which goes essentially to zero if the protons and neutrons are not touching each other. However, the coulomb repulsion between protons is relatively long range. Although only adjacent nuclear particles attract each other, all of the protons repel each other. Nuclear stability is achieved by interlacing more and more neutrons. For nuclei above lead ($Z = 82$), there are so many protons that complete stability is impossible.

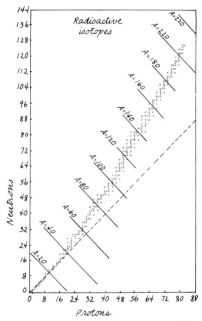

Number of neutrons and protons in stable nuclei.

The nuclear binding is tightest when proton-neutron pairs can be formed. If a proton and neutron combine to form a deuterium nucleus, their mass-energy is less than that of the original particles.

$$m_{\text{proton}} = 1.6724 \times 10^{-27} \text{ kg}$$

$$m_{\text{neutron}} = 1.6748 \times 10^{-27} \text{ kg}$$

$$\text{Sum of separate masses} = 3.3472 \times 10^{-27} \text{ kg}$$

$$m_{\text{deuteron}} = 3.3431 \times 10^{-27} \text{ kg}$$

$$\text{Mass difference} = 0.0041 \times 10^{-27} \text{ kg}$$

This lost mass is given off in the form of a high-energy photon (gamma ray) when the proton and neutron combine. The energy of the gamma ray, which is the negative binding energy of the deuteron, is

$$E = mc^2 = (4.1 \times 10^{-30} \text{ kg})(3.0 \times 10^8 \text{ m/s})^2$$

$$= 3.69 \times 10^{-13} \text{ J} = 2.3 \times 10^6 \text{ eV} = 2.3 \text{ MeV}$$

Question 23-17

Can we find the binding energy of a water molecule this way?

In this diagram we show the binding energy per nucleon (neutron or proton) as a function of the number of nucleons. The mass deficit for each

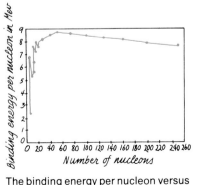

The binding energy per nucleon versus the number of nucleons.

element is calculated using the same method that we just used for deuterium. The deficit is then divided by the total number of nucleons in the nucleus of that element. Note that the binding energy per nucleon generally increases as the atomic number increases, although there are some special effects for certain combinations.

The general shape of the binding energy curve can be explained by assuming that the nuclear force is due to actual contact between nucleons. Greater binding can therefore be achieved by completely surrounding a nucleon with other nucleons. Surface nucleons are not so tightly bound as are interior ones. Increasing the total number of nucleons increases the volume-to-surface ratio. However, for the elements containing more than about 50 nucleons, the increasing coulomb repulsion reduces the total binding energy.

FUSION

If two protons and two neutrons could be combined into a helium nucleus, the mass defect would yield 28.4 MeV. One way to accomplish this might be to shoot one deuteron at another. To be sure, the mass defect of each deuteron is 2.3 MeV, but we would still end up with 23.8 MeV. Unfortunately for our human energy needs, it's very hard to get the deuterons to combine. Both are positively charged and so would have to collide with sufficient energy to overcome the repulsive electrostatic barrier.

Question 23-18

Why not just shoot a beam of deuterons into a block of deuterium ice?

Most of the attempts now being made to yield fusion energy depend on bringing deuterium or tritium together at high densities and at temperatures in the range of 100 million degrees. These are the conditions existing in the centers of stars where fusion of hydrogen into helium provides stellar energy. In the so-called hydrogen bomb, the fusing elements are compacted by the preliminary explosion of a uranium fission bomb.

FISSION

At the high end of the atomic numbers, energy can be gained by splitting up large nuclei into smaller ones. Of course, it would still take energy to break uranium into 235 protons and neutrons, but energy is yielded when ^{235}U breaks down into two smaller nuclei, such as barium and krypton. The binding energy per nucleon of uranium is about 7.6 MeV. For medium mass nuclei, the binding energy per nucleon is about 8.6 MeV. This difference of about 1 MeV per nucleon produces an average of 200 MeV for every uranium nucleus that is split into two smaller nuclei.

> **Question 23-19**
>
> On a human scale, 200 MeV is not much energy. How much energy is released if a mole of uranium undergoes fission? A mole of ^{235}U has a mass of 235 g, weighing about half a pound.

The heavy nuclei do not spontaneously split up, even though all that energy can be released by doing so. It takes a trigger to start the process. With ^{235}U, the capture of a low-energy neutron distorts the nucleus and throws it into oscillations that soon cause it to split in half (or sometimes into three pieces). In the process, two or three neutrons are emitted. If there is more ^{235}U surrounding the original fission site, the new neutrons can trigger more fission. The reaction is thus self-sustaining. With the right geometry of uranium and neutron absorbers, the process can be controlled to produce steady energy output in a reactor. With a different geometry, the process proceeds exponentially, producing a nuclear explosion.

The most common form of uranium, ^{238}U, can undergo fission, but only when bombarded by a high-energy nucleon. Consequently, it cannot produce a self-sustaining reaction.

RADIOACTIVITY

Nuclei with an imbalance of protons or neutrons can, under certain conditions, change into other nuclei. There are three main routes.

1. Some very heavy nuclei can emit a whole helium nucleus—a tightly bound combination of two protons and two neutrons. Historically this was called an alpha particle. Because of its double-positive charge, an alpha particle ionizes heavily, expending its kinetic energy in a relatively short path. Most alphas are emitted with energies of several MeV and can travel only a few centimeters in air.

When a nucleus emits an alpha, its atomic number drops by 2 and its nucleon number (or atomic mass) drops by 4. For instance, the decay of uranium-238 to thorium-234 is written:

$$^{238}_{92}\text{U} \rightarrow {}^{234}_{90}\text{Th} + {}^{4}_{2}\alpha$$

The half-lives for most alpha decays are in the range of billions of years. If the parent nucleus is unstable, why should it wait so long to decay? First, there is experimental evidence that large nuclei are to some extent composed of alpha units of four nucleons. The alpha is bound, however, just as if it were behind a high wall (an electrostatic one in this case). We might wonder if occasionally it gets enough energy to sail over the wall, even as a gas atom occasionally gets enough energy to escape from earth. However, all of the emitted alphas from a particular element have exactly the same kinetic energy, *and* the energy is considerably less than the height of the coulomb barrier required to get back into the nucleus. Apparently the alphas do not go over the wall. Instead, they go through it! According to the Heisenberg

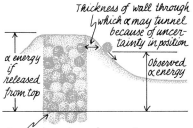

The quantum mechanical explanation of alpha radioactivity.

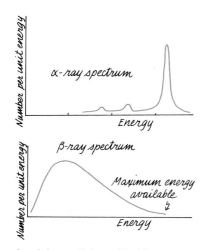

An alpha particle emitted from a particular nucleus must have one of several discrete energies. An electron emitted from a nucleus can have any energy up to a particular maximum.

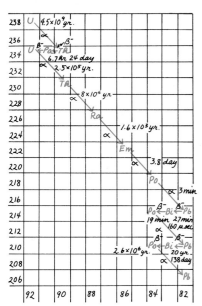

The radioactive decay sequence starting with ^{238}U.

uncertainty principle, there is a small probability that the position of the alpha will at some time be outside the nucleus. Once outside, it keeps going. The detailed theory explains the experimental facts perfectly.

2. Another way for a nucleus to change its ratio of neutrons to protons is to emit an electron, either negative or positive. The process is called *beta decay*. The prototype for beta decay is the decay of the neutron itself. The neutron, in free space, is unstable, decaying to a proton and electron with a half-life of about 11 minutes.

$$_0^1 n \rightarrow {}_1^1 p + {}_{-1}^0 e \; (+ \text{ anti-neutrino})$$

One of the best known beta decays is from the naturally occurring isotope, carbon-14. The decay has a half-life of only 5580 years, so any of it formed during the origin of the universe must have long since disappeared. It is created continuously, however, by the bombardment of nitrogen with neutrons, which are generated by cosmic rays in the upper atmosphere:

$$_7^{14} N + {}_0^1 n \rightarrow {}_6^{14} C + {}_1^1 H$$

The carbon-14 eventually decays according to the following scheme:

$$_6^{14} C \rightarrow {}_7^{14} N + {}_{-1}^0 e \; (+ \text{ anti-neutrino})$$

In some artificially radioactive elements, there are too many protons. In these cases, positive electrons (positrons) may be emitted. For instance,

$$_7^{13} N \rightarrow {}_6^{13} C + {}_{+1}^0 e \; (+ \text{ neutrino})$$

Although a definite amount of mass-energy difference exists between the mother and daughter nuclei of a beta decay, the electrons are emitted with varying energies up to the maximum. It turns out that in every beta decay, two particles are emitted. Along with the electron there is a neutrino—a particle with zero (or very small) mass, having no electric charge but with a spin of $\frac{1}{2}\hbar$, like the electron. The electron (beta particle) and the neutrino share the available energy.

3. When a nucleus undergoes a change, such as the emission of an alpha or a beta, it is frequently left in an excited, energetic state. When atoms are similarly excited, they emit light as they settle into their ground state. Nuclei also emit photons, although the energies are apt to be in the MeV range. These are called gamma rays.

RADIOACTIVE SEQUENCES

The radioactive heavy nuclei decay in series of reactions in which the alpha and beta decays approximately alternate. Each alpha decay reduces the atomic number by 2, and each negative beta decay increases it by 1. The series continues until a stable isotope of lead is reached. There are three main series. One starts with ^{238}U and ends at lead-206 (^{206}Pb). The sequence is shown in the diagram. Another series starts with ^{235}U and ends with ^{207}Pb. The third starts with thorium-232 (^{232}Th) and ends at ^{208}Pb. There is a similar series starting with neptunium-237 and ending with bismuth-209, but the half-life for neptunium-237 is so short that these nuclei no longer exist in nature.

OTHER PARTICLES

Let's take a census of the subatomic particles that we have met so far. They and others are organized in order of their masses in the chart on the next page. Down near the bottom of the chart is the familiar electron. One particle has no mass at all; photons travel at the speed of light and therefore cannot have rest mass. Neutrinos have rest mass less than 100 eV, and perhaps their mass is zero. Near the top of the chart are the nucleons—proton and neutron.

Note that even with these familiar particles there are novel complications. Each has an anti-particle, a particle with identical mass and spin but with opposite electric charge. Particles and anti-particles can be created in pairs, as long as energy and momentum are conserved. For instance, a gamma ray with an energy of 1.02 MeV or greater can create an electron pair. The rest mass of electron and positron is 0.51 MeV each. It might seem that neutrinos have so few non-zero attributes that there is nothing to be "anti" about. Nevertheless, the neutrino accompanying negative beta decay is experimentally different from its anti-particle that accompanies positive beta decay.

Actually, there are at least three electron-like particles with different masses. The muon was first detected in cosmic rays in the 1930s, although its nature was not understood until 10 years later. In every respect it is a heavy electron, having its own anti-muon and matching neutrinos. In 1978, a yet-heavier electron was found, dubbed the tau. The electrons and neutrinos seem to form a separate family of particles, called *leptons*, having no internal structure—or at least, none detected so far. Each member of the family has spin $\frac{1}{2}\hbar$. There is a conservation law within the family. No member can be created or destroyed without the cooperation of an appropriate anti-particle, such as one of the neutrinos.

The nucleons are the base of a large family of heavier particles called the *hyperons*—lambda, sigma, xi, omega, and so on. All are unstable, except the proton, and decay into the proton with half-lives in the 10^{-10} s range. (The neutron is an exception, which we will explain later. Its half-life is 11 minutes.) These particles have half-integral spins—$\frac{1}{2}\hbar$, or $\frac{3}{2}\hbar$, and so forth. They show internal structure and are currently pictured as composed of three members, each of a more fundamental group of particles called *quarks*. The theory based on quark arrangements has been highly successful in explaining the complicated spectrum of the many hyperons—their mass, charge, and spin relationships. Unfortunately, no quark has been detected in isolation (with the possible exception of one type of experiment). Always before, in explaining the nature of matter, experimenters soon produced and examined separately the proposed microstructure. In this case, although various theories attempt to explain the impossibility of extracting separate quarks, the situation is still delightfully confused.

There is another family of particles called *mesons*. These interact strongly with the hyperons and, to a first approximation, act as the nuclear glue holding nucleons together. They have integral spins of 0, 1, and so on, and are not subject to a conservation law of particles. As long as the appropriate energy and charge is available, mesons can be created or

Particles Stable Against Strong Nuclear Decay

class	name	particles	anti-particles	rest mass in Mev	half life in seconds	decay schemes	
BARYONS strongly interacting fermions (spin = half-integral)	Omega Hyperon	Ω^-	$\bar{\Omega}^+$	1676	$\sim 10^{-10}$	$\Omega^- \to \Lambda^0 + K^-$	
	Cascade Hyperon (Xi)	Ξ^0 Ξ^-	$\bar{\Xi}^+$ $\bar{\Xi}^0$	$\Xi^\pm \approx 1320$ $\Xi^0 \approx 1310$	$.9 \times 10^{-10}$ 1.0×10^{-10}	$\Xi^- \to \Lambda^0 + \pi^-$ $\Xi^0 \to \Lambda^0 + \pi^0$	
	Sigma Hyperon	Σ^+ Σ^0 Σ^-	$\bar{\Sigma}^-$ $\bar{\Sigma}^0$ $\bar{\Sigma}^+$	≈ 1190	$.6 \times 10^{-10}$ 1.2×10^{-10}	$\Sigma^+ \to p + \pi^0$ $\to n^0 + \pi^+$ $\Sigma^0 \to \Lambda^0 + \gamma$ $\Sigma^- \to n^0 + \pi^-$	(50%) (50%)
	Lambda Hyperon	Λ^0	$\bar{\Lambda}^0$	1115	1.7×10^{-10}	$\Lambda^0 \to p + \pi^-$ $\to n^0 + \pi^0$	(67%) (33%)
	Nucleon (proton-neutron)	p^+ n^0	$\bar{n}^0$ $\bar{p}^-$	939.5 938.2	$.7 \times 10^3$ Stable	$n \to p + e^- + \bar{\nu}$	
MESONS strongly interacting bosons (spin = 0)	η-Meson		η^0	548	$< 10^{-16}$	$\eta^0 \to \pi^+ + \pi^- + \pi^0$ $\to \pi^0 + e^+ + \nu$ $\to \pi^0 + \mu^+ + \nu$ $\to \mu^+ + \nu$ $\to \pi^+ + \pi^0$ $\to 2\pi^+ + 2\pi^-$	(5%) (5%) (64%) (19%) (6%) (2%)
	K-Meson	K^+ K^0 K^-	$\bar{K}^-$ $\bar{K}^0$ ($K^-=K^+$)	494 498	$.8 \times 10^{-8}$	$K^0_1 \to \pi^+ + \pi^-$ $\to 2\pi^0$ $K^0_1 \to \pi^+ + \pi^- + \pi^0$ $\to 3\pi^0$ $K^0_2 \to \pi^\pm + \mu^\mp + \nu$ $\to \pi^\pm + e^\mp + \nu$	($\approx 34\%$) ($\approx 16\%$) (7%) (19%) (24%)
		$\begin{Bmatrix} K^0 \\ \bar{K}^0 \end{Bmatrix}$			$.7 \times 10^{-10}$ 4×10^{-8}		
	π-Meson	π^+ π^0	π^-	140 135 140	1.8×10^{-8} $.7 \times 10^{-16}$ 1.8×10^{-8}	$\pi^- \to \mu^- + \bar{\nu}$ $\pi^0 \to \gamma + \gamma$ $(\pi^0 \to \gamma + e^+ + e^-$ 1%) $\pi^+ \to \mu^+ + \nu$ $(\pi^0 \to e^+ + \nu$ .01%)	
LEPTONS weakly interacting fermions (spin = ½ $\hbar$)	Tau	τ^-	τ^+	1807	$\sim 10^{-13}$	$\tau^\pm \to \mu^\pm + \nu + \bar{\nu}$ or $e^\pm + \nu + \bar{\nu}$	
	Muon	μ^-	μ^+	105.7	1.5×10^{-6}	$\mu^\pm \to e^\pm + \nu + \bar{\nu}$	
	Electron	e^-	e^+ (Positron)	.51	Stable		
	Neutrino-Tau	ν_τ	$\bar{\nu}_\tau$	0			
	Neutrino-Muon	ν_μ	$\bar{\nu}_\mu$	0			
	Neutrino-Electron	ν_e	$\bar{\nu}_e$	0			
MASSLESS BOSONS	Photon (spin = 1 $\hbar$)		γ	0	Stable		
	Graviton? (spin = 2 $\hbar$)			0	Stable	Not detected	

destroyed one at a time. In the quark theory, mesons are composed of a quark and an antiquark.

Electrons and neutrinos take no part in the strong nuclear interactions, but are subject to the Weak interaction as well as gravitation and electromagnetism. There are special conservation laws that must be obeyed by some of the four basic interactions, but not by others—although all must obey conservation of energy, momentum, and angular momentum. When a particle decays, the sequence must follow a path obeying all the conservation laws that apply to that interaction.

Consider, for example, the decay of a free neutron. Each of the three main interactions may be considered to be a possible decay route, as shown in the diagram. The decay will occur through whichever route allows the greatest energy difference and the smallest number of participants. Consider first the strong nuclear interaction. Although the neutron is more massive than the proton, it cannot simply transform into a proton, because that would violate charge conservation. There is not enough mass-energy to produce a proton and a negative meson. There is enough energy to produce a proton and electron, but the process cannot take place through the strong nuclear interaction. The electron is not influenced by that interaction. If the strong interaction were possible, the decay would happen in times of the order of 10^{-23} s.

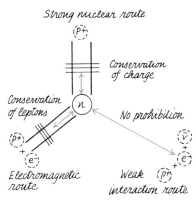

The forbidden and allowed decay routes of the neutron.

Since the strong interaction is forbidden, perhaps the neutron could decay through the electromagnetic interaction. It is weaker than the strong nuclear interaction by a factor of more than 100 and so the decay would take longer. Although the electron is influenced by electromagnetism, the neutrino is not. Because of the conservation of leptons, the neutron cannot decay simply to a proton and electron. Therefore, the electromagnetic route is blocked.

The remaining route is the Weak interaction. (Gravitation is weaker than the strong nuclear by a factor of 10^{39} and can be ignored for individual particle interactions.) The neutron, proton, electron, and neutrino are all subject to the Weak interaction, and there is sufficient energy to permit the decay. However, the energy difference between the neutron and the products is small, and the number of final particles is large (three). Furthermore, the Weak interaction is smaller than the strong nuclear interaction by a factor of 10^{13}. We should expect that it would take a long time for the participants to be arranged with the necessary combination of momenta so that the decay can take place. Indeed, the half-life is 11 minutes.

EPILOGUE

On the first page of this text we presented a logarithmic map of all the realms in the universe. One of the exciting features of our human condition is that we have learned how to explore all these realms using only a few principles and laws. There are four basic interactions; there are fewer than a dozen conservation laws. As we have seen, the motions of balls and pendulums turn out to be models for the motions of atoms and galaxies.

In this final chapter we have explored worlds within worlds. The chair you sit in, the water you drink, the air you breathe, and you yourself are all made of molecular units composed of groups of atoms. These in turn consist mostly of empty space populated by electrons and positive nuclei. The nucleons also have structure, acting like tiny molecules made up of combinations of quarks. However, we are obviously more than a simple collection of point objects. Organization is as real an entity as the particles. The ordered groupings are equivalent to information, and, in certain complex molecules, the organization has led to self-awareness and intelligence.

Although we have learned to probe the microworld and map the universe, we still do not know how or why the whole system began. We do not know why humans on this planet have developed the intelligence to learn these things. We do not know whether we are the only creatures in the universe who measure and try to comprehend it. Whether we are alone or not, and whether there is some great drama of creation, it seems peculiarly appropriate for humans to explore and understand their universe. Perhaps it is our proper role. But even if there is no drama, and even if we have no part, still the adventure is fun and noble in itself.

Answers to Questions

23-1: An energy of only 0.4 eV would not be enough to cause an electron to be emitted from tungsten, but that is only the average energy. The high-energy end of the distribution can give sufficiently energetic kicks to electrons frequently enough to make the white-hot filament a very copious electron emitter.

23-2: Here is one way to produce a beam of electrons with known velocity. If the electron is accelerated through a known potential difference, its kinetic energy becomes: $\frac{1}{2}mv^2 = eV$. Such a measurement would yield a value for v in terms of e/m: $v = \sqrt{2Ve/m}$. If the electron is then sent through a magnetic field, it travels in a circle with radius, r: $m = Ber/v = Ber/\sqrt{2Ve/m}$. Solve this equation for m in terms of B, r, V, and e.

23-3: Not only would the electrons spread out slightly vertically but, because they are charged particles moving in a magnetic field, they would be forced sideways in a direction perpendicular to their velocity and to the magnetic field.

23-4: $\lambda = c/f = (3 \times 10^8 \text{ m/s})/(3 \times 10^9 \text{ Hz}) = 1 \times 10^{-1}$ m = 10 cm. This is the wavelength used in microwave ovens.

23-5: The quantization law requires that the angular momentum of a system can change by only whole units of $\hbar$. If the intrinsic angular momentum is $1\hbar$, then the direction along an axis can change from in-line to perpendicular (a change of one unit from $+1\hbar$ to $0\hbar$ in the original direction), or from perpendicular to antiparallel with the original axis (a change of one unit from $0\hbar$ to $-1\hbar$).

If the intrinsic spin is $\frac{1}{2}\hbar$, then the angular momentum in a particular direction can change by $1\hbar$ only if the particle flips from parallel to antiparallel: $[(+\frac{1}{2}\hbar) - (-\frac{1}{2}\hbar)]$.

23-6: Not exactly. The quantum mechanics equations depend on the geometry, the nature of the particle, and the location of other particles. The solutions to these equations (which mathematically can have wavelike characteristics in space and time) are then combined or operated on to yield the *probability* of finding the particle in a particular position, or with a particular momentum. The probabilistic nature of these predictions corresponds with the actual experimental situation. Repeated measurements of the same atomic system usually yields a distribution of values, centered around a most probable value.

23-7: If a marble has a radius r_0, it has a volume: $V = \frac{4}{3}\pi r_0^3$. The volume of A marbles packed together would be $V_{tot} = KA\frac{4}{3}\pi r_0^3$, where K is the packing fraction for fitted spheres, a fraction close to 1. The radius of the whole nucleus would therefore be $r = K^{1/3} r_0 A^{1/3}$.

23-8:
$$r = \frac{1 \times (10^{-34})^2}{10^{-30}\, 10^{10}(10^{-19})^2} = 10^{-10}\,\text{m}$$

This is the order-of-magnitude size of any atom!

23-9:
$$E_{tot} = -\tfrac{1}{2}(10^{10})^2 \frac{10^{-30}(10^{-19})^4}{1(10^{-34})^2} \approx 10^{-18}\,\text{J}$$
$$\approx -10\,\text{eV}$$

23-10: The angular momentum of an electron in the first Bohr orbit is *exactly* $\hbar$. Since these orbital angular momenta are integral multiples of the basic unit, the uncertainty in angular momentum, ΔL, is zero. Therefore, $\Delta\theta = \infty$. There is complete uncertainty about the angular position, θ. This situation is incompatible with the model of an electron actually traveling from position to position in an orbit.

23-11: It is indeed difficult to calculate the interactions of more than two objects. However, complex groups can sometimes be considered, in first approximation, to consist of a main group that is motionless while a single other object moves. With the hydrogen molecule, for example, assume that the two protons are massive and stay fixed with respect to each other and the two electrons. The problem is thus reduced to calculating the behavior of two electrons moving around fixed centers. With sodium chloride, the inner electrons are largely shielded from what goes on at the atomic surface. In the first approximation, once the chlorine grabs the extra sodium electron, there is just a two-body interaction between ions.

23-12: Each band is divided into as many closely spaced energy levels as there are atoms in the crystal. A particular type of atom may provide two electrons each to roam the crystal. Each separate level can hold two electrons with their spins opposed.

23-13: The electrons in one material may have saturated the possible energy levels in that material and, therefore, not be able to gain any more energy in small amounts. However, a small amount of energy may be enough to move them into an unoccupied band in an adjacent material. If the electric field accelerates the electrons in that direction, they can gain energy.

23-14: No. Notice that the quantum of energy depends on the frequency. Infinitesimal amounts of radiation can be emitted or absorbed, but only at very low frequencies, corresponding to long wavelengths.

23-15:
$$E = hf = (6.6 \times 10^{-34}\,\text{J}\cdot\text{s})(7.5 \times 10^{14}\,\text{Hz})$$
$$= 4.95 \times 10^{-19}\,\text{J}$$
$$= 3.1\,\text{eV}$$

As we have seen, the valence electrons in the alkali metals are bound with energies in this region. Evidently, visible light photons can free electrons and cause chemical changes in material—people can be sunburned, colors bleach, photosynthesis occurs, and photography works.

23-16: Since the apparatus is occupied by photons for only 3×10^{-3} s out of each second, there is only about 0.3% chance that two photons will be there during the same time. It would seem that the chance for interference between photons, constructive or destructive, would be practically nil.

23-17: In theory, yes; in practice, no. The mass deficit is much too small to measure. Chemical binding energies are in the election-volt range; nuclear binding energies are larger by a factor of a million.

23-18: When a charged particle travels through matter, it loses energy to the atoms through which it passes. A direct hit on the nucleus of a target atom is rare, because the nucleus is so small. Most of the energy of the bombarding deuterons would be lost in ionizing the atoms in the deuterium ice.

23-19: $(6 \times 10^{23} \text{ nuclei})(200 \times 10^6 \text{ eV/nucleus})(1.6 \times 10^{-19} \text{ J/eV}) = 2 \times 10^{13}$ J

This is about the energy produced in 3 hours by the power plants at Niagara Falls.

Problems

1. How much energy can an ultraviolet photon deliver to an electron if its wavelength is 1×10^{-7} m?
2. In a field of 1×10^4 V/m, what is the diameter of a Millikan oil droplet that can just be supported if it carries one extra electron? (Assume that the density of oil is the same as that of water.)
3. If an electron is accelerated through a potential of 100,000 V and then sent into a magnetic field of 1000 G (1×10^{-1} T), what is the radius of curvature of its path?
4. A gamma ray with an energy of 2.6 MeV creates an electron pair. If the surplus energy is shared equally by electron and positron, what is the kinetic energy of each?
5. If the natural unit of angular momentum is $1\hbar$, how can an electron (or a proton) have a spin of $\tfrac{1}{2}\hbar$?
6. Show that the resonant frequency for flipping electrons in a magnetic field of 1000 G is about 3000 MHz.
7. What is the wavelength of electrons accelerated through 1000 V in an electron microscope? If this wavelength were the limiting factor (it is not), how small an object could be analyzed in such a microscope?
8. What would be the uncertainty of momentum of an electron if it were confined to a region as small as a nucleus that has a diameter of 10^{-14} m? What would be the possible energy in MeV of such an electron? (For energies much larger than the rest mass of a particle, the energy is approximately equal to pc, where $c = 3 \times 10^8$ m/s.)
9. For a very brief time, a π^0 (pi zero) meson can have as much mass-energy as a proton-anti-proton pair, since during a brief time interval the energy of any system is uncertain by an amount given by Heisenberg's principle. Thus during a brief time, energy need not be conserved. What is the longest time interval during which the π^0 can exist as a proton-anti-proton pair? The extra energy required is of the order of 2 GeV (2×10^9 eV).
10. What is the approximate radius of the nucleus and how many neutrons are there in ^{39}K (potassium), ^{56}Fe (iron), ^{209}Pb (lead)?
11. What is the velocity of the electron in the Bohr hydrogen atom for $n = 1$?
12. What is the difference in energy in electron volts between the two sodium lines: $\lambda_1 = 5.896 \times 10^{-7}$ m and $\lambda_2 = 5.890 \times 10^{-7}$ m?
13. Can red light with a wavelength of 7×10^{-7} m knock electrons out of a metal with a binding energy of 1 eV? If so, what is the maximum kinetic energy of the electron?

EPILOGUE 725

14. Explain in your own words and diagram the interference-of-light experiment when only one photon at a time is in the instrument.
15. Describe two phenomena most easily described by the wave model of light; describe two phenomena most easily described by a photon model of light. Explain why there is no controversy between the two models.
16. The binding energy for photoelectrons from a particular alkali metal is 3.0×10^{-19} J. (a) What is the shortest wavelength of light that will produce photoelectric emission? (b) What is the maximum electron kinetic energy if the wavelength of light is 4.4×10^{-7} m? (c) With the light of part (b), what retarding voltage will stop the current?
17. Explain in your own words and diagrams why the classical model of equipartition of energy, which predicted the distribution of energy between the oscillators in the wall of a black body and the enclosed radiation, led to the paradox of the "ultraviolet catastrophe."
18. List the wavelengths of the first 10 modes of standing waves that will fit into a one-dimensional pipe 1 m long. The waves must form nodes at both ends of the pipe. Note that there is no limit to the number of high-frequency modes.
19. Find the total binding energy in electron volts of $^{12}_{6}C$, which by definition has a mass of 12 atomic mass units (amu). What is the binding energy per nucleon? (Use the values of atomic constants on the inside of the back cover.)
20. The scattering of low-energy neutrons from crystals can be analyzed to reveal details of atomic structure. What are the momentum and energy of a neutron with a wavelength of 2×10^{-10} m (the atomic diameter)?
21. Assume that a proton is added to an existing nucleus that already has 100 protons. If the average spacing between the added proton and the rest of the positive charge is 1×10^{-14} m, how large is the electrostatic energy that detracts from the binding of the added proton? Express your answer is MeV.
22. How many alphas and how many betas are emitted as $^{238}_{92}U$ goes to $^{206}_{82}Pb$?
23. How many joules would be released by complete fission of 1 m^3 of $^{235}_{92}U$? The density of $^{235}_{92}U$ is 19 times that of water.
24. Would you expect that fission products are positive or negative beta emitters? Why?

APPENDIX

Conversions

Energy:
1.6×10^{-19} J = 1 eV   1 J = 6.24×10^{18} eV
4.18 J = 1 cal
3.6×10^6 J = 1 kWh = 0.86×10^6 cal
1 J = 9.48×10^{-4} Btu (British thermal unit)
1 kg = 9.0×10^{16} J/c^2
1 eV/molecule = 23 kcal/mole

Length:
1 m = 39.37 in. = 3.28 ft
1 in. = 2.54 cm
1 mi = 5280 ft = 1609 m = 1.6 km
1 light-year = 9.46×10^{15} m

Speed:
1 m/s = 3.28 ft/s = 3.6 km/h = 2.24 mi/h
88 ft/s = 60 mi/h = 96 km/h

Force:
1 N = 0.225 lb   1 lb = 4.45 N
The weight of 1 kg on the earth's surface is 9.8 N = 2.2 lb

Power:
1 kW = 1.34 horsepower   1 hp = 0.75 kW

Geometry

Surface area of sphere = $4\pi r^2$ *Volume of sphere* = $\frac{4}{3}\pi r^3$
Surface area of cylinder = $2\pi rL$ *Volume of cylinder* = $\pi r^2 L$
Circumference of circle = $2\pi r$ *Area of circle* = πr^2
Equation of straight line in two dimensions: $y = mx + b$

Greek Letters and Their Use in This Book

α	alpha	Sometimes used for angle
		Angular acceleration in rad/s^2
		Resistivity coefficient of temperature
β	beta	Sometimes used for angle
		Coefficient of volume expansion
γ	gamma	Ratio of specific heats of gases
		Surface tension
		Factor in Lorentz transformation
δ	delta (lowercase)	A small quantity
Δ	delta (capital)	The difference between two values of a quantity
		The change in a quantity
ϵ	epsilon	Permittivity

ζ	zeta	
η	eta	Viscosity
θ	theta	Usually used for angle
ι	iota	
κ	kappa	Dielectric constant
λ	lambda	Wavelength
		Linear charge density
μ	mu	Permeability
		Coefficient of friction
		Muon—a heavy electron
		Micro—prefix for 10^{-6}
ν	nu	Sometimes used for frequency
		Symbol for neutrino
ξ	xi	
o	omicron	
π	pi	Ratio of circumference to diameter of circle
		Pion—one of the mesons
ρ	rho	Radius to special point (e.g., to center of mass)
		Mass density
		Resistivity
σ	sigma (lowercase)	Area charge density
Σ	sigma (capital)	Summation
τ	tau	Torque
		Exponential time constant
υ	upsilon	
ϕ	phi (lowercase)	Usually used for angle
Φ	phi (capital)	Flux (number of lines of force)
χ	chi	Susceptibility
ψ	psi	Probability function in quantum mechanics
ω	omega (lowercase)	Angular frequency in rad/s
Ω	omega (capital)	Symbol for ohm (electrical resistance)

Other Special Letters

$\hbar = \dfrac{h}{2\pi}$ $\hbar$-bar is Planck's constant divided by 2π

$\mathcal{E}$ is used for emf, measured in volts

Approximations

$\sin\theta \approx \theta$ for small θ (θ in radians)

$\cos\theta \approx 1$ for small θ (θ in radians)

APPENDIX

$$\sin\theta = \theta - \frac{\theta^3}{3!} + \frac{\theta^5}{5!} - \cdots \quad (\theta \text{ in radians})$$

$$\cos\theta = 1 - \frac{\theta^2}{2!} + \frac{\theta^4}{4!} - \cdots \quad (\theta \text{ in radians})$$

$$e^x = 1 + x + \frac{x^2}{2!} + \frac{x^3}{3!} + \cdots$$

$$e^x \approx 1 + x \quad \text{for small } x$$

$$\ln(1+x) = x - \frac{x^2}{2} + \frac{x^3}{3} - \frac{x^4}{4} + \cdots \quad \text{for } |x| < 1$$

$$(1+x)^n = 1 + nx + \frac{n(n-1)}{2!}x^2 + \frac{n(n-1)(n-2)}{3!}x^3 + \cdots$$

This is an infinite series that converges if

$$\lim \left|\frac{a_{k+1}}{a_k}\right| < 1 \quad \text{as } k \to \infty$$

$$(a+b)^n = a^n + na^{n-1}b + \frac{n(n-1)}{2!}a^{n-2}b^2 + \cdots + b^n \quad \text{for } n \text{ a positive integer}$$

$$2^5 = 32 \quad \therefore \quad 2^{10} \approx 10^3$$

$$\sqrt{1+x} \approx 1 + \tfrac{1}{2}x \quad \text{for } x \text{ small compared with 1}$$

$$\pi \approx 3 \quad \pi^2 \approx 10$$

Trigonometry Relationships

$$\sin\theta = \frac{\text{opp}}{\text{hyp}} \quad \cos\theta = \frac{\text{adj}}{\text{hyp}} \quad \tan\theta = \frac{\text{opp}}{\text{adj}}$$

$$1 \text{ radian} = \frac{360}{2\pi} \text{ degrees} \approx 57°$$

For any θ, $\sin^2\theta + \cos^2\theta = 1$

$\sin 0° = \cos 90° = 0$

$\sin 30° = \cos 60° = \tfrac{1}{2} = 0.500$

$\sin 45° = \cos 45° = 1/\sqrt{2} = 0.707$

$\sin 60° = \cos 30° = \sqrt{3}/2 = 0.866$

$\sin 90° = \cos 0° = 1.000$

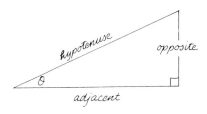

$$\sin(\theta \pm \phi) = \sin\theta\cos\phi \pm \cos\theta\sin\phi \quad \sin 2\theta = 2\sin\theta\cos\theta$$

$$\cos(\theta \pm \phi) = \cos\theta\cos\phi \mp \sin\theta\sin\phi \quad \cos 2\theta = \cos^2\theta - \sin^2\theta$$

$$\sin\tfrac{1}{2}\theta = \pm\sqrt{\frac{1-\cos\theta}{2}} \qquad \cos\tfrac{1}{2}\theta = \pm\sqrt{\frac{1+\cos\theta}{2}}$$

$\sin \theta + \sin \phi = 2 \sin \tfrac{1}{2}(\theta + \phi) \cos \tfrac{1}{2}(\theta - \phi)$

$\sin \theta - \sin \phi = 2 \cos \tfrac{1}{2}(\theta + \phi) \sin \tfrac{1}{2}(\theta - \phi)$

$\cos \theta + \cos \phi = 2 \cos \tfrac{1}{2}(\theta + \phi) \cos \tfrac{1}{2}(\theta - \phi)$

$\cos \theta - \cos \phi = -2 \sin \tfrac{1}{2}(\theta + \phi) \sin \tfrac{1}{2}(\theta - \phi)$

$\sin \theta \sin \phi = \tfrac{1}{2} \cos (\theta - \phi) - \tfrac{1}{2} \cos (\theta + \phi)$

$\cos \theta \cos \phi = \tfrac{1}{2} \cos (\theta - \phi) + \tfrac{1}{2} \cos (\theta + \phi)$

$\sin \theta \cos \phi = \tfrac{1}{2} \sin (\theta + \phi) + \tfrac{1}{2} \sin (\theta - \phi)$

Law of sines: $\dfrac{a}{\sin A} = \dfrac{b}{\sin B} = \dfrac{c}{\sin C}$

Law of cosines: $c^2 = a^2 + b^2 - 2ab \cos C$

Pythagorean theorem ($C = 90°$): $c^2 = a^2 + b^2$

Vectors

Components:

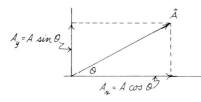

Scalar (dot) product: $\mathbf{A} \cdot \mathbf{B} = |A||B| \cos \theta$

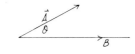

Vector (cross) product: $\mathbf{C} = \mathbf{A} \times \mathbf{B} = -\mathbf{B} \times \mathbf{A}$

$|C| = |A||B| \sin \theta$

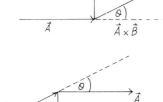

Algebra

Quadratic formula: $ax^2 + bx + c = 0$

$x = \dfrac{-b \pm \sqrt{b^2 - 4ac}}{2a}$

SI Base and Supplementary Units

	Quantity	Unit Name	Unit Symbol
SI base units	Length	meter	m
	Mass	kilogram	kg
	Time	second	s
	Electric current	ampere	A
	Thermodynamic temperature	Kelvin	K
	Amount of substance	mole	mol
	Luminous intensity	candela	cd
SI supplementary units	Plane angle	radian	rad
	Solid angle	steradian	sr

APPENDIX

SI Derived Units with Special Names

Quantity	SI Unit			
	Name	Symbol	Expression in Terms of Other Units	Expression in Terms of SI Base Units
Frequency	hertz	Hz		s^{-1}
Force	newton	N		$m \cdot kg \cdot s^{-2}$
Pressure, stress	pascal	Pa	N/m^2	$m^{-1} \cdot kg \cdot s^{-2}$
Energy, work, quantity of heat	joule	J	$N \cdot m$	$m^2 \cdot kg \cdot s^{-2}$
Power, radiant flux	watt	W	J/s	$m^2 \cdot kg \cdot s^{-3}$
Quantity of electricity, electric charge	coulomb	C	$A \cdot s$	$s \cdot A$
Electric potential, potential difference, electromotive force	volt	V	W/A	$m^2 \cdot kg \cdot s^{-3} \cdot A^{-1}$
Capacitance	farad	F	C/V	$m^{-2} \cdot kg^{-1} \cdot s^4 \cdot A^2$
Electric resistance	ohm	Ω	V/A	$m^2 \cdot kg \cdot s^{-3} \cdot A^{-2}$
Conductance	siemens	S	A/V	$m^{-2} \cdot kg^{-1} \cdot s^3 \cdot A^2$
Magnetic flux	weber	Wb	$V \cdot s$	$m^2 \cdot kg \cdot s^{-2} \cdot A^{-1}$
Magnetic flux density	tesla	T	Wb/m^2	$kg \cdot s^{-2} \cdot A^{-1}$
Inductance	henry	H	Wb/A	$m^2 \cdot kg \cdot s^{-2} \cdot A^{-2}$
Celsius temperature	degree Celsius	°C		K
Luminous flux	lumen	lm		$cd \cdot sr$
Illuminance	lux	lx	lm/m^2	$m^{-2} \cdot cd \cdot sr$
Activity (of a radionuclide)	becquerel	Bq		s^{-1}
Absorbed dose, specific energy imparted, kerma, absorbed dose index	gray	Gy	J/kg	$m^2 \cdot s^{-2}$
Dose equivalent, dose equivalent index	sievert	Sv	J/kg	$m^2 \cdot s^{-2}$

SI Prefixes

Factor	Prefix	Symbol	Factor	Prefix	Symbol
10^{18}	exa	E	10^{-1}	deci	d
10^{15}	peta	P	10^{-2}	centi	c
10^{12}	tera	T	10^{-3}	milli	m
10^{9}	giga	G	10^{-5}	micro	μ
10^{6}	mega	M	10^{-9}	nano	n
10^{3}	kilo	k	10^{-12}	pico	p
10^{2}	hecto	h	10^{-15}	femto	f
10^{1}	deka	da	10^{-18}	atto	a

INDEX

Absolute temperature, 275
Acceleration: angular (rotational), 116
  centripetal, 60
  constant, 36
  definition, 35
Achromat, 489
Action-reaction, 147
Adiabatic expansion, 292, 294, 322
Algebra formulas, 730
Alnico, 610
Alpha particle, 95, 717
Alternating current, 642
AM (amplitude modulation), 655, 673
Ameter, 566
Ampère's law, 580, 587
Ampere, unit of current, 543
Angle of incidence, 391, 462
Angle of minimum deviation, 469
Angle of reflection, 391, 462
Angle of refraction, 392, 462
Angular acceleration, α, 116
Angular frequency, ω, 116
Angular magnification, 487, 496
Angular momentum, 159
  atomic, 603, 689
  conservation, 162, 691
  direction, 167
  quantization, 169, 689
  vector nature, 160
Angular velocity, ω, 116
Antennas, 674, 675
Anti-node, 427
Anti-particles, 719
Approximations, 728
Arago's spot, 441
Archimedes' principle, 347
Atmospheric pressure, 72, 311
Atomic cesium clock, 11
Atomic masses, 305, front
    endpaper
Atomic model: Bohr, 161, 699
  first approximation, 305
Atomic nucleus, 699, 714
Atomic size, 2
Atomic structure, 607, 698, 704
Atwood's machine, 115
Avogadro's hypothesis, 276

Balmer series, 701
Band spectra, 705
Banking angle, 123
Barkhausen effect, 609
Barometer, 73, 346
Batteries, electrical, 563
Beats between waves, 411, 421
Bell, unit of intensity, 400
Bernoulli's equation, 358
β, ratio of speed to speed of light, 28
Beta decay, 718
Beta rays, 691
Betatron, 630
Bimetallic strips, 270
Binding energy, 208, 312, 715
Binoculars, 496
Biot-Savart law, 589
Black body radiation, 709
Blue sky, 670
Bohr model of atom, 161, 699
Boiling temperatures, table, 280
Boltzmann's constant, 276, 308, 331
Boyle's law, 276
Brewster's angle, 464
Brownian motion, 313
Bubble chamber tracks, 145, 688
Bugle notes, 432, 434
Bulk modulus, 75, 344, 384
Buoyant force, 348

Calorie, 284
Camera, 489
Capacitance, 528
Capacitive reactance, 645
Capacitor, charge decay, 619
Capacitors, 530
Capillary action, 355
Carbon^{14} dating, 13
Carnot cycle, 325
Cartesian diver, 349
Cathode rays, 686
Cavendish balance, 100
Celsius-Fahrenheit conversion, 272
Center of mass, 152
Center of mass coordinates, 231
Center of weight, 85

733

Centrifugal force, 215, 235
Centripetal acceleration, 60
Centripetal force, 118, 235
Charge of electron, 525, 686
Charging by induction, 509
Charles' law, 276
Chords, music, 436
Chromatic aberration, 489
Circular motion, 60, 116
Circular polarization of light, 170
Cloud chamber tracks, 141
Coefficient of friction, 103, 104
Coefficient of volume expansion, 278
Coherent waves, 420
Cohesion, 350
Collimators, 493
Collisions: elastic, 138, 182, 185
  equal mass particles, 184, 231
  inelastic, 141
Components, vector, 52
Conduction, thermal, 294
Conductivity in solids, 707
Conservation: angular momentum, 162
  energy, 188
  linear momentum, 137
  nuclear decay, 721
Constants: atomic, back endpaper
  general, front endpaper
  solar system, 7, 125, 165, back endpaper
Continuous spectra, 709
Convection, thermal, 294
Conversion units, 40, 727
Convex mirror, 480
Coriolis force, 235
coulomb, charge unit, 511
Coulomb's law, 510
Covalent, 705
Critical angle, refraction, 468
Cross product, 81, 160
Curie plot, magnetism, 606
Curved mirrors, 474
Cyclotron, 597

decibel, 400
Degree of freedom, 320
Densities, table, 305
Density of free electrons, 547
Density of gas, 310
Density measurement, 348
Depth of field, 491
Deuterium, 95, 714
Deuteron, 715
Diamagnetism, 605
Dielectric constants, 530
Diffraction, 442
  double slit, 448

  pin hole, 445
  single slit, 443
  x-rays, 452
Dimensional analysis, 41, 382
Dimensionality of variables, 40
Diode, I(v), 550
Dipole, 515
Dipole field, electric, 515
Dipole moment: electric, 515
  magnetic, 592, 600
Direct current, 542
Dispersion, 394, 410, 430, 469
Displacement, definition, 52
Distance measurements, astronomical, 4
Distortion: potential energy, 202, 211
  proportional to force, 70
Distribution of molecular energy, 309
Diverging lenses, 488
Doppler shift, 6, 410, 414
Dot product, 186, 516, 626
Drift velocity of electrons, 548
Dry cell, 565
Duality problem, 694, 712
DuLong and Petit rule, 289

Ear, 440
Earth's potential well, 206
Echoes, 377
Eddy currents, 631
Effective voltage in a-c, 643
Efficiency of heat engine, 326, 329
Einstein: Brownian motion, 313
  general theory of relativity, 258
  inertial-gravitational mass, 99
  photoelectric effect, 711
  special theory of relativity, 237
Elastic collisions, 138, 182, 185
Electrical capacitance, 528
Electric: cells, 543, 563
  charge, 533
  conductivity, 707
  current, 543
  field, see Field, electric
  flux, 658
  generators, 634
  ground, 569
  motors, 619, 633
  potential, 523, 528
  power, 544
  resistance, 551
Electrolytic cell, 543, 563
Electromagnetic frequencies, 388, 672
Electromagnetic radiation: energy, 665
  generation, 662
  momentum, 667

polarization, 669
pressure, 669
propagation, 471, 660
spectrum, 672
Electromotive force, emf (ℰ), 562, 624, 629
Electron: atomic periodicity, 692, 704
  charge, 525, 533, 686
  magnetic moment, 603, 688
  mass, 687
  microscope, 2
  mobility in wire, 548
  production, 686
  size, 693
  spin, 688
Electron volt, eV, 210
Electrostatic field, 513
Electrostatics, 504
Ellipsoidal mirror, 474
Energy density, magnetic, 639
Energy: binding, 715
  dipole, 601
  gravitational orbits, 209
  kinetic, 180
  mass equivalence, 252
  molecular formation, 318
  photon, 668, 673, 677, 679
  potential, 200
  rotational, 189, 677
Energy storage: capacitor, 531
  inductor, 638
Energy units: electron volt (eV), 210
  foot-pound, 217
  joule (J), 181
  kilowatt-hour (kWh), 537
Entropy, 330
Equal mass collisions, 184, 231
Equal tempered scale, 439
Equilibrium of forces, 78
Equilibrium, thermal, 284
Equilibrium of torques, 82
Equipartition of energy, 320, 710
Escape velocity, 208, 312
Eye, 492

Fahrenheit-Celsius conversion, 272
farad, 529
Faraday cage, 519
Faraday's homopolar generator, 634
Faraday's alw of induction, 618, 626
Fermat's principle, 472
Fermions, 693
Ferrites, 610, 669
Ferromagnetism, 606

Fiber optics, 468
Field: electric cylinder, 520
  electric dipole, 515
  electric inside sphere, 518
  electric parallel plates, 521
  electric point charge, 514
  electric sphere, 517
  electric in wire, 546
Field, electrostatic, 513
  in conductor, 519
Field, magnetic, 582
  dipole, 584, 592
  loop, 586, 591
  range of strength, 589
  solenoid, 586, 593
  straight wire, 585, 588
Field lines: electric, 516
  magnetic, 582
Filament, I(V), 549
First law of thermodynamics, 284
Fission, nuclear, 716
Fluid pressure, 344
Fluids: definition, 341
  solid, 366
Flux: electric, 658
  magnetic, 624
FM (frequency modulation), 655, 675
f-number of lens, 491
Focal point, 474, 476
Focusing forces on particle beams, 598
Foot-pound, 217
Force: buoyant, 348
  centrifugal, 215, 235
  centripetal, 118
  Coriolis, 235
  equilibrium, 66, 78
  gravitational, 97
  as interaction, 68
  magnetic on current (Lorentz), 594
  molecular, 213
  normal, 103
  static, 66-89
  unit, 67
  van der Waals, 705
Fourier analysis of waves, 429
Franklin, Benjamin, 506
Fraunhofer: diffraction, 442
  spectrum, 701
Free energy, surface tension, 352
Frequencies: of E-M waves, 388
  of oscillation, 129
  of rotation, 117
  of sound, 387, 389
Fresnel diffraction, 443
Friction, dry, 103

table of coefficients, 104
velocity dependent, 105
Functional dependence, 18
Fundamental frequency, 429
Fusion: heat energy of, 287, 318, 351
nuclear, 716

Galilean transformation, 225, 415
Galileo, 98, 272
Galvanometer, 566
Gamma rays, 681
Gases, kinetic theory, 307
Gas law, ideal, 277, 308
Gas pressure, 72, 275, 307, 311
Gas thermometer, 276
Gaussian pulse, 413
Gaussian surface, 519
gauss, magnetic field unit, 588
Gauss's law, 206, 516
Gears, 193
General theory of relativity, 258
Generators, electric, 634
Geometrical optics, definition, 460
Geometry formulas, 727
Grating, optical, 451
Gravitation: general theory of relativity, 258
Newton's law, 97
Gravitational constant, G, 97
g, 101
Gravitational orbits, 124
energy, 209
Gravitational potential energy, 202, 205
Greek letters, 727
Ground, electric, 569
Group velocity, 430
Growlers, 672

Hall effect, 599, 628
Harmonics, 429
Harmony, 435
Heat, definition, 285
Heat capacities, 286
Heat engine, 322
Heat of fusion, 287, 318
Heat of vaporization, 281, 287, 318
Heisenberg's uncertainty principle, 696
Helium: liquid, 365
nucleus, 95
spectra, 702
velocity of sound, 433
Henry, Joseph, 393
henry, unit of inductance, 636
Hero of Alexandria, 472
hertz, 378

Hertz's experiments, 671
Hole through earth, 206, 518
Hologram, 452
Hooke's law for springs, 70
Horsepower, 201, 544
Huygens' construction, 390
Hydraulic lever, 347
Hydrogen: isotopes, 714
specific heat, 320
spectra, 701
Hydrometer, 349
Hydrostatic paradox, 345
Hyperons, 719
Hysteresis, 609

Ideal gas law, 278, 308
Impedance: a-c, 646
wave, 396, 402
Impulse, 147, 148, 151
Inclined plane, 114, 194
Index of refraction, 465
table, 466
Inductance, 635
Induction, electrical, 509
Inductive reactance, 644
Inductor, charging, 637
Inertia, 95
Inertial reference frame, 232
Infrared, 678
Insulators, 707
Intensity: electromagnetic radiation, 665
sound, 398
Interactions, basic, 16, 504, 721
Interference, double slit, 446
Interference of waves, 440
Interferometer, Michelson's, 12, 239, 450
Internal energy, 285, 317
Internal resistance, 562
Invariant interval, 247
Ionic, 705
Iron, electron arrangement, 607
Irradiance, 665
Isothermal expansion, 292, 294, 322
Isotopes, 714

joule: definition, 181
size, 217
Joule, James, 181, 267

Kelvin temperature, 276, 328
Kepler's laws, 124, 163
Kilocalorie, 285
Kilocalories per mole, 218
Kilogram, 96, 99

Kilowatt-hour, 216, 557
Kinematic viscosity, 364
Kinetic energy: rotational, 189
  translational, 180
Kinetic theory of gases, 307
Kirchhoff's rules, 571, 620

Laser, 421, 446, 452
Latitude, 52
Lead storage battery, 565
Least time, 473
Lenses, 481
Lens-maker's equation, 483
Lenz's law, 624
Leptons, 719
Levers, 192
Lift, fluid, 361
Lightning, 508
Light: polarization, 170
  transmission model, 471
  velocity, 238, 661
  visible, 679
Light year, definition, 5
Linear equation, 31
Line spectra, 701
Local fields, 232
Longitude, 52
Longitudinal wave, 380
Lorentz force, 596, 629
Lorentz transformation, 244
L/R time constant, 638

Machines, simple, 191
Magnetic dipole, 592, 600
Magnetic domains, 608
Magnetic field: current in straight wire, 585, 588
  dipole, 584
  energy density, 639
  loop, 586, 591
  range of strength, 589
  solenoid, 586, 593
Magnetic flux, 624
Magnetic moment, electron, 603
Magnetic poles, 578
Magnetic potential energy, 203
Magnetic susceptibility, 605
Magnetism, 578
Magnification, 477, 485, 495, 496
Magnifying glass, 487
Mass: energy equivalence, 252
  gravitational, 97
  inertial, 96
  unit, 99
Maxwell-Boltzmann distribution, 309, 311, 331

Maxwell's equations, 658
Mechanical advantage, 192
Melting temperatures, table, 280
Meniscus, 354
Mercury thermometer, 269
Mesons, 145, 719
Meter, definition, 3
Michelson interferometer, 12, 239, 450, 713
Michelson-Morley experiment, 12, 239
Microscope, 494
Microwaves, 676
Middle A, 429
Millikan oil drop experiment, 533
Mirror: convex, 480
  curved, 474
  ellipsoidal, 474, 481
  parabolic, 474
  plane, 463
Mobility, electron, 548
Model for charge movement, 544
Model for gas molecules, 306
Model for light transmission, 471
Molar specific heat, 290, 321
Mole, 276
Molecular: beams, 310
  energy in gas, 309
  forces, 213
  potential wells, 210
  speeds in gas, 309, 312
Molecules, 704
Moment of inertia, 119
  table, 121
Momentum: angular, 159
  linear, 136
  vector nature, 141
Momentum conservation, 137, 143, 152
Monopole, 583
Motors, electric, 619, 633
Mu-metal, 610
Muon lifetime dilation, 244
Musical instruments, 431
Music chords and scales, 436

Natural frequency, 129, 378, 641
Neutrino, 692, 718
Neutrons, 699, 714, 719, 721
Newton's cooling law, 295
Newton's law of gravitation, 97
Newton's laws of motion: first, 78
  second, 96
  third, 146
Newton's rings, 449
Newton's second law for rotations, 119

Newton, unit of force, 67
   viscosity, 364
Nodes, 426, 447
Normal force, 103
Nuclear: binding energy, 715
   fission, 716
   fusion, 716
Nucleons, 719
Nucleus of atom, 699, 714
Null-reading devices, 570

Octave, 437
Oersted effect, 578
Ohm's law, 550
Orbits, gravitational, 124, 163
Oscillation, 125
Oxidation and reduction, 564

Pair production, 688
Palomar telescope, 446, 497
Parabolic mirror, 474
Parallel axis theorem, 171
Paramagnetism, 605
Paraxial rays, 475
Particle nature of light, 712
Particles, subatomic, 719
Pascal's vases, 346
Pauli exclusion principle, 603, 692
Pedal tone, 435
Pendulum: physical, 172
   simple, 129
Period: of oscillation, 128
   of rotation, 117
   of wave, 386
Periodicity of atomic electrons, 692, 704
Periodic table of elements, front endpaper
Permeability, 581
Permittivity, 523
Phase angle in a-c, 646
Phase changes, thermal, 280, 317
Phase diagram: carbon dioxide, 282
   water, 281
Phase of oscillations, 420
Phase vector, 645
Phase velocity, 430
Photoelectric effect, 680, 686, 711
Photon: energy, 668, 673, 677, 679
   momentum, 668
Physical optics, definition, 460
Physical pendulum, 172
Piano frequency chart, 389
Pin hole diffraction, 445
Pitch, musical, 377
Pitot tube, 359

Planck's constant, 169, 668, 677, 689, 697, 710
Plane mirror, 463
Planetary constants, 7, 125
Planetary orbits, 124
Platinum wire thermometer, 274
Poise, 363
Poiseuille formula, 365
Poisson's spot, 441
Polarization, circular, 170
Polarized light, 463, 656, 669
Polarizing filters, 670
Position: in two dimensions, 52
   $x(t)$, 30
Potential electric, 523
   cylinder, 527
   dipole, 528
   parallel plates, 527
   sphere, 526
   zero point, 205, 526
Potential energy, 200
   gravitational, 202
   magnets, 203
   negative, 205
   and restoring force, 211
   spring, 203
Potential wells: gravitational, 206
   molecular, 210
Potentiometer, 570
Power, 201, 216, 395, 544
   in a-c circuit, 647
   table for households, 558
Poynting vector, 666
Prandtl tube, 359
Precession, 167
   of earth, 168
Pressure: atmospheric, 72
   definition, 72
   fluids, 344
   gas, 72, 275, 307, 311
   negative, 356
   sound, 398
Principal rays, 477, 484
Prism, 469
Probability, molecular arrangement, 331
Projector, 494
proper time, 247
Protons, 699, 714, 719
Pulleys, 194

Quantization, angular momentum, 169, 689
Quantum mechanics, 545, 689, 694, 703, 706
Quarks, 719

INDEX

Radar, 676
Radian, definition, 3
Radiation as f(T), 710
Radiation, thermal, 296
Radioactive isotopes, natural, 13
Radioactivity, 717
Radius of gyration, j, 171
Radius of percussion, 174
Rainbow, 470
Ray optics, 462
Rays: paraxial, 475
  principal, 477
RC time constant, 621
Reactance, 644
Real image, 477
Recoil of projectile, 143
Red shift, 419
Reference frame: inertial, 232
  noninertial, 232
Reference frames, 224
Reflection, 390, 463
Refraction, 392, 462, 465
  index, 465, 466
Refrigerators, 330
Relativistic relationship of electromagnetism, 610
Relativity, special theory, 237
Resistance: electrical, 551
  internal, 562
Resistivity, 551, 552, 554
Resistors: construction, 557
  series and parallel, 559
Resolution of patterns, 446
Resolving power, 451
Resonance, 379, 675
Resonant frequency, 378, 641, 646
Resonant lines, 423
Restoring force and potential energy, 211
Resultant, vector, 53
Reversible process, 322
Reynolds number, 367
Rocket propulsion, 151
Rotational kinetic energy, 189
Rotational molecular energy, 677
Rumford, Count, 269

Satellites: energy, 209
  orbits, 124, 163
Scalar product, 186
Scalar quantities, definition, 54
Scales, music, 437
Schmidt lens, 498
Second law of thermodynamics, 327
Semiconductor, 708

Shear stress, 74, 341, 363
Simple harmonic motion, 130, 227, 378
Simple machines: gears, 193
  inclined plane, 194
  levers, 192
  pulleys, 194
  wheel and axle, 193
Single slit diffraction, 444
Siphon, 346
SI units, 730
Snell's law, 392, 465
Soap bubble, 342, 353, 449, cover
Solar system: angular momentum, 164
  constants, 7, 125, 165, back endpaper
  map, 9
Solid fluids, 366
Solid state, 706
Sound: frequency table, 387
  intensity, 398
  music, 431
  pressure, 398
  velocity, 385
Space colonies, 237, 265
Spark coil, 648
Specific heat, 287, 290, 318, 707
Spectra, 701, 702, 705, 709
Spectrometer, 451
Specular reflection, 463
Speed of light, 238, 661
Spherical aberration, 475
Spider webs, 354
Spin, 160, 545, 688
Spring constant, 70, 203, 378
Spring potential energy, 203
Stable isotopes, 715
Standard temperature and pressure, 277
Standing waves, 425
Stern-Gerlach atomic beam, 310, 689
Stokes' law, 367
Storage battery, 565
Strain, definition, 74
Streamlines, 357
Stress: definition, 74
  shear, 74, 341, 363
Strong nuclear interaction, 16, 504, 721
Subatomic particles, 719
Superconductivity, 555
Superfluids, 280, 365
Superposition of waves, 421
Surface tension, 350
Synchronization of clocks, 242

Telescope, 493, 495

Temperature: definition, 282, 328
  dependence of resistivity, 552, 554
  molecular energy, 308
  range of, table, 281
  scales, 271
Terminal velocity, 19, 44, 106
tesla, magnetic field unit, 588
Thermal conductivities, 295
Thermal equilibrium, 284
Thermal expansion, 279, 316
Thermal pollution, 329
Thermistor, 270
Thermocouple, 270
Thermodynamics: first law, 284
  second law, 327
  zeroth law, 283
Thermometer calibration, 271
Thermometers, 269
Thin film, light interference, 449
Thomson's cathode ray tube, 686
Thrust, of rocket, 152
Time: dilation, 241
  international unit, 9
  logarithmic scale, 8
  synchronization, 242
Time constant, 621, 638
Torques, 80, 119
Torque on wire loop, 599
Total internal reflection, 468
Trajectories, on flat earth, 56
Transformers, 647
Transverse wave, 380
Traveling waves, 413
Triangulation, astronomical, 4
Trigonometry formulas, 729
Triple point of water, 281
Tritium, 714
Turbulence, 366
TV, 655, 675
Twin paradox, 249

Ultraviolet, 680
Ultraviolet catastrophe, 710
Uncertainty principle, 696
Units, in equations, 40, 727
Unit test charge, 513
Unit test pole, 582
Universal gas constant, 277, 319
Universal gravitational constant, 97
Universe: age, 14
  logarithmic map, 1
Uranium-235, 715

Van Allen radiation belts, 599
Van de Graaff generator, 510
van der Waals' equation, 315

van der Waals' force, 705
Vector addition, 54, 77, 730
Vector multiplication: cross, 81, 160, 590, 730
  dot, 186, 516, 626, 730
Vector product, 81, 160, 590
Vector quantities, definition, 53
Velocity addition, relativistic, 248
Velocity: angular (rotational), 116
  charge flow, 547
  definition, 31
  drift of electrons, 548
  escape, 208
  fluid measurement, 359
  gas molecules, 309
  group, 430
  instantaneous, 33
  phase, 430
  light, 238, 661
  as slope of $x(t)$, 32
  sound, 385
  and speed, 28
  tangential, 117
  terminal, 19, 44, 106
  water ripples, 394
  wave, 381
    frequency dependence, 393
Venturi tube, 358
Violin string waves, 433
Virtual image, 464, 479
Viscosities, table, 364
Viscosity, 356, 362
  kinematic, 364
Visible light, 679
Volt, 524
Voltage, 548
Voltage divider, 569
Voltaic cells, 543, 563
Voltmeter, 566
Volume expansion coefficients, 279

Water: boiling point, 273
  freezing point, 273
  phase diagram, 281
  pressures and boiling points, 281
Watt, unit of power, 201, 216, 544
Wavelength, 387
Wave motion, 375
Wave nature of particles, 694
Waves: coherent, 420, 452
  impedance, 396, 402
  interference, 440
  power, 395
  reflection, 390, 402
  refraction, 392
  standing, 425

INDEX

traveling, 413
velocity, 381
  frequency dependence, 393, 410, 430, 469
Weak nuclear interaction, 16, 504, 721
Wheatstone bridge, 570
Wheel and axle, 193
Work, definition, 186

X-ray diffraction, 452
X-rays, 680

Young, interference of light, 447
Young's modulus, 75, 214

Zeroth law of thermodynamics, 283

Mass of alpha particle	6.6441×10^{-27} kg	3727.3 MeV
Atomic mass unit	1.661×10^{-27} kg	931.5 MeV
Unit electric charge	1.60×10^{-19} C	
r_{atom}	$\approx 1 \times 10^{-10}$ m	
Proton mass/electron mass	1836	